Geometry in History

S. G. Dani • Athanase Papadopoulos
Editors

Geometry in History

Editors
S. G. Dani
University of Mumbai & Department
of Atomic Energy
Centre for Excellence in Basic Sciences
Mumbai, Maharashtra, India

Athanase Papadopoulos
Institut de Recherche Mathématique
University of Strasbourg, CNRS
Strasbourg Cedex, France

ISBN 978-3-030-13611-6 ISBN 978-3-030-13609-3 (eBook)
https://doi.org/10.1007/978-3-030-13609-3

This Springer imprint is published by the registered company Springer Nature Switzerland AG.
The registered company address is: Gewerbestrasse 11, 6330 Cham, Switzerland

Preface

This volume consists of a collection of essays on geometry from a historical viewpoint, addressed to the general mathematical community interested in the history of ideas and their evolution.

In planning this book as editors we went with the conviction that writing on the basic geometrical concepts from a historical perspective is an essential element of our science literature, and that geometrico-historical surveys constitute an important ingredient, alongside purely mathematical and historical articles, for the development of the subject. We have also held, for many years, that we need such articles written by mathematicians, dealing with topics and ideas that they are directly engaged with and consider as fundamental, and it was our endeavour to put together an ensemble of articles of this variety on a broad spectrum of topics that constitute the general area of geometry. It actually turned out to be a nontrivial task to find colleagues capable and willing to write on the history of their field, but perseverance eventually led to this volume coming to fruition. We warmly thank all the authors of this volume for their contribution. We also thank all the referees, who shall stay anonymous, for their critical comments and constructive suggestions that helped substantially in improving the quality of the volume, and Elena Griniari for her editorial support.

The work on this book was completed during a stay of the second editor at the Yau Mathematical Center of Tsinghua University (Beijing).

Mumbai, India
Strasbourg, France
November 2018

S. G. Dani
Athanase Papadopoulos

Introduction

Comprehension of shapes has played a pivotal role, alongside of numbers, in the progress of civilizations, from the beginning. Significant engagement with shapes, or *geometry*,[1] is seen in the ancient cultures of Egypt, Mesopotamia, India, China, etc., from the very early times. A systematic approach to the subject, turning it into a discipline with an axiomatic foundation, was developed by the ancient Greeks, which served crucially as a basis to later rewritings and developments at the hands of Arab mathematicians during the Middle ages and in turn to the modern advent. The Greeks also addressed various philosophical issues associated with the subject that have been very influential in the later developments.

From the point of view of the subject of history of mathematics, there is a need for viewing the historical development of ideas of geometry as an integral whole. The present endeavour is seen by the editors as a limited attempt in that direction, focusing mainly on the historical antecedents of modern geometry and the internal relations within the latter.[2] In the overall context, the editors also felt the need to concentrate on bringing out the perspective of working mathematicians actively engaged with the ideas involved, in their respective areas, as against that of historians of mathematicians viewing developments in mathematics from the outside, a pursuit in which the issues involved and the flavour of the output are different from what we seek to explore in this project.

Mathematicians build upon the works of their predecessors, which they regularly reshape, refine and reinterpret (sometimes misinterpret). There are countless examples of ideas discovered concurrently and independently and of others that stayed in the dark until being rediscovered and used much later. This makes the history of mathematical ideas a living and intricate topic, and any attempt to say something

[1]Although the term *geometry* is etymologically associated with *earth measurement*, it was clear since the times of Plato that this field, from the moment it became mature, is more concerned with shape than with measurement.

[2]The present editors had organized a conference on the same theme, "Geometry in History," at the University of Strasbourg during 9–10 June 2015; the deliberations at the conference and the overall experience of the event have served as an inspirational precursor to bringing out this volume.

significant in this domain requires, on the one hand, a deep knowledge of the subject and also, on the other hand, a comprehensive vision of history. In the perception of the editors of this volume, the mathematical community is in acute need of articles presenting major geometrical ideas in a right historical perspective, paying attention also to the philosophical issues around the activity. This was a motivation for bringing out this volume, and we are happy to see that many of the essays in it have turned out to be a confluence of mathematics, history and philosophy, together with state-of-the-art mathematical research.

The book is divided into two parts. The first one, consisting of seven chapters, is concerned with topics that have roots in Greek antiquity and ramifications all through the history of mathematics. The second part, consisting of 12 chapters, treats more modern topics. We now describe briefly the themes dealt with in the individual chapters.

The opening essay by Stylianos Negrepontis is on a topic from the fourth century BCE, namely, Plato's theory of anthyphairesis, an idea that would serve as a precursor of continued fractions. It may be worthwhile to recall here that Plato was above all a mathematician and that his conceptualization of the world is deeply rooted in mathematics. The goal of this article is to explain the anthyphairetic nature of Plato's dialectics which is at the foundation of his theories of Ideas, of true Beings, of knowledge and of the distinction he draws between intelligible and sensible Beings. Negrepontis defends the thesis that the whole of Plato's philosophical system is based on the concept of periodic anthyphairesis. He also provides an explanation of Plato's praise of geometry and his criticism of its practice by the geometers. Plato criticized the axiomatic method, by which mathematicians rely on hypotheses that, according to him, have nothing to do with true knowledge. He was at the same time critical of the geometers' use of diagrams and of topics such as Eudoxus' theory of ratios (on which the theory of Dedekind cuts is based), of Archytas' theory of quadratic and cubic incommensurabilities and of the use of the notion of "geometric point" in the foundations of geometry. Instead, Plato claims that the method of Division and Collection—a philosophical expression of the periodic anthyphairesis—should be the only one to be used in the acquisition of all knowledge in geometry, in particular in the construction of numbers and of the straight line. At the same time, the author also proposes an explanation as to why geometry was so important to Plato.

The whole essay is based on a new analysis and a novel reading of several difficult dialogues of Plato, in particular the Parmenides, Sophist, Statesman, Timaeus and Phaedo.

The second chapter, written by Athanase Papadopoulos, is an exposition of René Thom's visionary ideas in his interpretation of the work of Aristotle—especially his treatises on biology—from a topological point of view. Thom was a dedicated reader of Aristotle. With his penetrating intuition as a mathematician, he gave a completely new explanation of some passages in the writings of the Stagirite, finding there the ideas of genericity, stratification, boundary, the Stokes formula and other topological notions. He completely adhered to Aristotle's theory of form which the latter expanded in his zoological treatises, and he highlighted the importance of these ideas in biology, and more particularly in embryology, namely, the idea of a form tending to its own realization. Thom's contribution to the interpretation of Aristotle's writings, an outcome of many years of reflection, constitutes a new link between science in Greek antiquity and modern mathematics.

In Chap. 3, Yuri Manin describes a continuous chain of ideas in mathematical thought that connects Greek antiquity to the modern times. He uses this chain as an argument for his thesis that the notion of "paradigm shifts", as an approach in the history of science promoted by the philosopher Thomas Kuhn (1922–1996), to the effect that radical changes in the objects of interest characterize the transitions from one epoch to another, does not apply to mathematics. On the contrary, says Manin, mathematical thought and the mathematically interesting questions evolve in a continuous manner. The ideas on which his argument is based involve the mathematics of space, time and periodicity, and he takes us through some episodes from the history of this subject, starting with Ptolemy's dynamical model of the solar system until the modern probabilistic models of elementary particles, Schrödinger's quantum mechanics amplitude interference and Feynman integrals, passing through the work of Fourier, and Mendeleev's table of chemical elements. In this description of the history, quantum mechanics, in Manin's words, becomes "a complexification of Ptolemy's epicycles".

In Chap. 4, Athanase Papadopoulos reviews the appearances of the notion of convexity in Greek antiquity, more specifically in the classical texts on mathematics and optics, in the writings of Aristotle and in art. While convexity was to turn into a mathematical field in the early twentieth century, at the hands of Minkowski, Carathéodory and others, this notion is found in mathematical works all the way from those of the Greeks. The thinking in many mathematical arguments of Euclid, Apollonius, Archimedes, Diocles and others is seen to be based on convexity considerations, even when the concept had not attained its maturity. In optics, the notion played an important role on account of convex and concave mirrors and lenses, the latter relating also to astronomy. Convexity also features in Aristotle's theories on a variety of topics and in ancient Greek art. This chapter puts the evolution of the notion of convexity in perspective.

Chapter 5, by Arkady Plotnitsky, is again concerned with the question of continuity of ideas in mathematics. Starting with the question of "what is a curve", and taking as a starting point the curves formed by images drawn 32,000 years ago on the curved walls of the Chauvet-Pont-d'Arc cave in the South of France, the author reflects on the evolution of mathematics, on the relation between mathematics and art and on the notion of modernity and the difference between "modernity" and "modernism", both in art and science. Modernism in mathematics, he argues, is essentially a conception characterized by the algebraization of spatiality while adhering to geometrical and topological *thinking*. The exposition wanders through the works of Fermat and of Descartes—where the notion of algebraization of geometry became essential—through a reflection on the role of experimentation in physics, and through the ideas of Heidegger on modern science as being essentially mathematical. A crisis concerning the "irrationality" of quantum mechanics and its "rationalization" by Werner Heisenberg is compared to the crisis of the incommensurable which traversed ancient Greek mathematics. The author also comments on John Tate's principle: "Think geometrically, prove algebraically", extending it to: "Think both intuitively geometrically and space-like geometrically, and prove algebraically", and taking examples in the modern works of Weyl, Weil, Grothendieck, Langlands and Tate.

The aim of Chap. 6, by Annette A'Campo and Athanase Papadopoulos, is to draw a continuous path from the theory of curves in Greek antiquity until the modern synthetic differential geometry of Busemann–Feller and Alexandrov. The route passes through the work of Huygens on evolutes, through that of Euler on curvature of surfaces embedded in 3-space and then through the developments made by the French school founded by Monge. These works, together with the background of the respective periods and of the authors concerned, are discussed extensively in this chapter. The evolution shows a return, in the twentieth century, to the fundamental methods of geometry that eschew the differential calculus.

Chapter 7, by Toshikazu Sunada, is a historical exposition of the development of geometry from ancient times until the modern period, with a view of this field as a tool for describing the shape of the universe. The account involves unavoidably questions of cosmology and of philosophy of space and time, and it takes us deep into the worlds of differential and projective geometry, topology and set theory, discussing notions like the infinite, infinitesimal, curvature, dimension and others.

The next two chapters are concerned with configuration theorems, that is, theorems of projective geometry whose statements involve finite sets of points and arrangements of lines.

Configuration theorems form a coherent subject which is again rooted in Greek antiquity, more precisely in the theory of conics, whose main founders are Apollonius, Pappus and Ptolemy. The subject continued to grow in the works of Pascal, Desargues, Brianchon, Poncelet, Steiner and others, and it has still a great impact on current research. A famous example of a configuration theorem is Pappus' theorem which concerns the alignment of three points: given two triples of aligned points A, B, C and a, b, c in the plane, the three intersection points of the three pairs of lines AB, Ba, Ac, Ca and Bc, Cb are also aligned. Another example is Pascal's theorem stating that if we take arbitrarily six points on a conic and if we join them pairwise so as to form a convex hexagon, then the three pairs of opposite sides of this hexagon meet in three points that lie on the same line. In the case where the conic degenerates to a pair of lines, Pascal's theorem reduces to Pappus' theorem. Thus, Pascal's theorem is a good example of a theorem discovered in the seventeenth century which is a generalization of a much earlier theorem.

In this volume, configuration theorems are considered in Chaps. 8 and 9.

Chapter 8, written by Victor Pambuccian and Celia Schacht, is concerned with the significance of Pappus' and Desargues' and other configuration theorems derived from them in the axiomatic foundation of geometry and with the interrelation of this topic with algebra and first-order logic. From a historical point of view, it has taken quite long for the wealth of significance of the theorems of Pappus and Desargues to be recognized. The recognition came only in the twentieth century, spurring with it also a flurry of activity in the general stream of configuration theorems, the genre they represent. The authors consider the article a story of "adventure through which their true importance became revealed".

Chapter 9, by Serge Tabachnikov, is a survey on the impact of configuration theorems and results inspired by them in modern dynamics, namely, in a series of iterated construction theorems motivated by configuration theorems, in the study of maps of the circle and self-similarity, in the study of actions of the modular group, in the theory of billiards in ellipses and in the study of caustics. Identities in the Lie algebra of motions of Euclidean and non-Euclidean geometries are interpreted as configuration theorems. The author also surveys the work of R. Schwarz on the pentagram map and the theory of skewers, recently developed by himself, which provides a setting for space analogues of plane configuration theorems.

In Chap. 10, Ken'ichi Ohshika shows how Poincaré's work on topology and geometry led him to the philosophy of science that he formulated in the various books he published. At the same time, he reviews some important points in Poincaré's work: the foundation of homology theory, the construction of 3-manifolds by gluing polyhedra or by using complex algebraic equations in three variables and the first results on what later became known as Morse theory, used by Poincaré in his classification of closed surfaces. Other questions raised by Poincaré that led to the development of topology are also mentioned in this chapter. Besides, Ohshika addresses the question as to why mathematicians of the stature of Poincaré are interested in the philosophy of science. This serves as an occasion for him to mention, briefly, the approaches of several philosophers in this field, including Kant, Frege, Husserl, Russell and Althusser. The latter talked about the "spontaneous philosophy of scientists" and at the same time insisted on the necessity of drawing a line separating science from ideology.

Chapter 11, by Alain Chenciner, is about the applications of the study of the dynamics of the iterates of a map obtained by perturbing the germ of the simplest map of the plane: a planar rotation at the origin. The author recalls how such a study led to the Andronov–Hopf–Neimark–Sacker bifurcation theory, concerned with invariant curves under a radial hypothesis of weak attraction (or repulsion) for generic diffeomorphisms with elliptic fixed points, and then to the Kolmogorov–Arnold–Moser (KAM) theory which deals with area-preserving maps. He brings out the relation with the so-called non-linear self-sustained oscillation theory of Lord Rayleigh and Van der Pol, and the theory of normal forms developed by Poincaré in his 1879 thesis in connection with the three-body problem, and he also mentions the relation with what he calls the "averaging of perturbations" technique, used by astronomers since the eighteenth century, which generalizes to the non-linear setting the Jordan normal form of a matrix. He concludes with a result of his own which interrelates several ideas presented earlier in the chapter.

Chapter 12, by François Laudenbach, starts with Gromov's h-principle, a principle that gives conditions under which a manifold carrying a geometric structure in a weak sense (e.g. an almost-complex structure, an almost symplectic structure, etc.) carries a genuine geometric structure (a complex structure, an almost symplectic structure, etc.,). At the same time, the author highlights a principle formulated by René Thom in 1959 which he calls a *homological h-principle*, and which is a precursor of the first h-principle that appeared in 1971 in a paper by Eliashberg and Gromov. Thom gave only a sketch of proof of his principle, a sketch which, as the author recalls, did not entirely convince the mathematical community. In his chapter, Laudenbach explains some combinatorial aspects of Thom's outline that may eventually lead to the completion of the proof of this theorem, highlighting the relations between this result and other results of various authors, including the sphere eversion by Smale and the Eliashberg–Mishachev holonomic approximation theorem. In this chapter, a special emphasis is laid on the relation between Thom's result and Thurston's jiggling lemma (1974), a result considered as an h-principle for foliations, which was crucial in the proof of Thurston's celebrated existence theorem in foliation theory.

The theme of Chap. 13, by Yakov Eliashberg, is flexibility and rigidity phenomena in symplectic geometry. The author starts by declaring that the study of these phenomena, as two general research directions, coexists in any mathematical field and that in symplectic topology these two directions are especially close to each other. This idea is further developed in the essay. The exposition starts with what the author calls "Poincaré's dream", namely, the fact that the qualitative properties of a Hamiltonian system are due to the existence of a symplectic form preserving its phase flow. Eliashberg states that this dream was realized in the 1960s, with the emergence of symplectic topology, and more particularly in the works of Arnold and Gromov. He makes a list of open problems formulated in the 1960s and 1970s that were foundational to some of the major work done in this field later. Several of these problems were solved by Gromov, thanks to his h-principle for contact and symplectic structures and his later introduction of J-holomorphic curves in symplectic topology. Eliashberg then discusses the current status of these problems, mentioning the relations with symplectic packing inequalities, Legendrian knots, Legendrian homology algebra and developments due to Donaldson, Bennequin, McDuff, Hofer and others.

Chapter 14 by William Goldman is a historical survey on the theory of locally homogeneous geometric structures. The theory was started by Charles Ehresmann, who laid its foundations in his 1936 paper *Sur les espaces localement homogènes* and in subsequent papers. In particular, Ehresmann introduced the notions of developing map, holonomy representation, normal structure and other related notions which led to what Goldman calls the "Thurston holonomy principle", or the "Ehresmann–Weil–Thurston holonomy principle". This principle establishes a relation between the classification of geometric structures on a manifold and the representation variety of its fundamental group into a Lie group, a topic of extensive activity today. Goldman talks about "Ehresmann's vision" which set the context for Thurston's geometrization program for 3-manifolds. He mentions the relation between the works of Poincaré, Lie, Klein and Elie Cartan and the later developments in the theories of discrete subgroups of Lie groups, complex projective structures, flat conformal manifolds and others.

Chapter 15 by Marc Chaperon is also closely related to the work of Ehresmann, this time in connection with that of Elie Cartan, on the development of differential geometry. While Cartan's contributions to the field include the concepts of frame bundle, canonical Pfaffian system, Cartan system (or canonical contact structure) and integral manifolds, those of Ehresmann include the theories of jets, fibre bundles, regular infinitesimal structures, connections, pseudogroups and groupoids. Ehresmann's space of jets is also mentioned in relation to the work of Thom, with Gromov's h-principle and with Smale's classification of immersions. The Cartan formula for differential forms leads to the so-called Poincaré lemma, to the first-order theory of partial differential equations, to the Gauss–Manin connection associated with a proper submersion and to the notion of monodromy.

Chapter 16, by Katsuhiro Shiohama and Bankteshwar Tiwari, is concerned with the comparison between Finsler and Riemannian geometries. The stress is on the asymmetry of the distance function associated with a Finsler manifold and in particular on the notions of reversibility constant, forward cut locus, forward injectivity radius, forward conjugate locus, forward pole, forward Busemann function, etc. Works in this domain by Blaschke, Whitehead, Myers, Rademacher, Busemann, Klingenberg, Berger, Omori, Weinstein and others are mentioned.

Chapters 17 and 18 are concerned with the topology of 3- and 4-manifolds, respectively.

Chapter 17, by Valerii Berestovskii, is centred on the work done in the twentieth century around the three-dimensional Poincaré conjecture. The geometrization conjecture and its proof by Perelman remain in the background, while the author reviews, instead, several topics related to the work done around the conjecture, including various characterizations of the 3-sphere that were formulated in the attempt to prove it, the relation with 4-manifolds, the group-theoretic equivalent forms of the conjecture, and recent works on the manifold recognition problem and the (simplicial) triangulation conjecture for topological manifolds of dimension at least five. The use of methods introduced by Busemann for dealing with questions closely related to the three-dimensional Poincaré conjecture is highlighted. The author emphasizes the fact that the history of the proof of the Poincaré conjecture and of other related conjectures once again demonstrates the unity of mathematics and at the same time the fruitfulness of ideas originating in mathematical physics.

Chapter 18, by Valentin Poénaru, is a survey of the important problems on four-dimensional manifolds. Two features of this dimension are highlighted: (1) the gap (the author refers to it as the "big abyss") between the differentiable and the topological categories, which is special to this dimension, exemplified by the fact that there are manifolds homeomorphic to four-dimensional Euclidean space but not diffeomorphic to it, and (2) the role played by mathematical physics. Several notions of special interest are highlighted, including the Whitehead manifold and Casson handles, which turned out to be fundamental in the theory. Works of several mathematicians, including Bing, Smale, Mazur, Milnor, Kervaire and Stallings, are mentioned. This is also the occasion for the author to talk about fundamental questions in the topology of manifolds that are solved in the particular dimension considered, and not in others. This is exemplified by the smooth four-dimensional Schoenflies conjecture which is still open. The chapter also mentions some episodes from the history of the proof of the Poincaré conjecture in higher dimensions.

The book ends with Chap. 19 which is an autobiographical account by Valentin Poénaru, of the period where he set out to be a mathematician. The author gives a vivid description of the problems he worked on, the people he met and others with whom he corresponded. The article gives us a glimpse of what the life of a mathematician was like in the Eastern part of Europe, at a time when the continent was divided by an iron curtain. A feeling emerges that mathematics is a space of liberty.

A word of conclusion is in order.

From its beginnings at the hands of the Ancients, the subject of geometry has now grown into a magnificent edifice of immense dimensions, and in this book, we have tried to trace out some of the representative developments and their interrelations, from a historical point of view, but from the perspective of the practitioners. The edifice, which is perpetually under construction, needs a large number of builders, many with specific skills, but the reader may notice that among these builders, a few names appear recurrently in the various chapters of this book: Poincaré, Ehresmann, Thom, Busemann, Alexandrov, Thurston and Gromov. Thus, while rooted in the broader objectives, our endeavour has also turned to be a celebration of the towering achievements in the area and of the major achievers.

This book project was undoubtedly circumscribed by the availability of expertise on various themes and of experts (known to us) willing to spare their time and effort to participate in such an endeavour. It is however hoped that enough ground has been covered towards the indicated objectives, so that this volume will serve as a meaningful step in building up further towards them.

Mumbai, India
Strasbourg, France
November 2018

S. G. Dani
Athanase Papadopoulos

Contents

Chapter 1
Plato on Geometry and the Geometers

Stelios Negrepontis

θεὰ σκέδασ' ἠέρα, εἴσατο δὲ χθών·
Odusseia, Book XIII, line 352
[G]oddess [Athena] dispersed the mist, and the land was recognized.

Abstract The present paper aspires to explain fully both the supreme importance of Geometry for Plato, and also the nature of the serious ongoing criticism that Plato (and the Academy) directs against the geometers, an explanation that has eluded modern scholars of Plato (since M. Ficino in the fifteenth century to our present day). In order to understand the criticism, it is necessary first to have a true understanding of the nature of Plato's philosophy. The most crucial concept in understanding Plato's philosophy, and essentially the only one, is the geometrical concept of periodic anthyphairesis of two magnitudes, say line segments, established with the Logos criterion (Sect. 1.2).

The Platonic true Being, the intelligible Platonic Idea, is a dyad of opposite parts in the philosophic analogue, Division and Logos-Collection, in close imitation of periodic anthyphairesis. Plato in effect isolates a method for acquiring full and complete knowledge, as it exists in a small but vital part of Geometry, namely in incommensurability by periodic anthyphairesis, and develops a general theory of knowledge, Division and Collection, of the Platonic Ideas, in close imitation to the complete acquisition of knowledge provided by the Logos criterion in periodic anthyphairesis.

The anthyphairetic nature of Platonic intelligible Beings was examined in detail by the author in earlier publications and is outlined in the present paper: the One of the second hypothesis of the *Parmenides* and its close relation to Zeno's arguments and paradoxes (outlined in Sect. 1.3) and the Division and Collection in the *Sophistes* and *Politicus*, where the genera and kinds in the Division are considered as hypotheses (outlined in Sect. 1.5). Furthermore, we establish in the present paper

S. Negrepontis (✉)
Department of Mathematics, Athens University, Athens, Greece
e-mail: snegrep@math.uoa.gr

S. G. Dani, A. Papadopoulos (eds.), *Geometry in History*,
https://doi.org/10.1007/978-3-030-13609-3_1

that the dialectics of the *Politeia* coincide with Division and Collection-Logos (Sect. 1.4), described also equivalently in analysis fashion as ascent from hypotheses to the "anhupotheton", the hypothesis-free (531d-e, 532a-b,d-e, 534b-d) (Sect. 1.6).

Without understanding the anthyphairetic nature of Plato's dialectics it is impossible to understand Plato's praise of Geometry and criticism of its practice by geometers.

Criticism I: the rejection by Plato of the indivisible geometric point in favor of the "indivisible line", according to Aristotle's *Metaphysics* 992a, 19–22, coincides with the rejection of the One of the first hypothesis as a true Being, a One similar to the geometric point, and the adoption of the One of the second hypothesis as a true Being, a One described by Zeno, Plato, and Xenocrates in terms akin to those of the indivisible line, in the *Parmenides* (Sect. 1.7).

Criticism II: the rejection by Plato of the use of hypotheses (namely basic definitions and postulates on lines, circles, angles in geometry, on units and numbers in Arithmetic) as principles, and not as stepping stones towards the true Being, the anhupotheton. The rejection of the axiomatic method on epistemological grounds, since hypotheses, namely definitions and postulates, are arbitrarily accepted and hence these, with all its consequences, cannot be known; knowledge, by Division and Collection, is achieved only when the generation of these hypotheses, the basic geometric (lines, circles, angles) and arithmetical (units, numbers) concepts and their Postulates, takes place within the Platonic true Beings, something possible because of their periodic anthyphairetic structure in the *Politeia*(510–511, 527a-b) (Sects. 1.8, 1.9, 1.11, and 1.12).

Praise of Geometry: Plato makes the extraordinary claim that the method of Division and Collection must be the method employed for the acquisition of true knowledge for all of Geometry (in place of Euclidean axiomatics). With this method he constructs the numbers, the straight line, and the circle (Sect. 1.10), and the three kinds of angle (Sect. 1.11).

Criticism III: the use of geometric diagrams is rejected by Plato, not simply because they are visible/sensible, but because they are sensible representations not provoking to the mind, as they should be, if they were represented as true sensibles, participating anthyphairetically in the intelligible by means of the receptacle/diakena, as presented analytically in the *Timaeus* 48a-58c and in preliminary manner in the *Politeia* 522e–525a. The geometers fall into this faulty use of geometric diagrams, as a result of their "dianoia" way of constructing their arguments (Criticism II) (Sects. 1.13 and 1.14).

Criticism IV: the rejection by Plato [1] of Archytas' non-anthyphairetic proofs of quadratic and cubic incommensurabilities (based on the arithmetical Book VIII of the *Elements* and eventually on the arithmetical Proposition VII.27 of the *Elements*), possibly expressed in the distinction between the eristic and dialectic way of going to infinity in the *Philebus* 16d-e, replacing Theaetetus' anthyphairetic proofs of quadratic incommensurabilities, [2] of Eudoxus' Dedekind-like theory of ratios of magnitudes (Book V of the *Elements*), expressed clearly in the second part of *Scholion In Euclidem* X.2, replacing Theaetetus' anthyphairetic one, and thus moving away from Plato's philosophy based on periodic anthyphairesis, and [3] of Archytas' non-anthyphaireic stereometric solution of the problem of the duplication of the cube, in the *Politeia* 527d-528e (Sect. 1.15).

All of Plato's criticisms of the geometers have a common strain: the practice of geometers is distancing away from periodic anthyphairesis and from Platonic true Beings, based on periodic anthyphairesis.

1.1 Introduction

According to a story of late antiquity there was an inscription in the entrance of the Academy proclaiming

> Let no one enter who is ignorant of geometry.[1]

The lateness of antiquity makes the authenticity of this story suspect; there is however a point in the *Politeia* where, speaking of his ideal city, Plato makes an essentially equivalent statement:

> "we must order in the strongest possible terms ("malista prostakteon") that the men of your Ideal City shall in no way ("medeni tropoi") neglect ("aphexontai") geometry". (*Politeia* 527c1-2)[2]

It is a strong understatement that the modern students of Plato (since Marsilio Ficino in the fifteenth century to our days) have not been able to explain why Geometry was so supremely important for Plato; and this cannot but be due to their failure to understand the real content of Plato's dialectics.

At the same time Plato is highly critical, mostly in the *Politeia*, of the way geometers employ and form their arguments about Geometry:

> [the geometers] do not deign to render a **Logos** ("logon... didonai") of [their hypotheses] to themselves or others, taking it for granted that they are obvious to everybody ("hos panti phaneron"). They take their start from these... (in 510c3-d5)
> this science [Geometry] is in direct contradiction ("pan tounantion echei") with the formation of arguments employed in it by its adepts ("tois en autei logois legomenois"). (in 527a1-8)

The present paper aspires to explain fully both the supreme importance of Geometry for Plato, and also the nature of the serious criticism that Plato directs against the geometers, the cause of their faulty practice, and the means he proposes to correct it and ascend to the intelligible Geometry.
[**Section 1.2** The Mathematics of Anthyphairesis.] The most crucial concept in understanding Plato's philosophy, and essentially the only one, is the geometrical

---

[1] Joannes Philoponus, *In Aristotelis libros de anima Commentaria* 15,117,26-27 Πυθαγόρειος δὲ ὁ Πλάτων, οὗ καὶ πρὸ τῆς διατριβῆς ἐπεγέγραπτο 'ἀγεωμέτρητος μὴ εἰσίτω'. Olympiodoros, *Prolegomena (tes Logikes)* 9,1 ἐπιγεγράφθαι ἐν τῷ τοῦ Πλάτωνος μουσείῳ 'ἀγεωμέτρητος μηδεὶς εἰσίτω'.

[2] "μάλιστα προστακτέον ὅπως οἱ ἐν τῇ καλλιπόλει σοι μηδενὶ τρόπῳ γεωμετρίας ἀφέξονται."

concept of periodic anthyphairesis of two magnitudes, say line segments.[3] The Mathematics of this concept, precursor of the modern theory of continued fractions, is closely connected with the great discovery of incommensurability by the Pythagoreans (who discovered the anthyphairesis of the diameter to the side of a square, and deduced incommensurability by employing Proposition X.2 of the *Elements*), and the subsequent great progress on the (palindromically) periodic anthyphairesis of quadratic irrationals by Theodorus and Theaetetus.[4] It is crucial to realize that the proofs of incommensurability of a to b, by establishing the Logos criterion for periodicity, in reality not only prove incommensurability but provide **full knowledge** of the ratio a to b.

Let us note that even though the method employed by the Pythagoreans is a contested topic in the History of Mathematics, there must be no doubt that the Pythagoreans proved the incommensurability by computing the anthyphairesis of the diameter to the side of a square, finding that its sequence of quotients is $[1, 2, 2, 2, \ldots]$, thus infinite, and employing Proposition X.2 of the *Elements*. In favor of such a method we can mention

(a) the knowledge by the Pythagoreans of the side and diameter numbers, the precursor of the convergents of square root of 2,
(b) the stories about Hippasus that involve his drowning into the sea, confused but clearly symbolizing the infinite nature of incommensurability, and
(c) the adoption of infinite and finite as the two basic principles of their philosophy; the infinite was essentially the infinity of the anthyphairesis, and the finite the preservation of the Gnomons, by which at a finite stage complete knowledge of (the anthyphairesis of) the diameter to the side is achieved.
(d) the fact that Zeno's arguments and paradoxes are based on periodic anthyphairesis and are related to the incommensurability of the diameter to the side of a square.

[**Sections 1.3–1.6**. Platonic Division and Collection, in the *Parmenides, Sophistes, Politicus* and *Politeia*, is a philosophic version of periodic anthyphairesis.] The whole of Plato's philosophical system is based on the geometrical concept of periodic anthyphairesis of two lines. In fact the Platonic true Being, the intelligible Platonic Idea is a dyad of opposite parts in the philosophic analogue, in imitation, as Plato himself clearly states,[5] of periodic anthyphairesis.

Plato's theory is a direct continuation of the philosophy of the Pythagoreans (and of Zeno's arguments, themselves concocted in order to support the new theory of

[3] A "magnitude" is used to render "megethos" in the *Elements*, meaning line, surface, or volume, and in opposition to 'number' ("arithmos"), which is always a natural number starting with 1 (or in fact with 2, as strictly speaking 1 is the unit generating all the numbers).

[4] Negrepontis [23].

[5] "ΣΩ. Ἴθι δή—καλῶς γὰρ ἄρτι ὑφηγήσω—**πειρῶ μιμούμενος** τὴν περὶ τῶν δυνάμεων ἀπόκρισιν, ὥσπερ ταύτας πολλὰς οὔσας ἑνὶ εἴδει περιέλαβες, οὕτω καὶ τὰς πολλὰς ἐπιστήμας ἑνὶ λόγῳ προσειπεῖν." *Theaetetus* 148d4-7.

his teacher, Parmenides, on the distinction between the way of Truth, realized by a Monad, a One, and the way of opinion, in which we live, borrowing heavily from the Pythagoreans).

Plato presents his theory (in practically all his dialogues, but in a more systematic way) in the second hypothesis of the *Parmenides* (outlined in Sect. 1.3),[6] in the dialogues *Sophistes* and *Politicus* (outlined in Sect. 1.5),[7] and in the *Politeia* (examined in Sects. 1.4 and 1.6).

[**Section 1.3**. The One of the second hypothesis in the *Parmenides* and Zeno's true Being, are similar philosophical versions of periodic anthyphairesis.] Parmenides was the first philosopher to propound the thesis that there are true unchanging monadic Beings, while our world is a changing world of mere opinion, the precursor of Plato's fundamental distinction between intelligible and sensible Beings. As we have gradually realized, Zeno accomplished a great synthesis of Pythagorean mathematics and Parmenidean philosophy, used the Pythagorean discoveries in service of a new form of the Parmenidean duality truth vs. opinion, in which infinite periodic anthyphairesis plays a key role, and has influenced Plato heavily, as can be seen from a comparison of Zeno's arguments and paradoxes with Plato's *Parmenides*.

In order to reach an understanding of Plato's criticism of the geometers and praise of a truly intelligible Geometry, it is necessary to reach first a clear understanding of the nature of the intelligible Platonic Beings.

This first step toward such an understanding is an outline of our study of the One of the second hypothesis in the *Parmenides* 142b-155e, the paradigmatical true Platonic Being. Our basic discovery is that the One becomes known by the method of Division and Collection, and that this method is the philosophic analogue of periodic anthyphairesis (a mathematical notion explained in Sect. 1.2): Division is infinite anthyphairetic division of the initial indefinite dyad One and Being, and Collection is the Logos criterion leading to periodicity, and the equalization of the parts One and Being. Progress in the study of Plato's philosophy has been greatly hampered by a lack of understanding of the anthyphairetic nature of Division and Collection.

[**Section 1.4**. Dialectics in the *Politeia* is Division and Collection.] We present the considerable evidence that the dialectics in the *Politeia* is, as in the second hypothesis of the *Parmenides*, exactly Division and Collection. Two descriptions of Division and Collection, (a) 531d9-e6, 532a5-b3, d8-e3 and (b) 534b3-d2, make this especially clear. To "provide Logos" is, as explained in Sect. 1.3, precisely to provide the philosophic analogue of the Logos Criterion for the establishment of periodicity in anthyphairesis. Thus the structure of the real intelligible Being in the *Politeia* coincides with the One of the second hypothesis in the *Parmenides*.

[**Section 1.5**. Genera of the Division and Collection in the *Sophistes* and *Politicus* as Hypotheses.] In the dialogues *Sophistes* and *Politicus* Plato carefully exhibits

---

[6]Negrepontis [24]. That Plato was greatly influenced by Zeno is attested by the comparison of Zeno's Fragments and paradoxes with Plato's *Parmenides*.

[7]Negrepontis [22, 23].

the method of Division and Collection, but in the abbreviated, in Zeno's tree-like dichotomy, assuming the form

$$G = a + b, \qquad (a = b + a_1 \text{ is omitted}),$$
$$b = a_1 + b_1, \qquad (a_1 = b_1 + a_2 \text{ is omitted}),$$
$$b_1 = a_2 + b_2, \qquad (a_2 = b_2 + a_3 \text{ is omitted}),$$

and so on, *ad infinitum.*

For completeness sake, we provide the first of these divisions, that of the Angler (in the beginning pages of the *Sophistes*). For details see [22, 23].

We also note that in these two dialogues Plato systematically treats the new parts-genera introduced in the division steps as "hypotheses". The systematic use of such a terminology might at first seem odd, until we realize that Plato, as Proclus[8] tells us, was extremely fond of the heuristic method of Analysis.

The abbreviated form of the Division makes it quite reasonable to think of Division as Analysis, and of the successive genera of the Division as higher hypotheses: starting from G, in analysis fashion we look for a higher hypothesis: we divide the initial genus G into a and b, and then decide that of the two a,b, the suitable higher hypothesis from which G follows is b, and so on, in quite the same way that geometrical analysis, such as the one by Pappus in the *Collection* 634,11–636,14, is described as a process proceeding from hypotheses to higher hypotheses.

The very abbreviated form of Division and Collection is precisely the Analysis in the direction of Division, starting with G, G follows from b, b follows from b_1, b_1 follows from $b_2, \ldots, b_{n-1}$ follows from b_n, ending into something known b_n exactly when the Logos criterion, say $a_n/b_n = a_1/b_1$, and Collection is achieved; and also the Synthesis and Deduction, in the direction opposite to Division, namely

$$b_n \longrightarrow b_{n-1} \longrightarrow \cdots \longrightarrow b_2 \longrightarrow b_1 \longrightarrow G.$$ [9]

Thus an advantage of the abbreviated form of the Division is to connect it with Analysis and Synthesis and to make reasonable the use of the terminology "hypotheses" for the parts-genera of the Division. This will be crucial for understanding the further description of Division and Collection as hypotheses-"anhupotheton" in the *Politeia*, examined in Sect. 1.6.

[**Section 1.6**. The Ascent from hypotheses to the anhupotheton is another description of Division and Collection.] According to the *Politeia* 510b6-9, 511b3-c2, 533c7-d3 passages, the ascent, a further terminology of analysis origin, from hypotheses to the "anhupotheton," the hypothesis-free, coincides with Division and Collection.

[8] *In Euclidem* 211,18–22.

[9] More details can be found in Negrepontis and Lamprinidis [26].

In order to achieve this identification, arguments are presented in favor of interpreting the phrase “tas hupotheseis anairousa” in 533c8 as “dividing the hypotheses” rather than some sort of “destroying the hypotheses”. Division of hypotheses makes sense in view of our remarks in Sect. 1.5, and is explicitly mentioned in the *Phaedo* 107b5-9. Thus the ascent to higher hypotheses is achieved by the division of hypotheses.

The aim of Division of hypotheses is to reach the “anhupotheton”. This unusual terminology can probably be understood as follows: hypothesis is identified with a part-genus of the infinite anthyphairesis of the initial dyad, and as such a hypothesis is divisible; the anhupotheton, the hypothesis-free is the division-free entity, the indivisible, that is obtained by the Logos and Collection. Thus “anhupotheton” has the same force as “indivisible”, precisely as the indivisible line, identified with the One of the second hypothesis in the *Parmenides*, as we will see in the next Sect. 1.7.

Thus hypotheses are related to Division, anhupotheton to Collection and Logos, and we have just another description of Division and Collection.

Summarising Sects. 1.3–1.6, Plato’s paramount concern in his philosophy is knowledge (“episteme”); just like a periodic anthyphairesis becomes fully known when the Logos Criterion, signifying the completion of an anthyphairetic period, a Platonic Being becomes fully known, by the method of Division and Collection by means of Logos, in close imitation of the geometric method. Thus Plato’s theory of Ideas, of true Beings is built in close imitation of the mathematics of periodic anthyphairesis, and full knowledge of true Beings, is achieved by the method of Division and Collection-Logos.

[**Sections 1.7–1.9**. Criticism I on the non-foundational role of the geometric point; Criticism II on the mistaken use of the hypotheses on numbers and figures.] Plato is not at all satisfied with the state of knowledge of the whole of Geometry, as it has developed by its practitioners, the geometers. The gist of his Criticism is against the axiomatisation of Geometry and Arithmetic for epistemological reasons. What geometers at the time of Plato regarded as the foundation of Geometry cannot be very different from the foundation presented in the *Elements*. The axiomatised foundation of Geometry, presented in Book I of the *Elements*, can be summarized in the basic Definitions of the geometric point, the straight line (segment), the circle, the three kind of angle, and the five Postulates, while the only foundation of Arithmetic, presented in Book VII of the *Elements*, consists in the Definitions of unit and number. Plato’s main concern was that the axiomatisation, consisting of the basic definitions and postulates, all accepted and assumed as true by the geometers, who build all their consequences in the form of propositions, in no way provides knowledge (“episteme”), since the hypotheses, on which everything else depends, are arbitrary and unknown.[10]

[10]Aristotle in *Analytics*, 72b5-73a5 mentions some unnamed, avoiding to name Plato explicitly, who have such an objection to axiomatisation (“Ενίοις μὲν οὖν διὰ τὸ δεῖν τὰ πρῶτα ἐπίστασθαι οὐ δοκεῖ ἐπιστήμη εἶναι”). Aristotle rejects this objection, because it would lead, by introducing higher and higher axioms, to a regress to infinity, something impossible (“ἀδύνατον γὰρ τὰ ἄπειρα διελθεῖν’). Cf. also “ἐὶ γὰρ εἰσὶν ἀρχαί, οὔτε πάντ’ ἀποδεικτὰ οὔτ’ εἰς ἄπειρον οἷόν τε βαδίζειν”

How can this be corrected? As we have already mentioned, the only method for the acquisition of knowledge that Plato regards as valid, Division and Collection, is an imitation of results in a small but important part of Geometry, the proofs of quadratic incommensurabilities developed by the Pythagoreans, Theodorus and Theaetetus. But what can be done for obtaining true knowledge of the Geometry as a whole? Plato's daring hypothesis is that the method of Division and Collection must somehow be applied to all of Geometry. But how can this be done? The argument roughly runs as follows:

Plato takes, in the *Politeia* 511c-d, as hypotheses (basic definitions and Postulates for Geometry and Arithmetic) the straight line and the circle, the three kinds of angle, and the numbers.

[**Section 1.7**. Criticism I: The geometric point, employed by the geometers, for the foundation of Geometry should be rejected, in favor of the indivisible line.] We cannot fail at this point to note the notable absence, among Plato's hypotheses, of the geometric point. According to Aristotle, in the *Metaphysics* 992a19-22, Plato, most surprisingly, rejects altogether the most emblematic concept in Geometry, that of the geometric point, from having any role in a true foundation of Geometry. Plato's rejection of the geometric point, going back to Zeno, essentially occurs in the *Parmenides*. The point, essentially the partless One of the first hypothesis in the *Parmenides* (137c-142a), is a rival One, failing to be a Platonic Being (141e4-142a2), to the One of the second hypothesis in the *Parmenides* (142b-155e), the paradigmatical true Platonic Being (155d3-5). Thus Plato's Geometry is without points, and is generated in a Platonic Being.

[**Section 1.8**. Criticism II: The hypotheses, on numbers in Arithmetic, and on lines, circles, and the three kinds of angle in Geometry that are taken as principles, should be replaced by dialectical hypotheses that are steps towards the anhupotheton.] The reason for Plato's difficulty with Geometry is epistemological, has to do with his concept of true knowledge. Plato, in 510c2-d3, criticizes the geometers/mathematicians, for employing the axiomatic method, a method that takes the hypotheses (in particular the definitions on numbers in Arithmetic and the definitions and postulates on lines circles, and the three kinds of angle in Geometry) as principles and is thus defective in knowledge (since these are arbitrarily accepted and not really known), and can only move downward in deductive fashion (so that all further "knowledge" depends on the arbitrary initial hypotheses and is thus problematic). But Plato goes much further, not simply criticizing but introducing a daring proposal for the acquisition of true knowledge in Mathematics; he

(a) isolates a small but vital part of Geometry, in which the acquisition of full and complete knowledge is possible, namely the Pythagorean and Theaetetean method of proving quadratic incommensurabilities, by establishing the Logos criterion and anthyphairetic periodicity (cf. Sect. 1.2);

84a32-33; "ἔτι αἱ ἀρχαὶ τῶν ἀποδείξεων ὁρισμοί, ὧν ὅτι οὐκ ἔσονται ἀποδείξεις δέδεικται πρότερον—ἢ ἔσονται αἱ ἀρχαὶ ἀποδεικταὶ καὶ τῶν ἀρχῶν ἀρχαί, καὶ τοῦτ' εἰς ἄπειρον βαδιεῖται, ἢ τὰ πρῶτα ὁρισμοὶ ἔσονται ἀναπόδεικτοι." 90b24-27.

(b) develops a philosophical theory of knowledge of the true Beings, the Platonic Ideas, the method of Division and Collection, in close imitation to the complete acquisition of knowledge provided for pairs of lines in quadratic incommensurability, by the Logos criterion and periodic anthyphairesis (cf. Sects. 1.3–1.6); and

(c) finally makes the extraordinary claim, expressed in the *Politeia* passages 511c3-d5 and 533b6-c6, that this method of Division and Collection must be the method employed for the acquisition of knowledge for all of Geometry and Mathematics. He thus sets in opposition the geometric hypothesis with the dialectical-philosophic hypothesis, and criticizes the geometers for failing to take as hypothesis a part-genus in the Division of a Platonic being, and thus to have an opposite part-genus, forming with it an anthyphairetic dyad, thereby being able to divide and generate higher hypotheses, and eventually receive Logos and periodicity, achieve Collection (by the self-similarity of periodicity) and full knowledge.

[**Section 1.9**. Criticism II, continued. The geometers are compelled to use hypotheses as principles because they need hypotheses to proceed to actions/geometric constructions.] The compulsion that geometers have to use hypotheses as principles is mentioned hand in hand with the role that the sensible diagrams play in Geometry in the 510b4-6, 511a3-b2, 511c6-8 passages, and this creates the impression that the use of these diagrams is the cause, the compulsion of the geometers to the wrong use of hypotheses. But in the final 527a1-b2 passage the compulsion that geometers have is explained by means of their need for intelligible/dianoia constructions in their deductive arguments (in this crucial passage visible diagrams are not mentioned). The following chain of causes and effects is then formed according to 527a1-b2: The geometers view geometry as a deductive science with proofs; proofs form their arguments with actions/geometric constructions; obtaining actions/geometric constructions compels the geometers to employ hypotheses as principles, namely compels the geometers to a "dianoia way" of forming their arguments. Thus the reason the geometers are compelled to proceed downwards, treating hypotheses as principles, and not as steps of ascent, lies with their thesis that Geometry is a deductive science, it is about proofs of propositions. A proof requires geometric constructions/actions, and actions depend on fixed geometric hypotheses, Definitions and Postulates, considered as principles. Thus the reason for the compulsion of the geometers to use hypotheses in a defective way, as principles, is purely intelligible and has nothing to do with the fact that the geometers employ visible/sensible diagrams (examined in Sects. 1.13 and 1.14). Roughly speaking, the geometers have adopted the axiomatic method because they think of Geometry as a deductive science, a science seeking proofs, deductions, for deductions and proofs constructions are needed, and constructions are themselves postulates (as Postulates I, II, III, V) or need postulates and definitions.

[**Section 1.10**. Praise of Geometry. Intelligible Foundation of Geometry. Straight lines are generated dialectically by Division, and circles by Collection.]

[**Section 1.11** Criticism II on the faulty use of the hypothesis of the three kinds of angle, and praise of the dialectic generation of angle, explained by Proclus.] Plato never explains what would be the dialectical treatment of the three kinds of angle, but Proclus does, in his illuminating comments in *In Euclidem* 131,3-134,7 and 188,20-189,12, referring specifically to the *Politeia* passage 510c2-d3. The true Being in which the intelligible three kinds of angle are generated is, not the general Platonic intelligible Being but, rather the "diameter itself to the side itself of a square", a true Being, since it is a dyad in periodic anthyphairesis. It appears that these three kinds of angle were understood dialectically by the Pythagoreans, earlier, before Plato's general theory, and explained in terms of the anthyphairetic approximants, the so called side and diameter numbers, in full agreement with the general approach. A pivotal role is played by the side and diameter numbers, their Pell equation

$$y^2 = 2x^2 \pm 1,$$

and Propositions II.12 and 13 of the *Elements*.
[**Section 1.12**. Criticism II+ for the axiomatic status of the Fifth Postulate is expressed by Proclus, who attempts unsuccessfully to derive it from a principle of finiteness, reminiscent of Eudoxus' principle, definition V.4 of the *Elements*.]
[**Sections 1.13 and 1.14**. Sensibles according to *Timaeus* 48a-58c, provoking and non-provoking perceptions in the *Politeia* 522e-525a, Plato's Criticism III on the use of geometric diagrams by the geometers].
[**Section 1.13**. Sensibles according to the *Timaeus* 48a-58c.] Plato's theory of the sensibles and their participation in the intelligibles, is presented mainly in the *Timaeus*, and uses the, ill-understood by students of Plato, receptacle/hollow space (48a-58c). This section is of independent interest, because we aim at a radically new interpretation of the receptacle, in terms of anthyphairesis.[11] In modern terms, anthyphairesis of a dyad of two line segments has its modern counterpart in the continued fraction of a real number, and a sensible, participating in an intelligible by essentially being an anthyphairetic approximation of it, corresponds to the modern convergent of that real number.

In the *Timaeus* theory, Plato, in order to express the participation of the sensibles to the **infinite** Division of the intelligible and at the same time to preserve his version of the traditional Greek physiology in terms of **only four** elements (earth, fire/air/water), a task that at first sight seems impossible, resorts to the, essentially mathematical, strategy of the receptacle/hollow space (the two identified as a result of our study, but we mostly refer to the hollow space). The idea is to express an anthyphairetic relation, say

$$a = kb + c, \qquad c < b,$$

[11]Negrepontis [21] and Negrepontis and Kalisperi [25].

equivalently, by the tight inequality $kb < a < (k+1)b$. In fact all steps of the anthyphairesis of a to b may equivalently be expressed by alternating tight anthyphairesis inequalities involving only the original magnitudes a and b, of the form

$$(T_{2n} - 1)a < S_{2n}b < T_{2n}a, \text{ or } S_{2n-1}b < T_{2n-1}a < (S_{2n-1} + 1)b,$$

where S_{2n-1}, S_{2n}, T_{2n-1}, T_{2n} are suitable natural numbers, tight in the sense that either a multiple of a or b is sandwiched between a multiple of b or a, respectively, and of its immediate successor, thus making the b's look like the a's or conversely. [**Section 1.14**. Plato's Criticism III on the use of geometric diagrams by the geometers 510d5-511a2, and the distinction in perceptions provoking and non-provoking to the mind 522e-525a.] The geometers employment of visible geometric diagrams for their study of Geometry, itself an intelligible activity (even if relegated to the lower kind of "dianoia"), might be thought the reason for Plato's criticism (Criticism III, as we call it) in 510b4-6, 511a3-b2, 511c6-8. But in fact, Plato is careful to distinguish between two kinds of perceptions, those provoking and those non-provoking to the mind (522e-525a), a distinction that cannot be really understood without obtaining first a full description of the sensibles as presented in the *Timaeus* 48a-58c (and studied in Sect. 1.13). But in fact, as far as I know, the provoking/non-provoking distinction of the perceptions has not been correlated before in the study of the role of the visible geometric diagrams.

The sensible according to the *Timaeus*, as analysed in Sect. 1.13, is a "confounded" dyad, say water with earth, in which one element of the dyad is the dominating and the other the dominated, assuming a form similar to the dominating; and the perceptions that are provoking to the mind are those that make clear this structure of a sensible, while the non-provoking are those that obscure this structure. It is clear that the visible geometric diagrams are non-provocative to the mind; the geometric representation of a line segment, is like the finger in 523c-524e, not as a confounded dyad, but as something isolated without suggesting its opposite.

This might give the impression that these diagrams, because of being non-provoking, are guilty of compelling the geometers to the use of hypotheses as principles. But the source of this compulsion has already been identified (in Sect. 1.9), mostly with the help of the passage 527a-b, with the need of the geometers to obtain actions/constructions in their proofs of propositions, and it is of a purely intelligible source and origin. Thus the geometric diagrams have nothing to do with the compulsion of the geometers to use hypotheses as principles.

The reason why the geometers employ visible geometric diagrams that are non-provoking to the mind is understood mostly from the passage 510d5-511a2. The geometers form their arguments **for the sake** of dianoia, namely for the sake of actions/constructions, but the arguments are **about** the visible diagrams. Thus the perception of these diagrams must be compatible, auxiliary to the dianoia reasoning of the geometers. As auxiliary to dianoia the visible diagrams cannot be perceptions provoking to the mind, because if they were, they would lead to the use of hypotheses as steps upwards, something that would go against the "dianoia" way of thinking to which the geometers are committed. Thus the diagrams used by "dianoia" thinking geometers, must definitely be non-provoking, perceptions (522e-525a)—as they are indeed.

In fact the way non-provoking perceptions are described as defective, namely as a perception in which the opposite perception is not suggested, in 522e-525a, is analogous to the way the hypotheses of the geometers are described as defective.

[**Section 1.15**. Criticism IV on the non-anthyphairetic, Dedekind-like theory of ratios of magnitudes by Eudoxus, the non-anthyphairetic proofs of quadratic and cubic incommensurabilities by Archytas, the duplication of the cube by Archytas.] There is a further criticism of the manner geometers study solid Geometry and in particular the duplication of the cube, in the *Politeia* 527d-528e, that fits well in the general framework of his criticism of (plane) Geometry. Here the geometers involved although not named are certainly Archytas and Eudoxus. Archytas succeeded in discovering elementary non-anthyphairetic proofs of the quadratic incommensurabilities, proved by Theaetetus by palindromically periodic anthyphairesis, a success certain to have had Plato's disapproval. Archytas' success prompted Eudoxus to obtain the theory of ratios of magnitudes, presented by Euclid in Book V of the *Elements*, in terms of Dedekind cuts, as we realize today, distancing himself from the anthyphairetic theory developed by Theaetetus and reported by Aristotle in *Topics* 158b-159a, a theory that certainly had Plato's strong disapproval. In turn Archytas used Eudoxus' theory to construct, by the Hippocratean method of the two means, a stereometric solution to the duplication of the cube. Plato, most probably under the false impression, that the anthyphairesis of a to b, in case $a^3/b^3 = 2/1$, is periodic, criticized Archytas for not finding a solution that would reveal this imagined periodicity.

1.2 Periodic Anthyphairesis

1.2.1 The Concept of Anthyphairesis

Let a, b be two natural numbers or magnitudes (e.g. line segments or volumes), with $a > b$.We measure the great a with the small b, and we find how many small b's are needed to cover the large a. There are then a natural number k_1, the quotient, and c_1, the remainder, such that

$$a = k_1 b + c_1, \text{ with } c_1 < b.$$

This is the first step of the anhyphairesis of a to b. If c_1 is zero, this is the end of the process. Otherwise a new pair of great and small is formed, namely the pair $b > c_1$.

Notice that the role of b is reversed: in the first step b is small, in the second step b is great. We repeat the process. There are a natural number k_2 and c_2, such that

$$b = k_2 c_1 + c_2, \text{ with } c_2 < c_1.$$

We continue in this way:

$$c_1 = k_3 c_2 + c_3, \text{ with } c_3 < c_2,$$

$$\dots$$

$$c_n = k_{n+2} c_{n+1} + c_{n+2}, \text{ with } c_{n+2} < c_{n+1}$$

$$\dots$$

Anthypharesis is the precursor of modern continued fractions.

1.2.2 Finite and Infinite Anthyphairesis

- If this process ends at some point (namely for some index n, c_n is zero, hence $c_{n-2} = k_n c_{n-1}$), then the anthyphairesis of a to b is finite. This is always the case if a, b are natural numbers, and the last non-zero remainder c_{n-1} is the greatest common measure-divisor of a and b (Propositions VII.1 and 2 of the *Elements*).
- If this process does not end, then the anthyphairesis of a to b is infinite. This can happen only if a and b are magnitudes. It is then proved that a and b do not have any common measure, that a and b are incommensurable:

Proposition (Proposition X.2 of the *Elements*) *If a, b are magnitudes, $a > b$, and the anthyphairesis of a to b is infinite, then a and b are incommensurable.*

Note that in this case there is an infinite, strictly decreasing sequence of remainders:

$$a > b > c_1 > c_2 > c_3 > \dots > c_n > c_{n+1} > \dots.$$

1.2.3 Periodic Anthyphairesis

The most important case of infinite anthyphairesis is the periodic one. An anthyphairesis is periodic if the sequence of quotients

$$k_1, k_2, k_3, \dots, k_n, k_{n+1}, \dots$$

of natural numbers is a periodic sequence of numbers.

An anthyphairesis, not necessarily infinite *a priori*, is recognized as periodic according to the following

'Logos' criterion *If there is an index n such that $a/b = c_n/c_{n+1}$ then the anthyphairesis of a to b is periodic.*

It then follows that

$$b/c_1 = c_{n+1}/c_{n+2},$$
$$c_1/c_2 = c_{n+2}/c_{n+3},$$
$$\dots,$$

and hence that the ratios of successive parts-remainders

$$a/b, b/c_1, c_1/c_2, c_2/c_3, \dots, c_n/c_{n+1}, \dots.$$

is a periodic sequence of ratios.

1.2.4 Pythagorean & Theaetetean Incommensurabilities; Theaetetus' Anthyphairetic Theory of Ratios of Lines Commensurable in Square

The great discovery by the Pythagoreans that made this theory possible was the incommensurability of the diameter to the side of a square. Although the method employed by the Pythagoreans is a contested topic in the History of Mathematics, there is strong evidence that the Pythagoreans proved the incommensurability by computing the anthyphairesis of the diameter to the side of a square, finding that its sequence of quotients is $[1, 2, 2, 2, \dots]$, thus infinite, and employing Proposition X.2 of the *Elements*.

Proposition (Pythagoreans) *If a, b are line segments such that b is the side of a square and a is the diameter (or diagonal, as we now call it) of the square, then*

(a) $a^2 = 2b^2$ (by the Pythagorean theorem),
(b) $a = b + c_1, b = 2c_1 + c_2$, and $b/c_1 = c_1/c_2$ (the Logos Criterion), and
(c) the anthyphairesis of the diameter a to the side b of the square is $Anth(a, b) = [1, 2, 2, 2, \dots]$ (infinite anthyphairesis),

hence (by Proposition X.2 of the Elements*) a, b are incommensurable.*

In favor of an anthyphairetic reconstruction we can mention

(a) the knowledge by the Pythagoreans of the side and diameter numbers, the precursor of the convergents of square root of 2,
(b) the stories about Hippasus that involve his drowning into the sea, confused but clearly symbolizing, according to Pappus, the infinite nature of incommensurability, and
(c) the adoption of infinite and finite as the two basic principles of their philosophy; the infinite was essentially the infinity of the anthyphairesis, and the finite the preservation of the Gnomons, by which at a finite stage complete knowledge of (the anthyphairesis of) the diameter to the side is achieved.

(d) the fact that Zeno's arguments and paradoxes are based on periodic anthyphairesis and relate to the incommensurability of the diameter to the side of a square.

1.2.5 Theaetetean Incommensurabilities and Anthyphairesis

Later, according to the *Theaetetus* 147d-148b, Theodorus proved that if $a^2 = Nb^2$, N a non-square number, and $3 \leq N \leq 17$, then the anthyphairesis of a to b is periodic; and, Theaetetus proved that if $a^2 = Nb^2$, and N is a non-square number, then the anthyphairesis of a to b is palindromically periodic, and thus by Preposition X.2 of the *Elements*, a and b are incommensurable. That the method is indeed anthyphairetic has been established in Negrepontis [23], by analyzing the philosophical imitation by Plato of Theaetetus' mathematical discoveries in the dialogues *Sophistes* and *Politicus*.

These discoveries greatly increased the class of incommensurable ratios and prompted the need for a theory of ratios of magnitudes. As Aristotle informs us in *Topics* 158b-159a, such a theory, based on the definition that

$$a/b = c/d \text{ iff } \mathrm{Anth}(a, b) = \mathrm{Anth}(c, d),$$

was developed; and there should be no doubt that the creator of the theory was Theaetetus. But what is not so well understood (cf. Knorr [14], Acerbi [2]), is that Theaetetus' theory was concerned not with general ratios but only with ratios whose squares are commensurable, or in fact rational ratios with respect to a given line (in the sense of Definition X.3 of the *Elements*). This can be seen from

(a) the fact that the anthyphairetic theory of ratios of magnitudes defined for all magnitudes needs Eudoxus condition, but the anthyphairetic theory of ratios of magnitudes defined for ratios commensurable in power only does not need Eudoxus condition, because of the anthyphairetic periodicity of these ratios, precisely by Theaetetus theorem on incommensurabilities,
(b) the fact that in the Theaetetean Book X of the Elements an "alogos" line (literally, a line without ratio) is defined as one incommensurable in square with respect to an assumed line (cf. Sect. 1.4.5), while, as can be proved, only quadratic ratios are needed in Book X, and
(c) the *Scholion In Euclidem* X.2, which sharply criticizes Eudoxus' general theory and differentiates it from the earlier theory of ratios only for quadratic ratios.

A forthcoming publication by Negrepontis and Protopapas [27] will deal with this subject.

Note. Since the Pythagoreans did not have a theory of ratios of magnitudes, the Pythagorean formulation of this definition would be in terms **not** of ratios, **but**, equivalently, in terms of square gnomons (appearing in Book II of the *Elements*). By definition, the dyad diameter a, side b of a square is a true Platonic being, since the condition $b/c_1 = c_1/c_2$ is the **Logos criterion** for anthyphairetic periodicity.

The modern concept, essentially equivalent to the ancient anthyphairesis, is the continued fraction of a (positive) real number. The theory was developed during sixteenth, seventeenth (Fermat, Wallis), eighteenth (Euler), and nineteenth century (Lagrange, Galois, Gauss). Fowler [10], in his book on *the Mathematics in Plato's Academy*, brings out the connection between modern continued fractions and ancient anthyphairesis.

1.3 The One of the Second Hypothesis in the *Parmenides*

The One of the second hypothesis in the *Parmenides* consists of a dyad, the One and the Being, in (a philosophic analogue of) periodic anthyphairesis. The dialectic meaning of Logos is the Logos criterion, the cause of periodic anthyphairesis.

According to the paper [24] of the present author, it is shown in the *Parmenides* 142d9-144e3 that the One of the second hypothesis in the *Parmenides* consists of two parts, the One and the Being, in a philosophic analogue of periodic anthyphairesis. We outline in this Section the results of [24].

1.3.1 The Infinite Anthyphairesis of the Dyad One and Being

In the first part, 142d9-143a3, it is shown that the dyad One Being is in infinite anthyphairetic division:

$$One = Being + One_1, Being > One_1,$$
$$Being = One_1 + Being_1, One_1 > Being_1,$$
$$\dots$$
$$One_n = Being_n + One_{n+1}, Being_n > One_{n+1},$$
$$Being_n = One_{n+1} + Being_{n+1}, One_{n+1} > Being_{n+1}$$
$$\dots$$

Thus there is then an infinite multitude of remainders-parts of the anthyphairetic division:

$$One > Being > One_1 > Being_1 > \dots > One_n > Being_n > \dots$$

1.3.2 The Logos Criterion and Anthyphairetic Periodicity

In the sequel, the passage 144c2-d4 is concerned with the presence of the One in the Being, setting the dilemma whether the presence of the One in each part of the

Being be the presence of a part of the One or of the whole One. It has similarities to the passage 130e4-131e7 in the Introduction of the *Parmenides*, posing the question (the dilemma of participation) whether the presence of the One, be as part or as whole, in each of the sensible that participates in the One.

The presence as "whole" is rejected in both 144c-d and 130e-131e, but the presence as "part" only in the 130e-131e. Because of this crucial difference, we are led to conclude that the "part" leg of the dilemma must have been found acceptable somewhere in the dialogue in between. And indeed the place discovered is 138a3-7, where it is stated that the presence in question is realized by a circle of "touchings" ("hapseis"). A "touching", as explained in the passage 148d5-149d7, occurs between parts of the dyad One, Being, successive in their order of generation, and touchings generate numbers according to the formula

$$\text{number of parts-units} = \text{touchings} + 1,$$

whose origin is clearly musical with terms-chords and musical intervals in place of parts-units and touchings, respectively. This correspondence suggests the interpretation that a "touching" in 138a3-7 is meant to be a ratio of consecutive parts. In view of the earlier statement that the division of the One to Being is anthyphairetic, this interpretation is tantamount to the statement that the anthyphairesis of One to Being satisfies the "logos" criterion and thus, in analogy to the mathematics of Sect. 1.2, finally the dyad One Being satisfies a philosophic analogue of periodic anthyphairesis.

The interpretation is strengthened by the realization that the *Philebus* 15a1-c3 sets a whole-part dilemma precisely as in the *Parmenides* 144c2-d4, and that this is followed by *Philebus* 15d4-5, a veritable replica of 138a3-7, with "logoi"-ratios in place of touchings.

1.3.3 Equalisation of Parts, in Consequence of Periodicity

The equalization of the One with the Being now proceeds smoothly, in the passage 144d4-e3 as follows:

$$\text{the number of parts of the One} = \text{logoi between successive parts (from One on)} + 1,$$

a finite number, since the "logoi" will be different till a complete circle-period of ratios is achieved, and there will be no other ratios. The periodicity then implies that

$$\text{the number of parts of the One} = \text{the number of parts of the Being},$$

on which equality the equalisation of One and Being rests.

1.3.4 Introduction of True Number, in Consequence of Equalisation

The equalisation of One and Being allows for the definition of a true number Two, namely a Two with equal(ised) units. The premature explanation given in 143c1-d5 in terms of logos, now becomes fully understandable. This definition of the Platonic number Two makes fully understandable the related comments by Aristotle in the *Metaphysics* (1081a14-15, 1081a23-25, 1083b30-32, 1091a24-29).

Note also that the highest number generated in the One is the number of ratios forming a complete period+1; no greater number can be formed. Thus a near paradox occurs in the One of the second hypothesis in the *Parmenides*: the parts of the One are infinite in multitude, while the parts of the One are finite in number. We will come back to this feature of the One in Sect. 1.4.

1.3.5 The Self-Similar Property of One and Many

The two basic features of the One of the second hypothesis are summarized by the statement that the One of the second hypothesis is both **One and Many**, in the sense that

- the One is divided anthyphairetically into Many, in fact an infinite multitude of, parts, and
- each part of the Many is equalized by the periodicity of the anthyphairesis, and thus the One of the second hypothesis is indeed a One in the sense of self-similarity, namely that each part is the same as the whole (**144e8-145a2**).

1.3.6 The One of the Second Hypothesis in the Parmenides Is the Paradigmatical True Intelligible Being

The One of the second hypothesis is declared, in 155d3-6, to be a true intelligible Being, a Platonic Idea. The One of the second hypothesis, although infinite in structure is nevertheless fully knowable, possesses "episteme" 155d6, since full knowledge of the One is achieved when a full period of the Logoi is completed.

1.3.7 The True Platonic Being Is Knowable by Name and Logos, Equivalently by Division and Collection

Plato refers in 155d8-e1 to the One and Many by "Name and Logos": by Name the Division into Many (names) is meant, and by Logos, the logos criterion for the periodicity in anthyphairesis is meant. Plato, by the expression "Name and Logos", refers, in the *Theaetetus* 201d8-202c5 and the *Sophistes* 218b5-c5, to the method of Division and Collection. Thus the One satisfies "Name and Logos", and becomes knowable, acquires "episteme", by the method of, Division and Collection.

1.3.8 *Properties of the One of the Second Hypothesis of the* Parmenides*: It Has Beginning, Middle and End; the Straight Line and the Circle Are Generated in It; It Is in Motion and at Rest*

Some consequences of One and Many are the following:

- The One has Beginning, Middle, End (145a4-b1), in the sense that the period of the One has Beginning, Middle, and End.
- The two figures, the straight line and the circle, defined in 137e1-4, are generated in the One (145b1-5).
  The straight is generated in the One by the infinite division, the circle by the anthyphairetic periodicity. The Platonic definition of the straight line and the generation of the two figures in the One will be explained later, in Sect. 1.9.
- The One is in the other ("en alloi"), and thus in infinite motion ("aei kineisthai"), and is in itself ("en heautoi"), and thus always at rest ("hestanai") (145e7-146a8). The One is in infinite motion by the Division, and is always at rest, because of self-similarity. The motion and rest of the One will play an important role later (Sects. 1.5 and 1.10.4).

1.4 The Dialectics of the *Politeia* Is Division and Logos-Collection; True Being in the *Politeia* Coincides with the One of the Second Hypothesis in the *Parmenides*

We present here the considerable evidence that the dialectics in the *Politeia* is Division and Collection, exactly as in the second hypothesis of the *Parmenides*. This step is necessary, since we intend to apply our interpretation of the *Parmenides* Division and Collection as periodic anthyphairesis, outlined in Sect. 1.3, to the *Politeia*.[12]

[12]Most Platonists fail to associate the *Politeia* dialectics with Division and Collection. Stenzel [31] had rejected the association, while Cornford [8] and Hare [12] do associate *Politeia* dialectics with Division and Collection; but again do not have an anthyphairetic interpretation of Division and Collection. Furthermore the anthyphairetic interpretation of Logos in the *Parmenides*, *Sophistes*-

1.4.1 Division According to Kinds vs. Division According to Name in the Politeia 454a4-9

> owing to their inability to study the subject under consideration by dividing according to kinds ("kat' eide diairoumenoi"), but pursueing the opposition to each other by dividing according to name itself ("kat' auto to onoma"), in eristic, not dialectic manner (*Politeia* 454a4-9)

The dialectic method is described in *Politeia* 454a4-9 by "Division according to kinds" ("kat' eide diairoumenoi"), set in opposition to the eristic method, which consists in proceeding with division according to name [only] ("kat'auto to onoma"). Division according to kinds clearly refers to the method of Division and Collection; in fact, in the *Sophistes* 252b1-6 and 253d1-e3, *Politicus* 285a7-b6, Division and Collection is repeatedly described as Division according to kinds (cf. Sect. 1.5). On the other hand, as we have seen, in the *Theaetetus* 201d8-202c5 and the *Sophistes* 218b5-c5, Division and Collection is equivalently described as "not only Name, but Logos as well". And in the *Philebus* 16c5-17a5, the dialectic method of Division and Collection is opposed to the eristic method of Division only.

Thus in the present *Politeia* passage, dialectic Division and Collection is opposed to eristic Division only.

1.4.2 Division and Logos in the Politeia

1.4.2.1 Dialectics as Knowledge of True Beings by Division into Kinds and Logos in the *Politeia* 531d9-e6, 532a5-b3 and d8-e3

> For you surely do not suppose that experts in these matters are dialecticians ('dialektikoi')? 'No, by Zeus,' he said, 'except a very few whom I have met.' "But have you ever supposed,' I said, 'that men who could not give and receive logos ("dounai te kai apodexasthai logon", 531e4-5) would ever know ("eisesthai") anything of the things we say must be known ("hon... dein eidenai", 531e5)?' 'No' is surely the answer to that too. (531d9-e6)

> In like manner, when anyone by dialectics ("dialegesthai") attempts by means of logos ("dia tou logou", 532a6-7) and apart from all perceptions of sense to find his way ("horman") to the true essence of each thing and does not desist till ("me apostei prin"[13]) he apprehends by thought itself the nature of the good in itself, he arrives at the end ("telei") of the intelligible. (532a5-b3)
> Tell me, then, what is the nature of this faculty of dialectic ("tou dialegesthai") ? Into what kinds has it been divided ("kata poia eide diesteken")? And what are its roads ("hodoi")? For it is these, it seems, that would bring us to the place where we may, so to speak, rest on the road ("hodou") and then come to the end ("telos") of our journeying ("poreias"). (532d8-e3).

Politicus and *Politeia* passages has not, to the best of my knowledge, been considered by any modern student of Plato.

[13]Cf. "he must not desist till ('me proaphistasthai prin', 285b1-2)" in the description of Division and Collection in the *Politicus* 285a4-b6.

From these two passages it follows that the knowledge of the intelligible Beings is achieved by Division into Kinds and Logos.

1.4.2.2 Division and Logos in the *Politeia* 534b3-d2

And do you not also give the name dialectician to the man who is able to receive the essential logos ("ton logon lambanonta", 534b3) of each true Being? And will you not say that the one who is unable to do this, in so far as he is incapable of rendering logos ("logon didona", 534b5) to himself and others, does not possess intelligence about the matter?

"How could I say that he does?" he replied. "And is not this true of the Good likewise[14]– that the man who is unable to define by means of logos ("diorisasthai toi logoi", 534b9) and subtracting from all other things ("apo ton alon panton aphelion") the idea of the Good, and who cannot, as it were in battle ("en machei"), running the gauntlet of all tests ("dia panton elegchon diexion"), and striving to test everything ("prothumoumenos elegchein") by essential reality and not by opinion, hold on his way through all this unwaveringly ("en pasi toutois aptoti diaporeuetai") by means of logos ("toi logoi", 534c3)—the man who lacks this power, you will say, does not really know ("eidenai", 534c4) the good itself or any particular good" (534b3-d2)

Thus knowledge of every idea, and of the idea of the good in particular, is obtained by subtracting the idea from all other things as if in battle, and holding on unwaveringly to logos.

Unmistakenly this is a description of obtaining knowledge by Division and Collection. The following table shows the great emphasis given in the passages shown (examined in Sect. 1.4.2) in the Logos and the resulting obtainment of knowledge of the true Being:

passage	Logos	knowledge
531d9-e6	531e4-5	531e5
532a5-b3	532a6-7	
533b6-c3		
534b3-d2	534b3, 534b5, 534b9, 534c3	534b9, 534c4

The passages should be compared with similar descriptions of Division and Collection in the *Sophistes* and the *Politicus*.

[14]In this passage the idea of the Good is, like any other true Being ("hosautos"), knowable by the method of Division and Collection. It is then difficult to reconcile its description as "epekeina ousias", beyond Being, in the *Politeia* 509b. Perhaps Plato regarded, apparently only in the *Politeia*, the Idea of the Good as the supreme true Being, in the senses of an intelligible analogue of the "apokatastatikos arithmos" for the visible universe, according to which there is a periodic restoration of the cosmic cycle (cf. Proclus, *Eis Politeian* 2, 15–19): when the (anthyphairetic) period of the Idea of the Good is completed, then the (anthyphairetic) period of every true Being will be completed as well, and there will be universal restoration.

let us divide in two ("schizontes dichei") the genus ("genos") we have taken up for discussion, and proceed always by way of the right-hand part of the thing divided ("meros tou tmethentos"), clinging close to the company to which the sophist belongs, until, having subtracted him of all the common kinds ("ta koina panta") and left him only his own ("oikeian") nature, we shall show him plainly first to ourselves and secondly to those who are most closely akin to the dialectic method. (*Sophistes* 264d10-265a1)
Where, then, shall we find the statesman's path? For we must find it, and, subtracting ("aphelontas") it from the others ("apo ton allon"), imprint upon it the seal of a single class (*Politicus* 258c1-3)
It is a very fine thing to separate ("diachorizein") the object of our search ("to zetoumenon") at once from the others ("apo ton allon"), if the separation can be made correctly (*Politicus* 262b2-3)
subtracting ("aphairountes") the Hellenic genus as one from all the others ("apo panton") (262d2-3)
cutting ("apotemnomenos") a myriad from all the others ("apo panton"), separating ("apochorizon") as one kind (262d7-e1)
We suspected a little while ago that although we might be outlining a sort of kingly shape we had not yet perfected an accurate portrait of the statesman, and could not do so until, by removing ("perielontes") those who crowd about him and contend with him for a share in his herdsmanship, we separated ("chorisantes") him from them ("ap' ekeinon") and made him stand forth alone and uncontaminated.(*Politicus* 268c5-10)
"A person might think that the definition of the art of weaving ("huphantiken") was adequate, not being able to realise that, although it has been separated ("apemeristhe") from many other kindred arts, nevertheless it has not yet been distinguished ("oupo dioristai") from the closely co–operative arts." *(Politicus* 280a8-b3)
"We shall certainly be undertaking a hard task in separating ("apochorizontes") this genus ("genos") from ("apo") the others ("ton allon") *(Politicus* 287d6-7)

1.4.3 Collection in One in the Politeia 531c9-d4, 537b8-c8

1.4.3.1 *Politeia* 531c9-d4

In 531c9-d4 it is stated that "if the investigation of all these studies goes far enough to bring out their community ("koinonia") and kinship ("suggeneia") with one another, and to infer their affinities ("oikeia"), then to busy ourselves with them contributes to our reaching the desired end, and the labor taken is not lost; but otherwise it is vain."

1.4.3.2 *Politeia* 537b8-c8

In 537b8-c8 it is stated that "they will be required to collect ("sunakteon") the studies which as children in their former education they pursued in a disordered manner ("cheden") into a comprehensive survey ("sunopsin") of their affinities ("oikeiotetos") with one another and with the nature of things."... "And it is also," said I, "the chief test of the dialectical nature and its opposite. For he who can view things comprehensively ("sunoptikos") is a dialectician; he who cannot, is not."

It is clear that both statements refer to the Collection of the Many into One. In the 537b8-c8 passage, "chuden" is contrasted with Collection; in the *Phaedrus* 264b3-c5, "chuden" is contrasted with the existence of middle and extremes—a consequence, as we saw (in Sect. 1.4), in the second hypothesis of the *Parmenides*, of Collection (*Parmenides* 145a8-b1). The Collection is described in the *Sophistes* 253d5-e7, *Politicus* 285a7-b6 in terms of "koinonein", in *Politicus* 285a7-b6 in terms of "oikeia", in *Politicus* 308c6, 311a1, *Phaedrus* 266b4 in terms of "sunagoge".

1.4.4 *Politeia 533c3-6: The True Being Has Beginning, Middle, and End*

The part 533c3-6 of the crucial passage 533b6-d3 (which will be discussed as a whole in Sect. 1.9) shows that the true Being of the *Politeia* has Beginning, Middle, and End, as the One of the second hypothesis in the *Parmenides* 145a8-b1.[15]

> "For where the beginning ("arche") is something that the geometer does not know, and the end ("teleute") and all that is in between ("metaxu") is interwoven ("sumpeplektai") is not truly known, what possibility is there that assent in such practice can ever become true knowledge ("episteme")?" "None," said he. (533c3-6)

As we saw in Sect. 1.3, this statement was a consequence of the Division and Collection, expressing the anthyphairetic periodicity. The passage connects the presence of Beginning, Middle, and End of the true Being, essentially a restatement of anthyphairetic periodicity, with knowledge of this Being.

1.4.5 *Geometric Irrational Lines vs. Dialectic Logos in 534d3-7*

> "But, surely," said I, "if you should ever nurture in fact your children whom you are now nurturing and educating in logos ("toi logoi"), you would not suffer them, I presume, to hold rule in the state, and determine the greatest matters, being themselves as irrational ("alogous") as the lines ("grammas"), so called in geometry." "Why, no," he said. (534d3-7)

This is, not just a silly linguistic quip, as some Platonists think but, a quite revealing passage that contrasts the philosophic 'logos' to the geometric "alogoi" lines. According to the definition, which is due to Theaetetus and appears in Book X of the *Elements*, a line a is "alogos" (literally, "ratio–less") with respect to a given ("protetheisa") line b if a^2 and b^2 are incommensurable. This terminology strongly suggests that the pre-Eudoxian, and almost certainly Theaetetean, theory of ratios, reported by Aristotle in *Topics* 158b–159a, was limited to rational lines only (cf. Sect. 1.2.5). In the dialogue *Theaetetus*, Theaetetus, upon hearing the lesson on quadratic incommesnsurabilities by Theodorus, had the idea and in fact succeeded in

[15] Cf. the comment in Sect. 1.8.3.2 about the *Phaedrus* 264b3-c5 passage.

proving the following quite substantial proposition: if two lines $a, b, a > b$, satisfy the condition

$$\langle a/b \text{ incommensurable, } a^2/b^2 \text{ commensurable}\rangle,$$

commensurability in power only for short, then the anthyphairesis of a to b is periodic, and in fact palindromically periodic (cf. Sect. 1.2.3). The comments of the anonymous *Scholion In Euclidem* X.2 are revealing.

In the subsequent two dialogues, *Sophistes* and *Politicus*, on the Platonic trilogy on Division and Collection and episteme (knowledge of true Beings) Plato is imitating Theaetetus' mathematical theory to dialectics. The imitation may be described briefly as follows: the philosophic analogue of a pair of lines a,b commensurable in power only is a Platonic Idea; the philosophic analogue of periodic anthyphairesis is the Platonic method of Division and Collection; and, the philosophic analogue of Theaetetus' proposition stating that a commensurable in power only pair of lines possesses periodic anthyphairesis is the fundamental statement in Platonic dialectics that a Platonic Idea possesses, and becomes knowable by Division and Collection. In fact the structure of the *Politicus*, coupled with the contents of Book X, leaves no doubt that Theaetetus had in fact proved not only periodicity but palindromic periodicity. We can now appreciate the contrast between philosophic dialectic logos and the geometric "alogos" lines. An "alogos" line is outside the realm of known anthyphairetic periodicity, and thus outside the geometry on which the Platonic method of Division and Collection is based, and is thus to be avoided, and not to be allowed that these "alogos" lines take control of the city ("ouk easai archontas en tei polei kurious ton megiston einai").

1.5 The Dialectic Meaning of "Hypothesis" in the Division and Collection of the *Sophistes* and *Politicus*. In the Abbreviated Form of the Division, the Parts-Genera Are Called "Hypotheses"

1.5.1 Division and Collection in the Sophistes and Politicus

The method of Division and Collection is the exclusive topic of the Platonic trilogy *Theaetetus-Sophistes-Politicus*. A study of these dialogues provides an independent confirmation that Division and Collection is a philosophic analogue of periodic anthyphairesis (cf. Negrepontis [22, 23], where older references by the author can be found). There is an interesting difference between the way a philosophic anthyphairesis is described in the *Parmenides* and in the *Sophistes-Politicus*. In the *Parmenides* 142d-143a all the steps of the anthyphairesis are given (say

$$B = a + b, a = b + a_1, b = a_1 + b_1, a_1 = b_1 + a_2, b_1 = a_2 + b_2, \text{ and so on,}$$

till the Logos criterion $a/b = a_k/b_k$ for some stage k) (cf. Sect. 1.3), but in the *Sophistes-Politicus*, the anthyphairesis appears in **abbreviated**, tree-like form (say

$$B = a + b, b = a_1 + b_1, b_1 = a_2 + b_2, \text{ and so on,}$$

till the Logos criterion $a/b = a_k/b_k$ for some stage k).

1.5.2 Example: The Division and Collection of the Platonic Being ⟨the Angler⟩ (Sophistes 218b-221c)

1.5.2.1 Division

In this Definition there are ten clearly described division steps (D1-D10); these are always binary divisions of a genus into two species opposite to each other, and in the immediately next step the one of these two species, which contains the final, under definition species of the art of Angling, is considered itself as the new genus and is further defined as before. These ten steps are summarized in a table as follows:

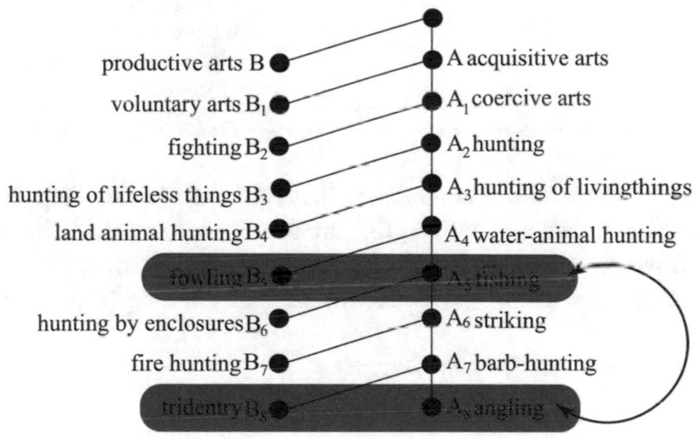

Table. Division and Collection for the Angler

1.5.2.2 Collection

Collection is achieved by Logos. In order to understand Logos, we must consider carefully Plato's comments for the Division steps D7 and D10.

These are his comments for step D10:

Plato's first comment for division step D10	
[220e3] that part which proceeds downward from above, because tridents are chiefly used in it, tridentry, I suppose	
	[221a2] proceeds in opposition from below upwards, being pulled up by twigs and rods

Plato's second comment for division step D10	
	[221b7] the part in which the blow is pulled from below upwards at an angle

It is clear, from the two comments of Plato for the division step D10, that the opposing species of tridentry and angling are described in a satisfactory way by the ratio from above downward to from below upwards. Schematically

tridentry/angling = from above downward barb-hunting/from below upwards barb-hunting.

Here are Plato's comment for step D7:

Plato's comment for division step D7	
	[221b5] that part which proceeds from below was fishing

It is clear, from Plato's comment for the division step D7, that the opposing species of fowling and fishing are described in a satisfactory way by the ratio from above downward to from below upwards. Schematically

fowling/fishing = from above downward water-animal hunting/from below upwards water-animal hunting.

The transitivity relation for the equality of ratios now implies that

tridentry/angling = fowling/fishing

clearly a philosophic version of the **Logos Criterion** for periodic anthyphairesis!

1.5.3 The Very Abbreviated Genus-Species Scheme of Division and Collection

The Division, as perceived by the soul, consists in the Abbreviated scheme, omitting, as mentioned in Sect. 1.5.1, the even-numbered steps, a scheme still rich enough to preserve intact the process of finding the Logos and thus achieving Collection.

Plato however presents a still further abbreviated form of the Division consisting in restricting attention only in the right-hand side of the given Division, going only from the genus to that species which will be further divided. Let us call this scheme the **Very Abbreviated** Genus-Species scheme of the Division and Collection. Thus for the Angler the Very Abbreviated Genus-Species scheme is the following:

> of the art
> as a whole half
> was acquisitive,
> and of the acquisitive half was coersive,
> and of the coercive half was hunting,
> and of hunting half was animal hunting,
> and of animal hunting half was half was water hunting,
> and of water hunting [half] was fishing,
> and of fishing half was striking,
> and of striking half was barb-hunting,
> and of barb-hunting [half] was angling. (*Sophistes* 221a7-c3)

A similar Very Abbreviated Genus-Species scheme has been given for the Sophist (*Sophistes* 268c5-d5).

Three basic things should be noted about the Very Abbreviated Genus-Species scheme:

(a) each entity in the Very Abbreviated scheme plays the role of a Genus to the immediately next entity which lays the role of a Species, hence each step is like a logical consequent followed by a logical antecedent; for example, in the case of the scheme for the Angler, a Genus-consequent is the art of hunting, the immediately next entity, the Species-antecedent, is the art of animal hunting, and, indeed, every "animal hunting", is certainly a "hunting". Hence every movement from an entity in the Very Abbreviated scheme is an inverse implication, while the inverse scheme is a chain of logical implications, and, thus, has the structure of a mathematical proof;

(b) the scheme is however something more that just the counter of a sequence of logical implications, since the steps in it, being determined by the Division process of a Platonic Being (the Angler in this case), are in natural order and succession; and,

(c) the Logos, present in the Abbreviated Division scheme, is **lost** in this scheme, since the successive difference of each genus or species is missing, and so the Very Abbreviated Genus-Species scheme does not have, by itself, the power to provide true knowledge, but, with proper dialectical ingenuity and heuristics, logos and knowledge may be recaptured.

The Very Abbreviated Genus-Species scheme plays a fundamental role in our Platonic-anthyphairetic interpretation of Analysis and Synthesis (cf. Sect. 1.6.3, 1.10.4, 1.10.5) and in the Platonic generation of the straight line in Zeno-like fashion (cf. Sect. 1.10.2).

1.5.4 The Parts Generated in the Division and Collection in the Sophistes and Politicus Are Called and Treated as Hypotheses

There are two distinct advantages in presenting the abbreviated, tree-like form of Division and Collection, of which one is the following[16]: it makes Division appear a process very similar to Analysis, as practiced by the geometers.

In fact, starting with genus G, in analysis fashion, we regard it as hypothesis and look for a higher hypothesis from which G will follow; we divide G into a and b, and then we decide that of a, b, the suitable higher hypothesis is b, and so on, till we reach something known, in this case the Logos criterion that turns the hypothesis into a true intelligible Being, in quite the same way that geometrical Analysis, such as the one described by Pappus in the *Collection*, is a process proceeding from hypotheses to higher hypotheses, till we reach something known to be (mathematically) true.

We will return to this point in Sect. 1.6.2 in connection with passage 511b3-c2. We should not be surprised, then, to find that the parts-species in every step of the Divisions in the *Sophistes-Politicus* are presented as hypotheses (with words such as "thesomen", "theteon", etc.). Exactly in this sense, in every Division step of the method of Division and Collection, what is being divided is always a hypothesis.

Here is a reasonably complete, but not exhaustive, list of the use of hypothesis in connection with the examples of Division and Collection.

1.5.4.1 Parts of the Division as Hypotheses in the *Sophistes*

shall we "thesomen" the angler as a man with an art or without an art (219a4-6)
in which species, acquisitive or productive, shall we "tithomen" the angler? (219d1-3)

[16]The second advantage of the abbreviated, tree-like Division will be discussed in Sect. 1.10.2 in connection with the generation of the intelligible line.

"thentas" the open species–part fighting and the secret one hunting (219e1-2)
which of the two is it "theteon" that the angler is just a man or a man with an art?, shall we "thesomen" that he is just a man or a sophist?, which art will it be "theteon" that he possesses? (221c-d)
but "thes" whatever you like, either "titheis" that there is no tame animal,. . . or (222b2–c2)
"tithentas" one species the competitive and the other the pugnacious (225a2–6)
"tithemenous" one species of the pugnacious violent (225a8–b2)
controversy "theteon" into two species (225b3–11)
that part of argumentation, which deals with business contracts. . . but carried on informally and without rules of art "theteon" a separate species (225b12–c10)
"themena", "theteon", "theteon" (227d13-228e5)
the sophist "theteon" in the species of juggler and imitator (235a)
I was uncertain in which of the two species, the likeness–making and the fantastic, the sophist "theteon" (236c-d)
to which of the two species, the likeness–making and the fantastic should we "thesomen" the sophist? (264c)
but I "theso" the species of things made by divine art and the species of things mad by human art) (265e)
I "tithemi" two species of production, each of which is twofold (266d)
and what division can we "thesomen" greater than that which separates knowledge and ignorance? (267b)
shall we "thesomen" one species the simple imitator and the other the dissembling imitator? (268a)
we "ethemen" the sophist in the ignorant species (268b-c).

1.5.4.2 Parts of the Division as Hypotheses in the *Politicus*

"theteon" him also among those who have a science, or not? (258b4)
"thesomen" that the statesman, king, master, and householder too, for that matter, are all one, to be grouped under one title, or shall we say that there are as many arts as names? (258e9)
Now to which of these two classes is the kingly man "theteon"? "thesomen" him to the art of judging, as a kind of spectator, or rather to the art of commanding, inasmuch as he is a ruler? (260c1–3)
"thentes" kings to the science of giving orders of one's own, disregarding all the rest and leaving to someone else the task of "thesthai" a name ? (260d11–e8)
"themenos" one name to all the rest (262e1)
"thomen" him out alone as herdsman and tender of the human herd, while countless others dispute his claim? (268b8–c3)
Well then, "thomen" the whole arts of fulling and mending be no part of the care and treatment of clothes, or shall we declare that these also are entirely included in the art of weaving? (281b5)
Will our definition of the art of weaving (I mean the part of it we selected) be satisfactory if we "tithomen" all the activities connected with woollen clothing be the noblest and the greatest? (281d1)
Just as in the previous case, you know, "etithemen' all the arts which furnished tools for weaving as contingent causes (287c6)
For all the arts furnish any implement, great or small, for the state, "theteon" as contingent causes; for without them neither state nor statesmanship could ever exist, and yet I do not suppose "thesomen" any of them as the work of the kingly art. (287d1-4)

And shall we not "thesomen" a sixth class of that which furnishes to all these the materials of which and in which all the arts we have mentioned fashion their works, a very various class, the offspring of many other arts? (288d4-5)
And property in food and all the things which, mingling parts of themselves with parts of the body, have any function of keeping it in health, we may say is the seventh class, and we "thesthai" it collectively our nourishment, unless we have some better name to give it. (288e8-289a3)
All this "hupotithentes" to the arts of husbandry, hunting, gymnastics, medicine, and cooking more properly than to that of statesmanship (289a3-5)
there was the primary possession, which ought in justice to "tethen" first, and after this the instrument, receptacle, vehicle, defence, plaything, nourishment (289a9-b2)
but to the rule of the many "etithemen" then only a single name, democracy; now, however, that "theteon" divided (302d3-5)
We, then, not "thesomen" the art of the generals statesmanship, since it is subservient (305a8)
Among all the parts we must look for those which we call excellent but "tithemen" in two opposite classes. (306c7-8)

Consequently, we will have no difficulty to accept the divisibility of hypotheses in the context of Division and Collection of an intelligible Being.

1.6 *Politeia* 510b6-9, 511b3-c2, 533c7-d3: The Ascent-Analysis from the Divisible Hypotheses to the Indivisible "anhupotheton" Coincides with the Method of Division and Collection

Our next step, after it has been shown that *Politeia* dialectics is Division and Collection, is to show (making use of the fact, noted in Sect. 1.5, that the "parts"—genera, generated in the steps of the abbreviated form of Division, in the *Sophistes* and *Politicus*, are called there "hypotheses") that the ascent from the hypotheses to the "anhupotheton" is just another description of Division and Logos-Collection. We will examine three passages: *Politeia* 510b6-9, 511b3-c2, and the most crucial 533c7-d3.

1.6.1 Politeia 510b6-9

"while there is another section in which it advances from an assumption ("ex hupotheseos") to a beginning or principle that transcends assumption ("ep' archen anhuotheton"), and in which it makes no use of the images employed by the other section, relying on ideas ("autois eidesi") only and progressing systematically through ideas ("di' auton")." (510b6-9)

This passage introduces the unfamiliar term "anhupotheton"—hypotheses-free. The intelligible proceeds from hypotheses to the hypothesis free, without any recourse to sensible images, but by means of kinds. It is reasonable to assume, by

Sects. 1.3, 1.4 and 1.5, that the intelligible hypotheses are the parts-genera of the true Being.

1.6.2 Politeia 511b3-c2

> "Understand then," said I, "that by the other section of the intelligible I mean that which Logos itself ("autos o logos") is in contact with ("haptetai") the power of dialectics ("dialegesthai"), treating its hypotheses not as beginnings ("archas") but truly as hypotheses, footings ("epibaseis"), and springboards ("hormas") so to speak, to enable it to rise to the "anhupotheton", which is the beginning ("archen") of the all, achieving contact ("hapsamenos") with it (511b3-7)
> again ("palin") having as parts the parts of it [the beginning], so to proceed downward ("katabainei")
> to the end ("epi teleuten"),
> making no use whatever of but only of pure ideas moving on through kinds ("eidesi autois") to kinds ("eis auta") and ending with kinds ("teleuta eis eide")." (511b7-c2)

This is an important and difficult passage; it consists of two parts. The first part (511b3-7) is a repetition of the first passage (510b6-9), but with the addition that it is "Logos itself" that treats the hypotheses-parts in such way that they ascend to the "anhupotheton". This strongly suggests that Plato is talking here about Division and Collection, and that anhupothteton is just another name for the intelligible Platonic being, and the One of the second hypothesis (*Parmenides* 145a8-b1).

The second part (511b7-c2) is still about the Logos. The Logos starting now from the beginning principle and proceeding in a purely intelligible manner, from kinds through kinds comes to the end. Thus Logos starts from the beginning, passes through the middle, and comes to the end. This is the beginning–middle–end of a full anthyphairetic period, a basic feature of the One of the second hypothesis.

This interpretation of the ascension of hypotheses to the anhupotheton, by means of Logos identifies it with Division and Collection.

1.6.3 Politeia 533c7-d3

The most crucial passage concerning the dialectics in the *Politeia* is passage 533c7-d3

> "Then," said I, "is not dialectics the only process of inquiry that advances in this manner, "anairousa"[17] the hypotheses, towards the principle itself in order to find confirmation there? And it is literally true that when the eye of the soul is sunk in the barbaric slough, dialectic gently draws it forth and leads it up ("anagei ano")." (533c7-d3)

We proceed to an interpretation of passage 533c7-d3

[17]We leave this crucial term untranslated for the moment, yet uncommitted as to its meaning, and discuss it below in this section.

1.6.3.1 "tas hupotheseis anairousa", 533c8: "anairesis" Is Used with the Meaning of "Division" in Kinds by Aristotle and Commentators and in Two Significant Passages by Proclus

Robinson [29, pp. 166–167] offered this explanation for "anairousa":

> And what is "destroying" hypotheses? The latter question is so difficult that many readers take the text for unsound, and emend.... What could dialectic possibly destroy with regard to all hypotheses? Why, precisely their hypothetical character, of course.

Robinson's interpretation was adopted by Annas [3, p.287]:

> Dialectic proceeds by "destroying the hypotheses". We have seen that this means destroying their hypothetical nature,

and reached the status of a consensus by Mueller [19, p.188]:

> Some later Platonists used this passage to belittle mathematics, and modern scholars have debated what Socrates could have in mind by destroying the hypotheses of mathematics. I think it is fair to say that there is now consensus that the only destruction Socrates has in mind is the destruction of the hypothetical character of mathematical hypotheses through subsumption under an unhypothetical starting point.

There has not been much progress since Robinson. In a recent work on the *Politeia*, Benson [5, Footnote 61, p. 256], writes:

> I here follow what Mueller (1992, 188) calls the "consensus that the only destruction Socrates has in mind is the destruction of the hypothetical character of mathematical hypotheses through subsumption under an unhypothetical starting point." See also, e.g., Robinson (1953, 161) and Annas (1981, 278). Against the consensus, see, e.g. McLarty (2005,128-29). Most recently, Bailey (2006, 125) offers a reading in keeping with the consensus but which avoids reading "anairousa" as "destroying."

Reeve [28, p. 219]:

> Dialectic thus becomes the philosopher's peculiar craft or science, since it is "the only method of inquiry that, doing away with hypotheses, journeys to the first principle itself in order to be made secure" (533c8–d1).

But in fact a good meaning of "anairesis" is subtraction-"division" (in kinds), as the following passages indicate. We start with some passages by Aristotle and Commentators.

> "diairesin e anairesin" (Aristotle's *Sophistici Elenchi* 183a11, 11-12),
> "anairesin"..."touto hos ephermeneutikon tes diaireseos" *(Scholia in Aristotelis Sophisticos Elenchos* 33),
> "hen anairesin autos diairesin ekalesen" (Ammonius, *Aristotelis librum de interpretatione commentaries* 27,1),
> "antanairesis" (Aristotle, *Topics* 158b33; Alexander, *In Topics* 545,17)

We note that here, the term used, "antanairesis" was the earlier name for "anthyphairesis", and thus "anairesis" here had the same force as "huphairesis", subtraction.

We continue with two Proclus passages:

> Persephone has that name ... because of dividing the kinds ("dia to diakrinein ta eide") and separating them from each other ("chorizein allelon"), since "phonos" hints at ("ainittomenou") "anairesis", ... (Proclus, *Commentary to Cratulus* 173,1-5).

The explanation of this revealing Proclus passage is as follows:

Plato, in the *Cratulus* 404c7-d1, tells us that the name of the godess Phersphone is fearful ("deinon"), at the same time showing that the goddess is wise ("sophen"). It is simple enough to understand why the name was a cause of fear to the people: the second part "phone" of the name "Persephone" is directly related to "phonos", meaning murder and death.

It is less clear why the name of "Persephone" shows that she is wise. Proclus gives a number of explanations, of which one is imaginative: we continue with "phonos"; the word "phonos" is hinting at the (more general) word "anairesis", a meaning of which is an action causing death, slaying, putting to death, practically synonymous with "phonos".

But now Proclus considers another meaning of "anairesis", namely 'diairesis', and the most wise "diairesis" is the Platonic "diairesis kat' eide", another name for Division and Collection. Thus Persephone, being an expert in philosophical Division and Collection, is wise. But even if a meaning of "anairesis" is division, why do we end up being reminded of "division in kinds"? The only place, where Plato might associate "anairesis" with Division and Collection, is at the *Politeia* 533c8, "tas hupotheseis anairousa". It is then clear that Proclus regards

- the word "anairesis" as a word that can serve as a description of the method of Division in kinds, and
- the phrase "tas hupotheseis anairousa", 533c8, as a phrase describing "Division of hypotheses".

> A second is the method of diairesis ("diairetike"), dividing ("diairousa") into its natural parts ("kat' arthra") the genus ("genos") proposed, and which [method of diairesis] provides a starting point ("aphormen") for the demonstration ("tei apodeixei") of the construction ("tes kataskeues")
> of the proposed [genus] by means of the division ("anaireseos") of the other [genera] (Proclus, *In Euclidem* 211, 23–27).

Thus, the method of Division according to species, to be exact according to "arthra", by dividing the given genus, provides a starting point ("aphormen") for the proof of the construction of the initial Genus, by the "anairesis" of the other (genera). Here "anairesis" of a genus has again the meaning of "division" of this genus. But the claim that this division of the genera provides a starting point for the demonstration of the construction of the initial genus is not clear and requires explanation.

Proclus here, clearly, adopts the abbreviated tree-like form of the Division and Collection:

$$B = a + b, b = a_1 + b_1, b_1 = a_2 + b_2, \text{ and so on,}$$

till the Logos criterion $a/b = a_k/b_k$ for some stage k.

The Logos turns the hypothesis b_k into something true, given, since it is a part in a true Being (remember that a true Being is a philosophic periodic anthyphairesis), and thus forms the starting point for the demonstration of the construction of the initial genus G.

The demonstration of the construction of the initial genus G runs as follows:

b_k is true, given. But b_k obviously implies b_{k-1}, since the part-hypothesis b_{k-1} is a greater part than b_{k-1}, and thus b_{k-1} is true, given, too.

Continuing in this way, we conclude that the genus G is true, given, and constructed. Of course the demonstration:

b_k given; if b_k given, then b_{k-1} given; hence b_{k-1} given; ..., hence G given,

is the synthesis of the analysis that results from the abbreviated Division in kinds:

let G be a hypothesis;

b_1 is a higher hypothesis, from which G follows;

b_2 is a higher hypothesis from which b_1 follows;

...

b_k, because of the Logos Criterion, is no longer a hypothesis,

but something true and given, an "anhupotheton".

Thus there is in every intelligible Being, intelligible analysis and intelligible synthesis. In this way Plato is able to subsume the mathematical method of proof from an antecedent to a consequent, and the mathematical method of analysis and synthesis within the context of the dialectical method of the (abbreviated, tree–like) diaeresis-division in an intelligible Being.

As we saw, in Sect. 1.6.2, this analysis and ascent is the content of the *Politeia* 511b3-c2 passage.

Cherniss [12, p. 418, Footnote 54], accepts the divisive meaning of "anaireseos" in Proclus, *In Euclidem,* pp. 211, 23–27, and interprets it as referring "to the separating-off" of the privative term in dichotomy, in agreement with our interpretation. But Cherniss' footnote is continued in relation to the appearance of the expression "tas hupotheseis anairousa" in the *Politeia* 533c8, and will return to it later, in Sect. 1.6.3.4.

1.6.3.2 "tas hupotheseis anairousa", 533c8: Is in Opposition of "tas hupotheseis akinetous", 533c2, Indicating Intelligible Motion and Division of the Hypotheses

Hypotheses, as practiced in Mathematics (geometry, arithmetic), cannot achieve true knowledge ("episteme"), because the geometers use hypotheses in the wrong way:

they keep the hypotheses unmoved,
and being unmoved these hypotheses cannot be provided with logos;
and because the hypotheses have not been provided with logos,
they do not form a true Being,
and the knowledge attained does not have beginning, middle, end,
and is thus not true knowledge.

By contrast the dialectic method is the only one that proceeds by "anairousa" the hypotheses (533c8), towards the first principle and true being.

The crucial word here is "anairousa". It is clear that "hypoheseis anairoumenai", is in contrast, in opposition to "hypotheseis akinetous", hypotheses unmoved, and thus "anairousa" must carry a meaning of motion, in fact intelligible motion. Thus it is natural to believe that "anairesis" is not a destruction but a motion of hypotheses. Now we have seen, in Sect. 1.3, that intelligible motion is in terms of the Division; and thus "anairousa" should be rendered in terms of division.

Thus from the two possible meanings, "dividing" vs. "destroying", of "anairousa", preferable is "dividing" them.

1.6.3.3 "tas hupotheseis anairousa", 533c8: In Agreement with "tas hupotheses dielete" in the *Phaedo* 107b7

We finally examine two key passages in the *Phaedo*, closely related to the *Politeia* 533b6-d3.

> and when you had to give "logos" of the hypothesis, you would give it in the same way by hypothesizing some other hypothesis which of the higher ones ("ton anothen") seemed best, and so on until you reached something adequate ("hikanon"). (*Phaedo* 101d5-e1)
> the first hypotheses ought to be more carefully examined, even though they seem to you to be certain. And if you divide them ("autas ... dielete") sufficiently ("hikanos") you will, as it seems to me, follow by "logos", so far as it is possible for man to do so. And if this is made clear ("saphes"), you will seek no farther. (*Phaedo* 107b5-9)

It is clear that both of these passages are about ascent, to something "adequate" in the *Phaedo*, to the "anhupotheton" in the *Politeia*.[18]

In the first passage we go to higher and higher hypotheses, till we achieve something adequate, in the second we divide the hypotheses till we achieve the

[18] Proclus, *In Parmeniden* 622,29-623,28 and 655,16-656,2, clearly identifies "hikanon" and "anhupotheton". "hikanos" in the *Sophistes* 221b2 also fits perfectly (cf. Sect. 1.5.2).

adequate; so it seems that we go from a hypothesis to a higher hypothesis, by division of the initial hypothesis in two, and from these two we choose the suitable, best one.

But the first passage also sets as the purpose of the procedure to give "logos" to the initial hypothesis, while the second also mentions the attainment of "logos". Thus we aim to divide the hypotheses, till we reach something adequate, at which point we will presumably have succeeded in giving "logos" to the hypothesis. The description can be described succinctly as Division and Logos, and is thus essentially a description of the method of Division and Collection, but where the parts of the indefinite dyad undergoing Division and Collection are now called "hypotheses".

But it can also be described as ascent to the adequate/anhupotheton. We reach the preliminary conclusion that the ascent to the anhupotheton coincides with the method of Division and Collection.

We summarise the arguments given in Sects. 1.6.3.2 and 1.6.3.3: 533c8 "tas hupotheseis anairousa" is habitually translated as "doing away with hypotheses", but that is not satisfactory. The greater passage 533b6-d7 contrasts geometry as practiced by geometers, who "leave the hypotheses at rest" ("hupotheseis... akinetous eosi"), thereby unable to provide "logos" ("me dunamenai logon didonai") and obtain true knowledge "idein"), vs. the dialectic method, the only one that "anairei" the hypotheses, and ascends to the "anhupotheton" principle, obtaining true knowledge ("episteme").

The "anairesis" of the hypotheses should, by Sect. 1.6.3.2, be contrasted to the hypotheses staying at rest, and this is hardly achieved by doing away with, destroying, them. The opposite of staying at rest is moving, and moving is certainly associated with Division.

In fact, as we saw in Sect. 1.6.3.3, the *Phaedo* 107b5-9 refers to the "adequate" ("hikanos") division of hypotheses, meaning till we obtain something adequate ("hikanon"), namely the anhupotheton, and providing logos and episteme corresponds to the first hypothesis.

1.6.3.4 In Conclusion

- "tas hupotheseis anairousa", 533c8, should be rendered as "dividing the hypotheses";[19]
- the passage 533c7-d3 consists in a description of Division and Collection, in the form Division ("anairousa") and Logos, where the Division of the true Being is

[19]Thus the second part of Cherniss [6, p. 418, Footnote 54], claiming that "tas hupotheseis anairousa" in the *Politeia* 533c8 is not a reference to division in kinds, is not correct; furthermore his claim, that Robinson's [29, p. 171]) notion that "Proclus seems to have understood division as belonging to the upward path." is mistaken, is also incorrect. Indeed, Proclus, in both *In Euclidem* 211, 23-27, and *Commentary to Cratulus* 173,1-5, identifies Division with "anairesis", and regards Division in the upward path.

the (binary, anthyphairetic) division of hypotheses into higher hypotheses; and the Logos, the logos of periodicity, by which Collection is achieved is a Logos given to a hypothesis; and,
- the passage 533c7-d3 is explicitly an ascent from the hypotheses to the anhupotheton 533d2-3.

Thus, our analysis of the passage *Politeia* 533c7-d3 shows that

- the "anhupotheton" is identical with the one of the second hypothesis in the *Parmenides*,[20]
- the Ascent from hypotheses to higher hypotheses is identical with the abbreviated Division of parts into more divisions, and
- the "Logos provided to the hypothesis" is the logos of periodicity and Collection.

This unusual terminology "anhupotheton" can be understood as follows: hypothesis is identified with a part–genus of the infinite anthyphairesis of the initial dyad, and as such a hypothesis is divisible; the anhupotheton, the hypothesis-free is the division-free entity, the indivisible, that is obtained by the Logos and Collection. Thus "anhupotheton" has the same force as "indivisible",[21] precisely as the indivisible line, identified with the One of the second hypothesis in the *Parmenides* (in the next Sect. 1.7).

1.7 Plato's Criticism I of the Geometers

The geometric point, the One of the first hypothesis of the *Parmenides* is rejected as a foundational concept of Geometry, in favor of the indivisible line, the One of the second hypothesis of the *Parmenides*.

The first definition in Book I of Euclid's *Elements* is that of a geometric point.

Def. I. A point is that which has no part. (Σημεῖόν ἐστιν, οὗ μέρος οὐθέν.)

The attentive reader of Plato will doubtlessly note that there is no mention of geometric point (although he talks about numbers, straight lines, circles, angles) in his dialogues. Aristotle in *Metaphysics* 992a19-22[22] states that Plato

(i) rejected the concept of the partless geometric point ("stigme") as a foundational concept of geometry, and
(ii) opted instead for the concept of the "indivisible line" ("atomous grammas").

---

[20]This identification is explicitly confirmed by Proclus, *In Parmeniden* 1033,32–35.

[21]Cf. ἀλλὰ γὰρ ἡμῖν ἔτι καὶ τοῦτο σκεπτέον, ἆρ' **ἄτομον** ἤδη ἐστὶ πᾶν ἤ τινα ἔχον διαίρεσιν ἀξίαν ἐπωνυμίας. *Sophistes* 229d5-6.

[22]ἔτι αἱ στιγμαὶ ἐκ τίνος ἐνυπάρξουσιν; τούτῳ μὲν οὖν τῷ γένει καὶ διεμάχετο Πλάτων ὡς ὄντι γεωμετρικῷ δόγματι, ἀλλ' ἐκάλει ἀρχὴν γραμμῆς—τοῦτο δὲ πολλάκις ἐτίθει—τὰς ἀτόμους γραμμάς.

It is very difficult to conceive of geometry without the concept of geometric point, but Plato suggests just that. Furthermore he replaces it with something mysterious, and seemingly contradictory, the indivisible line. We will try to make sense out of both, quite unexpected and difficult to comprehend, statements (i) and (ii), which in fact go back to Zeno.

The mystery is substantially illuminated when we realise that the point corresponds to the absolutely partless One of the first hypothesis of the *Parmenides*, and this is replaced by the indivisible line, which is identified with the One of the second hypothesis, as a true Being and as the true foundation of geometry.

1.7.1 The Geometric Point Is Rejected by Zeno and Plato as a True Being

First of all, we learn from Simplicius, *in Physics* 138,29-139,3, that Zeno, although convinced that true Being is some sort of One, was puzzled, mystified ("eporei") as to what sort of One would be suitable. He was nevertheless rejecting the point as the suitable One. At this point Simplicius cites Zeno's *Fragment B2* (139, 11–15) according to which something that has no magnitude ("megethos"), but is absolutely partless, cannot be a true Being. Now a similar rejection can be seen with the One of the first hypothesis in Plato's *Parmenides*; thus the One of the first hypothesis is absolutely partless (137c4-d3), like a geometric point, and is decidedly rejected as true Being in Plato's *Parmenides* (141e7-142a6).

Thus it appears that the rejection (i) must be understood as follows: Plato believes that the foundations of Geometry must rest on true Beings, and the point, not being a true Being, cannot serve as such foundation.

1.7.2 The Indivisible Line Is Accepted by Zeno, Plato and Xenocrates as a True Being

Next, Zeno, in his argument against the Many (assuming, for the purpose of contradiction, that each of the many sensibles satisfies properties of the true Being) does reveal the nature of Zeno's true Being. We outline the nature of Zeno's arguments.

1.7.2.1 [One and Many] Zeno's *Fragment B1* on True Being Agrees with the One Is Many, Namely the Infinite Anthyphairetic Division of the One Being and the Many Are One, Namely the Equalization of Parts in the One of the Second Hypothesis in the *Parmenides* 142d9-143a3

One half of Zeno's *Fragment B1* (Simplicius, *in Physics* 141, 2–8) is the following [in brackets the notation according to our interpretation]:

> it is necessary that each part [B] has magnitude and thickness ("pachos") and one part [b] is deficient ("apechein") with respect to the other part [a]. And concerning the part in excess ("prouchontos") [a] the same statement. Because this [a] too will have magnitude and some part [b] of it will be in excess ("proexei") [with respect to the other part a_1]. And the same holds if stated once ("hapax") and if stated for ever ("aei").
>
> Because there is no part [a_n, or b_n, resp.] that does not have a part in excess [b_n, or a_{n+1}, resp.], and there is no part [a_n, or b_n, resp.] which does not consist of one part and another part [$a_n = b_n + a_{n+1}$, or $b_n = a_{n+1} + b_{n+1}$, resp.].

Thus the following relations are produced

$$
\begin{aligned}
B &= a + b, b < a, \\
a &= b + a_1, a_1 < b, \\
&\dots \\
a_n &= b_n + a_{n+1}, a_{n+1} < b_n, \\
b_n &= a_{n+1} + b_{n+1}, b_{n+1} < a_{n+1} \\
&\dots
\end{aligned}
$$

and an infinite anthyphairetic division is described starting with a dyad $\langle a, b \rangle$ in a whole B. Thus, corresponding to the infinite anthyphairetic division of the One Being in the *Parmenides* 142d9-143a3, we have that Zeno's true Being consists of a dyad $\langle a, b \rangle$ in infinite dichotomic anthyphairesis (Simplicius, *eis Phusica* 140,34).

In the other half of *Fragment B1*, presumably using the corresponding *Fragment B3*, concludes that [in Zeno's true Being] "each part of the many is identical ("tauton") to itself and One" (Simplicius, *in Physics* 139, 18–19), thus the Many are One (in the self-similar sense that each of the many is equalized to the One). This is confirmed by *Parmenides* 129d8, and also by the Simplicius, *in Physics* 138,18–139,3 a passage, according to which Zeno's true Being

- is not only One (as the geometric point is),
- is not only Many (as the sensible are),
- is One (as Parmenides would wish), and Many (as Parmenides would not wish),

exactly as the One of the second hypothesis in the *Parmenides*. It follows that Zeno's true Being is One and Many in precisely the same way that Plato's One of the second hypothesis of the *Parmenides* is One and Many (*Parmenides* 144d4-e1, e1-3).

1.7.2.2 [Infinite and Finite] Zeno's *Fragment B3* on True Beings Essentially Coincides with the Infinity of the Multitude of the Parts of the One, and the Finiteness of the Number of Parts of the One in the Second Hypothesis of the *Parmenides* 144d4-e1

The infinity of the multitude of the parts in *Fragment B3* coincides essentially with the corresponding anthyphairetic infinity in *Fragment B1*. The crucial statement in part of Zeno's ***Fragment B3*** (Simplicius, *in Physics* 140,29-31) states:

> It is necessary ("anagke") that
> [the parts of one part] are as many ("tosauta") as ("hosa") [the parts of the other part], and neither more ("oute pleiona") nor less ("oute elattona") than they.
> And if [the parts of one part] are as many ("tosauta") as ("hosa") [the parts of the other part], then they would be finite ("peperasmena").

Thus Zeno's *Fragment B3* can be seen to be practically identical in wording to the *Parmenides* 144d4-e1:

> "And it is necessary ("anagke") that the divisible ("meriston") "[One] must be as many parts ("tosauta") as ("hosa") the parts [of the being]." "It is necessary ("anagke")." "Then what we said just now—that the Being was divided into in an infinite number ("pleista") of parts—was not true for it is not divided, you see, into any more ("pleio") parts than one, but, as it seems, into equal ("isa") as the one."

Thus, according to Zeno's *Fragment B3* (Simplicius, *eis Phusica* 140, 27), Zeno's true Being consists of a dyad a, b, such that the number of parts of a is equal to the number of parts of b.

It follows that Zeno's true Being is Infinite and Finite in precisely the same way that Plato's One of the second hypothesis of the *Parmenides* is Infinite and Finite (*Parmenides* 144d4-e1, e1-3).

1.7.2.3 The Identification of Zeno's True Being with Plato's True Being

On the basis of Sects. 1.7.2.1 and 1.7.2.2, Zeno's true Being has the same anthyphairetic structure with, and satisfies practically all the statements that hold, for the One of the second hypothesis, and thus we have no hesitation to identify Zeno's true Being with the One of the second hypothesis in the *Parmenides*, namely with Plato's true Being.

This identification is further confirmed by comparing Zeno's arguments and paradoxes with the properties of the One of the second hypotheseis in the *Parmenides*. According to Zeno's third paradox of motion (Aristotle' *Physics* 239b30) and the *Parmenides* 129e1, Zeno's true Being is in motion and at rest. The corresponding statement for Plato's true Being is in the *Parmenides*145e7-146a8. According to the *Parmenides* 127e3, 129b1-2, Zeno's true Being is similar and dissimilar, exactly as the One of the second hypothesis in the *Parmenides* 147c1-148d4, and this is essentially equivalent to saying that it is Infinite and Finite (*Parmenides* 144e8-145a4 and 158e1-159a4).

1.7.2.4 The Description of Plato's True Being as an Indivisible Line

Zeno's term "amegethes" ("outhen echei megethos"), without magnitude, in *Fragment B1* and Xenocrates' term "atomos", indivisible, line refer to the same entity. According to Simplicius, *in Physics* 140,6-18, 142,16-27, Xenocrates, being a "geometric and wise man", could not contradict the geometric principle and did not ignore the nature of magnitude [namely the infinite divisibility of magnitudes, "ten ep' apeiron tomen"]; by "indivisible line" he meant a line that, even though it was divisible *ad infinitum*, nevertheless was not divisible in kind ["oude toi eidei diairetton"], namely the number of kinds was finite.

According to Proclus, *eis Timaion* 2,245,23-246,7, by indivisible line Xenocrates did not mean the mathematical line but the ratio of the line showing true Being ("ton logon tes grammes ton ousiode").

Although Plato never explains or even really uses the term "indivisible", still, rather unexpectedly an interesting and revealing occurrence of the term comes up in a division step in the *Sophistes*, where it is being asked

> ἆρ' ἄτομον ἤδη ἐστὶ πᾶν ἤ τινα ἔχον διαίρεσιν ἀξίαν ἐπωνυμίας.
> whether it [the last kind] is already indivisible or still admits of division worthy of a name
> *Sophistes* 229d5-6[23]

It is clear that what Plato really asks at this point is whether we have reached the Logos criterion for establishing periodicity, at which point the ratio is repeated, and, although division is being continued *ad infinitum*, no new kinds are being produced.

Thus the near paradox that occurs in the "indivisible line", namely that the parts of the "indivisible line" are infinite in multitude, while the kinds ("eide") are only finite, practically coincides with the near paradox we have observed (in Sect. 1.3) in the One of the second hypothesis in the *Parmenides*, namely that the parts of the One are infinite in multitude, while the parts of the One are finite in number.

We have no difficulty to identify the "indivisible line" with the One of the second hypothesis in the *Parmenides*, possibly in a somewhat degraded form, as it might be thought of as a geometric, and not a purely dialectic entity.

The *Metaphysics* statement is now fully understandable: according to Plato, Geometry must be founded on what he regards as true Being, whose paradigm is the One of the second hypothesis in the *Parmenides*, and which essentially coincides with the indivisible line,[24] and not on the geometric point, which coincides with the One of the first hypothesis in the *Parmenides*. A byproduct of our analysis was the (anthyphairetic) interpretation of Zeno's three *Fragments B1, B2,B3*, and the close relation they have with the one of the second hypothesis in the *Parmenides*.

The detailed arguments of the present Sect. 1.7 can be found in [24].

---

[23]Cf. ἀλλὰ γὰρ ἡμῖν ἔτι καὶ τοῦτο σκεπτέον, ἆρ' **ἄτομον** ἤδη ἐστὶ πᾶν ἤ τινα ἔχον διαίρεσιν ἀξίαν ἐπωνυμίας. *Sophistes* 229d5-6.

[24]One may conjecture that the name "indivisible line" was given, by Platonic circles, so as to make the indivisible line (which according to Plato replaces the point in the foundation of Geometry) appear as closely as possible similar to the partless, indivisible point.

1.7.3 *Proclus' Compromise Between Plato and Euclid in* In Euclidem *85,1-96,15: Proclus' "Point" Is, in Fact, the Indivisible Line; and, the Flow ("rhesis") of the "Point" Is the Motion of the Indivisible Line*

Plato, as we have seen, had rejected the geometric point as having a foundational role in geometry, opting instead for the indivisible line. But Euclid in the *Elements* does take the point as having a foundational role. Proclus, a Platonist, in his *Commentary on Euclid's Elements* is then in a difficult situation. He must either criticize Plato, something impossible, or Euclid, something difficult. He then chooses a compromise: he comments on the point, as if it were the point introduced in the *Elements*, but in fact his comments are not about the partless point but about the indivisible line.

Proclus several times describes the point as a mixture of the Infinite and the Finite, e.g.

> the point ("semeion")... although it is being determined by the Finite ("peras"), it secretly ("kruphios") contains the Infinite ("apeiron") power, by virtue of which it generates all the line intervals ("diastemata"). (88,2-5)

This brings Proclus' conception of point much closer to the One of the second hypothesis, and thus to the indivisible line; and thus it brings Proclus, as expected, much closer to Plato's view about the true foundation of Geometry. It seems clear that for Proclus the geometric point is (a lower form of) the One of the second hypothesis.

The **geometric point**, being partless, is like the One of the first hypothesis in the *Parmenides*; and the One of the fist hypothesis is **not** in motion (139b2-3). (As mentioned in Sect. 1.4, the point and the partless One have been correlated by Zeno, Simplicius, and probably by Plato). Thus the partless geometric point cannot be in motion. But Proclus in 97,9-12 talks about the flow ("rhusis") of the point.

> but that which calls it the flowing of a point ("semeiou rhusin") appears to explain it in terms of its generative cause ("aitias gennetikes") and sets before us not line in general but the immaterial[25] ("aulon") line. (97, 9-12)

Since the indivisible point is for Proclus the Platonic indivisible line, and this is as we already analysed, a (degraded) One of the second hypothesis, and since the One of the second hypothesis is in motion 145e7-146a8, a motion that is a direct consequence of Division and Collection, it follows that the Proclean point does have a motion, the flow of point. This motion is intelligible, it is related to Division (and Collection), and it does generate, by Division, new intervals (as in the 88, 2-5 passage). It is instrumental in producing in the intelligible Being the first three Postulates of Euclidean geometry (in Sect. 1.10.4 below). For the third Postulate the periodic nature of the One is also needed.

[25]Morrow [20, p. 79], changes "aulon"–"immaterial" to its opposite "enulon"–"material", which involves a serious misunderstanding.

1.8 Plato's Criticism II: Hypotheses, Such as Numbers, Lines, Circles, Angles and Their Postulates, Are Employed by Mathematicians as Principles, Are Not Divided, Are Not Moved and Do Not Attain Logos (510c2-d3, 511c3-d5, 533b6-c6)

At this point, after Sect. 1.7, we have a general interpretation of ascent from intelligible hypotheses to the "anhupotheton", in terms of the Division and Collection of an intelligible Being, namely in terms of (the philosophical version of) periodic anthyphairesis. Plato's **criticism II** of the geometers, expressed in the three *Politeia* passages 510c2-d3, 511c3-d5, and 533b6-c6, sets in opposition geometric hypotheses with the dialectical-philosophic hypotheses: while the geometric hypotheses are about intelligible entities, nevertheless they are not employed as stepping stones to ascend to higher hypotheses and eventually to the "anhupotheton", they do not provide "Logos" for them, and therefore they cannot obtain true knowledge of them, but instead they leave them unmoved and are compelled to follow a descent from them. These Platonic claims will be fully explained in terms of our anthyphairetic interpretation.

1.8.1 Politeia 510c2-d3: Against the Geometers' Axiomatic Method of Mathematics, Because in This Method Logos Cannot Be Provided of Theses Hypotheses

> "For I think you are aware that students of geometry and reckoning ("logismous") and such subjects **first** set as **hypotheses** ("hupothemenoi") the odd and the even ("to te peritton kai to artion") and the figures ("ta schemata") and three kinds of angle ("gonion tritta eide") and other things akin to these in each branch of science, regard them as known ("hos eidotes"), and, treating them as **hypotheses** ("hupotheseis"), do not deign to render a **Logos** ("logon... didonai") of them to themselves or others, taking it for granted that they are obvious to everybody ("hos panti phaneron"). They take their start from these, and pursuing the inquiry from this point on consistently, conclude ("teleutosin") with that for the investigation of which they set out ("hormesosi")." "Certainly," he said, "I know that." (510c2-d3)

This passage (510c2-d3) is important, because in it Plato clarifies how he views the hypotheses of the geometers. He mentions three specific hypotheses, the odd and the even, the figures, and the three kinds of angle. We take it that Plato by "odd and even" means numbers, by "figures" the straight line and the circle, and of course by the "three kinds of angle" the acute, obtuse, and right angle. Usually scholars take these three hypotheses as meant to be just examples of geometric hypotheses, but we will argue that in fact they are much more, they represent in fact nothing less than the totality of the concepts needed for the foundation of geometry and arithmetic.

At this point it will help if we are reminded of the basic Definitions and Postulates appearing in Books I and VII of Euclid's *Elements*. By "basic" we mean the definitions that cannot be derived from earlier definitions. I suggest that the basic Definitions in the *Elements* are: the geometric point (Def. I.1), the straight line (Def. I.4.),[26] the circle (Def. I.15),[27] the right (Def. I.10), obtuse (Def. I.11), and acute angle (Def. I.12).[28]

Now if we take into account that, as we have seen in Sect. 1.7, the point has been replaced by the indivisible line in Plato's foundation of Geometry, all the other basic concepts of Euclidean geometry are included in the two geometric hypotheses that Plato mentions here. The same is true with Arithmetic, as here the only basic definitions are that of unit (Def. VII.1) and number (Def. VII.2).[29]

There have been heated arguments among scholars of Plato on whether his hypotheses are definitions or postulates. Plato by hypotheses cannot mean only Postulates, since the hypothesis about the odd and the even is a hypothesis about numbers, and there are no arithmetical Postulates. Moreover it is clear that the foundation of Geometry needs both Definitions and Postulates, and it is equally clear that the ascent from the hypotheses to the (dialectic) principle will have as a result to subsume the whole foundation of Geometry in Plato's dialectics, in Division and Logos-Collection. Thus there must be no question that by hypotheses Plato means all that is needed for the foundation of Geometry, and this must include both the basic geometric and arithmetic definitions and the geometric Postulates.[30]

Thus Plato criticizes the geometers for their axiomatic method, as it appears in the *Elements*, taking as arbitrary principles the basic definitions and postulates of geometry and arithmetic, rejects it, and instead proposes hypotheses that are parts-genera in the Division of a Platonic Being, and thus can proceed upward in the Division and Collection scheme of this Platonic Being and attain "logos".

---

[26]Def. I.4. A straight line is a line which lies evenly with the points on itself.

[27]Def. I.15. A circle is a plane figure contained by one line such that all the straight lines falling upon it from one point among those lying within the figure are equal to one another;

[28]Def. I.10. When a straight line set up on a straight line makes the adjacent angles equal to one another, each of the equal angles is right, and the straight line standing on the other is called a perpendicular to that on which it stands.
Def. I.11. An obtuse angle is an angle greater than a right.
Def. I. 12. An acute angle is an angle less than a right angle.

[29]Def. VII.1. An unit is that by virtue of which each of the things that exist is called one.
Def. VII.2. A number is a multitude composed of units.

[30]Post. I.1.- To draw a straight line from any point to any point.
Post. I.2. To produce a finite straight line continuously in a straight line.
Post. I.3. To describe a circle with any centre and distance.
Post. I.4. That all right angles are equal to one another.
Post. I.5. That, if a straight line falling on two straight lines make the interior angles on the same side less than two right angles, the two straight lines, if produced indefinitely, meet on that side on which are the angles less than the two right angles.
There are no arithmetic Postulates.
(Translation of basic definitions and Postulates following Heath [13]).

1.8.2 *Politeia 511c3-d5: Against the Geometers' Hypotheses, Because They Cannot Ascend to the Indivisible Anhupotheton*

"I understand," he said; "not fully, for it is no slight task that you appear to have in mind, but I do understand that you mean to distinguish the aspect of reality and the intelligible, which is contemplated by the power of dialectic, as something truer and more exact than the object of the so-called arts and sciences whose hypotheses are arbitrary starting-points ("hai hupotheseis archai").

And though it is true that those who contemplate them are **compelled** to the use of understanding and not of the senses ("dianoiai men **anagkazontai** alla me aisthesesin", 511c7), yet because they do not ascend to the beginning ("me ep' archen anelthontes") in the study of them but start from hypotheses ("ex hupotheseon") you do not think they possess true intelligence about them although the things themselves are intelligibles ("noeton onton") when apprehended in conjunction with a first principle ('meta arches'). And I think you call the mental habit of geometers and their like understanding ("dianoian") and not intelligence ("noun") because you regard understanding ("dianoian") as something intermediate between opinion and intelligence." (511c3-d5)

The passage 511c3-d5 states in a clear and detailed manner criticism II of the geometers by Plato, in that they use hypotheses as principles not as steps for higher hypotheses.

A notable statement here is that the geometers are compelled ("anagkazontai") to use the lower intelligible part and not their senses, when they study hypotheses as principles. The explanation of the use of "anagkazontai" (511c7), completed with three more uses of this term (510b5, 511a4, 527a6), is deferred for Sect. 1.9. In the present section we are content with the meaning of the use of hypothesis as starting principle, and not as stepping stone for higher hypotheses and the anhupotheton.

1.8.3 *Politeia 533b6-c6*

Passage 533b6-d3 is sandwiched between passages 531d9-e6, 532a5-b3 and d8-e3, a description of Division and Collection, as discussed in Sect. 1.4.2.1, and passage 534b3-d2, a description of Division and Collection, discussed in Sect. 1.4.2.2.

Thus, it is natural to consider that the in between passage 533b6-d3 is about Division and Collection, as well.

1.8.3.1 *Politeia* 533b6-c3: Against the Geometers' Hypotheses Because They Are Left Unmoved and 'Logos' Is Not Given of Them

the rest which we said did in some sort lay hold on true Being ("ontos")—geometry and the studies that accompany it—are, as we see, dreaming about true Being, but the clear waking vision of it is impossible for them as long as they leave the hypotheses ("hupothesesi") which they employ unmoved ("akinetous") and cannot give "logos" ("logon didonai") of them. (533b6-c3).

In the 533b6-c3 passage the mathematical hypotheses are said to be unable to move, namely to move upward by division to higher hypotheses, and also unable to be given Logos, the formation of ratio of a part with the immediately succeeding in its anthyphairetic generation part, which then satisfies the Logos criterion for anthyphairetic periodicity.

1.8.3.2 *Politeia* 533c3-6: Against the Geometers' Hypotheses Because, Not Having "Beginning, Between and End" These Hypotheses Are Not Known

> For where the beginning ("arche") is something that the geometer does not know, and the end ("teleute") and all that is in between ("metaxu") is interwoven ("sumpeplektai") is not truly known, what possibility is there that assent in such practice can ever become true knowledge ("episteme")?" "None," said he. (533c3-6)

The *Politeia* passage 533c3-6 points to an intelligible Being knowable by Division and Collection. This is so, because the passage should be compared to the *Parmenides* 145a8-b1 concerning the One of the second hypothesis:

> "Then the One, it appears, will have a beginning ("archen"), an end ("teleuten"), and a middle ("meson")." "It will.",

which was a direct consequence of the Division and Collection of the One (144e8-145a4), namely of periodic anthyphairesis; as we mentioned in Sect. 1.3, "beginning, middle–"metaxu", end" refers to the beginning, middle, and end of the period.

In *Phaedrus* 263e-264e, a dialogue whose dialectic is definitely Division and Collection, a truly living Being must have beginning, middle, end, unlike Lysias' speech, and for that criticized.

Thus the fact that true knowledge ("episteme"), unlike that knowledge supplied by the geometers (533c3-6), must possess Beginning–Middle–End is consistent with knowledge attained by means of Division and Collection, as with the One of the second hypothesis.

We note that according to 533b6-c6

(a) the geometers keep their hypothesis unmoved [by division of hypotheses] and therefore cannot achieve logos (533b6-c3),
(b) their knowledge does not have beginning, middle and end (533c3-6) [since they do not achieve logos and periodicity], and
(c) they are unable to ascend [to logos] by division of the hypotheses (533c6-d3).

1.8.4 The Nature of Plato's Criticism II Against the Axiomatisation of Mathematics

It is important to understand how the dialectic knowledge proposed by Plato replaces the mathematical knowledge. The axiomatic method of the mathematicians, criticized in the *Politeia* passages (examined in Sects. 1.8.1–1.8.3), accepts the truth of the hypotheses, definitions (in the sense that e.g. a straight line is a true and correct concept as defined) and postulates; once these are accepted, then everything follows, is deduced by pure logic, by the rules of inference. The weak point is that the definitions and postulates themselves are not known in any satisfactory sense, but in fact are arbitrary, hence in a sense nothing is known in Mathematics. Thus Plato rejects the axiomatic method of the geometers because this method is not founded on, and hence cannot lead to, knowledge.

Plato's method accepts as true Beings those dyads, mathematical or philosophic, that are in a periodic anthyphairesis. True Beings become known by the method of (anthyphairetic) Division and Collection—equalisation of parts by means of Logos, an analogue of the Logos criterion for periodic anthyphairesis. The hypotheses are simply the parts genera in the Division of this true Being and they are true and known only by receiving Logos (the ratio of the part—genus to its immediately next in generation is one in the finite, because of anthyphairetic periodicity, sequence of ratios). Thus, true knowledge is imparted upon the hypotheses, according to Plato, only by the method of Division and Collection (Sects. 1.3–1.6); hence Plato's insistence that hypotheses are not isolated, as in mathematics, but are parts of a true Being. Thus Plato believes that true knowledge may be obtained only within a true Being, namely a dyad satisfying the philosophic analogue of periodic anthyphairesis.

The three *Politeia* passages 510c2-d3, 511c3-d5, and 533b6-c3, considered in the present section, explain only the meaning of Plato's criticism II, that they use hypotheses as principles, not as steps toward "anhupotheton".

From these passages it is clear that the geometers, unlike the dialecticians, take the three fundamental hypotheses given in *Politeia* 510c2-d3 as starting principles, and, for this reason, move downwards. The geometers' hypotheses are not parts-genera of the Division of a true Being, but are considered in isolation, and thus cannot be divided, cannot give their place to higher hypotheses, and thus have no way to be provided with "logos", lose their hypothetical status, and become true and known. This is the problem of the geometrical hypotheses: they cannot be provided with "logos", the "logos" of anthyphairetic periodicity that, according to Plato, alone bestows its recipients with intelligible existence and knowledge. Without "logos" a hypothesis is just that, a hypothesis, an arbitrary, not itself not known and therefore not able to produce knowledge, starting principle. And since, according to the axiomatic method every proposition and construction in Mathematics is deduced from these initial arbitrary and unknown principles, all mathematical knowledge is open to doubt.

1.9 Criticism II: The Reason the Geometers Are Compelled to Proceed Downwards Treating Hypotheses as Principles

1.9.1 The Geometers Are Compelled to Use Hypotheses as Principles (510b4-6, 511a3-b2, 511c6-8)

On the other hand the geometer is **compelled** to consider hypotheses as starting principles, not as steps for ascending to higher hypotheses, eventually to the anhupotheton.[31]

> the soul is **compelled** ("**anagkazetai**", 510b5) to investigate starting from hypotheses ("ex hupotheseon"),
> by using as images ("hos eikosin") the things imitated in the former division, proceeding not up to a principle ("ep'archen") but down to a conclusion ("epi teleuten") (510b4-6)
> "This then is the class that I described as intelligible ("noeton eidos"),
> it is true, but with the reservation that
> the soul is **compelled** ("**anagkazomenen**", 511a4) to employ hypotheses ("hupothesesi") in the investigation of it, not proceeding to a principle ("ep' archen"), because of its inability to step out of ("ekbainein") and rise above ("iousan... anotero") its hypotheses ("ton hupotheseon"), and using as images or likenesses the very objects that are themselves copied and adumbrated by the class below them, and that in comparison with these latter are esteemed as clear and held in honor." (511a3-b2)
> those who contemplate them are **compelled** to use their understanding and not their senses ("dianoiai men **anagkazontai** alla me aisthesesin", 511c7)(511c6-8).[32]

1.9.2 The Reason That the Geometers Are Compelled to Employ Hypotheses as Principles Is That They Form Their Arguments for the Sake of Actions/Geometric Constructions (527a1-b2)

> "This at least," said I, "will not be disputed by those who have even a slight acquaintance with geometry, that
> this science is in direct contradiction ("pan tounantion echei") with the formation of arguments employed in it by its adepts ("tois en autei logois legomenois")."
> "How so?" he said.
> "Their language is most ludicrous ("mala geloios"), though *necessary* ("*anagkaios*", 527a6),
> for ("gar", 527a7)

[31] ψυχὴ ζητεῖν **ἀναγκάζεται** ἐξ ὑποθέσεων, οὐκ **ἐπ' ἀρχὴν** πορευομένη ἀλλ' ἐπὶ τελευτήν (510b4-6); ὑποθέσεσι δ' **ἀναγκαζομένην** ψυχὴν χρῆσθαι 511α4 περὶ τὴν ζήτησιν αὐτοῦ, οὐκ ἐπ' ἀρχὴν ἰοῦσαν, ὡς οὐ δυναμένην τῶν ὑποθέσεων ἀνωτέρω ἐκβαίνειν (511a3-b2).

[32] τὸ ὑπὸ τῶν τεχνῶν καλουμένων, αἷς αἱ ὑποθέσεις ἀρχαὶ καὶ **διανοίᾳ μὲν ἀναγκάζονται ἀλλὰ μὴ αἰσθήσεσιν** αὐτὰ θεᾶσθαι οἱ θεώμενοι 511c6-8, in the passage 511c3-d5, examined in Sect. 1.8.2.

they construct all their arguments ("pantas tous logous poioumenoi")
as if they were *acting* ("prattontes") and
for the sake of action ("praxeos heneka", 527a7).
For all their talk is of squaring ("tetragonizein") and extending ("parateinein") and adding ("prostithenai") and the like,
whereas in fact the entire study in mathematics ("mathema") is
for the sake of pure knowledge ("gnoseos heneka", 527b1)."
"That is absolutely true," he said. (527a8-b2)

Passage 527a1-b2 is crucial, because it is the only one that not simply describes, but explains **why** the geometers are compelled to employ hypotheses as principles. So we are justified in looking closely at the passage. At first Plato states that the geometers, when forming their arguments, are in opposition, in contradiction, with the nature of Geometry. Here what is meant no doubt is the often previously stated criticism that the geometers proceed downwards and not upwards as they should, in the opposite direction of the correct one, another way of stating that the geometers employ the hypotheses as principles and not as steps for proceeding upwards to the anhupotheton (527a2-4).

Next he states that the geometers are *compelled* into this "ridiculous" situation: "*anagkaios*", 527a6.

Next he sets to provide the reason why ("*gar*", 527a7) the geometers are compelled to proceed downwards. The reason is that the geometers form their arguments for the sake of action ("praxeos heneka", 527a7) and not for the sake of pure knowledge ("gnoseos heneka", 527b1), as they should. We thus have here a causal connection ("gar") between the compulsion of the geometers to employ hypotheses as principles (the essence of Criticism II) and the fact that the geometers form their arguments for the sake of actions. By pure knowledge we well understand that Plato means the knowledge of the logos of periodicity, the method of Division and Collection, as already explained. But we must look more carefully at the unusual term "praxis". Fortunately, Plato provides us with a rather detailed explanation of what exactly he means by "praxis" (527a6-9); he mention three examples of "actions": squaring (of a line),[33] adding (two lines),[34] extending (a line) to a greater line.[35] We realize that all three examples of action are geometric constructions. We then must come to the conclusion that by "action"-praxis-Plato means geometric construction.

Indeed constructions are basic in Euclidean geometry. In a typical Euclidean Proposition, the proof of the proposition is naturally divided into two parts: first are the required constructions on the visible diagram related to the proposition, next

[33] "tetragonizein" (squaring): possible by the hypothesis Proposition I.46.

[34] "prostithenai" (adding): possible by the hypotheses Postulate 2, Proposition I.2.

[35] "parateinein" (extending, stretching): possible by the hypothesis Postulate 5. The usual translation is "applying", presumably in the sense of application of areas (Propositions VI.28,29, where the term used is "parabalein"), presumably because Plato in *Meno* 87a5 is using "parateinein" in exactly this sense. But "parateinein" has the principal meaning of "stretching out along", "extend", cf. [17, p. 1327].

are the deductive steps, making use of these constructions, in which there is no need of the visible diagram. Thus Plato by actions, refers to the first part of the proof, in which the necessary geometrical constructions are set forth. The rest of the proof is "pure logic", rules of inference.

But these intelligible constructions/actions are possible only by employing the geometric hypotheses, the basic Definitions and Postulates of Euclidean Geometry. In mentioning the three specific actions of squaring, extending, adding, Plato might have in mind the (Pythagorean) Proposition II.10, perhaps the only proposition in the *Elements* where all these three actions appear. This diagram starts with a straight line, and is completed gradually with actions, such as addition of straight lines,[36] forming squares with given side,[37] (in Proposition II.10). extension of non-parallel (by Postulate 5) lines till they meet.[38]

Thus the reason the geometers are compelled to proceed downwards, treating hypotheses as principles, and not as steps of ascent, lies with the thesis of the geometers that Geometry is a deductive science, it is about proofs of propositions. A proof requires geometric constructions/actions, and actions depend on geometric hypotheses, Definitions and Postulates. Thus the reason of the compulsion of the geometers to use hypotheses in a defective way, as principles, is purely intelligible and has nothing to do with the fact that the geometers employ visible/sensible diagrams. Roughly speaking, the geometers have adopted the axiomatic method because they think of Geometry as a deductive science, for deductions and proofs constructions are needed, and constructions are themselves postulates or need postulates.

1.10 The Ascent from Intelligible Hypotheses to the Anhupotheton Generates Intelligible Number, Straight Line, and Circle (*Parmenides* 137e, 142b-145b, *Politeia* 522b8-526c8). Plato's Praise of Geometry (*Politeia* 527b3-c2)

At this point we have completed our interpretation of Platonic dialectics: Platonic Ideas, of which the One of the second hypothesis in the *Parmenides* is the pararadigm, are knowable by means of Division and Collection, which is a philosophic analogue of Logos criterion, periodic anthyphairesis and resulting equalization of parts (Sect. 1.3); while the ascent from hypotheses to the anhupotheton in the *Politeia* was found to be just another description of Division and

[36]"προστεθῇ", "τῇ προσκειμένῃ", "προσκείσθω" (in Proposition II.10).

[37]"tetragonizein" (squaring): "τὸ ἀπὸ τῆς ὅλης", "τὸ ἀπὸ τῆς προσκειμένης", "τὰ συναμφότερα τετράγωνα", "τὰ ἀπὸ τῶν ΑΔ, ΔΒ τετράγωνα", "ἀπὸ τῶν ΑΓ, ΓΔ τετραγώνων" (in Proposition II.10).

[38] "ἐκβαλλόμεναι", "ἐκβεβλήσθωσαν" (in Proposition II.10).

Collection, where the hypotheses are the parts-species of the Division, ascending by division to the anhupotheton, the Logos of periodicity and the resulting Collection (Sect. 1.5 and 1.6).

We are now well prepared to explain what Plato regards as the intelligible numbers, and the intelligible straight line and circle, and how they receive Logos. We will deal with the three kinds of angle in Sect. 1.11.

1.10.1 *Intelligible Numbers (Parmenides 142b1-145 and Politeia 522b8-526c8). The Mathematical Hypothesis of the Even and the Odd (Definitions VII.1, 2 of Unit and Number in the* Elements*), and the Ascent by Division and Collection from Hypotheses to Anhupotheton Leading to Intelligible Numbers (*Parmenides, Philebus *16c-17a and* Politeia *522b8-526c8)*

Dialectic numbers have been adequately treated in the second hypothesis in the *Parmenides* (outlined in Sect. 1.3). In order to generate the dialectic numbers we start with the indefinite dyad that exists in any intelligible true Being, the One and the Being, for the paradigmatical one in the *Parmenides*. The anthyphairetic Division produces an infinite multitude of parts, the ever higher hypotheses, ascending, after a finite number of stages, to the Logos criterion of anthyphairetic periodicity, the anhupotheton of the *Politeia*, and the true intelligible numbers, consisting of equalized units, and indivisible in the sense of Xenocrates, explained in Sect. 1.7.

A dialectic number has been defined as follows:

Definition Let

$$One > Being > One_1 > Being_1 > \ldots > One_{n-1} > Being_{n-1} > One_n > \ldots,$$

denote the infinite sequence of remainders of the anthyphairesis of the indefinite dyad One, Being of the one of the second hypothesis in the Parmenides, and let k be the least natural number satisfying the Logos criterion for the periodicity of the anthyphairesis of a to b:

$$One/Being = One_k/Being_k.$$

A dialectic (intelligible) number is a number consisting of units the parts of the initial segment

$$One > Being > One_1 > Being_1 > \ldots > One_{k-1} > Being_{k-1}.$$

The same ascent to the anhupotheton and the intelligible numbers has been described in the *Politeia* 522b-526c. There, the quest for the intelligible numbers starts from a lower starting point, from the sensible (provoking) images of the intelligibles, which will be described in Sect. 1.14 below. But once the soul realizes the ascending jump from the sensible provoking image to the intelligible indefinite dyad, the remaining steps towards the full ascent to the anhupotheton, in the *Politeia*, are exactly the same as those described in the second hypothesis in the *Parmenides*, with which we attain the logos of periodicity, the resulting equalization of parts, so that the Many are One, and the achievement of dialectic number, with equalized units. Here are the corresponding statements:

(a) The One is One and infinite in multitude

> The One, then, split up by Being, is many and infinite multitude (*Parmenides* 144e3-5).
> For we see the same thing at once as one and as an infinite multitude (*Politeia* 525a4-6).

(b) Units of number equal to each other

> [Being is divided] into the same finite number of parts as One, for Being is not wanting to the One, nor the One to Being, but being two equalised throughout (*Parmenides* 144d5-e3).
> My good friends, what numbers are these you are talking about, in which the one is such as you postulate, each unity equal to every other without the slightest difference (*Politeia* 526a1-4)

(c) If both make the number two, then each is One

> "And if things are two, must not each of them be one?" "Certainly" "Then since the units of these pairs are together two, each must be individually one." "That is clear" (*Parmenides* 143d1-4).
> "And if it appears to be two, each of the two is a distinct unit." "Yes." (*Politeia* 524b7-9).

Since the process in the *Politeia* 522b8-526c8, explicitly an ascent (sphodra ano agei, 525b11-d7), necessarily to the anhupotheton, coincides from a certain point on with the Division and Collection of the One of the second hypothesis in the *Parmenides*, we obtain a confirmation that:
the intelligible hypotheses of the *Politeia* coincide with the parts of the One Being, the ascent corresponds to the anthyphairetic division of the hypotheses-parts, and the anhupotheton is achieved by the logos of periodicity in anthyphairesis.

1.10.2 The Intelligible Straight Line Generated by the Division of the One of the Second Hypothesis in the Parmenides (Parmenides 137e, 142b-145b)

Plato defines the intelligible straight line in the

> "And the straight ("euthu"), again, is that of which the middle ("to meson") is in front of ("epiprosthen") both the extremes ("eschatoin")." "It is" (*Parmenides* 137e3-4).

The definition of the straight line can naively be thought as being inspired from sense experience, such as the eclipse of the sun, the impossibility of communication between sun and earth, being caused by the obstruction of the light from the sun to the earth by the intervening moon (Proclus, *In Euclidem* 109,25–110,4). But Plato's definition has a deeper intelligible meaning, based on a dialectical meaning of 'communication', the equalisation.

> In the progress from the causes ("apo ton aition") in the true Beings ("en tois ousi"), the middle ("ta mesa") become divisive ("diairetiika") of the mutual communication ("koinonias") of the extremes ("ton akron") (Proclus, *In Euclidem* 110,4–8).

In order that the beginning and the end be able to communicate and be equalized with each other, periodicity and circularity is needed. The absence of Logos and periodicity leaves us only with infinite anthyphairetic Division, in which no equalization and communication takes place, and this characterizes straightness.

The vague Definition I.4 of the *Elements*

> A straight line ("eutheia gramme") is a line which lies ("keitai") equally, evenly ("ex isou") in relation to the points on itself ("tois eph' heautois semeiois").

can be thought of as derivative from the Platonic one:
the straight line lies equally, evenly in relation to the points on it
exactly because
the end points in a straight line cannot communicate with each other (as the remarks in John Philoponus, Comments to Categories 13, 1, 154, 24–155, 8).

Plato shows, as we saw in Sect. 1.3, that the intelligible straight line is generated in the One of the second hypothesis in the *Parmenides* 145b1-5, as a consequence of the Division and Collection. Proclus' comments make it clear that the intelligible straight line is the product of the Infinite, namely of the infinite anthyphairetic division, in contrast to the circle that is the product of Collection, and anthyphairetic periodicity (cf. Proclus, *In Euclidem* 104,11-14; 107,11-15; 107,20-108,2).

The second advantage of the abbreviated Division, in the *Sophistes-Politicus* style, first mentioned in Sect. 1.5 (in connection with the description of the parts-species of the Division as hypotheses), is its use in the generation of the intelligible straight line.

The Division in the One of the second hypothesis in the *Parmenides*, namely the infinite anthyphairetic division of the dyad One Being, described in Sect. 1.3, abbreviated by considering only the odd stages of the anthyphairesis, induces a dichotomy scheme, as in the Divisions in the *Sophistes* and the *Politicus*, and this dichotomy scheme generates the intelligible straight line, as in Zeno's first paradox of motion, as described in Simplicius, *eis Phusica* 453,30–454,9, and outlined as follows:

Division, intelligible motion of the one of the second hypothesis	Generation of intelligible straight line
$Whole = One + Being$	We start with the whole B, which consists of two parts One and $Being$, $B = One + Being$
	we leave the part **One** undivided (corresponding to the distance covered by the runner after the first stage of Zeno's paradox)
$Being = One_1 + Being_1$	and recall the division of the part $Being$ in two parts One_1 and $Being_1$, $Being = One_1 + Being_1$;
	we leave the part One_1 undivided, and
	we add it to the previous undivided part One, $One + One_1$ (corresponding to the distance covered by the runner after the second stage of Zeno's paradox),
...	...
$Being_k = One_{k+1} + Being_{k+1}$	and recall the division of the part $Being_k$ in two parts One_{k+1} and $Being_{k+1}$, $Being_{k+1} = One_{k+1} + Being_{k+1}$;
	we leave the part One_{k+1} undivided
	we add it to the previous undivided parts $One + One_1 + \ldots + One_k + One_{k+1}$ (corresponding to the distance covered by the runner in the $(k+1)^{th}$ stage of Zeno's paradox),
...	...

The intelligible straight line is thus generated from the Division, by going from parts to "higher" parts, namely parts generated in higher stages of the anthyphairesis; and since, as we saw in Sect. 1.5, the intelligible hypotheses are precisely the parts in the anthyphairetic division of the One Being, the straight line is generated by going from hypotheses to higher hypotheses. The parts-hypotheses are in fact stepping stones ("epibaseis", 511b6) for higher parts-hypotheses. It is clear that the ascent is realized by division. The intelligible straight line is generated without reaching the Logos, Collection, and "anhupotheton".[39]

[39]Platonists seem to believe that the intelligible straight line is itself a true Being that can somehow be conceived as perfect straight line, without any imperfections of a sensible straight line; but, as we have found, the intelligible straight line is present in **every** intelligible true Being, and is the manifestation of its anthyphairetic Division.

1.10.3 *The First Three Postulates of Geometry in the Intelligible One*

The first three Postulates in the *Elements* are:

> Let the following be postulated:
> [Postulate I]. To draw a straight line from any point to any point.
> [Postulate II]. To produce a finite straight line continuously in a straight line.
> [Postulate III]. To describe a circle with any centre and distance.

Proclus, *In Euclidem* suggests that these Postulates follow from the definitions of the point, the straight line, and the circle, and the motion of a point:

> The drawing of a line from any point to any point [Postulate I] follows from the conception of the line as the flowing ("rhesin") of a point and of the straight line as its uniform ("homalen") and undeviating ("aparegkliton") flowing ("rhesin"). (185,8-12)
> And if we take a straight line as limited by a point ("semeioi peratoumenes") and similarly imagine its extremity ("to peras") as moving uniformly over the shortest route, the second postulate will have been established. (185,15-18)
> And if we think of a finite line as having one extremity moving about this stationary point, we shall have produced the third postulate. (185,19-21)

But as we have explained in Sect. 1.7, Proclus' point behaves like the indivisible line and the One of the second hypothesis in the *Parmenides*, and Proclus' motion and rest of a point like the motion and rest of the One of the second hypothesis, explained in Sect. 1.3, are really the Division and the self-similarity of the One of the second hypotheses, respectively.

We leave to the reader to explore the details on the manner in which Proclus establishes the validity of the first three Postulates from the definition and the motion of the point.

1.10.4 *The Intelligible Circle Is Generated by the Anthyphairetic Periodicity and Resulting Collection of the One of the Second Hypothesis in the* Parmenides

> "The round ("stroggulon"), of course, is that of which the extremes ("ta eschata") are everywhere equally distant ("ison apechei") from the middle ("apo tou mesou")." "Yes" (Proclus, *In Euclidem* 110,4-8).

The Definitions I.15,16 of the *Elements* is not substantially different from Plato's, except that Plato avoids the use of the term "kentron" (center), preferring "meson" (middle).

Plato shows, as we saw in Sect. 1.3, that the intelligible circle is generated in the one of the second hypothesis in the *Parmenides* 145b1-5, as a consequence of the anthyphairetic periodicity and resulting Collection. Proclus makes it perfectly clear that the intelligible circle is indeed produced from the Division and Collection. We emphasise that the intelligible circle is exactly a consequence of the anthyphairetic periodicity.[40]

> Let us once more ascend from these details to the contemplation of their paradigms. Let us think among them of the center, with its unitary, indivisible, and steadfast superiority in every instance; the distances from the center [rays], as the ways in which the One issues forth as far as possible into infinite plularity [Division of One into Many]; and the circumference of the circle, as the element through which, in the return to the center by the things that have gone forth from it, the multitude of powers are collected ("elissestai") into their own union ("henosin"), all pressing toward it and desiring activity around it. [Collection of Many into One]. (Proclus, *In Euclidem* 153,12-22, translation based on Morrow [20, pp. 121–122])

Thus, the center of the intelligible circle is the intelligible One, not a point of course but still an indivisible entity, the rays of the intelligible circle are the intelligible straight lines generated by the infinite anthyphairetic Division-motion of the One, and the circumference of the intelligible circle generated by anthyphairetic periodicity.

Thus, the same initial indefinite dyad One Being generates, by anthyphairetic Division-ascent the parts-hypotheses, and, as we saw in Sect. 1.10.2, the straight line, while the Logos of Periodicity and Collection—equalisation lead to ascent to the "anhupotheton".

If we consider the equalization that results from periodicity, then we obtain the units for the intelligible numbers, as we saw in Sect. 1.10.1, but if consider just periodicity itself, then we obtain the intelligible circle.

Platonic analysis and synthesis, in relation to Aristotelian division-definition-analysis and synthesis-deduction, in the *Analytica Protera*, and *Hustera*, and the *Nicomachean Ethics*,, and Pappus' theoretical account of theoreic and problematic geometric analysis and synthesis, in the *Collection*, has been studied in Negrepontis-Lamprinidis [26] and Lamprinidis [15]. A rough outline of the intelligible ascent and descent has been given in Sect. 1.6.3, in connection with the Proclus, *In Euclidem* 211, 23-27 passage. We will expend no space to deal with it in more detail.

Note that in this passage the ascent is from hypotheses to higher hypotheses, and the ascent and descent is by kinds ("eide"); thus, the hypotheses are "eide", confirming our interpretation of intelligible hypotheses as parts-kinds in the Division, given in Sect. 1.5.

[40]Platonists seem to believe that the intelligible circle is itself a true Being that can somehow be conceived as perfect circle, without any imperfections of a sensible circle (cf. Annas [3, p. 251]: "mathematicians talk about circles and lines, not about the physical diagrams that illustrate them, nor about the unique Form of Circle and Form of Line"); but, as we have found, the intelligible circle is in every intelligible true being, and is the manifestation of its anthyphairetic periodicity and Collection.

1.10.5 Plato's Praise of Geometry (527b3-c2)

We have explained so far that the intelligible structure of true Beings generates basic mathematical notions at a purely intelligible level. The sole structure is Division and Collection, and, as we have seen, is modeled after periodic anthyphairesis. In such a true Being we have seen that the intelligible numbers, the intelligible straight line and circle (with the first three Postulates), and intelligible analysis and synthesis-demonstration are generated (and, as we will see below in Sect. 1.11, the intelligible angles of three kinds are also generated).

It is then quite natural that Plato, in the *Politeia* 527b-c, expresses his admiration for Geometry, calling it a subject concerned with true Beings and true knowledge. The last statement is strongly reminiscent of the legendary "motto" that was supposed to stand at the entrance of the Academy:

> "And must we not agree on a further point?" "What?" "That it is the knowledge ("gnoseos") about the eternally existent ("tou aei ontos"), and not about something which at some time comes into being and passes away." "That is readily admitted," he said, "for geometrical knowledge ("geometrike gnosis") is about the eternally existent ("tou aei ontos")." "Then, my good friend, it would tend to draw ("holkon") the soul to truth, and would be productive ("apergastikon") of a philosophic attitude of mind ("philosophou dianoias"), directing upward the faculties that now wrongly are turned earthward." "Nothing is surer," he said. "Then", said I, "we must order in the strongest possible terms ("malista prostakteon') that the men of your Ideal City shall in no way ('medeni tropoi') neglect ('aphexontai') geometry". (527b3-c2)

1.11 Proclus' Completion of Criticism II and Praise of the Pythagoreans for Their Success to Derive the Three Kinds of Angles Dialectically

Plato does not explain at some point what would be the dialectical treatment of the three kinds of angle, he has mentioned in the *Politeia* 510c2-d3. But fortunately Proclus *In Euclidem* 131,3-134,7 and 188,20-189,12 has quite illuminating comments, indicating that the dialectical treatment of the three kind of angle in terms of the side and diameter numbers, their satisfaction of the Pell equation $y^2=2x^2\pm 1$, and Propositions II.12 and 13 in the *Elements*, is an early Pythagorean discovery.

1.11.1 The Three Kinds of Angles in the Elements

The Euclidean definitions of the three kinds of angles Definitions 10–12 in the first Book of the *Elements* define the right, obtuse and acute angle, respectively.

> Def. I.10. When a straight line set up on a straight line makes the adjacent angles equal to one another, each of the equal angles is a right angle....

Def. I.11. An obtuse angle is an angle greater than a right angle.
Def. I.12. An acute angle is an angle less than a right angle.

1.11.2 *Politeia 510c2-6: The Geometers Do Not 'Provide Logos' for the Hypothesis 'the Three Kinds of Angles'*

Plato, in the *Politeia* 510c2-d3 criticizes, as we have seen in Sect. 1.8, these definitions as being 'from hypothesis' ('ex hupotheseos'), according to the practice of the geometers, and there is need to ascend from hypotheses to the "anhupotheton" and thus to be able to 'provide Logos' to the hypothesis. As we have seen in Sect. 1.6, the ascent will be realized by Division and Collection of a true Being.

1.11.3 *Proclus: The Pythagoreans 'Provide Logos' for the Three Kinds of Angles*

Proclus' illuminating comments in *In Euclidem* 131,3-134,7 explain how this happens with the special case of the hypotheses of the three angles.

> These are the three kinds of angle that Socrates in the *Politeia* [510c2-6] says are accepted as hypotheses by geometers,... the right, the obtuse, and the acute....
> Most geometers are unable to provide "logos" for this classification but take it as a hypothesis that there are three angles; and if we demand a cause [for this classification], they deny that we have a right to ask it of them. (Proclus, *In Euclidem* 131, 17-21)

We have seen that the dialectical-Platonic expression "to provide Logos" refers to the Logos criterion with which periodic anthyphairesis is established.

Proclus further explains that the Pythagoreans are able to 'provide Logos' to this hypothesis by employing their principles of Infinite and Finite. (The Pythagoreans would not use the Logos terminology, but rather its equivalent precursor, the preservation of Gnomons).

> But the Pythagoreans, who refer the solution of this triple distinction to **first principles**, have no difficulty in giving the causes of this difference among rectilinear angles... For one of their principles is constituted by **the finite** ('peras') which is the cause ("aitia") of definiteness ("horou") and self-identity ("tautotetos") of all things that have come into completion, of equality ("isotetos") and of everything in the better of the two columns of contraries;
> their other principle is the **infinite** ("apeiron") which produces progression to infinity ("ep' apeiron proodon"), increase and diminution ("auxesin kai meiosin"), inequality ("anisoteta"), and every sort of difference ("heteroteta") among the things it generates, and in general is the head of the inferior column. (Proclus, *In Euclidem* 131, 21-132, 6)

The Pythagorean principle of the Infinite coincides with the infinite anthyphairesis and Division, and the principle of the Finite with the preservation by means of Gnomons of the same equation of application of areas in excess, the precursor of the Logos criterion, and Collection.

The Pythagoreans provide Logos by appealing to the two principles of the Infinite and the Finite. Such an appeal is essentially an appeal, in Plato's language, to Division and Collection. But an appeal to Division and Collection has meaning only in connection with a dyad in periodic anthyphairesis, geometrical or philosophical. In fact Plato, in the *Politeia* 510d5-e1, refers to the square itself and the diameter itself ("tou tetragonou autou... kai diametrou autes") right next to 510c2-6; indeed the dyad <diameter, side of a square> is in periodic anthyphairesis, and thus, according to our analysis of intelligible Platonic Beings in Sects. 1.3–1.6, qualifies as a true intelligible Platonic Being. For the Pythagoreans, the only entity on which they employed these principles would be the diameter to the side of the square. For this entity the Logos of periodicity has been described in Sect. 1.2.4. Thus we conclude that Proclus in these comments has in mind the diameter to the side of a square.

1.11.4 Proclus Connects the Dyad <Acute, Obtuse Angle> with the Infinite and the Right Angle with the Finite

But next Proclus, *In Euclidem* 132, 8-17, connects the principle of the Infinite, namely the Division of the diameter to the side of a square, with the acute and obtuse angles, and the principle of the Finite, namely the Logos and Collection of the diameter to the side of a square, with the right angle:

> The idea which proceeds from **the finite** ("peratos") should produce the right angle ("orthen gonian"), one ("mian"), ruled by equality ("isoteti") and similarity ("homoioteti") to every other right angle, always determinate ("horismenen") and staying the same ("ten auten hestosan"), not admitting of either growth ("auxesin") or diminution ("meiosin");
> whereas the idea that comes from the **infinite** ("apeirias"), being second in rank and dyadic ("duadikos"), reveals a pair of angles divided about the right angle by inequality ("anisoteti") according to the greater and smaller ("kata to meizon kai elasson"), and subject to infinite motion ("aperanton kinesin") according to the more and less ("kata to mallon kai hetton"), the one becoming obtuse ("amblunomenes") more and less, the other becoming acute ("oxunomenes"). (Proclus, *In Euclidem* 132, 8-17)

Thus Proclus connects the dyad < acute, obtuse angle> with the anthyphairetic Division of the diameter to the side of a square, and the right angle with Logos and Collection. These connections might appear to make little sense and call for an explanation. But we will show now that they make, indeed, perfect sense.

1.11.5 Side and Diameter Numbers, the Ancient Convergents, and the Solution of Pell's Equation for $N = 2$

Instrumental for our interpretation is the sequence, Pythagorean in origin, of side and diameter numbers. It is known that the Pythagoreans, following the discovery of the incommensurability of the diameter to the side of a square, outlined in Sect. 1.2, proceeded with the introduction and study of the side and diameter numbers, the ancient precursors of the convergents of the square root of 2, and, based on them, with the solution of Pell's equations

$$x^2 = 2y^2 + 1 \text{ and } x^2 = 2y^2 - 1.$$

These important discoveries are recounted by the neo-pythagoreans Theon Smyrneus, *eis Platona* 43,1-45,8 and Iamblichus, *eis Nikomachon* 91,4-93,7, and by Proclus, *eis Politeian* 2,24,16-25,13;2,27,1-29,4.

1.11.5.1 Anthyphairetic Definition of the Side and Diameter Numbers

Definition For every natural number n, the nth side p_n and the nth diameter q_n number are the unique natural numbers such that

$$\text{Anth}(q_n, p_n) = [1, 2, 2, 2, \dots (n-1 \text{ times})]$$

(namely equal to the initial segment of length n of the infinite anthyphiresis of the diameter to the side of a square), and q_n, p_n are relatively prime.

Since $\text{Anth}(q_n, p_n) = [1, 2, 2, \dots, 2(n-1 \text{ times})]$ equals to the initial sequence of n elements of the sequence of quotients of the anthyphairesis of the diameter to the side of a square, the anthyphairetic Division of the diameter to the side of a square up to stage n can be described equivalently by the side and diameter numbers, the ancient convergents, of level n.

Thus the infinite anthyphairetic Division

$$\text{Anth}(\text{ diameter, side of a square}) = [1, 2, 2, 2, \dots]$$

is described equivalently by the whole approximating sequence (q_n, p_n) of side and diameter numbers.

1.11.5.2 Recursive Definition of the Side and Diameter Numbers

The definition of the side and diameter numbers is preserved by Nicomachus, Iamblichus, and Proclus, in an equivalent recursive form as follows:

$$p_1 = 1 = q_1,\ p_{n+1} = p_n + q_n,\ q_{n+1} = q_n + 2p_n \text{ for every natural number } n.$$

The recursive definition is a consequence of the anthyphairetic definition. We omit the rather straightforward proof of this implication.

1.11.5.3 Pell Property of the Side and Diameter Numbers

The recursive definition together with Proposition II.10 of the has been used by the Pythagoreans for an inductive proof of the following

Proposition (The Fundamental Pell Property of Side and Diameter Numbers)

$$q_n^2 = \begin{cases} 2p_n^2 - 1 \text{ if } n \text{ is odd, and} \\ 2p_n^2 + 1 \text{ if } n \text{ is even.} \end{cases}$$

It has been strongly contested by scholars whether these neo-pythagorean authors really describe a strict (inductive) proof. (Freudenthal [11], van der Waerden [34] argued in favor, Fowler [10], Unguru [33, p. 286], Acerbi [1, p.72]) against). The present author has novel arguments in favor of the inductive character of this proof. Details can be found in Chapter 8 of the book by S. Negrepontis and V. Farmaki, *History of ancient Greek Mathematics*, Vol. I, 2019 (in Greek).

1.11.6 Pell Property and the Dyad of Acute and Obtuse Angles

The Pell property is instrumental for the dialectical treatment of the three kinds of angles.

Definition For any natural number n let p_n, q_n denote the side and diameter numbers of level n. We form the isosceles triangle $A_nB_nC_n$, approximating the rectangular triangle, with sides $A_nB_n = A_nC_n = p_n$, and $B_nC_n = q_n$, and we let h_n denote the angle $B_nA_nC_n$, approximating the right angle.

Proposition *The angle h_n is acute if n is odd, and obtuse if n is even.*

Proof By the Pell property of side and diameter number for n odd and even number, and the Propositions II.12 and 13 of the *Elements* applied to the triangle $A_nB_nC_n$.

1.11.7 Dialectical Definition of the Three Kinds of Angle

Now Proclus' comments become well understood in the context of the Division and Collection of the diameter to the side of a square. Indeed

- the acute and obtuse angles are higher and higher hypotheses generated in the infinite anthyphairetic Division of the diameter to the side, while
- the ascent to the right angle is realized by the Logos criterion of periodicity and Collection of the diameter to the side.

It follows from the approximating nature of the sequence of side and diameters numbers that we can define and determine all acute and obtuse angles by the side and diameter numbers.

Since the infinite anthyphairetic Division of the diameter to the side determines the side and diameter numbers, it follows that the infinite anthyphairetic Division of the diameter to the side determines all acute and obtuse angles.

Definition (Dialectical Definitions of Acute And Obtuse Angle) An angle h is *acute* if and only if angle $h <$ angle h_n for some odd number n, and an angle h is *obtuse* if and only if angle $h >$ angle h_n for some even number n.

The relation of the right angle with the principle of the Finite and the Logos Criterion. The first two steps of the anthyphairesis of the diameter a to the side b are

$$a = b + c_1, b = 2c_1 + c_2, \text{ hence } c_1 = a - b, c_2 = b - 2c_1 = 3b - 2a.$$

The Logos criterion $b/c_1 = c_1/c_2$, equivalently, the condition $b.c_2 = c_1{}^2$, is attained, since the condition is equivalent to $a^2 = 2b^2$, defining a right angle isosceles triangle by Pythagoras theorem and heir converses (Propositions I.47, II.12 and 13 of the *Elements*).

Thus Proclus' illuminating comments in *In Euclidem* 131,3-134,7 lead unmistakenly to the conclusion that the ascent of the status of the three kind of angles is realized not in any true Being in general, as was the case with numbers and line and circle, but in specific true Being diameter itself and the side itself.

1.11.8 The Fourth Postulate and Its Dialectical Content

We will consider the Fourth Postulate in Euclid's *Elements*.

Postulate IV. And that all right angles are equal to one another.

It is remarkable that Proclus, just before these philosophic comments, in 188,20-189,12, produces a proof of the Fourth Postulate, perfectly valid in fact, and agreeing in essence with the proof given in the Hilbert axiomatisation of Euclidean Geometry, concluding with the comment "this proof has been given by other commentators and required no great study."

It is clear then that the value of the Fourth Postulate is mostly philosophical, simply strengthening the dialectic status of the right angle.

> This postulate also shows that the rightness of the angles is akin to the equality, as the acuteness and obtuseness are akin to the inequality. The rightness is in the same column with the equality, for both of them belong under the Finite... But the acuteness and obtuseness are akin to the inequality...; for all of them are the offspring ("ekgonoi") of the Infinite. (Proclus, *In Euclidem* 191, 5-11)

Thus, according to Proclus, for the Pythagoreans and Plato, the **equality** of the right angles in the Fourth Postulate is a consequence of the principle of the Finite, namely of the Logos and Collection in the (intelligible Being) diameter to the side, by which the parts-hypotheses of the anthyphairesis of diameter to side are equalized. This is the only meaning of the dialectic equality.

1.12 Completion of Criticism II by Proclus: The Geometers' Failure to Derive the Fifth Postulate from a Platonic Principle of Finiteness

The stubborn belief by the ancient Platonists, expressed epigrammatically by Proclus

> This [the Fifth Postulate] ought to be struck altogether ("pantelos diagraphein") from the postulates (191, 21-22)

on the derivability of the Fifth Postulate (Proclus, *In Euclidem* 191,16-193,9), a postulate of Pythagorean origin,[41] appeared to rest on the general conviction that all Geometry was under the umbrella of Platonic philosophy.

It was realised that the fifth postulate is a principle of the finitisation of infinity. In fact Proclus, *In Euclidem* 394,8-400,16 describes the paradoxical nature of Proposition I.35, indicating that there is a similarity between Platonic Beings and Proposition I.35, or in fact of Propositions III.21,31, in the sense that they both exhibit a finitisation of infinity. From a mathematical point of view, it is clear that, while both the Fifth Postulate and periodic anthyphairesis, the basis for Plato's dialectics, are finitisations of infinity, nevertheless they are quite different sort of finitisations, e.g. there is no indication of periodicity as the cause of the Fifth Postulate.

Still Platonists, like Ptolemy or Proclus, thought that it would be possible to deduce in some sense the Fifth Postulate from some principle of the Finite. More specifically, Proposition I.30, shows that "parallelism" is "similarity of position"

[41] Since the Pythagoreans used it in the proof of Proposition I.32 of the *Elements*, according to Eudemus in Proclus, *In Euclidem* 379,1-18 (in contrapositive form), and in the proof of Proposition II.10 of the *Elements*, according to Proclus, *eis Politeian* 2.24-28 (in direct form).

(373,5-23) and that the Fifth Postulate is for parallelism what the Enallax property is for analogy (357,9-16). The attempted proof of the Fifth Postulate by Ptolemy (365,5-368,26) relies on some kind of alternation, reminiscent of the Enallax property, while Proclus'attempted proof, on an appeal to Aristotle's philosophic principle of the finite, proceeds in the same way that the Enallax property for magnitudes is proved in Book V (Proposition V.16), by an appeal to the Eudoxian principle (Definition V. 4 and Proposition V.8).

But in reality there was never a satisfactory account of the Fifth Postulate from the Platonic principles of Infinite and Finite, equivalently, the method of Division and Collection. The Fifth postulate has always been Achilles' heel of Plato's approach to Geometry.

1.13 The Sensibles in Plato's *Timaeus* 48a2-58c4

Back to Plato's dialectics. We wish to gain an essential understanding of the way Plato views the relation of Geometry with the sensible figures and diagrams geometers use in their proofs (to be studied in Sect. 1.14). For this it is first necessary to obtain a deep and correct understanding of the way Plato views the sensibles. There are some indications about them in various dialogues but the main source for Plato's sensibles is the *Timaeus* 48a-58c, where he introduces the receptacle and the hollow spaces. The present section has independent interest in that it outlines an original interpretation of the sensibles and the receptacle/hollow space. The basic ideas for the anthyphairetic interpretation of the sensibles, by means of the *Timaeus*' "diakena"-receptacle have been outlined in initial form in Negrepontis [21]; in detailed form they will appear in a forthcoming work by Negrepontis and Kalisperi [25].

1.13.1 The True Opinion of an Intelligible Being Is Identified with an Initial Finite Segment of the Anthyphairesis of the Intelligible Being

The sensibles participating in an intelligible Being are routinely described as copies, images, incomplete approximations of the intelligible in which they participate. But our interpretation of the nature of the intelligible Being and its knowledge by means of Division and Collection in terms of (anthyphairetic) Division and Collection (by means of Logos, establishing periodic anthyphairesis) makes possible a deeper and precise understanding.

Equivalent descriptions of Division and Collection-Logos are

(a) Name and Logos (*Parmenides* 155d8-e1, *Theaetetus* 201e2-202b5, *Sophistes* 218c1-5, 221a7-b2, 268c5-d5),

(b) true opinion ("alethes doxa") and Logos (*Symposium* 202a2-10, 209e5-211d1, *Meno* 97e2-98c4 (where instead of "logos", the equivalent expression "logismos aitias" is employed), *Theaetetus* 201c8-d3, 202b8-d7, *Timaeus* 27d5-28a5, 51d3-e6).

Furthermore, true opinion is described as

- alogos (*Symposium* 202a6, *Theaetetus* 201d1, *Timaeus* 28a3, 51e4), and
- finite in nature (*Sophistes* 264b1, *Scholia In Euclidem* X.2, 1-6).

It now follows, from our anthyphairetic interpretation of Division and Collection, that *a true opinion of an intelligible Being is the knowledge provided by some first finite steps of the anthyphairetic Division of the intelligible Being.*[42]

1.13.2 The Knowledge of a Sensible Participating in an Intelligible Platonic Being Is Identified with a True Opinion of the Intelligible Being

Once the basic nature of an intelligible being has been clarified (in Sects. 1.3–1.6), the sensibles participating in an intelligible being, the philosophic analogue of periodic anthyphairesis, there seems to be little choice for the sensibles, being described as copies, images, incomplete approximations of the intelligible to which they participate, but to be, in one way or another, the philosophic analogue of the anthyphairetic finite initial approximations, the modern "convergents", of the full infinite periodic anthyphairesis of the intelligibles. The model of the sensibles is the sequence of side and diameter numbers, indeed an initial segment of the full anthyphairesis of the intelligible diameter to the side of a square.

In view of our remarks in Sect. 1.13.1, this is confirmed

- in the *Politeia* 475e6-480a13 especially 479d3-10 ("doxaston" 478a11, b2, b3, e3, 479d7), and used in the analogy of the Divided Line 509d1-510b1 ("doxaston" 510a9), repeated briefly in the *Timaeus* 29c and
- in the *Timaeus* 27d5-28a4 (and 28b8-c2) ("doxaston" 28a3), and 51d3-52a7, on the need of the receptacle,

because in these passages the knowledge of a sensible that participates in an intelligible Platonic being is squarely equated with a true opinion of that intelligible Being.

It follows that the knowledge of a sensible that participates in an intelligible Platonic being is equated with the knowledge provided by some first finite steps of the anthyphairetic Division of the intelligible Being.

[42]Details are given in Negrepontis [22, Section 7].

The relation of participation of a sensible in an intelligible entity will, then, be completely analogous to the relation between a pair of diameter and side numbers to the diameter to the side of a square.[43]

1.13.3 *The Geometrisation of the Sensibles in the* Timaeus

1.13.3.1 The Association of the Four Elements of the Sensible World, Earth, Fire, Air, Water, with the Canonical Polyhedra Cube, Pyramid, 8hedron, 20hedron of Book XIII of the *Elements*, Respectively (*Timaeus 55d-56c*)

We now turn to the *Timaeus*.

In, the model of the perceptible world, as presented in the *Timaeus*, Plato, following the Greek physiology tradition, conceives of the sensible world as consisting of four elements, earth, water, air, fire. But to this tradition, he introduces a bold and unexpected geometrisation, exploiting the theory of regular solids, most likely completed not long before the writing of the *Timaeus* by Theaetetus, in the form presented in Book XIII of the *Elements*; according to it earth consists of cubes, fire of pyramids, air of octahedra, water of icosahedra (55d-56c).

1.13.3.2 The Fundamental Role of the Two Basic Triangles a and b for the Surfaces of the Four Polyhedral (*Timaeus* 53c4-54b5)

The most remarkable and in fact crucial consequence of this Platonic geometrisation is that the three of the four polyhedra involved, namely fire-pyramid, air-octahedron, water-icosahedra have their surface consisting entirely of equilateral triangles, while the fourth element, earth, has its surface consisting of squares.

Furthermore, a square consists of two rectangular isosceles triangles, called hereafter *triangle a*, and an equilateral triangle consists of two equal rectangular non-isosceles triangles, called hereafter *triangle b* (53c4-54b5). The importance of this consequence in the Platonic theory of sensibles will be analysed below.

[43] and in modern terms, analogous to the relation between the **convergents** of a continued fraction and the continued fraction itself.

1.13.4 The Incommensurability Between the Areas of the Triangles a and b, and Their Canonical Comparison by Equating Their Hypotenuses

1.13.4.1 The Incommensurability of the Areas of the Triangles *a* and *b*

Plato states, in 56d6-e7, that

> fire, air, water are freely exchangeable at the rational rate 4:8:20, namely at the rates corresponding to the respective numbers of triangles b covering the surface of each of these three solids,

while, in 54b6-c5,[44] and 56d1-6[45] that

> it is impossible that earth will ever turn into one of the other three elements.

It is natural to assume, and we do assume, that the impossibility of exchange of earth into the other three elements is explainable, as well, in terms of the surfaces of the solids corresponding to these surfaces, namely in terms of the impossibility of exchange of triangle a into triangle b.

Let us first state precisely the impossibility of the exchange of triangle a into triangle b: it means that any multiple Ma of the area of a cannot be equal to any multiple Nb of the area of b. Thus there are no numbers M and N such that the two areas are equal $Ma = Nb$. But we realize that this statement is precisely the statement that the area of the triangle a is **incommensurable** to the area of the triangle b.

Now triangle b is the constituent of the pyramid, and in 54d5-55a4 the construction of a pyramid from triangle b is outlined. This construction is closely related with the construction in Proposition XIII.13, where it is also shown that the pyramid is comprehended, inscribed in a given sphere, and that the square on the diameter of the sphere is one and a half times the (square) on the side of the pyramid.

Proposition [XIII.13]

Πυραμίδα συστήσασθαι καὶ σφαίρᾳ περιλαβεῖν τῇ δοθείσῃ καὶ δεῖξαι, ὅτι ἡ τῆς σφαίρας διάμετρος δυνάμει ἡμιολία ἐστὶ τῆς πλευρᾶς τῆς πυραμίδος.

---

[44] τὰ γὰρ τέτταρα γένη δι' ἀλλήλων εἰς ἄλληλα ἐφαίνετο πάντα γένεσιν ἔχειν, -οὐκ ὀρθῶς φανταζόμενα· γίγνεται μὲν γὰρ ἐκ τῶν τριγώνων ὧν προῃρήμεθα γένη τέτταρα, **τρία μὲν** ἐξ ἑνὸς τοῦ τὰς πλευρὰς **ἀνίσους** ἔχοντος, **τὸ δὲ τέταρτον** ἓν μόνον ἐκ τοῦ **ἰσοσκελοῦς** τριγώνου συναρμοσθέν.
οὔκουν δυνατὰ πάντα εἰς ἄλληλα διαλυόμενα ἐκ πολλῶν σμικρῶν ὀλίγα μεγάλα καὶ τοὐναντίον γίγνεσθαι, **τὰ δὲ τρία** οἷόν τε·(*Timaeus* 54b6-c5).

[45] γῆ μὲν συντυγχάνουσα πυρὶ διαλυθεῖσά τε ὑπὸ τῆς ὀξύτητος αὐτοῦ φέροιτ' ἄν, εἴτ' ἐν αὐτῷ πυρὶ λυθεῖσα εἴτ' ἐν ἀέρος εἴτ' ἐν ὕδατος ὄγκῳ τύχοι, μέχριπερ ἂν **αὐτῆς** πῃ συντυχόντα τὰ μέρη, πάλιν συναρμοσθέντα αὐτὰ αὑτοῖς, **γῆ** γένοιτο—**οὐ γὰρ εἰς ἄλλο γε εἶδος ἔλθοι ποτ' ἄν** (*Timaeus* 56d1-6).

To construct a pyramid, to comprehend it in a given sphere, and to prove that the square on the diameter of the sphere is one and a half times the square on the side of the pyramid.

Thus in a given sphere of diameter D, a pyramid is constructed with side p, comprehended in the given sphere, and $D^2 = (3/2)p^2$. The hypotenuse of the triangle b constituent to the constructed pyramid is the side p.

In the same way the triangle a is the constituent of the cube, and in 55b3-c4 the construction of a cube from triangle a is outlined. This construction is closely related with the construction in Proposition XIII.15, where it is also shown that the cube, like the pyramid ("ἧ καὶ τὴν πυραμίδα"), is comprehended, inscribed in a given sphere, and that the square on the diameter of the sphere is three times the (square) on the side of the cube.

Proposition [XIII.15]

Κύβον συστήσασθαι καὶ σφαίρᾳ περιλαβεῖν, ἧ καὶ τὴν πυραμίδα, καὶ δεῖξαι, ὅτι ἡ τῆς σφαίρας διάμετρος δυνάμει τριπλασίων ἐστὶ τῆς τοῦ κύβου πλευρᾶς.

To construct a cube and comprehend it in a sphere, like the pyramid; and to prove that the square on the diameter of the sphere is triple of the square on the side of the cube.

Thus in a given sphere of diameter D, a cube is constructed with side k, comprehended in the given sphere, and $D^2 = 3k^2$. The hypotenuse of the triangle a constituent to the constructed cube is the diameter q of the square with side k.

This suggests the way of the "**canonical**" comparison of the areas of the triangles a and b. Triangle a will be compared to triangle b, when triangle a is considered as constituent part of a cube and triangle b is considered as constituent part of a pyramid, AND these two solids, the cube and the pyramid are comprehended by the same sphere, as in fact in Proposition XIII.18.

We obtain, as a consequence, the following

Proposition *The hypotenuse p of the triangle b, namely the side of the pyramid, and the hypotenuse of the triangle a, namely the diameter q of the square side of the cube, when the pyramid and the cube are comprehended* ***in the same sphere****, are equal, $p = q$.*

Proof Indeed, in this common sphere of diameter say D, the side p of the pyramid satisfies, by Proposition XIII.13, $p^2 = 2/3D^2$, and the side k of the cube satisfies, by Proposition XIII.15, $k^2 = 1/3D^2$. This allows for the canonical comparison of triangle a and triangle b. The hypotenuse of the triangle b is the side p. The hypotenuse q of the triangle a is the diameter of the square side, hence satisfies $q^2 = 2k^2$. Thus we have

$$p^2 = 2/3D^2 = 2(1/3D^2) = 2k^2 = q^2.$$

Hence $p = q$.

We are then fully justified in comparing the areas of triangles *a* and *b* by identifying their hypotenuses.[46]

1.13.4.2 Comparison of the Areas of the Triangles *a* and *b*

Let us assume for convenience that the common hypotenuse has length 2. The surface of the earth-cube consists of squares, equivalently of rectangular isosceles triangles (with sides in ratio $\sqrt{2}, \sqrt{2}, 2$, and area 1), while the surface of the fire-pyramid consists of equilateral triangles, equivalently of rectangular scalene triangles of type *b* (with sides in ratio 1, $\sqrt{3}$, 2, and area $\frac{\sqrt{3}}{2}$).

1.13.4.3 The Anthyphairesis of Ma to Nb, for Any Numbers M,N, Is Periodic

The relation of any conglomeration of earth, consisting say of M elements, and having total surface area M.1, to any conglomeration of fire, consisting say of N elements, and having total surface area $N.\frac{\sqrt{3}}{2}$ is, thus, always incommensurable (*ibid.* 54b-d, 56b-e); and, from what was said above, we expect that the interaction of earth and fire will be anthyphairetic, and in fact following the anthyphairesis of M to $N.\frac{\sqrt{3}}{2}$.

The simplest (M=N=1) is

$$\text{Anth}(\text{area of triangle a, area of triangle b}) = \text{Anth}(2, \sqrt{3}) = \left[1, \text{period}(6, 2)\right],$$

but these anthyphaireses get quite complicated, e.g.

$$\text{Anth}(\text{area of triangles 4a, area of triangles 17b}) = \text{Anth}(8, 17\sqrt{3}) =$$
$$\Big[3, \text{period}\Big(1, 2, 7, 1, 1, 1, 3, 35, 1, 28, 2, 8, 1, 1, 3, 6, 1, 1, 5, \mathbf{7},$$
$$5, 1, 1, 6, 3, 1, 1, 8, 2, 28, 1, 35, 3, 1, 1, 1, 7, 2, 1, \mathbf{6}\Big)\Big].$$

In fact there is an infinite variety of combinations, Ma to Nb, as noted in 57c7-d6 but always remaining commensurable in power, hence, according to Theaetetus' theorem, with palindromically periodic anthyphairesis.[47]

---

[46]In retrospect we realize that the reason why Plato takes the half-square triangle a, and the half isopleuron triangle b, is for the purpose of having two right angled triangles with equal hypotenuses.

[47]Negrepontis [23].

On the other hand, the interactions between any two of the three fire, air, water is of the form Mb to Nb for any numbers M,N, and is thus governed by finite anthyphairesis.

So according to Plato's *Timaeus* the really different elements in the material world are just two, the earth on one side, whose surface elements consist of triangles of the type a, and the fire-air-water on the other, whose surface elements consist of triangles of the type b, a dyad of opposites.

1.13.5 The Problem of Participation of the Sensibles in the Intelligibles

On the one hand, we expect that the sensibles are governed by their participation in the intelligible. In the *Timaeus* 51b6-52a7, and also in the *Philebus* 29a6-d6, Plato expressly introduces the intelligible fire and earth, and states that the sensible fire and earth participate in the intelligible fire and earth.[48] According to our analysis of intelligible Beings, the intelligible Being that shapes the material world consists of a dyad of opposites (opposite line segments), the intelligible Fire B and Earth A, in periodic anthyphairesis. It is clear that the anthyphairesis of the intelligible A to B must be equal to the anthyphairesis of the area of the triangle a to the area of the triangle b. Thus, the Platonic idea in which the four elements participate is the idea with the dyad

$$\langle \text{ ideal earth A, ideal fire B}\rangle,$$

with A, B line segments, such that the ratio

$$A/B = \text{ ratio area of triangle } a/ \text{ area of triangle } b.$$

(Such lines A,B can be constructed by choosing a "protetheisa" line, say r, use Proposition I.44 of the *Elements* to find lines A, B such that $a = rA, b = rB$, and finally use the pivotal Proposition VI.1 of the *Elements*, to conclude that $A/B = a/b$. Then, according to the Theaetetus approach to the theory of ratios of magnitudes, reported by Aristotle in the *Topics* 158b-159a, the anthyphairesis of A to B equals the anthyphairesis of a to b, hence (parindromically) periodic; thus the dyad A, B is a Platonic idea). The sensibles participate into the intelligible, as explained in Sect. 1.13.2, by means of an **initial segment of the full infinite periodic intelligible anthyphairesis.**

[48] ἔστιν τι **πῦρ αὐτὸ ἐφ' ἑαυτοῦ** καὶ πάντα περὶ ὧν ἀεὶ λέγομεν οὕτως **αὐτὰ καθ' αὑτὰ ὄντα** ἕκαστα. (*Timaeus* 51b6-52a7).

On the other hand, in the model of the physical world with only four, and for Plato, in fact, essentially only two, elements, actual **anthyphairesis** in the form of successive divisions is immediately seen to be **impossible**: for, if a is earth-cube and b is fire-pyramid, then the successive remainders in their anthyphairetic intercourse, in fact infinitely many because of incommensurability, will necessarily correspond to ever new elements, beyond the initial ones. Thus if

$$a = k_1 b + a_1, \text{ with } a_1 < b,$$

is the first step in the anthyphairetic division of a to b, the remainder a_1 is a new element, different from earth a, and different from fire/air/water b.

Plato specifically rejects the possibility that the world is made up of an infinite multitude of elements, or "kosmous" (55c7-d6).[49]

Thus the triangles a and b cannot interact by anthyphairetic division.

[The triangles a and b on the other hand are divisible in only one way, that does not change their shape: the isosceles rectangular triangle of type a is divisible in two smaller isosceles rectangular triangle also of type a [dichotomy] (55b4-7), and the non-isosceles rectangular triangle of type b is divisible in three smaller non isosceles rectangular triangle also of type b [trichotomy] (54d5-e3).]

At this point we face a seemingly impossible situation:
the traditional Greek theory, to which as we saw Plato subscribes, that all sensibles are produced from (only) the four elements (55c7-d8) appears to be incompatible and at odds with the anthyphairetic interpretation of the sensible world, which, as a consequence of our anthyphairetic interpretation of the participation of the sensibles in the intelligible, confirmed by the incommensurability of earth to fire (56d5-6) requires the generation of an infinity of new parts.

1.13.6 *The Purpose of the Receptacle/Hollow Space Is to Transform the Intelligible Anthyphairetic Division into Equivalent Tight Inequalities (***Timaeus** *56c8-57c6, 57d7-58c4)*

Mathematically there is a way out of this seemingly impossible situation: we can replace the infinitely many cuts and divisions, ever producing new elements, of the anthyphairetic Division by an equivalent sequence of tight double inequalities

[49] cf. Proclus, *eis Timaion* 2,49,29-50,2 on the meaning of "kosmous":
καὶ ταῦτα τῷ τε Πλάτωνι συμφωνότατά ἐστι, νῦν μὲν λέγοντι τὸν οὐρανὸν ἐκ τῶν τεττάρων εἶναι στοιχείων ἀναλογίᾳ συνδεδεμένων καὶ ὅλον τὸν κόσμον συνεστάναι, μικρὸν δὲ ὕστερον τὰ πέντε σχήματα πλάττοντι καὶ πέντε κόσμους ἀποκαλοῦντι (ταῦτα γὰρ καὶ πέμπτην οὐσίαν τῷ οὐρανῷ δίδωσι καὶ τὴν τῶν στοιχείων εἰσάγει τετρακτύν) καὶ τῇ ἀληθείᾳ συνερχόμενα.

expressible in terms of **only** the two initial elements, intelligible earth A and intelligible fire B. Plato's philosophical strategem of the receptacle/hollow space in the *Timaeus* serves precisely this purpose and function: the intelligible anthyphairetic Division of A to B is turned into a sequence of tight inequalities (involving triangles a and b), occurring in the receptacle/hollow space and generating the sensible entity.

We have expressed the proposition mostly in the language of the *Timaeus*[50] in order to indicate the close agreement of Plato's description of the receptacle/hollow space with the mathematical content of the proposition.

Proposition (Fundamental Proposition of Receptacle/Hollow Space) (Timaeus *56c8-57c6, 57d7-58c4). There is a sequence of natural numbers S_n, T_n, such that every sensible entity is a confounded*[51] *dyad of opposites, resulting from the entrance and compression*[52] *of a conglomeration of the small triangles b's into the hollow spaces of a conglomeration of the large triangles a's, formed alternatingly*

either

for every odd stage $2n-1$ of the intelligible anthyphairesis by a tight[53]*/compressed inequality of the type*

$$\boxed{S_{2n-1}b < T_{2n-1}a} < (S_{2n-1}+1)\,b,$$

not leaving even one single empty hollow space[54] *b, sandwitching the conglomeration of the triangle a's by the conglomeration of the triangle b's, with the*

[50]The revolution ("periodos") of the All, since it comprehends the Kinds, tightens ("sfiggei") them all, seeing that it is circular ("kukloteres") and tends naturally to come together to itself; and thus it suffers not even a single ("outhemian") hollow space ("kenoteta") to be left. Wherefore, fire most of all has permeated ("dielekuthen") all things, and in a second degree air, as it is by nature second in fineness; and so with the rest; for those that have the largest constituent parts have the largest hollow space left in their construction, and those that have the smallest the least. Thus the tightening of the compression ("pileseos sunodos") forces together the small bodies into the hollow space ("diakena") of the large. Therefore, when small bodies are placed beside large, and the smaller disintegrate ("diakrinein") the larger while the larger unite ("sugkrinei"') the smaller. (58a4-b7).

[51]Μέγα μὴν καὶ ὄψις καὶ σμικρὸν ἑώρα, φαμέν, ἀλλ' οὐ κεχωρισμένον ἀλλὰ συγκεχυμένον τι. *Politeia* 524c3-4.

[52]ἡ δὴ **τῆς πιλήσεως** σύνοδος τὰ σμικρὰ εἰς τὰ τῶν μεγάλων **διάκενα συνωθεῖ**. 58b4-5
"pilesis" 58b4-5 compression LSJ 1404,
"sunothein" 53a6, 58b5, force together, compress LSJ 1730.

[53]ἡ τοῦ παντὸς περίοδος, ἐπειδὴ συμπεριέλαβεν τὰ γένη, κυκλοτερὴς οὖσα καὶ πρὸς αὑτὴν πεφυκυῖα βούλεσθαι συνιέναι, **σφίγγει** πάντα 58a4-7
"sphiggein" 58a7 bind tight, tighten LSJ 1741.

[54]κενὴν χώραν οὐδεμίαν ἐᾷ λείπεσθαι, 58a7. Note that our rendering of "outhemia" is not as "and thus it suffers **no** void place to be left" (as translated by W.R.M. Lamb. Cambridge, MA, Harvard University Press; London, William Heinemann Ltd. 1925), but "and thus it suffers **not even one** single hollow space to be left", treating the hollow space of B or of A, as unit. Consider a similar meaning of "outhen" in the *Parmenides* 144c3.

b's winning, while the a's are defeated, forcing the a's to become dissimilar to themselves and similar to them,[55] *thus disintegrating them,*[56] *and stay with the b's,*[57]

or

for every even stage 2n of the intelligible anthyphairesis by a tight/compressed inequality of the type

$$(T_{2n-1})\, a < \boxed{S_{2n} b < T_{2n} a},$$

not leaving even one single empty hollow space a, sandwitching the conglomeration of the triangle b's by the conglomeration of the triangle a's, with the a's winning, while the b's are defeated, forcing the b's to become dissimilar to themselves and similar to them, thus uniting them,[58] *and stay with the a's.*

Thus, even though the dyad ⟨triangle a, triangle b⟩ cannot satisfy the intelligible anthyphairesis of the dyad $\langle A, B\rangle$, for the reasons explained in Sect. 1.13.5, it still does satisfy the equivalent tight anthyphairetic inequalities, and participation is rescued.

Proof Every sensible entity participates in the intelligible entity with dyad $\langle A, B\rangle$. Let the intelligible dyad $\langle A, B\rangle$ have (infinite periodic) anthyphairesis

$$\begin{aligned}
A &= k_0 B + A_1, \quad A_1 < B, \\
B &= m_0 A_1 + B_1, \quad B_1 < A_1, \\
&\dots \\
A_n &= k_n B_n + A_{n+1}, \quad A_{n+1} < B_n, \\
B_n &= m_n A_{n+1} + B_{n+1}, \quad B_{n+1} < A_{n+1} \\
&\dots
\end{aligned}$$

Claim. There is a double sequence S_n, T_n of numbers, such that the first $2n - 1$ intelligible anhtyphairetic relations are equivalent to the tight/compressed inequality of **receptacle/hollow space**

$$S_{2n-1} b < T_{2n-1} a < (S_{2n-1} + 1)\, b;$$

[55] νικηθέντα... ὅμοιον τῷ κρατήσαντι γενόμενον, they are defeated and, instead, assume one form similar to the victorious Kind 57b6-7; τὰ δὲ ἀνομοιούμενα ἑκάστοτε ἑαυτοῖς, ἄλλοις δὲ ὁμοιούμενα, those corpuscles which from time to time become dissimilar to themselves and similar to others 57c3-4.

[56] τῶν ἐλαττόνων τὰ μείζονα **διακρινόντων**, 58b6-7.

[57] αὐτοῦ σύνοικον μείνῃ, stay dwelling therewith as a united family 57b7.

[58] τῶν δὲ μειζόνων ἐκεῖνα [elattona] **συγκρινόντων** 58b7.

and the first $2n$ intelligible anhtyphairetic relations are equivalent to the tight/compressed inequality of **receptacle/hollow space**

$$(T_{2n} - 1)\,a < S_{2n}b < T_{2n}a.$$

Proof of the claim. We indicate the first two steps of the proof.

The first intelligible anthyphairetic relation

$$A = k_0 B + A_1, \quad A_1 < B,$$

is obviously equivalent to a **tight** inequality

$$S_0 B < T_0 A < (S_0 + 1)\,B, \text{ with } S_0 = k_0, T_0 = 1,$$

involving only the initial indefinite dyad A, B, and with no presence of the remainder A_1. Since $A/B = a/b$, hence $\text{Anth}(A, B) = \text{Anth}(a, b)$, we obtain the tight/compressed inequality of **receptacle/hollow space**

$$S_0 b < T_0 a < (S_0 + 1)\,b$$

The first two intelligible anthyphairetic relations

$$A = k_0 B + A_1, \quad A_1 < B,$$
$$B = m_0 A_1 + B_1, \quad B_1 < A_1$$

[are expressed equivalently as $S_0 B < T_0 A < (S_0 + 1)\,B$, and $k_1 A_1 < B < (k_1 + 1)\,A_1$, and upon substituting A_1 by its equal $A - k_1 B$, and performing elementary manipulations we obtain that] are equivalent to a **tight** inequality

$$(T_1 - 1)\,A < S_1 B < T_1 A, \text{ with } T_1 = m_0\Big((m_0 + 1)k_0 + 1\Big) - 1,$$
$$S_1 = (m_0 k_0 + 1)\Big((m_0 + 1)k_0 + 1\Big).$$

Since $A/B = a/b$, hence $\text{Anth}(A, B) = \text{Anth}(a, b)$, we obtain the tight/compressed inequality of **receptacle/hollow space**

$$(T_1 - 1)a < S_1 b < T_1 a. \qquad \square$$

A similar transformation from division to equivalent double inequalities, involving only the two elements a and b, is possible inductively for the anthyphairetic relations of any order. We omit the details of the interesting computations involved.

An excellent example of the fundamental proposition is in the formation of the flesh and the bone; both flesh (74c5-d20) and bone (73e1-5) are generated by mixing earth vs. fire/water, but with different outcomes in the mixture.

> Wherefore, with this intent, our modeler mixed and blended together **water** and **fire** and **earth**, and compounding a ferment of acid and salt mixed it in therewith, and thus molded **flesh full of sap** and **soft**. (*Timaeus* 74c5-d2)

The flesh is produced by, and consists of, an anthyphairetic dyad of opposites, earth vs. fire/water; the winner of this mixture is apparently the soft and watery conglomeration forming a tight inequality of the form

$$Sb < Ta < (S + 1)b,$$

where S many elements b of water/fire have entered *into* T many elements a of earth, maximally, that is anthyphairetically, namely not allowing emptiness even by one hollow space of water. Thus the mixture assumes the form of the winner, according to the sandwiched tight inequality, a form that is soft and watery ("ἔγχυμον καὶ μαλακὴν").

> And **bone** he compounded in this wise. Having sifted **earth** till it was pure and smooth, he kneaded it and moistened it with marrow; then he placed it in **fire**, and after that dipped it in **water**, and from this back to **fire**, and once again in **water**; and by thus transferring it many times from the one element to the other he made it so that it was **soluble by neither**. (*Timaeus* 73e1-5)

The bone is produced by, and consists of, an anthyphairetic dyad of opposites, earth vs. fire/water; the winner of this mixture is apparently the insoluble hard conglomeration forming a double inequality of the form

$$(T - 1)a < Sb < Ta,$$

where S elements b of water/fire have entered into T elements a of earth, maximally, that is anthyphairetically, namely not allowing emptiness even by one hollow space of earth. Thus the mixture assumes the form of the winner, according to the sandwiched tight inequality, a form that is hard and insoluble.

1.13.7 The Sensibles Participating in the Intelligible Platonic Being F Are Sometimes F and Sometimes Not-F

The anthyphairetic description of the intelligibles in the second hypothesis of the *Parmenides* coupled with the description of the sensibles that participate in an intelligible in the *Timaeus* in terms of anthyphairetic inequalities make clear the repeated claim of Plato that the sensibles participating in say the idea of the beautiful are sometimes beautiful B and sometimes not-beautiful not-B (*Symposium* 210e2-211b5,[59] *Politeia* 479a5-b8[60]).

Indeed the idea of the Beautiful consists of a dyad $\langle B, \text{not } B\rangle$ in periodic anthyphairesis, while the sensibles participating in the idea are alternately either of the form

$$S.\text{ not } B < T.B < (S+1).\text{ not } B$$

or of the form

$$(T-1).B < S.\text{ not } B < T.B.$$

The first form of sensible appears as not-Beautiful, while the second form of sensible appears as beautiful. Paradigmatical case is the side and diameter numbers.

1.13.8 The Identification of the "Receptacle" ("hupodoche", 48e-53c) with the "Hollow Space"

At this point it becomes evident that the earlier and rather mysterious concept of "receptacle" ("hupodoche", 48e-53c), coincides with the hollow space ("diakena").

> and that the Nurse of Becoming (τιθήνην), being liquefied (ὑγραινομένην) and ignified (πυρουμένην) and receiving also the forms of earth and of air, and submitting (πάσχουσαν) to all the other affections (πάθη) which accompany these, but owing to [the Receptacle]

---

[59] *Symposium* 210e2-211b5.
ὃς γὰρ ἂν μέχρι ἐνταῦθα πρὸς τὰ ἐρωτικὰ παιδαγωγηθῇ, θεώμενος ἐφεξῆς τε καὶ ὀρθῶς τὰ καλά, πρὸς τέλος ἤδη ἰὼν τῶν ἐρωτικῶν ἐξαίφνης κατόψεταί τι θαυμαστὸν τὴν φύσιν καλόν, τοῦτο ἐκεῖνο, ὦ Σώκρατες, οὗ δὴ ἕνεκεν καὶ οἱ ἔμπροσθεν πάντες πόνοι ἦσαν, [1] πρῶτον μὲν ἀεὶ ὂν καὶ οὔτε γιγνόμενον οὔτε ἀπολλύμενον, [2] οὔτε αὐξανόμενον οὔτε φθίνον, [3] ἔπειτα οὐτῇ μὲν καλόν, τῇ δ' αἰσχρόν, [4] οὐδὲ τοτὲ μέν, τοτὲ δὲ οὔ, [5] οὐδὲ πρὸς μὲν τὸ καλόν, πρὸς δὲ τὸ αἰσχρόν, [6] οὐδ' ἔνθα μὲν καλόν, ἔνθα δὲ αἰσχρόν, [7] ὡς τισὶ μὲν ὂν καλόν, τισὶ δὲ αἰσχρόν· [8] οὐδ' αὖ φαντασθήσεται αὐτῷ τὸ καλὸν οἷον πρόσωπόν τι οὐδὲ χεῖρες οὐδὲ ἄλλο οὐδὲν ὧν σῶμα μετέχει, [9] οὐδέ τις λόγος οὐδέ τις ἐπιστήμη, [10] οὐδέ που ὂν ἐν ἑτέρῳ τινι, οἷον ἐν ζώῳ ἢ ἐν γῇ ἢ ἐν οὐρανῷ ἢ ἔν τῳ ἄλλῳ, ἀλλ' αὐτὸ καθ' αὑτὸ μεθ' αὑτοῦ μονοειδὲς ἀεὶ ὄν, τὰ δὲ ἄλλα πάντα καλὰ ἐκείνου μετέχοντα τρόπον τινὰ τοιοῦτον, οἷον [1] γιγνομένων τε τῶν ἄλλων καὶ ἀπολλυμένων μηδὲν ἐκεῖνο [11] μήτε τι πλέον μήτε ἔλαττον γίγνεσθαι [12] μηδὲ πάσχειν μηδέν.

[60] *Politeia* 479a5-b8.
Τούτων γὰρ δή, ὦ ἄριστε, φήσομεν, [1] τῶν πολλῶν καλῶν μῶν τι ἔστιν ὃ οὐκ αἰσχρὸν φανήσεται; [2] καὶ τῶν δικαίων, ὃ οὐκ ἄδικον; [3] καὶ τῶν ὁσίων, ὃ οὐκ ἀνόσιον; Οὔκ, ἀλλ' ἀνάγκη, ἔφη, [1] καὶ καλά πως αὐτὰ καὶ αἰσχρὰ φανῆναι, καὶ ὅσα ἄλλα ἐρωτᾷς. Τί δὲ [4] τὰ πολλὰ διπλάσια; ἧττόν τι ἡμίσεα ἢ διπλάσια φαίνεται; Οὐδέν. [5] Καὶ μεγάλα δὴ καὶ σμικρὰ [6] καὶ κοῦφα καὶ βαρέα μή τι μᾶλλον ἃ ἂν φήσωμεν, ταῦτα προσρηθήσεται ἢ τἀναντία;Οὔκ, ἀλλ' ἀεί, ἔφη, ἕκαστον ἀμφοτέρων ἕξεται.

being filled quite full (ἐμπίμπλασθαι) with powers that are neither similar (μήθ' ὁμοίων δυνάμεων) nor balanced (μήτε ἰσορρόπων), is not balanced (ἰσορροπεῖν) in not even one of her parts (κατ' οὐδὲν αὐτῆς, 52e3) but sways (ταλαντουμένην) unequable (ἀνωμάλως) everywhere (πάντῃ) and is herself moved (σείεσθαι) by these [powers] (52d4-e5)

In this passage, the receptacle is described as "formless" (like gold 50a5-c2, "ekmageion" 50c2-e4), as being moved—changed by the triangles that enter, and as itself moving the triangles (like "plokanon" 52e5-53a1), depending on the outcome of the 'battle'. We do not need here the detailed argument for this fundamental identification; it will appear in the forthcoming paper [25].

1.14 Plato's Criticism III of the Geometers: The Geometers Employ Sensible Geometric Figures That Are Non-provoking to the Mind (510d5-511a2, 522e5-525a2)

1.14.1 Plato's Criticism III of the Geometers, Employment of Visible Diagrams in Their Study of Geometry

The geometers employment of visible geometric diagrams for their study of Geometry, itself an intelligible activity (even if relegated to the lower kind of "dianoia"), might be thought the reason, in view of Plato's low regard for visible, sensible entities, for his criticism (Criticism III, as we call it) in 510b4-6, 511a3-b2, 511c6-8. But in fact, Plato, in the subsequent passage 522e-525a, is careful to distinguish between two kinds of perceptions, those provoking and those non-provoking to the mind, a distinction that cannot be really understood without obtaining first a full description of the sensibles as presented in the *Timaeus* 48a-58c (and studied in Sect. 1.13). So it becomes immediately clear that Plato is not rejecting all perceptions, and that the use of sensible images by the geometers is not automatically something negative for Plato.

Thus before determining the nature of Plato's criticism about the use of visible diagrams, it is best to look into this distinction into provoking and non-provoking perceptions.

In fact, as we shall see, Criticism III has been misunderstood by modern Platonists,

(a) because they have failed to connect criticism III with the division of the representation of sensibles to provoking and non-provoking the mind to ascent in the *Politeia* 522e5-525a2[61]; and,

[61]Platonists in general failed to realize the significance of provoking perceptions in connection with the geometers reliance on sensible images, e.g. Annas [3] writes in p. 278: "Mathematics has two defects compared with dialectic. It relies on visible diagrams, and it does not question its assumptions. Plato never indicates that, or how, these two defects are connected".

(b) because they failed to obtain the true meaning of the sensibles and of the participation of the sensibles in the intelligibles in Plato's philosophy, as presented in the *Timaeus* 48a-58e (examined in Sect. 1.13).

1.14.2 *Politeia 522e5-523b7: Perceptions Provoking and Non-provoking the Mind to Ascent*

Plato in the *Politeia* 522e5-524d6 is drawing a distinction of perceptions, dividing them in

- perceptions that provoke ("parakalounta") the soul to ascend to the intelligible (that draw the mind to essence and reality ("ton pros noesin agonton", "helktikoi pros ousian"), that are conducive ("agoga") to such ascent), and
- perceptions that are non-provoking.

In the introductory section 522e5-523b7, the distinction is described thus:

> it is one of those studies which we are seeking that naturally conduce to the awakening of thought ("ton pros noesin agonton'), but that no one makes the right use of it, though it really does tend to draw the mind to essence and reality ("helktikoi pros ousian"). (523a1-3)
> the things I distinguish in my mind as being or not being conducive ("agoga te... kai me") to our purpose (523a6)
> if you can discern that some reports of our perceptions do not provoke thought to reconsideration ("ou parakalounta ten noesin eis episkepsin") because the judgment of them by sensation seems adequate. (523a10-b4)

1.14.3 *The Perception of a Sensible Is Provoking If It Can Be Considered at the Same Time as Its Opposite, and Non Provoking, If It Cannot Be So Considered (523b8-525a2)*

Plato then proceeds to explain this distinction. The introductory section is followed by more detailed sections, one on non-provoking perceptions 523b8-e7:

> The perceptions that do not provoke thought ("ou papakalounta") are those that do not at the same time step out ("me ekbainei") to an opposite ("enantian") perception.
> Those that step out ("ta d' ekbainonta") I set down as provocatives ("parakalounta"), when the perception no more manifests one thing than its opposite ("to enantion"). (523b9-c3)
> For in none of these cases is the soul of most men impelled to question the reason and to ask what in the world is a finger, since the faculty of sight never signifies to it at the same time ("hama") that the finger is the opposite ("tounantion") of a finger. (523d3-6)

and another on provoking perceptions 523e7-524c12:

> In the first place, the sensation that is set over ("tetagmene") the hard is of necessity set over ("tetachthai") also to the soft, and it reports to the soul that the same thing ("tauton") is both hard and soft to its perception.""It is so," he said. "Then," said I, "is not this again a case

where the soul must be at a loss as to what significance for it the sensation of hardness has, if the sense reports the same thing ("to auto") as also soft? And, similarly, as to what the sensation of light and heavy means by light and heavy, if it reports the heavy as light, and the light as heavy?" "Yes, indeed," he said, "these communications to the soul are strange ("atopoi") and invite reconsideration ("episkepseos deomenai")." (524a1-b2)

The description ends with a concluding section 524c13-d6:

defining as provocative ("parakletika") things that impinge ("empiptei") upon the senses together ("hama") with their opposites ("tois enantiois"). (524d2-6)

Example of the non-provoking type, is the perception of a finger, as finger, since there is nothing to suggest that the finger can also be seen as non-finger. The provoking ("parakletika") type is a perception coupled with its opposite perception, such as the smallness of fingers with the greatness of fingers, or thickness with thinness, or softness and hardness, or heaviness with lightness. Thus it is provoking, if we see three fingers the small, the middle, and the large, because the middle finger is at the same time perceived as great (with respect to the small) and as small (with respect to the great). A feature of provocative perceptions, by no means fully explained, is described as follows:

"Sight too saw the great and the small, we say, not separated ("achoristos") but confounded. ("sugkechumene") "Is not that so?" "Yes." (524c3-5)

Furthermore perception provoking to ascent to the intellect, as described in 524e2-525a2.

But if some opposition ("enantioma") is always seen coincidentally with it, so that it no more appears to be one than the opposite ("tounantion"), there would forthwith be need of something to judge between them, and
it would **compel** ("anagkazoit' an") the soul to be at a loss and to inquire, by arousing thought in itself ("kinousa en heaute ten ennoian"), and to ask, whatever then is the one as such, and thus the study of unity will be one of the studies that guide ("agogon") and convert ("metastreptikon") the soul to the contemplation of true Being.(524e2-525a2).

1.14.4 The Perceptions That Are Provoking to the Mind, as Described in the Politeia *522e-525a, Coincide with the True Representations of Sensibles, as Described in the* Timaeus *48a-58e*

At this point we will compare the provoking and non-provoking representations of sensibles given in the *Politeia* (Sects. 1.14.2–1.14.3) with the nature of the sensibles, as described in the *Timaeus*. In fact this is the reason we have inserted Sect. 1.13 on the nature and the structure of the sensibles in the *Timaeus*.

It is clear that a sensible, if it is to ascend to the intelligible, cannot but ascend to the intelligible to which it participates; and the participation of the sensibles to the intelligible is, as we have seen in Sects. 1.13.1–1.13.2, the participation of a finite initial anthyphairesis of the intelligible dyad to the whole infinite anthyphairesis.

But why does Plato single out the property of a sensible able to provoke ascent of the soul to the intelligible to be the almost contradictory compresence of a property with its opposite in a confounded sort of dyad?

The intelligible is, as we have seen in Sects. 1.3–1.6, a dyad of opposites in periodic anthyphairesis, and by means of periodicity, the two elements of the dyad, are equalized, "separated" in the sense that each of the parts is a unit (by the equalisation) and together they form the intelligible number two, exactly as with the paradigmatical true Being, the One of the second hypothesis in the *Parmenides*, outlined in Sect. 1.3. The intelligible dyad in the *Timaeus* is $\langle$ earth, water $\rangle$, and the periodic anthyphairesis is between a number of triangles of type a to a number of triangles of type b.

According to the interpretation of the sensibles in the *Timaeus*, outlined in Sect. 1.13, a sensible (participating in this intelligible) is an anthyphairetic dyad, a "confounded" dyad, with anthyphairesis a finite approximating initial segment of the infinite anthyphairesis of the intelligible indefinite dyad $\langle$ earth, water $\rangle$. A sensible is an anthyphairetic dyad with finite anthyphairesis, in the form of one of two double inequalities.

A typical sensible is a mixture of water and earth, in the sense that the parts of water Sb have entered and filled the hollow spaces/diakena of earth Ta. In case this results in "sugkrisis"

$$(T-1)a < Sb < Ta$$

this mixture will take the solid form of earth, and in case it results in "diakrisis"

$$Sb < Ta < (S+1)b,$$

the mixture takes the fluid form of water. Thus, this true sensible entity is "a confounded" dyad (since the diakena of earth are filled with the parts of water), and not a separated, namely intelligible, dyad, as described in 524c3-4. Furthermore a true sensible is perceived simultaneously as A and not A, as described in 523b8-525a2.

Indeed, this entity is both earth and not earth (or both water and not water); and also, it is earth in the present stage that may be transformed into water in the next stage (or conversely), by passing from sugkrisis to diakrisis (or conversely), which happens when a sensible which is described by a $2n$-th anthyphairetic approximation is transformed into the $(2n+1)$-th anthyphairetic approximation (or from the $(2n-1)$-th to the $2n$-th, resp.). Thus we have the true explanation of why a sensible entity is always "such as" earth, or "such as" water (and not "this" earth or "this" water), as described in the *Timaeus* 49c7-50b5. The question was discussed by Cherniss in [7], but not answered in a satisfactory manner. The anthyphairetic approximations of the sequence of side and diameter numbers to the side and diameter of a square, and the corresponding approximations of the right angle by an infinite sequence of alternating approximations of acute and obtuse angles, as studied in Sect. 1.11, mathematically equivalent forms of describing a sensible entity, are confounded dyads, and of the form both A and not A.

We can thus see that a representation, a perception, of a sensible is provoking the mind towards the intelligible, according to 522e-525a, if it represents truthfully and faithfully a sensible, as described according to the *Timaeus* 48a-58e.

1.14.5 The Visible Geometric Diagrams Are Not-Provoking the Mind Upwards

Now the natural question, in relation to Plato's Criticism III, appears to be: do the visible geometric diagrams constitute provoking or non-provoking representations of sensibles?

The provoking sensible is, in some way, present with its opposite (a straight line with a greater straight line, or conversely, an acute angle with its opposite obtuse or conversely), or equivalently as a confounded dyad, in terms of tight inequalities, leaving not a single hollow space, as described in the *Timaeus*, participating in the Division scheme of a Platonic Being.

But in the geometric diagrams a straight line is represented as one, isolated line segment between joining two points; and an acute angle is presented as an isolated angle, without its opposite obtuse.These representations are very much like the one, isolated finger, in Plato's discussion in 523b8-524d6 (cf. Sect. 1.13.3), and thus they qualify as non-provocative representations of the sensibles, representations that do not provoke toward true intelligible knowledge.

The distinction of perceptions in provoking and non provoking the soul to ascent brings the blame for the failure of the geometers to ascend to the intelligible to the geometers themselves: it is not the geometers' fault that they are using visible diagrams, but it is their fault that, they are not using visible diagrams and perceptions correctly (in fact, "no one is using them correctly", 523a1-3), that they are using non-provoking instead of provoking visible diagrams.

1.14.6 The Representations of the Sensibles That Are Not Provoking Towards Ascent to the Intelligible Are Analogous to the Geometric Hypotheses That Do Not Ascend to the Anhupotheton

In fact the way non-provoking perceptions are described, as a perception in which the opposite perception is not suggested, in 522e-525a, is analogous to the way the hypotheses of the geometers are described. What makes a hypothesis geometric and non-dialectical is that it is isolated, without an opposite, forming an anthyphairetic dyad, as in the definitions of the *Sophistes* or *Politicus*; and, what makes a representation of a sensible non-provoking the mind upwards is the almost contradictory compresence of an opposite, forming an anthyphairetic dyad.

1.14.7 The Non-provoking Representation of Sensibles in the Geometric Diagrams Is Not the Cause for the Compulsion of the Geometers to Employ Hypotheses in a Dianoia Non-Dialectical Way

This close analogy between non-provoking representations of sensibles and non-dialectic hypotheses might now give the impression that the visible geometric diagrams, because of being non-provoking, are guilty of compelling the geometers to the use of hypotheses as principles. But the source of this compulsion has already been identified (in Sect. 1.9), mostly with the help of the passage 527a-b, with the need of the geometers to obtain actions/constructions in their proofs of propositions, and it is of a purely intelligible source and origin. Thus the geometric diagrams have nothing to do with the compulsion of the geometers to use hypotheses as principles.

1.14.8 The Geometers Form Their Arguments for the Sake of Intelligible Beings (Studied by Means of Dianoia), but About the Visible Diagrams of These Beings (510d5-511a2)

The 510d5-511a2 passage is important because in it Plato defines carefully the relation of the intelligible dianoia type hypotheses and the related actions/constructions role with the sensible/visible diagrams play in Geometry.

> "And do you not also know that they [the geometers] further make use of the visible forms ("tois horomenois eidesi") and construct their arguments ("tous logous poiountai") about ("peri") them,
> though they are arguing by means of "dianoia" ("dianooumenoi")
> not about them [visible forms]
> but about those things of which they are a likeness,
> constructing their arguments ("tous logous poioumenoi")
> for the sake ("heneka") of the square itself ("tou tetragonou autou") and the diagonal itself ("tes diametrou autes"), and
> not for the sake ("heneka") of the image of it which they draw?
> And so in all cases.
> The very things which they mould and draw, which have shadows and images of themselves in water, these things they treat in their turn as only images ("eikosin chromenoi"), but what they really seek is to get sight of those intelligible beings, which can be seen only by means of "dianoia"." "True," he said. (510d5-511a2)

Thus, the geometers construct their arguments ("tous logous poiountai", "tous logous poioumenoi") **about** ("peri") these visible diagrams, even though these arguments are **for the sake of** ("heneka") intelligible real, Platonic Beings, such as the square itself and the diameter itself, understood though by the (imperfect) method of "dianoia", namely by taking hypotheses as principles. The present passage, examined together with 527a1-b2 (analysed in Sect. 1.9.2), in which there is no mention of sensibles, but where it is stated that the geometers are compelled to employ hypotheses in a defective way because they form their arguments for the sake of geometric constructions.

Thus the visible representations of the diagrams must be compatible, auxiliary to the dianoia reasoning of the geometers. As auxiliary to dianoia the visible diagrams cannot be perceptions provoking the mind upwards, because if they were, they would lead to the use of hypotheses as steps upwards, something that would go against the "dianoia" way of thinking to which the geometers are committed. Thus the diagrams used by "dianoia" thinking geometers, must definitely be non-provoking, perceptions (522e-525a)—as they are indeed.

1.14.9 The Geometric Diagrams Are Non-Provoking the Mind, Because the Geometers Form Their Arguments by Means of Dianoia and the Diagrams Are in Accordance with "dianoia" Way of Thinking

Plato's non-provoking representations of geometric entities in the geometric diagrams is thus seen as a consequence rather than as a cause of the use by the geometers of hypotheses that are principles, and of a geometry that moves down, deductively. The non-provoking visible geometric diagrams are used by the geometers because they are in close accord with the use of geometric hypotheses, conceived in "dianoia" non-dialectic way of thinking, not conversely.

1.15 Criticism IV of the Academy: Criticism of Eudoxus' Theory of Ratios of Magnitudes (*Scholion In Euclidem X.2*) and Archytas' Duplication of the Cube (527d-528e)

Our interpretation of Plato's view of (plane) Geometry, and its relation to philosophical true Beings and Ideas, sheds considerable light on Plato's view on Stereometry and his criticism on the solution of the problem of the Duplication of the Cube provided by Archytas, in the *Politeia* 527d-528e.

Inspired from music and epimoric ratios, Archytas was the first to conceive of arithmetical, non-anthyphairetic proofs of quadratic incommensurabilities, first shown in terms of periodic anthyphairesis by Theaetetus (as mentioned in Sect. 1.2). His proofs were based on Proposition VII.27 (if a, b are relatively prime numbers, then a^2, b^2 are relatively prime as well), and their derivative Proposition VIII.14, on the existence of an arithmetical mean proportional. One feature of these proofs is that they readily generalize to cubic incommensurabilities, since by VII.27, if a, b are relatively prime numbers, then a^3, b^3 are relatively prime as well and Proposition VIII.14 holds for **two arithmetical mean proportionals**. Despite the mistaken claim in the *Theaetetus* 148b2 ("καὶ περὶ τὰ στερεὰ ἄλλο τοιοῦτον", "and similarly in the case of solids"), Theaetetus' anthyphairetic method of proof on quadratic incommensurabilities does not extend to cubic incommensurabilities.

Since historically the development of a theory of ratios of magnitudes was closely associated with incommensurabilities, the discovery by Archytas of non-anthyphairetic proofs of incommensurabilities, and not only quadratic, but cubic as well, not covered by Theaetetus' theory of ratios of magnitudes (mentioned in Sect. 1.2), made necessary and pressed forward for a **non-anthyphairetic theory of ratios of magnitudes**. This is precisely the feat of Eudoxus, exposed in Book V of the *Elements*, who realized that instead of determining the ratio of two magnitudes by its position between the odd below and the even above arithmetical ratios of the convergents, one can simply pass, by the grace of Eudoxus-Archimedean condition, to Dedekind cuts.

Next, the new theory of ratios by Eudoxus made possible Archytas' most ingenuous duplication of the cube by means of constructing two mean proportionals (as initially conceived by Hippocrates). It is generally not realized that this construction must come AFTER a general theory of ratios of magnitudes has been given, that Theaetetus' theory was inadequate to deal with cubic ratios, and thus necessarily after Eudoxus' theory. Thus for a given line b, he constructed line a, such that $a^3 = 2b^3$, by constructing two mean proportionals lines m and a between $2b$ and b, namely $2b/m = m/a = a/b$; then, indeed,

$$a^3/b^3 = (2b/m).(m/a).(a/b) = 2b/b = 2/1.$$

By Proposition VI.14 of the *Elements* (based on Eudoxus' fundamental proposition V.9 and on Proposition VI.1), m and a must be constructed so that both $m^2 = 2ab$ (parabola) and $ma = 2b^2$ (hyperbola) are satisfied, the point of intersection of two conic sections.

While the non-anthyphairetic proofs of the quadratic incommensurabilities could well be considered as inferior to the anthyphairetic (in that they do not provide full knowledge of the ratio involved), and thus would not cause any great concern to the mathematical and philosophical supporters of an anthyphairetic approach (Plato's annoyance is likely expressed in the distinction between the eristic, not producing numbers and thus not making use of periodic anthyphairesis and logos, and the dialectic way of attaining infinity (namely by means of periodic anthyphairesis) in the *Philebus* 16d-e), Eudoxus' theory and Archytas' duplication of the cube were certainly *casus belli* for Plato. The annoyance of the Academy with Eudoxus' new and general theory of ratios can be seen in *Scholion In Euclidem* X.2. In the second part of this *Scholion*, it is lamented that if Eudoxus theory of ratios of magnitudes (Book V of the *Elements*) is adopted, then the difference between alogon and rheton, that is fundamental in the earlier Theaetetean theory of magnitudes, is obliterated and destroyed.

> Εἰ πάντα τὰ μεγέθη τὰ πεπερασμένα δύναται πολλαπλασιαζόμενα ἀλλήλων ὑπερέχειν (τοῦτο δὲ ἦν τὸ λόγον ἔχειν,ὡς ἐν τῷ πέμπτῳ μεμαθήκαμεν), τίς μηχανὴ τὴν τῶν ἀλόγων ἐπεισφέρειν διαφοράν;

As we have mentioned in Sect. 1.2.4, Theaetetus' anthyphairetic theory of ratios of magnitudes was only for a limited class of magnitudes, related to lines commensurable in square, and certainly not for general magnitudes.

Plato in the *Politeia* 527d-528e is generally considered to have in mind the problem of the duplication of the cube.

> The right way is next in order after the second dimension to take the third (μετὰ δευτέραν **αὔξην τρίτην** λαμβάνειν). This, I suppose, is the dimension of cubes and of everything that has depth." (ἔστι δέ που τοῦτο **περὶ τὴν τῶν κύβων αὔξην** καὶ **τὸ βάθους μετέχον**) 528b2-3.

The duplication of the cube is then criticized because the seekers of such duplication are arrogant, and do not know the great value that logoi, presumably the logoi leading to anthyphairetic periodicity, would have in relation to this problem.

> Seekers in this field would be too arrogant
> (οι περὶ ταῦτα ζητητικοὶ **μεγαλοφρονούμενοι**) 528c1
>
> the seekers, not knowing why "logoi" would be useful
> (ὑπὸ δὲ τῶν ζητούντων **λόγον οὐκ ἐχόντων** καθ' **ὅτι χρήσιμα**) 528c5-6

Plutarch, *Quaestiones Convivales* 718 B8-F4, deals at length with the matter, and in what appears to be the main criticism, he states[62]:

> [geometry], according to Philolaus, is the chief and principal of all, and does bring back and turn the mind, as it were, purged and gently loosened from sense. For this reason **Plato** himself also reproached **Eudoxus, Archytas, Menaechmus** and their followers for trying to lead away **the problem of the duplication of a solid** into constructions that use instruments and that are mechanical, just as if they were trying to obtain **the two mean proportionals apart from "logos"** in whatever way it was practicable. For by this means all that was good in geometry would be lost and corrupted, it falling back again to sensible things, and not rising upward and considering immaterial and immortal images, in which God being versed is always God.

Plato must have believed, or hoped, that the anthyphairesis of a, b, for $a^3 = 2b^3$, would be periodic, and thus "have logos" of periodicity, as is the case of the duplication of the square, with $a^2 = 2b^2$, and thus constitute a dyad of a true Platonic Being, since $\mathrm{Anth}(a, b) = [1, 2, 2, 2, \ldots]$.[63] Under this mistaken,

---

62 μάλιστα δὲ γεωμετρία κατὰ τὸν Φιλόλαον ἀρχὴ καὶ μητρόπολις οὖσα τῶν ἄλλων ἐπανάγει καὶ στρέφει τὴν διάνοιαν, οἷον ἐκκαθαιρομένην καὶ ἀπολυομένην ἀτρέμα τῆς αἰσθήσεως. διὸ καὶ **Πλάτων αὐτὸς ἐμέμψατο τοὺς περὶ Εὔδοξον καὶ Ἀρχύταν καὶ Μέναιχμον** εἰς ὀργανικὰς καὶ μηχανικὰς κατασκευὰς **τὸν τοῦ στερεοῦ διπλασιασμὸν** ἀπάγειν ἐπιχειροῦντας, ὥσπερ πειρωμένους **δίχα λόγου δύο μέσας ἀνὰ λόγον**, ᾗ παρείκοι, λαβεῖν. ἀπόλλυσθαι γὰρ οὕτω καὶ διαφθείρεσθαι τὸ γεωμετρίας ἀγαθὸν αὖθις ἐπὶ τὰ αἰσθητὰ παλινδρομούσης καὶ **μὴ φερομένης** ἄνω μηδ' ἀντιλαμβανομένης τῶν ἀιδίων καὶ ἀσωμάτων εἰκόνων, πρὸς αἷσπερ ὢν ὁ θεὸς ἀεὶ θεός ἐστιν (718E4-F4).

63 It is not realized by students of the *Meno*, that in the *Meno* 82a7-86c3 the recollection (ἀνάμνησις) achieved is the recollection, i.e. repetition, of the logos of periodicity of the duplication of the square, and not simply the Pythagorean theorem for the orthogonal isosceles triangle, resulting in the complete knowledge (ἐπιστήμη) of the duplication ratio.

as it turned out, belief, the duplication of the cube by Archytas could not be considered satisfactory at all by Plato, since by his construction Archytas did not prove periodicity of anthyphairesis and did not provide the full knowledge of the true Being involved. It is clear, in view of our analysis that the reproach of Plato, echoed in Plutarch, is that Eudoxus and Archytas are “trying to obtain the two mean proportionals” necessary for the duplication of the cube “apart from logos” (“δίχα λόγου”), namely apart from the logos of periodicity in anthyphairesis, the only one bestowing true existence, and therefore “not rising upward”.

Little [16] prefers the reading “δι’ ἀλόγου” in place of the reading “δίχα λόγου”, and adds the following Footnote in p. 272:

> ὥσπερ πειρωμένους δι’ ἀλόγου δύο μέσας ἀνὰ λόγον, ᾗ παρείκοι, λαβεῖν’ The δι’ ἀλόγου is hard to translate and may not even be what Plutarch originally wrote. This specific phrase has a rather large number of textual issues as evidenced by the variant readings discussed in the Loeb Classical Library/Perseus version of the Plutarch text. At least one editor has suggested that the whole phrase should be omitted from the text!.

However, “δίχα λόγου” (“apart from logos”) and “δι′ ἀλόγου” (“by means of something lacking logos”) are practically synonymous, and both fit well with our anthyphairetic interpretation.

1.16 Epilogue. The Un-Platonic Victory of Axiomatization: From Euclid to Peano and Hilbert

It is clear that Plato thought that Geometry and in fact all Mathematics flowed from the true Platonic Beings, and thus Geometry was subservient to Platonic philosophy and dialectics. It is ironic that the vehicle, through which this superiority was claimed, was a purely mathematical concept, periodic anthyphairesis, elevated to the status of supreme philosophic principle.

Euclid, a geometer of Platonic inclination, followed a middle ground between mathematics and Platonic philosophy, providing definitions for the point, the straight line, and the arithmetical unit that were philosophically meaningful but mathematically useless. Surprisingly for a Platonist his treatment of anthyphairesis is rudimentary in Book X and fully suppressed in Book V.

Peano in 1889 and Hilbert in 1899 provided axiomatic foundation of Arithmetic and of Euclidean Geometry, respectively, free from any philosophic supervision, formalizing the emancipation of modern era Mathematics. Hilbert achieved this by refusing to give definitions of the basic concepts (point, line, plane) and of their basic relations (incidence, betweeness, application). Hilbert’s approach served as the model for the axiomatisation of set theory by Zermelo-Fraenkel, in principle comprising all Mathematics. Godel, a professed Platonist, with his incompleteness theorem in 1931, brought fourth some crucial shortcomings of Peano’s axiomatisation, but these were not at all in line with Plato’s.

Translation of Ancient Sources

† The translation of the ancient passages is based on the following works, with modifications by the author
Euclid, *Elements*–T. L. Heath, *The thirteen books of Euclid's elements*. Cambridge: Cambridge University Press. (First edition 1908, Second Edition 1926, Reprinted by Dover Publications, New York, 1956).
Plato, *Phaedo*–H. N. Fowler; Introduction by W.R.M. Lamb. Cambridge, MA, Harvard University Press; London, William Heinemann Ltd. 1966.
Plato, *Politeia*–P. Shorey; Cambridge, MA, Harvard University Press; London, William Heinemann Ltd. 1969.
Plato, *Parmenides, Theaetetus, Sophistes, Politicus*–H. N. Fowler; Cambridge, MA, Harvard University Press; London, William Heinemann, 1921.
Plato, *Timaeus*–W.R.M. Lamb. Cambridge, MA, Harvard University Press; London, William Heinemann Ltd. 1925.
Plutarch, *Quaestiones Convivales- Moralia*, Volume IX, Table-talk, Books 7-9. Dialogue on Love, ed. E. L. Minar, Jr.,F. H. Sandbach, W. C. Helmbold, with an English translation, Loeb Classical Library 425, Harvard University Press, Mass., 1961.
Proclus, *In Euclidem*–G. R. Morrow, *Proclus A Commentary on the First Book of Euclid's Elements*. Translated, with Introduction and Notes, Princeton University Press, Princeton, 1970.
Simplicius, *On Aristotle's Physics 1.3-4*–P. Huby and C.C.W. Taylor, Bloomsbury, London, 2011.

References

1. F. Acerbi, Plato: Parmenides 149a7-c3. A Proof by Complete Induction?, *Arch. Hist. Exact Sci.* 55 (2000) 57–76.
2. F. Acerbi, Drowning by Multiples: Remarks on the Fifth Book of Euclid's Elements, with Special Emphasis on Prop. 8, *Archive for History of Exact Sciences*, 57 (2003) 175-242.
3. J. Annas, *An Introduction to Plato's Republic*, Clarendon Press, Oxford, 1981.
4. D. T. J. Bailey, Plato and Aristotle on the Unhypothetical, *Oxford Studies in Ancient Philosophy* 30 (2006), 101–126.
5. H. H. Benson, *Clitophon's Challenge, Dialectic in Plato's Meno, Phaedo, and Republic*, Oxford University Press, 2015.
6. H. Cherniss, Plato as a mathematician, *The Review of Metaphysics* 4 (1951), 395–425.
7. H. Cherniss, A Much Misread Passage of the Timaeus (Timaeus 49 C 7-50 B 5), *The American Journal of Philology* 75, (1954) 113-130.
8. F. M. Cornford, *Plato's Theory of Knowledge*, Routledge & Kegan Paul LTD, London, 1935.
9. V. Farmaki, S. Negrepontis, *Axioms in Search of a definition*, Abstract, HPM 2008 History and Pedagogy of Mathematics, 14 - 18 July 2008, Mexico City, México.
10. D. Fowler, *The Mathematics of Plato's Academy, a new reconstruction*, Second edition, Clarendon Press, Oxford, 1999.

11. H.Freudenthal, Zur Geschichte der vollstandigen Induktion, *Archives Internationales d'Histoire des Sciences*, 6 (1953), 17–37.
12. R. M. Hare, Plato and the Mathematicians, in R. Bambrough (ed.), *New Essays on Plato and Aristotle*, London, Routledge and Kegan Paul, 1965, pp. 21-38.
13. T. L. Heath, *The thirteen books of Euclid's Elements*, Cambridge University Press, Cambridge, 1926.
14. W.R. Knorr, *The Evolution of Euclidean Elements: A Study of the Theory of Incommensurable Magnitudes and its Significance of Early Greek Geometry*, Reidel, Dordrecht, 1975.
15. D. Lamprinidis, *The Anthyphairetic Nature of Plato's Mathemata,* Doctoral Dissertation, Athens University, 2015 [in Greek].
16. J. B. Little, A Mathematician Reads Plutarch: Plato's Criticism of Geometers of His Time, *Journal of Humanistic Mathematics* 7 (2017) 269-293.
17. H. G. Liddell, R.Scott, Revised-Augmented by R.S. Jones, *Greek-English Lexicon*, Clarendon Press, Oxford University, 1996.
18. C. McLarty, 'Mathematical Platonism' Versus Gathering the Dead: What Socrates Teaches Glaucon, *Philosophia Mathematica* 13 (2005), pp. 113–134.
19. I. Mueller, Mathematical Method and Philosophical Truth, in *The Cambridge Companion to Plato*, edited by R. Kraut, Cambridge University Press, Cambridge, 1992, pp. 170–199.
20. G.R. Morrow, *Translation of Proclus Commentary on the First Book of Euclid's Elements, with Introduction and Notes*, Princeton University Press, 1970.
21. S. Negrepontis, The Anthyphairetic Nature of the Platonic Principles of Infinite and Finite, *Proceedings of the 4th Mediterranean Conference on Mathematics Education*, 28-30 January 2005, Palermo, Italy, pp. 3–27.
22. S. Negrepontis, Plato's theory of knowledge of Forms by Division and Collection in the Sophistes is a philosophic analogue of periodic anthyphairesis (and modern continued fractions),`arXiv:1207.2950`, 2012; to appear in *Proceedings of the Colloquium on "La Demonstration de l' Anliquite a I'age classique*, June 3-6, 2008, Paris, ed. by R. Rashed.
23. S. Negrepontis, *The Anthyphairetic Revolutions of the Platonic Ideas*, `arXiv:1405.4186`, 2014; *Revolutions and Continuity in Greek. Mathematics*, ed. M. Sialaros, Walter de Gruyter, 2018, pp. 335-381.
24. S. Negrepontis, The periodic anthyphairetic nature of the One of the Second Hypothesis in Plato's Parmenides, *Proceedings of the Conference Mathématiques et musique : des Grecs à Euler*, 10-11 September, 2015, Strasbourg, ed. X. Hascher and A. Papadopoulos, Hermann, Paris (to appear).
25. S. Negrepontis & D. Kalisperi, *Timaeus' receptacle and hollow space*, in preparation.
26. S. Negrepontis & D. Lamprinidis, History and Epistemology in Mathematics education, *Proceedings of the 5th European Summer School ESU5*, (ed) E. Barbin, N. Stehlikova, C. Tzanakis, Vydavatelsky Servis, Plzen, 2008, pp. 501–511.
27. S. Negrepontis and D. Protopapas, *Theaetetus' theory of ratios of magnitudes*, in preparation.
28. C. D. C. Reeve, Blindness and reorientation: education and the acquisition of knowledge in the Republic, in *Plato's Republic A Critical Guide*, edited by M. L. McPherran, Cambridge University Press, 2010, pp. 209–228,
29. R. Robinson, *Plato's Earlier Dialectic*, 2nd ed., Clarendon Press, Oxford, 1953.
30. W. D. Ross, *Plato's Theory of Ideas,* Oxford: Clarendon Press, 1951.
31. J. Stenzel, *Plato's Method of Dialectic*, Translation D.J. Allan, Arno Press, New York; 1973.
32. C.W.W. Taylor, Plato and the Mathematicians, An Examination of Professor Hare's views, *Philosophical Quarterly* 17 (1967), 193–203.
33. S. Unguru, Greek Mathematics and Mathematical Induction, *Physis*, 28 (1991), 273–289.
34. B. L. van der Waerden, *Science Awakening*, Noordhoff, Groningen, 1954.

Chapter 2
Topology and Biology: From Aristotle to Thom

Athanase Papadopoulos

Abstract René Thom discovered several refined topological notions in the writings of Aristotle, especially the biological ones. More generally, he understood that some assertions made by philosophers from Greek antiquity have a definite topological content, even if they were stated more than two and a half millennia before the field of topology was born. He adhered completely to Aristotle's theory of form which the latter developed especially in his biological treatises. Thom emphasized the importance of these ideas in biology, and more particularly in embryology, namely, the idea of a form tending to its own realization. In this article, we expand on these ideas of Thom. At the same time, we highlight some major ideas in the works of Aristotle and Thom in biology and we comment on their conceptions of mathematics and more generally of science.

AMS Classification: 03A05, 01-02, 01A20, 34-02, 34-03, 54-03, 00A30, 92B99

2.1 Introduction

Bourbaki, in his *Éléments d'histoire des mathématiques*, at the beginning of the chapter on topological spaces, writes [16, p. 175]:

> The notions of limit and continuity date back to antiquity; one cannot make their complete history without studying systematically, from this point of view, not only the mathematicians, but also the Greek philosophers, and in particular Aristotle, and without tracing the evolution of these ideas through the mathematics of the Renaissance and the

A. Papadopoulos (✉)
Université de Strasbourg and CNRS, Strasbourg Cedex, France
e-mail: athanase.papadopoulos@math.unistra.fr

S. G. Dani, A. Papadopoulos (eds.), *Geometry in History*,
https://doi.org/10.1007/978-3-030-13609-3_2

> beginning of differential and integral calculus. Such a study, which would be interesting to carry on, would have to go much beyond the frame of the present note.[1]

We notice that in this quote, the reference is to "Greek philosophers" rather than "Greek mathematicians." One may recall in this respect that at that time, mathematics was closely related to philosophy, even though there was a clear distinction between the two subjects. Many mathematicians were often philosophers and vice-versa, and mathematics was taught at the philosophical schools. A prominent example is Plato who, in his Academy, was essentially teaching mathematics. We may recall here a note by Aristoxenus, the famous fourth century BCE music theorist who, reporting on the teaching of Plato, writes that the latter used to entice his audience by announcing lectures on the Good or similar philosophical topics, but that eventually the lectures turned out to be on geometry, number theory or astronomy, with the conclusion that "the Good is One" [15, p. 98]. Aristotle was educated at Plato's academy, where he spent 20 years, learning mathematics in the pure Pythagorean tradition, and he was one of the best students there—Plato used to call him "the Brain of the School". This is important to keep in mind, since Aristotle's mathematical side, which is one of the main topic of the present article, is usually underestimated.

Over half a century since Bourbaki made the statement we quoted, a systematic investigation of traces of topology in the works of the philosophers of Greek antiquity has not been conducted yet. One problem for realizing such a program is that the ancient Greek texts that survive are generally analyzed and commented on by specialists in philosophy or ancient languages with little knowledge of mathematics, especially when it comes to topology, a field where detecting the important ideas requires a familiarity with this topic. In particular, it often happens in this context that the translation of terms conveying topological ideas does not render faithfully their content; this is the well-known problem of mathematical texts translated by non-mathematicians.

Thom offers a rare, even unique, and important exception. He spent a great deal of his time reading (in Greek) the works of Aristotle and detecting the topological ideas they contain. He also pointed out a number of mistakes in the existing translations and proposed more precise ones. In his 1991 article *Matière, forme et catastrophes* [48, p. 367], he declares that his past as a mathematician and initiator of catastrophe theory probably confers upon him a certain capacity of noticing in the work of Aristotle aspects that a traditional philosophical training would have put aside. He writes in his article *Les intuitions topologiques primordiales de l'aristotélisme* [45] (1988) that the earliest terms of Greek topology available

[1] "Les notions de limite et de continuité remontent à l'antiquité ; on ne saurait en faire une histoire complète sans étudier systématiquement de ce point de vue, non seulement les mathématiciens, mais aussi les philosophes grecs et en particulier Aristote, ni non plus sans poursuivre l'évolution de ces idées à travers les mathématiques de la Renaissance et les débuts du calcul différentiel et intégral. Une telle étude, qu'il serait certes intéressant d'entreprendre, dépasserait de beaucoup le cadre de cette note." [For all the translations that are mine, I have included the French text in footnote.]

in the Greek literature are contained in Parmenides' poem, *On Nature*, and he notes, in particular, that the word *sunekhes* (συνεχές) which appears there is usually translated by *continuous*, which is a mistake, because this word, in that context, rather refers to (arcwise) connectedness. This is not only a matter of terminology; it is part of a quest for a comprehension of what are exactly the topological notions that are present in the literature of Ancient Greece. In order to make such a comment, and, especially given that no precise topological terminology existed in antiquity, a definite understanding of the topological ideas and intuitions involved in the writings of that epoch, together with the precise terminology of modern topology, is required. There are many other examples of statements of the same kind by Thom, and we shall note a few of them later in our text. Thanks to Thom's translation and interpretation of Aristotle and of other Greek philosophers, certain passages that were obscure suddenly become clearer and more understandable for a mathematician. This is one of the themes that we develop in this article.

In several of his works, Thom expressed his veneration for the philosophers of Greek antiquity. He started quoting them extensively in his books and articles that he wrote in the beginning of the 1960s, a time when his mathematical work became part of a broader inquiry that included biology and philosophy. In the first book he published, *Stabilité structurelle et morphogenèse : Essai d'une théorie générale des modèles*, written in the period 1964–1968, he writes [38, p. 9][2]:

> If I have quoted the aphorisms of Heraclitus at the beginning of some chapters, the reason is that nothing else could be better adapted to this type of study. In fact, all the basic intuitive ideas of morphogenesis can be found in Heraclitus: all that I have done is to place these in a geometric and dynamic framework that will make them someday accessible to quantitative analysis. The "solemn, unadorned words,"[3] like those of the sibyl that have sounded without faltering throughout the centuries, deserve this distant echo.

Aristotle, to whom Thom referred constantly in his later works, placed biology and morphology, the science of form, at the basis of almost every field of human knowledge, including physics, psychology, law and politics. Topology is involved, often implicitly but also explicitly, at several places in his writings. Bourbaki, in the passage cited at the beginning of this article, talks about limits and continuity, but Thom highlighted many other fundamental topological questions in the works of the Stagirite. Among them is the central problem of topology, namely, that of reconstructing the global from the local. He noted that Aristotle saw that this problem, which he considered mainly from a morphological point of view, is solved in nature, especially in the vegetal world, where one can reconstruct a plant from a piece of it. Even in the animal world, where this phenomenon is exceptional—in general, one cannot resurrect an animal from one of its parts—he saw that if we

[2]We quote from the English translation, *Structural stability and morphogenesis. An outline of a general theory of models* published in 1975.

[3]Thom refers here to Heraclitus, Fragment No. 12: "And the Sibyl with raving mouth, uttering words solemn, unadorned, and unsweetened, reaches with her voice a thousand years because of the god in her." (Transl. A. Fairbanks, In: The First Philosophers of Greece, Scribner, 1898).

amputate certain mollusks, crustaceans and insects of one of their organs, they can regenerate it. Furthermore, Aristotle understood that it is possible to split certain eggs to obtain several individuals.

In the foreword to his *Esquisse d'une sémiophysique*, a book published in 1988 and which carries the subtitle *Physique aristotélicienne et théorie des catastrophes*,[4] Thom writes [47, p. viii][5]: "It was only quite recently, almost by chance, that I discovered the work of Aristotle. It was fascinating reading, almost from the start." More than a source of inspiration, Thom found in the writings of the Stagirite a confirmation of his own ideas. On several occasions, he expressed his surprise at seeing in these writings statements that he himself had come up with, in slightly different terms, without being aware of their existence. He declares in his 1991 article *Matière, forme et catastrophes* [48, p. 367]: "In the last years, in view of my writing of the *Semiophysics*, I was led to go more thoroughly into the knowledge of the work of the Stagirite."[6] He found in that work the idea that, because we cannot understand everything in nature, we should concentrate our efforts on stable and generic phenomena. Aristotle formulated the difference between a generic and a non-generic phenomenon as a difference between a "natural phenomenon" and an "accident." Interpreting this idea (and related ones) in terms of modern topology is one of the epistemological contributions of Thom. He writes in his *Esquisse* [47, p. 12] (p. ix of the English translation):

> [...] If I add that I found in Aristotle the concept of genericity (ὡς ἐπὶ τὸ πολύ), the idea of stratification as it might be glimpsed in the decomposition of the organism into homeomerous and anhomeomerous parts by Aristotle the biologist, and the idea of the breaking down of the genus into species as images of bifurcation, it will be agreed that there was matter for some astonishment.

The notion of "homeomerous" (ὁμοιομερής), to which Thom refers in this passage, occurs in Aristotle's *History of animals*, his *Parts of animals*, and in his other zoological writings. It contains the idea of self-similarity that appears in modern topology. The Aristotelian view of mathematics, shared by Thom, tells us in particular that mathematics emerges from our daily concepts, rather than being some Platonist ideal realm. Among the other topological ideas that Thom found in Aristotle's writings, we mention the notions of open and closed set, of cobordism, and the Stokes formula. We shall quote the explicit passages below.

Thom published several articles on topology in Aristotle, including *Les intuitions topologiques primordiales de l'aristotélisme* (*The primary intuitions of Aristotelianism*) [45] (1988) which is an expanded version of a section in Chapter 7 of the *Esquisse* [43], *Matière, forme et catastrophes* (*Matter, form and catastrophes*) [48] (1991), an article published in the proceedings of a conference held at the UNESCO

[4]We quote from the English translation published in 1990 under the title *Semio Physics: A Sketch*, with the subtitle *Aristotelian Physics and Catastrophe Theory.*

[5]Page numbers refer to the English translation.

[6]Ces dernières années, en vue de la rédaction de ma *Sémiophysique*, j'ai dû entrer plus profondément dans la connaissance de l'œuvre du Stagirite.

headquarters in Paris at the occasion of the 23rd centenary of the Philosopher, *Aristote topologue* (Aristotle as a topologist) [53] (1999) and others. It results from Thom's articles that even though Aristotle did not build any topological theory in the purely mathematical sense, topological intuitions are found throughout his works. Thom writes in [45, p. 395]: "We shall not find these intuitions in explicit theses and constructions of the theory. We shall find them mostly in some 'small sentences' that illuminate the whole corpus with their brilliant concision."[7]

Thom considered that mathematics is a language that is adequate for philosophy. He expressed this in several papers, and in particular in the article *Logos phenix* [40] and others that are reprinted in the collection *Modèles mathématiques de la morphogenèse* [41]. We shall elaborate on this below, but let us note right away that beyond the discussion that we conduct here concerning topological notions in philosophy, we are aware of the fact that the question of how much philosophical notions carrying a mathematical name (infinity, limit, continuity, etc.) differ from their mathematical counterparts is debatable. Indeed, one may argue that mathematics needs precise terms and formal definitions. Such a position is expressed by Plotnitsky in his essay published in the present volume [31], in which he writes:

> What made topology a mathematical discipline is that one can associate algebraic structures (initially numbers, eventually groups and other abstract algebraic structures, such as rings) to the architecture of spatial objects that are invariant under continuous transformations, independently of their geometrical properties.

Plotnitsky also quotes Hermann Weyl from his book *The continuum* [57], where the latter writes that the concepts offered to us by the intuitive notion of continuum cannot be identified with those that mathematics presents to us. Thom had a different point of view. For him, mathematics and our intuition of the real world are strongly intermingled. In the article *Logos phenix* [41, p. 292], he writes: "How can we explain that mathematics represents the real world? The answer, I think, is offered to us by the intuition of the continuous."[8] For him, it is the geometrical notion of continuous that gives a meaning to beings that would have needed infinitely many actions. He mentions the paradox of Achilles and the tortoise that allows us to give a meaning to the infinite sum 1/2+1/4+1/8+... He writes that here, "the infinite becomes seizable in action."[9] Conversely, he says, the introduction of a continuous underlying substrate allows us to explain the significant—non-trivial—character of many mathematical theorems [41, p. 293].

Beyond philosophy, Thom used the mathematics that he developed in order to express natural phenomena. On several occasions, he insisted on the fact that the

---

[7]Ces intuitions, on ne les trouvera pas dans les thèses et les constructions explicites de la théorie. On les trouvera surtout dans quelques "petites phrases", qui illuminent tout le corpus de leur éclatante concision.

[8]Comment expliquer que les mathématiques représentent le réel ? La réponse, je crois, nous est offerte par l'intuition du continu.

[9]L'infini devient saisissable en acte.

mathematical concepts need not be formalized or rigorously expressed in order to exist. He recalls in [14] that "rigor" is a Latin word that reminds him of the sentence *rigor mortis* (the rigor of a dead body). He writes that rigor is "a very unnecessary quality in mathematical thinking." In his article [44], he declares that anything which is rigorous is not significant.[10]

Our aim in the next few pages is to expand on these ideas. A collection of essays on Thom and his work appeared in print recently, [30].

2.2 D'Arcy Thompson

Before talking more thoroughly about Aristotle and Thom, I would like to say a few words on the mathematician, biologist and philosopher who, in many ways, stands between them, namely, D'Arcy Thompson.[11] On several occasions, Thom referred to the latter's book *On growth and form* (1917), a frequently quoted work whose main object is the existence of mathematical models for growth and form in animal and vegetable biology at the level of cells, tissues and other parts of a living organism. D'Arcy Thompson emphasized the striking analogies between the mathematical patterns that describe all these parts, searching for general laws based on these patterns. This was new, compared to the contemporary approach to biology which relied on comparative anatomy and which was influenced by Darwin's theory of evolution.

In reading *On growth and form*, Thom had the same reaction he had when he read Aristotle: he was amazed by the richness of the ideas expressed in this book, and by their closeness to his own ideas. In turn, Thompson had a boundless admiration for Aristotle who had placed biology at the center of his investigations. In an article titled *On Aristotle as biologist*, Thompson calls the latter "the great biologist of Antiquity, who is *maestro di color che sanno*,[12] in the science as in so many other departments of knowledge." [54, p. 11] Likewise, Thom, in his articles *Les intuitions topologiques primordiales de l'aristotélisme* and *Matière, forme et catastrophes*,

[10]Tout ce qui est rigoureux est insignifiant.

[11]D'Arcy Wentworth Thompson (1860–1948) was a Scottish biologist with a profound passion for mathematics and for Greek science and philosophy. He was also an accomplished writer and his *magnum opus*, *On growth and form*, is an authentic literary piece. He is known as the first who found a relation between the Fibonacci sequence and some logarithmic spiral structures in the animal and vegetable life (mollusk shells, ruminant horns, etc.), but this is only a minor achievement compared to the rest of his work, and in particular, the development of the general field of mathematical morphology, which had an enormous impact on mathematicians and on biologists. D'Arcy Thompson's writings were influential on several twentieth century thinkers including C. H. Waddington, Alan Turing, Claude Lévi-Strauss and Le Corbusier and on artists such as Richard Hamilton and Eduardo Paolozzi. D'Arcy Thompson translated Aristotle's *History of animals*, a translation which is still authoritative.

[12]The quote is from Dante's *Inferno*, in which Aristotle is mentioned: "I saw the master of those who know."

refers to Aristotle as "the Master" ("le Maître"). Thompson writes in *On growth and form* [55, p. 15]:

> [Aristotle] recognized great problems of biology that are still ours today, problems of heredity, of sex, of nutrition and growth, of adaptation, of the struggle for existence, of the orderly sequence of Nature's plan. Above all he was a student of Life itself. If he was a learned anatomist, a great student of the dead, still more was he a lover of the living. Evermore his world is in movement. The seed is growing, the heart is beating, the frame breathing. The ways and habits of living things must be known: how they work and play, love and hate, feed and procreate, rear and tend their young; whether they dwell solitary, or in more and more organized companies and societies. All such things appeal to his imagination and his diligence.

Before Thom, D'Arcy Thompson foresaw the importance of topology in biology. He writes, in the same treatise [55, p. 609 ff]:

> [...] for in this study of a segmenting egg we are on the verge of a subject adumbrated by Leibniz, studied more deeply by Euler, and greatly developed of recent years. [...] Topological analysis seems somewhat superfluous here; but it may come into use some day to describe and classify such complicated, and diagnostic, patterns as are seen in the wings of a butterfly or a fly.

D'Arcy Thompson refers to the Aristotelian precept of studying the generic and to discard the accident as a mathematician's approach. He writes [54, p. 1032]: " [...] we must learn from the mathematician to eliminate and to discard; to keep the type in mind and leave the single case, with all its accidents, alone; and to find in this sacrifice of what matters little and conservation of what matters much one of the particular excellences of the method of mathematics."[13]

The question of the relation between the local and the global, which is dear to topologists, is addressed in *On growth and form* in several ways. D'Arcy Thompson refers to Aristotle and to other nineteenth century biologists. He writes (p. 1019):

> The biologist, as well as the philosopher, learns to recognize that the whole is not merely the sum of its parts. It is this, and much more than this. For it is not a bundle of parts but an organization of parts, of parts in their mutual arrangement, fitting one with another, in what Aristotle calls "a single and indivisible principle of unity"; and this is no merely metaphysical conception, but is in biology the fundamental truth which lies at the basis of Geoffroy's (or Goethe's) law of "compensation," or "balancement of growth."

It is of course not coincidental that Thompson mentions Goethe and Geoffroy-Saint-Hilaire and it is well worth to recall here that Goethe was also a passionate student of biology, in particular, of form and morphology. He wrote an essay on the evolution of plants based on their form. To him is generally attributed the first use of the word "metamorphosis" in botanics—even though the concept was present in Aristotle, who studied the metamorphosis of butterflies, gnats and other insects.

Isidore Geoffroy Saint-Hilaire, the other naturalist to whom D'Arcy Thompson refers, wrote an important treatise in three volumes titled *Histoire naturelle générale*

[13] D'Arcy Thompson refers to W. H. Young's address at the 1928 Bologna ICM (1928), titled "The mathematical method and its limitations."

des règnes organiques (General natural history of organic worlds) in which he carries out a classification of animals that is still used today. In commenting on Aristotle, Geoffroy writes in this treatise [22, vol. I, p. 19 ss.]:

> He is, in every branch of human knowledge, like a master who develops it alone. He reaches, he extends, the limit of all sciences, and at the same time he penetrates their intimate depths. From this point of view, Aristotle is an absolutely unique exception in the history of human thought, and if something here is amazing, it is not the fact that this exception is unique, but that there exists one, as such a meeting of faculties and of knowledge is surprising for anyone who wants to notice it psychologically. [...] Among his multiple treatises, the *History of animals* and the *Parts* are the main monuments of his genius.[14]

The biologist Thomas Lecuit, newly appointed professor at the Collège de France, gave his first course, in the year 2017–2018, on the problem of morphogenesis. He started his first lecture by a tribute to D'Arcy Thompson, emphasizing the latter's contribution to the problem of understanding the diversity of forms, the mathematical patterns underlying them and their transformations, and highlighting the continuing importance of his 100 years old book *On growth and form*. The book is now among the few most important books that formed the basis of modern science. In 2017, a 1-week workshop dedicated to that book was held at the Lorentz Center (Leiden) and the Institute for Advanced Studies (University of Amsterdam). The talks were on the impact of this work on mathematics, biology, philosophy, art and engineering.

D'Arcy Thompson addressed at the same time biologists and mathematicians. He writes, in the Epilogue of *On growth and form*: "While I have sought to shew the naturalist how a few mathematical concepts and dynamical principles may help and guide him, I have tried to shew the mathematician a field of his labour—a field which few have entered and no man has explored."

Let me end this section by emphasizing the fact that *On growth and form* is not only a scholarly book, but also a magnificent literary piece. D'Arcy Thompson himself was proud of his literary gifts. In an obituary notice, Clifford Dobell quotes him saying the following: "To spin words and make pretty sentences is my one talent, and I must make the best of it. And I am fallen on an age when not one man in 20,000 knows good English from bad, and not one in 50,000 thinks the difference of any importance. [...] The little gift of writing English which I possess, and try to cultivate and use, is, speaking honestly and seriously, the one thing I am a bit proud and vain of—the one and only thing." [20, p. 612]

[14] Il est, dans chaque branche du savoir humain, comme un maître qui la cultiverait seule; il atteint, il recule les limites de toutes les sciences, et il en pénètre en même temps les profondeurs intimes. Aristote est, à ce point de vue, une exception absolument unique dans l'histoire de l'esprit humain, et si quelque chose doit nous étonner ici, ce n'est pas qu'elle soit restée unique, c'est qu'il en existe une : tant une semblable réunion de facultés et de connaissances est surprenante pour qui veut s'en rendre compte psychologiquement. [...] Entre ses nombreux traités, les deux monuments principaux de son génie sont l'*Histoire des animaux* et *le Traité des parties*.

2.3 Aristotle, Mathematician and Topologist

Aristotle is very poorly known as a mathematician, a situation which is unfair. Although he did not write any mathematical treatise (or, rather, there is no indication that such a treatise existed), Aristotle had an enormous influence on mathematics, by his thorough treatment of first principles, axioms, postulates and other foundational notions of geometry, by his classification of the various kinds of logical reasonings, his reflections on the use of motion (that is, what we call today isometries) in axioms and in proofs, and on the consequences of that use, etc. There are may ways in which Aristotle preceded Euclid, not in compiling a list of axioms, but in his profound vision and thoughts on the axiomatic approach to geometry. We refer in particular to his discussions of first principles in his *Posterior analytics*,[15] to his insistence in the *Metaphysics*[16] on the fact that a demonstrative science is based on axioms that are not provable, to his discussion of the *reductio ad absurdum* reasoning in the same work,[17] and there are many other ideas on the foundations of mathematics in his work that one could mention. Besides that, several mathematical propositions are spread throughout his works. In particular, we find results related to all the fundamental problems of mathematics of that epoch: the parallel problem in the *Prior analytics*,[18] the incommensurability of the diagonal of a square[19] and the squaring of the circle (by means of the squaring of lunules) in the same treatise,[20] etc. There is an axiom in the foundations of geometry that was given the name *Aristotle's axiom*; see Greenberg's article [23]. In Book III of *On the heavens*, talking about form, one of his favorite topics, and commenting on a passage of Plato's *Timaeus* in which regular polyhedra are associated with the four sublunar elements (earth, water, air, fire), Aristotle states that there are exactly three regular figures that tile the plane, namely, the equilateral triangle, the square and the regular hexagon, and that in space, there are only two: the pyramid and the cube.[21] It is possible that this passage is the oldest surviving written document in which this theorem is stated.[22] One may also consider Aristotle's *Problems*,[23] a treatise in 38 books, assembled by themes, one of the longest of the Aristotelian corpus. It consists of a list of commented (open) problems, of the kind mathematicians are used to edit,

[15] *Posterior analytics*, [4] 74b5 and 76a31-77a4.

[16] *Metaphysics* [8] 997a10.

[17] *Posterior analytics*, [4] 85a16ff.

[18] *Prior analytics* [2] 65a4–9, 66a11-15.

[19] *Prior analytics* [2] 65b16-21.

[20] *Prior analytics* [2] 69a20-5.

[21] *On the heavens* [7] 306b1-5.

[22] The reason why Aristotle makes this statement here is not completely clear, but it is reasonable to assume that it is because Plato, in the *Timaeus* conjectured that the elementary particles of the four elements have the form of the regular polyhedra he associated to them; hence the question raised by Aristotle concerning the tiling of space using regular polyhedra.

[23] The translations below are from [5].

with the difference that Aristotle's problems concern not only mathematics but all subjects of human knowledge: psychology, biology, physics, acoustics, medicine, ethics, physiology, etc. In total, there are 890 problems. They generally start with the words "Why is it that..." For instance, Problem 48 of Book XXVI asks: *Why is it that the winds are cold, although they are due to movement caused by heat?* Problem 19 of Book XXXI asks: *Why is it that when we keep our gaze fixed on objects of other colours our vision deteriorates, whereas it improves if we gaze intently on yellow and green objects, such as herbs and the like?* Problem 20 of the same book asks: *Why is it that we see other things better with both eyes, but we can judge of the straightness of lines of writing better with one eye?* Some of the problems concern mathematics. For instance, Problem 3 of Book XV asks: *Why do all men, Barbarians and Greek alike, count up to 10 and not up to any other number, saying for example, 2, 3, 4, 5 and then repeating then, "one-five", "two-five", just as they say eleven, twelve?* Problem 5 of the same book starts with the question: *Why is it that, although the sun moves with uniform motion, yet the increase and decrease of the shadows is not the same in any equal period of time?* Book XVI is dedicated to "inanimate things." Problems 1 and 2 in that book concern floating bubbles. Problem 2 asks: *Why are bubbles hemispherical?*[24] Problem 5 asks: *Why is it that a cylinder, when it is set in motion, travels straight and describes straight lines with the circles in which it terminates, whereas a cone revolves in a circle, its apex remaining still, and describes a circle with the circle in which it terminates?* Problem 5 of the same book concerns the traces of oblique sections of a cylinder rolling on a plane. Problem 6 concerns a property of straight lines: *Why is it that the section of a rolled book, which is flat, if you cut it parallel to the base becomes straight when unrolled, but if it is cut obliquely becomes crooked?*

Another mathematical topic discussed in detail in Aristotle's works is that of infinity, for which the Greeks, at least since Anaximander (sixth century BCE) had a name, *apeiron* (ἄπειρον), meaning "boundless." This notion, together with that of limit, is discussed in the *Categories*, the *Physics*, the *Metaphysics*, *On the heavens*, and in other writings. In the *Physics*,[25] Aristotle mentions the two occurrences of the infinite in mathematics: the infinitely large, where, he says, "every magnitude is surpassed" and the infinitely small, where "every assigned magnitude is surpassed

[24]D'Arcy Thompson was also fascinated by the questions of form and transformation of floating bubbles, in relation with the question of growth of a living cell submitted to a fluid pressure (cf. *On growth and form*, Chapters V–VII). On p. 351, he talks about "the peculiar beauty of a soap-bubble, solitary or in collocation [..] The resulting form is in such a case so pure and simple that we come to look on it as well-nigh a mathematical abstraction." On p. 468, he writes: "Bubbles have many beautiful properties besides the more obvious ones. For instance, a floating bubble is always part of a sphere, but never more than a hemisphere; in fact it is always rather less, and a very small bubble is considerably less than a hemisphere. Again, as we blow up a bubble, its thickness varies inversely as the square of its diameter; the bubble becomes a 150 times thinner as it grows from an inch in a diameter to a foot." Later in the same chapter, Thompson talks about clustered bubbles (p. 485). He quotes Plateau on soap-bubble shapes.

[25]*Physics*, [1] 201a-b.

in the direction of smallness." He disliked the "unbounded infinite" and his interest lay in the second kind. He writes in the same passage:

> Our account does not rob the mathematicians of their science, by disproving the actual existence of the infinite in the direction of increase, in the sense of the untraversable. In point of fact they do not need the infinite and do not use it. They postulate only that the finite straight line may be produced as far as they wish.

This is a reference to the occurrence of the infinitely large in the axioms of geometry (predating Euclid). It is interesting to note that Descartes considered that only the infinitely small appears in mathematics. The infinitely large, in his conception, belongs to metaphysics only.[26]

Aristotle, in his discussion of infinity, makes a clear distinction between the cases where infinity is attained or not. In a passage of the *Physics*,[27] he provides a list of the various senses in which the word "infinite" is used: (1) infinity incapable of being gone through; (2) infinity capable of being gone through having no termination; (3) infinity that "scarcely admits of being gone through"; (4) infinity that "naturally admits of being gone through, but is not actually gone through or does not actually reach an end." He discusses the possibility for an infinite body to be simple infinite or compound infinite.[28] In the same passage, he talks about form, which, he says, "contains matter and the infinite." There is also a mention of infinite series in the *Physics*.[29]

One should also talk about the mathematical notion of continuity in the writings of the Philosopher.

In the *Categories*, Aristotle starts by classifying quantities into discrete or continuous. [30] He declares that some quantities are such that "each part of the whole has a relative position to the other parts; others have within them no such relation of part to part," a clear reference to topology. As examples of discrete quantities, he mentions number (the integers) and speech. As examples of continuous quantities, he gives lines, surfaces, solids, time and place (a further reference to topology). He explains at length, using the mathematical language at his disposal, why the set of integers is discrete. According to this description, two arbitrary integers "have no common boundary, but are always separate." He declares that the same holds for speech: "there is no common boundary at which the syllables join, but each is separate and distinct from the rest." This is a way of saying that each integer (respectively each syllable) is isolated from the others, the modern formulation of discreteness. Aristotle writes that a line is a continuous quantity "for it is possible to find a common boundary at which its parts join." He declares that space is a continuous quantity "because the parts of a solid occupy a certain space, and these have a common boundary; it follows that the parts of space also, which are occupied

[26]See the comments by R. Rashed in his article *Descartes et l'infiniment petit* [33].

[27]*Physics*, [1] 204a.

[28]*Physics*, [1] 204b10.

[29]*Physics* [1] 206a25-206b13.

[30]*Categories* [10] 4b20.

by the parts of the solid, have the same common boundary as the parts of the solid." Time, he writes, is also a continuous quantity, "for its parts have a common boundary." The notion of boundary is omnipresent in this discussion. Thom was fascinated by this fact and he emphasized it in his writings. We shall discuss this in more detail later in this paper.

In another passage of the *Categories*, Aristotle discusses position and relative position, notions that apply to both discrete and continuous quantities: "Quantities consist either of parts which bear a relative position each to each, or of parts which do not."[31] Among the quantities of the former kind, he mentions the line, the plane, the solid space, for which one may state "what is the position of each part and what sort of parts are contiguous." On the contrary, he says, the parts of the integers do not have any relative position each to each, or a particular position, and it is impossible to state what parts of them are contiguous. The parts of time, even though the latter is a continuous quantity, do not have position, because, he says, "none of them has an abiding existence, and that which does not abide can hardly have position." Rather, he says, such parts have a *relative order*, like for number, and the same holds for speech: "None of its parts has an abiding existence: when once a syllable is pronounced, it is not possible to retain it, so that, naturally, as the parts do not abide, they cannot have position." In this and in other passages of Aristotle's work, motion and the passage of time are intermingled with spatiality. It is interesting to see that Hermann Weyl, in his book *Space, time and matter*, also insisted on the importance of the relation between, on the one hand, motion, and, on the other hand, space, time and matter [58, p. 1]: "It is in the composite idea of *motion* that these three fundamental conceptions enter into intimate relationship."

Talking about position, we come to the important notion of place.

Several Greek mathematicians insisted on the difference between space and place, and Aristotle was their main representative. They used the word khôra (χώρα) for the former and topos (τόπος) for the latter. Again, Thom regards Aristotle's discussion from a mathematician's point of view, and he considers it as readily leading to a topological mathematization, although it does not use any notational apparatus (which, at that time, was nonexistent) or the technical language of topology to which we are used today. In the *Physics*, Aristotle gives the following characteristics of place[32]: (1) Place is what contains that of which it is the place. (2) Place is no part of the thing. (3) The immediate place of a thing is neither less nor greater than the thing. (4) Place can be left behind by the thing and is separable. (5) All place admits of the distinction of up and down, and each of the bodies is naturally carried to its appropriate place and rests there, and this makes the place either up or down.

This makes Aristotle's concept of place close to our mathematical notion of boundary. Furthermore, it confers to the notion of place the status of a "relative" notion: a place is defined in terms of boundary, and the boundary is also the

[31] Categories [10] 5a10.

[32] *Physics*, [1] 211a ff.

boundary of something else. At the same time, this is not too far from the notion of "relative position" that was formalized later on in Galilean mechanics. All this is an indication of the fact that the notion of place in Aristotle's writings is a broad subject, and there is a wide literature on it. Aristotle himself considered place as a difficult subject. He writes in the same passage:

> We ought to try to make our investigations such as will render an account of place, and will not only solve the difficulties connected with it, but will also show that the attributes supposed to belong to it do really belong to it, and further will make clear the cause of the trouble and of the difficulties about it.

Aristotle then introduces a dynamical aspect in his analysis of place. He discusses motion and states that locomotion and the phenomena of increase and diminution involve a variation of place. The notion of boundary is omnipresent in this discussion. In the same passage of the *Physics*, he mentions the notion of "inner surface" which surrounds a body, an expression on which Thom insisted, as we shall see later. Aristotle writes:

> When what surrounds, then, is not separate from the thing, but is in continuity with it, the thing is said to be in what surrounds it, not in the sense of in place, but as a part in a whole. But when the thing is separate or in contact, it is immediately "in" the inner surface of the surrounding body, and this surface is neither a part of what is in it nor yet greater than its extension, but equal to it; for the extremities of things which touch are coincident.

Regarding place and its relation to boundary, we mention another passage from the *Physics*[33]: "Place is [...] the boundary of the containing body at which it is in contact with the contained body. (By the contained body is meant what can be moved by way of locomotion)." Thom commented on this passage on several occasions. In his paper *Aristote topologue* [53], he discusses the relation, in Aristotle's writings, between topos and *eschata* (ἔσχατα), that is, the limits, or extreme boundaries. We shall elaborate on this below.

The overall discussion of place in Aristotle's work, and its relation with shape and boundary is involved, and the various translations of the relevant passages in his writings often differ substantially from each other and depend on the understanding of the translator. This not a surprise, and the reader may imagine the difficulties in talking in a precise manner about a notion of boundary that does not use the language of modern topology. Aristotle emphasized the difficulty of defining place. In the *Physics*, he writes:

> Place is thought to be something important and hard to grasp, both because the matter and the shape present themselves along with it, and because the displacement of the body that is moved takes place in a stationary container, for it seems possible that there should be an interval which is other than the bodies which are moved. [...] Hence we conclude that *the innermost motionless boundary of what contains is place.* [...] Place is thought to be a kind of surface, and as it were a vessel, i.e. a container of the thing. Further; place is coincident with the thing, for boundaries are coincident with the bounded.[34]

[33] *Physics*, [1] 211b5-9.

[34] *Physics* [1] 212a20.

Place is also strongly related to form. Aristotle states that form is the boundary of the thing whereas place is the boundary of the body that contains the thing. Thom interpreted the texts where Aristotle makes a distinction between "the boundary of a thing" and "the boundary of the body that contains it"—a structure of "double ring" of the eschata, as he describes it—using homological considerations. More precisely, he saw there a version of the Stokes formula. We shall review this in Sect. 2.4 below.

Let us quote a well-known text which belongs to the Pythagorean literature, written by Eudemus of Rhodes (fourth century BCE), in which the latter quotes Archytas of Tarentum. This text shows the kind of questions on space and place that the Greek philosophers before Aristotle addressed, e.g., whether space is bounded or not, whether "outer space" exists, and the paradoxes to which this existence leads (see [25, p. 541]):

> "But Archytas," as Eudemus says, "used to propound the argument in this way: 'If I arrived at the outermost edge of the heaven [that is to say at the fixed heaven], could I extend my hand or staff into what is outside or not?' It would be paradoxical not to be able to extend it. But if I extend it, what is outside will be either body or place. It doesn't matter which, as we will learn. So then he will always go forward in the same fashion to the limit that is supposed in each case and will ask the same question, and if there will always be something else to which his staff [extends], it is clear that it is also unlimited. And if it is a body, what was proposed has been demonstrated. If it is place, place is that in which body is or could be, but what is potential must be regarded as really existing in the case of eternal things, and thus there would be unlimited body and space." (Eudemus, Fr. 65 Wehrli, Simplicius, In Ar. Phys. iii 4; 541)

Let us now pass to other aspects of mathematics in Aristotle. We mentioned in the introduction his notion of homeomerous. This is used on several occasions in his works, and especially in his zoological treatises. He introduces it at the beginning of the *History of animals* [12], where he talks about simple and complex parts. A part is simple, he says, if, when divided, one recovers parts that have the same form as the original part, otherwise, it is complex. For instance, a face is subdivided into eyes, a nose, a mouth, cheeks, etc., but not into faces. Thus, a face is a complex (anhomeomorous) part of the body. On the other hand, blood, bone, nerves, flesh, etc. are simple (homeomorous) parts because subdividing them gives blood, bone, nerves, flesh, etc. Anhomeomorous parts in turn are composed of homeomerous parts: for instance, a hand is constituted of flesh, nerves and bones. The classification goes on. Among the parts, some are called "members." These are the parts which form a complete whole but also contain distinct parts: for example, a head, a leg, a hand, a chest, etc. Aristotle also makes a distinction between parts responsible of "act", which in general are anhomeomorous (like the hand) and the others, which are homeomorous (like blood) and which he considers as "potential parts." In D'Arcy Thompson's translation, the word "homeomerous" is translated by "uniform with itself," and sometimes by "homogèneous." We mention that the term homeomoerous was also used in geometry, to denote curves that are self-similar in the sense that any part of them can be moved to coincide with any other part. Proclus, the fifth century mathematician, philosopher and historian of mathematics, discusses this notion in his *Commentary on the First Book of Euclid's Elements*. He gives as an example of a homeomerous curve the cylindrical helix, attributing the definition to Apollonius in his book titled *On the screw*, a work which does not survive. Among the curves of similar shape that are not homeomerous, he gives the example of Archimedes' (planar) spiral, the conical helix, and the spherical helix (see p. 95 of ver Eecke's edition of his *Commentary on the First Book of Euclid's Elements* [32]). In the same work, Proclus attributes to Geminus a result stating that there are only three homeomerous curves: the straight line, the circle and the cylindrical helix [32, p. 102].

Aristotle was a mathematician in his unrelenting desire for making exhaustive classifications and of finding *structure* in phenomena: after all, he introduced his *Categories* for that purpose. In the *History of animals*, like in several other treatises he wrote, the sense of detail is dizzying. He writes (D'Arcy Thompson's translation)[35]:

> Of the substances that are composed of parts uniform (or homogeneous) with themselves, some are soft and moist, others are dry and solid. The soft and moist are such either absolutely or so long as they are in their natural conditions, as, for instance, blood, serum, lard, suet, marrow, sperm, gall, milk in such as have it flesh and the like; and also, in a different way, the superfluities, as phlegm and the excretions of the belly and the bladder. The dry and solid are such as sinew, skin, vein, hair, bone, gristle, nail, horn (a term which as applied to the part involves an ambiguity, since the whole also by virtue of its form is designated horn), and such parts as present an analogy to these.

Thom, in Chapter 7 of his *Esquisse*, comments on Aristotle's methods of classification developed in the *Parts of animals*, relating it to his own work as a topologist. He recalls that Aristotle

> attacks therein the Platonic method of *dichotomy*, suggesting in its place an interrogative method for taking the substrate into consideration. Thus, if we propose to attain a definition characterizing the "essence" of an animal, we should not, says Aristotle, pose series of questions bearing on "functionally independent" characteristics. For example, "Is it a winged or a terrestrial animal?" or "Is it a wild animal or a tame one?" Such a battery of questions bearing on semantic fields—genera—unrelated one to the other, can be used in an arbitrary order. The questionnaire may indeed lead to a definition of sorts, but it will be a purely artificial one. It would be more rational to have a questionnaire with a *tree* structure, its ramification corresponding to the substrate. For instance, after the question: "Is the animal terrestrial?", if the answer is yes, we will ask, "Does the animal have legs?" If the answer is again yes, we then ask, "Is the foot all in one (solid), or cloven, or does it bear digits?" Thus we will reach a definition which is at the same time a description of the organism in question. Whence a better grasp of its essence in its phenomenal aspect. Aristotle observes, for instance, that if one poses a dilemma bearing on a private opposition, presence of A, absence of A, the natural posterity of the absence of A in the question-tree is empty. In a way the tree of this questionnaire is the reflection of a dynamic inside the substrate. It is the dynamic of the blowing-up of the centre of the body (the soul), unique in potentiality, into a multitude of part souls in actuality. In a model of the catastrophe type, it is the "unfolding" dynamics.

---

[35] *History of animals* [12] 487a.

It may be fitting to cite also the following passage from the *History of animals*, on the mysteries of embryology, and more precisely on the relation between the local and the global that plays a central role in this domain[36]:

> The fact is that animals, if they be subjected to a modification in minute organs, are liable to immense modifications in their general configuration. This phenomenon may be observed in the case of gelded animals: only a minute organ of the animal is mutilated, and the creature passes from the male to the female form. We may infer, then, that if in the primary conformation of the embryo an infinitesimally minute but absolutely essential organ sustain a change of magnitude one way or the other, the animal will in one case turn to male and in the other to female; and also that, if the said organ be obliterated altogether, the animal will be of neither one sex nor the other. And so by the occurrence of modification in minute organs it comes to pass that one animal is terrestrial and another aquatic, in both senses of these terms. And, again, some animals are amphibious whilst other animals are not amphibious, owing to the circumstance that in their conformation while in the embryonic condition there got intermixed into them some portion of the matter of which their subsequent food is constituted; for, as was said above, what is in conformity with nature is to every single animal pleasant and agreeable.

To close this section, I would like to say a few more words on Aristotle as a scientist, from Thom's point of view, and in particular, regarding the (naive) opposition that is usually made between Aristotelian and Galilean science, claiming that the mathematization of nature, as well as the so-called "experimental method" started with Galileo and other moderns, and not before, and in any case, not with Aristotle.

The lack of "mathematization" in Aristotle's physics, together with the absence of an "experimental method" are due in great part to the lack of measurements, and this was intentional. In fact, Aristotle, like Thom, was interested in the qualitative aspects of phenomena, and not the quantitative. This is a mathematician's approach. We may quote Thurston: "Ultimately, what we seek when we study mathematics is a qualitative understanding." [56, p. 74] Furthermore, like Thom after him, Aristotle was reluctant to think in terms of a "useful science": he was the kind of scholar who was satisfied with doing science for the pleasure of the intellect. Thom's vision of Aristotelian science was completely different from the commonly accepted one, and the fact that Aristotle was a proponent of the qualitative vs. the quantitative was in line with his own conception of science. Let us quote him, from his article *Aristote et l'avènement de la science moderne : la rupture galiléenne* (Aristotle and the advent of modern science: the Galilean break), published in 1991 [49, p. 489]:

> I would be tempted to say that one can see frequently enough a somewhat paternalistic attitude of condescension regarding Aristotle in the mouth of contemporary scholars. I think that this attitude is not justified. It is usually claimed that the Galilean epistemological break brought to science a radical progress, annihilating the fundamental concepts of Aristotelian physics. But here too, one must rather see the effect of this brutal transformation as a scientific revolution in the sense of Kuhn, that is, a change in paradigms, where the problems solved by the ancient theory stop being objects of interest and disappear from the speculative landscape. The new theory produces new problems, which it can solve more

[36] *History of animals* [12] 589b30 ff.

> or less happily, but above all, it leads to an occultation of all the ancient problematics which nevertheless continues its underground journey under the clothing of new techniques and new formalisms.
>
> I think that very precisely, maybe since ten years, one can see in modern science the reappearance of a certain number of themes and methods that are close to the Aristotelian doctrine, and I personally welcome this kind of resurgence that I will try to describe for you. What I will talk to you about may not belong as much to Aristotle, of whom, by the way, I have a poor knowledge, than to certain recent evolutions of science that recall, I hope without excessive optimism, certain fundamental ideas of Aristotelian physics and metaphysics.[37]

In his development of these ideas, Thom talks about Aristotelian logic, closely linked to his ontology, making science appear as a "logical language", forming an "isomorphic image of natural behavior." He also talks about the "mathematization of nature" that the so-called Galilean break brought as a hiatus between the mathematical and the common languages, and about the disinterest in the Aristotelian notion of formal causality that characterizes modern science and which would have been so useful in embryology. He gives examples from Aristotelian mechanics, in particular his concept of time, which he links to ideas on thermodynamics, entropy and Boltzmann's H-Theorem which describes the tendency of an isolated ideal gas system towards a thermodynamical equilibrium state.

Even if Thom's point of view is debatable, it has the advantage of making history richer and giving it a due complexity.

As a final remark on mathematics in Aristotle, I would like to recall his reluctance to accept the Pythagorean theories of symbolism in numbers, which he expressed at several places in his writings.[38] Incidentally, mathematicians tend to refer to Plato rather than to Aristotle, because the former's conception of the world gives a more prominent place to mathematics. But whereas for Plato, mathematics has an abstract nature and is dissociated from the real world, for Aristotle, it is connected with

[37] Je serais tenté de dire qu'une attitude de condescendance un peu paternaliste se remarque assez fréquemment dans la bouche des savants contemporains à l'égard d'Aristote. Et je pense que cette attitude n'est pas justifiée. Il est courant de dire que la rupture épistémologique galiléenne a amené en science un progrès radical, réduisant à néant les concepts fondamentaux de la physique aristotélicienne. Mais là aussi, il faut voir l'effet de cette transformation brutale plutôt à la manière d'une révolution scientifique au sens de Kuhn, c'est-à-dire comme un changement de paradigmes : les problèmes résolus par l'ancienne théorie cessant d'être objet d'intérêt et disparaissant du champ spéculatif. La nouvelle théorie dégage des problèmes neufs, qu'elle peut résoudre avec plus ou moins de bonheur, mais surtout elle conduit à occulter toute la problématique ancienne qui n'en poursuit pas moins son cheminement souterrain sous l'habillement des nouvelles techniques et des nouveaux formalismes.

Je pense que très précisément, depuis peut-être une dizaine d'années, on voit réapparaître dans la science moderne un certain nombre de thèmes et de méthodes proches de la doctrine aristotélicien, et je salue quant à moi cette espèce de résurgence que je vais essayer de vous décrire. Ce dont je vous parlerai ce n'est donc peut-être pas tant d'Aristote, que je connais mal d'ailleurs, que de certaines évolutions récentes de la science qui me semblent évoquer, sans optimisme excessif, j'espère, certaines idées fondamentales de la physique et de la métaphysique aristotélicienne.

[38] Cf. for instance *Metaphysics* [8] 1080b16-22.

nature. The latter's point of view is consistent with the one of Thom. In his paper *Logos phenix* published in the book *Modèles mathématiques de la morphogenèse* [41] which we already mentioned, he writes: "What remains in me of a professional mathematician can hardly accept that mathematics is only a pointless construction without any connection to reality."[39]

We shall talk more about Thom's view on Aristotle in the next sections.

2.4 Thom on Aristotle

Chapter 5 of Thom's *Esquisse* is titled *The general plan of animal organization* and is in the lineage of the zoological treatises of Aristotle, expressed in the language of modern topology. Thom writes in the introduction: "This presentation might be called an essay in transcendental anatomy, by which I mean that animal organisation will be considered here only from the topologists' abstract point of view." He then writes: "We shall be concerned with ideal animals, stylized images of existing animals, leaving aside all considerations of quantitative size and biochemical composition, to retain only those inter-organic relations that have a topological and functional character." In Section B of the same chapter, Thom returns to Aristotle's notion of homeomerous and anhomeomerous, in relation with the stratification of an animal's organism, formulating in a modern topological language the condition for two organism to have the same organisation.

An organisms, in Thom's words, is a three-dimensional ball O equipped with a stratification, which is finite if we decide to neglect too fine details. For example, when considering the vascular system, Thom stops at the details which may be seen with the naked eye: arterioles and veinlets. He writes: "We will thus avoid introducing fractal morphologies which would take us outside the mathematical schema of stratification." Seen from this point of view, the homeomerous parts are the strata of dimension three (blood, flesh, the inside of bones...), the two-dimensional strata are the membranes: skin, mucous membrane, periosteum, intestinal wall, walls of the blood vessels, articulation surfaces, etc., the one-dimensional strata are the nerves: vessel axes, hair, etc., and the zero-dimensional strata are the points of junction between the one-dimensional strata or the punctual singularities: corners of the lips, ends of hair, etc.

[39]Ce qui reste en moi du mathématicien professionnel admet difficilement que la mathématique ne soit qu'une construction gratuite dépourvue de toute attache au réel.

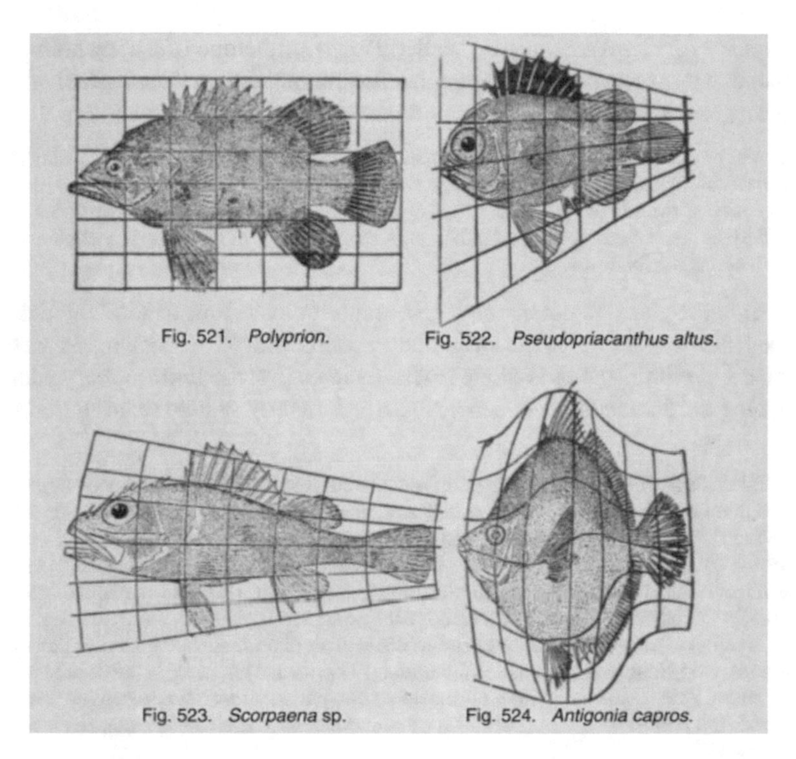

Homeomorphisms preserving stratifications, from D'Arcy Thompson's *On growth and form*

Thom says that two organisms O and O' "have the same organisation" if there exists a homeomorphism $h : O \to O'$ preserving this stratification. He claims that such a formalism generalizes D'Arcy Thompson's famous diagrams and makes them more precise. The reference here is to Thompson's sketches from *On growth and Form* which describe passages between various species of fishes using homeomorphisms that preserve zero-dimensional, one-dimensional, two-dimensional and three-dimensional strata.

Thom developed his ideas on the stratification of the animal body in the series of lectures given in 1988 at the Solignac Abbey, a medieval monastery in the Limousin (South of France). The lectures are titled *Structure et fonction en biologie aristotélicienne* (*Structure and function is Aristotelian biology*) and the lecture notes are available [46]. On p. 7 of these notes, he addresses the question of when two animals have isomorphic stratifications, and he uses for that the notion of isotopy between stratified spaces: two sets E_1, E_2 have isotopic stratifications if there exists a stratification of the product $E \times [0, 1]$ such that the canonical projection $p : E \times [0, 1] \to [0, 1]$ is of rank one on every stratum of E, with $E_1 = p^{-1}(1)$ and $E_2 = p^{-1}(2)$. He considers that this notion is implicitly used by Aristotle in his classification of the animals, insisting on the fact that the latter neglected all the quantitative differences and was only interested in the qualitative ones. In the same passage, he recalls that D'Arcy Thompson, who translated Aristotle's *History of Animals*, acknowledged that he found there the idea of his diagrams.

Chapter 7 of Thom's *Esquisse* is called *Perspectives in Aristotelian biology*. The first part concerns topology and bears the title *The primordial topological intuitions of Aristotelianism: Aristotle and the continuum*. It starts as follows [47, p. 165]:

> We shall present here those intuitions which we believe sub-tend all Aristotelianism. They are ideas that are never explicitly developed by the author, but which—to my mind—are the framework of the whole architecture of his system. We come across these ideas formulated "by the way" as it were, condensed into a few small sentences that light up the whole corpus with their bright concision.

Thom highlights and comments on several citations from Aristotle which show that the latter was aware of the basic notions of topology. He declares in particular that a careful reading of Aristotle's *Physics* shows that the Philosopher understood the topological distinction between a closed and an open set. He writes [47, pp. 167–168]:

> Careful reading of the *Physica* leaves little doubt but that [Aristotle] had indeed perceived this difference. "It is a whole and limited; not, however, by itself, but by something other than itself"[40] could hardly be interpreted except in terms of a bounded open set. In the same vein, the affirmation: "The extremities of a body and of its envelope are the same"[41] can be identified, if the envelope is of a negligible thickness, with the well-known axiom of general topology: "Closure of closure is closure itself" expressed by Kuratowski at the beginning of this century. This allows the Stagirite to distinguish two infinites: the great infinite that envelops everything and the small infinite that is bounded. This latter is the infinite of the continuum, able to take an infinity of divisions (into parts that are themselves continuous). Whence the definition he proposes: "The infinite has an intrinsic substrate, the sensible continuum."[42]

In his 1988 article *Les intuitions topologiques primordiales de l'aristotélisme* [45] and in his 1991 article *Matière, forme et catastrophes* [48], Thom returns to these matters, explaining that the modern topological distinction between an open and a closed set is expressed in Aristotle's philosophical distinctions between form and matter and between actuality and potentiality. He gives an explanation for the difference between the notion of bounded and unbounded open set in Aristotle's philosophical system: the former may exist as a substrate of being whereas the latter cannot [45, p. 396]. He highlights the presence of the notion of boundary in formulae such as: "Form is the boundary of matter," [45, p. 398] and "Actuality is the boundary of potentiality" ([45, p. 399] and [48, p. 380]). He recalls that the paradigmatic substance for Aristotle is the living being, which is nothing else than a ball in Euclidean space whose boundary is a sphere (provided, Thom says, one neglects the necessary physiological orifices), that is, a closed surface *without boundary*. Shapeless matter is enveloped by form—*eidos*—in the same way as the boundary of a bronze statue defines its shape. The boundary of a living organism is its skin, and its "interior" exists only as a potentiality. A homeomerous part of an animal has generally a boundary structure constituted by anhomeomerous

[40] *Physics* [1] 207a24-35.

[41] *Physics* [1] 211b12.

[42] *Physics* [1] 208a.

parts. Thus, the substrate of a homeomerous part is not closed.[43] Mathematics, philosophy and biology are intermingled in this interpretation of a living organism, following Aristotle, with formulae like: "The opposition homeomerous-anhomeomerous is a 'representation' (a homomorphic image) of the metaphysical opposition: potentiality-act. As the anhomeomerous is part of the boundary of a homeomerous of one dimension higher, we recover a case of the application of *act as boundary of the potentiality*."[44]

Thom refers to Chapter 16 of Book Z of the *Metaphysics*, interpreting a sentence of the *Metaphysics* on entelechy (ἐντελέχεια) (the "creative principle", by which being passes from potentiality to action): "Entelechy separates", as the opposition between an open and a closed set, more precisely, by the fact that "a closed segment is 'actual' as opposed to the semi-open interval which exists only as a potentiality."[45] He writes again, in the same passage [48, p. 381]: "An open substrate characterizes potential entity. A closed substrate is required for the acting being."[46]

In a passage of the *Esquisse*, Thom talks about Aristotle as "the philosopher of the continuous" [47, p. viii], and he considers that his chief merit was that he was "the only one who thought in terms of the continuous" for hundreds, may be thousands of years. In his article *Logos phenix*, Thom writes: "How can we explain that mathematics represents the real? The response, I think, is offered to us by the intuition of the continuous. [...] The introduction of an underlying continuous substrate allows us henceforth to explain the significant—non-trivial—character of several mathematical theorems."[47] [41, p. 292ff].

In the book *Prédire n'est pas expliquer* [50, pp. 81–82], turning to the distinction that the Greek philosophers made between the discrete and the continuous, he declares: "For me, the fundamental aporia of mathematics is certainly the opposition between the discrete and the continuous. And this aporia at the same time dominates all thought. [...] The origin of scientific thinking, we find it in the apories of Zeno of Elea: the story of Achilles and the tortoise. Here, we find the crucial opposition between the discontinuous and the continuous."[48] Thom says that this aporia, or

[43]Une partie homéomère a en général un bord constitué d'anhoméomères ; ainsi le substrat d'un homéomère n'est pas—en général—un ensemble fermé au sens de la topologie moderne [45, p. 398]

[44]L'opposition homéomère-anhoméomère est une "représentation" (une image homomorphe) de l'opposition métaphysique : puissance-acte. Comme l'anhoméomère est partie du bord d'un homéomère de dimension plus grande, on retrouve ainsi un cas d'application de l'*acte bord de la puissance* [45, p. 400].

[45]Le segment fermé est "actuel" par opposition à l'intervalle semi-ouvert qui n'est qu'en puissance.

[46]Un substrat ouvert caractérise l'entité en puissance. Un substrat fermé est requis pour l'être en acte.

[47]Comment expliquer que les mathématiques représentent le réel ? La réponse, je crois, nous est offerte par l'intuition du continu. [...] L'introduction d'un substrat continu sous-jacent permet dès lors de s'expliquer le caractère signifiant—non trivial—de bien des théorèmes de la mathématiques.

[48]Pour moi, l'aporie fondamentale de la mathématique est bien dans l'opposition discret-continu. Et cette aporie domine en même temps toute la pensée. [...] L'origine de la pensée scientifique, on

fundamental opposition, which he also calls the "foundational contradiction"[49] received illusory solutions,[50] in particular, the one which pretends to generate the continuous out of the discrete: Thom refers here to the mathematical generation of the real numbers out of the rationals, by the two competing and essentially equivalent methods: Cauchy sequences and Dedekind cuts, whose highly non-constructive character makes them, from his point of view, illusory and fantasmic, despite the fact that they are useful from a purely mathematical (technical) point of view. In reality the discussion concerns the philosophical status of the concept of continuity. Thom declares in the book, *Prédire n'est pas expliquer* [50], concerning this side of the concept of continuity and the distinction between the continuous and the discrete:

> The continuous is in some way the universal substrate of thought, and in particular of mathematical thought. But we cannot think of anything in an effective way without having something like the discrete in the continuous flow of mental processes: there are words, sentences, etc. The *logos*, discourse, is always discrete; these are words, coming in with a certain order, but they are discrete words. And the discrete immediately calls down the quantitative. There are points: we can count them; there are words in a sentence: we can classify them quantitatively by the grammatical function they occupy in a sentence. However there is an undeniable multiplicity."[51]

On the same subject, in his article *Logos phenix*, Thom writes [41, p. 294]:

> Meaning is always tied to the attribution of a place of spatial nature to an expression formally encoded. There should always be, in any meaningful message, a discontinuous component tied to the generative mechanisms of language—to symbols—, and a continuous component, a substrate, in which the continuous component cuts out a place.[52]

In the foreword to the *Esquisse*, Thom writes that Aristotle's geometrical insight was founded uniquely on an intuition of the continuous, where a segment of a straight line is not made out of points but of sub-segments. Neither Dedekind nor Cantor, he says, have taken that road. He writes that the single isolated point (the one we consider when we take a point O on the $x'x$-axis), exists only "potentially."

---

la trouve dans les apories de Zénon d'Élée : l'histoire d'Achille et de la tortue. Il y a là l'opposition cruciale entre discontinu et continu.

[49]Thom talks of a "contradiction de base"; cf. his paper *Un panorama des mathématiques* [51].

[50]Thom talks in [51] of "fantasmatic" solutions.

[51]Le continu est en quelque sorte le substrat universel de la pensée, et de la pensée mathématique en particulier. Mais on en peut rien penser de manière effective sans avoir quelque chose comme le discret dans ce déroulement continu de processus mentaux : il y a des mots, il y a des phrases, etc. Le *logos*, le discours, c'est toujours du discret ; ce sont des mots entrant dans une certaine succession, mais des mots discrets. Et le discret appelle immédiatement le quantitatif. Il y a des points : on les compte ; il a des mots dans une phrase : on peut les classer quantitativement par la fonction grammaticale qu'ils occupent dans la phrase, mais il n'empêche qu'il y a une incontestable multiplicité.

[52]Le sens est toujours lié à l'attribution d'une place de nature spatiale à une expression formellement codée. Il y aurait toujours, dans tout message signifiant, une composante discontinue liée aux mécanismes génératifs de la langue—aux symboles—, et une composante continue, un substrat, dans lequel la composante continue découpe une place.

Such a point, according to him, aspires to actuality by duplicating itself into two points O_1, O_2, O_1 adhering to the left, O_2 to the right. "These two points then being *distinct* even though they are *together* (ἅμα), the two half-segments so limited attain full existence, being in actuality." [47, p. viii] Thom refers for that to the *Metaphysics* 139a3-7.[53] The question is also that of knowing what expressions like "is made of" or "are part of", applied to points and lines, mean. In his Solignac 1988 notes, quoting a passage from Aristotle's *Parts of animals* in which the latter compares an animal vascular system to a garden irrigation ditch,[54] Thom notes that the Philosopher goes as far as to say that blood is not part of the organism,[55] because it is dense there, and that this does not belong to the definition of "being part of" [46, p. 7].

In the same foreword to his *Esquisse*, Thom recalls that Aristotle considered that the underlying substrate of both matter and form is continuous [47, p. viii]:

> I knew of course that the hylomorphic schema—of which I make use in the catastrophe formalism—originated in the Stagirite's work. But I was unaware of the essential fact that Aristotle had attempted in his *Physics* to construct a world theory based not on numbers but on continuity. He had thus (at least partly) realized something I have always dreamed of doing—the development of mathematics of the continuous, which would take the notion of continuum as point of departure, without (if possible) any evocation of the intrinsic generativity of numbers.

On the same subject, Thom recalls in the *Esquisse* that Aristotle's decision to quit Plato's Academy is due to a disagreement with his master concerning the notion of continuity. He explains this in a long passage [47, p. 166]:

> The Ancients knew that the moving point generates a curve, that a moving curve generates a surfaces, and that the movement of a surface generates a volume. It seems that the aging Plato—or his epigones—considered this generation to be of the type of discrete generativity, that of the sequence of natural integers. So the point, which is a pure "zero", could not serve as a base of this construction—whence the necessity of "thickening" the point into a "unsecable length" (ἄτομος γραμμή), which was the generating principle of the straight line (ἀρχὴ γραμμῆς). The *Timaeus*' demiurge could then use this unsecable length to construct the polygons and polyhedrons which constitute the elements. It is odd to note that this kind of hypothesis still haunts our contemporary physicists; the elementary length (10^{-33} cm) below which space no longer has a physical meaning, or that absolute spatial dimension given by the confinement of quarks in nuclear physics, are so many absolute "lengths" associated with physical agents. Why did Aristotle reject this sort of hypothesis? No doubt because he held number generativity in disregard. His revolt against Plato is that of the topologist against the arithmetic imperialism, that of the apostle of the qualitative against the quantitative. Aristotle basically postulates the notion of *continuity* (συνεχές), and it is in the name of the divisibility of the continuum that he refuses the "indivisible lines." *A priori*

[53]In a footnote, Thom refers to a passage in Dieudonné's *Pour l'Honneur de l'esprit humain* [18] as a "fine example of modern incomprehension of the Aristotelian point of view." In p. 229 of this book, Dieudonné criticizes Aristotle for his view of infinity, on the basis of the passage 231a21-232a22 of the *Physics* where he discusses points on a straight line, describing his reasoning as an example of "mental confusion."

[54]*Parts of animals* [13] 668a10-13.

[55]*Parts of animals* [13] 636b21.

> this is a paradoxical position. Indeed, Aristotle never admitted the existence of space in the sense in which we have considered it since Descartes. We know why: his substantialist metaphysics required that extent is made a predicate of the substance (the *topos*); in no way could substance, matter, be a predicate of space. For Aristotle space is never generated by some intrinsic generative mechanism as our Cartesian space is generated by the $\mathbb{R}^3$ additive group of translations; at the most it is the place of some entity (*ousia*), for it is never empty. This decision to relegate space to a kind of total ostracism led him, by a singular rebound, to multiply the kinds of matter. Each time of change (μεταβολή), each genus (γένος), requires a specific matter. But all these matters have one thing in common: they are continua (συνεχές); in this sense they all have the character of spatial extension.

Needless to say, the question of the discreteness or continuity of the ultimate constitution of nature has been at the edge of philosophical thought, at least since the fifth century BC. One thinks of Empedocles, who developed the theory of the four original elements (earth, water, air and fire), assuming the existence of indivisibles, and of Leucippus and Democritus (who was the latter's disciple), who shaped the oldest known theories of stable atoms of various sizes living in an otherwise infinite vacuum. Although these thinkers (like many others) founded schools and had followers, their atomistic theories were soon overshadowed by the aura of Aristotle and Plato whose theories postulated a continuous substrate for matter.[56] One may also mention here Anaxagoras, the fifth century BCE philosopher who was known to be Socrates' teacher, who is described by Clerk Maxwell as the thinker to whom "we are indebted for the most important service to the atomic theory, which, after its statement by Democritus, remained to be done." Maxwell explains this assertion by adding that "Anaxagoras, in fact, stated a theory which so exactly contradicts the atomic theory of Democritus that the truth or falsehood of the one theory implies the falsehood or truth of the other. The question of the existence or non-existence of atoms cannot be presented to us with greater clearness than in the alternative theories of these two philosophers" [27]. In the same article, Maxwell talks about the notion of homeomerous in Anaxagoras' thinking. He writes in particular: "The essence of the doctrine of Anaxagoras is that the parts of a body are in all respects similar to the whole. It was therefore called the doctrine of Homoiomereia." We note by the way that Maxwell refers to the Greek philosophers in several of his works.

Atomism, in Europe, as an active school of thought, was revived only by the end of the Middle Ages. Among the mathematicians from the modern period, we mention Riemann who addressed again in a dramatic way the question of discreteness of continuity in his Habilitation lecture *Über die Hypothesen, welche der Geometrie zu Grunde liegen* (On the hypotheses that lie at the bases of geometry) [34] (1854) (see also the discussion in [29]), and Hermann Weyl, who continued Riemann's tradition.

[56]Diogenes Laërtius, in his *Lives and Opinions of Eminent Philosophers*, writes: "Aristoxenus, in his Historical Commentaries, says that Plato wished to burn all the writings of Democritus that he was able to collect; but that Amyclas and Cleinias, the Pythagoreans, prevented him, as it would do no good; for that copies of his books were already in many hands. And it is plain that that was the case; for Plato, who mentions nearly all the ancient philosophers, nowhere speaks of Democritus, not even in those passages where he has occasion to contradict his theories" [19, §38].

Aristotle, in a passage of the *Topics*,[57] quotes Plato's definition of a straight line. The latter disliked the notion of point, considering that it has no geometric meaning. To compensate for this, he used the notion of "unsecable", that is, "indivisible" length. At several places in his writings, Aristotle discusses the relation between a point and a line. For him, a line is a continuous object and as such it cannot be neither a collection of points nor a collection of indivisible objects of any kind. A point may only be the start, or the end, or a division point of a line, but it is not a magnitude. The question of the nonexistence of indivisibles is so important for him that it is treated in several of his writings.[58] There exists a treatise called *On indivisible lines*, belonging to the Peripatetic school (may be to Aristotle), in which the author criticizes item by item the arguments of the disciples of indivisible lines [9].

In his treatise *On the heavens*, Book III, Chapter 1, Aristotle discusses the analogous question in higher dimensions, that is, the impossibility of a surface to be a collection of lines, and of a solid to be a collection of surfaces, unless, he says, we change the axioms of mathematics, and he adds that this is not advisable. Talking of a change in the axioms in such a setting is a typical attitude of Aristotle acting as a mathematician.

Nikolai Luzin, the founder of the famous Moscow school of topology and function theory (1920s), at the beginning of his book *Leçons sur les ensembles analytiques et leurs applications* [26] writes: "The goal of set theory is to solve the following question of highest importance: can we consider the linear extent in an atomistic manner as a set of points, a question which by the way is not new, and goes back to the Eleats."[59]

Entering into the question of whether a geometric line, or, more generally, a geometric body, is constituted of its points, and if yes, in what sense this is so, leads us deep into considerations which several philosophers of Ancient Greece have thoroughly considered (we mentioned Plato and Aristotle, but these questions were extensively studied before them, especially by Zeno of Elea). One may think of the axiomatization of the real line based on the rational numbers, realized in the nineteenth century by Cantor, Dedekind and others, which, as we recalled, was described by Thom as "illusory" and "fantasmic". Let me quote here a sentence by Plotnitsky in the present volume [31] which expresses the current viewpoint: "In sum, we do not, and even cannot, know how a continuous line, straight or curved (which does not matter topologically), is spatially constituted by its points, but we have algebra to address this question, and have a proof that the answer is rigorously undecidable."

In quoting Plato while talking about Aristotle, a general comment is in order, regarding the relation between the two philosophers and on their respective attitudes

[57]*Topics*, [3] 148b27.

[58]*Physics* [1] 215b12-22, 220a1-21 et 231b6, *On generation and corruption* [11] 317a11, *On the heavens* [7] III. 1, 299a10 ff.; there are other passages.

[59]Le but de la théorie des ensembles est de résoudre la question de la plus haute importance: si l'on peut considérer ou non l'étendue linéaire d'une manière atomistique comme un ensemble de points, question d'ailleurs peu nouvelle et remontant aux éléates.

toward mathematics and natural science. The easiest way to deal with this subject is to oppose the points of view of the two philosophers concerning many topics. This is the general trend that is followed by many modern specialists of Aristotle or of Plato, and the result is a magnification of the differences between the two philosophers. But despite the fact that Plato and Aristotle had different points of view on the nature of mathematical objects, there are important principles on which they both agreed, and some apparent conflicts between them or contradictions between their thoughts deserve to be treated with the most careful consideration. Thom writes, in a note on p. 186 of his *Esquisse*: "The relations between Plato and Aristotle constitute one of the *topoi* of philosophical erudites. [...] My own position on the question is that of an autodidact."[60] There is an essay, written by Aristotle (or his school), called Περὶ ἰδεῶν (*Peri ideōn*, On ideas), of which a fragment survives, in which the Stagirite criticizes some arguments of his master on forms. The reader interested in the question of how Aristotle understood Plato's ideas on form should consult the critical edition of excerpts of this work, accompanied by thorough commentaries, published by Gail Fine [21].

In an appendix to the present article, Stelios Negrepontis collected some notes on the attitudes of Plato and Aristotle toward some of the questions that we address in this paper.

Let us return to the notion of boundary.

It is interesting to see that this notion is included in Euclid's *Elements* among the elementary notions, at the same level as "point", "line", "angle," etc. The boundary, there, is what defines a figure. Definition 14 of Book I says: "A figure is that which is contained by any boundary or boundaries [24]." This may be put in parallel with Thom's idea, following Aristotle, that a form is defined by its boundary. Euclid also uses the notion of boundary when he talks about the measure of angles (not only rectilinear angles). Proclus, in his *Commentary on the First Book of Euclid's Elements*, writes that the notion of boundary belongs to the origin of geometry since this science originated in the need to measure areas of pieces of land. Aristotle, in the *Physics*, says that a body may be defined as being "bounded by a surface."[61]

The notion of form, since the origin of geometry, is closely related to the notion of boundary, and it is not surprising to see that the mathematical notion of boundary which was essential in Thom's mathematical work, is also central in his philosophical thought. He declares in the interview *La théorie des catastrophes* conducted in 1992 [52]:

> All the unity of my work is centered at the notion of boundary, since the notion of cobordism is only one of its generalizations. The notion of boundary seems to me the more important today since I am interested in Aristotelian metaphysics. For Aristotle, the boundary is an

[60] Thom gives, as references to this subject, the books [35] by Robin and [17] by Cherniss.

[61] *Physics* [1] 204b.

individualisation principle. The marble statue is matter, in the block from where the sculptor extracted it, but it is its boundary which defines its form.[62]

Thom also talks about the unity of his work in the series of interviews *Prédire n'est pas expliquer*. He declares (1991, [50], pp. 20–21):

Truly, there exists a real unity in my reflections. I can see it only today, after I pondered a lot about it, at the philosophical level. And this unity, I find it in the notion of boundary. That of cobordism is related to it. The notion of boundary is all the more important since I was immersed into Aristotelian metaphysics. For Aristotle, a being, in general, is what is here, separated. It possesses a boundary, it is separated from the ambient space. In other words, the boundary of an object is its form. A concept has also a boundary, viz. the definition of that object. On the other hand, this idea that the boundary defines the object is not completely exact for a topologist. It is only true in the usual space. But the fact remains that, starting from this notion of boundary, I developed a few mathematical theories that were useful to me; then I looked into the applications, that is, on the possibilities of sending a space into another one, in a continuous manner. From here, I was led to study cusps and folds, objects that have a mathematical meaning.[63]

Let us return to the Stokes formula. It establishes a precise relation between a domain and its boundary. In his article *Aristote topologue* [53] (1999), Thom writes that one can interpret a passage of Aristotle's *Physics* in which he talks about the "minimal limit of the enveloping body"[64] as the homological Stokes formula: $d \circ d = 0$, the dual of the usual Stokes formula, concerning differential forms and expressing the fact that the boundary of the boundary is empty. "This formula, he writes, essentially expresses the closed character of a human being. Because if there is boundary, then there is blood loss, with a threat to life. Hence the role of the operator $d^2 = 0$ from homological algebra as an *ontology detector*, and its profound biological interpretation."[65] This identification of the Stokes formula

[62]Toute l'unité de mon travail tourne autour de la notion de bord, car la notion de cobordisme n'en est qu'une généralisation. La notion de bord me paraît d'autant plus importante aujourd'hui que je m'intéresse à la métaphysique aristotélicienne. Pour Aristote, le bord est principe d'individuation. La statue de marbre est matière dans le bloc d'où le sculpteur l'a tirée, mais c'est son bord qui définit sa forme.

[63]En vérité, il existe une réelle unité dans ma réflexion. Je ne la perçois qu'aujourd'hui, après y avoir beaucoup réfléchi, sur le plan philosophique. Et cette unité, je la trouve dans cette notion de bord. Celle de cobordisme lui est liée. [...] La notion de bord est d'autant plus importante que j'ai plongé dans la métaphysique aristotélicienne. Pour Aristote, un être, en général, c'est ce qui est là, séparé. Il possède un bord, il est séparé de l'espace ambiant. En somme, le bord de la chose, c'est sa forme. Le concept, lui aussi, a un bord : c'est la définition de ce concept. Cette idée que le bord définit la chose n'est d'ailleurs pas tout à fait exacte pour un topologue. Ce n'est vrai que dans l'espace usuel. Il reste que, partant de cette notion de bord, j'ai développé quelques théories mathématiques qui m'ont servi ; puis je me suis penché sur les applications, c'est-à-dire sur les possibilités d'envoyer un espace dans un autre, de manière continue. J'en ai été amené à étudier les fronces et les plis, objets qui ont une signification mathématique.

[64]*Physics* [1] 211b11.

[65]La formule explique essentiellement le caractère clos de l'être vivant. Car s'il y a un bord, il y a perte de sang, avec menace pour la vie. D'où le rôle de *détecteur d'ontologie* qu'est l'opérateur bord $d^2 = 0$ de l'algèbre homologique et de sa profonde interprétation biologique.

with a formula in Aristotle was already carried out in his 1991 article *Matière, forme et catastrophes* [48, p. 381]. In the series of interviews *Prédire n'est pas expliquer* (1991), he also talks about the homological Stokes formula $d^2 = 0$ and his biological interpretation: "The boundary of the boundary is empty; this is the great axiom of topology and of differential geometry in mathematics, but it is an expression of the spacial integrity of the boundary of the organism."[66] [50, p. 111] We recall incidentally that the Stokes formula is the basic tool in the proof of the theorem stating that the characteristic numbers of two cobordant manifolds, computed from the tangent bundles, coincide, and that the great theorem of Thom, the one for which he was awarded the Fields medal, is the converse of this one.

We should also talk about the relation between the local and the global in biology, and in particular, the problems of "local implies global" type.

Thom was fascinated by the relation between the local and the global. He declares in his 1992 interview on catastrophe theory [52] that the big problem of biology is the relation between the local and the global, that this is a philosophical problem which has to do with "extent" and that, at the same time, it is the object of topology. Catastrophe theory has something to say about the common features of the evolution of the form of a wave, of a cloud, of a living cell, of a fish and of any other living being, but also about the question of how does morphogenesis—the birth of form—affect the development of a form. This is the biological counterpart of the question of how the local implies the global. Thom, in the interview, recalls that topology is essentially the study of the ways that make a relation between a given local property and a global property to be found, or conversely: given a global property of a space, to find its local properties, around each point. He concludes by saying that there is a profound methodological unity between topology and biology.

In mathematics, Thom considered that the basic theorems of analysis (and they are very few, he says, "may be five or six") are concerned with the relation between the local and the global [51, p. 187]. One example of a passage "global $\Rightarrow$ local" is the implicit function theorem where, from a set defined by a non-singular equation of the form $F(x_1, \ldots, x_n) = 0$, one deduces local properties of the space of solutions. Another example is the Taylor formula, which gives a more precise equation valid in the neighborhood of a point, out of a global equation. Regarding the passage "local $\Rightarrow$ global," Thom discusses in his paper [51] several occurrences of the notion of analytic continuation, including the flat universal deformations of germs intrinsically defined by local algebras that are used in catastrophe theory.

Talking about Aristotle in a mathematical context, one is tempted to say a few words about the logic he founded, the so-called Aristotelian logic. This is different from the abstract mathematical logic—a nineteenth century invention. Thom had his own ideas on the matter, and, needless to say, if the question of which among the two logics is more suitable to science may be raised, Thom's preference goes

[66]Le bord du bord est vide ; c'est le grand axiome de la topologie, de la géométrie différentielle en mathématiques, mais cela exprime l'intégrité spatiale du bord de l'organisme.

to Aristotelian logic. We leave to him the final word of this section, from his article *Aristote et l'avènement de la science moderne*, [49, p. 489]:

> Aristotle had perfectly understood that there is no pertinent logic without an ontology which serves for it as foundations. In other words, if a logic may serve to describe in an efficient way certain aspects of the real world, if logical deduction is a reflexion of the behavior of real phenomena, then, that logic must necessarily be connected with the reality of the external world. And indeed, it is very clear that Aristotle's logic was one with his physics. [...] I will say in the most formal way that the so-called progress of logic, realized since the appearance in the 19th century of formal logic with Boole were in fact Pyrrhic progresses, in the sense that what we gained from the point of view of rigor, we lost it from the point of view of pertinence. Logic wanted to be separate from any ontology, and, for that, it became a gratuitous construction, in some way modeled on mathematics, but such an orientation is even less justified than in the case of mathematics.[67]

2.5 On Form

Thom was thoroughly involved in questions of morphogenesis. In his article *Matière, forme et catastrophes*, he recalls that it was in 1978 that for the first time he made the connection between catastrophe theory and Aristotle's hylemorphism theory. In the foreword to his *Esquisse*, he writes [47, p. VIII]: "I knew of course that the hylomorphic schema—of which I make use in the catastrophe formalism—originated in the Stagirite's work." Aristotle's theory of hylemorphism is discussed thoroughly in the *Esquisse*. According to that theory, every being (whether it is an object or a living being) is composed in an inseparable way of a matter (*hylé*, ὕλη)[68] and form (*morphê*, μορφή). Matter, from this point of view, is a potentiality, a substrate awaiting to receive form in order to become a substance—the substance of being, or being itself. In the *Metaphysics*, we can read: "I call form the essential being and the primary substance of a thing."[69] In the treatise *On the soul*, Aristotle states that the soul is the *form* of a human being.[70] In the *Physics*, he writes that

[67] Aristote avait parfaitement compris qu'il n'y a pas de logique pertinente sans une ontologie qui lui sert de fondement. Autrement dit, si une logique peut servir à décrire efficacement certains aspects du réel, si la déduction logique est un reflet du comportement des phénomènes réels, eh bien, c'est que la logique doit nécessairement avoir un rapport avec la réalité du monde extérieur. Et il est bien clair en effet que la logique d'Aristote faisait corps avec sa physique. [...] Je serai tout à fait formel en disant que les prétendus progrès de la logique, réalisés depuis l'apparition de la logique formelle avec Boole au XIXe siècle, ont été en fait des progrès à la Pyrrhus, en ce sens que ce qu'on a gagné du point de vue de la rigueur, on l'a perdu du point de vue de la pertinence. La logique a voulu se séparer de toute ontologie et, de ce fait, elle est devenue une construction gratuite, un peu sur le même modèle que les mathématiques, mais une telle orientation est encore moins motivée que dans le cas des mathématiques.

[68] The word ὕλη, before Aristotle, designated shapeless wood, and the introduction of this word in philosophy is due to Aristotle himself.

[69] *Metaphysics* [8] 1032b1-2.

[70] *On the soul*, [6] 412a11.

the φύσις (nature) of a body is its form,[71] and that flesh and bone do not exist by nature until they have acquired their form.[72] Matter and form also make the difference between the domain of interest of a physicist and that of a mathematician. The former, according to Aristotle, studies matter and form,[73] whereas the latter is only concerned with form.[74] In the *Metaphysics*, he writes that mathematical objects constitute a class of things intermediate between forms and sensibles.[75] In another passage of the *Physics*,[76] he writes that "matter desires form as much as a female desires a male." Form, according to him, is *what contains*, and it may even contain the infinite: "For the matter and the infinite are contained inside what contains them, while it is form which contains."[77] It appears clearly in these writings that Aristotle did not conceive form as a self-contained entity which lives without matter.

The paper *L'explication des formes spatiales : réductionnisme ou platonisme* (The explanation of spacial forms: reductionism or platonism) (1980) [43] by Thom concerns the notion of form and its classification. Thom tried there to give a mathematical basis to phenomenological concepts. In this setting, the substrate of a morphology is the four-dimensional Euclidean space, and a *form* is then a closed subset of space-time[78] up to a certain equivalence relation, and one of the fundamental problems in morphology is to make precise, from the mathematical point of view, this equivalence relation. In biology, Thom talks about a "congruence in the sense of D'Arcy Thompson" [55]. The last chapter of D'Arcy Thompson's book *On growth of form* is titled *the theory of transformations, or the comparison of related forms* and it contains sketches of that congruence, in various animal and human settings. The relation satisfies certain metrical constraints which, Thom says, "are generally impossible to formalize." This is, he says, the problem that biometrics has to solve: for instance, to characterize a certain bone of a given animal species. The problem is obviously open, but Thom adds that often, some subtle psychological mechanisms of form recognition will allow one to decide almost immediately whether two objects have the same form or not.

[71] *Physics* [1] 193a30.

[72] *Physics* [1] 193b.

[73] *Metaphysics* [8] 1037a16-17 and *Physics* [1] 194a15.

[74] *Posterior analytics* [4], 79a13.

[75] *Metaphysics* [8] 1059b9.

[76] *Physics* [1] 192a24.

[77] *Physics* [1] 207a35.

[78] Thom speaks of space-time in the classical sense, that is, a four-dimensional space whose first three coordinates represent space and the fourth one time. This is not the space-time that is used in the theory of relativity.

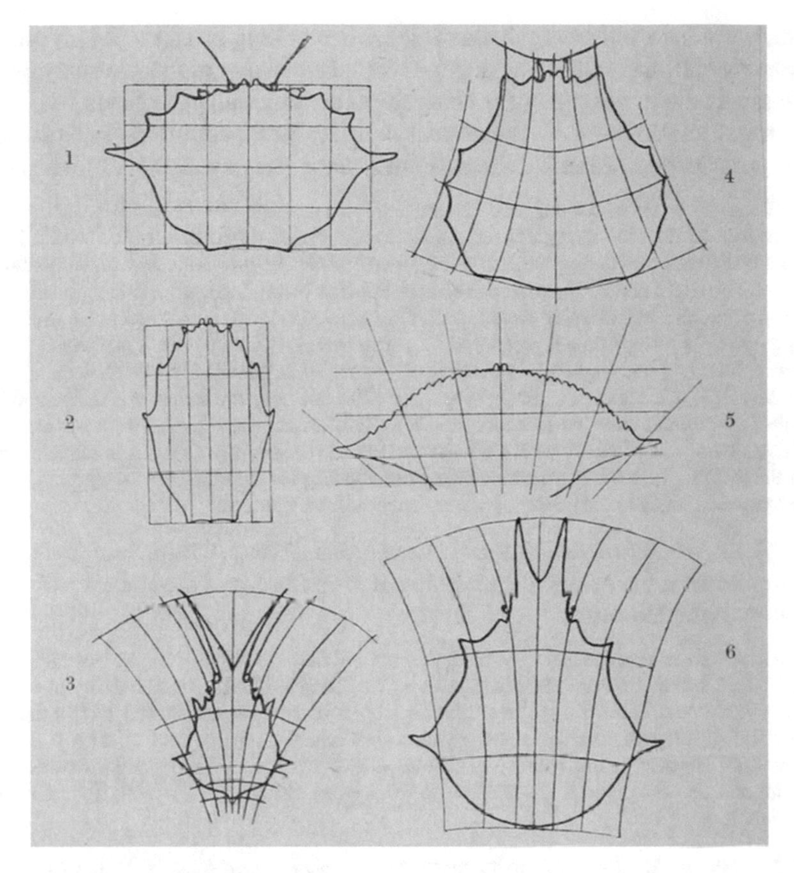

Examples of D'Arcy Thompson congruences, from his book *On growth and form*: transformations between carapaces of various crabs

Like biology, linguistics was, for Thom, part of the general theory of forms. In the introduction of his 1980 Solignac course [42], he writes that biology is a morphological discipline, concerned with form, and topology, as a branch of mathematics which involves the study of form, is at the basis of theoretical biology. From his point of view, there are two steps in a morphological discipline: the classification—giving names to the various forms, the identification of stable forms, etc.—and, after that, the theorization, namely, building a theory which is "generative" in the sense that it confers to certain forms (or aggregates of form) a certain power of determining other forms which are close to them. This program, says Thom, is partially realized in linguistics, which he considers as a morphological discipline. In 1971, he published an essay on the subject, *Topologie et linguistique*[79] [37] in which he develops a general theory of linguistics based on topology and where the accent is on morphology, again in a pure Aristotelian tradition. For

[79]This article was published 5 years later in Russia, with an introduction by Yuri Manin.

Thom, a sentence, a phrase, whether it is written or oral, is a form. More than that, morphology is what unifies language—a complicated process whose study pertains at the same time to physiology, psychology, sociology, and other fields.

In his article *Aristote et l'avènement de la science moderne* (1991), discussing the relation between Aristotle and modern science, Thom writes [49, p. 491ff]:

> I belong to those who think that the hylemorphic schema is still valid, because it is equivalent to the classifying role of concept in the verbal description of the world. [...] I am convinced that during the last years, in several disciplines, there appeared situations that can be explained by the presence of local fields or forms and that absolutely justify the old Aristotelian hylemorphic model, according to which nature is in some sense captured by form. Of course, I do not pretend the fact that here, Aristotelian form, the "*eidos*", was a being that had nothing mathematical. It was an entity that carried its own "*energeia*", its activity, and it is clear that, for Aristotle, form did not have the status of a mathematical object that would have led him to a certain form of Platonism. The fact remains that the Aristotelian "*eidos*" has a certain efficient virtue which, anyway, one has to explain, and in the theories of modern science which I am alluding to, the efficiency of the "*eidos*" is expressed in mathematical terms, for instance using structural stability.[80]

The book *Structural stability and morphogenesis* starts with a *Program* in which Thom presents the problem of succession of forms as one of the central problems of human thought. He writes:

> Whatever is the ultimate nature of reality (assuming that this expression has a meaning), it is indisputable that our universe is not chaos. We perceive beings, objects, things to which we give names. These beings or things are forms or structures endowed with a degree of stability; they take up some part of space and last for some period of time. [...] we must concede that the universe we see is a ceaseless creation, evolution, and destruction of forms and that the purpose of science is to foresee this change of form, and, if possible, to explain it.

In his article *Aristote et l'avènement de la science moderne*, [49] Thom declares that since the advent of Galilean physics, which emphasizes motion in a world in which there is no place for generation and corruption,[81] considerations on form disappeared from physics, even though morphology is present in biology. Modern

---

[80]Je suis de ceux qui croient que le schème hylémorphique garde toute sa valeur, car il est l'équivalent du rôle classificateur du concept dans la description verbale du monde. [...] Je suis convaincu que ces dernières années ont vu dans un assez grand nombre de disciplines réapparaître des situations qu'on peut expliquer par la présence de champs locaux ou de formes et qui justifient tout à fait à mon avis le vieux modèle hylémorphique d'Aristote, selon lequel la matière en quelque sorte est capturée par la forme. Bien entendu, je ne me dissimule pas qu'ici la forme aristotélicienne, l'"*eidos*", était un être qui n'avait rien de mathématique. C'était une entité qui portait en elle son "*energeia*", son activité, et il est clair que, pour Aristote, la forme n'avait pas un statut de caractère mathématique qui l'aurait obligé à une certaine forme de platonisme. Mais il n'en demeure pas moins que l'"*eidos*" aristotélicien a une certaine vertu efficace qu'il faut expliquer de toute façon, et dans les théories de la science moderne auxquelles je fais allusion, l'efficace de l'"*eidos*" s'exprime en termes mathématiques, par la théorie de la stabilité structurelle par exemple.

[81]This is also a reference to Aristotle's *On generation and corruption* [11].

science, he says, is characterized by the disappearance of this central notion of form, which played a central role in Aristotle's ontology [49, p. 491].[82]

Ovid's *Metamorphoses* starts with the line: "I want to speak about bodies changed into new forms [28]." In Book I, the author recounts how gods ended the status of primal chaos, a "raw confused mass, nothing but inert matter, badly combined, discordant atoms of things, confused in the one place." In Hesiod's *Theogony*, chaos is the name of the first of the primordial deities. Then come, in that order, Gaia (Earth), Tartarus (at the same time a place, the Deep abyss), and Eros (love).

Chapter 2 of *Structural stability and morphogenesis* is titled *Form and structural stability*. It starts with the question: *what is form?* Thom declares that the answer is beyond his task. One may mention here D'arcy Thompson, who writes in *On growth and form* [55, p. 1032]: "In a very large part of morphology, our essential task lies in the comparison of related forms rather than in the precise definition of each."

From the purely mathematical point of view, saying that a form is a geometric figure may be a good start, but then one has to agree on what is a "geometric figure." Aristotle, in *On the soul*, [6] 414b19, considers it useless to try to define a "figure", saying that it is a "sort of magnitude" (425a18). We already noted that Euclid defines a figure as "that which is contained by any boundary or boundaries" (Definition 14 of Book I of the *Elements*). Thus, we are led again to the notion of boundary. In the same trend, Plato, defines a figure as the boundary of a solid (*Meno* 76A). Furthermore, one has to introduce an equivalence relation between forms—for instance, it is natural to assume that two figures in the plane which differ by a translation "have the same form"—and, technically speaking, this should also depend on what kind of geometry we are talking about: Euclidean, projective, etc. Topological equivalence (homeomorphism) is certainly too weak, and metric equivalence (isometry) too restrictive. For example, we would naturally consider that a homothety preserves form. But Thom notes that there are instances where a square drawn in a plane such that two of its sides are horizontal (and the other two vertical) does not have the same "form" as a square placed in such a way that its sides make an angle of 45° with the horizontal. The development of the theory of forms depends on the use that one wants to make of it, and for Thom, the main use of this theory is in biology. This is indeed the main subject of his book *Structural stability and morphogenesis*. Topology is an adequate language for describing spaces of forms and Thom's tools are the notions of stability, bifurcation, attractor, singularity, envelope, etc. He considers equivariant Hausdorff metrics on spaces of form.

In the article *Structuralism and biology* [39], which was published the same year as the French version of *Structural stability and morphogenesis*, Thom writes that the foundations of a structure requires a precise lexicon of elementary chreods and

[82]Le monde de Galilée est un monde de mouvement, mais où génération et corruption n'ont point de place, d'où la disparition quasi totale en physique moderne des considérations de forme ; il n'y a pas de morphologie inanimée. Bien entendu, en biologie, il y a par contre de la morphologie. Mais alors, il n'y a plus de mathématique, au moins en tant qu'instrument de déduction. C'est cette disparition de la notion centrale de forme qui caractérise la science moderne, alors que cette notion jouait un rôle central dans l'ontologie d'Aristote.

the introduction of the notion of conditional chreod, and that catastrophe theory gives the mathematical models for such structures.

One of Thom's aims in his book *Structural stability and morphogenesis* was to introduce in biology the language of differential topology, in particular basic notions such as differentiable manifolds, vector fields, genericity, transversality, universal unfolding, etc. In the introduction, he mentions two predecessors in this domain, D'Arcy Thompson whom we already mentioned on several occasion, and C. H. Waddington, whose concepts of "chreod" and "epigenetic landscape" played a germinal part in Thom's work.[83] The image represented here is extracted from Waddington's *Strategy of the Genes* (1957) [59] where the gene is represented as a small ball rolling in a golf field. The idea expressed by this representation is that a tiny change in the initial conditions leads to drastic changes in the path the rolling ball will take. Waddington's main idea was that the development of a cell or an embryo does not depend only on its origin, but also on the landscape that surrounds it. In his *Esquisse* (1988), Thom returns to Waddington's epigenetic landscape,

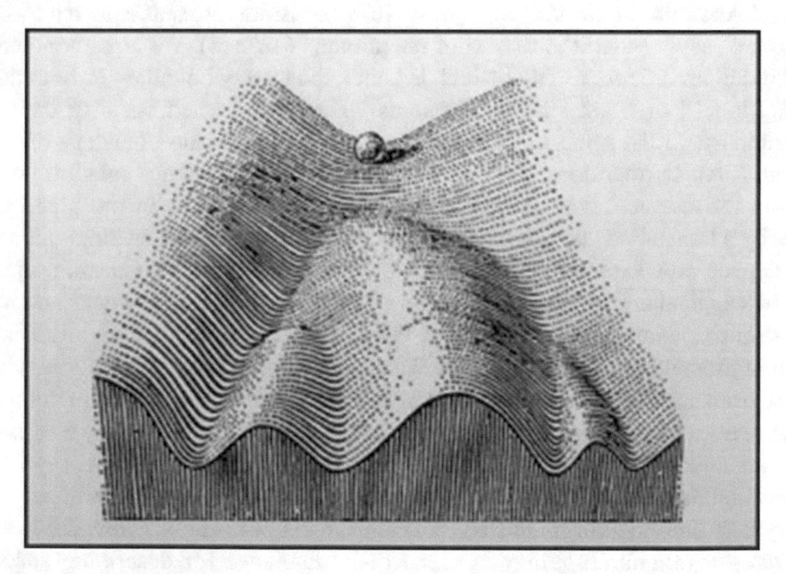

An epigenetic landscape, from C. H. Waddington, *The Strategy of the Genes*, George Allen & Unwin, 1957, p. 29

[83]Conrad Hal Waddington (1905–1975), to whom Thom refers, was a well-known biologist, working on developmental biology, that is, the study of growth and development of living organisms. The term "chreod" which he introduced in this field (from the Greek χρή, which means "it is necessary to" and ὁδός, which means "way") designates the transformations underwent by a cell during its development, until it finds its place as part of the organism. During this development, the cell is subject to an incredible amount of forces exerted on it from its environment to which it is permanently adjusting. An "epigenetic landscape" is a representation of a succession of differentiation phenomena that a cell undergoes by hills and valleys.

describing it as a "metaphor which played a primordial role in catastrophe theory." [47, p. 19]

Waddington wrote two prefaces to Thom's *Structural stability and morphogenesis*, one for the first French edition, and another one for the English edition. He writes in the first one [38, p. xix]:

> I cannot claim to understand all of it; I think that only a relatively few expert topologists will be able to follow all his mathematical details, and they may find themselves less at home in some of the biology. I can, however, grasp sufficient of the topological concepts and logic to realise that this is a very important contribution to the philosophy of science and to theoretical general biology in particular. [...] Thom has tried to show, in detail and with precision, just how the global regularities with which biology deals can be envisaged as structures within a many-dimensional space. He not only has shown how such ideas as chreods, the epigenetic landscape, and switching points, which previously were expressed only in the unsophisticated language of biology, can be formulated more adequately in terms such as vector fields, attractors, catastrophes, and the like; going much further than this, he develops many highly original ideas, both strictly mathematical ones within the field of topology, and applications of these to very many aspects of biology and of other sciences. [...] It would be quite wrong to give the impression that Thom's book is exclusively devoted to biology. The subjects mentioned in his title, *Structural stability and morphogenesis*, have a much wider reference; and he related his topological system of thought to physical and indeed to general philosophical problems. [...] In biology, Thom not only uses topological modes of thought to provide formal definitions of concepts and a logical framework by which they can be related; he also makes a bold attempt at a direct comparison between topological structures within four-dimensional space-time, such as catastrophe hypersurfaces, and the physical structures found in developing embryos. [...] As this branch of science [theoretical biology] gathers momentum, it will never in the future be able to neglect the topological approach of which Thom has been the first significant advocate.

Another mathematician who could have been invoked in the preceding pages is Leonardo da Vinci, who is the model—probably the supreme model—for a rare scientist/artist combination. Leonardo is also the prototype of a scholar who spent all his life learning. He was a theoretician of form. In his notebooks, he captured the dynamics of form in nature and in various situations. At an advanced age, he became thoroughly involved in biology, in particular, in exploring the ideas of birth and beginning of life. His approach to form in biology, like those of Goethe and D'Arcy Thompson after him, was greatly motivated by an aesthetic awareness and appreciation of nature. He introduced some of the first known theories on the fetus, and one of his notebooks is entirely dedicated to embryology. He had personal ideas on the role of the umbilical cord and he developed theories on the nutritional and respiratory aspects of the embryo, as well as on its rate of change in form during the various phases of its growth.

Leonardo was famous for taking a long time for the execution of the works that the various patrons ordered to him, and, as a matter of fact, he was blamed for spending more time on studying mathematics than on painting. Gabriel Séailles, his well-known nineteenth century biographer, in his book *Léonard de Vinci, l'artiste et le savant : 1452–1519 : essai de biographie psychologique* [36], quotes a letter from the Reverend Petrus de Nuvolaria to Isabelle d'Esté, Duchess of Milan, who was a leading figure of the Italian Renaissance, in which he says about Leonardo:

"His mathematical studies were, for him, the cause of such a disgust for painting that he barely stands holding a brush."[84] Séailles also quotes Sabba da Castiglione, a writer and humanist who was his contemporary, who writes in his memoirs: "Instead of dedicating himself to painting, he gave himself fully to the study of geometry, architecture and anatomy."

Leonardo was a dedicated reader of Aristotle. The Renaissance was, in great part, a return to the Greek authors and in this sense, Thom was, like Leonardo, the prototype of a Renaissance man. Not only he participated in the renewed interest in Aristotle's work, but he shed a new light on it, helping us better understanding his biology and his mathematics.

Acknowledgements I am grateful to Marie-Pascale Hautefeuille who read several versions of this paper and made corrections, to Stelios Negrepontis who read an early version and shared with me his thoughts on Plato that are included in the appendix that follows this paper, and to Arkady Plotnitsky who made extremely helpful comments on an early version. Part of this paper was written during a stay at the Yau Mathematical Sciences Centre of Tsinghua University (Beijing).

References

1. Aristotle, The Physics, transl. P. Hardie and R. K. Gaye, In: *The works of Aristotle translated into English*, ed. W. D. Ross and J. A. Smith, Oxford, Clarendon Press, 1930.
2. Aristotle, Prior analytics, transl. A. J. Jenkinson, In: *The works of Aristotle translated into English*, ed. W. D. Ross and J. A. Smith, Oxford, Clarendon Press, 1928.
3. Aristotle, The Topics, transl. A. W. Pickard-Cambridge, In: *The works of Aristotle translated into English*, ed. W. D. Ross and J. A. Smith, Oxford, Clarendon Press, 1928.
4. Aristotle, Posterior analytics, transl. G. R. G. Mure, In: *The works of Aristotle translated into English*, ed. W. D. Ross and J. A. Smith, Oxford, Clarendon Press, 1928.
5. Aristotle, The Problems, transl. E. S. Forster, In: *The works of Aristotle translated into English*, ed. W. D. Ross, Oxford, Clarendon Press, 1927.
6. Aristotle, On the soul, transl. J. A. Smith, In: *The works of Aristotle translated into English*, ed. W. D. Ross and J. Smith, Oxford, Clarendon Press, 1927.
7. Aristotle, On the heavens, transl. J. L. Stocks, In: *The works of Aristotle translated into English*, ed. W. D. Ross and J. A. Smith, Oxford, Clarendon Press, 1922.
8. Aristotle, Metaphysics, transl. W. D. Ross, In: *The works of Aristotle translated into English*, ed. W. D. Ross and J. A. Smith, Oxford, Clarendon Press, 1908, 2d. ed. 1928.
9. Aristotle, On Indivisible Lines, In: Opuscula, transl. E. M. Edghill, T. Loveday, E. S. Forster, L. D. Dowdall and H. H. Joachim, In: *The works of Aristotle translated into English*, ed. W. D. Ross and J. A. Smith, Clarendon Press, 1913.
10. Aristotle, The Categories, transl. E. M. Edghill, In: *The works of Aristotle translated into English*, ed. W. D. Ross and J. A. Smith, Clarendon Press, 1930.
11. Aristotle, On generation and corruption, transl. H. H. Joachim, In: *The works of Aristotle translated into English*, ed. W. D. Ross, Clarendon Press, 1930.
12. Aristotle, History of animals, transl. D'Arcy Wentworth Thompson, In: *The works of Aristotle translated into English*, ed. W. D. Ross and J. A. Smith, Clarendon Press, 1910.

[84]Ses études mathématiques l'ont à ce point dégoûté de la peinture, qu'il supporte à peine de prendre une brosse.

13. Aristotle, On the parts of animals, transl. W. Ogle, In: *The works of Aristotle translated into English*, ed. W. D. Ross and J. A. Smith, Clarendon Press, 1912.
14. M. Atiyah et al., Responses to: A. Jaffe and F. Quinn, Theoretical mathematics: Toward a cultural synthesis of mathematics and theoretical physics, (*Bull. Amer. Math. Soc.* (N.S.) 29, 1993, no 1, pp. 1–13), *Bull. Amer. Math. Soc.* (N.S.), 30, 1994, no 2, pp. 178–207.
15. A. Bellis, *Aristoxène de Tarente et Aristote, Le traité d'harmonique*, Paris, Klincksieck, coll. Études et commentaires, 1986.
16. N. Bourbaki, *Éléments d'histoire des mathématiques*, Hermann, Paris, 1974 (New printing, 1984, 2007), 3rd ed. (1st ed. 1960).
17. F. Cherniss, *Aristotle's criticism of Plato and the Academy*, Johns Hopkins University Press, Baltimore, 1949.
18. J. Dieudonné, *Pour l'Honneur de l'esprit humain : les mathématiques aujourd'hui*, Hachette, Paris, 1987.
19. Diogenes Laertius, *The lives and opinions of eminent philosophers*, transl. by C. D. Yonge, Henry G. Bohn, London, 1853.
20. C. Dobell, D'Arcy Wentworth Thompson. 1860–1948, *Obituary Notices of Fellows of the Royal Society*, 6 (1949) (18), pp. 599–617.
21. G. Fine, On ideas: *Aristotle criticism of Plato's theory of forms*, Clarendon Press, Oxford, 1993.
22. I. Geoffroy Saint-Hilaire, *Histoire naturelle générale des règnes organiques*, 3 volumes, Masson, Paris, 1854.
23. M. J. Greenberg, Aristotle's axiom in the foundations of geometry, *Journal of Geometry*, October 1988, Volume 33, Issue 1–2, pp. 53–57.
24. T. L. Heath, *The thirteen books of Euclid's* Elements, Cambridge University Press, 1908, Reprint, Dover.
25. C. Huffman, *Archytas of Tarentum: Pythagorean, Philosopher and Mathematician King*. Cambridge University Press, Cambridge, 2005.
26. N. N. Luzin, *Leçons sur les ensembles analytiques et leurs applications*, Gauthier-Villars, Paris, 1930.
27. J. C. Maxwell, Molecules, *Nature*, 8 (September 1873), pp. 437–441.
28. Ovid, *The Metamorphoses*, translation A. S. Kline, Charlottesville, University of Virginia Electronic Library, 2000.
29. A. Papadopoulos, Physics in Riemann's mathematical papers, In: *From Riemann to differential geometry and relativity* (L. Ji, A. Papadopoulos and S. Yamada, ed.) Berlin: Springer, pp. 151–207.
30. A. Papadopoulos (ed.) *René Thom, Portait mathématique et philosophique*, CNRS Éditions, Paris, 2018, 460 pages.
31. A. Plotnitsky, On the Concept of Curve: Geometry and Algebra, from Mathematical Modernity to Mathematical Modernism. In: *Geometry in history* (S. G. Dani and A. Papadopoulos, ed.), Springer Verlag, 2019, pp. 153–212.
32. Proclus de Lycie, *Les Commentaires sur le premier livre des Eléments d'Euclide*. Traduits pour la première fois du grec en français avec une introduction et des notes par Paul Ver Eecke. Desclée de Brouwer, Bruges, 1948.
33. R. Rashed, Descartes et l'infiniment grand, *Boll. Stor. Sci. Mat.* 33, No. 1, 151–169 (2013).
34. B. Riemann, Über die Hypothesen, welche der Geometrie zu Grunde liegen, published by R. Dedekind, after Riemann's death, in Abhandlungen der Königlichen Gesellschaft der Wissenschaften zu Göttingen, Vol. 13, 1867.
35. L. Robin, *La théorie des idées et des nombres*, Félix Alcan, Paris, 1908.
36. G. Séailles, *Léonard de Vinci, l'artiste et le savant : 1452–1519 : essai de biographie psychologique*, Perrin, Paris, 1892.
37. R. Thom, Topologie et linguistique, In: *Essays in Topology and Related Topics*, mémoires dédiés à Georges de Rham, Springer, 1971, pp. 226–248.

38. R. Thom, *Stabilité structurelle et morphogénèse : Essai d'une théorie générale des modèles.* Mathematical Physics Monograph Series. Reading, Mass.: W. A. Benjamin, Inc. Advanced Book Program, 1972. English translation: *Structural stability and morphogenesis. An outline of a general theory of models.* Transl. by D. H. Fowler, with a Foreword by C. H. Waddington, Reading, Mass.: W. A. Benjamin, Inc., 1975.
39. R. Thom, Structuralism and biology, In: *Towards a theoretical biology IV*, ed. C. H. Waddington, Univ. of Edinburgh Press, 1972, pp. 68–82.
40. R. Thom, Logos phenix, *Critique*, 387–388, 1979, p. 790–800.
41. R. Thom, *Modèles mathématiques de la morphogenèse*, 1e édition, Paris, 10/18 UGE, 1974, nouvelle édition, Paris, Christian Bourgois, 1980.
42. R. Thom, École d'automne de biologie théorique, Abbaye de Solignac (Haute-Vienne), 22 septembre – 10 octobre 1980, publication du CNRS.
43. R. Thom, L'explication des formes spatiales : réductionnisme ou platonisme, In: *La morphogénèse de la biologie aux mathématiques*, Actes de trois colloques organisés par l'École Pratique des Hautes Études, Maloine S. A. ed., coll. Recherches Interdisciplinaires, Paris, 1980, pp. 71–79.
44. R. Thom, La science malgré tout, Organum vol. XVII, Encyclopedia Universalis, L'homme et son savoir (1980), pp. 5–11.
45. R. Thom, Les intuitions topologiques primordiales de l'aristotélisme, *Revue Thomiste*, juillet-septembre 1988, No. 3, tome 88, pp. 393–409.
46. R. Thom, Structure et fonction en biologie aristotélicienne, In: *Biologie théorique*, Solignac, 1988, Paris, éditions du CNRS, 1993, publié aussi dans *Apologie du logos*, Paris, Hachette, 1990, pp. 247–266.
47. R. Thom, *Esquisse d'une sémiophysique : Physique aristotélicienne et théorie des catastrophes*, Paris, InterEditions, 1988. English translation by V. Meyer, *Semio Physics: A Sketch. Aristotelian Physics and Catastrophe Theory*, Addison-Wesley, 1990.
48. R. Thom, Matière, forme et catastrophes, In: *Penser avec Aristote*, Dir. M. A. Sinaceur, Paris, Érès, Toulouse, 1991, pp. 367–398.
49. R. Thom, Aristote et l'avènement de la science moderne : la rupture galiléenne. In: *Penser avec Aristote* (M. A. Sinaceur, ed.), Érès, Toulouse, 1991, pp. 489–494.
50. R. Thom, *Prédire n'est pas expliquer : Entretiens avec Emile Noël*, Champs, Editions Eshel coll. La Question, Paris 1991.
51. R. Thom, Un panorama des mathématiques, In: *1830–1930: A century of geometry*, (L. Boi, D. Flament, J.-M. Salanskis, ed.), Springer Verlag, Berlin, Heidelberg, New York, 1992, pp.184–191.
52. R. Thom, *La théorie des catastrophes*, INA-Éditions ESHEL, 1992. Réalisation Francis Bouchet.
53. R. Thom, Aristote topologue, *Revue de synthèse*, 4e s. no. 1, janv.–mars 1999, pp. 39–47.
54. D'Arcy Thompson, On Aristotle as biologist: With a Prooemion on Herbert Spencer, Being the Herbert Spencer Lecture Delivered Before the University of Oxford, on February 14, 1913, Oxford, Clarendon Press, 1913.
55. D'Arcy Thompson, *On growth and Form*, Cambridge University Press, 1st. ed., 1917.
56. W. P. Thurston, *Three-dimensional geometry and topology* Vol. 1, edited by S. Levy, Princeton University Press, Princeton, N.J., 1997.
57. H. Weyl, *The Continuum: A critical examination of the foundation of analysis*, translation of *Das Kontinuum, kritische Untersuchungen über die Grundlagen der Analysis*, 1st edition German, Leipzig, Weit & Comp., 1918, tr. S. Pollard and T. Bole, Originally published by Thomas Jefferson University Press, Kirksville, MO, 1987, reprint Dover, Mineola, NY, 1994.
58. H. Weyl, *Raum Zeit Materie*, 1921, several revised editions, English translation by H. L. Bose, Space—Time—Matter, First edition 1922, New Edition, 1950, Dover.
59. C. H. Waddington, *The Strategy of the Genes. A Discussion of Some Aspects of Theoretical Biology*, George Allen & Unwin, London, 1957.

Appendix to the article Topology and biology: From Aristotle to Thom by A. Papadopoulos

Stelios Negrepontis

I would like to make a few comments on three subjects that Thom addresses in his writings: (1) indivisible (unsecable) lines; (2) homeomerous and non-homeomeomerous; (3) topos and chora.

1. Zeno and Plato: Geometric Point and Indivisible Line Parmenides hypothesized the existence of true entities, unchangeable monads, higher than the sensible changing multiple entities. The nature of these monads was not clear from the description Parmenides gave in his poem. Parmenides had two students, Zeno and Melissus, who adopted completely diverging interpretations of these monads. For Melissus, a monad is partless, like a geometric point. But Zeno, influenced by the Pythagoreans, accepted as monad the self-similar one that results from the infinite anthyphairesis/continued fraction of diameter to the side of a square (square root of two). This is a One in a self-similar sense. This monad was adopted by Plato as well, improved by the periodic nature of more general quadratic incommensurabilities, proved in the meantime by Theodorus and Theaetetus. This is in essence the Platonic Idea.

Later, Xenocrates called the Platonic Idea an indivisible line. But an indivisible line is not literally indivisible as Thom seems to think. It is a line like every other geometric line, infinitely divisible. It is indivisible only in the following sense: $a = k_1 b + c_1, b = k_1 c_1 + c_2, c_1 = k_3 c_2 + c_3, \ldots, c_n = k_{n+2} c_{n+1} + c_{n+2}, \ldots$ If this anthyphairesis is periodic then there will be an m so that $a/b = c_m/c_{m+1}$. After stage m, the division continues but does not produce new logoi and new species, because of the repetition of the logoi by periodicity.

Thus, Zeno and Plato were totally committed to infinite divisibility and in this sense to the continuum.

The divisions of Plato in the *Sophistes* and the *Politicus* that are attacked by Aristotle are imitations of the above periodic anthyphairesis, and not finite Linneus type biological divisions.

2. Self-similarity (Homeomerous and Non-homeomeomerous) The difference between homeomerous and non-homeomeomerous parts is of central interest in Plato's philosophy, since he treats a Platonic Idea, which is an imitation of a dyad of lines in periodic anthyphairesis, as a self similar One, monad. One such recurring monad for Plato is Virtue (arête). This, being a Platonic Idea consists of two initial opposite parts, bravery (andreia) and prudence (sophrosune).

Now the high point in the dialogue *Protagoras* is when Socrates asks Protagoras (329d): How do you think these two are parts of virtue, in the sense that nose, eyes, mouth, ears are parts of face, or in the sense of gold? Protagoras gives the wrong answer, not realizing the need for self-similarity.

That the Oneness of a true intelligible Being in Plato's philosophy is the self-similar nature of a dyad in periodic anthyphairesis is argued in my papers [2, 3].

3. Topos and Chora: The Boundary Determines the Content What Aristotle says about topos and chora is related to Plato's idea of a receptacle/hollow space in the *Timaeus*, occurring after Timaeus 48. There, in Timaeus 48-63, the four elements: earth, fire, air, water are correlated (not identified) with the cube (hexahedron), pyramid (tetrahedron), octahedron, icosahedron. But all interconnection, interaction, motion and change in nature occurs by the interaction (anthyphairesis) between their respective boundaries. Thus, Plato, here, is again the first one to insist on the fact that form-boundary determines content. (He does so of course for his own reasons, because only by considering the bounding surfaces he can get essentially a dyad, instead of four elements, and in fact a dyadic periodic anthyphairesis.)

An outline of the importance of the form of the boundary of the canonical polyhedra in the *Timaeus* can be found in Section 12 of my paper [1] in this volume.

Needless to say, Thom's mathematical discoveries in Aristotle are most fascinating. One can nevertheless regret that Thom, who was a true philosopher-mathematician and a thorough reader of Aristotle, probably relied, for what concerns Plato, on scholars that did not have a real understanding of the latter's writings. Incidentally, D'Arcy Thompson had written, back in 1928, a paper *Excess and Defect; or the little more and the little less*, in the journal *Mind*, that was perhaps to find, together with the Plato scholar A. E. Taylor, traces of anthyphairesis in the *Epinomis*—something that they never followed up.

References

1. S. Negrepontis, Plato on Geometry and the geometers, In: *Geometry in history* (S.G. Dani and A. Papadopoulos, ed.), Cham: Springer, 2019. https://doi.org/10.1007/978-3-030-13609-3_1.
2. S. Negrepontis, *The Anthyphairetic Revolutions of the Platonic Ideas*, arXiv:1405.4186, 2014; *Revolutions and Continuity in Greek. Mathematics*, ed. M. Sialaros, Walter de Gruyter, 2018, pp. 335-381.
3. S. Negrepontis, The periodic anthyphairetic nature of the One of the Second Hypothesis in Plato's Parmenides, *Proceedings of the Conference Mathématiques et musique : des Grecs à Euler*, 10-11 September, 2015, Strasbourg, ed. X. Hascher and A. Papadopoulos, Hermann, Paris (to appear).

Chapter 3
Time and Periodicity from Ptolemy to Schrödinger: Paradigm Shifts vs Continuity in History of Mathematics

Yuri I. Manin

To Bob Penner, cordially

Abstract I briefly consider the Kuhnian notion of "paradigm shifts" applied to the history of mathematics and argue that the succession and intergenerational continuity of mathematical thought was undeservedly neglected in the historical studies. To this end, I focus on the history of mathematical theory of time and periodicity, from Ptolemy's epicycles to Schrödinger's quantum amplitudes interference and contemporary cosmological models.

AMS 2010 Mathematics Subject Classification: 01A99

3.1 Introduction

In his influential treatise [9], Thomas Kuhn developed an approach to the history of natural science(s) based upon the assumption that this history can be naturally subdivided into periods. According to Kuhn, the transitions from one period to the next one (called "revolutions") are characterised by a radical change of the basic assumptions, experimental and observational practices, and acceptable types of argumentation. Any such set of assumptions is shared by the learned community during each development phase of "normal science", and its change is called a "paradigm shift".

Kuhn himself was reluctant about extending this view to the history, philosophy, religion, and much of the social science(s). He believed that they are formed rather by a "tradition of claims, counterclaims, and debates over fundamentals."

Y. I. Manin (✉)
Max Planck Institute for Mathematics, Bonn, Germany
e-mail: manin@mpim-bonn.mpg.de

S. G. Dani, A. Papadopoulos (eds.), *Geometry in History*,
https://doi.org/10.1007/978-3-030-13609-3_3

The motivation of this brief essay was a desire to discuss the applicability of the Kuhnian view on history of mathematics. I argue that the succession and intergenerational continuity of mathematical thought was undeservedly neglected in the science studies. To this end, I focus on the history of the mathematical theory of time and periodicity, from Ptolemy's epicycles to Schrödinger's quantum amplitudes interference and Feynman integrals.

According to the concise description in [2], my essay lands somewhere in the uncharted territory between History of Science and Science Studies. Kuhn's book originated Science Studies "as a self-conscious field of inquiry" [2, p. 801]. Hence this article belongs to it. But it focuses on the intrinsic continuity and the peculiarities of forms of historical legacy in understanding space, time, and periodicity that, for many historians, might be completely outside their fields of vision.

If one rejects, as I do here, the assumption about (this particular flow of) history as a sequence of revolutions, then the idea of paradigm shifts cannot claim anymore its leading role.

I accept here the more general viewpoint that Mathematics has a position mediating, or bridging, daily life, common sense, philosophy, and physics. Those fragments of mathematical knowledge that can become subjected to "reality tests" part are more sensitive to respective "revolutions" or "paradigm shifts", whereas those parts that are closer to "pure mathematics" show rather a kind of continuous development as is argued in this paper.

Also, some light on my position can be thrown by comparison of the history of developing knowledge on the scale of civilisations (cf. [5], [14], [15], [17]) with the history of development of cognition in the individual brain of a growing human being (cf. [12]).

oritur sol et occidit et ad locum suum revertitur
lustrans universa in circuitu pergit spiritus et in circulos suos revertitur
VULGATA CLEMENTINA, Ecclesiastes 1:5–1:6

Acknowledgements A project of this paper was conceived during a brief stimulating conversation of Yu. M. with Lorraine Daston about the role of Kuhn's "paradigm shifts" doctrine in the history of applied (or rather, "applicable") mathematics in a broad sense of this word.

Andreea S. Calude provided informative data about the prehistory of the cognitive behaviour of early *Homo Sapiens.*

When this paper was already written, Matilde Marcolli drew my attention to the article [4] which contains a very careful and sensible survey of a considerable part of the same historical background (but excluding quantum mechanics, cosmology and my remarks on "computational mentality").

I am very grateful to them for inspiring communication.

3.2 Brief Summary and Plan of Exposition

I will start with a few words about notions and formulas summarising some basic *mathematical* tools used in the *contemporary* discussions of time and periodicity.

Fundamental is the fact that these tools are subdivided into two complementary parts: geometric ones involving *space/time intuition* (as in Euclid's *Elements*) and algebraic/calculus ones involving formulae and computations and generally having linguistic character. Arguably (cf. [10] and references therein) this is one reflection of the general dynamic patterns of interactions between right and left brain.

Start with an Euclidean plane **P** endowed with Euclidean metric. Then a choice of a point 0, of a line L passing through it, and of its orientation, determines an identification of the set of all points of this line with the set of real numbers **R**: this is its "coordinatisation". Call it the x-line and denote now by L_x.

Choose now another line passing through the same point, oriented, and orthogonal to the x-line; call it y-line L_y. Now we can construct a "coordinatisation" of the whole plane **P** i.e., the identification of the set of points of **P** with the set of ordered pairs of real numbers $\mathbf{R}^2$.

At this point, we can start describing various figures, actors of plane Euclidean geometry, by equations and inequalities between various algebraic expressions involving x and y. So, for example, a circle of radius r whose center is a point (x_0, y_0), is the set of all points (x, y) whose coordinates satisfy the equation $(x - x_0)^2 + (y - y_0)^2 = r^2$.

We may call it *Cartesian picture of geometry.*

Similarly, using three coordinates one gets an algebraic picture of Euclidean geometry of space; passing to four coordinates, with time axis added to three space axes, we get the scene for Newtonian mechanics. But some fragments of this scene were already recognisable in the world pictures going back to the times of Archimedes and Ptolemy, as the celebrated Antikythera mechanism modeling the movement of heavens and relating them to the chronological dating of historical events [7].

Arguably, one important contribution of history of "periodicity" to mathematics was the crystallisation of the notions of "definition", initially emerging as secondary to the notions of "axiom" and "theorem" as in Euclid's "Elements".

The less obvious one was a "reification" of the idea of symmetry: statements and proofs of most theorems of Euclidean geometry are not dependent on the choice of origin of coordinates and therefore invariant with respect to parallel shifts of the whole space, and also with respect to rotations, conserving angles. Thanks to this, one can introduce "Cartesian" coordinates also on the space of all Euclidean symmetries of a Euclidean plane/space.

Finally, I must mention that, using the terminology of one of the schools of Science Studies, when I briefly quote and/or interpret mathematical intuition and historical data, I appeal mostly to *"ethnomathematics in the European context"* leaving aside many interesting achievements and inputs that came from Eastern,

Chinese and other regions of the global world. For a much more complete and balanced treatment, see e.g. [18].

3.3 Mathematics and Physics of Periodicity

3.3.1 *Antiquity: Euclidean Geometry, Ptolemy's Epicycles, Antikythera Mechanism*

This book I bought in Venice for one ducat in the year 1507
Albrecht Dürer inscription in Euclid's book
from his library

Before starting the central themes of our discussion, I must say explicitly that accumulation and intergenerational transmission of knowledge, became possible only at a certain stage of development of human language(s), and somewhat later, of written languages.

Moreover, as I argued in [11], pp. 159–167 and 169–189, the most important new functions of emerging language consisted *not* in the transmission of concrete information about "here and now" ("in this grove a deer is grazing"), but rather in creation of *"spaces of possibilities"*. Gods, heavens and netherworlds powerfully influenced human's collective behaviour, even if they could never be located here and now.

Since the concept of *here and now* itself later entered physical theories as coordinate origin, it would be interesting to trace its history as far back in time as possible. I am grateful to Andreea Calude who informed me that deep reconstruction (to about $15 \cdot 10^3$ years back from now) seemingly recovers old common Indoeuropean roots for "now" but not for "here", cf. [13]. Perhaps, a psychologically motivated substitution for "here" was furnished by very old ("ultraconserved") words for "I" and "you".

Passing now to the real origins of modern scientific knowledge about the Solar System and the Universe in the Greco-Roman and Hellenistic worlds, we see that its foundations were laid between 300 BCE and 200 CE and connected in particular with the names of Euclid, Archimedes, and Ptolemy. The history of "here and now", however, must alert us to the tracing also of the background history of the development of various new "languages of science", of translations and mutual interactions between these languages, and their intergenerational functioning.

Euclid of Alexandria conjecturally lived and worked during about 325 BCE–270 BCE in the south Mediterranean Greek colonial city. He created the richest and at that time logically perfect axiomatic description of two- and three-dimensional spaces with metric and their symmetry groups that were made explicit only many centuries later when the language of coordinates was created and one could speak about geometry using languages of algebra/calculus etc. Still, perception of Euclid's "Elements" as *the* foundational, almost *sacral* treatise survived till the nineteenth century: in particular new editions and translations of his *Elements* after spreading of printing were second only to the Bible.

See a very remarkable book [1] by Oliver Byrne, *"surveyor of her Majesty's settlements in the Falkland Islands and author of numerous mathematical works"*, where he keeps texts of all his geometric chapters but rewrites all of Euclid's definitions (axioms), statements and proofs in pseudo-algebraic formulas in which letters $a, d, c, \ldots, x, y, z$ that traditionally (for us) serve as notations for variables, constants, functions etc. are replaced by coloured pictures of angles, triangles et al.

Claudius Ptolemy conjecturally was born about 85 CE in Egypt and died about 165 CE in Alexandria, Egypt. His greatest achievement described in the "Almagest" is a dynamical model of the Solar System. This model is geocentric. This is justified by the fact that all our observation of planets and Sun are made from the Earth. It represents the visible movements of the planets and the sun as complex combination of uniform circular motions along *epicycles*, whose centres also move uniformly along their "secondary" epicycles, and finally various centres themselves are cleverly displaced from their expected ideal positions.

We do not know much about the computational devices that were used in antiquity in order to make Ptolemy's model and other models of observable periodicities such as lunar phases quantitatively comparable with observations. However, one remarkable archaeological discovery was made in 1900 when a group of sponger fishers from Greece during of bad weather anchored their boats near the island Antikythera and while they were diving discovered at a depth of 42 m an ancient shipwreck. Besides bronze and marble statues, it contained a very corroded lump of bronze. All these remnants were transferred to the National Archaeological Museum in Athens, and after several decades of sophisticated studies and reconstructions, a general consensus arose summarised in [16] as follows:

> The Antikythera mechanism is an ancient astronomical calculator that contains a lunisolar calendar, predicts eclipses, and indicates the moon's position and phase. Its use of multiple dials and interlocking gears eerily foreshadows modern computing concepts from the fields of digital design, programming, and software engineering.

For a description of continuing disagreements about details of the reconstruction, see [3].

Digression 1: The Number π In the history of geometric models of periodicity, the number π plays a crucial role. Since Babylonian and Egyptian times, π was considered ("defined") as the ratio of the length of a circle to its diameter that can be measured in the same way as other physical constants are measured. So in order to get an (approximate) value of π, one can first say, draw a circle using compasses, and then measure its length using a string. Independence of the result on the diameter is also an experimental fact which very naturally appears during land surveying. Finally, the approximate values are always rational numbers, or rather, *names* of some rational numbers, that can be transferred by means acceptable in the relevant culture: see a very expressive account by Ph. E. B. Jourdain [8] written at the beginning of the twentieth century.

Arguably, the first modern approach to π was found by Archimedes (about 287–212 BCE). This approach consisted in approximating π from below by the values of perimeters of inscribed regular n-gons (diameter is for simplicity taken as unit of length). Manageable and fast converging formulas for consecutive approximations are obtained by passing from an N-gon to $2N$-gon etc.

3.3.2 Fourier Sums and Fourier Integrals: Epicycle Calculus

As we reminded in Sect. 3.2, after choosing orthogonal coordinates and scale identifying an Euclidean plane with $\mathbf{R}^2$, one can describe the circle of radius r_0 with centre (x_0, y_0) as the set of points (x, y) such that $(x - x_0)^2 + (y - y_0)^2 = r_0^2$. The variables change $x = r_0(x_0 + \sin 2\pi t)$, $y = r_0(y_0 + \cos 2\pi t)$ describes then the movement of a point along this circle, with angular velocity one, if t is interpreted as time flow. Replacing t by $v_0 t$ we can choose another velocity.

In turn, we can put in the formulae above $x_0 = r_1(x_1 + \sin 2\pi v_1 t)$, $y_0 = r_1(y_1 + \cos 2\pi v_1 t)$, in order to make the centre (x_0, y_0) move along another circle with uniform angular velocity, etc. We get thus an analytic description of Ptolemy's picture, or rather its projection on a coordinate plane in our space, which can be complemented by projections on other planes.

In order to use it for computational purposes, we must input the observable values of (x_i, y_i) and v_i, $i = 0, 1, \ldots$, for, say, planetary movements. The Antikythera mechanism served as a replacement of these formulae for which the language was not yet invented and developed. This language in its modern form and the analytic machinery were introduced only in the eighteenth to nineteenth centuries: i.e. Fourier sums/series $\sum_i (a_i \sin it + b_i \cos it)$ and more sophisticated Fourier integrals were initiated by Jean-Baptiste Joseph Fourier (1768–1830).

Joseph Fourier had a long and complicated social and political career, starting with education in the Convent of St. Mark, and including service in the local Revolutionary Committee during the French Revolution, imprisonment during the Terror time, travels with Napoleon to Egypt, and office of the Prefect of the Department of Isère (where Joseph Fourier was born).

Returning briefly to Fourier's mathematics, I would like to stress also an analogy with Archimedes legacy, namely, observational astronomy and mathematics of his "Psammit" ("The Sand Reckoner"). Archimedes wanted to estimate the size of the observable universe giving an estimate of the number of grains of sand needed to fill it. Among other difficulties he had to overcome, was the absence of language (system of notation) for very large (in principle, as large as one wishes) integers. He solved it by introducing inductively powers of 10, so that any next power might be equal to the biggest number, defined at the previous step.

Digression 2: The Number e and "Computational Mentality" The famous Euler number $e = 2.7182818284590\ldots$ and his series

$$e^x = 1 + \sum_{n=1}^{\infty} \frac{x^n}{n!}$$

were only the last steps of a convoluted history, with decisive contributions due to John Napier (1550–1617, Scotland), Henry Briggs (1561–1630, England), Abraham de Moivre (1667–1754, France), among others, and finally Leonhard Euler (1707, Basel, Switzerland–1783, St Petersburg, Russia).

As already with Archimedes, and later with the Masters of the Antikythera mechanism, one of the great motivations of the studies in this domain was the necessity to devise practical tools for computations with big numbers and/or numbers whose decimal notation included many digits before/after the decimal point/comma: this is what I call here "computational mentality". This ancient urge morphed now into such ideas as "Artificial intellect" and general identification of the activity of neural nets with computations.

So, for example, Briggs logarithm tables allowing to efficiently replace (approximate) multiplications by additions consisted essentially in the tables of numbers $10^7 \cdot (1 - 10^{-7})^N$, $N = 1, 2, 3, \ldots 10^7$. The future Euler's number e was hidden here as a result of passing to the limit $e^{-1} = \lim(1 - N^{-1})^N$, $N \to \infty$, which Briggs never made explicit. However, the way Napier approached logarithms included approximate calculations of logarithms of the function sin, which may be considered as the premonition of the Euler formula $e^{ix} = \cos x + i \sin x$ that later played the key role in mathematical foundations of quantum mechanics.

3.3.3 *Quantum Amplitudes and Their Interference*

As we have seen, the basic scientific meta-notions of "observations" and "mathematical models" explaining and predicting results of observations, go back to deep antiquity. The total body of scientific knowledge accumulated since then, was enriched during the nineteenth and twentieth centuries also by recognition that "scientific laws", that is, the central parts of mathematical models explaining more or less directly the results of observations, are *qualitatively different* at various space/time scales: see a comprehensive survey [6], in particular, the expressive table on pp. 100–101.

A breakthrough in understanding physics at the very large scale Universe (cosmology) was related to Einstein's general relativity (or gravity) theory, whereas on the very small scale, the respective breakthrough came with quantum mechanics and later quantum field theory. Bridging these two ways of understanding Nature still remains one of the main challenges for modern science.

One can argue that an "observable" bridge between these two scales is the existence and cognitive activities of *Homo Sapiens* on our Earth (and possibly elsewhere), but the discussion of the current stage of "observations" and "explanations" in biology would have taken us too far away from the subject of this short essay; cf. [12]. Anyway, the key idea of scientific observation includes some understanding of how a subject of human scale can interact with objects of cosmic/micro scales.

Studying the small scale physics unavoidably involved the necessity of working out mathematical models of probabilistic behaviour of elementary particles that was observed and justified in multiple experiments. It was preceded by a remarkable cognitive passage: from the observable properties of chemical reactions to Mendeleev's intellectual construct of the Periodic Table to the images of atoms of the Chemical Elements as analogs of the Solar System with nucleus for Sun and electrons orbiting like planets. This cognitive passage might be compared with the evolution of astronomy from antiquity to Copernicus, Galileo and Newton.

When experimental methods were developed for working quantitatively with unstable (radioactive) atoms, small groups of electrons, etc., a new theoretical challenge emerged: observable data involved random, probabilistic behaviour, but the already well developed mathematical tools for describing randomness did not work correctly in the microworld!

The emergent quantum mechanics postulated that "probabilities" of classical statistics, expressed by real numbers between 0 and 1, must be replaced in the microworld by *probability amplitudes*, whose values are *complex numbers* that after some normalisation become complex numbers lying in the complex plane on the circle of radius 1 and centre 0. It must be then explained how to pass from the hidden quantum mechanical picture to the observable classical statistical picture. Many different paths along this thorny way were discovered in the 20s of the twentieth century, in particular, in classical works of Werner Heisenberg, Wolfgang Pauli, Erwin Schrödinger, et al. One way of looking at "quantization" of the simplest classical system, point-like body in space, is this. Any classical trajectory of this system is a curve in the space of pairs *(position, momentum)*. Here position and momentum can be considered as Cartesian coordinate triples whose values, of course, pairwise commute. To the contrary, in the quantum mechanical space, the commutator between position and momentum is not zero. This can be envisioned as a replacement of possible classical trajectories of such a system by their wave functions which are not localised. Probabilistic data occur when one adds, say, one more point-like body and/or interaction with a macroscopic environment created by an experimenter.

Mathematical descriptions of all this are multiple and all represent a drastic break with "high school", or "layman", intuition. One remarkable example of pedagogical difficulties of quantum physics can be glimpsed in the famous Lectures on Quantum Mechanics by the great Richard Feynman.

In our context, the most essential is the fact that quantum interaction in the simplest cases of quantum mechanics is described via Fourier sums, series, and integrals in (finite dimensional) *complex* spaces endowed with *Hermitean* metrics, in place of real Euclidean space with real metric.

The quantum mechanical amplitudes are given by Fourier sums or series of the form $\sum_n a_n e^{it}$ where a_n are complex numbers, and t is time, whereas probabilities in classical statistic descriptions are given by the similar sums with real a_n, and it replaced by inverse (also real) temperature $-1/T$.

In this sense, quantum mechanics is a complexification of Ptolemy's epicycles.

In the currently acceptable picture, our evolving Universe can be dissected into "space sections" corresponding to the values of global *cosmological* time (e.g. in the so called Bianchi cosmological models) to each of which a specific temperature of background cosmic radiation can be ascribed. Going back in time, our Universe becomes hotter, so that at the moment of the Big Bang (time = 0) its temperature becomes infinite. This provides a highly romantic interpretation of the correspondence $-1/T \leftrightarrow it$.

References

1. O. Byrne. *The first six books of the Elements of Euclid in which coloured diagrams and symbols are used instead of letters for the greater ease of learners, By Oliver Byrne surveyor of her Majesty's settlements in the Falkland Islands and author of numerous mathematical works. London William Pickering 1847*. Facsimile publication by Taschen.
2. L. Daston. Science Studies and the History of Science. *Critical Inquiry*, Vol. 35, No. 4, pp. 798–813. U. of Chicago Press Journals, Summer of 2009. https://doi.org/10.1086/599584
3. T. Freeth. The Antikythera mechanism. *Mediterranean Archaeology and Archaeometry*, vol. 2, no. 1, pp. 21–35.
4. G. Gallavotti. Quasi periodic motions from Hipparchus to Kolmogorov. *Rendiconti Acc. dei Lincei, Matematica e Applicazioni.*, vol. 12, 2001, pp. 125–152.
5. St. Greenblatt. *The rise and fall of Adam and Eve.* W. W. Norton & Company, 2017.
6. G. 't Hooft, St. Vandoren. *Time in powers of ten. Natural Phenomena and Their Timescales.* World Scientific, 2014.
7. A. Jones. *A portable cosmos: revealing the Antikythera mechanism, scientific wonder of the ancient world.* Oxford UP, 2017.
8. Ph. E. B. Jourdain. *The Nature of Mathematics.* Reproduced in the anthology "The World of Mathematics" by James R. Newman, vol. 1, Simon and Schuster, NY 1956.
9. Th. Kuhn. *The Structure of Scientific Revolutions (2nd, enlarged ed.).* University of Chicago Press, 1970. ISBN 0-226-45804-0.
10. Yu. I. Manin. De Novo Artistic Activity, Origins of Logograms, and Mathematical Intuition. In: *Art in the Life of Mathematicians*, Ed. Anna Kepes Szemerédi, AMS, 2015, pp. 187–208.
11. Yu. Manin. *Mathematics as Metaphor. Selected Essays, with Foreword by Freeman Dyson.* American Math. Society, 2007.
12. D. Yu. Manin, Yu. I. Manin. Cognitive networks: brains, internet, and civilizations. In: *Humanizing Mathematics and its Philosophy*, ed. by Bh. Sriraman, Springer International Publishing AG, 2017, pp. 85–96. https://doi.org/10.1007/978-3-319-61231-7_9. arXiv:1709.03114
13. M. Pagel, Q. D. Atkinson, A. S. Calude, A. Meade. *Ultraconserved words point to deep language ancestry across Eurasia.* www.pnas.org/cgi/doi/10.1073/pnas.1218726110
14. D. Park. *The How and the Why. An essay on the origins and development of physical theory.* Princeton UP, New Jersey, 1990.
15. K. Simonyi. *A Cultural History of Physics. Translated by D. Kramer.* CRC Press, Boca Raton, London, New York, 2012.

16. D. Spinellis. The Antikythera Mechanism: a Computer Science Perspective. *Computer*, May 2008, pp. 22–27.
17. G. Woolf. Orrery and Claw. *London Review of Books*, 18 Nov. 2010, pp. 34–35.
18. D. Wootton. *The Invention of Science: A New History of the Scientific Revolution.* Penguin, 1976, 784 pp.

Chapter 4
Convexity in Greek Antiquity

Athanase Papadopoulos

Abstract We consider several appearances of the notion of convexity in Greek antiquity, more specifically in mathematics and optics, in the writings of Aristotle, and in art.

AMS Classification: 01-02, 01A20, 34-02, 34-03, 54-03, 92B99

4.1 Introduction

The mathematical idea of convexity was known in ancient Greece. It is present in the works on geometry and on optics of Euclid (c. 325–270 BCE), Archimedes (c. 287–212 BCE), Apollonius of Perga (c. 262–190 BCE), Heron of Alexandria (c. 10–70 CE), Ptolemy (c. 100–160 CE), and other mathematicians. This concept evolved slowly until the modern period where progress was made by Kepler, Descartes and Euler, and convexity became gradually a property at the basis of several geometric results. For instance, Euler, in his memoir *De linea brevissima in superficie quacunque duo quaelibet puncta jungente* (Concerning the shortest line on any surface by which any two points can be joined together) (1732), gave the differential equation satisfied by a geodesic joining two points on a differentiable convex surface.

Around the beginning of the twentieth century, convexity acquired the status of a mathematical field, with works of Minkowski, Carathéodory, Steinitz, Fenchel, Jessen, Alexandrov, Busemann, Pogorelov and others.

My purpose in this note is to indicate some instances where the notion of convexity appears in the writings of the Greek mathematicians and philosophers of antiquity. The account is not chronological, because I wanted to start with convexity

A. Papadopoulos (✉)
Université de Strasbourg and CNRS, Strasbourg Cedex, France
e-mail: athanase.papadopoulos@math.unistra.fr

S. G. Dani, A. Papadopoulos (eds.), *Geometry in History*,
https://doi.org/10.1007/978-3-030-13609-3_4

in the purely mathematical works, before talking about this notion in philosophy, architecture, etc.

The present account of the history of convexity is very different from the existing surveys on the subject. For sources concerning the modern period, I refer the reader to the survey by Fenchel [12].

4.2 Geometry

Euclid uses convexity in the *Elements*, although he does not give any precise definition of this notion. In Proposition 8 of Book III, the words *concave* and *convex* sides of a circumference appear, and Euclid regards them as understood. Propositions 36 and 37 of the same book also involve convexity: Euclid talks about lines falling on the convex circumference. The constructibility of certain convex regular polygons is extensively studied in Book IV of the same work. Books XI, XII and XIII are dedicated to the construction and properties of convex regular polyhedra. The word "convex" is not used there to describe a property of these polyhedra, but Euclid relies extensively on the existence of a circumscribed sphere, which (in addition to the other properties that these polyhedra satisfy) implies that the polyhedra are convex. Such a sphere is also used by him for addressing the construction question: to construct the edge length of a face of a regular polyhedron in terms of the radius of the circumscribed sphere.

There is a rather long passage on the construction and the properties of the regular convex polyhedra derived from their plane faces in Plato's *Timaeus*,[1] written about half-a-century before Euclid's *Elements* appeared, and it is commonly admitted that Plato learned this theory from Theaetetus, a mathematician who was like him a student of the Pythagorean geometer Theodorus of Cyrene. Actually, Plato in the *Timaeus*, was mainly interested in the construction of four out of the five regular polyhedra that he assigned to the four elements of nature, namely, the tetrahedron, the octahedron, the icosahedron and the cube. In particular, he shows how the faces of these polyhedra decompose into known (constructible) triangles, he computes angles between faces, etc. [23, p. 210ff]. The regular convex polyhedra were part of the teaching of the early Pythagoreans (cf. [16]). All this was discussed at length by several authors and commentators, see in particular Heath's notes in his edition of the *Elements* [14] and Cornford's comments in his edition of the *Timaeus*.

Apollonius, in the *Conics*, uses the notion of convexity, in particular in Book IV where he studies intersections of conics. For instance, he proves that two conics intersect in at most two points "if their convexities are not in the same direction" (Proposition 30).[2] Proposition 35 of the same book concerns the tangency of conics

[1] *Timaeus*, [23], 53C-55C.

[2] *The Conics*, Book IV, [3, p. 172].

with convexities in opposite directions.[3] Proposition 37 concerns the intersection of a hyperbola with another conic, with convexities in opposite directions,[4] and there are several other examples where convexity is involved. We note that conics themselves are convex—they bound convex regions of the plane.

In the works of Archimedes, considerations on convexity are not limited to conics but they concern arbitrary curves and surfaces. Right at the beginning of his treatise *On the sphere and the cylinder* [13, p. 2], Archimedes introduces the general notion of convexity. The first definition affirms the existence of "bent lines in the plane which either lie wholly on the same side of the straight line joining their extremities, or have no part of them on the other side."[5] Definition 2 is that of a concave curve:

> I apply the term *concave in the same direction* to a line such that, if any two points on it are taken, either all the straight lines connecting the points fall on the same side of the line, or some fall on one and the same side while others fall on the line itself, but none on the other side.

Definitions 3 and 4 are the two-dimensional analogues of Definitions 1 and 2. Definition 3 concerns the existence of surfaces with boundary[6] whose boundaries are contained in a plane and such that "they will either be wholly on the same side of the plane containing their extremities,[7] or have no part of them on the other side." Definition 4 is about concave surfaces, and it is an adaptation of the one concerning concave curves:

> I apply the term *concave in the same direction* to surfaces such that, if any two points on them are taken, the straight lines connecting the points either all fall on the same side of the surface, or some fall on one and the same side of it while some fall upon it, but none on the other side.

After the first definitions, Archimedes makes a few *assumptions*, the first one being that among all lines having the same extremities, the straight line is the shortest. The second assumption is a comparison between the lengths of two concave curves in the plane having the same endpoints, with concavities in the same direction, and such that one is contained in the convex region bounded by the other curve and the line joining its endpoints:

> Of other lines in a plane and having the same extremities, [any two] such are unequal whenever both are concave in the same direction and one of them is either wholly included between the other and the straight line which has the same extremities with it, or is partly included by, and is partly common with, the other; and that [line] which is included is the lesser.

[3] *The Conics*, Book IV, [3, p. 178].

[4] *The Conics*, Book IV, [3, p. 183].

[5] The quotations of Archimedes are from Heath's translation [13].

[6] "Terminated surfaces", in Heath's translation.

[7] In this and the next quotes, since we follow Heath's translation, we are using the word "extremities", although the word "boundary" would have been closer to what we intend in modern geometry.

The third and fourth assumptions are analogues of the first two in respect of area instead of length. Assumption 3 says that among all surfaces that have the same extremities and such that these extremities are in a plane, the plane is the least in area. Assumption 4 is more involved:

> Of other surfaces with the same extremities, the extremities being in a plane, [any two] such are unequal whenever both are concave in the same direction and one surface is either wholly included between the other and the plane which has the same extremities with it, or is partly included by, and partly common with, the other; and that which is included is the lesser [in area].

Almost all of Book I (44 propositions) of Archimedes' treatise *On the sphere and the cylinder* uses in some way or another the notion of convexity. Several among these propositions are dedicated to inequalities concerning length and area under convexity assumptions. For instance, we find there inequalities on the length of polygonal figures inscribed in convex figures. Archimedes proves the crucial fact that the length of a convex curve is equal to the limit of polygonal paths approximating it, and similar propositions concerning area and volume, in particular for figures inscribed in or circumscribed to a circle or a sphere. His booklet *On measurement of a circle* is on the same subject. His treatise *On the equilibrium of planes* is another work in which convexity is used in a fundamental way. Postulate 7 of Book I of that work says that "in any figure whose perimeter is concave in one and the same direction, the center of gravity must be within the figure."

Another discovery of Archimedes involving the notion of convexity is his list of thirteen polyhedra that are now called semi-regular (Archimedean) convex polyhedra. These are convex polyhedra whose faces are regular polygons of a not necessarily unique type but admitting a symmetry group which is transitive on the set of vertices.[8] The faces of such a polyhedron may be a mixture of equilateral triangles and squares, or of equilateral triangles and regular pentagons, or of regular pentagons and regular hexagons, etc. It turns out that the Archimedean polyhedra are finite in number (there are essentially thirteen of them, if one excludes the regular ones). Archimedes' work on this subject does not survive but Pappus, in Book V of the *Collection*, reports on this topic, and he says there that Archimedes was the first to give the list of 13 semi-regular polyhedra [19, pp. 272–273].

Proclus (411–485) uses the notion of convexity at several places of his *Commentary on the first book of Euclid's Elements*; see e.g. his commentary on Definitions IV, VIII and XIX in which he discusses the angle made by two circles, depending on the relative convexities of the circles [25, pp. 97, 115, and 141].

Ptolemy, in Book I of his *Almagest*, establishes a necessary and sufficient condition for a convex quadrilateral to be inscribed in a circle in terms of a single relation between the lengths of the sides and those of the diagonals of that quadrilateral.[9] The relation, known as *Ptomely's relation*, has always been very useful in geometry.

[8] We are using modern terminology.

[9] Ptolemy's proof with the reference to Heiberg's edition is quoted in Heath's edition of Euclid [14, Vol. 2 p. 225].

4.3 Mirrors and Optics

Convex and concave mirrors are traditionally associated with imagination and phantasies, because they distort images. There are many examples of visual illusions and deceptions caused by convex and concave mirrors. Plato, in the *Republic* (in particular, in the well-known cave passage),[10] uses this as an illustration of his view that reality is very different from sensible experience. According to him, reason, and especially mathematics, allows us to see the real and intelligible world of which otherwise we see only distorted shadows.

The Roman writer Pliny the elder (first c. CE) at several places of his *Natural history* refers to the concept of concave or convex surface. In a passage on mirrors, he writes [24, Vol. VI, p. 126]:

> Mirrors, too, have been invented to reflect monstrous forms; those, for instance, which have been consecrated in the Temple at Smyrna. This, however, all results from the configuration given to the metal; and it makes all the difference whether the surface has a concave form like the section of a drinking cup, or whether it is [convex] like a Thracian buckler; whether it is depressed in the middle or elevated; whether the surface has a direction transversely or obliquely; or whether it runs horizontally or vertically; the peculiar configuration of the surface which receives the shadows, causing them to undergo corresponding distortions: for, in fact, the image is nothing else but the shadow of the object collected upon the bright surface of the metal.

Regarding surfaces receiving shadows, let me also mention the sundials used in Greek antiquity that have the form of a convex surface (see Fig. 4.1). The curve traced by the shadow of the extremity of a bar exposed to the sun has an interesting mathematical theory. The seventeenth-century mathematician Philippe de la Hire, in his treatise titled *Gnomonique ou l'art de tracer des cadrans ou horloges solaires sur toutes les surfaces, par différentes pratiques, avec les démonstrations géométriques de toutes les opérations* (Gnomonics, or the art of tracing sundials over all kind of surfaces by different methods, with geometrical proofs of all the operations) [10], conjectured that the theory of conic sections originated in the practical observations of sundials. Otto Neugebauer, in his paper *The Astronomical Origin of the Theory of Conic Sections* [18], made the same conjecture. This is also discussed in the article [1] in the present volume.

With mirrors, we enter into the realm of optics, where convexity is used in an essential way. The propagation properties of light rays, including their reflection and their refraction properties on convex and concave mirrors, were studied extensively by the Greek mathematicians. Catoptrics, the science of mirrors, was considered as a mathematical topic and was closely related to the theory of conic sections. (The Greek word katoptron, κάτοπτρον means mirror.) Since ancient times, the study

[10] Actually, in the cave passage ([22], Book VII, 514a–521d), not only images are distorted because the walls are not planar, but also one sees only shadows, apparent contours. Thom, in his *Esquisse d'une sémiophysique* ([34, p. 218] of the English translation) sees there the mathematical problem of reconstructing figures from their apparent contours.

Fig. 4.1 A Greek sundial with convex plate, from Ai Khanoum (Afghanistan), third to second c. BCE

in this field involved, not only plane mirrors, but curved ones as well, concave or convex. A few books on catoptrics dating from Greek antiquity have reached us, some only in Arabic or Latin translation. There is a treatise with a possible attribution to Euclid, compiled and amplified by Theon of Alexandria (fourth c. CE), containing reflection laws for convex and concave mirrors; cf. [11]. In particular, the author studies there the position of the focus of a concave mirror, that is, the point where sun rays concentrate after reflection, so as to produce fire.

The geometry of mirrors is related to conic sections. Book III of Apollonius' *Conics* addresses the question of reflection properties of these curves. A treatise by Heron of Alexandria which survives in the form of fragments is concerned with the laws of reflection on plane, concave and convex mirrors and their applications. In another treatise on catoptrics, attributed to Ptolemy, whose third, fourth and fifth Books survive, the author studies the reflection properties on plane, spherical convex and spherical concave mirrors. A book titled *Catoptrics* by Archimedes does not survive but is quoted by later authors, notably by Theon of Alexandria in his *Commentary on Ptolemy's Almagest* (I.3). One should also mention the work of Diocles (third to second c. BCE) which survives in the form of fragments, citations by other authors, and translations and commentaries by Arabic mathematicians. Diocles was a contemporary of Apollonius and his work on optics is inseparable from the theory of conics. To him is attributed the first investigation of the focal property of the parabola. (Heath in his edition of the *Conics* [2, p. 114] notes that Apollonius never used or mentioned the focus of a parabola.) Diocles studied mirrors having the form of pieces of spheres, paraboloids of revolution, ellipsoids of revolution, and other surfaces. His work was edited by Rashed in [26], a book containing critical editions of Arabic translations of Greek texts on the theory of burning mirrors, in particular those by Diocles on elliptical and parabolic burning mirrors.[11] The questions of finding the various shapes that a mirror can take in order to concentrate sun rays onto a point and produce fire at that point, and conversely, given a mirror, to find (possibly) a point where sun rays reflected on that mirror concentrate and produce fire, are recurrent in the Greek treatises on optics. There are proposition in Euclid's *Catoptrics* [11] dealing with burning mirrors. The introductory chapter of Lejeune's *Recherches sur la catoptrique grecque d'après les sources antiques et médiévales* (Researches on Greek catoptrics following antique and medieval sources) [17] contains an interesting brief history of this subject.

Optics is related to astronomy, in particular because of lenses. Convex lenses were already used in Greek antiquity to explore the heavens. There is a famous passage of the *Life of Pythagoras* written by Iamblichus, the Syrian neo-Pythagorean mathematician of the third century CE, in which the author recounts that Pythagoras, at the moment he made his famous discovery of the relation between ratios of integers and musical intervals, was pondering on the necessity of finding a device which would be useful for the ear in the same manner as the dioptre is useful for the sight [16, p. 62]. Dioptres are a kind of glasses used to observe the celestial bodies.

4.4 Billiards in Convex Domains

A question raised by Ptolemy is known since the Renaissance as *Alhazen's problem.*[12] This problem, in its generalized form, concerns reflection in a convex mirror, and, in the modern terminology, it can be regarded as a problem concerning

[11]The Latin word *focus* means fireplace, which led to the expression "burning mirror."

[12]The name refers to Ibn al-Haytham, the Arab scholar from the Middle Ages known in the Latin world as Alhazen, a deformation of the name "Al-Haytham." Ibn al-Haytham is especially famous for his treatise on Optics (*Kitāb al-manāzir*), in seven books (about 1400 pages long), which was translated into Latin at the beginning of the thirteenth century, and which was influential on Johannes Kepler, Galileo Galilei, Christiaan Huygens and René Descartes, among others. An important part of what survives from his work in geometry and optics was translated and edited by Rashed [27, 28]. Ibn al-Haytham is the author of an "intromission" theory of vision saying that it is the result of light rays penetrating our eyes, contradicting the theories held by Euclid and Ptolemy who considered, on the contrary, that vision is the result of light rays emanating from the eye ("extramission" theory). It is possible though that Euclid, as a mathematician, adhered to the theory where visual perception is caused by light rays traveling along straight lines emitted from the eye that strike the objects seen, in order to develop his mathematical theory of optics as an application of Euclidean geometry. This also explains the fact that Euclid's optics does not include any physiological theory of vision, nor any physical theory of colors, etc. Needless to say, besides this rough classification into an intromission theory and an extramission theory of light, there is a large amount of highly sophisticated and complex theories of vision and of light that were developed by Greek authors, which were related to the various philosophical schools of thought, and at the same time to the mathematical theories that were being developed.

trajectories in a convex billiard table. In its original form, the problem asks for the following: given a circle and two points that both lie outside or inside this circle, to construct a point on the circle such that the two lines joining the given points to that point on the circle make equal angles with the normal to the circle at the constructed point. Ibn al-Haytham,[13] in his *Kitāb al-Manāzir*, found a geometrical solution of the problem using conic sections and asked for an algebraic solution. He referred to Ptolemy while writing on his problem, and in fact, a large part of his work on optics was motivated by Ptolemy's work on this subject, which he criticized at several points. A. I. Sabra published an edition of Ibn al-Haytham's *Optics* [30], and wrote a paper containing an account of six lemmas used by Ibn al-Haytham in his work on the problem [29, 31–33].

A number of prominent mathematicians worked on Alhazen's problem. We mention in particular Christiaan Huygens, who wrote several articles and notes on it; they are published in Volumes XX and XXII of his *Complete Works* edition [15]. J. A. Vollgraff, the editor of Volume XXII of these works, writes on p. 647: "At the beginning of 1669, we can see Huygens absorbed by mathematics. He was busy with Alhazen's problem. This is one of the problems of which he always strived to find, using conic sections, the most elegant solution."[14] Volume XX of the *Complete Works* contains a text read by Huygens in 1669 or 1670 at the Royal Academy of Sciences of Paris on this problem titled *Problema Alhaseni* (p. 265). There is also a note titled *Construction d'un Problème d'Optique, qui est la XXXIX*e *Proposition du Livre V d' Alhazen, et la XXII*e *du Livre VI de Vitellion* (Construction of a problem on optics, which is Proposition XXXIX of Book V of Alhazen and Proposition XXII of Book VI of Vitellion)[15] and a note (p. 330) titled *Problema Alhazeni ad inveniendum in superficie speculi sphaerici punctum reflexionis* (Alhazen's problem on finding the reflection point on the surface of a spherical mirror), in the same volume. Vol. XXII of the *Complete Works* [15] contains an article dating from 1673 titled *Constructio et demonstratio ad omnes casus Problematis Alhazeni de puncto reflexionis* (Construction and proof of all the cases of Alhazen's reflection point problem). There are other pieces related to Alhazen's problem in Huygens' complete works. Among the other mathematicians who worked on this problem, we mention the Marquis de l'Hôpital and Isaac Barrow. Leonardo da Vinci conceived a mechanical device to solve the problem. Talking about Leonardo, let us note that he also conceived devices to draw conics; cf. [20].

[13]See Footnote 12.

[14]Au commencement de 1669 nous voyons Huygens absorbé par la mathématique. Il s'occupa du problème d'Alhazen. C'est là un des problèmes dont il a toujours eu l'ambition de trouver, par les sections coniques, la solution la plus élégante.

[15]Vitellion is the name of a thirteenth-century mathematician who edited works of Alhazen on optics.

4.5 Aristotle

Aristotle mentions convex and concave surfaces at several places in his writings, usually in his explanations of metaphysical ideas, and generally as an illustration of the fact that the same object may have two very different appearances, depending on the way one looks at it; like the circle, which may seem concave or convex, depending on the side from which one sees it.

In the *Nicomachean Ethics* [6],[16] talking about the soul, Aristotle discusses the fact that it has a part which is rational and a part which is irrational. He asks whether this distinction into two parts is comparable to the distinction between the parts of a body or of anything divisible into parts, or whether these two parts are "by nature inseparable, like the convex and concave parts in the circumference of a circle." The latter response is, according to him, the correct one, because the soul is one, and the fact that it has rational and irrational behaviors are different phases of the same thing.

Convexity is also mentioned in the *Meteorology* [5]. Here, Aristotle explains the origin of rivers and springs. He writes[17]:

> For mountains and high ground, suspended over the country like a saturated sponge, make the water ooze out and trickle together in minute quantities but in many places. They receive a great deal of water falling as rain (for it makes no difference whether a spongy receptacle is concave and turned up or convex and turned down: in either case it will contain the same volume of matter) and, they also cool the vapour that rises and condense it back into water.

Chapter 9 of Aristotle's *Physics* [8] is concerned with the existence of void, and it is an occasion for the Philosopher to discuss actuality and potentiality, in various instances. Convexity and concavity are again used as metaphorical entities. He writes[18]:

> For as the same matter becomes hot from being cold, and cold from being hot, because it was potentially both, so too from hot it can become more hot, though nothing in the matter has become hot that was not hot when the thing was less hot; just as, if the arc or curvature of a greater circle becomes that of a smaller—whether it remains the same or becomes a different curve—convexity has not come to exist in anything that was not convex but straight.

Thus, he says, something which becomes cold and hot was potentially cold or hot, like a thing which is convex: it may become more convex, because it was potentially convex, but it cannot become straight. Chapter 13 of the same treatise is concerned with the meaning of different words related to *time*. Aristotle writes[19]:

> Since the "now" is an end and a beginning of time, not of the same time however, but the end of that which is past and the beginning of that which is to come, it follows that, as the

[16] *Nicomachean Ethics* [6], 1102a-30.

[17] *Meteorology* [5], 350a10.

[18] *Physics* [8], 217a30-b5.

[19] *Physics* [8], 222b1.

circle has its convexity and its concavity, in a sense, in the same thing, so time is always at a beginning and at an end.

Thus, again like a circle, depending on the side from which one looks at it, may seem concave or convex, the term "now", depending on the side from which we look at it, may be the beginning or the end of time.

Besides talking about the convex and the concave as two sides of the same thing, Aristotle liked to give the example of the convex and the concave, while he talked about opposites. In the *Mechanical problems* [9], talking about the lever, and the fact that with this device, a small force can move a large weight, he writes[20]:

> The original cause of all such phenomena is the circle; and this is natural, for it is in no way strange that something remarkable should result from something more remarkable, and the most remarkable fact is the combination of opposites with each other. The circle is made up of such opposites, for to begin with it is composed both of the moving and of the stationary, which are by nature opposite to each other.[...] an opposition of the kind appears, the concave and the convex. These differ from each other in the same way as the great and small; for the mean between these latter is the equal, and between the former is the straight line.

Aristotle's fascination for the circle is always present in his works. In another passage of the same treatise, he writes[21]:

> [the concave and the convex] before they could pass to either of the extremes, so also the line must become straight either when it changes from convex to concave, or by the reverse process becomes a convex curve. This, then, is one peculiarity of the circle, and a second is that it moves simultaneously in opposite directions; for it moves simultaneously forwards and backwards, and the radius which describes it behaves in the same way; for from whatever point it begins, it returns again to the same point; and as it moves continuously the last point again becomes the first in such a way that it is evidently changed from its first position.

Book V of Aristotle's *Problems* [7] is titled *Problems connected with fatigue*. Problem 11 of that book asks: "Why is it more fatiguing to lie down on a flat than on a concave surface? Is it for the same reason that it is more fatiguing to lie on a convex than on a flat surface?" As usual in Aristotle's *Problems*, the question is followed by comments and partial answers, some of them maybe due to Aristotle, and others presumably written by students of the Peripatetic school. The comments on this problem include a discussion on the pressure exerted on a convex line, saying that it is greater than that exerted on a straight or concave line. They start with:

> For the weight being concentrated in one place in the sitting or reclining position causes pain owing to the pressure. This is more the case on a convex than on a straight surface, and more on a straight than on a concave; for our body assumes curved rather than straight lines, and in such circumstances concave surfaces give more points of contact than flat surfaces. For this reason also couches and seats which yield to pressure are less fatiguing than those which do not do so.

[20] *Mechanical problems* [9], 847b.

[21] *Mechanical problems* [9], 848a.

Book XXXI of the same work is titled *Problems connected with the eyes*, and Problem 25 of that book involves a discussion of convexity in relation with vision. The question is: "Why is it that though both a short-sighted and an old man are affected by weakness of the eyes, the former places an object, if he wishes to see it, near the eye, while the latter holds it at a distance?" In the comments, we read:

> The short-sighted man can see the object but cannot proceed to distinguish which parts of the thing at which he is looking are concave and which convex, but he is deceived on these points. Now concavity and convexity are distinguished by means of the light which they reflect; so at a distance the short-sighted man cannot discern how the light falls on the object seen; but near at hand the incidence of light can be more easily perceived.

The treatise *On the Gait of Animals* is a major biological treatise of Aristotle, and it concerns motion and the comparison of the various ways of motion for animals (including human beings). In Chapter 1, Aristotle writes[22]:

> Why do man and bird, though both bipeds, have an opposite curvature of the legs? For man bends his legs convexly, a bird has his bent concavely; again, man bends his arms and legs in opposite directions, for he has his arms bent convexly, but his legs concavely. And a viviparous quadruped bends his limbs in opposite directions to a man's, and in opposite directions to one another; for he has his forelegs bent convexly, his hind legs concavely.

In Chapter 13 of the same treatise, we read[23]:

> There are four modes of flexion if we take the combinations in pairs. Fore and hind may bend either both backwards, as the figures marked A, or in the opposite way both forwards, as in B, or in converse ways and not in the same direction, as in C where the fore bend forwards and the hind bend backwards, or as in D, the opposite way to C, where the convexities are turned towards one another and the concavities outwards.

4.6 Architecture

The Parthenon, Erechteum and Theseum columns, and more generally, Doric columns, are not straight but convex. Several explanations for this fact have been given, but none of them is definitive. F. C. Penrose published a book titled *An investigation of the principles of Athenian architecture; or, the results of a survey conducted chiefly with reference to the optical refinements exhibited in the construction of the ancient buildings at Athens* [21], whose object he describes as (p. 22)

> the investigation of various delicate curves, which form the principal architectural lines of certain of the Greek buildings of the best period; which lines, in ordinary architecture, are (or are intended to be) straight. In the course of our inquiries we shall perhaps be enabled in some degree to extend and correct our views of the geometry and mathematics of the ancients, by establishing the nature of the curves employed [...] The most important curves

[22] *On the Gait of Animals* [4], 704a15.

[23] *On the Gait of Animals* [4], 712a.

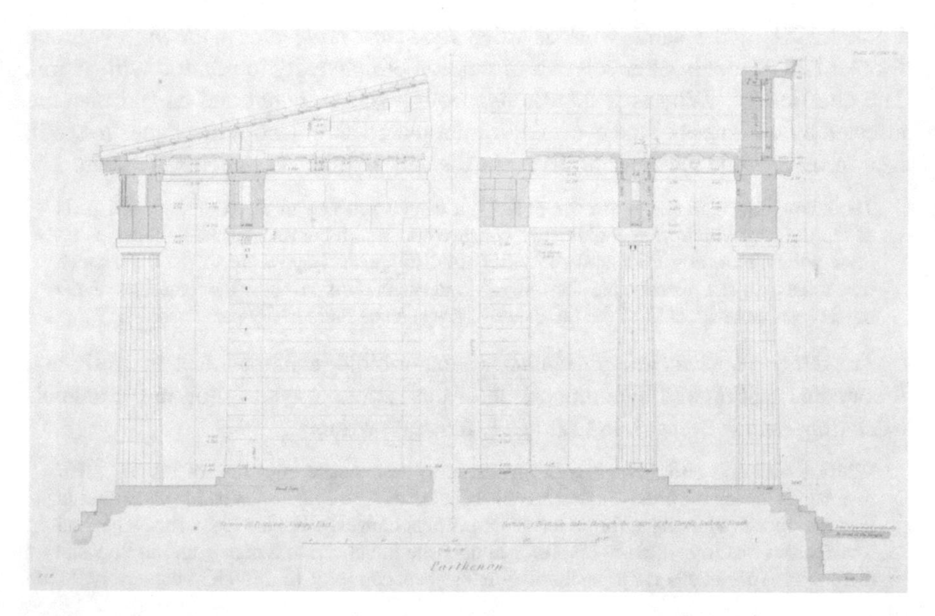

Fig. 4.2 A drawing of the Parthenon, from F. C. Penrose's *Investigation of the Principles of Athenian Architecture*

> in point and extent are those which form the horizontal lines of the buildings where they occur; such as the edges of the steps and the lines of the entablature, which are usually understood to be straight level lines, but in the steps of the Parthenon and some other of the best examples of Greek doric, are convex curves [. . .].

For instance, the columns of the Parthenon (see Fig. 4.2) are "in wonderful agreement at all points" with a piece of a parabola [21, p. 41].

Penrose says that the first mention of the curvature properties in Greek architecture was made by the Roman historian of architecture Vitruvius. Referring to him again on p. 39 of his essay [21], Penrose writes that this phenomenon, called *entasis* (from a Greek word meaning to stretch a line, or to bend a bow), is the

> well-known increment or swelling given to a column in the middle parts of the shaft for the purpose of correcting a disagreeable optical illusion, which is found to give an attenuated appearance to columns formed with straight sides, and to cause their outlines to seem concave instead of straight. The fact is almost universally recognized by attentive observers, though it may be difficult to assign a conclusive reason why it should be so.

One possible explanation which Penrose gives is "simply an imitation of the practice of Nature in giving almost invariably a convex outline of the limbs of animals" (p. 116) and "of the trunks and branches of trees" (p. 105).

This leads us to the question of architecture and art as an imitation of Nature, whose lines are seldom straight, and sometimes intricately curved. Curved are also the roads of poetic creation. The first lines of Canto I of Dante's *Inferno* read:

Midway upon the journey of our life
I found myself within a forest dark,
For the straightforward pathway had been lost.[24]

The roads of mathematical discovery are even more curved; they are twisted, and very lengthy.

References

1. A. A'Campo-Neuen and A. Papadopoulos, A path from curvature to convexity. In: *Geometry in history* (S. G. Dani and A. Papadopoulos, ed.) Springer Verlag, 2019.
2. Apollonius of Perga, *Treatise on conic sections*, ed. T. L. Heath, Cambridge University Press, 1896.
3. *Apollonius: Les Coniques*, tome 2: Livre IV, commentaire historique et mathématique, R. Rashed (ed.) Walter de Gruyter, 2009.
4. Aristotle, On the Gait of Animals, transl. A. S. L. Farquharson, In: *The works of Aristotle translated into English*, Oxford, Clarendon Press, Oxford, 1908.
5. Aristotle, Meteorology, translated by E. W. Webster In: *The works of Aristotle translated into English*, vol. III, ed. W. D. Ross, Oxford, Clarendon Press, 1923.
6. Aristotle, Nicomachean Ethics, transl. W. D. Ross, In: *The works of Aristotle translated into English*, vol. VII, Oxford, Clarendon Press, 1925.
7. Aristotle, The Problems, transl. E. S. Forster, In: *The works of Aristotle translated into English*, vol. VII, ed. W. D. Ross, Oxford, Clarendon Press, 1927.
8. Aristotle, The Physics, transl. P. Hardie and R. K. Gaye, In: *The works of Aristotle translated into English*, vol. II, ed. W. D. Ross and J. A. Smith, Oxford, Clarendon Press, 1930.
9. Aristotle, Mechanical problems, In: *The works of Aristotle in Twenty-three volumes*, vol. XIV, ed. W. S. Hett, The Loeb Classical Library, Cambridge (Mass.) and London, 1936.
10. Ph. de la Hire, Gnomonique ou l'art de tracer des cadrans ou horloges solaires sur toutes les surfaces, par différentes pratiques, avec les démonstrations géométriques de toutes les opérations, Paris Estienne Michallet, 1682.
11. Euclide, l'Optique et la catoptrique, trad., introduction et notes par P. Ver Eecke, Desclée de Brouwer, Bruges, 1938. Nouveau tirage, Librairie Albert Blanchard, Pari, 1959.
12. W. Fenchel, Convexity through the ages. In: *Convexity and its Applications* (P. M. Gruber and J. M. Wills, ed.), pp. 120–130, Birkhäuser, Basel, 1983.
13. T. L. Heath (ed.), *The works of Archimedes*, Cambridge University Press, 1897, Reprint, Dover, 2002.
14. T. L. Heath, *The thirteen books of Euclid's Elements*, 3 volumes, Cambridge University Press, 1908, Reprint, Dover.
15. C. Huygens, *Œuvres complètes*, 22 volumes, Société Hollandaise des Sciences, La Haye, 1888–1950.
16. Iamblichus, *Life of Pythagoras*, transl. T. Taylor, Watkins, London, 1818.
17. A. Lejeune, Recherches sur la catoptrique grecque d'après les sources antiques et médiévales, Académie Royale de Belgique, Bruxelles, 1957.

[24] *Nel mezzo del cammin di nostra vita*
mi ritrovai per una selva oscura
ché la diritta via era smarrita.

18. O. Neugebauer, The astronomical origin of the theory of conic sections, *Proc. Am. Phil. Soc.* 92, No. 3 (1948), pp. 136–138.
19. Pappus d'Alexandrie. La collection mathématique ; œuvre traduite pour la première fois du grec en français par Paul Ver Eecke. Paris, Bruges, Desclée de Brouwer, 1933.
20. C. Pedretti, Studi Vinciani: Documenti, Analisi e inediti leonardeschi, Librairie E. Droz, Paris, 1957.
21. F. C. Penrose, *An Investigation of the Principles of Athenian Architecture; or, The Results of a Survey Conducted Chiefly with Reference to the Optical Refinements Exhibited in the Construction of the Ancient Buildings at Athens*, Macmillan, London, 1888.
22. Plato, *the Republic*, transl. P. Shorey, Harvard University Press, 1935.
23. Plato, *the Timaeus*, translation with commentary by F. M. Cornford, Kegan Paul, Trench, Trubner, London, 1937.
24. Pliny, *Natural history*, tr. J. Bostock, Cambridge, 1857.
25. Proclus de Lycie, Les Commentaires sur le premier livre des Éléments d'Euclide. Traduits pour la première fois du grec en français avec une introduction et des notes par Paul Ver Eecke. Desclée de Brouwer, Bruges, 1948.
26. R. Rashed, Les catoptriciens grecs. I. Les miroirs ardents, Collection des Universités de France, Les Belles Lettres Paris, 2000.
27. R. Rashed, *Ibn al-Haytham, New spherical geometry and astronomy, A History of Arabic sciences and mathematics*, vol. 4, Routledge, Tayler and Francis, London and New York, 2014.
28. R. Rashed, *Ibn al-Haytham's Geometrical Methods and the Philosophy of Mathematics*, vol. 5, Routledge, Taylor and Francis, London and New York, 2018.
29. A. I. Sabra, Ibn al-Haytham's lemmas for solving "Alhazen's Problem." *Arch. Hist. Exact Sci.* 26 (1982), pp. 299–324.
30. A. I. Sabra (ed.), *The optics of Ibn al-Haytham, Books I–III: On Direct vision*. Translated and edited by A. I. Sabra, 2 volumes, London, Warburg Institute, University of London, 1989.
31. G. Simon, Roger Bacon et Kepler lecteurs d'Alhazen. *Arch. Internat. Hist. Sci.* 51 (2001), no. 146, pp. 38–54.
32. A. M. Smith (ed.), Ptolemy's Theory of Visual Perception: An English Translation of the Optics, with Introduction and Commentary, *Trans. Am. Phil. Soc.* Vol. 86, part 2, American Philosophical Society, Philadelphia, 1996.
33. A. M. Smith, Alhacen's Theory of Visual Perception: A Critical Edition, with English Translation and Commentary, of the First Three Books of Alhacen's *De Aspectibus*, the Medieval Latin Version of Ibn al-Haytham's Kitāb al-Manāẓir, *Transactions of the American Philosophical Society*, Philadelphia, American Philosophical Society, 2001.
34. R. Thom, *Esquisse d'une sémiophysique : Physique aristotélicienne et théorie des catastrophes*, Paris, InterEditions, 1988. English translation by V. Meyer, *Semio Physics: A Sketch. Aristotelian Physics and Catastrophe Theory*, Addison-Wesley, 1990.

Chapter 5
On the Concept of Curve: Geometry and Algebra, from Mathematical Modernity to Mathematical Modernism

Arkady Plotnitsky

Abstract We consider the concept of curve in the context of the transition from mathematical "modernity" to mathematical "modernism," the transition defined, the article argues, by the movement from the primacy of geometrical to the primacy of algebraic thinking. The article also explores the ontological and epistemological aspects of this transition and the connections between modernist mathematics and modernist physics, especially quantum theory, in this set of contexts.

AMS Classification: 00A30, 01A65, 01A60, 81P05

5.1 Prologue

The frescoes of the Chauvet-Pont-d'Arc Cave in southern France, painted roughly 32,000 years ago, and the subject of Werner Herzog's 3-D documentary, *The Cave of Forgotten Dreams* (2010), are remarkable not only because of the extraordinary richness and quality of their paintings, or how well they are preserved, or, closer to my subject here, their prehistorical images of curves, which are found in other, some earlier, cave paintings, but also and especially because these curves, delineating animal figures (there is only one human figure), are drawn on the intricately curved surfaces of the cave. This unfolding curved-surface imagery compelled Herzog to use 3-D technology for his film. The resulting cinematography, the temporal image of curves and curved surfaces, the curved image of time (Herzog's theme as well) is remarkable phenomenologically, aesthetically, and, for a geometer, mathematically. The access to the cave, discovered in 1994, is severely limited, and Herzog was lucky to get permission to enter and film in it. But one could imagine what the likes of Gauss, Lobachevsky, Riemann and Poincaré (who spent a lifetime thinking about curves on surfaces) would have thought if they had had a chance to see the cave,

---

A. Plotnitsky (✉)
Literature, Theory, and Culture Program, Purdue University, West Lafayette, IN, USA
e-mail: plotnits@purdue.edu

S. G. Dani, A. Papadopoulos (eds.), *Geometry in History*,
https://doi.org/10.1007/978-3-030-13609-3_5

which seems, by its very existence, to challenge Euclid, nearly our contemporary on this 30,000-year-old time scale. Almost, but not quite! One needs to know and, first, to invent a great deal of mathematics to think of this challenge, and the mathematics as developed, along with science, philosophy, and art, by the ancient Greeks brings them and us together as contemporaries.

There is an immense intervening history between these paintings and us, a history that has erased beyond recovery most of the thinking that created them, and whatever comments one can make concerning this thinking are bound to be conjectural. It is unlikely that mathematics existed at the time, although this may depend on how one understands what mathematics is. What may be said with more confidence is that human thinking, thanks to the neurological structure of the human brain, had by then (50,000–100,000 years ago is a current rough estimate for when this structure emerged) a component that led to the rise of mathematical thinking and eventually to mathematics itself. Concerning the longer prehistory one can only invent evolutionary fables, akin to those concerning the origins of thinking, consciousness, language, logic, music, or art, which may be plausible and useful but are unlikely to ever be confirmed.[1] Art is one endeavor where we might be close to the cave painters in the Chauvet Cave and elsewhere. But then, is art possible without some mathematical thinking, or mathematics without some artistic thinking, or thinking in general without either, or without philosophical thinking? This is doubtful, as my argument in this article will suggest, without, however, making a definitive claim to that effect, which may not be possible given where our understanding (neurological, psychological, or philosophical) of the nature of thinking stands now.

5.2 From Mathematical Modernity to Mathematical Modernism

While, inevitably, invoking earlier developments, in particular ancient Greek mathematics, the history I address here begins with mathematics at the rise of *modernity*, especially the work of Fermat and Descartes, crucial to our concept of curve as well, but having a much greater significance for all subsequent mathematics. One can assess the character of mathematical thinking then more reliably because this character is close to and has decisively shaped the character of our mathematical thinking now. I then move to Riemann and, finally, to mathematical *modernism*, which emerged sometime around 1900, shaped by the preceding development of nineteenth-century mathematics, roughly, from K. F. Gauss on, with Riemann as the most crucial figure of this development and, in the present view, conceptually, already of modernism. Thus, even this shorter trajectory, curve, of the idea of curve

[1]G. Tomlinson's book on the prehistory of music, with the revealing title "A Million Years of Music," confirms this view, even if sometimes against its own grain [64].

will be sketched by way of a discrete set of points, which would hopefully allow the reader to envision or surmise the curve or curves (for there are, again, more than one of them) connecting the dots. This *uncircumventable circumstance* notwithstanding (the metaphor of a curve invades this sentence, too), I do aim to offer a thesis concerning the idea of curve in mathematics and a thesis concerning mathematics itself, as defined by mathematical modernism as fundamentally algebraic, and to offer a historical and conceptual argument supporting this thesis.

I shall more properly outline my key concepts in the next section, merely sketching them in preliminary terms here, beginning with modernity and modernism. Modernity is a well-established broad cultural category, defined by a set of cultural transformations, "revolutions," that extends, roughly, from the sixteenth century to our own time. The rise of modernity has been commonly associated with the concept of Renaissance, especially in dealing with its cultural aspects, such as philosophy, mathematics and science, and literature and art. We are more cautious in using the rubric of Renaissance now, and prefer to speak of the early modern period, implying a greater continuity with the preceding as well as subsequent history. This caution is justified. On the other hand, in the present context, invoking the Renaissance is not out of place either, as referring to the rebirth of the ways of thinking, not the least the mathematical thinking, of ancient Greece. The rise of modernity was in part shaped by new mathematics, such as analytic geometry, algebra, and calculus, and new, experimental-mathematical sciences of nature, beginning with the physics of Kepler, Descartes, Galileo and Newton, a set of developments often referred to as the Scientific Revolution.

By contrast, modernism is a well-established, even if not uniquely defined, denomination only when applied to literature and art, while a recent and infrequent denomination when applied to mathematics or science, a denomination, moreover, commonly borrowed from its use in literature and art, as by H. Mehrtens and J. Gray, my main references here [26, 44]. *Historically*, both phenomena, modernist literature or art and modernist mathematics or science, are commonly understood as belonging to the same period, roughly from the 1890s to the 1940s, but as in various ways extending to and, certainly, continuing to impact our own time. Both forms of modernism are considered to be defined by major transformations in their respective domains, and there is a consensus that significant changes in both domains did take place during that time. However, the complex and multifaceted nature of these transformations makes it difficult to conclusively ascertain their nature and causality. Some of these transformational effects had multiple causes, and conversely, some of these causes combine to produce single effects. It is hardly surprising, then, that, *conceptually*, the thinking concerning modernism in any field is diverse and, in each case, only partially reflects the nature of modernism in a given field or the relationships, on modernist lines, between different fields, even between modernist mathematics and modernist physics.

This is an unavoidable limitation, and it cannot be circumvented by the conception of mathematical modernism to be offered in this article, which intersects with other such conceptions, but, to the best of my knowledge, does not coincide with any of them. I can only argue that this type of mathematical thinking emerged during the historical period in question, but not that it exhausts what can be termed mathematical modernism, let alone capture the development of mathematics during that period. Not all of this mathematics was modernist by any definition I am familiar with. Indeed, it cannot be captured by any single definition, any more than mathematics in general. While I find the term mathematical modernism useful and historically justified, more important are key conceptual formations that, I argue, decisively, even if not uniquely, characterize the mathematics or science that emerged during that period. I am, however, ready to admit that these formations could be given other denominations. One cannot hope for a unique name here (any more than in general), which is a good thing, because new names open new trajectories of thought.

Mathematical modernism will be primarily understood here as mathematical thinking that gives mathematics a *fundamentally algebraic character*. By way of broad preliminary definitions, I understand algebra as the mathematical formalization of the relationships between symbols, arithmetic as the mathematical practice dealing specifically with numbers, geometry as the mathematical formalization of spatiality, especially (although not exclusively) in terms of measurement, and topology as the mathematization of the structure of spatial or spatial-like objects apart from measurements, specifically in terms of continuity and discreteness. The corresponding mathematical fields are algebra, number theory, geometry, and topology. Analysis deals with the questions of limit, and related concepts such as continuity (where it intersects with topology), differentiation, integration, measurement, and so forth. There are multiple intersections between these fields, and there are numerous subfields and fields, like arithmetic algebraic geometry, that branched off these basic fields.

Defining algebra as the mathematical formalization of the relationships between symbols makes it part of all mathematics, at least all modern mathematics. Ancient Greek geometry was grounded, at least expressly, in arithmetic, although one might detect elements of symbolism there as well, especially at later stages of its development, certainly by the time of Diaphantus, sometimes called "the father of algebra." Geometrical and topological mathematical objects always have algebraic components as part of their structure, while algebraic objects may, but need not have geometrical or topological components. Two other, narrower or field-specific, considerations of algebra are important for my argument as well. The first, standing at the origins of algebra as a mathematical discipline, is that of algebra as the study of algebraic (polynomial) equations, is important also because all equations are in effect forms of algebra, which includes equations associated with calculus and then differential equations, crucial to the history of mathematics from mathematical modernity to mathematical modernism. The second is that defined by algebraic structures, such as groups or associative algebras (groups, especially symmetry groups, are also crucial to geometry and topology). These two senses of algebra bring into the landscape of mathematical modernity and then modernism, and the transition from one to the other, the figure of Galois who was the first to connect these two senses of algebra, which he did in a radically revolutionary way (also by introducing the concept of group). Galois is, arguably, the most notable figure, next to Riemann, in this history and is an even earlier (proto)modernist than Riemann was, although the limits of this article will allow me to comment on Galois only in passing. All these aspects of algebra are part of the algebraization defining mathematical modernism and the concepts of curve that come with it.

What gives the present conception of mathematical modernism its bite is that it applies *fundamentally*, rather than merely operationally, across modernist mathematics: It defines not only fields, such as analysis or mathematical logic that, while not disciplinarily classified as algebra, are governed by structures that are algebraic, but also fields like geometry and topology, that, while having technical algebraic aspects, are conceptually and disciplinarily juxtaposed to algebra. According to the present view, it is not only a matter of having an algebraic component as part of the mathematical structure of their objects but also and primarily a matter of defining these objects algebraically. Without aiming thus to contain the nature of mathematics (which is impossible in any event), one might say that the following three elements are always found in mathematics: concepts, structures, and logic, each generally more rigorously formalized than when they are found elsewhere, especially when mathematics is not used, the way it is used in physics, for example. While, however, structures and logic always entail one or another form of algebra, this is not necessarily so in the case of concepts, which may be strictly geometrical or topological. My argument here is that mathematical modernism brings algebra into the architecture of mathematical concepts, including those found in geometry or topology, even though algebra in turn accommodates the disciplinary demands of these fields.

An emblematic case is algebraic topology (a revealing denomination in itself), a field important for my argument here on several accounts, especially given its significance for algebraic geometry and Grothendieck's work.[2] Algebraic topology does have an earlier history (extending from Leibniz and, more expressly, Euler) preceding the rise of the discipline as such with Riemann, Poincaré, and others. This history, however, is not comparable to that of geometry from the ancient Greeks on, until modernism. By contrast, from modernism on, geometry and topology developed equally and in interaction with each other, and with differential topology, a field that emerged along with algebraic topology. What makes algebraic topology a mathematical discipline is that one can associate algebraic structures (initially numbers, eventually groups and other abstract algebraic structures, such as rings) to the architecture of spatial objects that are invariant under continuous transformations, independently of their geometrical properties, such as those associated, directly or implicitly, with measurements. This makes topology topo-*logy* vs. geo-*metry*. By the same token, in retrospect, topology is almost inherently categorical. It relates, functorially, the objects of topological and algebraic categories, a form of algebraic

[2] I will be less concerned with general or point-set topology, which has a different and much longer history, extending, arguably, to the ancient Greek thinking, although my claim concerning the modernist algebraization of mathematics could still be made in this case. See A. Papadopoulos' contribution to this volume for an illuminating discussion of the topological aspects of Aristotle's philosophy, via Thom's engagement with Aristotle [48].

thinking that is one of the culminating conceptions of mathematical modernism, for example and in particular, in Grothendieck's algebraic geometry.

This is not to say that the spatial (geometrical or topological) character of mathematical objects defined in modernist geometry and topology in terms of algebra disappears. It remains important at least on two counts, both of which are, however, consistent with my argument. First, the algebra defining these objects has a special form that may be called "spatial algebra."[3] Spatial algebra arises from algebraic structures that mathematically define geometrical or topological objects and reflects their proximity to $\mathbb{R}^3$ and mathematical spatial objects there, that are close to our phenomenal intuition and the geometry and topology associated with this intuition. This proximity may be, and commonly is, left behind in rigorous mathematical definitions and treatments of such objects, beginning with $\mathbb{R}^3$ itself. The same type of algebra may also be used to define mathematical objects that are no longer available to our phenomenal intuition, apart from using the latter to create heuristic metaphorical images of such space. Among characteristic examples of such objects are a projective space (a set of lines through the origin of a vector space, such as $\mathbb{R}^2$ in the case of the projective plane, with projective curves defined algebraically, as algebraic varieties) and an infinite-dimensional Hilbert space (the points of which are typically square-integrable functions or infinite series, although a Euclidean space of any dimension is a Hilbert space, too). In sum, spatial algebra is an algebraization of spatiality that makes it rigorously mathematical, topologically or geometrically, as opposed to something that is phenomenally intuitive or is defined philosophically, even in the case of spatial objects in $\mathbb{R}^3$. As such, it also allows us to define spatial-algebraic objects across a broad mathematical spectrum, and by doing so to extend the fields of topology and geometry.

At the same time, and this is the second count on which mathematical objects defined by spatial algebra retain their connections to geometrical and topological thinking, analogies with $\mathbb{R}^3$ continue to remain useful and even indispensable. Such analogies may be rigorous (and specifically algebraic) or metaphorical, with both types sometimes used jointly. Thus, the analogues of the Pythagorean theorem or parallelogram law in Euclidean geometry, which holds in infinite-dimensional Hilbert spaces over either $\mathbb{R}$ or $\mathbb{C}$, are important, including in applications to physics, especially quantum theory, the mathematical formalism of which is based in Hilbert spaces (of both finite and infinite dimensions) over $\mathbb{C}$. More generally, our thinking concerning geometrical and topological objects is not entirely translatable into algebra. This was well understood by D. Hilbert in his axiomatization of Euclidean geometry, even though this axiomatization had a spatial-algebraic character, in the

[3]Finding a good term poses difficulties because such, perhaps more suitable, terms as "geometric algebra" and "algebraic geometry," are already in use for designating, respectively, the Clifford algebra over a vector space with a quadratic form and the study of algebraic varieties, defined as the solutions of systems of polynomial equations. This object and this field, however, equally exemplify the modernist algebraization of mathematics.

present sense, including in establishing an algebraic model (the field $\mathbb{C}$) of his system of axioms in order to prove its consistency [34]. According to D. Reed:

> [A]fter a chapter in which [Hilbert] provides himself with more tools like geometry and algebra [in this following Descartes], he goes on to demonstrate in a truly spectacular way:
> (*) a "theory of plane area can be derived from the axioms" (but not a theory of volume);
> (*) Desargues' theorem, which states that if two triangles are situated in a plane so that pairs of corresponding sides are parallel then the lines joining the corresponding sides pass through one and the same point or are parallel, expresses a criterion for a "plane" geometry to form part of "space" geometry; and
> (*) Pascal's theorem, which states that if A, B, C, and A^1, B^1, C^1 are two sets of points on two intersecting lines and if AB^1 is parallel to BC^1 and AA^1 is parallel to CC^1 then BA^1 is parallel to CB^1, is dependent in a very specific way on the so-called Axiom of Archimedes.
> None of these statements can be given a simple unequivocal expression in the realm of algebra even though models from "analytic geometry" are used in the demonstrations. In other words, while algebra is useful as a tool in the demonstration of geometrical statements it is not useful in formulating the statements themselves. [59, pp. 33–34]

Reed is right in arguing for the significance of geometrical thinking and expression in mathematics. On the other hand, his claim concerning algebra as not being useful in formulating geometrical statements is an over-simplification, whether as a general claim or as reflecting Hilbert's thinking, even in Euclidean geometry, where our geometrical intuition is more applicable and where certain proofs, such as many of those supplied by the Elements could be geometrical [22].[4] Thus, as Hilbert was well aware, a more natural setting for Desargues' and Pascal's theorems is projective geometry, which these theorems helped to usher in, in a setting, however, that we cannot visualize and that is spatial-algebraic. In other words, making a symmetrical assessment, while (Euclidean) geometrical and topological intuitions are helpful and even irreducible, spatial algebra and, thus, algebra itself, at least since Fermat and Descartes, or Diaphantus, if not Euclid, is irreducible in turn even in topology and geometry.

One can get further insight into this situation by considering a related principle due to J. Tate, whose thinking bridged number theory and algebraic geometry in highly original and profound ways: "Think geometrically, prove algebraically." It was introduced in the book (co-authored with J. Silverman) on "the rational

[4]This may remain true in low-dimensional geometry or topology. I would argue, however, that spatial algebra is still irreducible there because one commonly converts topological operations into algebraic ones. This conversion in low dimensions was essential to the origin of algebraic topology. On the other hand, the recent development of low-dimensional topology, following, among others, W. Thurston's work, from the 1970s on, is a more geometrically oriented trend that, to some degree, counters the twentieth-century modernist algebraic trends and returns to Riemann's and Poincaré's topological thinking, but only to a degree, because the algebraic structures associated with these objects remain crucial. Some of the most powerful (modernist) algebraic tools of algebraic topology and algebraic geometry have been used and sometimes developed during this more geometrical stage of the field. These areas have important connections to quantum field theory and then string theory, as in E. Witten's work, which, especially in quantum field theory, are fundamentally algebraic, in part by virtue of their probabilistic nature.

points of elliptic curves," a context that is more expressly modernist as far as the algebraization of the geometrical is concerned and as such is more illuminating in the present context. The title-phrase combines algebra ("rational points") and geometry ("curves"), and implies that geometry, at least beyond that of $\mathbb{R}^3$ and even there, requires algebra to be mathematically rigorous. According to Silverman and Tate:

> It is also possible to look at polynomial equations and their solutions in rings and fields other than $\mathbb{Z}$ or $\mathbb{Q}$ or $\mathbb{R}$ or $\mathbb{C}$. For example, one might look at polynomials with coefficients in the finite field $\mathbb{F}_p$ with p elements and ask for solutions whose coordinates are also in the field $\mathbb{F}_p$. You may worry about your geometric intuition in situations like this. How can one visualize points and curves and directions in A^2 when the points of A^2 are pairs (x, y) with $x, y \in \mathbb{F}_p$? There are two answers to this question. The first and most reassuring one is that you can continue to think of the usual Euclidean plane, i.e., $\mathbb{R}^2$, and most of your geometric intuition concerning points and curves will still be true when you switch to coordinates in $\mathbb{F}_p$. The second and more practical answer is that the affine and projective planes and affine and projective curves are defined algebraically in terms of ordered pairs (r, s) or homogeneous triples $[a, b, c]$ without any reference to geometry. So, in proving things one can work algebraically using coordinates, without worrying at all about geometrical intuitions. We might summarize this general philosophy as: *Think geometrically, prove algebraically*. [62, p. 277]

Affine and projective planes and curves, no longer available to our phenomenal intuition, can in principle be defined without any reference to ordinary language and concepts. The latter are more difficult and perhaps impossible to avoid in geometry, at least in the kind of intuitive geometry Silverman and Tate refer to, rather than what I call spatial algebra, which, I argue, ultimately defines (nearly) all geometry rigorously. Even in these more intuitively accessible cases, we still think algebraically, too, by using spatial algebra, if with the help of geometrical intuitions, except, as noted, possibly in dealing with low-dimensional topological and geometrical objects, where more immediately spatial (topological and geometrical) arguments could be used more rigorously. It is also true that a mathematician can develop and use intuition in dealing with discrete geometries as such, say, that of the Fano plane of order 2, which has the smallest number of points and lines (seven each). However, beyond the fact that they occur in the two-dimensional regular plane, the diagrammatic representations of even the Fano plane are still difficult to think of as other than spatial algebra, in this case, combinatorial in character. While useful and even indispensable, our Euclidean intuitions are limited even when we deal with algebraic curves in the Euclidean plane, let alone in considering something like a Riemann surface as a curve over $\mathbb{C}$, or curves and other objects of finite or projective geometries, abstract algebraic varieties, Hilbert spaces, the spaces of noncommutative geometry, or geometric groups, a great example of the extension of spatial algebra to conventionally algebraic objects.

Paul Dirac, recognized as the greatest algebraic virtuoso among the founding figures of quantum theory, was, nevertheless, reportedly fond of referring to geometrical thinking in quantum mechanics and quantum field theory, both mathematically based in Hilbert spaces, of both finite and infinite dimensions over $\mathbb{C}$, and the algebras of Hermitian operators there (e.g., [23]). It is difficult to surmise, especially just from reported statements, what Dirac, famous for his laconic style, exactly had in mind. If, however, one is to judge by his writings, they appear to suggest that at stake are the algebraic properties and relations, and methods of investigations they suggest, modeled on those found in geometrical objects, defined by algebraic structures, in short, spatial algebra, as just explained, working with which was part of Dirac's algebraic virtuosity. Indicatively, notwithstanding his insistence on the role of geometry in Dirac's quantum-theoretical thinking, O. Darigold's analysis of this thinking shows the significance of algebra there [16]. Thus, as he says, "roughly, Dirac's quantum mechanics could be said to be to ordinary mechanics what noncommutative geometry is to intuitive geometry" [16, p. 307]. Noncommutative geometry, however, the invention of which was in part inspired by quantum mechanics, is a form of spatial algebra ([13, p. 38]; [53, pp. 112–113]). One encounters similar appeals to geometrical thinking in referring to transfers of geometrical methods and techniques to spatially algebraic or just algebraic objects (thus making them spatially algebraic), such as in dealing with groups and group representations in quantum mechanics, initially developed in a more geometrical context beginning with S. Lie and F. Klein or in using the idea of metrics in geometric group theory.

In what sense, then, apart from being defined by spatial algebra, may such spaces be seen as spaces, in particular, as relates to our phenomenal intuition, including visualization? The subject is complex and it is far from sufficiently explored in cognitive psychology and related fields, an extensive research during recent decades notwithstanding, including as concerns cultural or technological (digital technology in particular) factors affecting our spatial thinking. It would be difficult to make any definitive claims here. It does appear, however, that, these factors notwithstanding, our three-dimensional phenomenal intuition is shared by us cognitively and even neurologically in shaping our sense of spatiality. Part of this sense appears to be Euclidean, insofar as it corresponds to what is embodied in $\mathbb{R}^3$ (again, a mathematical concept), keeping in mind that the idea of empty space, apart from bodies of one kind or another defining or framing it, is an extrapolation, because we cannot have such a conception phenomenally or, as Leibniz argued against Newton, physically. We can have a mathematical conception of space itself. To what degree our phenomenal spatiality is Euclidean remains an open question, for example, in dealing with the visual perception of extent and perspective (e.g., [21, 25, 63]). It is nearly certain, however, that, when we visualize spatial-algebraically defined objects or even more conventional geometrical spaces or geometries once the number of dimensions is more than three, we visualize only three- (or even two-) dimensional configurations and supplement them by algebraic structures and intuitions. R. Feynman instructively explained this process in describing visual intuition in thinking in quantum theory, as cited in S. Schweber [61, pp. 465–466]. Obviously, such anecdotal evidence is not sufficient for any definitive claim. It appears, however, to be in accord with the current neurological and cognitive-psychological research, as just mentioned, which suggests the dependence of our spatial intuition, including visualization, on two- and three-dimensional phenomenal intuition. This was arguably why Kant thought of this intuition, which he saw as that of Euclidean three-dimensional spatiality, as given to us a priori. That this intuition is entirely Euclidean or that it is given to us a priori, rather than developed by experience (we would now say, neurologically), may be and has been challenged. On the other hand, its three-dimensional character appears to be reasonably certain.

As Tate must have been aware, mathematical thinking concerning geometrical and topological objects cannot be reduced to our naïve Euclidean intuitions, even though it may not be possible or desirable to exclude them. Silverman and Tate's example from differential calculus (given to further illustrate their philosophy of thinking geometrically but proving algebraically), that of finding a tangent line to a curve, confirms this point [62, pp. 377–378]. The invention of calculus, an essentially algebraic form of mathematics, was not so much about proving algebraically, as the standard of proof then was geometry. Newton, was compelled to present his mechanics in terms of geometry rather than calculus in his 1687 *Principia*, in part, as he explained, to assure a geometrical demonstration of his findings, also in the direct sense of showing something by means of phenomenal visualization, rather than in terms of the algebra of calculus [45]. Calculus was about *thinking* algebraically, as was especially manifested in Leibniz's version, rather than about rigorous proofs.

Calculus was a decisive development in understanding the geometry of curves as continuous objects, a major rethinking of the nature of curve and curvature, with, artistically and culturally, deep connections to the Baroque, the style, or more accurately, the mode of thought, defined by the ideas of curve and inflection, with Leibniz being the defining philosopher of the Baroque as well as, in his case correlatively, the coinventor of calculus [17, 51]. "Inflection is the ideal genetic element of the variable curve," G. Deleuze says in *The Fold: Leibniz and the Baroque* [17, p. 15]. Baroque thinking was also thinking in terms of infinite variations of curves, reminding one of moduli spaces of curves, yet another of Riemann's major discoveries. For the moment, one might argue that it is not in fact possible to understand the concept of continuous curve mathematically apart from calculus or some form of proto-calculus (as in Archimedes, for example), and the subject, accordingly, should have been given more consideration here. However, even in enabling this understanding, calculus was a new form of algebra, as is, again, especially manifested in Leibniz's version of it, but found in Newton as well. Fermat, the founding father of the study of elliptic curves (which led him to his famous "last" theorem), played a key role in this history, too, even if he fell short of inventing calculus. Mathematical modernism, I argue, is ultimately defined by thinking in terms of algebra rather than in terms of continuity, even in thinking of continuity itself, for example, and in particular, in considering differentiable objects, differentiable manifolds, as we define them, following Riemann. The field known as "differential algebra" (introduced in the 1950s) is another confirmation of this modernist view in one of its later incarnations, and, as it may be argued to have a Leibnizian genealogy, the connection, via algebra, between modern and modernist mathematics. An earlier modernist example of this connection was "symbolic differentiation" for Hilbert space operators (infinite-dimensional matrices) in quantum mechanics by M. Born and P. Jordan [9, 52, p. 121].

My conception of modernist mathematics as an algebraization of mathematics, even in the case of topology and geometry, is an extension of Tate's principle. This extension retains its second part but modifies the first: "Think both Intuitively Geometrically and Spatially-Geometrically: Prove Algebraically." In this form the principle could sometimes apply in algebra as well, which has benefited from the introduction of spatial-algebraic objects from Fermat and Descartes to Grothendieck and beyond, for example, in the case of geometric group theory, the study of which was founded by M. Gromov, one of the more intuitive contemporary geometers, on this type of principle. But then proving something is thinking, too, as Tate would surely admit.

The history of mathematical thinking concerning curves or straight lines (a special class of curves) is part of the origin, if not a unique origin, of this philosophy "Think both Intuitively Geometrically and Spatially-Geometrically: Prove Algebraically," which, in modernity, begins with Fermat and Descartes. Their thinking and work, which overshadow Silverman and Tate's passage just cited, bridge modern and modernist mathematics and physics, from the birth of modern mathematics. Silverman and Tate's statement that "in proving things one can work algebraically using coordinates, without worrying at all about geometrical intuitions" could have been made by Descartes, and it was one of his main points in his analytic geometry. The concept of an elliptic curve, especially when considered in its overall conceptual architecture, presented in their book, is strictly modernist, as is in fact is all algebraic geometry. This concept has other modernist dimensions, for example, by virtue of its Riemannian genealogy as (a) belonging to the theory of functions of a complex variable; (b) as, for each such a curve, both a two-dimensional topological (real) manifold and a one-dimensional complex manifold, to Riemann's theory of manifolds, central to the history of modernist geometry, and (c) as, topologically, a torus, a figure at the origin of topology, as a mathematical discipline. Both (b) and (c) manifest the modernist algebraization of geometry and topology, via spatial algebra, expressly, but it is found in (a), too, even if in a more oblique and subtler way. It might be added that a major part of Grothendieck's work in algebraic geometry, his theory of étale cohomology, discussed below, originates in Riemann's ideas of a covering space over a Riemann surface, one of Riemann's several great inventions. Algebraic curves, beginning with elliptic curves (the simplest abelian varieties) were the objects for which étale cohomology groups were established first, by an elegant calculation, exemplary of the mathematical technologies to which modernism gave rise [4, 5]).

Although their manifestation in Silverman and Tate's passage cited here is particularly notable because of echoing Descartes, these historical connections to the rise of modern algebra and analytic geometry (which algebraic geometry brings to its, for now, ultimate form) are not surprising. As all conceptions or undertakings, no matter how innovative, the modernist algebraization of mathematics has a history. While more prominent in the nineteenth century, the history of algebraization, at the very least, again, by way of practice, although it has, especially with Descartes, deep philosophical roots as well, begins with the mathematics at the outset of modernity, such as that of Fermat and Descartes.[5] Geometry was then still more dominant than algebra, and it had continued to be dominant for quite a while, even though this dominance diminished with the shift, often noted, of interest from geometry to algebra and number theory from around the time of Gauss, a key figure in this shift. Gauss' work was also central to the development of geometry during the same period and a major influence on Riemann's thinking concerning geometry, which, however, only testify to the rising significance of the relationships between algebra and geometry during the period leading to modernism. In any event, the possibility of making geometry algebraic (in either sense, that of algebraic geometry and the present one) entered mathematical thinking with Fermat and Descartes.

In some respects, the view of mathematics as fundamentally algebraic returns to *a* Pythagorean view of mathematics, which is not the same as *the* Pythagoreans' view, which was more arithmetical, although arithmetic is a form of algebra in its broad modern sense (the mathematical field-specificity of arithmetic or number theory, say, from Gauss on, is a separate issue). Geometry was of course a key part of Pythagorean mathematics. For one thing, it appears that these were the Pythagoreans (who exactly, is conjectural) who discovered the existence of incommensurable magnitudes by considering the diagonal of the square and thus in geometry, in effect by means of what I call spatial algebra, or proto-spatial algebra. (The "irrationals" in our algebraic language and, with it, our sense, are borrowed from Latin, and not Greek.) This discovery led to the crisis of ancient Greek mathematics. According to Heath's commentary: The "discovery of incommensurability must have necessitated a great recasting of the whole fabric of elementary geometry, pending the discovery of the general theory of proportion applicable to incommensurable as well as to commensurable magnitudes" ("Introductory Note," [22, v. 3, p. 1]; [53, pp. 416–417]). Thus, the *history* of mathematical modernism defined by algebra is very old, possibly as old as mathematics itself. On the other hand, thus combining, as history often does, the continuous and the discontinuous along different lines, the specific form this definition takes in modernism is a break with the past.

At stake, thus, is the rethinking of the very nature and practice of mathematics by making algebra a fundamental part of it, including topology and geometry, even in the cases of mathematicians whose thinking has a strong geometrical or topological orientation, such as Riemann, who figures centrally in this history. Riemann is, arguably, a unique case of a modernist combination of geometrical, topological, and algebraic thinking, further combined with real and complex analysis, and with number theory (the ζ-function and the distribution of primes) added to the mix, even though, as might be expected, various aspects of this Riemannian synthesis are found in the work on his predecessors, such as Gauss, Cauchy, Abel, and Dirichlet. I am referring not only to the multifaceted character of Riemann's work and his contributions to the interrelationships of these diverse fields in his work, but also and primarily to the significance of these interrelationships for modernism, which could,

[5]Descartes' *La Géométrie* was originally published as an appendix to his *Discourse on Method*, and it was part of a vast philosophical agenda that encompassed mathematics [18].

nevertheless, still be defined, *in these relationships*, in terms of the algebraization of mathematics. This situation makes Riemann's position in the history of modernism more complex, especially because of strong geometrical and topological dimensions of his thinking that resist algebraization, without, as I shall argue, diminishing his importance in this history but instead reflecting the complexity of this history and of modernism itself.

I shall further argue that this algebraization was often accompanied or even codefined by three additional, often interrelated, features, which are, along with the algebraization of the mathematics used, equally found in modernist physics. It is possible to define modernism in mathematics and physics by the presence of all four features. Doing so, however, would narrow modernism too much, as against seeing it in terms of modernist algebraization of mathematics, possibly accompanied by some or all of these additional features. These features are as follows.

The first feature, which gradually emerged throughout the nineteenth century, with Gauss, Abel, and Galois, as notable early examples, was a movement toward the independence and self-determination of mathematics as a field, especially its independence from physics and, with it, from the representation of natural objects. This feature has been seen as central to mathematical modernism by commentators who used the rubric and even defined it accordingly by Merthens and, following him, Gray [26, 44]. As will be seen presently, however, modernist mathematics, either in this or the present (algebraically oriented) definition, acquired a new, nonrepresentational, role in physics with quantum theory. This feature was closely related to the development of algebra, beginning with Gauss, Abel, and Galois; new, sometimes related, areas of analysis, such as the theory of elliptic functions; and then projective and finite geometries, in part following Riemann's work. Riemann's own thinking, as that of his teacher Gauss, retained close connections to physics, testifying to the complex nature of this history. As I explain below, this independence is also related to the independence of mathematics' from ordinary language and concepts, with which algebra could dispense more easily than geometry. This independence becomes crucial for modernist physics as well, especially quantum theory, which is essentially algebraic in character, in contrast to more geometrically oriented classical physics and relativity.

The second feature, discussed most explicitly in the final section of this article, is the role of technological thinking, in this case in considering mathematical technology in mathematics itself and in physics (where the use of mathematics is technological), in contrast to the dominance of ontological or realist thinking, defined by claims concerning *how* what exists or is claimed to exist actually exists. The "ontological" and "realist" are not always seen as the same, but their shared aspects allow these terms to be used interchangeably in the present context. On the other hand, the nouns "ontology" and "reality" will be used differently, because, as I shall explain presently, "reality" may be defined as disallowing ontology or realism.

The third feature, the emphasis on which, arguably, distinguishes most the present understanding of mathematical modernism from other concepts of mathematical modernism, is a radical form of epistemology linked to and in part enabled by the combination of other modernist trends: the modernist algebraization of mathematics, a movement toward the independence of mathematics from physics, and, especially, a shift from more ontological to more technological thinking, in the case of quantum theory (in the present interpretation) to the point of abandoning or even precluding ontological thinking altogether. As will be seen, ontological thinking (in this case concerning the ontology of thought rather than matter) retains a greater role in mathematics itself. In physics this epistemology, again, extends to the point of placing the ultimate constitution of reality (referring, roughly, to what exists) beyond a representation or even beyond conception, and thus beyond ontology, referring, as just noted, to such a representation or at least conception of the constitution of reality or existence, rather than merely to the fact that something exists. In this view, quantum objects or something in nature that compels us to think of quantum objects is assumed to exist, while no representation or even conception of what they ultimately are or how they exist is possible. That does not preclude thinking and knowledge in quantum theory or elsewhere, along with and in part enabled by surface-level ontologies (physical, mathematical, conceptual, and so forth) which enable this thinking and knowledge. On the other hand, any knowledge or even conception concerning and thus any ontology of the ultimate nature of reality is precluded. Thinking and knowledge would concern certain surface levels of reality, surface ontologies. Indeed, the unknowable or even unthinkable ultimate nature of reality is inferred from the effects it has at these surface levels. Importantly, however, this conception of reality as that which is beyond thought is still the product of thought. This conception is, moreover, interpretive in nature and the justification for assuming it is practical, and applicable only insofar as things stand now rather than is theoretically guaranteed to be true.

This epistemology, too, can be traced to Riemann's Habilitation lecture of 1854, "The Hypotheses That Lie at the Foundations of Geometry" [60], one of the founding works of mathematical modernism, as Riemann's view of the foundations of geometry is a radical reconceptualization of mathematics, pursued, correlatively, in his other works as well, such as that on the concept of a Riemann surface. In his Habilitation lecture, Riemann uses a remarkable phrase "a reality underlying space" [60, p. 33]. This phrase implies, on Kantian lines, that this reality may not be spatial in the sense of our usual phenomenal sense of spatiality: it could be discrete, for example. I am not contending that Riemann saw this reality as beyond representation (discrete or continuous, flat or curved, or three- or more-dimensional, all of which possibilities he entertained), let alone conception, any more than did Kant, a key figure in this history. While, in defining his epistemology by distinguishing noumena or things-in-themselves, as objects, and phenomena or representations appearing in our thought, Kant places things-in-themselves beyond representation or knowledge, he allows that a conception of them could be formed and, if logical, accepted for practical reasons, and even in principle be true, although this truth cannot be guaranteed [35, p. 115]. In its most radical form, the modernist epistemology, as defined in this article, in principle excludes that such as a conception can be formed, keeping in mind the qualifications just noted to the effect that this conception of reality as that which is beyond conception is still human and is only practically justified.

Both Riemann and then Einstein appear to have thought that an adequate mathematical representation of the ultimate nature of physical reality, a conception ideally close to the truth of nature, is, in principle, possible, as deduced from our experience and knowledge. This, for example, would allow one to conclude, against Kant's view, the geometry of the space is not Euclidean and its physics is not Newtonian, although of all people Kant might have been more open to this view than others, given new mathematics and science. A similar, more mathematically grounded, view was found in Heisenberg's later works. Heisenberg argued there that "the 'thing-in-itself' is for the atomic physicist, if he uses this concept at all, finally a mathematical structure; but this structure is—contrary to Kant—indirectly deduced from experience" [33, p. 91]. Kant's view was more complex and more open. It is impossible to know what Kant would have thought if he'd had been confronted with quantum physics, or, again, non-Euclidean geometry or relativity. If anything, his epistemology is closer to the one advocated here than just about any modern philosopher, apart from Nietzsche.

In any event, neither Riemann nor Einstein thought that the ultimate constitution of physical reality could be beyond conception altogether. This is the position adopted here in view of quantum mechanics and following Heisenberg (in his early work, as opposed to his later thinking) and N. Bohr, although neither might have assumed that quantum objects and behavior are beyond conception rather than only beyond representation and knowledge.[6] As I explain below, however, Heisenberg's and Bohr's positions are still different from that of Kant concerning phenomena vs. things-in-themselves, in this case, defining phenomena as what is observed in measuring instruments and objects as quantum objects, which cannot be observed, as effects they have on measuring instruments by interacting with the latter. Bohr, it could be noted in passing, was influenced in his interpretation of quantum mechanics in terms of what he called complementarity (a mutually exclusive nature of certain experiments we can perform and, correlatively, certain concepts we can use) by the concept of a Riemann surface as a way of dealing with multivalued functions of a complex variable [50, pp. 235–238].

Heisenberg's and Bohr's epistemology arises in part in view of the algebraic rather than, as in classical physics or relativity, geometrical, relationship between the mathematics of a physical theory and physical reality in its ultimate constitution, assumed by theory. This algebraic relationship between a (mathematical) physical theory and physical reality was no longer representational, because, in Bohr's words, "In contrast to ordinary mechanics, the new quantum mechanics does not deal with a [geometrical] space-time description of the motion of atomic particles," while, nevertheless, providing probabilistic or statistical predictions that are fully in accord with the available experimental evidence [8, v. 1, p. 48]. Eventually, Bohr argued, more radically, that "in quantum mechanics, we are not dealing with an arbitrary renunciation of a more detailed analysis of atomic [quantum] phenomena, but with a recognition that such an analysis is *in principle excluded*" [8, v. 2, p. 62]. The

[6]For a detail discussion of the subject, see the companion article by the present author [56].

true meaning of this statement is brought out by Bohr's view of the (irreducible) difference, following but ultimately reaching beyond Kant, between quantum *phenomena*, defined by what is observed in measuring instruments, and quantum *objects*, responsible for these phenomena, as effects of the interactions between quantum objects and measuring instruments, effects manifested in the measuring instruments. Bohr's statement, then, means that there could be no analysis that would allow us to represent, physically or mathematically, quantum objects and behavior, although Bohr might not have thought that they are beyond conception, which is an *interpretation* adopted here. It is, again, important that at stake here are interpretations, those of the type, indeed two types, just defined, amidst still other interpretations (some of which are realist) of quantum theory, and not the ultimate truth of nature, which we do not know and may never know or even conceive of, concerning which this article makes no claims.

It is worth noting that probability theory is fundamentally algebraic, as is, accordingly, its use in physics or elsewhere. Indeed, that probability theory is defined by the role of events, either in the real world or some model world, makes it akin to physics and its use of mathematics, in the case of quantum theory, in nonrealist interpretations, in effect making the latter a form of probability theory. The origin of probability theory, in the work of G. Cardano, B. Pascal and Fermat (who thus makes yet another appearance in the history of algebra) coincides with the emergence of algebra, as part of the rise of modernity. As I. Hacking persuasively argued in explaining why the theory emerged in the seventeenth century rather than earlier, some form of algebra was necessary for probability theory [28]. Quantum mechanics, however, at least, again, in nonrealist interpretations, reshaped, the relationships between the algebra of probability and the algebra of theoretical physics, as against previous uses of probability, for example, in classical statistical physics. There the relationships between them is underlain by a geometrical picture of the behavior of the individual constituents of the systems considered, assumed to follow the (causal) laws of classical mechanics. By contrast, as became apparent beginning with M. Planck's discovery of quantum phenomena in 1900, even elementary individual quantum objects and the events they give rise to had to be treated probabilistically. One needed to find a new theory to make correct probabilistic or statistical predictions concerning them. Heisenberg was able to accomplish this task with quantum mechanics, which only predicted the probabilities of what was observed in measuring instruments, considered as quantum phenomena, without representing the behavior of *quantum objects*, even the elementary ones, an imperative that had previously defined fundamental physics, including relativity.[7] This mathematics, never previously used in physics, was

[7]That, again, does not exclude either realist or causal interpretations of quantum mechanics or alternative theories of this behavior that are realist or causal. The so-called Bohmian mechanics is one example of such an alternative theory. Unlike quantum mechanics, however, Bohmian mechanics expressly violates the requirement of locality, which entered physics with relativity theory and which dictates that the instantaneous transmission of physical influences between spatially separated systems is forbidden.

essentially (Heisenberg did not initially use these terms) that of infinite-dimensional Hilbert spaces over $\mathbb{C}$, a modernist concept.

As I shall discuss later in this article, related epistemological considerations are relevant in considering modernist mathematics itself, as became apparent beginning at least with G. Cantor's set theory and became more pronounced in subsequent developments, such as those leading to K. Gödel's incompleteness theorems and P. Cohen's proof of the undecidability of Cantor's continuum hypothesis. In mathematics, moreover, it may not be possible to speak of the ultimate nature of reality, however inconceivable, as existing independently of thought, in the way one is able to do in quantum physics. There one might more readily assume the ultimate reality of matter that exists independently of us, say, as something that has existed before we were here and will continue to exist when we are no longer here, even if any conception concerning this reality or the impossibility of forming such a conception is a product of our thought and thus can only exist insofar as we exist.[8] On the other hand, while one might easily accept what we think of as real in our thought, assuming the existence of a single nonmaterial reality existing independently of our individual thinking is a more complicated matter. This is not to say that this type of assumption has not been made in mathematics, philosophy, or art, from Parmenides and Plato to the mathematical Platonism of the twentieth century (which was important to the project of the independence of mathematics and to mathematical modernism), with numerous Platonisms, whether so named or not, between them or after mathematical Platonism. Not many of them, certainly not twentieth-century mathematical Platonism, are the same as Plato's own Platonism.

The concept of curve, as it emerged in modernist mathematics, is, I argue here, exemplary, in some respects even uniquely exemplary, of the modernist situation outlined here, beginning with the modernist extension of the view of Riemann surfaces as curves over $\mathbb{C}$, which is only possible if one thinks of them spatial-algebraically. The mathematics of complex numbers was, especially, again, in and following Gauss' work, itself a crucial part of the history that eventually brought modern mathematics to modernist mathematics. This mathematics, too, is traceable to the origin of modern algebra, in considering the roots of polynomial equations, essentially related to the algebra of curves. However, the view that something (topologically) two-dimensional is a curve is essentially modernist. But then, as noted, in modernist mathematics, even something topologically zero-dimensional may be a curve, a situation anticipated by Riemann as well in considering discrete manifolds in his Habilitation lecture.

Between his work on Riemann surfaces, which are both (differentiable) manifolds and, while topologically two-dimensional, curves, and his ideas concerning the foundations of geometry, which properly grounded non-Euclidean, *curved*,

[8]The so-called many-worlds interpretation of quantum mechanics, which aimed to resolve some of the paradoxes of the theory in a realist and causal way, does not affect this point, because this kind of material reality is still retained within each world involved, and there are no connections between these worlds.

geometries, Riemann becomes a key figure for our understanding of the idea of curve, as he is for many developments of modernist mathematics. Riemann's thinking figures in quantum theory, too, by virtue of his introduction of the idea of an infinite-dimensional manifold, of which Hilbert spaces are examples. Riemann is the highest point of the arc, a curve, from Fermat and Descartes to our own time, via A. Weil, Grothendieck, and their followers, making the work of each of these figures, in Nietzsche's phrase, the mathematical "philosophy of the future," a subtitle of his 1888 *Beyond Good and Evil: A Prelude to a Philosophy of the Future* [46]. As most of Nietzsche's works, it belongs to the time around the rise of modernism, which it influenced in philosophy, and literature and art, or even in mathematics, as in the case of F. Hausdorff [26, p. 222], and physics, as, likely, in the case of Bohr [54, p. 116].

5.3 Fundamentals of Mathematical Modernism

I would like now to establish more firmly the key concepts that ground my view of mathematical modernism, as sketched in the Introduction, beginning with modernity and modernism. "Modernity" is customarily seen, and will be seen here, as a broad cultural category. It refers to the period of Western culture extending from about the sixteenth century to our own time: we are still modern, although during the last 50 years or so, modernity entered a new stage, sometimes known as postmodernity, defined by the rise of digital information technology.[9] Modernity is defined by several interrelated transformations, sometimes known as revolutions, although each took a while. Among them are scientific (defined by the new cosmological thinking, beginning with the Copernican heliocentric view of the Solar system, and the introduction by Descartes, Galileo, and others, of the mathematical-experimental science of nature); industrial or, more broadly, technological (defined by the transition to the primary role of machines in industrial production and beyond); philosophical-psychological (defined by the rise of the concept of the individual human self, beginning with Descartes' concept of the Cogito); economic (defined by the rise of capitalism); and political (defined by the rise of Western democracies).

One might add to this standard list, in which Descartes figures prominently already, the mathematical revolution, which is rarely expressly discussed as such, although it figures in discussions of the rise of modernity as part of the scientific revolution and, occasionally, because of the invention of calculus and then probability theory, both seen as defined by a modern way of thinking. The rise of algebra was, however, equally important in this revolution and conceptually fundamental because algebra was also crucial to the discoveries and developments of calculus and probability theory, in which calculus came to play a major role as

[9]Thus, postmodernity was also epistemologically shaped by certain developments in mathematics and science, most of which are modernist in the present sense (e.g., [37]).

well. Algebra was the defining aspect of modern mathematics and physics, although geometry remained dominant for a quite while in both and has never, including in modernism, entirely lost its independence and importance. Thus, while the laws of classical mechanics, embodied in its equations, are algebraic (all equations are), they are grounded in a geometrical picture of the world, including the curved motion of classical bodies, such as, paradigmatically, planets moving around the Sun, although analytic geometry or algebraic laws of classical mechanics added algebra to this geometry. This type of modern geometrical thinking will continue to define physics, including Einstein's relativity (although it does have modernist aspects as well), until quantum mechanics and its modernist algebraic approach, introduced by Heisenberg.

As I noted at the outset, in contrast to modernity, "modernism" has been primarily used as an aesthetic category, referring to certain developments in literature and art in the first half of the twentieth century, from roughly the 1900s on, represented by such figures as Stéphane Mallarmé, W. B. Yeats, Ezra Pound, James Joyce, Franz Kafka, Reiner M. Rilke, Virginia Woolf, and Jorge Luis Borges in literature; Pablo Picasso, Wassily Kandinsky, and Paul Klee, in art; and Arnold Schoenberg and Igor Stravinsky in music. On occasion, it has been applied to the philosophy of, roughly, the same period, such as that of Nietzsche, Bergson, Husserl, and Heidegger. Gray considers Husserl in the context of the foundations of mathematics and mentions Nietzsche because of Hausdorff's interest in him [26, p. 222], but he does not discuss modernism in philosophy. The denomination has rarely been used in considering mathematics and physics, or science, as opposed to "modern," used frequently, but with different periodizations. In mathematics, "modern" tends to refer to the mathematics that had emerged in the nineteenth century, with the likes of Gauss, Abel, Cauchy, and Galois, and then developed into the twentieth century, thus overlapping with modernist mathematics in the present definition. In fact, the term "modern algebra" was introduced, referring essentially to abstract algebra (presented axiomatically), as late as 1930 by van der Waerden in his influential book under this title, based on the lectures given by Emil Artin and E. Noether [65]. In physics it refers to all mathematical-experimental physics, from Galileo and Descartes on, which is fitting because this physics emerged along with and shaped the rise of modernity as a cultural formation, as just explained, making it fundamentally scientific. After the discovery of relativity and quantum theory, the term "classical physics" was adopted for the preceding physics, still considered modern, by virtue of its mathematical-experimental character. The present article, by contrast, uses the designation modern for the mathematics emerging at the same time. If modernity is scientific, it is also because it is mathematical. As Heidegger argued in commenting on Galileo and Descartes, "modern science is experimental because of its mathematical project" [29, p. 93]. Thus, it was the concept of the second-degree curve that supported and even defined the experimental basis of physics and astronomy, in Kepler, Galileo, and Descartes, who gave these mathematics "coordinates," the concept central to all modern and then modernist physics.

Using of the term "modernism" in considering, historically and conceptually, mathematics and science is, as noted, still quite infrequent. Two most prominent examples, mentioned from the outset of this article, are H. Mehrtens' 1990 *Moderne Sprache, Mathematik: Eine Geschichte des Streits um die Grundlagen der Disziplin und des Subjekts formaler Systeme* [44] and, in part following Mehrtens' book (but also departing from it in several key respects), Gray's 2007 *Plato's Ghost: The Modernist Transformation of Mathematics* [26]. Gray's conception of modernism covers developments in topology, set theory, abstract algebra, mathematical logic, and foundations of geometry that had reached their modernist stage around 1900, focusing most on geometry, with Hilbert's Foundations of Geometry as his conceptual center in this regard, and on logical, especially set-theoretical, foundations of mathematics, where Hilbert again, figures centrally. In sum, for Gray, the most representative figure of mathematical modernism is Hilbert. By contrast, in the present view of modernism (defined differently), it is Riemann, while still by and large respecting the chronology of modernism adopted by Gray, a chronology contemporaneous with the rise of literary or artistic modernism.[10]

Gray briefly comments on literary and artistic modernism, and his title comes, not coincidentally, from that of a poem by Yeats, one of the major modernist poets. Gray only minimally considers these connections (e.g., [26, p. 185]). He prefers to focus on mathematics and the philosophy of mathematics. He could in my view

[10]The modernist aspects of Riemann's work, equally in Gray's definition of modernism, pose difficulties for Gray, because Riemann preceded modernism by several decades [26, p. 5]. It is not a problem for the present argument, firstly, because the present view of modernism is different, and, secondly, because modernism is seen here as more continuous with modern mathematics from Fermat and Descartes on, a longer history in which Riemann's work is a decisive juncture. This continuity is recognized by Gray, but it seems to worry him because it disturbs the stricter chronology he considers. The present view emphasizes, in part following G. W. F. Hegel, the conceptual over the chronological, even in historical considerations. Gray, in addition, appears to see the axiomatic, not central for Riemann (in contrast to Hilbert), rather than the conceptual, as more characteristic of modernism. In the present view, modernism is more about concepts and their history than about the chronology of events or developments, such as those associated with the spreading of modernist thinking or practices. This chronology cannot of course be disregarded, but a concept or a form of practice in a given field can precede a chronologically defined state of this field, with which this concept or practice would be in accord. This accord is not an "anticipation" but a determinate quality of a concept or a form of practice. Riemann's concepts and practice are modernist, in the present (or, with some differences, Gray's) definition, and a similar claim could be made, helped by his revolutionary algebraic thinking, concerning Galois. The degree or even the existence of such an accord, or to what degree this accord reflects the understanding of this concept by its inventor, is a matter of interpretation, which could be contested. Riemann's thinking has complexities when it comes to the role of algebra there because of the topological and geometrical aspects of his thinking, which often take the center stage, while algebra, when still present, appears in a supporting role. This is, however, only so in a more narrow or technical sense, as opposed to the broader sense assumed here as defining modernism. Riemann's work, as noted, is defined by the joint workings of geometry, topology, algebra, and analysis in his mathematics, added by philosophical and physical, aspects of his thinking. Hilbert made major contributions in all these areas as well (apart from topology), but one does find the same type of fusion of different fields dealing with a given subject that one finds in Riemann, as in the case of Riemann surfaces.

have given more attention to physics, especially quantum theory, which he by and large bypasses. (Gray does comment on relativity.) Unfortunately, especially given the role of the image of the curve in modernist art, such as that of Klee (e.g., [17, pp. 14–15]), I cannot address the connections between modernist mathematics or sciences and modernist literature and art in detail either. I would, however, argue, more strongly than Gray, for the validity of these connections in sharing some of the key conceptual features. This view follows Bohr, who said, in speaking of quantum theory: "We are not dealing here with more or less vague analogies, but with an investigation of the conditions for the proper use of our conceptual means shared by different fields" [8, v. 2, p. 2]. It is not merely a matter of traffic, for example, metaphorical, between fields, but of parallel situations in each that justifies the use of the term modernism in considering them. Indeed, this article explores this type of parallel between modernist mathematics and quantum theory, mathematically equally defined by the role of algebra in them.[11] That does not of course mean that the specificity of each field, such as that of mathematics vs. physics, or that of either vs. that of literature and art (or that of literature vs. that of art) is dissolved even in considering such parallel situations, let alone in general. Such parallels often give new dimensions to this specificity, for example, as I argue, in the case of modernist mathematics and modernist physics in bringing out the fundamentally algebraic character of both.

Although Gray's concept of mathematical modernism is different from the one adopted here, there are relationships between them. These relationships are complex and considering them in detail would be difficult. While Gray offers a discussion of modernist algebra (which would of course be impossible to avoid), he does not address, except occasionally and mostly by implication, the modernist algebraization of mathematics, including geometry and topology. In fact, some key developments in modernist algebra, too, are not given by Gray the attention they deserve, such as Noether's work in algebra, one of the great examples of mathematical modernism, central to more abstract developments of algebraic topology (as in H. Hopf's work) and a bridge between R. Dedekind and Grothendieck, helpfully discussed by C. McLarty [42]. Gray also largely bypasses epistemological considerations central to the present analysis. Gray acknowledges the connections in mathematical modernism in the case of relativity [26, p. 324, n. 28]. But he misses nonrealist thinking found in quantum theory, which connects physical reality in its ultimate constitution with mathematics without recourse to realism. In fairness, related epistemological aspects of modernism are suggested by Gray in the context

[11]The nature of these connections and, in part correlatively, the effectiveness of using the term modernism, specifically by Mehrtens and Gray, have been questioned, for example, by S. Feferman [24] and L. Corry [15]. While both articles (that of Feferman is a review of Gray's book) make valid points, I don't find them especially convincing on either count, in part because their engagement with modernist art is extremely limited and because neither considers the epistemological dimensions of modernism, which are, in my view, important in addressing these connections. For an instructive counter argument to Mehrtens, challenging his historical claims, specifically those concerning F. Klein, see [6].

of Cantor's set theory and logical foundations of mathematics, especially Cantor's continuum hypothesis and Gödel's theorems. There is no discussion of quantum theory either.

On the other hand, the project of the *independence* and *self-determination* of mathematics is central to Gray, as it was to Mehrtens (whose views I shall put aside). This trend had, as I said, been gradually emerging throughout the nineteenth century. This independence is especially manifested as an independence from physics or, more generally, from considering mathematical objects as idealizations from natural objects, the type of idealization that was central to physics and its use of mathematics from Descartes and Galileo on. As I argue here, however, quantum theory, by its algebraic nature, also established new, nonrepresentational, relationships between mathematics and physics, and thus mathematics and nature, by using modernist mathematics. It was, echoing the literary parallel, the end of realism and the beginning of modernism, and not only echoing because similar relationships between representation and reality emerged in literary or artistic modernism. Quantum mechanics did not diminish faith in the classical ideal. Einstein or E. Schrödinger, the coinventor of quantum mechanics (and Einstein, too, made momentous contributions to quantum theory), never relinquished the hope that this ideal would be eventually restored to fundamental physics. Their uncompromising positions have served as inspirations for many others who share this hope, in fact a majority among physicists and philosophers alike. Einstein won this *philosophical* part of his debate with Bohr. *Physics* is a different matter. The question, which was the main question in the Bohr-Einstein debate, is whether nature would allow us a return to realism. While Einstein thought that it should, Bohr thought that it *might not*, which is not the same as it never will. As our fundamental theories are manifestly incomplete, especially given that of quantum field theory, our best theory of the fundamental forces of nature (electromagnetism, the weak force, and the strong forces) apart from gravity, and general relativity, our best theory of gravity, are inconsistent with each other, the question remains open, and the debate concerning it continues with undiminished intensity.

By contrast, mathematical realism and, especially, mathematical Platonism (a modernist development, which is, as I said, only superficially related to Plato's thought) has been important for the project of the independence of mathematics. This project had been developing as part of *modern* mathematics, but by 1900, with the rise of mathematical modernism, it reached the stage of breaking with connections representing or idealizing natural objects in all areas of mathematics, notably in geometry, making it "profoundly counterintuitive." "This realization," Gray contends, "marks a break with all philosophy of mathematics that present mathematical objects as idealizations from natural ones: it is characteristic of modernism" [26, p. 20].

The history of realization is much longer and, to some degree or in some of its aspects, it began even with the emergence of mathematics itself, including geometry, but it was certainly quite advanced by 1800 or thereabout, with non-Euclidean geometry as part of it (e.g., [27]). Indeed, this history may also be seen as that of divorcing mathematical concepts from our general phenomenal intuition as well, culminating in modernism and more characteristic of it (than a break with the view of mathematical objects as idealization from natural ones, although there are connections between these two breaks). In this divorce, algebra, including spatial algebra, has, I argue, played a major role. This divorce was stressed by H. Weyl, himself a major figure of mathematical modernism. Weyl made his point in his 1918 book, *The Continuum* [69], following closely his 1913 book on Riemann surfaces [73], and followed even more closely by his 1918 classic on the mathematics of relativity, *Space Time Matter* [68].[12] All three books were linked by their shared modernist problematic, as defined here, keeping in mind that Weyl's own position was realist, which was, however, common among mathematical modernists. The idea of curve, too, was equally crucial in all three contexts—Riemann surfaces, the continuum, and, again, via Riemann, Einstein's general relativity. According to Weyl: "the conceptual world of mathematics is so foreign to what the intuitive continuum presents to us that the demand for coincidence between the two must be dismissed as absurd" [69, p. 108]. "Coincidence" is not the same as "relations," which, as noted above, are unavoidable, at least insofar as it is difficult to think of continuity, spatially, apart from one or another phenomenal intuition of it. Even algebra involves general phenomenal intuition, even a spatial one, for example, in considering matrices as arrangements of symbols, which was crucial to Heisenberg's discovery of quantum mechanics, in the course of which he reinvented matrix algebra through so arranging certain mathematical elements involved [52, pp. 30–31]. On the other hand, it is entirely possible to define a given continuum, such as that of a line or curve, algebraically. This situation emerged with Cantor's introduction of set theory, and the multiplicity of infinities, the infinity of infinities, there, and his continuum hypothesis, and then the discovery of esoteric objects, such as Peano's curve, and related developments leading to Gödel's incompleteness theorem, and finally Cohen's proof of the undecidability of the continuum hypothesis, which brought new, ultimately irresolvable, complexity to the idea of continuum.[13] In sum, we do not, and even cannot, know how a continuous line, straight or curved (which does not matter topologically), is spatially constituted by its points, but we have algebra to address this question, and have a proof that the answer is rigorously undecidable. I shall further address the philosophical underpinning of this situation in the final section of this article. This history and related modernist developments, such as the concept of dimension (which Cantor's rethinking of the concepts of continuum required) is extensively considered by Gray, confirming Weyl's point, made as part of his own important contribution to this history in *The Continuum*.

Weyl's point concerning the conceptual world of mathematics as unavailable to our general phenomenal intuition or, by implication, ordinary language, exceeds the question of the continuum, and pertains to most of modernist mathematics

[12]Weyl's classic book had undergone several editions, some of them with significant revisions. I cite here the last edition.

[13]Intriguingly, Cohen ultimately thought that the hypothesis was likely to be false [12, p. 151].

or physics, such as relativity and quantum theory, which was even able to take advantage of this divorce between mathematics and our general phenomenal intuition, or a representation of natural objects. Weyl added to his statement cited above: "Nevertheless, those abstract schemata supplied us by mathematics must underlie the exact sciences of domains of objects in which continua play a role" [69, p. 108]. This comment was undoubtedly made with Einstein's relativity in mind, as Weyl's next book, *Space Time Matter*, was already in the works [68]. While his careful formulation implies the representational role of such schemata, it has a modernist twist given that those abstract schemata are, mathematically, divorced to one degree or another from our common phenomenal intuition, which entails the use of algebra, in the case of Einstein's general relativity, based in Riemannian geometry of differentiable manifolds as a form of spatial algebra.

This type of divorce, I argue, has been equally at work in modernist mathematics and modernist physics, and, while prepared by previous developments, such as analytical mechanics and Maxwell's electromagnetism, it reaches its modernist stage with relativity, beginning with special relativity, and especially quantum mechanics. Both used modernist mathematics, respectively, that of Riemannian geometry (Minkowski's spacetime of special relativity is a pseudo-Riemannian manifold) and that of infinite-dimensional Hilbert spaces, equally mathematically divorced, as abstract continuous schemata, from our general phenomenal intuition. Relativity still did this in a realist way, as the break from our phenomenal intuition does not entail a divorce from realism or ontology, because the latter could be mathematical. In fact, this ontology has been mathematical in all modern physics from Descartes and Galileo on, even when it is supplemented by or, in classical mechanics, originates in our phenomenal intuition. In addition, as noted above, one could still use one's phenomenal, including geometrical, intuition heuristically, to help our thinking, or, to return to Tate's principle, one could still think geometrically, as well as spatial-algebraically, while proving things or (which is not the same) making more rigorous arguments and calculations in physics, algebraically. Still, relativity entailed a radical departure from classical physics. For one thing, as Weyl was of course aware, the relativistic law of addition of velocities (defined by the Lorentz transformation) in special relativity, $s = (v + u)/(1 + \frac{1}{c^2}vu)$, for collinear motion ($c$ is the speed of light in a vacuum), runs contrary to any intuitive (geometrical) representation of motion that we can have. This concept of motion is, thus, no longer a mathematical refinement of a daily concept of motion in the way the classical concept of motion is. Relativity was the first physical theory that defeated our ability to form a phenomenal conception of an elementary physical process. But it still allowed for a mathematical and conceptual representation of physical reality.

Quantum mechanics, by contrast, only used mathematics for providing probabilistic predictions concerning the outcomes of quantum experiments, quantum events, without providing a representation or even conception of the processes responsible for these events, in which case geometrical intuitions are of no help to us at all. At most we can have spatial algebra. This gives an entirely new role to abstract continuous schemata, such as those of Hilbert spaces, in physics, that of predicting, probabilistically, the outcome of irreducibly discrete events. These predictions are, moreover, made possible by rules *added* to the formalism rather than being part of it, such as Born's rule, which relates, essentially by using complex conjugation, complex quantities of formalism to real numbers corresponding to the probabilities of quantum events. Basically, one takes the square moduli of the eigenvalues of the operators associated with quantum variables, such as position, momentum, or energy (or equivalently, multiply these eigenvalues by their complex conjugates), which gives one real numbers, corresponding, once suitably normalized, to the probabilities of observed events, associated with the corresponding measurements. The standard rule for adding the probabilities of alternative outcomes is changed to adding the corresponding amplitudes and deriving the final probability by squaring the modulus of the sum. The algebra of probabilities changes!

The modernist situation just outlined bears significantly on the question of language in quantum theory and in mathematics itself. It is helpful to briefly consider first this role in mathematics, beginning with geometry, where ordinary language or ordinary concepts have always played a greater role than in algebra, beginning with Euclid's *Elements* and its very first definition: "A point is that which has no part" [22, v. 1, p. 15]. At the same time, there is a movement, enabled by and enabling mathematics, away from ordinary language and concepts, because what makes "points," or "lines," part of geometry as mathematics are not their definition but the relationships between and among them in Euclidean geometry, a fact on which Hilbert capitalized two millenia later. Descartes' "geometry," as presented in *La Géométrie*, offers an important contrast to both Euclid and Hilbert alike. It is not axiomatic but "problematic," as well as, correlatively, algebraic. It primarily deals with problems (some of which may be theorems in the usual sense), thus nearly erasing Euclid's distinction, which is difficult to sustain, between problems and theorems. There is an affinity with Riemann in this regard. But there is also a major difference. Riemann thinks in terms of concepts [55]. Descartes thinks in terms of equations, and points and lines are understood accordingly. Modern algebraic geometry will eventually bring Descartes and Riemann together, with Grothendieck's work as the culmination of this history.

While not axiomatic, Descartes' thinking suggests the possibility of a different, more algebraic, axiomatization, which was part of the project of mathematical modernism (both in the present and in Gray's definition), as manifested in Hilbert's *Foundations of Geometry*, first published in 1899. Hilbert's often cited earlier remark, apparently made in 1891, offers an intriguing angle: "One must be able to say at all times—instead of points, straight lines, and planes—tables, chairs, and beer mugs" [71, p. 635]. What Hilbert exactly had in mind is not entirely certain and has been interpreted in a variety of ways. Without attempting to give it a definitive interpretation, my reading would be as follows, in accordance with Weyl's point concerning the conceptual world of mathematics as foreign to that of our general phenomenal intuition. Hilbert uses his example only to indicate that both sets are that of connected entities, and that one should properly speak of neither "points, straight lines, and planes," as geometry did from Euclid on, nor "tables, chairs, and beer mugs," nor anything else referred to by means of ordinary language, but instead use algebraic symbols and algebraic relationships between them, without referring to any objects in the world represented by ordinary language.[14] Ordinary language, however, still plays an important role and, arguably, cannot be entirely dispensed with. While we can replace points, straight lines, and planes, and relationships between them with symbols, it may be difficult for our thought, at least our unconscious thought, but perhaps even our conscious thought, to replace them with tables, chairs, and beer mugs. As explained earlier, a decade later, Hilbert's *Foundations of Geometry* tells us as much. Still, with modernism, mathematics and physics break with ordinary language or thinking more deliberately and radically, as expressed in Weyl's 1918 remark just considered. In physics, this break, although gradually emerging earlier as well, becomes pronounced with quantum mechanics, where it became complete in considering quantum objects and behavior, in nonrealist interpretations.

According to Heisenberg, "it is not surprising that our language [or concepts] should be incapable of describing the processes occurring within atoms, for . . . it was invented to describe the experiences of daily life, and these consist only of processes involving exceedingly large numbers of atoms. It is very difficult to modify our language so that it will be able to describe these atomic processes, for words can only describe things of which we can form mental pictures, and this ability, too, is a result of daily experience" [32, p. 11]. Words can do more, including make the statement that tells us that words cannot describe the processes occurring within atoms, which, however, does not undermine Heisenberg's main point. It follows that, while classical physics, at least classical mechanics, may rely on, and was born from, a mathematical refinement of our daily phenomenal intuition, concepts, and language, atomic physics can no longer do so. However, as Heisenberg realized in his discovery of quantum mechanics, it can still use mathematics. As just discussed in considering Weyl's argument, relativity and the preceding quantum theory, or even some developments of classical physics have, with the help of mathematics, already broken, at least in part, with our daily intuition and concepts. Heisenberg clearly realized this. As he said, following the passage just cited: "Fortunately, mathematics is not subject to this limitation, and it has been possible to invent a mathematical scheme—the quantum theory [e.g., quantum mechanics]—which seems entirely adequate for the treatment of atomic processes" [32, p. 11]. Mathematics allows one to circumvent the limits of our phenomenal representational intuition, also involving visualization, sometimes used, including by Bohr, to translate the German word for intuition, *Anschaulichkeit*. "Visualization" and its avatars are often invoked by Bohr, by way of this translation, in considering quantum objects and behavior, as being beyond our capacity to phenomenally represent them (e.g., [8, v. 1, pp. 51, 98–100, 108, v. 2, p. 59]). Ultimately, Bohr came to see quantum objects and behavior as being beyond any representation, if not conception (a view adopted here), including a mathematical

[14]I borrow the juxtaposition between Hilbert's remark and Euclid's definition of a point from G. E. Martin [40, p. 140], who, however, only states this juxtaposition without interpreting it.

one, a view adopted by Heisenberg at the time of the comments just cited. By contrast, as I noted, in his later thinking Heisenberg appears to be more open to the possibility of a mathematical representation of the ultimate structure of matter, still, however, in the absence of a physical representation of it, a form of strictly mathematical realism or Platonism. (e.g., [33, pp. 91, 147–166]).

That Heisenberg *found* a mathematical scheme that could predict the data in question was as fortunate as that mathematics is free of this limitation, for, as just noted and as Heisenberg must have realized, this freedom is also at work in relativity or even classical physics, beginning at least with Lagrange's and Hamilton's analytical mechanics. It is true that matrix algebra was introduced in mathematics before Heisenberg, who was, again, unaware of it and reinvented it, although the unbounded infinite matrices that he used were not previously studied in mathematics and were given a proper mathematical treatment by M. Born and P. Jordan later [9]. But, even if Heisenberg had been familiar with it, his scheme would still have needed to be invented as a mathematical model dealing with quantum phenomena. Heisenberg discovered that this was possible to do in terms of probabilistic or statistical predictions in the absence of any representation or even conception of quantum objects and their behavior. Indeed, mathematics now becomes primary in an even more fundamental sense than in its previous use in physics. This is because, given that we have no help from physical concepts, mathematics is our only means to develop the formalism we need. Quantum physics does contain an irreducible nonmathematical remainder because no mathematics can apply to quantum objects and behavior. But then, nothing else, physics or philosophy, for example, could apply either. Heisenberg's key physical intuition was that there could be no physical intuition that could apply to quantum objects and processes, while one could use mathematics to predict the outcomes of experiments, thus redefining the relationships between mathematics and physics.

This redefinition was grounded in the primacy of algebra, moreover, not only as against classical physics and relativity but also as against the preceding quantum theory, specifically, Bohr's 1913 atomic theory, initially, as that of the hydrogen atom. The theory retained a geometrical, orbital, representation of electrons' motion in so-called stationary states, even though it renounced any mechanical conception of transitions between such states. It had its Keplerian, "Harmonia-Mundi," appeal (Bohr's orbits were elliptical, too) in defining the ultimate microscopic constitution of nature. Developed by Bohr and others to apply to more complex atoms, the theory had major successes over the next decade. However, it ran into formidable problems and proved to be inadequate as a fundamental theory of atomic constitution. To rectify the situation Heisenberg made an extraordinary move, unanticipated at the time, because nearly everyone was expecting a return to a more geometrical picture partially abandoned by Bohr's theory. Against these expectations, in Heisenberg's scheme there were no orbits anymore but only states of quantum objects, states, moreover, never accessible as such and hence not available to a theoretical representation, but only manifested in their effects on measuring instruments. This, however, still allowed his theory to predict the probabilities of what can be observed in quantum experiments, which became the core of Heisenberg's approach.

Even before his paper announcing his discovery was published [31], Heisenberg explained this "more suitable" concept as follows: "What I really like in this scheme is that one can really reduce *all interactions between atoms and the external world ... to transition probabilities*" between quantum measurements (Heisenberg, Letter to Kronig, 5 June 1925; cited in [43, v. 2, p. 242]; emphasis added). It was Heisenberg's renunciation of any geometrical representation of quantum objects and behavior, thus replacing the geometry of curves with the algebra of probabilities, that led him to the discovery of quantum mechanics.

As he says at the outset of his paper: "in quantum theory it has not been possible to associate the electron with a point in space, *considered as a function of time*, by means of observable quantities. However, even in quantum theory it is possible to ascribe to an electron the emission of radiation" [31, p. 263; emphasis added]. The effect of such an emission could be observed in a measuring instrument and its occurrence can be assigned probability or (if the experiment is repeated many time) statistics. My emphasis reflects the fact that, in principle, a measurement could associate an electron with a point in space, but not by linking this association to a function of time representing the continuous motion of this electron, in the way it is possible in classical mechanics. If one adopts a nonrealist interpretation, one cannot assign any properties to quantum objects themselves, not even single out such properties, such as that of having a position, rather than only certain joint ones, which are precluded by the uncertainty relations. One could only assign physical properties to the measuring instruments involved. On the other hand, Heisenberg's approach put into question the privileged *position* that the position variable had previously occupied in physics. Heisenberg described his next task as follows, which shows the genealogy of his derivation in Bohr's atomic theory:

> In order to characterize this radiation we first need the frequencies which appear as functions of two variables. In quantum theory these functions are in the form:
>
> $$\nu(n, n-\alpha) = 1/h\{W(n) - W(n-\alpha)\}$$
>
> and in classical theory in the form
>
> $$\nu(n, \alpha) = \alpha\nu(n) = \alpha/h(dW/dn)$$
>
> [31, p. 263]

This difference leads to a difference between classical and quantum theories as regards the combination relations for frequencies, which, in the quantum case, correspond to the Rydberg-Ritz combination rules, again, reflecting, in Heisenberg's words, "the discrepancy between the calculated orbital frequency of the electrons and the frequency of the emitted radiation." However, "in order to complete the description of radiation [in correspondence, by the mathematical correspondence principle, with the classical Fourier representation of motion] it is necessary to have not only frequencies but also the amplitudes" [31, p. 263]. On the one hand, then, by the correspondence principle, the new, quantum-mechanical equations must formally contain amplitudes, as well as frequencies. On the other hand, these amplitudes could no longer serve their classical physical function (as part of a continuous representation of motion) and are instead related to discrete transitions between stationary states. (Nor ultimately do frequencies because of the non-classical character of the Rydberg-Ritz combination rules.) In Heisenberg's theory and in quantum theory since then, these "amplitudes" are no longer amplitudes of physical motions, which makes the name "amplitude" itself an artificial, symbolic term. Linear superposition in quantum mechanics is of a fundamentally different nature from any superposition found in the classical wave theory. In nonrealist interpretations, this superposition is not even physical: it is only mathematical. In classical physics this mathematics represents physical processes; in quantum mechanics it does not. Amplitudes are instead linked to the probabilities of transitions between stationary states: they are what we now call probability amplitudes. The corresponding probabilities are derived, from Heisenberg's matrices, by a form of Born's rule for this limited case. (Technically, one needs to use the probability density functions, but this does not affect the main point in question.) One can literally see here a conversion of the classical continuous geometrical picture of oscillation or wave propagation, as defined by frequencies and amplitudes, into the algebra of probabilities of transitions between discrete quantum events.

Algebra, in part as the spatial algebra of Hilbert spaces, was the mathematical technology of predictions concerning the outcomes of quantum experiments, eventually, with quantum field theory, in high energy (relativistic) quantum regimes, in the absence of mathematical ontology of the ultimate reality, defined by the quantum constitution of nature, an ontology found in relativity or classical physics before it. Quantum electrodynamics is the best experimentally confirmed physical theory ever. It was the triumph of "the Heisenberg [algebraic] method," as Einstein characterized it in 1936, while still skeptical about its future, a decade of major successes of quantum mechanics notwithstanding. Even apart from the fact that Einstein's unwavering discontent with quantum mechanics and his debate with Bohr concerning it were a decade long by then as well, Einstein's assessment of Heisenberg's algebraic method was hardly unexpected given Einstein's preference for realism and geometry. As he said: "[P]erhaps the success of the Heisenberg method points to a purely algebraic method of description of nature, that is, to the elimination of continuous functions from physics. Then, however, we must give up, in principle, the space-time continuum [at the ultimate level of reality]. It is not unimaginable that human ingenuity will some day find methods which will make it possible to proceed along such a path. At present however, such a program looks like an attempt to breathe in empty space" [20, p. 378]. For some, by contrast, beginning with Bohr, Heisenberg's method was more like breathing fresh mountain air. The theory has been extraordinarily successful and remains our standard theory of quantum phenomena in both low and high-energy quantum regimes, governed by quantum mechanics and quantum field theory respectively.

A few qualifications are in order, however. First of all, one must keep in mind the complexity of this algebra, which involves objects that are not, in general, discontinuous, although certain key elements involved are no longer continuous functions, such as those used in classical physics, and are replaced by Hilbert-space operators (over $\mathbb{C}$). Some continuous functions are retained, because the Hilbert spaces involved are those of such functions, considered as infinite-dimensional vectors in dealing with continuous variables such as position and momentum, keeping in mind that these variables themselves are represented by operators. These functions, vectors, are those of complex (rather than, as in classical physics, real) variables and the vector spaces that they comprise, or associated objects such as operator algebras, have special properties, such as, most crucially, noncommutativity. These vectors, of which Schrödinger's wave function is the most famous example, play an essential role in calculating (via Born's rule) the probabilities of the outcomes of quantum experiments. In fact, given that they deal with Hilbert spaces, quantum mechanics and quantum field theory involve mathematical objects whose continuity is denser than that of regular continua such as the (real number) spacetime continuum of classical physics or relativity. In contrast to these theories, however, the continuous and differential mathematics used in quantum theory, along with the discontinuous algebraic one, relates, in terms of probabilistic predictions, to the physical discontinuity defining quantum phenomena, which are discrete in relation to each other, while, at least in nonrealist interpretations, quantum objects and their behavior are not given any physical or mathematical representation or even conceptions—continuous or discontinuous.

Thus, as Bohr was the first to fully realize, Heisenberg's algebraic method brings about a radical change of our understanding of the nature of physical reality, an understanding ultimately depriving us not of reality but of realism, which was, for Einstein, the most unpalatable implication of Heisenberg's method. In saying that "we must give up, in principle, the space-time continuum," Einstein must have had in mind the spacetime continuum in representing, by means of the corresponding theory, the ultimate reality considered, and possibly in attributing the spacetime continuum to this reality, something, defining his geometrical philosophy of physics (embodied in general relativity), that Einstein was extremely reluctant to give up. The idea that this reality may ultimately be discrete had been around for quite a while by then. In particular, it was, as noted, proposed by Riemann as early as 1854, speaking of "the reality underlying space" [60, p. 33]. It was Riemann's concept of continuous (actually, differentiable) manifolds and Riemannian geometry this concept defined that grounded Einstein's general relativity and his view of the ultimate nature of physical reality as the spacetime continuum, threatened by quantum theory. The idea of the discrete nature of ultimate reality has acquired new currency in view of quantum mechanics and quantum field theory, as advocated by, among others, Heisenberg in the 1930s, and is still around. In the present view, the ultimate nature of physical reality is beyond representation and even conception (neither Bohr nor Heisenberg might, again, have been ready to go that far) and, as such, may not be seen as either continuous or discontinuous. Discreteness only pertains to quantum phenomena, observed in measuring instruments, while continuity has no physical significance at all. It is only a feature of the formalism of quantum mechanics, which at the same time relates to discrete phenomena by predicting the probabilities or statistics of their occurrence.

While Kant's philosophy may be seen as an important precursor to this epistemology, beginning with the difference between objects and phenomena as its basis, in its stronger form, which places the ultimate nature of reality beyond conception (rather than only beyond representation or knowledge), this epistemology is manifestly more radical than that of Kant. This because, as I explained, in Kant's epistemology, noumena or objects as things-in-themselves are, while unknowable, still in principle conceivable and that conception might even be true, even though there is no guarantee that it is true [35, p. 115]. Even if Bohr adopted a weaker view, which only precludes a representation of quantum objects and behavior, it is still more radical than that of Kant, because, while a conception of quantum objects and behavior is in principle possible, it cannot be unambiguously used in considering quantum phenomena, at least as things stand now. I am not saying that the stronger view is physically necessary, but only that it is interpretively possible. There does not appear to be experimental data that would compel one to prefer either view. These views are, however, different philosophically because they reflect different limits that nature allows our thought in reaching its ultimate constitution. "As things stand now" is an important qualification, equally applicable to the strong view adopted here, even though it might appear otherwise, given that this view precludes any conception of the ultimate reality not only now but also ever, by placing it beyond thought altogether. This qualification still applies because a return to realism is possible, either on experimental or theoretical grounds even for those who hold this view. This return may take place because quantum theory, as currently constituted, may be replaced by an alternative theory that allows for or requires a realist interpretation, or because either the weak or the strong nonrealist view in question may become obsolete, even for those who hold this view, with quantum theory in place in its present form. It is also possible, however, that this view, in either the weak or strong version, will remain part of our future fundamental theories.

It is reasonable to assume that something "happens" or "changes," for example, that an electron changes its quantum state in an atom, say, from one energy level to another, between observations that then register this change. But, if one adopted the present interpretation, one could do so only if one keeps in mind the provisional nature of such words as "happen," "change," or "atom," which are ultimately inapplicable in this case, as are any other words or concepts. Quantum objects are defined by their capacity to create certain specific effects observed in measuring instruments and changes in what is so observed from one measurement to the other, changes described in language with the help of mathematics, without allowing one to represent or even conceive of what they are or how they change. According to Heisenberg:

> There is no description of what happens to the system between the initial observation and the next measurement. . . . The demand to "describe what happens" in the quantum-theoretical process between two successive observations is a contradiction in adjecto, since the word "describe" [or "represent"] refers to the use of classical concepts, while these concepts cannot be applied in the space between the observations; they can only be applied at the points of observation. [33, pp. 47, 145]

The same, it follows, must apply to the word "happen" or any word we use, and we must use words and concepts associated with them, even when we try to restrict ourselves to mathematics as much as possible. There can be no physics without language, but quantum physics imposes new limitations on using it. Heisenberg adds later in the book: "But the problem of language is really serious. We wish to speak in some way about the structure of the atoms and not only about 'facts'—the latter being, for instance, the black spots on a photographic plate or the water droplets in a cloud chamber. But we cannot speak about the atoms in ordinary language" [33, pp. 178–179]. Nor, by the same token, can we use, in referring to the atoms, ordinary concepts, from which our language is not dissociable, or for that matter philosophical or physical concepts. Heisenberg's statements still leave space for the possibility of representing "the structure of atoms" and thus the ultimate constitution of matter mathematically, without providing a physical description of this constitution. Indeed, as I said, this was the position adopted by Heisenberg by the time of these statements [33, pp. 91, 147–166]. At the time of his discovery of quantum mechanics, he saw the quantum-mechanical formalism strictly as the means of providing probabilistic predictions of the outcomes of quantum experiments. Physically, it was only assumed that "it [was] possible to ascribe to an electron the emission of radiation [a photon] [the effect of which could be observed in a measuring instrument]," without providing any physical mechanism for this emission [31, p. 263].

Language remains unavoidable and helpful in mathematics and physics alike. In physics, this significance of language is more immediate, as Bohr, again, observed on many occasions. Thus, he said: "[W]e must recognize above all that, even when the phenomena transcend the scope of classical physical theories, the account of the experimental arrangement and the recording of observations must be given in plain language, suitably supplemented by technical physical terminology. This is a clear logical demand, since the very word 'experiment' refers to a situation where we can tell others what we have done and what we have learned" [8, v. 2, p. 72; emphasis added]. This also ensures the objective and (objectively) verifiable nature of our measurements or predictions, just as in classical physics. The fundamental difference in this regard between classical and quantum physics is that in quantum physics, we deal with objects, quantum objects, which cannot be observed or represented, in contradistinction to quantum phenomena, defined by what is observed in measuring instruments as the impact of unobservable quantum objects. This difference in principle exists in classical mechanics as well, just as it does in our observations of the world, as was realized by Kant, who introduced his epistemology in the wake of Newton, whose mechanics was crucial to Kant, along with and correlatively to Euclidean geometry. There, however, as Bohr noted on the same occasion, the interference of observation "may be neglected," which is no longer the case in quantum physics [8, v. 2, p. 72].[15] Thus, paradigmatically, we can

[15]Classical statistical physics introduces certain complications here, which are, however not essential because the behavior of individual constituents of the systems considered there is

observe how planets move along the curves of their orbits, without our observational process having any effect. Not so in quantum mechanics. Nobody has ever observed, at least thus far, an electron or photon as such, in motion or at rest, to the degree that either concept ultimately applies to them, or any quantum objects, qua quantum objects, no matter how large (and some could be quite large). It is only possible to observe traces, such as spots on photographic plates, left by their interactions with measuring instruments. Hence, Bohr invokes "the essential ambiguity involved in a reference to physical attributes of [quantum] objects when dealing with phenomena where no sharp distinction can be made between the behavior of the objects themselves and their interaction with the measuring instruments" [8, v. 2, p. 61]. It follows that any meaningful ("unambiguous") representations or even conception of quantum objects and their independent behavior is "*in principle* excluded" [8, v. 2, p. 62]. On the other hand, each such trace in measuring instruments or a specific configuration of such traces can be treated as a permanent record, which can be discussed, communicated, and so forth. In this sense, such traces or our predictions concerning them are, again, as objective as they are in classical physics or relativity, except that quantum records are only verifiable as probabilistic or statistical records in all quantum physics, which is only the case in classical statistical physics. Classical mechanics or relativity give ideally exact predictions, which are not possible in quantum mechanics, because identically prepared quantum experiments in general lead to different outcomes. Only the statistics of multiple identically prepared experiments are repeatable. It would be difficult, if not impossible, to do science without being able to reproduce at least the statistical data and thus to verify the prediction of a given theory, which is possible in quantum physics.

Bohr's qualification, "plain language, suitably supplemented by technical physical terminology," introduces an additional subtlety, which extends to the mathematics of quantum theory and to mathematics itself. In the latter case, however, Bohr's formulation may be reversed to "technical terminology, suitably supplemented by plain language," although it may be a matter of balance, especially when philosophical considerations are involved. Thus, Riemann's Habilitation lecture famously contains only one real formula, which did not prevent it from decisively shaping the subsequent history of geometry, dominated, especially from modernism on, by technical, sometimes nearly impenetrably technical, algebraic treatments.

Consider his defining concept, that of manifold—*Mannigfaltigkeit*. Riemann's German is important. Although the term "*Mannigfaltigkeit*" was not uncommon in German philosophical literature, including in Leibniz and Kant, it is worth noting that the German word for the Trinity is "*Dreifaltigkeit*," thus, etymologically, suggesting a kind of "three-folded-ness," which could not have been missed by Riemann, or, for that matter, Leibniz and Kant. It is the "folded-ness" that is of the main significance here in shaping Riemann's concept philosophically. English

governed by the deterministic laws of classical mechanics. In quantum mechanics, even elementary individual objects (the so-called elementary particles) can only be handled probabilistically, and in the present view, their behavior is beyond representation or even conception.

"manifold" picks it up, as does French "*multiplicité*" [*pli*] which was initially used to translate Riemann's term, but is no longer, being replaced by *variété* (English "variety" is used for algebraic varieties), perhaps because, unlike German *Mannigfaltigkeit* and English *manifold*, it also refers to multiplicity in general.[16] Different general (or philosophical) concepts implied by terminological fluctuations of these terms do shape their mathematical choices and uses. These concepts add important dimensions to our understanding of these choices or their intellectual and cultural significance in a given case, such as that of Riemann's concept of *Mannigfaltigkeit* [55]. On the other hand, a mathematical definition of a manifold allows us to dispense with these connections or, again, from its connection to intuitive geometrical thinking, and also extend this concept in mathematics or in physics. This is something Riemann's lecture gives us as well, even though some of this mathematics is still expressed verbally, which would be quite uncommon now and has been uncommon for quite a while, uncommon but not entirely absent.

Thus, one finds this type of approach in Poincaré's work, as in parts of his series of papers on the curves defined by differentials published in the 1880s and related work [57], which also led him to the so-called qualitative theory of differential equations.[17] Poincaré's strategy in these papers was also novel (and exhibited a contrast to or even a reversal of algebraic modernist trends emerging at the time) in that, in Gray's words, it was to consider "the solutions as curves, not as functions, and to consider the global behavior of these curves" [26, p. 254]. Gray adds: "Two kinds of topological thinking entered this early work: the algebraic topological ideas of the genus of a surface and the recognition that many surfaces are characterized by their genus alone; and the point-set topological idea of everywhere dense and perfect sets, which though not original with Poincaré, are put to novel uses" [26, p. 254]. Among many remarkable outcomes of this thinking was Poincaré's analysis of curves and flows on a torus, an elliptic curve, if considered over $\mathbb{C}$. Poincaré's work is a chapter in its own right in the modernist history of the concept of curve. His "conventionalism" in physics is also important for the history of modernist

---

[16]See [3, pp. 523–524] on Grothendieck's use of the term "multiplicity," which is, on the one hand, specific (close to what is now called "orbifold"), and on the other hand, is clearly chosen to convey the multiple, plural nature of the objects considered. This is also true concerning Riemann's concept of manifold. I would argue that Riemann and Grothendieck share thinking in terms of multiplicities as their primary mathematical philosophy, a modernist trend that is especially pronounced in their thinking. As will be seen, this philosophy, manifested already in Grothendieck's early work in functional analysis, drives his use of sheaves and category theory (both concepts of the multiple), and then his concept of topos. Nothing is ever single. Everything is always positioned in relation to a multiplicity, is "sociological," and is defined and studied as such, which is itself a trend characteristic of modernism.

[17]While, the concept of "qualitative" is of much interest in the context of this article, it would require a separate treatment. I might note, however, that, while the qualitative could be juxtaposed to the quantitative, it has more complex relationships with the algebraic, which is not the same as the quantitative, just as the geometrical is not the same the qualitative. Still the genus of a surface, which is a number and thus is quantitative, is important in a qualitative approach to its topology or geometry. See note 4 above.

epistemology, from relativity to quantum mechanics, even though his position was ultimately realist. I cannot unfortunately address either subject within my scope.

Poincaré, however, joins Fermat, Descartes, Gauss, Riemann, and Hilbert in reminding us that curves are still curves, and while they may and even must be replaced by or rather translated into algebra, their geometry never quite leaves our thought and our work and exposition of mathematics. Riemann makes an extraordinary use of this situation in his work, again, nearly unique, even next to other great figures just mentioned, in mixing geometry, topology, analysis, and algebra with each other and all of them with philosophical concepts, general phenomenal intuition, and the power of language, in turn intermixed as well. It is, I might add, not a matter of inventing evocative metaphors, but rather of using these multiple, manifold, means for creating new mathematical and physical (and sometimes philosophical) concepts, such as that of *Mannigfaltigkeit* [55, pp. 341–342].

I close this section with a more general point central for my argument concerning the modernist transformation of mathematics and, in part via this mathematics, modernist physics, into essential algebraic mathematical theories, keeping in mind other components of this transformation to which this qualification equally pertains. As discussed from the outset, we still depend on and are helped by a more conventional geometrical or topological thinking in modernist, fundamentally algebraic, thinking; general phenomenal intuition; ordinary language and concepts, or other general aspects of human thinking or cognition, such as narrative, for example.[18] On the other hand, these aspects of our thinking may also become limitations in mathematics and physics alike, and, as Heisenberg argued, in quantum mechanics, modernist mathematics frees us from these limitations, or at least gives us more freedom from them. Technically, so does all mathematics, geometry and topology included, vis-à-vis other components just listed, but algebra and, with it, modernist mathematics extends this freedom. Making a curve an algebraic equation, as in Fermat and Descartes; extending the concept of curve in mathematics to include (topologically speaking) surfaces by making them curves over $\mathbb{C}$, as (at least in

[18]On narrative in mathematics, see [19]. Of particular interest in the present context, as part of the history leading to the modernist algebraization of mathematics, is B. Mazur's contribution there, which offers a discussion of L. Kronecker's "dream, vision, and mathematics" in "Visions, Dreams, and Mathematics" [41]. It might be added that Kronecker's "dream, vision, and mathematics," also decisively shaped those full-fledged modernist ideas of Weil. It may also be connected to Grothendieck's work. See the article by A'Campo et al for a suggestion concerning this possibility, as part of a much broader network, opened by Grothendieck's work on Galois theory ("the absolute Galois group"), which confirms Galois' work as a key juncture of the trajectory leading from modernity to modernism in mathematics [2, p. 405, also n. 12]. These themes could be conceptually linked to quantum field theory, via M. Kontsevitch's work on the "Cosmic Galois Group" (Cartier 2001), noted below (note 19). The article by A'Campo et al is also notable for a remarkable narrative trajectory of Grothendieck's work it traces. This confirms the role of narrative as part of mathematics itself and the philosophy of mathematics rather than only of the history of mathematics, a key theme of Mazur's and other articles in [19]. The present author's contribution to this volume deals with the epistemology of narrative, along the lines of this article [53].

effect) in Riemann, or even more so making a curve, or even just a point, a topos in Grothendieck; and using infinite-dimensional Hilbert spaces as the predictive mathematical technology of quantum mechanics (in the absence of a representation of quantum objects and behavior, which would depend on physical concepts) in Heisenberg, are all examples of taking advantage of this freedom. This freedom may not be complete, but it makes possible pursuits of previously insurmountable tasks.

5.4 Curves from Modernity to Modernism: Three Cases

5.4.1 Curves as Algebra: Descartes/Fermat/Diophantus

I shall now discuss three cases shaping the idea of curve, and geometry and mathematics in general from modernity to modernism. For the reasons explained in the Introduction, I leave aside most of the relevant earlier history and begin with Fermat and Descartes. Then I move to Riemann, and finally, to Weil and Grothendieck, with Riemann still as a key background figure.

Fermat's work is both remarkable and seminal historically, also in influencing Descartes' work and the development of calculus, and, of course, especially in view of his famous, "last," theorem, the study of algebraic and specifically elliptic curves. The deeper mathematical nature of elliptic curves was ultimately revealed by unexpected connections, via the Taniyama-Shimura conjecture (now the "modularity theorem") and related developments, to Fermat's last theorem, which enabled Wiles' proof of the theorem, as a consequence of the modularity theorem for semistable elliptic curves, which he proved as well. These connections could not of course have been anticipated by Fermat. On the other hand, his ideas concerning elliptic curves remain relevant, and are a powerful manifestation of the algebraization and number-theorization of the geometrical ideas then emerging. Thus, according to Weil, first commenting on Fermat's last theorem and Fermat's famous remark "that he discovered a truly remarkable proof for [it] 'which this margin is too narrow to hold,' " and then on Fermat's study of elliptic curves:

> How could he have guessed that he was writing for eternity? We know his proof for biquadrates . . . ; he may well have constructed a proof for cubes, similar to the one which Euler discovered in 1753 . . . ; he frequently repeated those two statements . . . , but never the more general one. For a brief moment perhaps, and perhaps in his younger days . . . , he must have deluded himself into thinking that he had the principle of a general proof; what he had in mind on that day can never be known.
>
> On the other hand, what we possess of his methods for dealing with curves of genus 1 is remarkably coherent; it is still the foundation for the modern theory of such curves. It naturally falls into two parts; the first one, directly inspired by Diophantus, may conveniently be termed a method of ascent, in contrast with the descent which is rightly regarded as Fermat's own. Our information about the latter, while leaving no doubt about its general features, is quite scanty in comparison with Fermat's testimony about the former . . . .

> and the abundant (and indeed superabundant) material collected by Billy in the *Inventum Novum*.
>
> In modern terms, the "ascent" is nothing else than a method of deriving new solutions for the equations of a curve of genus 1. What was new here was of course not the principle of the method: it has been applied quite systematically by Diophantus... and, as such, referred to, by Fermat as well as by Billy, as "*methodus vulgaris*." The novelty consisted in the vastly extended use which Fermat made of it, giving him at least a partial equivalent of what we would obtain by the systematic use of the group theoretical properties of the rational points on a standard cubic. Obviously Fermat was quite proud of it; writing for himself on the margins of his Diophantus, he calls it "*nostro invention*," and again, writing to Billy: "it has astonished the greatest experts." [67, pp. 104–105]

It still does, which was Weil's point as well. The epoch of algebraization and spatial algebraization of elliptic curves and of mathematical curves in general had commenced, with the arc from Fermat to Wiles through many points of modern and then modernist algebraic geometry. Weil's own work was one these points, even a trajectory of its own, leading him to his rethinking of algebraic geometry, which had a momentous impact on Grothendieck, who, however, also radically transformed it in turn. It would, again, be more accurate to speak of a network of trajectories, manifested in Wiles' proof, which brings together so many of them. It is hard, however, to abandon the metaphor of a curve when dealing with the history of the idea of curve itself.

Descartes took full advantage of this algebraization and gave it its modern coordinate form, still very much in use, thus, as I said, making his project of analytic geometry an intimation of modernist thinking in mathematics and physics at the heart of modernity. This project has its history, too, as part of the history of algebra, especially the concept of equation that, as we just saw, emerged in ancient Greek mathematics, especially with Diophantus (around the third century CE), whose ideas were, again, central to Fermat. Analytic geometry, however, by expressly making geometry algebra, gave mathematics its, in effect, independence of physics and of material nature, thus, along with the work of Descartes' contemporary fellow algebraists, again, in particular Fermat, initiating mathematical modernism within modernity.

In the simplest possible terms, analytic geometry did so because the equation corresponding to a curve, say, $X^2 - 1 = 0$ for the corresponding parabola, could be studied as an algebraic object, independently of its geometrical representation or its connection to physics, which eventually enabled us to define curves even over finite fields and thus as discrete objects, as considered above. A curve becomes, in its composition, defined by its equation, divested from its representational geometrical counterpart. It no longer geometrically idealizes the reality exterior to it. It only represents itself, is its own ontology, akin to a *line* of poetry. The equation, algebra, is the poetry of the curve, confirming and amplifying a separation of a mathematical curve from any curve found in the world, which defines all mathematics. When we say in mathematics, "consider a curve *X*," we separate it from every curved object in the world, in the way poetry separates its words and ideas from those denoted by ordinary language and the world they represent, as A. Badiou noted in commenting on Mallarmé's theory of poetry, based in this separation [7, p. 47]. This poetry of algebra can define a discrete curve, or can make a curve a surface, or a surface a curve, give it an even more complex spatial algebraic architecture, or, to continue with my artistic metaphor, an ever-more complex composition, such as that of a moduli space, the Teichmüller space (also the Teichmüller *curve*), Grothendieck's or Hilbert's scheme and representable-functors, ... we are as yet far from exhausting the limits of this "poetry" of Riemann surfaces/curves (e.g., [1]).

5.4.2 Curves as Surfaces, Surfaces as Curves: Riemann/Riemann/Riemann

The idea of a Riemann surface is one of Riemann's (many) great contributions to modern and eventually modernist mathematics. In Papadopoulos' cogent account:

> In his doctoral dissertation, Riemann introduced Riemann surfaces as ramified coverings of the complex plane or of the Riemann sphere. He further developed his ideas on this topic in his paper on Abelian functions. This work was motivated in particular by problems posed by multi-valued functions $w(z)$ of a complex variable z defined by algebraic equations of the form
>
> $$f(w, z) = 0,$$
>
> where f is a two-variable polynomial in w and z.
>
> Cauchy, long before Riemann, dealt with such functions by performing what he called "cuts" in the complex plane, in order to obtain surfaces (the complement of the cuts) on which the various determinations of the multi-valued functions are defined. Instead, Riemann assigned to a multi-valued function a surface which is a ramified covering of the plane and which becomes a domain of definition of the function such that this function, defined on this new domain, becomes single-valued (or "uniform"). Riemann's theory also applies to transcendental functions. He also considered ramified coverings of surfaces that are not the plane. [47, p. 240]

The idea of a Riemann surface gains much additional depth and richness when considered along with, and in terms of, Riemann's concept of manifold, his other great invention, introduced, around the same time, in his Habilitation lecture [60]. Riemann did not do so himself, although he undoubtedly realized that Riemann surfaces were manifolds, and they have likely been part of the genealogy of the concept of manifold. Riemann's surfaces were first expressly defined as manifolds by Weyl in *The Concept of a Riemann Surface* [73]. Understanding the concept of a Riemann surface as a complex curve is helped by this perspective. It is an intriguing question whether Riemann himself thought of them as curves, but it would not be surprising if he had. Weyl undoubtedly did, although the point does not figure significantly in his book, focused on the "surface" nature of Riemann surfaces, defined, however, in spatial-algebraic terms. This may be surprising. But then, Weyl was not an algebraic geometer. The work of É. Picard, a key figure in the history of algebraic geometry would be more exemplary in considering this aspect of Riemann's concept [47, 49]. However, that a Riemann surface (with which a family of algebraic curves could be associated) is a manifold is crucial for "making it" both a surface and a curve.

Weyl's argumentation leading him to his definition is an application of a principle very much akin to Tate's "Think geometrically, prove algebraically" or its extension here, "Think both Intuitively Geometrically and Spatially-Geometrically: Prove Algebraically." It is also a manifestation of the spirit of Riemann's thinking in which, as noted earlier, geometry and algebra, indeed geometry, topology, algebra, and analysis, come together in a complex mixture of the rigorous and the intuitive, algebraic and spatial-algebraic, mathematical and physical, mathematical and philosophical, and so forth. His work on his ζ-function and number theory could be brought into this mix as well. According to Weyl:

> It was pointed out ... that one's intuitive grasp of an analytic form [an analytic function to which a countable number of irregular elements have been added] is greatly enhanced if one represents each element of the form by a point on a surface F in space in such a way that the representative points cover F simply and so that every analytic chain of elements of the form becomes a continuous curve on F. To be sure, from a purely objective point of view, the problem of finding a surface to represent the analytic form in this visual way may be rejected as nonpertinent; for in essence, three-dimensional space has nothing to do with analytic forms, and one appeals to it not on logical-mathematical grounds, but because it is closely associated with our sense perception. To satisfy our desire for pictures and analogies in this fashion by forcing inessential representation of objects instead of taking them as they are could be called an anthropomorphism contrary to scientific principles. [73, p. 16]

I note in passing a criticism, apparent here, of the logicist philosophy of mathematics, which theorized mathematics as an extension of logic and, championed by, among others, Bertrand Russell, was in vogue at the time. This is, however, a separate subject. Weyl will now proceed, again, in the spirit of Riemann, to his definition of a two-dimensional manifold and eventually Riemann's surface, *intrinsically*, rather than in relation to its ambient three-dimensional space. Riemann was building on Gauss' ideas concerning the curvature of a surface and his, as he called it, "*theorema egregium*," which states that the curvature of a surface, which he defined as well, was intrinsic to the surface. It is also this concept and the corresponding spatial algebra that enables one to define a Riemann surface as a curve over C. This intrinsic and abstract, spatial-algebraic, view of a Riemann surface was often forgotten by Riemann's followers, especially at earlier stages of the history of using Riemann's concept. According to Papadopoulos, who in part follows Klein's assessment:

> Riemann not only considered Riemann surfaces as associated with individual multi-valued functions or with meromorphic functions in general, but he also considered them as objects in themselves, on which function theory can be developed in the same way as the classical theory of functions is developed on the complex plane. Riemann's existence theorem for meromorphic functions with specified singularities on a Riemann surface is also an important factor in this setting of abstract Riemann surfaces. Riemann conceived the idea of an abstract Riemann surface, but his immediate followers did not. During several decades after Riemann, mathematicians (analysts and geometers) perceived Riemann surfaces as objects embedded in three-space, with self-intersections, instead of thinking of them abstractly. They tried to build branched covers by gluing together pieces of the complex

> plane cut along some families of curves, to obtain surfaces with self-intersections embedded in three-space. [47, p. 242]

According to Weyl (whom Papadopoulos cites): "Thus, the concept 'two-dimensional manifold' or 'surface' will not be associated with points in three-dimensional space; rather it will be a much more general abstract idea," in effect a spatial-algebraic one in the present definition, and thus is modernist. Weyl's position concerning the nature of mathematical reality is a different matter. As is clear from his philosophical writings (e.g., [72]), Weyl was ultimately a realist (albeit not a Platonist) in mathematics and physics alike, his major contribution to quantum mechanics notwithstanding, contributions also dealing with the role of group theory there, yet another modernist trend in mathematics and physics alike [70]. This aspect of the situation is, however, secondary for the moment, although one might still ask whether if considering a given Riemann surface as either a (topologically) real two-dimensional surface or a curve over C, deal with the same mathematical object. Weyl continued as follows:

> If any set of objects (which will play the role of points) is given and a continuous coherence between them, similar to that in the plane, is defined we shall speak of a two-dimensional manifold. Since all ideas of continuity may be reduced to the concept of neighborhood, two things are necessary to specify a two-dimensional manifold:
>
> (1) to state what entities are the "points" of the manifold;
> (2) to define the concept of "neighborhood." [73, pp. 16–17]

One hears here an echo, deliberate or not, of Hilbert's "tables, chairs, and beer mugs," for "points, straight lines, and planes," mentioned above. In the present view, this means one should define entities, such as points, lines, neighborhoods, by using algebraic symbols and algebraic relationships between them, without referring to any objects in the world represented by ordinary language, even if still using this language, as the concept of a Riemann surface as a curve and then its avatars such as Gromov's concept of a pseudoholomorphic curve (a smooth map from a Riemann surface into an almost complex manifold) exemplify. Its connections to our phenomenal sense of surface are primarily, if not entirely, intuitive, when it comes to the idea of continuity, for example, as defined by Weyl here, in terms of the concept of neighborhood. In any event, a Riemann surface is certainly not a curve in any phenomenal sense. As defined by Weyl, in a pretty much standard way, the concept of manifold is a spatial-algebraic one in the present definition. Weyl's more technical definition, again, pretty much standard, given next, and then his analysis of Riemann surfaces only amplified this point.

This multifaceted nature of Riemann surfaces equally and often jointly defined the history of complex analysis, the main initial motivation for Riemann's introduction of the concept of a Riemann surface, and the history of algebraic curves, both building on this concept, and other developments, for example, in abstract algebra and number theory, including Riemann's work on the ζ-function and the distribution of primes.[19] All these developments were unfolding towards modernism during the period between Riemann and Weyl, whose book initiated the (modernist) treatment of the concept of a Riemann surface that defines our understanding of the concept. This history explains my triple subtitle, "Riemann/Riemann/Riemann"—the Riemann of the concept of manifold, the Riemann of the concept of Riemann surfaces, the Riemann of complex analysis. A few more Riemanns could be added. This multiple and entangled history shaped algebraic geometry, eventually leading to the work of A. Weil, Grothendieck, M. Artin, J. Tate, and others, ultimately extending mathematical modernism to our own time.

5.4.3 *Curves as Discrete Manifolds: Grothendieck/Weil/Riemann*

One of the great examples of this extension is the concept of algebraic curve or algebraic variety in general over a finite field and the study of such objects by the standard tools of algebraic topology, in particular homotopy and cohomology theories, which have previously proven to be effective tools, technologies, for the study of complex algebraic varieties.

The origin of this project goes back to Weil, a key figure of the later stage of mathematical modernism, especially in bringing together algebra, geometry, and number theory, in which he was a true heir of Fermat (and he probably saw himself as one), as well as of Kronecker (in this case, Weil certainly saw himself as one). Riemann is still a key figure in the history leading to Weil's work in algebraic geometry, first of all, again, in view of his concepts of a Riemann surface and a covering space, but also the least by virtue of introducing the concept of a discrete manifold in his Habilitation lecture. (Riemann, thus, was instrumental in the history of both discrete and infinite-dimensional spaces of modernism.) G. Fano, one of the founders of finite geometry, belonged to the Italian school of geometry (1880s–1930s), contemporary with and an important part of the history of mathematical

[19]For an extensive historical account of the history of complex function theory, only mentioned in passing here, see [10], which considers at length most key developments conjoining geometry and complex analysis, from Cauchy to Riemann and then of Riemann's work [10, pp. 189–213, 259–342]. Intriguingly, the algebra of quantum field theory found the way to use Riemann's algebraic work, his work and his hypothesis concerning the ζ-function (one of the greatest, if not the greatest, of yet unsolved problems of mathematics). The ζ-function plays an important role in certain versions of higher-level quantum field theory. See P. Cartier's discussion, which introduces an intriguing idea of the "Cosmic Galois group" [11] and A. Connes and M. Marcoli's book [14], which explores the role of Riemann's differential geometry in this context. The latter is a long and technical work in noncommutative geometry, which uses Grothendieck's motive cohomology theory, but see p. 10 for an important definition of "the Riemann-Hilbert correspondence." This is yet another testimony to the fact that much of modernism in mathematics and even in physics takes place along the trajectory or again, a network of trajectories between Riemann and Grothendieck. See note 16.

modernism (Fano was a student of F. Enriques), strongly influenced by Riemann's thought. Representatives of the Italian school (quite a few of them, even counting only major figures) made major contributions to many areas of geometry, especially algebraic geometry, which formed an important part of the very rich and complex (modernist) history, leading to Weil's work under discussion. Weil suggested that a cohomology theory for algebraic varieties over finite fields, now known as Weil cohomologies, could be developed, by analogy with the corresponding theories for complex algebraic varieties or topological manifolds in general. Weil's motivation was a set of conjectures (these go back to Gauss), known as the Weil conjectures, concerning the so-called local ζ-functions, which are the generating functions derived from counting numbers of points on algebraic varieties over finite fields. These conjectures, Weil thought, could be attacked by means of a proper cohomology theory, although he did not propose such a theory himself.

In order to be able to do so, one needed, first, a proper topology, which was nontrivial because the objects in question are topologically discrete. A more "native" topology that could be algebraically defined by them, known as Zariski's topology, did not work, because it had too few open sets. The decisive ideas came from Grothendieck, helped by the sheaf-cohomology theory and category theory, known as "cohomological algebra," by then the standard technology of algebraic topology. Using these tools, a hallmark of Grothendieck's thinking throughout his career, and his previous concepts, such as that of "scheme," eventually led him to topos theory, arguably the culminating example of spatial algebra, and étale cohomology, as a viable candidate for Weil's cohomology, which it had quickly proven to be. By using it, Grothendieck (with Artin and J.-L. Verdier) and P. Deligne (his student) were able to prove Weil's conjectures, and then Deligne, who previously proved the Riemann hypothesis conjecture (considered the most difficult one), found and proved a generalization of Weil's conjectures. Grothendieck's key, extraordinary, insight, also extending what I call here spatial algebra in a radically new direction, was to generalize, in terms of category theory, the concept of "open set," beyond a subset of the algebraic variety, which was possible because the concept of sheaf and of the cohomology of sheaves could be defined by any category, rather than only that of open sets of a given space. Étale cohomology is defined by this type of replacement, specifically by using the category of étale mappings of an algebraic variety, which become "open subsets" of the finite unbranched covering spaces of the variety, a vast and radical generalization of Riemann's concept of a covering space. Grothendieck was also building on some ideas of J.-P. Serre. Part of the origin of this generalization was the fact that the fundamental group of a topological space, say, again, a Riemann surface, could be defined in two ways: it can either be defined more geometrically, as a group of the sets of equivalence classes of the sets of all loops at a given point, with the equivalence relation given by homotopy (itself an example of the history of the idea of curve); or it can be defined even more algebraically, as a group of transpositions of covering spaces. In this second, algebraic, definition, the fundamental group is analogous to the Galois group of the algebraic closure of a field. Serre was the first to consider for finite fields, importantly for Grothendieck's work on étale cohomology, a concept that, thus, has its genealogy in both Galois' and Riemann's thought. (The connection has been established in the case of Riemann surfaces long before then [e.g., [73, p. 58]].) Grothendieck's concept of étale mappings gives a sufficient number of additional open sets to define adequate cohomology groups for some coefficients, for algebraic varieties over finite fields. In the case of complex varieties, one recovers the standard cohomology groups (with coefficients in any constructible sheaf).

Some of the most elegant calculations concern algebraic curves over algebraically closed fields, beginning with elliptic ones [4, 5]. These calculations are also important because they are the initial step in calculating étale cohomology groups for other algebraic varieties by using the standard means of algebraic topology, such as spectral sequences of a fibration. My main point at the moment is that (spatial) algebra makes algebraic curves over such finite fields fully mathematically analogous to standard algebraic curves, beginning, again, with elliptic curves, as studied by Fermat, again, a major inspiration for Weil.

Now, the category of étale mapping is a topos, a concept that is, for now, the most abstract form of what I call here spatial algebra. Although, as became apparent later, étale cohomologies could be defined for most practical uses in simpler settings, the concept of topos remains crucial, especially in the present context, because it can be seen as the concept of a covering space over a Riemann surface converted into the (spatial-algebraic) concept of topology of the surface itself, and then generalized to any algebraic variety. The concept of topos also came to play a major role in mathematical logic, a major development of mathematical modernism, thus bringing it together with the modernist problematic considered here. The subject cannot, however, be addressed here, except by noting that mathematical logic is already an example of modernist algebraization of mathematics, with radical epistemological implications concerning the nature of mathematical reality, or the impossibility of such a concept. On the other hand, Grothendieck's use of his topoi in algebraic geometry is essentially ontological rather than logical, although his overall philosophical position concerning the nature of mathematical reality remains somewhat unclear, for example, whether it conforms or not to mathematical Platonism, and the subject will be put aside here as well. In any event, it does not appear that Grothendieck was ever thinking of his topoi or in general in terms of breaking with the ontological view of mathematics. My main focus here is the mathematical technologies that the concept of topos, whatever its ontological status, enables, such as étale cohomology. Such technologies may suggest a possible break with the possibility of the ultimate ontological description of mathematical reality, again, assuming that any ultimate reality, say, again, of the type considered in physics, is even possible in mathematics.

It would not be possible here to present topos theory in its proper abstractness and rigor, prohibitive even for those trained in the field of algebraic geometry. The essential philosophical nature of the concept, briefly indicated above, may, however, be sketched in somewhat greater detail, as an example of both a rich mathematical concept in its own terms and of the modernist problematic in question here. First, very informally, consider the following way of endowing a space with a structure, generalizing the definition of topological space in terms of open subsets, as mentioned above. One begins with an arbitrarily chosen space, X, potentially any given space, which may initially be left unspecified in terms of its properties and structure. What would be specified are the relationships between spaces applicable to X, such as mapping or covering one or a portion of one, by another. This structure is the arrow structure $Y \rightarrow X$ of category theory, where X is the space under consideration and the arrow designates the relationship(s) in question. One can also generalize the notion of neighborhood or of an open subspace of (the topology of) a topological space in this way, by defining it as a relation between a given point and space (a generalized neighborhood or open subspace) associated with it. This procedure enables one to specify a given space not in terms of its intrinsic structure (e.g., a set of points with relations among them) but "sociologically," throughout its relationships with other spaces of the same category, say, that of algebraic varieties over a finite field of characteristic p [38, p. 7]. Some among such spaces may play a special role in defining the initial space, X, and algebraic structures, such as homotopy and cohomology, as Riemann in effect realized in the case of covering spaces over Riemann surfaces, which, as I explained, was one of the inspirations for Grothendieck's concept of topos and more specifically of an étale topos.

To make this a bit more rigorous (albeit still quite informal), I shall briefly sketch the key ideas of category theory. It was introduced as part of cohomology theory in algebraic topology in 1940 and, as I said, later extensively used by Grothendieck in his approach to cohomological algebra and algebraic geometry, eventually leading him to the concept of topos.[20] Category theory considers multiplicities (which need not be sets) of mathematical objects conforming to a given concept, such as the category of differential manifolds or that of algebraic varieties, and the arrows or morphisms, the mappings between these objects that preserve this structure. Studying morphisms allows one to learn about the individual objects involved, often to learn more than we would by considering them only or primarily individually. In a certain sense, in his Habilitation lecture, Riemann already thinks categorically. He does not start with a Euclidean space. Instead, the latter is just one specifiable object of a large categorical multiplicity, here that of the category of differential or, more narrowly, Riemannian manifolds, an object marked by a particularly simple way we can measure the distance between any two points. Categories themselves may be viewed as such objects, and in this case one speaks of "functors" rather than "morphisms." Topology relates topological or geometrical objects, such as manifolds, to algebraic ones, especially, as in the case of homotopy and cohomology groups, introduced by Poincaré. Thus, in contrast to geometry (which relates its spaces to algebraic aspects of measurement), topology, almost by its nature, deals

[20]One of his important, but rarely considered, contributions is his work on Teichmüller space, the genealogy of which originates in Riemann's moduli problem, powerfully recast by Grothendieck in his framework. Especially pertinent in the present context is the idea of a "Teichmüller curve" and then Grothendieck's recasting of it, another manifestly modernist incarnation of the idea of curve, via Riemann. Conversely, the theory provided an important case for Grothendieck to use his new technology. Étale cohomology came next. This is yet another modernist trajectory extending from Riemann and Grothendieck. For an excellent account, see A'Campo et al. [1].

with functors between categories of topological objects, such as manifolds, and categories of algebraic objects, such as groups.

A topos in Grothendieck's sense is a category of spaces and arrows over a given space, used especially for the purpose of allowing one to define richer algebraic structures associated with this space, as explained above. There are additional conditions such categories must satisfy, but this is not essential at the moment. To give one of the simplest examples, for any topological space S, the category of sheaves on S is a topos. The concept of topos is, however, very general and extends far beyond spatial mathematical objects (thus, the category of finite sets is a topos); indeed, it replaces the latter with a more algebraic structure of categorical and topos-theoretical relationships between objects. On the other hand, it derives from the properties of and (arrow-like) categorical relationships between properly topological objects. The conditions, mentioned above, that categories that form topoi must satisfy have to do with these connections.

Beyond enabling the establishing of a new cohomology theory for algebraic varieties, as considered above, topos theory allows for such esoteric constructions as nontrivial or nonpunctual single-point "spaces" or, conversely, spaces (topoi) without points (first constructed by Deligne), sometimes slyly referred to by mathematicians as "pointless topology." Philosophically, this notion is far from pointless, especially if considered within the overall topos-theoretical framework. In particular, it amplifies a Riemannian idea that "space," defined by its relation to other spaces, is a more primary object than a "point" or, again, a "set of points." Space becomes a Leibnizean, "monadological" concept, insofar as points in such a space (when it has points) may themselves be seen as a kind of monad, thus also giving a nontrivial structure to single-point spaces. These monads are certain elemental but structured entities, spaces, rather than structureless entities (classical points), or at least as entities defined by (spatial) structures associated with and defining them [1]. Naturally, my appeal to monads is qualified and metaphorical. Leibniz's monads are elemental souls, the atoms of soul-ness, as it were. One might, however, also say, getting a bit more mileage from the metaphor, that the space thus associated with a given point is the soul of this point, which defines its nature. In other words, not all points are alike insofar as the mathematical (and possibly philosophical) nature of a given point may depend on the nature or structure of the space or topos to which it belongs or with which it is associated in the way just described. This approach gives a much richer architecture to spaces with multiple points, and one might see (with caution) such spaces as analogous to Leibniz's universe composed by monads. It also allows for different (mathematical) universes associated with a given space, possibly a single-point one, in which case a monad and a universe would coincide. Grothendieck's topoi are possible universes, possible worlds, or com-possible worlds in Leibniz's sense, without assuming, like Leibniz (in dealing with the physical world), the existence of only one of them, the best possible.

One might also think of this ontology as an assembly of surface ontologies (Grothendieck's concept of topos is, again, ontological, rather than logical, as in his logical followers), in the absence of any ultimate ontology, or even, as against physics, any ultimate mathematical or otherwise mental reality, thus connecting on modernist lines the multiple and the unthinkable. Topoi are multiple universes, defined ontologically, in the absence of a single ultimate reality underlying them; they are investigated by means of technologies such as cohomologies or homotopies (which can be defined for them as well). I shall not consider Grothendieck's topos as such from this perspective, which, again, does not appear to be Grothendieck's own. Instead, I shall discuss next the ontological and epistemological architecture of modernism mathematics more generally in relation to the concept of technology, conceived broadly so as to include the means by which mathematics studied itself; and I shall briefly comment on topos theory in this context. Quantum theory will, yet again, serve as a convenient bridge, in this case as much because of the differences as the similarities between physics and mathematics.[21]

5.5 Mathematical Modernism Between Ontologies and Technologies

While, roughly speaking, technology is a means of doing something, enabling us to get "from here to there," as it were, the concept of mathematical technology that I adopt extends more specifically the concept of "experimental technology" in modern, post-Galilean, physics, defined, as explained, by its jointly experimental and mathematical character. I note, first, that experimental technology is a broader concept than that of measuring instruments, with which it is most commonly associated in physics. It would, for example, involve devices that make it possible to use the measuring instruments, a point that, as will be seen, bears on the concept of technology in mathematics. Thus, the experimental technology of quantum physics, from Geissler tubes and Ruhmkorff coils of the nineteenth century to the Large Hadron Collider of our time, enables us to understand how nature works at the ultimate level of its constitution. In the present interpretation, this technology allows us to know the effects this constitution produces on measuring devices (described, along with these effects themselves, by classical physics), without allowing us to represent or even conceive of the character of this constitution. The character of these effects is, however, sufficient for creating theories, defined by their *mathematical technologies*, such as quantum mechanics and quantum field theory, that can predict these effects. Thus, quantum physics is only about the relationships between mathematical and experimental technologies used, vis-à-vis classical physics or relativity, or mathematics itself. All mathematics used in quantum physics is technology; in mathematics, or in classical physics or relativity, some mathematics is also used ontologically. Quantum objects themselves are not technology; they are a form of reality that technology helps us to discover, understand, work with, and so forth, but in this case, at least in the present

[21]The discussion to follow is partly adopted from [54, pp. 265–274]. My argument here is essentially different, however.

interpretation, without assuming or even precluding any ontological representation of this reality. Quantum objects can of course become part of technology, beginning with the quantum parts of measuring instruments through which the latter interact with quantum objects, or as parts of devices we use elsewhere, such as lasers, electronic equipment, MRI machines, and so forth.

In the wake of Heisenberg's discovery and Born and Jordan's work in 1925 [9, 31], Bohr commented as follows:

> In contrast to ordinary mechanics, *the new quantum mechanics does not deal with a space-time description of the motion of atomic particles*. It operates with manifolds of quantities which replace the harmonic oscillating components of the motion and symbolize the possibilities of transitions between stationary states [manifested in measuring instruments]. These quantities satisfy certain relations which take the place of the mechanical equations of motion and the quantization rules [of the preceding quantum theory]. ...
>
> It will interest mathematical circles that the mathematical instruments created by the higher algebra play an essential part in *the rational formulation* of the new quantum mechanics. Thus, the general proofs of the conservation theorems in Heisenberg's theory carried out by Born and Jordan are based on the use of the theory of matrices, which go back to Cayley and were developed especially by Hermite. It is to be hoped that a new era of mutual stimulation of mechanics and mathematics has commenced. To the physicists it will at first seem deplorable that in atomic problems we have apparently met with such a limitation of our usual means of visualization. This regret will, however, have to give way to thankfulness that mathematics in this field, too, presents us with the tools to prepare the way for further progress. [8, v. 1, pp. 48, 51; emphasis added]

Bohr's appeal to "the *rational* formulation of the new quantum mechanics" merits a brief digression, especially in conjunction with his several invocations of the "irrationality" inherent in quantum mechanics, a point often misunderstood. The "irrationality" invoked here and elsewhere in Bohr's writings is not any "irrationality" of quantum mechanics, which Bohr, again, sees as a "rational" theory [8, v. 1, p. 48]. Bohr's invocation of "irrationality" is based on an analogy with irrational numbers, reinforced perhaps by the apparently irreducible role of complex numbers and specifically the square root of -1, i (an irrational magnitude in the literal sense because it cannot be presented as a ratio of two integers) in quantum mechanics, or quantum field theory. It is part of the history of the relationships between algebra (initially arithmetic) and geometry, from the ancient Greeks on. As noted earlier, the ancient Greeks, who discovered the (real) irrationals, could not find an arithmetical, as opposed to geometrical, form of representing them. The Greek terms were "alogon" and "areton," which may be translated as "incommensurable" and "incomprehensible," the latter especially fitting in referring to quantum objects and processes. The problem was only resolved, by essentially modernist mathematical means (algebra played a major role), in the nineteenth century, after more than 2000 years of effort, with Dedekind and others, albeit, in view of the undecidability of Cantor's continuum hypothesis, perhaps only resolved as ultimately unresolvable. It remains to be seen whether quantum mechanical "irrationality" will ever be resolved by discovering a way to mathematically or otherwise represent quantum objects and processes. As thing stand now, quantum mechanics is a rational theory of something that is irrational in the sense of being inaccessible to a rational representation or even to thinking itself. In other words, at stake is a replacement of a *rational representational* theory, classical mechanics, with a *rational probabilistically or statistically predictive* theory. This replacement is the rational quantum mechanics introduced by Heisenberg, a reversal of what happened in the crisis of the incommensurable in ancient Greek mathematics, which compelled it to move from arithmetic to geometry.

Heisenberg's thinking revolutionized the practice of theoretical physics and redefined experimental physics or reflected what the practice of experimental physics had in effect become in dealing with quantum phenomena. The practice of experimental physics no longer consists of tracking what happens or what would have happened independently of our experimental technology, but in creating new configurations of this technology, which allows us to observe effects of quantum objects and behavior manifested in this technology.[22] This practice reflects the fact that what happens is unavoidably defined by what kinds of experiments we perform, and how we affect quantum objects, rather than by tracking their independent behavior, although their independent behavior does contribute to what happens. The practice of theoretical physics no longer consists in offering an idealized mathematical representation of quantum objects and their behavior, but in developing mathematical technology that is able to predict, in general (in accordance with what obtains in experiments) probabilistically, the outcomes of always discrete quantum events, observed in the corresponding configurations of experimental technology.

Taking advantage of and bringing together two meanings of the word "experiment" (as a test and as an attempt at an innovative creation), one might say that the practice of quantum physics is the first practice of physics that is both, jointly, fundamentally experimental and fundamentally mathematical. That need not mean that this practice has no history; quite the contrary, creative experimentation has always been crucial to mathematics and science, as the work of all key figures discussed in this article demonstrates. Galileo and Newton, are two great examples in classical physics: they were experimentalists, both in the conventional sense (also inventors of new experimental technologies, new telescopes in particular) and, in their experimental and theoretical thinking alike, in the sense under discussion at the moment. Nevertheless, this experimentation acquires a new form with quantum mechanics and then extends to higher level quantum theories, and, as just explained, a new understanding of the nature of experimental physics. The practice of quantum physics is *fundamentally* experimental because, as just explained, we no longer track, as we do in classical physics or relativity, the independent behavior of the systems considered, and thus track what happens in any event, by however ingenious

[22]I qualify by "unavoidably" because we can sometimes define by an experiment what will happen in classical physics, say, by rolling a ball on a smooth surface, as Galileo did in considering inertia. In this case, however, we can then observe the ensuing process without affecting it. This is not so in quantum physics, because any new observation essentially interferes with the quantum object under investigation and defines a new experiment and a new course of events. Only some observations do in classical physics.

experiments. We *define* what will happen in the experiments we perform, by how we experiment with nature by means of our experimental technology.

By the same token, quantum physics is fundamentally mathematical, because its mathematical formalism is equally not in the service of tracking, by way of a mathematical representation, what would have happened anyhow, which tracking would shape the formalism accordingly, but is in the service of predictions required by experiments. Indeed, quantum theory experiments with mathematics itself, more so and more fundamentally than does classical physics or relativity. This is because quantum theorists invent, in the way Heisenberg did, effective mathematical schemes of whatever kind and however far they may be from our general phenomenal intuition, rather than proceeding by refining mathematically our phenomenal representations of nature, which limits us in classical physics or even (to some degree) in relativity. One's choice of a mathematical scheme becomes relatively arbitrary insofar as one need not provide any representational physical justification for it, but only need to justify this scheme by its capacity to make correct predictions for the data in question. It is true that in Heisenberg's original work the formalism of quantum mechanics extended (via the correspondence principle) from the representationally justified formalism of classical mechanics. Heisenberg and then other founders of the theory (such as Born and Jordan, or Dirac) borrowed the equations of classical mechanics. However, they replaced the variables used in these equations with Hilbert-space operators, thus using modernist mathematics, which was no longer justified by their representational capacity but, in Heisenberg's words, by "the agreement of their predictions with the experiment" [32, p. 108]. One's mathematical experimentation may, thus, be physically motivated, but it is not determined by representational considerations, the freedom from which also liberates one's mathematical creativity. Rather than with the equations of classical mechanics, one could have started directly with Hilbert-spaces and derived the necessary formalism by certain postulates, as was done by von Neumann in his classical book, admittedly, with quantum mechanics already in place [66]. Other versions of the formalism, such as the C*-algebra version and, more recently, the category-theory version are products of this type of mathematical experimentation. It is true that all these versions have thus far been essentially mathematically equivalent, and in particular, the role of complex numbers appears to be unavoidable. It is difficult, however, to be entirely certain that this will remain the case in the future, even if no change is necessary because of new experimental data. The invention of quantum theory was essentially modernist in its epistemology and its spirit of creative experimentation (which it shared with contemporary modernist literature and art) alike, as well as in its use of modernist mathematics. Heisenberg was the Kandinsky of physics.

One could indeed think of the technological functioning of mathematics even in mathematics itself: certain mathematical instruments, such as homotopy or cohomology groups, are technologies akin to measuring instruments in physics, with the role of reality taken in each case by the corresponding topological space. According to J.-P. Marquis, who borrows his conception of mathematical technology from quantum physics "they provide information about the corresponding topological space. . . . [T]hey are epistemologically radically from . . . transformation [symmetry] groups of a space. They do not act on anything. The purpose of these geometric devices is to classify spaces by their different *homotopy [or cohomology] types*." By contrast, fibrations, for example, important for using homotopy and cohomology groups, including, as noted, in étale cohomologies, are not "measuring instruments," but rather "devices that make it possible to apply measuring instruments [such as cohomology and homotopy groups] and other devices" [39, p. 259]. The key point here is that the invention and use of mathematical technologies is crucial for mathematics, in modernism from Riemann's concept of the genus of a Riemann surface to Grothendieck's invention of étale cohomology, with the whole history of algebraic topology between them, keeping in mind that any technology can and is eventually likely to become obsolete, as Marquis notes [39, p. 259]. In quantum theory, all mathematics used is technology (vs. classical physics or relativity where it can also be used ontologically) and it can become obsolete, too, as that of classical physics became in quantum physics. Ontologies can become obsolete, too, such as, at least for some, that of set theory, replaced by category theory, which redefines, for example, the concept of topological space. On the other hand, the concept of physical reality is unlikely to go away any time soon. (The name may change, and "matter" has sometimes been used instead.) Could the same be said about some form of mathematical reality? I would like to offer a view that suggests that it is possible to answer this question in the negative, thus fundamentally differentiating mathematical and physical reality.

First, I note that, in parallel with the experimental and mathematical technology used in quantum physics, the mathematical technology in mathematics may not only be used to help us to represent mathematical reality (although it may be used in this way, too) but also to enable us to experiment with this reality, without representing it. In mathematics, moreover, where all our ontologies and technologies are mental (although they can be embodied and communicated materially), one need not assume the independent reality, shared or not, of the type assumed, as material reality, in physics, beyond representation or even conception as the ultimate character of this reality may be assumed to be, as it is in quantum theory in nonrealist interpretations. Technically, it follows that, if this reality is beyond representation or even conception, it is not possible to rigorously claim that this reality as such is single any more than multiple. The "sameness" of this reality is itself an effect ascertainable by our measuring instruments, which, however, compel us to assume that we deal with the same types of quantum objects (electrons, photons, and so forth) and their composites, regardless where and when we perform our experiments. In any event, mathematical realities are always multiple, a circumstance of which, as we have seen, Grothendieck's topos theory takes advantage, as it also experiments with and even creates them. In fact, as I shall now suggest, mathematical realities always belong to individual human thinking, although they may be related to each other. Indeed, they always are so related, to one degree or another, which prevents them from being purely subjective. On the other hand, they can acquire a great degree of objectivity because they can be (re)constructed, for example in checking mathematical proofs. This situation is parallel to that of quantum physics, too, where, as explained earlier, this objectivity is provided by quantum phenomena, observed in measuring instrument, while, however, the ultimate material reality is still assumed.

The nature of mathematical reality, representable or not, has been debated since Plato, whose ghost still overshadows this debate in modernist mathematics, as Gray rightly suggests, and the differences between his and the present view of modernism do not affect this point [26]. Mathematical Platonism assumes the existence of mathematical reality, whether representable by our mathematical concept or not (there are different views on this point within mathematical Platonism), as independent of our thinking. I shall not enter these debates, including those concerning mathematical Platonism, apart from noting that the questions of the domain, "location," of this reality have, in my view, never been adequately answered. It is difficult to think of that which is not material and yet is outside human thought, unless it is divine, which, however, is not a common assumption among those who subscribe to mathematical Platonism. In any event, I only assume here the reality of human thought, thus, generally, different for each of us, as the only domain in relation to which one can speak of the reality of mathematical objects and concepts, in juxtaposition to the material reality of nature in physics. It is possible to assume that the Platonist mathematical reality is a potentiality—the same, even if multiply branched, potentiality—in principle realizable by human thought. I shall comment on this possibility presently.

What could be claimed to exist, ontologically, in our thought without much controversy are mathematical specifications, from strict definitions to partial and indirect characterizations (implying more complete or direct future specifications). Such specifications would involve concepts, structures, logical propositions, or still others elements, which could be geometrical or topological, as well as algebraic. In this respect, there is no difference between geometrical (or topological) and algebraic specifications. All such specifications can, at least in principle, be expressed and presented in language, verbally or in writing, visually, digitally for example, in other words technologically. While algebra helps our mathematical writing (to paraphrase Tate, "think geometrically, write algebraically!"), digital technology helps our geometrical specifications and expression. The computer-generated images of chaos theory are most famous, but low-dimensional topology and geometry have been similarly helped by digital technology.

The question, then, is whether anything else exists in our thought beyond such specifications and the local ontologies they define, at least in our conscious thought, because we can unconsciously think of other properties of a given object or field, which either may eventually be made conscious or possibly never become conscious and thus known. This qualification does not, however, change the nature of the question, because one could either claim that the unconscious could still only contain such specifications or that the object or broader reality in question, as different or exceeding such specifications, somehow exists there. In other words, essentially the same alternative remains in place. On the other hand, some unconscious specifications may never become conscious. It is quite possible that some mathematics, even very great mathematics, or for that matter poetry, never left the unconscious, and there are accounts by mathematicians or poets of dreaming of mathematics or poetry which could not be remembered as what it actually was, although sometimes it can. Only our memory of dreams is conscious, but never dreams themselves. By the same token, whatever is in our unconscious can at some point enter our consciousness: if the Chauvet Cave is the cave of dreams, it is because the consciousness of those who painted them realized their previously unconscious thinking, even though some parts of these painting were, undoubtedly, still manifestations of the unconscious. It is beyond doubt also that our unconscious does a great deal of mathematical thinking, as it does most of our thinking in general, and some of it may never be realized by our conscious thought. Sometimes, not uncommonly in the logical foundations of mathematics (Hilbert held this view), the consistency of a given definition of a mathematical entity is identified with the existence of the corresponding object, a form of mathematical Platonism, if this existence is assumed to be possible outside human thought. Either way, this view poses difficulties given Gödel's incompleteness theorems or even Cantor's set theory and its paradoxes. Our mathematical specifications must of course be logically consistent.

My assumption here is that nothing mathematical actually exists in thought beyond what can be thus specified, perhaps, again, in one's unconscious. This differentiates the situation from that of quantum physics. While quantum phenomena or quantum theories are specified in the same sense (and quantum theories are mathematical in the first place), one assumes, by a decision of thought concerning one's interpretation of quantum phenomena, the existence of the ultimate physical reality, which is beyond representation or even conception and thus specification. I leave aside for the moment whether something nonmathematical can exist in thought apart from any specification, although the position I take here compels me to answer in the negative in this case as well. In the present view, only physical matter in its ultimate constitution exists in this way. This assumption has been challenged as well, with Plato as the most famous ancient case and Bishop Berkeley as the most famous modern case, and is occasionally revived, as a possibility, in the context of quantum theory, but it is still a common assumption. As just noted, it is quite possible that there are (mentally) real things that exist in our unconscious that will never become conscious. It is equally possible, however, that they will enter our consciousness at one point or another. The ultimate constitution of nature, in this interpretation of quantum physics, is not assumed to ever become available, as things stand now. This does not of course preclude that such specifications cannot be made more complete or modified by new concepts, structures, and logical propositions, which would change the objects or concept in question, as say, a Riemann surface, as it developed during over, by now, a long period. However, in the present view, it is no longer possible to see such changes as referring to the same mathematical object (which can, again, be a broad and multiple entity), approached by our evolving concepts. Instead, they create new objects or concepts. Thus, new classes of Riemann surfaces are created by each modification of the concept. There are no Riemann surfaces as such, existing by themselves and in themselves, at any given point of time; there is only what we can think or say about them at a given point of time. By contrast, as things stand now, nothing can be in principle improved in our understanding of the ultimate constitution of matter. We can only improve our understanding and, by using mathematics and new experimental technology, our predictions of the effects on this constitution manifested in measuring instruments.

The present view of mathematical reality has thus a constructivist flavor, in part following Kant's view of mathematics as the synthetic construction of mathematical proposition and concepts by thought [35].[23] According to Badiou, "mathematics is a thought," part of the ontology of thought, and for Badiou this ontology is mathematical, an argument he makes via modernist mathematics, from Cantor's set theory to Grothendieck's topos theory [7, p. 45]. Without addressing Badiou's argument itself (different from the one offered here), I take his thesis literally in the sense that mathematics is only what can be thought, created by thought and then expressed, communicated, and so forth, thus also in accord with the Greek meaning of *máthēma* as that which can be known and learned, or taught.

It is of course not uncommon to encounter a situation in which a mathematical entity (again, possibly a large and multiple one) that cannot be given, now or possibly ever, an adequate mathematical specification, and is only specified partially mathematically or more fully otherwise. It may, for example, be specified as a phenomenal object or set of objects by means of philosophical concepts, but that can nevertheless be consistently related to, indirectly, and by means of a more properly specified mathematical concept or set of concepts. The latter concepts may, then, function as mathematical technologies which enable one to work with and, to the degree possible, understand this entity, as fibrations or homotopy and cohomology group allow us to understand better and more properly specify the corresponding topological spaces. These technologies are crucial and, while found in all mathematics, their persistent use, in part, against, relying on ontology, is characteristic of mathematical modernism, because of the persistence of the situations of the type just described. In the present view, however, any such entity can only be seen as existing or real if it is sufficiently specified in some way: in terms of phenomenal intuition or philosophical concepts, perhaps partially supplemented by mathematical concepts or structures. It cannot be assigned reality beyond such a specification.

In fact, as we know, in view of Gödel's incompleteness theorems, mathematics, at least if it is rich enough to contain arithmetic, cannot completely represent itself: it cannot mathematically formalize all of its concepts, propositions, or structures, and ultimately itself so as to guarantee its consistency. But it does not necessarily follow that the corresponding unspecifiable reality exists, although one can make this assumption, as Gödel ultimately did on Platonist lines, claiming that there is, at least for now, no human means, mathematical or other, to specify this reality. It only follows that it is impossible to prove that all possible specifications, within

[23] I do not refer by this statement to the trend known as "constructivism" in the foundational philosophy of mathematics, from intuitionism on, relevant as it may be, in part given Kant's influence. I use the term "constructivist" more generally.

any such system, are consistent. (Gödel's theorems do allow that the system can in fact be proven to be inconsistent.) What may be inconceivable is why this is the case, the reality that is responsible for it, which is, however, not a mathematical or even meta-mathematical question, any more than the question why our interaction with nature by means of quantum physics enables us to make correct probabilistic predictions. These questions belong to the biological and specifically neurological nature of our thought, although they may not ultimately be answerable by biology or neuroscience either.

One could speak in considering such as yet unspecified mathematical objects or concepts in terms of a hypothetical potentiality, defined by the assumption that a mathematical object or concept of a certain type could or should exist. Such potentialities are, moreover, only partially, probabilistically, determined by what is sometimes called plausible reasoning, very important in mathematical thinking, as rightly argued by Polya [58]. There are different and possibly incompatible way in which this potentially may be become reality. Consider, paradigmatically, thinking of the equation $X^2 + 1 = 0$ and complex numbers. While this equation (which may be safely assumed to exist in our thought as a mathematical entity) had no real solution, one could have and some had envisioned that it should have a solution and that a mathematical entity or a multiple of such entities, a new type of number, should exist. This hypothesis came to be realized, also literally, insofar, as complex numbers eventually became a mathematical reality. In the present view, however, they were not a mathematical reality before they were correspondingly specified in somebody's thought, say, by the time of Gauss, who was crucial in allowing complex numbers to become a mathematical reality, each time one thinks of them, but in the present view, not otherwise.

The present view, thus, precludes the assumption of an independent mathematical reality. This assumption, again, commonly defines reality in physics, even if this reality is assumed to be beyond representation or conception, as in quantum theory, thus, consistently with the present view of physical reality, as opposed to mathematical Platonism or other positions that claim the existence of mathematical reality independent of human thought, which is in conflict with the present view of mathematical reality. In sum, in the present view, in mathematics all reality is constructed, and this construction may, ontologically, involve multiple "mathematics," as Grothendieck's topos-theoretical ontology shows. This multiplicity is also a consequence of Gödel's undecidability, as exemplified by Cantor's continuum hypothesis, mentioned above. This hypothesis was crucial not only for the question of continuity but also for the question of Cantor's hierarchical order of infinities (the infinity of which was one of his discoveries) and thus for the whole edifice of Cantor's set theory. The hypothesis was proven undecidable by Cohen. It follows, however, that one can extend classical arithmetic in two ways by considering Cantor's hypothesis as either true or false, that is, by assuming either that there is no such intermediate infinity or that there is. This allows one, by decisions of thought, to extend arithmetic into mutually incompatible systems that one can construct, ultimately infinitely many such systems, because each on them will contain at least one undecidable proposition. It is, as noted, in principle possible to assume that all such possible constructions form a single, if multiply branching, potentiality, ultimately realizable in principle. I shall not assess this view except by noting that even if one adopted it strictly in this form, one would still only allow for a vast constructible mathematical potentiality rather than independent mathematical reality. Would this view be *in practice* equivalent to mathematical Platonism? While it may be in practice, the difference in principle would remain important, both in general and because that it would be impossible to assume that, being infinite, this potentiality could ever be realized. In practice, mathematics, again, creates new mathematical realities and, with them, new mathematical potentialities all the time, quite apart from any undecidable propositions. This process will only end when mathematics is no longer with us, and one day it might not be, although curves are likely to remain with us as long as we are around. By contrast, in quantum physics in nonrealist interpretations, the ultimate reality is assumed to exist as unconstructible or (as this view is still constructivist), *constructed* as unconstructible. But this unconstructible physical reality may be related to by means of constructed mathematical realities, such as that of Hilbert-spaces mathematics, again, meaning by a Hilbert space what we can think about or use and objectively share, rather than an independently existent mathematical object.

Thus, along with all realism in physics, the present view radically breaks with all Platonism in mathematics, especially with mathematical Platonism, but arguably with any form of Platonism hitherto. As I said, not all Platonism in mathematics is mathematical Platonism: that of Plato is not. Some forms of realism in physics are, again, forms of Platonism, too, as are, for example, some versions (known as ontological) of the so-called structural realism, according to which mathematical structures are the only reality [36]. As I indicated, Heisenberg, in his later thinking was inclined to this type of view, as against the time of his creation of quantum mechanics [33, pp. 91, 147–166].

While, however, breaking with Platonism, even Plato's own, the modernist thinking considered here in mathematics and physics does retain something, perhaps the most important thing, from Plato—from the *spirit* of Plato—rather than the *ghost* of Plato, intimately linked as these two words, spirit and ghost, are. This thinking retains the essential role of the movement of thought, something as crucial to Plato as to mathematical modernism, however anti-Platonist the latter may become. Heisenberg (whose father was a classicist) was reading Plato's *Timaeus* in the course of his discovery of quantum mechanics, in which he in effect reinvented Hilbert spaces over $\mathbb{C}$, a double, physical and mathematical, modernism [43, v. 2, pp. 11–14]. Some of Plato's thinking, led Heisenberg to his invention of a new mathematical technology in physics, under radically non-Platonist, epistemological assumptions. (Heisenberg, again, adopted a more Platonist view in his later thinking.) That this technology already existed in mathematics does not diminish the significance of this mathematical invention, especially given that Heisenberg used infinite unbounded matrices, never considered previously. The work of the mathematical figures considered here, from Fermat and Descartes to Riemann and from Riemann to Grothendieck and beyond, to split for a moment (but only for a moment) modernity and modernism, was shaped by the spirit of the movement of thought, the spirit that connects modernity and modernism, in mathematics and science, as it does in philosophy and art.

5.6 Conclusion

I close on a philosophical and artistic note by citing Heidegger's conclusion in "The Question Concerning Technology":

> There was a time when it was not technology alone that bore the name techne. Once that revealing that brings forth truth into the splendor of radiant appearing also was called techne.
>
> Once there was a time when the bringing forth of the true into the beautiful was called techne. And the poiesis of the fine arts also was called techne.
>
> In Greece, at the outset of the destining of the West, the arts soared to the supreme height of the revealing granted them. . . . And art was simply called techne. It was a single, manifold revealing. It was . . . , promos, i.e., yielding to the holding-sway and the safekeeping of truth.
>
> The arts were not derived from the artistic. Art works were not enjoyed aesthetically. Art was not a sector of cultural activity.
>
> What, then, was art—perhaps only for that brief but magnificent time? Why did art bear the modest name techne? Because it was a revealing that brought forth and hither, and therefore belonged within poiesis. It was finally that revealing which holds complete sway in all the fine arts, in poetry, and in everything poetical that obtained poiesis as its proper name. . . .
>
> Whether art may be granted this highest possibility of its essence in the midst of the extreme danger [of modern technology], no one can tell. Yet we can be astounded. Before what? Before this other possibility: that the frenziedness of technology may entrench itself everywhere to such an extent that someday, throughout everything technological, the essence of technology may come to presence in the coming-to-pass of truth.
>
> Because the essence of technology is nothing technological, essential reflection upon technology and decisive confrontation with it must happen in a realm that is, on the one hand, akin to the essence of technology and, on the other, fundamentally different from it.
>
> Such a realm is art. But certainly only if reflection on art, for its part, does not shut its eyes to the constellation of truth after which we are questioning.
>
> Thus questioning, we bear witness to the crisis that in our sheer preoccupation with technology we do not yet experience the coming to presence of technology, that in our sheer aesthetic-mindedness we no longer guard and preserve the coming to presence of art. Yet the more questioningly we ponder the essence of technology, the more mysterious the essence of art becomes. [30, pp. 34–35]

I would argue that modernist mathematics, in its more expressly technological aspects and in general, and physics, where in quantum theory all mathematics used is a technology, are *techne* in a sense *close* to that Heidegger wants to give this term here. The reason that I see them as close rather than the same is that Heidegger would allow that the ultimate reality could be accessed by what he saw as the true thought, which he saw as artistic or poetic thought in the sense of this passage. In contrast to some (including some modernist) poetry and art, he sees modernist technology (in its conventional sense) and modernist mathematics and science, including, one might plausibly surmise, as it is understood here, as a form of forgetting rather than approaching techne as art found in ancient Greek thinking.

It is not even clear that he would grant this to the ancient Greek mathematics, much as he admired ancient Greek, especially pre-Socratic, thought, in philosophy and poetry, and he sees the forgetting of the thought in question as beginning with Socrates and Plato. Admittedly, Heidegger's position is complex, especially insofar as how artistic thought can do this remains "mysterious," the mystery that appears to be deepened by our attempts to understand the essence of our, modern and modernist, technology. Nevertheless, Heidegger allows at least the possibility of *thinking* [*Denken*] (his preferred term) this truth, even if not representing it. I would contend, however, that modernist epistemology, even when, in its most radical form, it places the ultimate nature of reality beyond thought itself in physics or rejects the existence of such a (single) ultimate reality in mathematics altogether, does not preclude thought from reaching "the supreme height of the revealing granted them," albeit "creation" might be a better word than "revealing," if there is no ultimate reality that can be revealed. Even if it exists, as in physics, it still cannot be revealed, and in mathematics, again, everything is created, constructed. Coming together of techne and truth is still possible under these conditions and is perhaps not possible otherwise, regardless of one's aspirations for how far our thought can reach. We cannot dispense with truth. What changes are the relationships between truth and reality, and both concepts themselves, while realism and the corresponding concepts of truth still apply and are indispensable at surface levels. *Techne* and truth do come together under these conditions.

This, I have argued here, is precisely what happens in the thought of Riemann, Hilbert, Weyl, Weil, and Grothendieck, and those who followed them in mathematics, or their predecessors, from Fermat and Descartes, or the thought of those who used mathematics in physics, from Kepler and Galileo to Einstein and Heisenberg, and beyond, the Platonist or realist aspirations of many, even most, of these figures notwithstanding. Their thought continues, in mathematics or physics, not the least when it comes to the idea of curve, even when a curve is a surface, the project of the painters of curves of the Chauvet Cave, the cave of dreams, no longer forgotten. The discovery of the cave gave these dreams back to us, and these dreams are about much more than curves, just as modernist art, such as that of Klee, or the modernist mathematics of curves are so much more.

Perhaps, however, our history has kept these dreams alive all along by keeping alive the creative nature of our thought, dreams that we began dreaming well before the frescoes of the Chauvet Cave were painted. Some form of mathematical thinking, just as some form of artistic or philosophical thinking, must have always been part of our history as thinking beings and our dreams, in either sense. The history that at some point gave (we may never know how!) our brain the capacity to have these dreams is immeasurably longer, ultimately as long as the history of life or even the Universe itself, in which, at some point, life has emerged.

Acknowledgements I am grateful to Athanase Papadopoulos for reading the original draft of the article and helpful suggestions for improving it, and for productive exchanges on the subjects considered here.

References

1. N. A'Campo, L. Ji, A. Papadopoulos "On Grothendieck's construction of the Teichmüller space," in A. Papadopoulos (ed.) *Handbook of Teichmüller Theory, Volume VI* (European Mathematical Society, Zürich, 2016), pp. 35–70.
2. N. A'Campo, L. Ji, A. Papadopoulos, "Action of the absolute Galois Group," in A. Papadopoulos (ed.) *Handbook of Teichmüller Theory, Volume VI* (European Mathematical Society, Zürich, 2016), pp. 397–436.
3. N. A'Campo, L. Ji, A. Papadopoulos, "On Grothendieck's tame topology," in Papadopoulos, A. (ed.) *Handbook of Teichmüller Theory, Volume VI* (European Mathematical Society, Zürich, 2016), pp. 521–536.
4. M. A. Artin, *Grothendieck Topology* (Harvard Department of Mathematics, Cambridge, MA, 1962).
5. M. A. Artin, Grothendieck, J-L. Verdier, *Séminaire de Géométrie Algébrique du Bois Marie, 1963–1964, Théorie des topos et cohomologie étale des schémas - (SGA 4), vol. 1* (Springer, Berlin/New York, 1972).
6. A. Badiou, *Briefings on Existence* (SUNY Press, Albany, NY, 2006). Translated by N. Madarasz.
7. J. Bair, P. Blaszczyk, P. Heinig, M. Katz, J. P. Schafermeyer, D. Sherry, "Klein vs. Mehrtens: Restoring the Reputation of a Great Modern," arXiv[math.HO]1803.02193v.1 (2008).
8. N. Bohr, *The Philosophical Writings of Niels Bohr*, 3 vols. (Ox Bow Press, Woodbridge, CT, 1987).
9. M. Born, P. Jordan, "Zur Quantenmechanik" *Zeitschrift für Physik* 34, 858–888 (1925).
10. U. Bottazzini, J. Gray, *Hidden Harmony—Geometric Fantasies: The Rise of Complex Function Theory* (Springer, Berlin, 2013).
11. P. Cartier, (2001) "Mad Days Work: From Grothendieck to Connes and Kontsevitch, The Evolution of Concepts of Space and Symmetry," *Bulletin of the American Mathematical Society* 38 (4), 389–408 (2001).
12. P. Cohen, *Set Theory and the Continuum Hypothesis* (Dover, Mineola, NY, 2008).
13. A. Connes, *Noncommutative Geometry* (Academic Press, San Diego, CA, 1994). Translated by S. K. Berberian.
14. A. Connes, M. Marcoli, *Noncommutative Geometry, Quantum Fields, and Motives* (American Mathematical Society, Hindustan Book Agency, 2007).
15. L. Corry, "How Useful is the Term 'Modernism' for Understanding the History of Early Twentieth-Century Mathematics?" in M. Epple, F. Müller (eds.) *Science as a Cultural Practice: Modernism in Science* (de Gruyter, Berlin, 2019).
16. O. Darigold, *From c-Numbers to q-Numbers: The Classical Analogy in the History of Quantum Theory* (University of California Press, Berkeley, CA, 1993).
17. G. Deleuze, *The Fold: Leibniz and the Baroque* (Continuum, New York 2005). Translated by T. Conley.
18. Descartes, *Discourse on Method, Optics, Geometry, and Meteorology* (Hackett, Indianapolis, IN, 2001). Translated by P. Olscamp.
19. A. Doxiadis, B. Mazur (eds.) *Circles Disturbed: The Interplay of Mathematics and Narrative* (Princeton University Press, Princeton, NJ, 2012).
20. Einstein, "Physics and reality," *Journal of the Franklin Institute* 221, 349–382 (1936).
21. C. Erkelens, "The Perceptual Structure of Visual Space," *i-Perception* 6 https://doi.org/10.1177/2041669515613672 (2015).
22. Euclid, *Thirteen Books of the Elements*, 3 vols, T. L. Heath (ed.) (Dover, Mineola, NY, 1989).
23. G. Farmelo, G. "Dirac's Hidden Geometry", *Nature* 437, 323 (2005).
24. S. Feferman, "Modernism in Mathematics," *American Scientist* 97 (5), 417 (2009).
25. J. Foley, N. Ribeiro-Filho, J. Da Silva, "Visual Perception of extend and the geometry of visual space," *Vision Research* 44, 147–156 (2004).

26. J. Gray, *Plato's Ghost: The Modernist Transformation of Mathematics* (Princeton University Press, Princeton, NJ, 2007).
27. M. Gromov, "Spaces and Questions," in N. Alon, J. Bourgain, A. Connes, M. Gromov, V. Milman (eds.), *Visions in Mathematics* (Borkhäuser, Boston, MA, 2010), pp. 118–161.
28. I. Hacking, *The Emergence of Probability: A Philosophical Study of Early Ideas about Probability, Induction and Statistical Inference* (2nd edition), Cambridge University Press, Cambridge, UK, 2006).
29. Heidegger, *What is a Thing?* (Gateway, South Bend, IN, 1967). Translated by W. B. Barton and V. Deutsch (1967).
30. Heidegger, *The Question Concerning Technology, and Other Essay* (Harper, New York, 2004).
31. Heisenberg, "Quantum-theoretical re-interpretation of kinematical and mechanical relations" (1925), in B. L. Van der Waerden (ed.) *Sources of Quantum Mechanics, Dover*, Mineola, NY, 1968), pp. 261–277.
32. Heisenberg, *The Physical Principles of the Quantum Theory* (Dover, Mineola, NY, 1930, rpt. 1949). Translated by K. Eckhart and F. C. Hoyt.
33. Heisenberg, *Physics and Philosophy: The Revolution in Modern Science* (Harper & Row, New York, 1962).
34. D. Hilbert, *Foundations of Geometry* (Open Court, La Salle, IL, 1999). Translated L. Unger and P. Bernays.
35. Kant, (1997) *Critique of Pure Reason* (Cambridge, UK: Cambridge University Press, 1997). Translated by P. Guyer and A. W. Wood.
36. J. Ladyman, "Structural Realism," *The Stanford Encyclopedia of Philosophy* (Winter 2016 Edition), Zalta, E. N. (ed.), URL = https://plato.stanford.edu/archives/win2016/entries/structural-realism/ (2016).
37. J-F. Lyotard, *The Postmodern Condition: A Report on Knowledge* (University of Minnesota Press, Minneapolis, MN, 1984). Translated by G. Bennington and B. Massumi.
38. Y. Manin, "Georg Cantor and His Heritage," in *Algebraic Geometry: Methods, Relations, and Applications: Collected Papers dedicated to the memory of Andrei Nikolaevich Tyurin*. Proc. Steklov Inst. Math. vol. 246 (2004), pp. 195–203. Also published in: Mathematics as Metaphor: Selected Essays of Yuri I. Manin, American Mathematical Society, Providence, R.I., 2007, pp. 45–54.
39. J-P. Marquis, "A Path to Epistemology of Mathematics: Homotopy Theory," in: J. Ferreir, G. Gray (eds.) *The Architecture of Modern Mathematics: Essays in History and Philosophy* (Oxford University Press, Oxford, 2006), pp. 239–260.
40. G. E. Martin, *The Foundations of Geometry and the Non-Euclidean Plane* (Springer, Berlin, 1975).
41. B. Mazur, "Visions, Dreams, and Mathematics," in A. Doxiadis, B. Mazur (eds.) *Circles Disturbed: The Interplay of Mathematics and Narrative* (Princeton University Press, Princeton, NJ, 2012), pp. 183–210.
42. C. McLarty, "Emmy Noether's set-theoretical topology: From Dedekind to the rise of functors," in J. Ferreir, G. Gray (eds.) *The Architecture of Modern Mathematics: Essays in History and Philosophy* (Oxford University Press, Oxford, 2006), pp. 211–236.
43. J. Mehra, H. Rechenberg, *The Historical Development of Quantum Theory*, 6 vols. (Springer, Berlin, 2001).
44. H. Mehrtens, *Moderne Sprache, Mathematik: Eine Geschichte des Streits um die Grundlagen der Disziplin und des Subjekts formaler Systeme* (Suhrcampf, Frankfurt am Main, 1990).
45. Newton, Sir I., *The Principia: Mathematical Principles of Natural Philosophy* (University of California Press; Berkeley, CA, 1999). Translated by I. B. Cohen and A. Whitman.
46. Nietzsche, *Beyond Good and Evil: A Prelude to a Philosophy of the Future* (Vintage, New York, 1966). Translated by W. Kaufmann.
47. A. Papadopoulos, "Riemann Surfaces: Reception by the French School," in L. Ji, A. Papadopoulos, S. Yamada (eds.) *From Riemann to Differential Geometry and Relativity* (Springer, Berlin, 2017), pp. 237–294.

48. A. Papadopulos, "Topology and Biology: From Aristotle to Thom," S.G. Dani, A. Papadopoulos (eds.) *Geometry in History* (Springer, Cham, 2019). https://doi.org/10.1007/978-3-030-13609-3_2.
49. É. Picard, *Traité d'analyse*, 3 volumes (Gauthier-Villars, Paris, 1891, 1893, 1896).
50. A. Plotnitsky, *The Knowable and the Unknowable: Modern Science, Nonclassical Thought, and the "Two Culture"* (University of Michigan Press, Ann Arbor, MI, 2002).
51. A. Plotnitsky, "Algebras, Geometries, and Topologies of the Fold: Deleuze, Derrida, and Quasi-Mathematical Thinking, with Leibniz and Mallarmé," in P. Patton, J. Protevi (eds.) *Between Deleuze and Derrida* (New York: Continuum, 2003).
52. A. Plotnitsky, *Epistemology and Probability: Bohr, Heisenberg, Schrödinger and the Nature of Quantum-Theoretical Thinking* (Springer, New York, 2009).
53. A. Plotnitsky, "Adventures of the Diagonal: Non-Euclidean Mathematics and Narrative," in A. Doxiadis, B. Mazur (eds.) *Circles Disturbed: The Interplay of Mathematics and Narrative* (Princeton University Press, Princeton, NJ, 2012), pp. 407–446.
54. A. Plotnitsky, *The Principles of Quantum Theory, from Planck's Quanta to the Higgs Boson: The Nature of Quantum Reality and the Spirit of Copenhagen* (Springer/Nature, New York, 2016).
55. A. Plotnitsky, "'Comprehending the Connection of Things:' Bernhard Riemann and the Architecture of Mathematical Concepts," in L. Ji, A. Papadopoulos, S. Yamada (eds.) *From Riemann to Differential Geometry and Relativity* (Springer, Berlin, 2017), pp. 329–363.
56. A. Plotnitsky, "The Heisenberg Method": Geometry, Algebra, and Probability in Quantum Theory, *Entropy* 20 (9), 656–702 (2018).
57. H. Poincaré, *Oeuvres de Henri Poincaré*, vol. 1, P. Appel, J. Drach (eds.) (Gautier-Villars, Paris, 1928).
58. G. Polya, *Mathematics and Plausible Reasoning, Volume 1: Induction and Analogy in Mathematics* (Princeton, University Press, Princeton, NJ, 1990).
59. D. Reed, *Figures of Thought: Mathematics and Mathematical Texts* (Rutledge, London, 1995).
60. Riemann, "On the Hypotheses That Lie at the Foundations of Geometry" (1854) in P. Pesic (ed.), *Beyond Geometry: Classic Papers from Riemann to Einstein* (Dover, Mineola, NY, 2007) pp. 23–40.
61. S. S. Schweber, *QED and the Men Who Made It: Dyson, Feynman, Schwinger, and Tomonaga* (Princeton University Press, Princeton, NJ, 1994).
62. J. Silverman, J. Tate, *Rational Points on Elliptic Curves* (Springer, Heidelberg/New York, 2015).
63. P. Suppes, "Is Visual Space Euclidean?" *Synthese* 35, 397–421 (1977).
64. G. Tomlinson, *Million Years of Music: The Emergence of Human Modernity* (MIT Press, Cambridge, MA, 2015).
65. B. L. Van der Waerden, *Moderne Algebra. Teil I* (Springer, Berlin, 1930).
66. J. Von Neumann, *Mathematical Foundations of Quantum Mechanics* (Princeton University Press, Princeton, NJ, 1932, rpt. 1983). Translated by R. T. Beyer.
67. A. Weil, *Number Theory: An Approach through History from Hammurapi to Legendre* (Springer, Berlin, 2001).
68. H. Weyl, *Space Time Matter* (Dover, Mineola, NY, 1952). Translated by H. L. Brose.
69. H. Weyl, *The Continuum: A Critical Examination of the Foundation of Analysis* (Dover, Mineola, NY, 1928, rpt. 1994). Translated by S. Pollard and T. Bole.
70. H. Weyl, *Theory of Groups and Quantum Mechanics* (Dover, Mineola, NY, 1928, rpt. 1984).
71. H. Weyl, "David Hilbert and his mathematical Work," *Bulletin of the American Mathematical Society* 50, 612–654 (1944).
72. H. Weyl, *Philosophy of mathematics and Natural Science* (Princeton University Press, Princeton, NJ, 2009)
73. H. Weyl, *The Concept of a Riemann Surface* (Dover, Mineola, NY, 2013). Translated by G. L. MacLane.

Chapter 6
From Euclid to Riemann and Beyond: How to Describe the Shape of the Universe

Toshikazu Sunada

Dedicated to the memory of Marcel Berger (14 April 1927–15 October 2016).

The purpose of this essay is to trace the historical development of geometry while focusing on how we acquired mathematical tools for describing the "shape of the universe." More specifically, our aim is to consider, without a claim to completeness, the origin of Riemannian geometry, which is indispensable to the description of the space of the universe as a "generalized curved space."

But what is the meaning of "shape of the universe"? The reader who has never encountered such an issue might say that this is a pointless question. It is surely hard to conceive of the universe as a geometric figure such as a plane or a sphere sitting in space for which we have vocabulary to describe its shape. For instance, we usually say that a plane is "flat" and "infinite," and a sphere is "round" and "finite." But in what way is it possible to make use of such phrases for the universe? Behind this inescapable question is the fact that the universe is not necessarily the ordinary 3D (3-dimensional) space where the traditional *synthetic geometry*—based on a property of parallels which turns out to underlie the "flatness" of space—is practised. Indeed, as Einstein's theory of *general relativity* (1915) claims, the universe is possibly "curved" by gravitational effects. (To be exact, we need to handle 4D *curved space-times*; but for simplicity we do not take the "time" into consideration, and hence treat the "static" universe or the universe at any instant of time unless otherwise stated. We shall also disregard possible "singularities" caused by "black holes.")

An obvious problem still remains to be grappled with, however. Even if we assent to the view that the universe is a sort of geometric figure, it is impossible for us to look out over the universe all at once because we are strictly confined in it. How can

T. Sunada (✉)
Meiji Institute for the Advanced Study of Mathematical Sciences, Meiji University, Nakano-ku, Tokyo, Japan
e-mail: sunada@meiji.ac.jp

S. G. Dani, A. Papadopoulos (eds.), *Geometry in History*,
https://doi.org/10.1007/978-3-030-13609-3_6

we tell the shape of the universe despite that? Before Albert Einstein (1879–1955) created his theory, mathematics had already climbed such a height as to be capable to attack this issue. In this respect, Gauss and Riemann are the names we must, first and foremost, refer to as mathematicians who intensively investigated curved surfaces and spaces with the grand vision that their observations have opened up an entirely new horizon to *cosmology*. In particular, Riemann's work, which completely recast 3000 years of geometry executed in "space as an a priori entity," played an absolutely decisive role when Einstein established the theory of general relativity.

Gauss and Riemann were, of course, were not the first to be involved in cosmology. Throughout history, especially from ancient Greece to Renaissance Europe, mathematicians were, more often than not, astronomers at the same time, and hence the links between mathematics and cosmology are ancient, if not in the modern sense. Meanwhile, the venerable history of cosmology (and cosmogony) overlaps in large with the history of human thought, from a reflection on primitive religious concepts to an all-embracing understanding of the world order by dint of reason. It is therefore legitimate to lead off this essay with a rough sketch of philosophical and theological aspects of cosmology in the past, while especially focusing on the image of the universe held by scientists (see Koyré [16] for a detailed account). In the course of our historical account, the reader will see how cosmology removed its religious guise through a long process of secularization and was finally established on a firm mathematical base. To be specific, Kepler, Galileo, Fermat, Descartes, Newton, Leibniz, Euler, Lagrange, and Laplace are on a short list of central figures who plowed directly or indirectly the way to the mathematization of cosmology.

The late nineteenth century occupies a special position in the history of mathematics. It was in this period that the autonomous progress of mathematics was getting apparent more than before. This is particularly the case after the notion of *set* was introduced by Cantor. His theory—in concord with the theory of topological spaces—allowed to bring in an entirely new concept of abstract space, with which one may talk not only about *(in)finiteness* of the universe in an *intrinsic manner*, an issue inherited since classical antiquity, but also about a global aspect of the universe despite that we human beings are confined to a very tiny and negligible planet in immeasurable space. In all these revolutions, Riemann's theory (in tandem with his embryonic research of *topology*) became encompassed in a broader, more adequate theory, and eventually led, with a wealth of new ideas and methodology, to modern geometry.

It is not too much to say that geometry as such (and mathematics in general) is part of our cultural heritage because of the profound manifestations of the internal dynamism of human thought provoked by a "sense of wonder" (the words Aristotle used in the context of philosophy). Actually, its advances described in this essay may provide an indicator that reveals how human thought has progressed over a long period of history.

In this essay, I do not, deliberately, engage myself in the cutting-edge topics of differential geometry which, combined with topology, analysis, and algebraic geometry, have been highly cultivated since the latter half of the twentieth century.

I also do not touch on the *extrinsic* study of manifolds, i.e., the theory of configurations existing in space—no doubt an equally significant theme in modern differential geometry. In this sense, the bulk of my historical commentary might be quite a bit biased towards a narrow range of geometry. The reader interested in a history of geometry (and of mathematics overall) should consult Katz [14].

Acknowledgements I wish to thank my colleague Jim Elwood whose suggestions were very helpful for improvement of the first short draft of this essay. I am grateful to S.G. Dani, Athanase Papadopoulos, Ken'ichi Ohshika, and Polly Wee Sy for the careful reading and for all their invaluable comments and suggestions to the enlarged version.

6.1 Ancient Models of the Universe

Since the inception of civilization that emerged in a number of far-flung places around the globe like ancient Egypt, Mesopotamia, ancient India, ancient China, and so on, mankind has struggled to understand the universe and especially how the world came to be as it is.[1] Such attempts are seen in mythical tales about the birth of the world. "Chaos", "water", and the like, were thought to be the fundamental entities in its beginning that was to grow gradually into the present state. For example, the epic of *Atrahasis* written about 1800 BCE contains a creation myth about the Mesopotamian gods *Enki* (god of water), *Anu* (god of sky), and *Enlil* (god of wind). The Chinese myth in the *Three Five Historic Records* (the third century CE) tells us that the universe in the beginning was like a big egg, inside of which was darkness, chaos.[2]

The question of the origin is tied to the question of future. Hindu cosmology has a unique feature in this respect. In contrast to the didactics in monotheistic Christianity describing the end of the world as a single event (with the Last Judgement) in history, the *Rig veda*, one of the oldest extant texts in Indo-European language composed between 1500 BCE–1200 BCE, alleges that our cosmos experiences a creation-destruction cycle almost endlessly (the view celebrated much later by Nietzsche). Furthermore, some literature (e.g. the *Bhagavata Purana* composed between the eighth and tenth century CE or as early as the sixth century CE) mentions the "multiverse" (infinitely many universes), which resuscitated as the modern astronomical theory of the *parallel universes*.

In the meantime, thinkers in the colonial towns in Asia Minor, Magna Graecia, and mainland Greece, cultivated a love for systematizing phenomena on a rational basis, as opposed to supernatural explanations typified by mythology and folklore

[1]To know the birth and evolution of the universe is "the problem of problems" at the present day as seen in the *Big Bang theory*, the most prevailing hypothesis of the birth.

[2]The term "chaos" (χάος) is rooted in the poem *Theogony*, a major source on Greek mythology, by Hesiod (ca. 750 BCE–ca. 650 BCE). The epic poet describes Chaos as the primeval emptiness of the universe.

in which the cult of Olympian gods and goddesses are wrapped. Many of them were not only concerned with fundamental issues arising from everyday life, as represented by Socrates (ca. 470 BCE–399 BCE), but also labored to mathematically understand multifarious phenomena and to construct an orderly system. They appreciated purity, universality, a certainty and an elegance of mathematics, the characteristics that all other forms of knowledge do not possess. Legend has it that Plato (ca. 428 BCE–ca. 348 BCE) engraved the phrase "Let no one ignorant of geometry enter here" at the entrance of the *Academeia* (Ἀκαδημία) he founded in ca. 387 BCE in an outskirt of Athens. Whether or not this is historically real, Academeia indubitably put great emphasis on mathematics as a prerequisite of philosophy.[3]

Eventually, Greek philosophers began to speculate about the structure of the universe by deploying geometric apparatus. Exemplary is the *spherical model*, with the earth at the center, proposed by Plato himself, and his two former students Eudoxus of Cnidus (ca. 408 BCE–ca. 355 BCE) and Aristotle from Stagira (384 BCE–322 BCE).[4] In his dialogue *Timaeus*, Plato explored cosmogonical issues, and deliberated the nature of the physical world and human beings. He referred to the *demiurge* (δημιουργός) as the creator of the world who chose a *round sphere* as the most appropriate shape that embraces within itself all the shapes there are. Meanwhile, Eudoxus proposed a sophisticated system of homocentric spheres rotating about different axes through the center of the Earth, as an answer to his mentor's question how to reduce the apparent motions of heavenly bodies to uniform circular motions. This was stated in the treatise *On velocities*—now lost, but Aristotle knew about it. Eudoxus may have regarded his system simply as an abstract geometrical model; Aristotle took it to be a description of the real world, and organized it into a kind of fixed hierarchy, conjoining with his metaphysical principle (*Metaphysics*, XII).

Aristotle's spherical model (Fig. 6.1) was refined later to a sophisticated geocentric theory by the Alexandrian astronomer Ptolemy (ca. 100 CE–ca. 170 CE). The latter's vision of the universe was set forth in his mathematico-astronomical treatise *Almagest*, and had been accepted for more than 1200 years in Western Europe and the Islamic world until Copernicus' heliocentric theory emerged (Sect. 6.3) because his theory succeeded, up to a point, in describing the apparent motions of the sun,

[3]Among Plato's thirty-five extant dialogues, there are quite a few in which the characters (Socrates in particular) discuss mathematical knowledge in one form or another; say, *Hippias major*, *Meno*, *Parmenides*, *Theaetetus*, *Republic*, *Laws*, *Timaeus*, *Philebus*. In the *Republic*, Book VII, Socrates says, "Geometry is fully intellectual and a study leading to the good because it deals with absolute, unchanging truths, so that it should be part of education."

[4]Pythagoras of Samos (ca. 580 BCE–ca. 500 BCE) was the first to declare that the earth is a sphere, and that the universe has a soul and intelligence. Plato was a devout *Pythagorean*, originally meaning the member of a mathematico-religious community created by Pythagoras (ca. 530 BCE) in Croton, Southern Italy.

Fig. 6.1 Aristotle's model (P. Apianus, *Cosmographicus liber*, 1524)

moon, and moon, and planets.[5] What is notable in his system is the use of numerous *epicycles* (ἐπίκυκλος), where an epicycle is a small circle along which a planet is assumed to move, while each epicycle in turn moves along a larger circle (*deferent*).[6] This schema—a coinage of the Greek doctrine—was first set out by Apollonius of Perga (ca. 262 BCE–ca. 190 BCE), and developed further by Hipparchus of Nicaea (ca. 190 BCE–ca. 120 BCE).

[5] The title "Almagest" was derived from the Arabic name meaning "greatest." The original Greek title is *Mathematike Syntaxis* (Μαθηματικὴ Σύνταξις). It was rendered into Latin by Gerard of Cremona (ca. 1114–1187) from the Arabic version found in Toledo (1175), and became very widely known in Western Europe before the Renaissance.

[6] The reasonable accuracy of Ptolemy's system results from that an "almost periodic" motion (in the sense of H. A. Bohr)—a presumable nature of planetary motions—can be represented to any desired degree of approximation by a superposition of circular motions.

At all events, an intriguing (and arguable) feature of their model is the idea of unreachable "outermost sphere." In Aristotle's concept (*Physics*, VIII, 6), it is the domain of the Prime Mover (τὸ πρώτη ἀκίνητον), a variant of Plato's notion in his cosmological argument unfolded in the *Timaeus*, which caused the outermost sphere to rotate at a constant angular velocity.

We shall come back to the spherical model in Sect. 6.3 after discussing a relevant issue, and recount at some length how this peculiar model had an effect on philosophical and religious aspects of cosmology.

6.2 What Is Infinity?

Besides the kinematical nature of celestial bodies, what seems to lie in the background of Aristotle's view about the universe is his philosophical thought about *infinity*. Actually the issue of infinity was a favorite subject for Greek philosophers, dating back to the pre-Socratic period.[7]

Deliberating over his predecessors' vision, Aristotle distinguished between *actual* and *potential* infinity. Briefly speaking, actual infinity is "things" that are completed and definite, thereby being *transcendental* in nature, while potential infinity is "things" that continue without terminating; more and more elements can be always added, but with no recognizable ending point, thus being what can be somehow corroborated within the scope of the capacity of human deed or thought. For a variety of possible reasons, Aristotle rejects actual infinity, claiming that only potential infinity exists, and captures space as something infinitely divisible into parts that are again infinitely divisible, and so on (*Physics*, III). His doctrine, which had satisfied nearly all scholars for a long time,[8] was employed by himself to find a way out of *Zeno's paradoxes* , especially the paradox (παράδοξος) of Achilles and the and the Tortoise.[9] This, in a rehashed form, says, "*The fastest runner [Achilles]*

[7]Anaximander of Miletus (ca. 610 BCE–ca. 546 BCE) was the first who contemplated about infinity, and employed the word *apeiron* (ἄπειρον meaning "unlimited") to explain all natural phenomena in the world. Anaxagoras of Clazomenae (ca. 510 BCE–ca. 428 BCE) wrote the book *About Nature*, in which he says, "All things were together, infinite in number." Assuming the infinite divisibility of matter, he avers, "There is no smallest among the small and no largest among the large, but always something still smaller and something still larger."

[8]G. W. F. Hegel, a pivotal figure of German *idealism*, defended Aristotle's perspective on the infinite in his *Wissenschaft der Logik* (1812–1816), though he used the terms "true (absolute) infinity" and "spurious infinity (*schlecht Unendlichkeit*)" instead.

[9] Zeno of Elea (ca. 490 BCE–ca. 430 BCE) was a student of Parmenides (ca. 515 BCE–ca. 450 BCE) who contended that the true reality is absolutely unitary, unchanging, eternal, "the one." To vindicate his teacher's tenet, Zeno offered the four arguable paradoxes "The Dichotomy," "Achilles and the Tortoise," "The Arrow," and "The Stadium."

During the Warring States period (476 BCE–221 BCE) in China, the *School of Names* cultivated a philosophy similar to Parmenides'. Hui Shi (ca. 380 BCE–ca. 305 BCE) belonging to this school says, "Ultimate greatness has no exterior, ultimate smallness has no interior."

in a race can never overtake the slowest [tortoise], because the pursuer must first get to the point whence the pursued started, so that the slowest must always hold a lead" (*Physics*, VI; 9, 239b15). To quote Aristotle's reaction in response to the quibble, Zeno's argument exploits an ambiguity in the nature of 'infinity' because Zeno seems to insist that Achilles cannot *complete* an infinite number of his actions (getting the point where the tortoise was); that is, 'complete' is the word for actual infinity, but 'infinite number of actions' is the phrase for potential infinity.

Aristotle's resolute rejection of actual infinity seems in part to come from the fact that mathematicians of those days had no adequate manner to treat continuous magnitude and could do, all in all, quite well without actual infinity. This is distinctively seen in the statement in Euclid's *Elements*, Book IX, Prop. 20, about the infinitude of prime numbers, which deftly asserts, "Prime numbers are more than any assigned multitude of prime numbers."[10]

On the other hand, it appears that Archimedes of Syracuse (ca. 287 BCE–ca. 212 BCE) had a prescient view of infinity, as adumbrated in the *Archimedes Palimpsest*,[11] a tenth-century Byzantine Greek copy housed at the Metochion of the Holy Sepulcher in Jerusalem. To our astonishment, in the 174-pages text (specifically in the *Method*, Prop. 14), he duly compared two infinite collections of certain geometric objects by means of a *one-to-one correspondence* (henceforth OTOC); see [21]. This is surely related to the concept of actual infinity that has been revived in nineteenth century (Sect. 6.18).

Archimedes was also a master of the *method of exhaustion* (or the method of double contradiction) which originated, however incomplete, with Antiphon the Sophist (ca. 480 BCE–ca. 411 BCE) and Bryson of Heraclea (born ca. 450 BCE), and exploited by Eudoxus to avoid flaws that may happen when we treat infinity in a naive way. Such a flaw is found in the claim by the two originators; they contended that it is possible to construct, with compass and straightedge, a square with the same area as a given circle C. Their argument (criticized roundly by Aristotle in the *Posterior Analytics* I, 9, 75b40) is as follows: From the correct fact that such a construction is possible for a given polygon (Euclid's *Elements*, Book II, Prop. 14), they elicited the incorrect consequence that the same is true for C on the grounds that the regular polygon with 2^n edges inscribed in C *eventually coincides with* C as we let n increase endlessly (needless to say, a passage to a limit does not necessarily preserve the given properties).[12]

[10]Euclid does not seem to entirely refrain from the use of non-potential infinity. In the *Elements*, Book X, Def. 3, he says, "there exist straight lines *infinite in multitude* which are commensurable with a given one" [7, Book X, p. 10].

[11]In 1906, J. L. Heiberg, the leading authority on Archimedes, confirmed that the palimpsest, overwritten with a Christian religious text by thirteenth-century monks, included the *Method of Mechanical Theorems*, one of Archimedes' lost works by that time. Our knowledge of Archimedes was greatly enriched by this fabulous discovery.

[12]The approach taken by Antiphon and Bryson, however incorrect, was appropriately adopted by Archimedes, who obtained $223/71 < \pi < 22/7$ using two regular polygons of 96 sides inscribed and circumscribed to a circle (*Measurement of the Circle*).

Incidentally, the problem Antiphon and Bryson challenged is designated "squaring the circle", one of the three big problems on constructions in Greek geometry; the other problems are "doubling the cube" and "trisecting the angle." Anaxagoras is the first who worked on squaring the circle, while Hippocrates of Chios (ca. 470 BCE–ca. 410 BCE) squared a *lune*. This held out hope to square the circle for a time, but all attempts met with failure. It was in 1882 that this problem came to a conclusion; Ferdinand von Lindemann (1852–1939) proved the impossibility of such a construction by showing that $\pi = 3.141592\cdots$ is a *transcendental number*; i.e., π is not a solution of an algebraic equation with integral coefficients (1882). Doubling the cube and trisecting the angle are also impossible as P. L. Wantzel showed (1837).

In the background of the method of exhaustion is the premise that a given quantity α (not necessarily numerical) can be made smaller than another one given beforehand by successively halving A. This, in modern terms, says that for any quantity β, there exists a natural number n with $\alpha/2^n < \beta$; thus the premise goes along well with Anaxagoras' view of "limitless smallness"[13] and is, though restricted to very special situations, regarded as a harbinger of the *predicate calculus* in modern logic—the branch of logic that deliberately deals with quantified statements such as "there exists an x such that $\cdots$" or "for any x, $\cdots$"—and particularly the ϵ-δ argument invented for the rigorous treatment of *limits* in the nineteenth century which adequately avoids endless processes (Remark 19.1). A variant of this premise is: "Given two quantities, one can find a multiple of either which will exceed the other." This is what we call the *Axiom of Archimedes* since it was explicitly formulated by Archimedes in his work *On the Sphere and Cylinder*.[14] Eudoxus and Archimedes combined these premises with *reductio ad absurdum* (proof by contradiction) in a judicious manner,[15] and established various results on area and volume by highly sophisticated arguments. Noteworthy is that, in his computations of ratios of areas or volumes of two figures, Archimedes made use of *infinitesimals* in a way similar to the one in integral calculus at the early stage (e.g. *Quadrature of the Parabola*).

[13]The word "exhaustion" for this reasoning was first used by Grégoire de Saint-Vincent (1584–1667); *Opus Geometricum Quadraturae Circuli et Sectionum Conti*, 1647, p. 739.

[14]The *Elements*, Book V, Def. 4 states, "Magnitudes (μέγεθος) are said to have a ratio to one another *which can, when multiplied, exceed one another*." A similar statement appears in Aristotle's *Physics* VIII, 10, 266b 2. The name "Axiom of Archimedes" was given by O. Stolz in 1883, and was adopt by David Hilbert (1862–1943) in a modern treatment of Euclidean geometry (Remark 10.1 (2)).

[15]Reductio ad absurdum is a reasoning in the Eleatics philosophy initiated by Parmenides. Incidentally, the Chinese word for "contradiction" is "máodùn," literally "spear-shield," stemming from an anecdote in the *Han Feizi*, an ancient Chinese text attributed to the political philosopher Han Fei (ca. 280 BCE–ca. 233 BCE). The story goes as follows. A dealer of spears and shields advertised that one of his spears could pierce any shield, and at the same time said that one of his shields could defend from all spear attacks. Then one customer queried the dealer what happens when the shield and spear he mentioned would be used in a fight.

Some 1500 years later, the issue of infinity was examined by two scholars. Thomas Bradwardine (ca. 1290–ca. 1349)—a key figure in the *Oxford Calculators* (a group of mathematics-oriented thinkers associated with Merton College)—employed the principle of OTOC in discussing the aspects of infinite. His observation is construed, in modern terms, as "an infinite subset could *be equal* to its proper subset" (*Tractatus de continuo*, 1328–1335). This is a polemic work directed against atomistic thinkers in his time such as N. d'Autrécourt who, following the classical atomic concept, considered that matter and space were all made up of indivisible atoms, as opposed to Aristotle's doctrine. After a while, Nicole Oresme (1320–1382), a significant scholar of the later Middle Ages, elaborated the thought that the collection of odd numbers is *not smaller than* that of natural numbers because it is possible to count the odd numbers by the natural numbers (*Physics Commentary*, around 1345).[16]

Meanwhile, the magic of infinity has bewildered Galileo Galilei (1564–1642). In his *Discorsi e Dimostrazioni Matematiche Intorno a Due Nuove Scienze* (1638), written during house arrest as the result of the Inquisition, he says, in a similar vein to Oresme's claim, "Even though the *number* of squares should be less than that of all natural numbers, there appears to exist as many squares as natural numbers because any natural number corresponds to its square, and any square corresponds to its square root" (*Galileo's paradox*).

Remark 2.1

(1) The principle of OTOC seems to have its roots deep in a fundamental faculty of human brain. Thinking back on how the species acquired "natural numbers" when they did not yet have any clue about numerals to count things, we are led to the speculation that they relied on the principle of OTOC. More specifically, ancient people are supposed to check whether their cattle put out to pasture returned safely to their shed, without any loss, by drawing on OTOC with some identical things such as sticks of twigs prepared in advance. They extended the same manner in counting things in various aggregations. This experience over many generations presumably made people aware that there is "something in common" behind, not depending on things, and they eventually got a way of identifying as an entity all aggregations among which there are OTOCs. This entity is nothing but a natural number.[17]

Once reached this stage, it did not take much time for people to come up with assigning symbols and names, and evolving primitive arithmetic through empirical fumbling, especially the *division algorithm*, if not in the way we express it today. In fact, numerical notation systems allowing us to represent

---

[16]Oresme was in favor of the spherical model of the universe, but took a noncommittal attitude regarding the Aristotelian theory of the stationary Earth and a rotating sphere of the fixed stars (*Livre du ciel et du monde* and *Questiones super De celo, Questiones de spera*).

[17]A. N. Whitehead said, "The first man who noticed the analogy between a group of seven fishes and a group of 7 days made a notable advance in the history of thought" (*Science and the Modern World*, 1929).

numbers with a few symbols evolved from the division algorithm. For instance, lining up rod-like symbols as |, ||, |||, ||||, |||||, ||||||, $\cdots$ is the most primitive way to represent numerals, as seen in the first few of the Babylonian and Chinese numerals. In the *quinary system*, we split up a given $||| \cdots |$ into several groups each of which consists of five rods ||||| together with a group consisting of rods less than five (this is an infantile form of the division algorithm). We then replace each ||||| by a new symbol, e.g. $\top$; thus, for example, |||||||||||| = ||||| ||||| || is exhibited as $\top\ \top$ ||. We do the same for the symbol $\top$, and continue this procedure (this story is, of course, considerably simplified).

(2) The *decimal system*, contrived in India between the first and fourth centuries, made it possible to express every number by finitely many symbols $0, 1, \ldots, 9$.[18] This epochal format amplified arithmetic considerably, and was brought through the Islamic world to Europe in the tenth century. The names to be mentioned are Muhammad ibn Mūsā al-Khwārizmī (ca. 780 CE–ca. 850 CE), Gerbert of Aurillac (ca. 946 CE–1003), and Leonardo Fibonacci (ca. 1170 –ca. 1250). In ca. 825 CE, al-Khwārizmī wrote a treatise on the decimal system. His Arabic text (now lost) was translated into Latin with the title *Algoritmi de numero Indorum*, most likely by Adelard of Bath (ca. 1080–ca. 1152). Gerbert is said to be the first to introduce the decimal system in Europe (probably without the numeral zero). Fibonacci, best known by the sequence with his name, travelled extensively around the Mediterranean coast, and assimilated plenty of knowledge including Islamic mathematics. His *Liber Abaci* (1202) popularized the decimal system in Europe.

The decimal notation—practically convenient and theoretically being of avail—is firmly planted in our brain as a mental image of numbers. □

6.3 Is the Universe Infinite or Finite?

Let us turn to the issue of the universe. Were Aristotle's spherical model correct, there would be its "outside." So the question at once arises as to what the outside means after all. Is it something substantial or just speculative fabrication? Defending this ridiculous image of the universe, Aristotle explained away by saying, "there is neither place, nor void outside the heavens by reason that the heaven does not exist inside another thing" (*On the Heavens*, I, 9).

Aristotle's model continued to be employed even in the time when the Christian tenet dominated the scholarly world in Europe. Especially, influenced by Thomas Aquinas (1225–1274), who incorporated extensive Aristotelian philosophy through-

---

[18] The explicit use of zero as a symbol more than a placeholder was made much later. In India, zero was initially represented by a point. The first record of the use of the symbol "0" is dated in 876 (inscribed on a stone at the Chaturbhuja Temple). The historical process leading to the term "zero" is as follows: "sunya" in Hindu meaning *emptiness* → "sifr" (cipher) in Arabic → "zephirum" in Latin used in 1202 by Fibonacci → "zero" in Italian (ca. 1600).

out his own theology, the scientific substratum in Christianity was synthesized with the Aristotelian physics.[19] After a while, the Florentine poet Dante Alighieri (1265–1321) reinforced Aristotle's idea of "outside" in his unfinished work *Convivio*. He says, "in the supreme edifice of the universe, all the world is included, and beyond which is nothing; and it is not in space, but was formed solely in the Primal Mind." He further offered an imaginal vision of an intriguing macrocosm in order to turn down the queer consequence of the Biblical concept of ascent to Heaven and descent to Hell which connotes that Hell is the center of the spherical universe (Sect. 6.21).

An exception is Nicolaus Cusanus (Nicholas of Cusa, 1401–1464), a first-class scientist of his time. He alleges, refuting the prevalent outlook of the world, "The universe is not finite in the sense of physically unboundedness since there is no special center in the universe, and hence the outermost sphere cannot be a boundary." At the same time, he argued finiteness of the universe, by which he meant to say "privatively infinite" because "the world cannot be conceived of as finite, albeit it is not infinite" (*De Docta ignorantia*; 1440). This rather contradictory dictum elaborated in the ground-breaking book is a consequence of his analysis of conceivability and a parallelism between the universe and God.

A 100 years later, a dramatic turnabout took place. The heliocentric theory proposed by Nicolaus Copernicus (1473–1543) in his *De revolutionibus orbium coelestium* brings out the apparent retrograde motion of planets better than Ptolemy's theory. He already got his ideas—relying largely on Arabic astronomy typified by al-Battānī[20] (ca. 858 CE–929 CE)—some time before 1541 (*Commentariolus*), but resisted, in spite of his friends' persuasion, to make his theory public since he was afraid of being a target of contempt. It was in 1543, just before his death, that his work (with a preface which puts the accent on the hypothetical nature of the contents) was brought out.

Now, what did Copernicus think about the size of the universe? His model of the universe is spherical with the outermost consisting of motionless, fixed stars; thus being not much different from Aristotle's model in this respect. Meanwhile, Giordano Bruno (1548–1600) argued against the outermost sphere (while accepting Copernicus' theory), reasoning that the infinite power of God would not have produced a finite creation. He was quite explicit in his belief that there is no special place as the center of the universe (*De l'infinito universo et mondi*, 1584); thus indicating that the universe is endless, limitless, and *homogeneous*. His outspoken views, containing Hermetic elements, scandalized Catholics and Protestants alike.[21]

---

[19]In his masterwork *Summa Theologiae*, Aquinas asserts that God is perfect, complete, and embodies actual infinity. To reconcile his assertion with Aristotle's doctrine, he says, "God is an infinite that has bounds."

[20]His work *Kitāb az-Zīj* (Book of Astronomical Tables), translated into Latin as *De Motu Stellarum* by Plato Tiburtinus in 1116, was quoted by Tycho Brahe, Kepler and Galileo.

[21]Some modern scholars consider him as a magus, not a pioneer of science because *Hermeticism* is an ancient spiritual and magical tradition. Others, however, have argued that the magical world-view in Hermeticism was a necessary precursor to the *Scientific Revolution*.

Consequently, he was condemned as an impenitent heretic and eventually burned alive at the stake after 8 years' solitary confinement.

Bruno's view was partly shared by his contemporary Thomas Digges (1546–1595). He was the first to expound the Copernican system in English. In an appendix to a new edition of his father's book *A Prognostication everlasting*, Digges discarded Copernicus' notion of a fixed shell of immovable stars, presuming infinitely many stars at varying distances (1576)—a more tenable reasoning than Bruno's. What should be thought over here is that the religious atmosphere in England in his time was different from that in the Continent because of the English Reformation that started in the reign of Henry VIII.

Interrupting the chronological account, we shall go back to ancient Greece, where we find forerunners of Cusanus, Copernicus, and Bruno. Among them, Archytas of Tarentum (428 BCE–347 BCE) is a precursor of Bruno. He inferred that any place in our space looks the same, and hence no boundary can exist; otherwise we would have a completely different sight at a boundary point. He thus concluded infiniteness of the universe.[22] The Pythagorean Philolaus (ca. 470 BCE–ca. 385 BCE) relinquished the geocentric model, saying that the Earth, Sun, and stars revolve around an *unseen central fire*. As alluded to in Archimedes' work *Sand Reckoner*, Aristarchus of Samos (ca. 310 BCE–ca. 230 BCE)—apparently influenced by Philolaus—had speculated that the sun is at the center of the solar system for the reason that the geocentric theory is against his conclusion that the sun is much bigger than the moon. In his work *On the Sizes and Distances*, he figured out the distances and sizes of the sun and the moon, under the assumption that the moon receives light from the sun. In the *Sand Reckoner*, Archimedes himself harbored the ambition to estimate the size of the universe in the wake of Aristarchus' heliocentric spherical model. To this end, he proposed a peculiar number system to remedy the inadequacies of the Greek one and expressed the number of sand grains filling a cosmological sphere. The presupposition he made is that the ratio of the diameter of the universe to the diameter of the orbit of the earth around the sun equals the ratio of the diameter of the orbit of the earth around the sun to the diameter of the earth.

Aristarchus' heliocentric model was espoused by Seleucus of Seleucia, a Hellenistic astronomer (born ca. 190 BCE), who developed a method to compute planetary positions. In the end, however, the Greek heliocentrism had been long forgotten. Even Hipparchus who undertook to find a more accurate distance between the sun and the moon took a step backwards.

Now in passing, we shall make special mention of Alexandria founded at the mouth of the Nile in 332 BCE by Alexander the Great, after his conquest of Egypt. It was Ptolemy I, the founder of the Ptolemaic dynasty, who raised the city to a center of *Hellenistic culture*, an offspring of Greek culture that flourished around the

[22]Anaximander, belonging to the previous generation, conceived a mechanical model of the world based on a non-mythological explanatory hypothesis, and alleged that the Earth has a disk shape and is floating very still in the center of the infinite, not supported by anything, while Anaxagoras portrayed the sun as a *mass of blazing metal*, and postulated that Mind (νοῦς, Nous) was the initiating and governing principle of the cosmos (κόσμος).

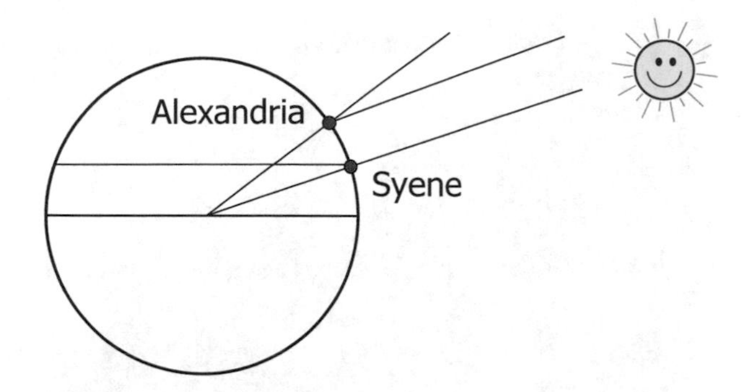

Fig. 6.2 The circumference of the earth

Mediterranean after the decline of Athens. He and his son Ptolemy II—both held academic activities in high esteem—established the Great Museum (Μουσεῖον), where many thinkers and scientists from the Mediterranean world studied and collaborated with each other. Archimedes stayed there sometime in adolescence.

Prominent among them is Eratosthenes of Cyrene (ca. 276 BCE–ca. 195 BCE), the third chief librarian appointed by Ptolemy II and famous for an algorithm for finding all prime numbers.[23] Pushing further the view about the spherical earth, he adroitly calculated the meridian of the earth. The outcome was 46,620 km, about 16% greater than the actual value. His work *On the measurement of the earth* was lost, but the book *On the circular motions of the celestial bodies* by Cleomedes (died ca. 489 CE) explains Eratosthenes' deduction relying on a property of parallels and the observation that at noon on the summer solstice, the sun casts no shadow in Syene (now Aswan), while it casts a shadow on one fiftieth of a circle (7.2°) in Alexandria on the same degree of longitude as Syene (Fig. 6.2). His resulting value is deduced from the distance between the two cities (5040 stades = 925 km), which he inferred from the number of days that caravans require for travelling between them.

The Ptolemaic dynasty lasted until the Roman conquest of Egypt in 30 BCE, but even afterwards Alexandria maintained its position as the center of scientific activity (see Sect. 6.11).

Finishing the excursion into ancient times, we shall turn back to the later stage of the Renaissance, the age when science was progressively separated from theology, and scientists came to direct their attention to scientific evidence rather than to theological accounts (thus science was becoming an occasional annoyance to the religious authorities). Representative in this period (religiously tumultuous times) are the Catholic Galileo and the Protestant Johannes Kepler (1571–1630); both gave a final death blow to Aristotelian/Ptolemaic theory.

[23] Archimedes' *Method* mentioned in Footnote 11 takes the form of a letter to Eratosthenes.

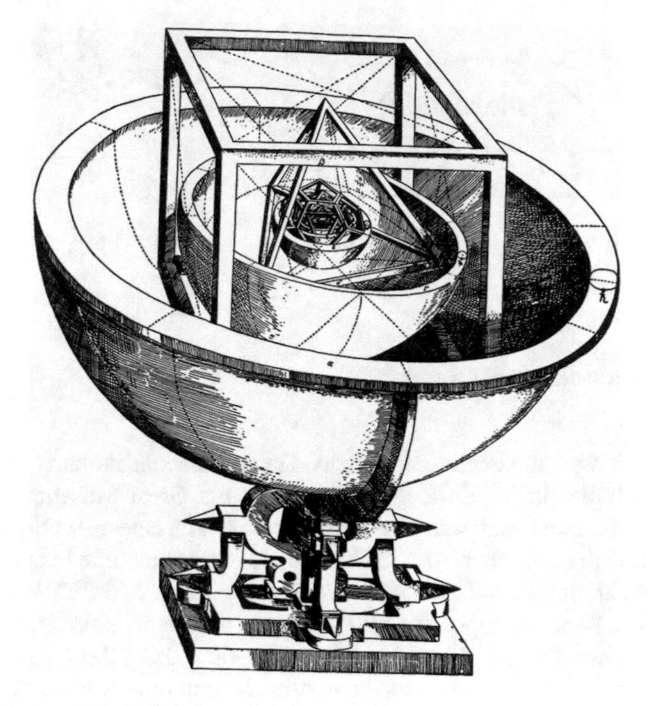

Fig. 6.3 Kepler's model of the Solar system

Kepler discovered the *three laws of planetary motion*, based on the data of Mars' motion recorded by Tycho Brahe (1546–1601).[24] His first and second laws were elaborated in the *Astronomia nova* (1609). The first law asserts that the orbit of a planet is elliptical in shape with the center of the sun being located at one focus. The second law says that a line segment joining the sun and a planet sweeps out equal areas in equal intervals of time. The third law stated in the *Epitome Astronomiae Copernicanae* (1617–1621) and the *Harmonices Mundi* (1619) maintains that the square of the orbital period of a planet is proportional to the cube of the semi-major axis of its orbit.

Before this epoch-making discovery, however, he attempted to explain the distances in the Solar system by means of "regular convex polyhedra" inscribing and circumscribed by spheres, where the six spheres separating those solids correspond to Saturn, Jupiter, Mars, Earth, Venus, and Mercury (*Mysterium Cosmographicum*, 1596; Fig. 6.3). He also rejected the infiniteness of the universe, on the basis of an astronomical speculation on the one hand, and on the traditional scholastic doctrine

[24]Tycho Brahe, the last of the major naked-eye astronomers, rejected the heliocentricism. His model of the universe is a combination of the Ptolemaic and the Copernican systems, in which the planets revolved around the Sun, which in turn moved around the stationary Earth.

on the other (for this reason, Kepler is portrayed as the last astronomer of the Renaissance, and not the first of the new age).

Galileo was not affected by Aristotelian prejudice that an experiment was an interference with the natural course of Nature. Without performing any experiment at all, medieval successors of Aristotle insisted, for instance, that a projectile is pushed along by the force they called "impetus," and if the impetus is expended, the object should fall straight to the ground. At the apogee of his scientific career, Galileo had conducted experiments on projectile motion, and observed that their trajectories are always parabolic (1604–1608). He further investigated the relation between the distance d an object (say, a ball) falls and the time t that passes during the fall, and found the formula $d = \frac{1}{2}gt^2$ (in modern terms), where $g = 9.8\ \mathrm{m/s^2}$, the *gravitational acceleration* (*Discorsi*). The crux of this formula is that the drop distance does not depend on the mass of a falling object, contrary to Aristotle's prediction (*Physics*, IV, 8, 215a25).[25]

Galileo turned his eyes on the universe. With his handmade telescope, he made various astronomical observations and confirmed that four moons orbited Jupiter. This prompted him to defend the heliocentricism, for provided that Aristotle were right about all things orbiting earth, these moons could not exist (*Sidereus Nuncius*, 1610). As an inevitable consequence, his Copernican view led to the condemnation at the Inquisition (1616, 1633). After being forced to racant, he was thrown into a more thorny position than Kepler, and had to avoid dangerous issues that may provoke the Church. Regarding the size of the universe, he denied the existence of the celestial sphere as the limit, while he was reluctant to say definitely that the universe is infinite, because of censorship by the Church. In a letter to F. Liceti on February 10, 1640, Galileo says, "the question about the size of the universe is beyond human knowledge; it can be only answered by the *Bible* and a divine revelation."

Next to Kepler and Galileo are Descartes and Pascal, the intellectual heroes of seventeenth-century France, who laid the starting point of the Enlightenment with their inquiries into truth and the limits of reason.

René Descartes (1596–1650), whose natural philosophy agrees in broad lines with Galileo's one, defended the heliocentrism, by saying that it is much simpler and distinct, but hesitated to make his opinion public upon the news of the Inquisition's conviction of Galileo (1633). As regards the size of the universe, he maintained at first that the universe is finite; but later became ambivalent. In a letter in 1649 to the rationalist theologian H. More of Cambridge, he conceded, after a long dispute, that the universe must have infinite expanse because one cannot think of the limit; for if the limit exists, one cannot help but think of the outside of space which must be the same as our space.

While, for Descartes, the *self* is prior and independent of any knowledge of the world, Blaise Pascal (1623–1662) pronounced that the order is reversed; that is,

[25]The *mean speed theorem* due to the Oxford Calculators (nothing but the formula for the area of a *trapezoid* from today's view) is a precursor of "the law of falling bodies."

knowledge of the world is a prerequisite to knowledge of the self. In the *Pensées* (1670), in comparison with the disproportion between our justice and God's, he writes, "Unity added to infinity does not increase it at all [....]: the finite is annihilated in the presence of the infinite and becomes pure nothingness" (fragment 418). Further, he says, "Nature is an infinite sphere of which the center is everywhere and the circumference nowhere" (fragment 199).[26]

As described hitherto, the posture to pay attention to the universe is a steady tradition of European culture. In its background, philosophy was nurtured in the bosom of cosmogony, and could not be separated from religion because of the intimacy between them. Especially after the rise of Christianity, Western scholars had to be confronted, at the risk of their life in the worst case, with God as a creator. Even after the tension between science and religion was eased, scientists could not be entirely free from God, whatever His image is. Such milieu led quite naturally to probing questions about our universe in return.

In summary, "whether the universe is finite or infinite," whatever it means, is an esoteric issue in religion, metaphysics, and astronomy in Europe that had intrigued and baffled mankind since dim antiquity. What matters most of all is whether the outside is necessary when we talk about finiteness of the universe.

6.4 From Descartes to Newton and Leibniz

Intuition for space surrounding us was the driving force for people to puff up their image of the universe. Needless to say, from the ancient times to the Middle Ages and even the early modern times, the mental image that nearly all people had is, if not being particularly conscious about it, the one described by *Euclidean space*, the model of our space named after the Alexandrian Euclid (Εὐκλείδης; ca. 300 BCE; see Sect. 6.9 for the details). Even the advocates of the spherical model of the universe imagined Euclidean space as the entity embracing all. Synthetic geometry executed on this model is what we call *Euclidean geometry*. It started with a collection of geometrical results acquired in Egypt and Mesopotamia by empirical investigations or experience of land surveys and constructions of magnificent and imposing structures, and had been systemized, as the search of universals, through the efforts of Greek geometers.

Putting it briefly, Euclidean space is *homogeneous* in the sense that there is no special place, and it is *isotropic*; i.e., there is no special direction.[27] These features are not expressly indicated in Greek geometry, but are guaranteed by a property

[26]This sentence, reminiscent of Cusanus' view, may be from the *Liber XXIV philosophorum* (Book of the 24 Philosophers), an influential philosophical and theological medieval text, usually attributed to Hermes Trismegistus, the purported author of the *Corpus Hermeticum*.

[27]Euclidean space holds one more significant aspect expressed by a property of parallels, but its true meaning had not been comprehended for quite a while; see Sect. 6.10.

of *congruence*[28]; namely any geometric figure can be rotated and moved to an arbitrary place while keeping its shape and size. What should be pinpointed here is that it was not until the nineteenth century that people explicitly conceptualized Euclidean space (or space as a mathematical object which fits in with our spatial intuition). Until then, *geometry* meant only Euclidean geometry, and the space where geometry is performed was considered as an a priori entity, or what amounts to the same thing, our space—a place of storage in which objects are recognized, and a place of manufacture in which objects are constructed—was not an entity for which we investigate whether our understanding of it is right or wrong, thereby all the propositions of geometry being considered "absolute truth."

As stated above, Euclidean geometry had been a lofty edifice for nearly 2000 years that nobody could break down. Only the appearance changed when Descartes invented the so-called *algebraic method*. This epoch-making method is elucidated in his *La Géométrie* [5], one of the three essays attached to the philosophical and autobiographical treatise *Discours de la méthode pour bien conduire sa raison, et chercher la vérité dans les sciences* (1637).[29]

Descartes' prime concern was, though he was trained in religion with the still-authoritarian nature, to find principles that one can know as true without any scruples. In the self-imposed search for *certainty*, he linked philosophy with science, and had the confidence that certainty could be found in the mathematical proofs having the apodictic character. Upon his emphasis on lucid methodology and dissatisfaction with the ancient arcane method, as already glimpsed in his *Regulae ad Directionem Ingenii* (1628), he attempted to create "universal mathematics" by bridging the gap between arithmetic and geometry that used to be thought of as different terrains. Indeed, the Greeks definitely distinguished geometric quantities from numerical values, and even thought that length, area, and volume belong to different categories, thus lumping them together makes no sense. Under such shackles (and being devoid of symbolic algebra), the "equality," "addition/subtraction," and the "large/small relation" for two figures in the same category were defined by means of geometric operations.[30] Specifically, they considered that two polygons (resp.

---

[28] Prop. 4 in the *Elements*, Book I, is the first of the congruence propositions.

[29] *La Géométrie* consists of Book I (*Des problèmes qu'on peut construire sans y employer que des cercles et des lignes droites*), Book II (*De la nature des lignes courbes*), and Book III (*De la construction des problèmes solides ou plus que solides*). The other two essays are *La Dioptrique* and *Les Météores*.

[30] The Greeks had difficulty to handle irrationals in their arithmetic and were forced to replace *algebraic* manipulations by geometric ones [22]. Actually, geometry had been thought of as far more general than *arithmetic* as seen in Aristotle's words, "We cannot prove geometric truths by arithmetic" (*Posterior Analytics*, I, 7; see also Plato's *Philebus*, 56d).

The word "algebra," stemmed from the Arabic title *al-Kitāb al-mukhtasar fi hisāb al-jabr walmuqābala* (The Compendious Book on Calculation by Completion and Balancing) of the book written approximately 830 CE by al-Khwārizmī (ca. 780–ca. 850) (Remark 2.1 (2)), was first imported into Europe in the early Middle Ages as a medical term, meaning "the joining together of what is broken." In addition, *algorithm*, meaning a process to be followed in calculations, is a transliteration of his surname al-Khwārizmī.

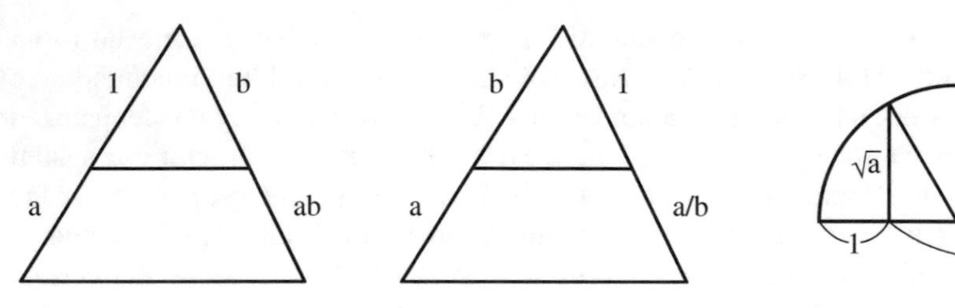

Fig. 6.4 Operations of segments

polyhedra) are "equal" if they are *scissors-congruent*; i.e., if the first can be cut into finitely many polygonal (resp. polyhedral) pieces that can be reassembled to yield the second.

Remark 4.1 Any two polygons with the same numerical area are scissors-congruent as shown independently by W. Wallace in 1807, Farkas Bolyai in 1832, and P. Gerwien in 1833. Gauss questioned whether this is the case for polyhedra in two letters to his former student C. L. Gerling dated 8 and 17 April, 1844 (*Werke*, VIII, 241–42). In 1900, Hilbert put Gauss's question as the third problem in his list of the 23 open problems at the second ICM. M. W. Dehn, a student of Hilbert, found two tetrahedra with the same volume, but non-scissors-congruent (1901); see [37]. □

In his essay *La Géométrie*, Descartes made public that all kinds of geometric magnitude can be unified by representing them as line segments, and introduced the fundamental rules of calculation in this framework.[31] His idea is to fix a unit length, with which he defines addition, subtraction, multiplication, division, and the extraction of roots for segments, appealing to the proportional relations of similar triangles and the Pythagorean Theorem, as indicated in Fig. 6.4. Then he adopts a *single axis* to represent these operations; thereby replacing the Greek *geometric algebra* by "numerical" algebra. On this basic format, he handles various problems for a class of algebraic curves. Priding himself on his invention, he says, in a fond familiar letter dated November 1643 to Elisabeth (Princess Palatine of Bohemia), that extra effort to make up the constructions and proofs by means of Euclid's theorems is no more than a poor excuse for self-congratulation of petty geometers because such effort does not require any scientific mind.

Afterwards, Franciscus van Schooten (1615–1660), who met Descartes in Leiden (1632) and read the still-unpublished *La Géométrie*, translated the French text into Latin, and endeavored to disseminate the algebraic method to the scientific community (1649). In the second edition (1659–1661), he added annotations and transformed Descartes' approach into a systematic theory, which made it far more accessible to a large readership.

[31] A letter dated March 26, 1619 to I. Beeckman indicates that Descartes already held a rough scheme at the age of 22. Pierre de Fermat (1601–1665) argued a similar idea in his *Ad locos plano et solidos isagoge* (1636), though not published in his life time.

Descartes' method, along with algebraic notations dating back to François Viète (1540–1603) who took a momentous step towards modern algebra,[32] was inherited as *analytic geometry* afterwards. With this progression, geometric figures were to be transplanted to algebraic objects in the coordinate plane $\mathbb{R}^2$ or the coordinate space $\mathbb{R}^3$.[33] What is more, this new discipline was integrally connected with the calculus initiated by Issac Newton (1642–1726) and Gottfried Wilhelm von Leibniz (1646–1716), independently and almost simultaneously, which is to be a vital necessity when we talk about the shape of our universe.

Newton learnt much from the *Exercitationes mathematicae libri quinque* (1657) by van Schooten and *Clavis Mathematicae* (1631) by William Oughtred (1574–1660) when he was a student of Trinity College, Cambridge. "Fluxion" is Newton's underlying term in his differential calculus, meaning the instantaneous rate of change of a *fluent*, a varying (flowing) quantity. His idea, to which he was led by personal communication with Issac Barrow (1630–1677) and also by Wallis' book *Arithmetica infinitorum* (1656),[34] is stated in two manuscripts; one is of October 1666 written when he was evacuated in a neighborhood of his family home at Woolsthorpe during the Great Plague, and another is the *Tractatus De Methodis Serierum et Fluxionum* (1671). In his calculus, Newton made use of power series in a systematic way.

Meanwhile, Leibniz almost completed calculus while staying in Paris (1673–1676) as a diplomat of the Electorate of Mainz in order to get Germany back on its feet from the exhaustion caused by the 30 Years' War. Although the aspired end of diplomacy was not attained, he had an opportunity to meet Christiaan Huygens (1629–1695), a leading scientist of his time, who happened to be invited by Louis XIV as a founding member of the Académie des Sciences. Very helpful for him was the suggestion by Huygens to read Pascal's *Lettre de Monsieur Dettonville*··· (1658). Further, his perusal of Descartes' work was an assistance in strengthening the basis of his thought. Around the time when Leibniz returned to Germany, he improved the presentation of calculus, and brought it out as two papers; *Nova methodus pro maximis et minimis*, and *De geometria recondita et analysi indivisibilium atque infinitorum*.[35]

Their calculus, though still something of a mystery (Remark 15.1 (2)), provided a powerful tool which enabled to handle more complicated geometric figures

---

[32] Viète, *In artem analyticem isagoge* (1591).

[33] It was Leibniz who first introduced the "coordinate system" in its present sense (*De linea ex lineis numero infinitis ordinatim ductis inter se concurrentibus formata, easque omnes tangente, ac de novo in ea re Analysis infinitorum usu*, Acta Eruditorum, **11** (1692), 168–171). Incidentally, Oresme used constructions similar to rectangular coordinates in the *Questiones super geometriam Euclidis* and *Tractatus de configurationibus qualitatum et motuum* (ca. 1370), but there is no evidence of linking algebra and geometry.

[34] John Wallis (1616–1703) propagated Descartes' idea in Great Britain through his work.

[35] His papers were published in Acta Eruditorum, **3** (1684), 467–473 and **5** (1686), 292–300. In the second paper, he states, "With this idea, geometry will make a far more greater strides than Viète's and Descartes'."

than the ones that Greek geometers treated in an *ad hoc* manner.[36] Indeed, differential calculus provides a unified recipe to find tangents of a general curve (say, transcendental curves Descartes did not deal with), and integral calculus (called the "inverse tangent problem" by Leibniz) allows more latitude in calculating the area of a general figure without any ingenious trick. Paramountly important is the discovery that tangent (a local concept) and area (a global concept) are linked through *the fundamental theorem of calculus* (FTC). [37]

There are other lines of evolution of Descartes' method and its offspring. Analytic geometry paved the way for *higher-dimensional* geometry (Sect. 6.14), which was to be incorporated into Riemann's theory of curved spaces (Sect. 6.15). *Algebraic geometry* is regarded as the ultimate incarnation of Cartesian geometry, which came of age from the late nineteenth century to the early twentieth century. Descartes' method also prompted to set up mathematical theories in a strictly logical manner, not relying on geometry, but relying on arithmetic wherein concepts involved are entirely framed in the language of real numbers or systems of such numbers (Sect. 6.16). This signifies in some sense that mathematicians eventually become aware of *mathematical reality*—a sort of Plato's world of *Forms* (εἶδος or its cognates)—distinguished from *physical reality* or the empirical world.

Remark 4.2

(1) Symbols and symbolization has played a significant role in the history of mathematics. Prior to Viète, the symbols $+$, $-$ were used by H. Grammateus in *Ayn new Kunstlich Buech* (1518). Actually, these symbols already appeared in 1489 to indicate surplus and deficit for the mercantile purpose. In 1525, the symbol $\sqrt{\cdot}$ was invented by C. Rudolff, a student of Grammateus. Subsequently, R. Recorde adopted the equal sign $=$ in 1557. After Viète, Oughtred innovated the symbol $\times$ for multiplication in his *Clavis Mathematicae* (1631). In the same year, the book *Artis analyticae praxis ad aequationes algebraicas resolvendas* by Thomas Harriot (ca. 1560–1621), which left a great mark on the history of symbolic algebras, was brought out posthumously, wherein the symbols $<$ and $>$ representing the magnitude relation appear. For all of this, the usage of symbols in Descartes' time was almost in agreement with that of today.
(2) When integral calculus was still in its nascent stage, Kepler computed the volumes of solids of revolution.[38] After a while, Bonaventura Cavalieri (1598–1647), a disciple of Galileo, proposed a naive but a systematic method, known as *Cavalieri's principle*, to find areas and volumes of general figures, which

[36]In the *Elements*, Book III, Def. 2, Euclid says, "a straight line is said to touch a circle which, meeting the circle and being produced, does not cut the circle." Curiously, in the *Metaphysics*, III, 998a, Aristotle reports that Protagoras (ca. 490 BCE–ca. 420 BCE) argued against geometry that a straight line cannot be perceived to touch a circle at only one point.

[37]I. Barrow, who strongly opposed to Descartes' method in contrast with Wallis, knew the FTC in a geometric form (Prop. 11, Lecture 10 of his *Lectiones Geometricae* delivered in 1664–1666). A transparent proof was given by Leibniz in his *De geometria*, 1686.

[38]*Nova stereometria doliorum vinariorum* (1615).

is thought of as intermediating between the Greek quadrature and integral calculus.[39] The linchpin of his innovation is the notion of *indivisibles*; he regards, for instance, a plane region as being composed of an infinite number of parallel lines, each considered to be an infinitesimally thin rectangle.

(3) Although calculus initiated by Newton and Leibniz was the same in essence, their styles differed in a crucial way. The difference was reflected in the notation they used. For instance, Leibniz invented the symbol dx/dt, while Newton employed the dot notation $\dot{x}$. This difference gave rise, as often said, to the fact that the British mathematics under a baneful influence of Newton's authority had fallen seriously behind the Continental counterpart. They were jolted by the *Mécanique Céleste* by Pierre-Simon Laplace (1749–1827) which was affiliated with the "Leibnizian school." John Playfair (1748–1819) says, "We will venture to say that the number of those in this island who can read the *Mécanique Céleste* with any tolerable facility is small indeed" (1808).

(4) Until the nineteenth century through a pre-classical theory implicit in the work of Newton, Leibniz and their successors, real numbers had been grasped as points on a straight line—the vision that dates back to Descartes and is credited to J. Wallis—or "something" (without telling what they are) approximated by rational numbers.[40]

Thus calculus had been built on a fragile base. In a letter to his benefactor dated March 29, 1826, Niels Henrik Abel (1802–1829) writes, "the tremendous obscurity one undoubtedly finds in analysis today. It lacks all plan and system [$\cdots$] The worst of it is, it has never been treated with rigor. There are very few theorems in advanced analysis that have been demonstrated with complete rigor" (Remark 19.1).

An entrenched foundation to calculus was furnished by Richard Dedekind (1831–1916). He constructed the real number system by appealing to what we now call the method of *Dedekind cuts*, partitions of the rational numbers into two sets such that all numbers in one set are smaller than all numbers in the other (1872).[41] To say the least, this understanding heralds the emancipation of analysis from geometry.

Long ago, Eudoxus developed an idea similar to Dedekind's in essence. Before that, Greek geometers had tacitly assumed that two magnitudes α, β are always commensurable (σύμμετρος); i.e., there exist two natural numbers m, n such that $m\alpha = n\beta$. Thus, with the discovery of incommensurable magnitudes which results from the Pythagorean Theorem, the issue arose as to how to define equality of two ratios. The impeccable definition attributed to Eudoxus is : "Magnitudes are said to be in the same ratio, the first to the second and the

---

[39] *Geometria indivisibilibus continuorum nova quadam ratione promota* (1635, 1653).

[40] Simon Stevin (1548–1620) renovated the notation for decimal fractions to make all computations easier, which contributed to comprehension of the nature of real numbers. In the 35-page booklet *De Thiende* (1585), he expressed 27.847 by 27⓪8①4②7③ for instance.

[41] In the same year, Cantor arrived at another definition of the real numbers (Footnote 147).

third to the fourth, when, if any equimultiples whatever are taken of the first and third, and any equimultiples whatever of the second and fourth, the former equimultiples alike exceed, are alike equal to, or alike fall short of, the latter equimultiples respectively taken in corresponding order" [7, Book V, p. 114]. In modern terminology, this is expressed as "Given four magnitudes $\alpha, \beta, \gamma, \delta$, the two ratios $\alpha{:}\beta$ and $\gamma{:}\delta$ are said to be the same if for all natural numbers m, n, it be the case that according as $m\alpha \gtreqless n\beta$, so also is $m\gamma \gtreqless n\delta$."

Related to the theory of proportions is the *anthyphairesis* (ἀνθυφαίρεσις), a method to find the ratio of two magnitudes. A special case is *Euclid's algorithm* used to compute the greatest common divisors of two numbers (*Elements*, Book VII, Prop. 2; see [20]). Specifically, for two magnitudes $\alpha > \beta$, it proceeds as $\alpha = n_1\beta + \gamma_1$ $(0 < \gamma_1 < \beta)$, $\beta = n_2\gamma_1 + \gamma_2$ $(0 < \gamma_2 < \gamma_1)$, $\gamma_1 = n_3\gamma_2 + \gamma_3$ $(0 < \gamma_3 < \gamma_2)$, $\gamma_2 = n_4\gamma_3 + \gamma_4$ $(0 < \gamma_4 < \gamma_3)$, $\cdots$ (Book X, Prop. 2). Here, α and β are commensurable if and only if this process terminates after a finite number of steps (i.e., $\gamma_n = 0$ for some n). This idea was handed down to us as the technique to obtain *continued-fraction expansions* afterward. Namely, for positive real numbers α, β, eliminating γ_i in the above, we have

$$\frac{\alpha}{\beta} = n_1 + \frac{1}{n_2+}\,\frac{1}{n_3+}\,\frac{1}{n_4+}\cdots \qquad (= [n_1, n_2, n_3, n_4 \ldots]). \qquad \square$$

6.5 A New Approach in Classical Geometry

Leaving Euclidean geometry aside for a moment, we shall touch on *projective geometry*, the embryonic subject that came up during the Renaissance period and had matured in the nineteenth century. This new approach is exclusively concerned with quantity-independent properties such as "three points are on a line" (*collinearity*) and "three lines intersect at a point" (*concurrency*) that are invariant under *projective transformations* (the left of Fig. 6.5).[42]

Projective geometry has a historical link to (linear) *perspective*, an innovative skill in drawing contrived by Filippo Brunelleschi (1377–1446), which allowed Renaissance artists to portray a thing in space and landscapes as someone actually might see them.[43] The theoretical aspect of perspective was investigated by Leon Battista Alberti (1404–1472) and Piero della Francesca (1416–1492). In the preface of his book *Della Pittura* (1435) dedicated to Brunelleschi, Alberti emphatically

[42] Imitating this wording, one may say that Euclidean geometry is a geometry that treats invariant properties under congruence transformations. Such a perspective is ascribed to Christian Felix Klein (1849–1925), who promulgated his scheme in the booklet *Vergleichende Betrachtungen über neuere geometrische Forschungen* ("Erlangen Program" in short) in 1872.

[43] One of the earliest to have used perspective is Masaccio. He limned the *San Giovenale Triptych* (1422) based on the principles he learned from Brunelleschi, which emblematizes the transition from medieval mysticism to the Renaissance spirit.

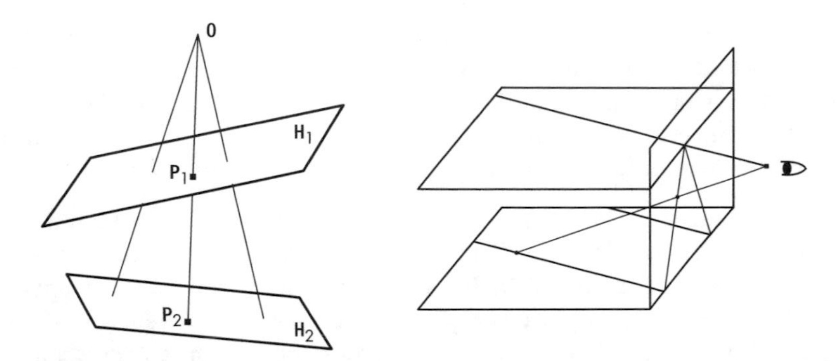

Fig. 6.5 Perspective

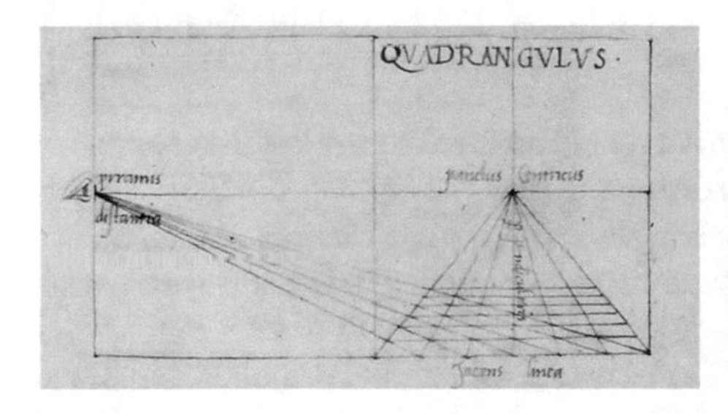

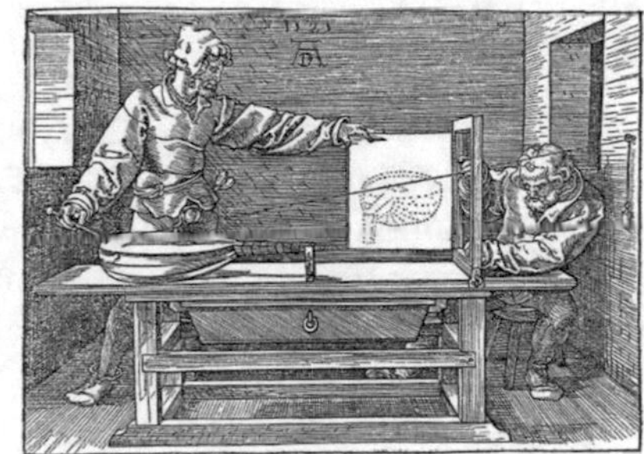

Fig. 6.6 Alberti's drawing in *Della Pictura* and Dürer's illustration

writes, "I used to marvel and at the same time to grieve that so many excellent and superior arts and sciences from our most vigorous antique past could now seem lacking and almost wholly lost. ··· Since this work [due to Brunelleschi] seems impossible of execution in our time, if I judge rightly, it was probably unknown and unthought of among the Ancients" (translated by J. R. Spencer). Piero's book *De Prospectiva Pingendi* (1475) exemplifies the effective symbiosis of geometry and art; he actually had profound knowledge of Greek geometry and made a transcription of a Latin translation of Archimedes' work.

The "Renaissance Man" Leonardo da Vinci (1452–1519)—stimulated by Alberti's book—fully deserves his reputation as a true master of perspective. His technique is particularly seen in his study for *A Magis adoratur* (ca. 1481). He says, "Perspective is nothing else than seeing a place or objects behind a plane of glass, quite transparent, on the surface of which the objects behind the glass are to be drawn" [30]. Albrecht Dürer (1471–1528), a weighty figure of the Northern Renaissance who shared da Vinci's pursuit of art, discussed in his work *Underweysung der Messung* (1525) an assortment of mechanisms for drawing in perspective from models (the right of Fig. 6.6). This work, in which he touched on "Doubling the cube," is the first on advanced mathematics in German.

Subsequently, Federico Commandino (1506–1575) published the work entitled *Commentarius in planisphaerium Ptolemaei* (1558). This, a commentary on Ptolemy's work *Planisphaerium*, includes an account of the *stereographic projection* of the celestial sphere, and is a work on perspective from a mathematical viewpoint (concurrently, he translated several works of ancient scholars).

Remark 5.1 Leonardo drew the illustrations of the regular polyhedra in the book *De divina proportione* (1509) by Fra Luca Bartolomeo de Pacioli (ca. 1447–1517), a pupil of Piero. The theme of the book is mathematical and artistic proportions, especially the *golden ratio* (divine proportion)—the positive solution $(1+\sqrt{5})/2 = [1, 1, 1, \ldots]$ of the equation $x^2 = x + 1$—and its application in architecture. The golden ratio appears in the *Elements*, Book II, Prop. 11; Book IV, Prop. 10–11; Book XIII, Prop. 1–6, 8–11, 16–18,[44] and is commonly represented by the Greek letter ϕ after the sculptor Pheidias (Φειδίας, ca. 480 BCE–ca. 430 BCE) who is said to have employed it in his work. □

The origin of projective geometry can be traced back to the work of Apollonius of Perga on *conic sections* and Pappus of Alexandria (ca. 290 CE–ca. 350 CE).[45] In his *Collection*, Book VII, where Pappus enunciated his *hexagon theorem*, he made use of a concept equivalent to what we call now the *cross-ratio*,[46] which was to enrich projective geometry from the quantitative side since it is invariant under projective transformations. Here the cross-ratio of collinear points A, B, C and D is defined as $[A, B, C, D] = (AC \cdot BD)/(BC \cdot AD)$, where each of the distances is signed according to a consistent orientation of the line.[47]

Projective geometry as a solid discipline was essentially initiated by Girard Desargues (1591–1661), a coeval of Descartes. In 1639, motivated by a practical purpose pertinent to perspective, he published the *Brouillon project d'une atteinte aux événemens des rencontres du Cone avec un Plan*.[48] Pascal took a strong interest in Desargues' work, and studied conic sections in depth at the age of 16. Philippe

[44]In Book VI, Def. 3, Euclid says, "A straight line is said to have been cut in *extreme and mean ratio* (ἄκρος καὶ μέσος λόγος) when, as the whole line is to the greater segment, so is the greater to the less." In today's terms, "the segment AB is cut at C in extreme and mean ratio when $AB{:}AC = AC{:}CB$." Note that AB/AC equals the golden ratio. Kepler, who proved that ϕ is the limit of the ratio of consecutive numbers in the Fibonacci sequence, described it as "a fundamental tool for God in creating the universe" (*Mysterium Cosmographicum*).

[45]The first to have studied conic sections is Menaechmus (ca. 380 BCE–ca. 320 BCE), who used them to solve "doubling the cube." The names *ellipse* (ἔλλειψις), *parabola* (παραβολη) and *hyperbola* (ὑπερβολη) for conic curves were introduced by Apollonius.

[46][Editor's note] The notion of cross ratio was already known to Menelaus of Alexandria (c. 70–c. 140); see [28, p. 340ff].

[47]The cross-ratio of real numbers x_1, x_2, x_3, x_4 is defined by $[x_1, x_2, x_3, x_4] := (x_3 - x_1)(x_4 - x_2)/(x_3 - x_2)(x_4 - x_1)$. The projective invariance of cross-ratios amounts to the identity $[T(x_1), T(x_2), T(x_3), T(x_4)] = [x_1, x_2, x_3, x_4]$, where $T(x) = (ax+b)/(cx+d)$ $(ad - bc \neq 0)$.

[48]The celebrated Desargues' theorem showed up in an appendix of the book *Exemple de l'une des manières universelles du S.G.D.L. touchant la pratique de la perspective* published in 1648 by his friend A. Bosse.

de La Hire (1640–1718) was affected by Desargues' work, too. He is most noted for the work *Sectiones Conicae in novem libros distributatae* (1685). However, partly because of the spread of Descartes's method among the mathematical community in those days and because of his peculiar style of writing, his work remained unrecognized until it was republished in 1864 by Noël-Germinal Poudra. Without being aware of Desargues' work, Jean-Victor Poncelet (1788–1867) laid a firm foundation for projective geometry in his masterpiece *Traité des propriétés projectives des figures* (1822).

In any case, projective geometry falls within classical geometry at this moment; but it has an interesting hallmark in view of infiniteness. Guidobaldi del Monte (1505–1607) in the *Perspectivae Libri VI* (1600) and Kepler in his *Astronomiae pars optica* (1604) proposed the idea of *points at infinity*. Subsequently, having in mind the "horizon" in perspective drawing (Fig. 6.5), Desargues appends points at infinity to the plane as idealized limiting points at the "end," and argues that two parallel lines intersect at a point at infinity. Later, the "plane *plus* points at infinity" came to be termed the *projective plane* (Sect. 6.16).

6.6 Euclid's Legacy in Physics and Philosophy

Not surprisingly, Euclidean geometry predominated in physics for a long time. Salient is Newton who adopted Euclidean geometry as the base of his grand work *Philosophiae Naturalis Principia Mathematica* (1687, 1713, 1726; briefly called the *Principia*), in which he formulated the *law of inertia*,[49] the *law of motion*, the *law of action-reaction* and the *law of universal gravitation*.

It was in 1666, exactly the same year when he worked out calculus, that Newton found a clue leading to the law of universal gravitation. Hagiography has it that he was inspired, in a stroke of genius, to formulate the law while watching the fall of an apple from a tree. Leaving aside whether this is a fact or not, all we can definitely say is, he concluded that the force acting on objects on the ground (say, apples) acts on the moon as well; he thus found the extraordinarily significant connection between the terrestrial and celestial which had been thought of as being independent of each other.

To go further with his subsequent observation, let O be the center of the earth, and let P be the position of the moon. We denote by v and a the speed and acceleration of P, respectively. The acceleration is directed towards O, and $a = v^2/R$, where R is the distance between O and P. If the orbital period of the moon is T, then $vT = 2\pi R$, so that $a = 4\pi^2 R/T^2$. Next, applying Kepler's third

[49]The law of inertia was essentially discovered by Galileo during the first decade of the seventeenth century though he did not understand the law in the general way (the term "inertia" was first introduced by Kepler in his *Epitome Astronomiae Copernicanae*). The general formulation of the law was devised by Galileo's pupils and Descartes (the *Principles of Philosophy*, 1644).

law to a circular motion, we have $R^3 = cT^2$ ($c > 0$). Hence $a = 4\pi^2 cR^{-2}$ and $ma = 4\pi^2 mcR^{-2}$, which is the force acting on P in view of the law of motion, and the law of universal gravitation follows. Conversely, based on his laws, Newton derived Kepler's laws through far-reaching deductions.[50]

Remark 6.1 We let $\boldsymbol{x}(t) = \big(x(t), y(t), z(t)\big) \in \mathbb{R}^3$ be the position of a point mass (particle) in motion at time t. The law of motion is expressed as $m\ddot{\boldsymbol{x}} = \boldsymbol{F}$, where m is the *inertial* mass of the particle, and $\boldsymbol{F}$ stands for a force. The law of universal gravitation says that a point mass fixed at the origin O attracts the point mass at $\boldsymbol{x}$ by the force $\boldsymbol{F} = -GMm\|\boldsymbol{x}\|^{-3}\boldsymbol{x}$,[51] where $G = 6.67408 \times 10^{-11}\,\mathrm{m}^3\,\mathrm{kg}^{-1}\,\mathrm{s}^{-2}$ is the *gravitational constant*, and M is the *gravitational mass* at the origin. What should be stressed is, as stated in the opening paragraph of the *Principia*, that the inertial mass coincides with the gravitational mass under a suitable system of units (this is by no means self-evident), so that the Newtonian equation is expressed as $\ddot{\boldsymbol{x}} = -GM\|\boldsymbol{x}\|^{-3}\boldsymbol{x}$, namely, the acceleration does not depend on the mass m as Galileo observed for a falling object. Kepler's laws of planetary motion are deduced from this equation. □

The Newton's cardinal importance in the history of cosmology lies in his understanding of physical space. According to him, Euclidean space is an "absolute space," with which one can say that an object must be either in a state of absolute rest or moving at some absolute speed. To justify this, he assumed that the fixed stars can be a basis of "inertial frame." His *bucket experiment*,[52] elucidated in the *Scholium* to Book 1 of the *Principia*, is an attempt to support the existence of absolute motion.

Setting aside the theoretical matters, what is peculiar about the *Principia* is that Newton adhered to the Euclidean style, and used laboriously the traditional theory of proportions in Book V of the *Elements* and the results in Apollonius' *Conics* even though he was well acquainted with Descartes' work, which was, as a matter of course, more appropriate for the presentation. He testified, writing about himself in the third person, "By the help of the *Analysis*, Mr. Newton found out most of Propositions of his *Principia Philosophiae*: but because the Ancients for making things certain admitted nothing into Geometry before it was demonstrated synthetically, he demonstrated the Propositions synthetically, that the System of Heavens might be founded upon good Geometry. And this makes it now difficult

[50]Newton tried to reconfirm his theory of gravitation by explaining the motion of the moon observed by J. Flamsteed, but the 3-body problem for the Moon, Earth and Sun turned out be too much complicated to accomplish his goal (he planned to carry a desired result as a centerpiece in the new edition of the *Principia*).

[51]Throughout, $\|\cdot\|$ and $\langle\cdot,\cdot\rangle$ stand for *norm* and *inner product*, respectively.

[52]When a bucket of water hung by a long cord is twisted and released, the surface of water, initially being flat, is eventually distorted into a paraboloid-like shape by the effect of centrifugal force. This shape shows that the water is rotating as well, despite the fact that the water is at rest relative to the bucket. From this, Newton concluded that the force applied to water does not depend on the relative motion between the bucket and the water, but it results from the absolute rotation of water in the stationary absolute space (Footnote 187).

for unskillful Men to see the Analysis by which those Propositions were found out" (Phil. Trans. R. Soc., **29** (1715), 173–224).

Apart from the style of the *Principia*, this epoch-making work (and its offspring) was "the theory of everything" in the centuries to follow as far as the classical description of the world is concerned. Indeed, Newton's laws seemed to tap all the secrets of nature, and hence to be the last word in physics. It was in the nineteenth century that physical phenomena inexplicable by the Newtonian mechanics were discovered one after another. A notable example is electromagnetic phenomena which are in discord with his mechanics at the fundamental level; see Remark 17.2 (4). Moreover various physicochemical phenomena called for entirely new explanations as well, and eventually led to quantum physics.

Newton was a pious Unitarian. His theological thought had him say: "Space is God's boundless uniform sensorium."[53] This gratuitous comment to his own comprehensive theory provoked a criticism from Leibniz, who said that God does not need a *sense organ* to perceive objects. Moreover, on the basis of "the principle of sufficient reason" and "the principle of the identity of indiscernibles," Leibniz claimed that space is merely relations between objects, thereby no absolute location in space, and that time is order of succession. His thought was unfolded in a series of long letters between 1715 and 1716 to a friend of Newton, S. Clarke.[54] Having said that, however, the existence of God is an issue which Leibniz could not sidestep. To be specific, his principle of sufficient reason made him assert that nothing happens without a reason, and that all reasons are *ex hypothesi* God's reasons. One may ask, for instance, "Why would God have created the universe here, rather than somewhere else?" That is, when God created the universe, He had an infinite number of choices. According to Leibniz, He would choose the best one among different possible worlds. As will be explained later (Sect. 6.12), his insistence, though having a strong theological inclination, is relevant to a fundamental physical principle.

At all events, the period that begins with Newton and Leibniz corresponds to the commencement of the close relationship between mathematics and physics. From then on, both disciplines have securely influenced each other.

Immanuel Kant (1724–1804), who stimulated the birth of German idealism, was influenced by the rationalist philosophy represented by Descartes and Leibniz on the one hand, and troubled by Hume's thoroughgoing skepticism on the other.[55] He felt the need to rebuild metaphysics to argue against Hume's view. Being

[53] The medieval tradition had so much effect on Newton that he was deeply involved in alchemy and regarded the universe as a cryptogram set by God. As J. M. Keynes says, he was not the first of the age of reason, but the last of the magicians (*Newton the Man*, 1947).

[54] Clarke (1717), *A Collection of Papers, which passed between the late Learned Mr. Leibniz, and Dr. Clarke, In the Years 1715 and 1716*.

[55] The empiricist David Hume is a successor of F. Bacon, T. Hobbes, J. Locke, and G. Berkeley. He says that an orderly universe does not necessarily prove the existence of God (*An Enquiry Concerning Human Understanding*, 1748). Hobbes, a critic of Aristotle, described the world as mere "matter in motion," maybe the most colorless depiction of the universe since the ancient atomists, but he did not abandon God in his cosmology (*Leviathan*, 1651).

also dissatisfied with the state of affairs surrounding metaphysics, in contrast to the scientific model cultivated in his days, Kant strove to lay his philosophical foundation of reason and judgement on secure grounds.

He was an enthusiast of Newtonian physics framed on Euclidean geometry, and asked himself, "Are space and time real existences? Are they only determinations or relations of things as Leibniz insists?" His inquiry brought him to the conclusion that Euclidean geometry is the inevitable necessity of thought and *inherent in nature* because our space (as an "absolute" entity) is a "sacrosanct" framework for all and any experience. He also held that space (as a "relative" entity) is the "subjective constitution of the mind" (see Sect. 6.22), He stressed further that mathematical propositions are formal descriptions of the a priori structures of space and time.[56]

In the work *Allgemeine Naturgeschichte und Theorie des Himmels* (1755), Kant discussed infiniteness of the universe. He believed that the universe is infinite, because God is the "infinite being," and creates the universe in proportion to his power (recall Bruno's view). In turn, he argues that it is not philosophically possible to decide whether the universe is infinite or finite; that is, we are incapable of perceiving so large distance, because the mind is finite.[57]

As Kant emphasized in the *Kritik der reinen Vernunft* (1781), abstract speculation must be pursued without losing touch with reality grasped by intuition; otherwise the outcome could be empty. This is not least the case for scientific knowledge. Yet intuition about our space turned out to be a clinging constraint on us; it was not easy to free ourselves from it.

6.7 Gauss: Intrinsic Description of the Universe

In Christianity, God is portrayed to be omnipresent, yet His whereabouts are unknown. Nonetheless, the laity usually personify God (if not a person literally), and believe that He "dwells" outside the universe (or *empyrean*). Imagining such a deity is not altogether extravagant from the view that we will take up in investigating a model of the universe though we do not posit the existence of God. What is available in place of God is, of course, *mathematics*.[58] Thus, we pose the question, "Is it mathematically possible to tell how the universe is curved without mentioning any outside of the universe?"

[56] Affected by the science in his time, Kant developed further the *nebular hypothesis* proposed by E. Swedenborg in 1734 (this hypothesis that the solar system condensed from a cloud of rotating gas was discussed later by Laplace in 1796).

[57] Hegel dismissed the claim that the universe extends infinitely.

[58] Robert Grosseteste (ca. 1175–1253) in medieval Oxford says, "mathematics is the most supreme of all sciences since every natural science ultimately depends on it" (*De Luce*, 1225). He maintained, by the way, that stars and planets in a set of nested spheres around the earth were formed by crystallization of matter after the birth of the universe in an explosion.

The answer is "Yes." To explain why, let us take a look at smooth *surfaces* as 2D toy models of the universe.[59] That is, we human beings are supposed to be confined in a surface. Here space in which the surface is located is thought of as the "outside," and is called the *ambient space*. A person in the outside plays the role of God incarnate. In our story, we allow him to exist for the time being; that is, we investigate surfaces in an *extrinsic* manner. At the final stage, we are disposed to remove him (and even the outside) from the scene, and to let the inhabitants in the surface find a way to understand his universe. In other words, what we shall do is, starting from an extrinsic study of surfaces, to look for some geometric concepts with which one can talk about how surfaces are curved without the assistance of the outside. Such concepts will be called *intrinsic*, one of the most vital key terms in this essay.

A few words about intrinsicness. Imagine that an inhabitant confined in a surface wants to examine the structure of his universe. The only method he holds is, like us, measuring the distance between two points and the angle between two directions. In our universe, we use the *light ray* to this end, which in a uniform medium propagates along shortest curves (straight lines) in space; a very special case of *Fermat's principle of least time* saying that the path taken between two points by a light ray is the path which can be traversed in the least time.[60] Hence shortest curves on a surface are regarded as something similar to paths traced by light rays (see Sect. 6.12), and it is reasonable to say that a quantity (or a concept) attached to a surface is "intrinsic" if it is derived from distance and angle measured by using shortest curves.

The progenitor of the intrinsic theory of surfaces is Johann Carl Friedrich Gauss (1777–1855). In 1828, he brought out the memoir *Disquisitiones generales circa superficies curvas*, where he formulated *bona fide* intrinsic curvature, which came to be called *Gaussian curvature* later.

Just for the comparison purpose, let us primarily look at the curvature of plane curves. Subsequent to the pioneering studies by Kepler (*Opera*, vol. 2, p. 175) and Huygens (1653–4), Newton used his calculus to compute curvature.[61] His approach is to compare a curve with *uniformly curved figures*; i.e., circles or straight lines. To be exact, given a point p on a smooth curve C, he singles out the circle (or the straight line) C_0 that has the highest possible order of contact with C at p, and then defines the curvature of C at p to be the reciprocal of the radius of C_0 (when C_0 is a straight line, the curvature is defined to be 0); thus the smaller the circle C_0 is, the more curved C is. To simplify his computation, take a coordinate system (x, y) such that the x-axis is the tangent line of C at p, and express C around p as the graph

[59]Throughout, "smooth" means "infinite differentiable."

[60]This principle was stated in a letter (January 1, 1662) to M. C. de la Chambre. Employing it, he deduced the *law of refraction*. The first discoverer of the law is Harriot (July 1601), but he died before publishing. W. Snellius rediscovered the law (1621). Descartes derived the law in *La Dioptrique* under a wrong setup; actually he believed that light's propagation was instantaneous. The finite speed of light was demonstrated by O. Rømer in 1676.

[61]*Tractus de methodis serierum et fluxionum* (1670–1671).

associated with a smooth function $y = f(x)$ with $f(0) = f'(0) = 0$ and $f''(0) \geq 0$. Compare $f(x) = \frac{1}{2}f''(0)x^2 + \frac{1}{3!}f'''(\theta x)x^3$ $(0 < \theta < 1)$ (Taylor's theorem) with the function $f_0(x) = R - \sqrt{R^2 - x^2} = \frac{1}{2}R^{-1}x^2 + \cdots$ corresponding to the circle C_0 of radius R, tangent to the x-axis at $x = 0$. Thus C_0 has the second order of contact with C at $(0, 0)$ if and only if $R = 1/f''(0)$. Therefore the curvature is equal to $f''(0)$.

Remark 7.1 For an *oriented curve* C (a curve with a consistent direction defined along the curve), we select a coordinate system (x, y) such that the direction of x-axis coincides with that of C at p. Then $f''(0)$ is possibly negative. We call $\kappa(p) := f''(0)$ the *signed curvature* of C at p. If $c : [a, b] \longrightarrow \mathbb{R}^2$ is a parameterization of C with $\|\dot{c}\| \equiv 1$, then $\ddot{c}$ is perpendicular to $\dot{c}$, and $\ddot{c}(s) = \kappa(c(s))\boldsymbol{n}(s)$, where $\boldsymbol{n}(s)$ is the unit vector obtained by the 90° counterclockwise rotation of $\dot{c}(s)$. Writing $c(s) = (x(s), y(s))$, we have $\boldsymbol{n}(s) = (-\dot{y}(s), \dot{x}(s))$, and $\kappa(c(s)) = \langle \ddot{c}(s), \boldsymbol{n}(s) \rangle = \dot{x}(s)\ddot{y}(s) - \dot{y}(s)\ddot{x}(s)$. □

In the eighteenth century, Continental mathematicians polished calculus as an effectual instrument to handle diverse problems in sciences. Leonhard Euler (1707–1783) was among those who helped to brings calculus to completion and employed it as a tool for the extrinsic geometry of surfaces. His particular interest was the curvature of the curve obtained by cutting a surface with the plane spanned by a tangent line and the normal line. This study brought him to the notion of *principal curvature*.[62] Then in 1795, Gaspard Monge (1746–1818) published the book *Application de l'analyse à la géométrie*, an early influential work on differential geometry containing a refinement of Euler's work. However neither Euler nor Monge arrived at arrived at intrinsic curvature. Here came Gauss, who found how to formulate it with absolute confidence in its significance.

What comes to mind when defining the curvature of a surface S at p is the use of a coordinate system (x, y, z) with origin p such that the xy-plane is tangent to S at p,[63] with which S is locally expressed as a graph of a smooth function $f(x, y)$ defined around $(x, y) = (0, 0)$ with $f(0, 0) = f_x(0, 0) = f_y(0, 0) = 0$, where $f_x = \partial f/\partial x$ and $f_y = \partial f/\partial y$. Then we obtain $f(x, y) = \frac{1}{2}(ax^2 + 2bxy + cy^2) + r(x, y)$, where $a = f_{xx}(0, 0)$, $b = f_{xy}(0, 0)$, $c = f_{yy}(0, 0)$, and $r(x, y)$ is of higher degree than the second. The shape of the graph (and hence the shape of S around p) is roughly determined by the quantity $ac - b^2$; it is paraboloid-like (resp. hyperboloid-like) if $ac - b^2 > 0$ (resp. $ac - b^2 < 0$) (Fig. 6.7). Given all this, we define the curvature $K_S(p)$ to be $ac - b^2$, which is independent of the choice of a coordinate system. For example, $K_S \equiv R^{-2}$ for the sphere of radius R.

To account the exact meaning of intrinsicness of the Gaussian curvature, we consider another surface S', and let Φ be a one-to-one correspondence from S to S' preserving the distance of two points and the angle between two directions (such

---

[62] *Recherches sur la courbure des surfaces*, Mém. Acad. R. Sci. Berlin, **16** (1767), 119–143.

[63] Gauss's original definition uses the map (*Gauss map*) from S to the unit sphere which associates to each point on S its oriented unit normal vector (*Disquisitiones generales*, Art. 6).

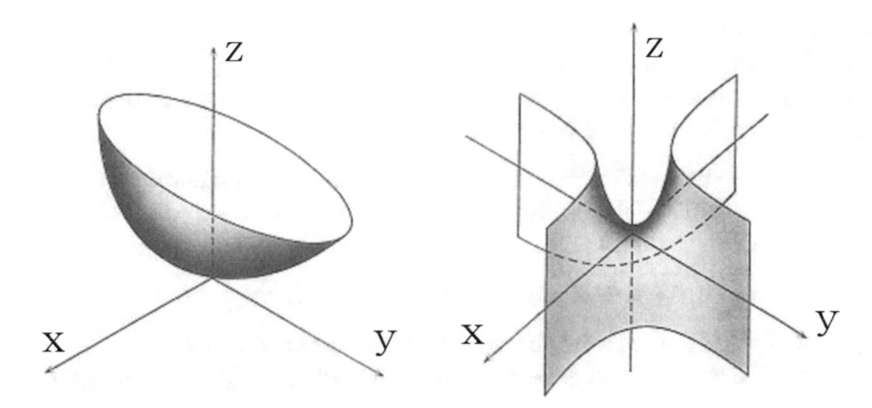

Fig. 6.7 Paraboloid and Hyperboloid

Φ is called an *isometry* in the present-day terms; see Sect. 6.20). If we would adopt the approach mentioned above on intrinsicness, Gauss's outcome could be rephrased as "$K_{S'}(\Phi(p)) = K_S(p)$." In particular, $K_S = 0$ for a cylindrical surface S, since S, which is ostensibly curved, is deformed without distortion to a strip in the plane. This example articulately signifies that the Gaussian curvature could be possibly defined without the aid of the outside.

Executing lengthy computations, Gauss confirmed the intrinsicness of K_S (see Sect. 6.17).[64] Since the goal is more than a pleasing astonishment, Gauss named his outcome "Theorema Egregium (remarkable theorem)" (*Disquisitiones generales*, Art. 12). Actually, it is no exaggeration to say that this theorem is a breakthrough not only in geometry, but also in cosmology. The following is his own words (1825) unfolded in a letter to the P. A. Hansen:

This research is deeply entwined with much else, I would like to say, with metaphysics of space and I find it difficult to shake off the consequences of this, such as for instance the true metaphysics of negative or imaginary quantities (*Werke*, XII, p. 8. See [3]).[65]

Although the Theorema Egregium has a deep metaphysical nature as Gauss put it, his incipient motivation came from the practical activity pertaining to the *geodetic survey* of Hanover to link up with the existing Danish grid that started in 1818 and

[64]Crucial in his argument is the commutativity of partial differentiation. Euler and A. C. Clairaut knew it, but the first correct proof was given by K. H. A. Schwarz in 1873.

[65]Even in the eighteenth century, there were a few who did not accept negative number. For example, F. Maseres said, "[negative numbers] darken the very whole doctrines of the equations and make dark of the things which are in their nature excessively obvious and simple" (1758). Although limited to the practical use, negative numbers already appeared in the *Jiuzhang Suanshu* (The Nine Chapters on the Mathematical Art) written in the period of the Han Dynasty (202 BCE–8 CE). The reason why the Chinese accepted negative numbers is, in all likelihood, that they had the concept of *yin* (negative or dark) and *yang* (positive or bright). Negative numbers were in use India at least from about the 7th century.

continued for 8 years. He traveled criss-cross in order to cover the whole country. Fueled by this arduous activity, he brought out many papers on the geometry of curves and surfaces from about 1820 onwards. The Theorema Egregium was a culmination of his intensive study in this field.

Remark 7.2 In her study of elasticity (1831), Sophie Germain introduced the *mean curvature* $H_S(p) := \frac{1}{2}(a + c)$, which appears in a characterization of *minimal surfaces*; that is, the surface area of S is locally minimized if and only if $H_S \equiv 0$. Putting $A = \begin{pmatrix} a & b \\ b & c \end{pmatrix}$, we have $K_S(p) = \det A$ and $H_S(p) = \frac{1}{2}\mathrm{tr}\, A$. The eigenvalues $\kappa_1(p)$, $\kappa_2(p)$ of A are called the *principal curvatures* of S. Obviously $K_S(p) = \kappa_1(p)\kappa_2(p)$ (*Disquisitiones generales*, Art. 8) and $H_S(p) = \frac{1}{2}(\kappa_1(p) + \kappa_2(p))$. □

6.8 Number Systems

It is off the subject, but while looking at the background behind Gauss's opinion about complex numbers quoted above, we shall prepare some material related to what we shall describe later (Sect. 6.21).

Gerolamo Cardano (1501–1576), who stood out above the rest at that time as an algebraist, acknowledged the existence of complex numbers in his *Ars Magna* (1545) that contains the first printed solutions to cubic and quartic equations. He observed that his formula may possibly involve complex numbers even when applied to a cubic equation possessing only real solutions.[66] Yet, he did not understand the nature of complex numbers and even thought that they are useless. Subsequently, Rafael Bombelli (1526–1572) took down the rules for multiplication of complex numbers (*binomio*) in his Book I of *L'Algebra*, while Descartes used the term "imaginary number" (*nombre imaginaire*) with the meaning of contempt. Wallis insisted that imaginary numbers are not unuseful and absurd when properly understood by using a geometric model just like negative numbers (*Algebra*, Vol. II, Chap. LXVI, 1673). Meanwhile, Euler persuaded himself that there is an advantage for the use of imaginary numbers (1751), and introduced the symbol i for the imaginary unit (1755). It was around 1740 that he found the earth-shaking formula $e^{\sqrt{-1}\theta} = \cos\theta + \sqrt{-1}\sin\theta$.

In 1799, Gauss gained his doctorate in absentia from the University of Helmstedt. The subject of his thesis is *the fundamental theorem of algebra*,[67] which, with a long history of inspiring many mathematicians, says that every polynomial equation

[66]For example, his formula applied to the cubic equation $x^3 - 15x - 4 (= (x-4)(x+2+\sqrt{3})(x+2-\sqrt{3})) = 0$ yields a solution $x = \sqrt[3]{2+\sqrt{-124}} + \sqrt[3]{2-\sqrt{-124}}$.

[67]*Demonstratio nova theorematis omnem functionem algebraicam rationalem integram unius variabilis in factores reales primi vel secundi gradus resolvi posse*. However, his proof had a gap. He gave three other proofs (1816, 1849) (see Sect. 6.21 for a "topological" proof).

with complex coefficients has at least one complex root.[68] As the title of his thesis indicates, he shied away from the use of imaginary numbers, notwithstanding that Gauss already had a tangible image. In a letter to F. W. Bessel, dated December 18, 1811, he communicated, "first I would like to ask anyone who wishes to introduce a new function into analysis to explain whether he wishes it to be applied merely to real quantities, and regard imaginary values of the argument only as an appendage, or whether he agrees with my thesis that in the realm of quantities the imaginaries $a+b\sqrt{-1}$ have to be accorded equal rights with the reals. Here it is not a question of practical value; analysis is for me an independent science, which would suffer serious loss of beauty and completeness, and would have constantly to impose very tiresome restrictions on truths which would hold generally otherwise, if these imaginary quantities were to be neglected..." (*Werke*, X, pp. 366–367; see [15]).

In 1833, William Rowan Hamilton (1805–1865) conceived the idea to express a complex number $a+b\sqrt{-1}$ by the point (a,b) in the coordinate plane $\mathbb{R}^2$,[69] which motivated him to construct 3D "numbers" with arithmetical operations similar to complex numbers, but the quest was a dead end. Alternatively, he discovered *quaternions*, the 4D numbers at the expense of commutativity of multiplication (1843). Specifically, the multiplication is defined by

$$(a_1,b_1,c_1,d_1)\cdot(a_2,b_2,c_2,d_2)=(a_1a_2-b_1b_2-c_1c_2-d_1d_2,\ a_1b_2+b_1a_2$$
$$+c_1d_2-d_1c_2,\ a_1c_2-b_1d_2+c_1a_2+d_1b_2,\ a_1d_2+b_1c_2-c_1b_2+d_1a_2). \quad (6.8.1)$$

In Hamilton's notations, $i=(0,1,0,0)$, $j=(0,0,1,0)$, $k=(0,0,0,1)$, so that quaternions are represented in the form: $a+bi+cj+dk\ \ (a,b,c,d\in\mathbb{R})$. A fact deserving special mention is that putting $\|\alpha\|=\sqrt{a^2+b^2+c^2+d^2}$ for $\alpha=(a,b,c,d)$, we have $\|\alpha\cdot\beta\|=\|\alpha\|\|\beta\|$, the identity similar to $|zw|=|z||w|$ for complex numbers, which agrees with the *four-square identity* discovered by Euler in 1748 during his investigation on sums of four squares (see Remark 14.1).

It may be said, incidentally, that Gauss had already discovered quaternions in 1818 (*Mutationen des Raumes*, *Werke*, VIII, 357–361). What he observed is,

[68] Albert Girard (1595–1632) was the first to suggest the fundamental theorem of algebra (FTA) (*Invention Nouvelle en l'Algèbre*, 1629). Meanwhile, Leibniz claimed that $x^4+1=0$ affords a counterexample for the FTA (1702). He assumed that the square root of i should be a more complicated "imaginary" entity, not able to be expressed as $a+bi$; but the truth is that $x=\pm\frac{1}{\sqrt{2}}(1+i)$ is a square root of i. Euler (1742) and D'Alembert (1746) counted on the FTA when they dealt with indefinite integrals of rational functions.

[69] Gauss had been in possession of the geometric representation of complex numbers since 1796. In 1797, C. Wessel presented a memoir to the Copenhagen Academy of Sciences in which he announced the same idea, but it did not attract attention. In 1806, J. R. Argand went public with the same formulation.

though his original expression is slightly different, that the rotation group SO(3) is parameterized by (a, b, c, d) with $a^2+b^2+c^2+d^2=1$ as

$$A(a,b,c,d) = \begin{pmatrix} a^2-b^2-c^2+d^2 & -2(ab+cd) & 2(bd-ac) \\ 2(ab-cd) & a^2-b^2+c^2-d^2 & -2(ad+bc) \\ 2(ac+bd) & 2(ad-bc) & a^2+b^2-c^2-d^2 \end{pmatrix},$$

and $A(a_1, b_1, c_1, d_1)A(a_2, b_2, c_2, d_2) = A\big((a_1, b_1, c_1, d_1) \cdot (a_2, b_2, c_2, d_2)\big)$. Note here that quaternions $\alpha = a + bi + cj + dk$ with $\|\alpha\| = 1$ form a group, which is identified with the *spin group* Spin(3), pertinent to the quantum version of *angular momentum* that describes *internal degrees of freedom* of electrons. Additionally, the kernel of the homomorphism $\alpha \mapsto A(\alpha)$ is a homomorphism of Spin(3) onto SO(3) whose kernel is $\{\pm 1\}$ (Remark 21.1).

Remark 8.1 Euler discovered his formula while studying differential equations with constant coefficients (1748). Previously, de Moivre derived a formula that later brought on what we now call *de Moivre's theorem* (1707), a precursor of Euler's formula. Subsequently, R. Cotes had given the formula $ix = \log(\cos x + i \sin x)$ in 1714, but overlooked that the *complex logarithm* assumes infinitely many values, differing by multiples of $2\pi i$, as found by Euler (1746).[70]

Before the problem of the complex logarithm was settled by Euler, there were arguments between Leibniz and Johann Bernoulli (1667–1748) in 1712 and then between Euler and Jean-Baptiste le Rond d'Alembert (1717–1783) around 1746 about $\log x$ for a negative x. Leibniz says that $\log(-1)$ does not exist, while Bernoulli alleges that $\log(-x) = \log x$. In the above-mentioned letter to Bessel in 1811, Gauss renewed the question about the *multivaluedness* of the complex logarithm, representing $\log z$ as the complex line integral $\int_c 1/z\, dz$, where c is a curve in $\mathbb{C}\backslash\{0\}$ joining 1 and z (see Sect. 6.21). He further referred to what we now call *Cauchy's integral theorem*.[71] □

---

[70]The *logarithmic function* originated from the computational demands of the late sixteenth century in observational astronomy, long-distance navigation, and geodesy. The main contributors are John Napier (1550–1617), Joost Bürgi, Henry Briggs, Oughtred, and Kepler. The term *logarithm*, literally "ratio-number" from λόγος and ἀριθμός, was coined by Napier in his pamphlet *Mirifici Logarithmorum Canonis Descriptio* (1614). Kepler studied Napier's pamphlet in 1619, and published the logarithmic tables *Chilias Logarithmoria* (1624), which he used in his calculations of the *Tabulae Rudolphinae Astronomicae* (1627), a star catalog and planetary tables based on the observations of Tycho Brahe.

[71]Augustin-Louis Cauchy defined complex line integrals in 1825 and established his theorem in 1851.

6.9 Euclid's Elements

Gauss is one of the discoverers of *non-Euclidean geometry* (a geometry consistently built under the assumption that there are more than one line through a point which do not meet a given line), but he did not publish his work and only disclosed it in letters to his friends.[72] His hesitation in part came from the atmosphere in those days; that is, his new geometry was entirely against the predominant Kantianism. He wanted to avoid controversial issues,[73] and requested his friends not to make it public.

The other discoverers of the new geometry are Nikolai Ivanovich Lobachevsky (1792–1856) and János Bolyai (1802–1860).[74] In 1826, Lobachevsky stated publicly that the new geometry exists.[75] Fourteen years later, as this research did not draw attention, he issued a small book in German containing a summary of his work to appeal to the mathematical community,[76] and then put out his final work *Pangéométrie ou précis de géométrie fondée sur une théorie générale et rigoureuse des parallèles*, a year before his death (1855); see [17]. Meanwhile, in a break with tradition, Bolyai independently began his head-on approach to the non-Euclidean properties as early as 1823. In a letter to his father Farkas Bolyai, on November 3, 1823, János says with confidence, "I discovered a whole new world out of nothing," and made public his discovery in a 26-pages appendix *Scientiam spatii absolute veram exhibens* to his father's book.[77]

Gauss was a perfectionist by all accounts. He did not publish many outcomes, fearing that they were never perfect enough; his motto was "*pauca sed matura*—few, but ripe." After his death it was discovered that many results credited to others had been already worked out by him (remember Cauchy's integral theorem; Remark 8.1). The extraordinariness of Gauss may be highlighted by the joke: *Suppose you discovered something new. If Gauss would be still alive and would browse your result, then he would say, "Ah, that's in my paper," and surely would take an unpublished article out of a drawer in his desk.* This was no joke for János Bolyai. Gauss received a copy of his father's book in 1832, and endorsed the discovery in the response. To János' disappointment, his cool reply reads, "The entire contents of your son's work coincides almost exactly with my own meditation, which has occupied my mind for from thirty to thirty-five years"

[72]For instance, his discovery was mentioned in a letter to F. A. Taurinus dated November 8, 1824 (*Werke*, VIII, p. 186).

[73]In a letter to Bessel dated January 27, 1829, Gauss writes, "I fear the cry of the Boetians [known as vulgarians] if I were to voice my views" (*Werke*, VIII, p. 200).

[74]Strictly speaking, both Lobachevsky and Bolyai discovered the 3D geometry, while Gauss dealt only with the 2D case (Sect. 6.11).

[75]Lobachevsky, *Exposition succincte des principes de la géométrie avec une démonstration rigoureuse du théorème des parallèles*;*O nachalakh geometrii*, Kazanskii Vestnik, (1829–1830).

[76]*Geometrischen Untersuchungen zur Theorie der Parallellinien* (1840).

[77]*Tentamen juventutem studiosam in elementa Matheseos Purae* (1832).

(*Werke*, VIII, p. 220). So much disappointed, János wholly withdrew from scientific activity, though Gauss wrote to Gerling (February 14, 1832), "I consider the young geometer Bolyai a genius of the first rank" (see [4]). As regards Lobachevsky, it was not until 1841 that his work came to the notice of Gauss, who looked over the *Geometrischen Untersuchungen* by chance. Gauss was very impressed with it, and praised Lobachevsky as a clever mathematician in a letter to his friend J. F. Encke (February 1841; *Werke*, VIII, p. 232).[78]

Now, what is known about Euclid and his *Elements*? It is popularly thought that Euclid was attached to Plato's Academeia or at least was affiliated with it, and that he was temporarily a member of the Great Museum during the reign of Ptolemy I. Truth to tell, very little is known about his life. Archimedes, whose life, too, overlapped with the reign of Ptolemy I, mentioned briefly Euclid in his *On the Sphere and the Cylinder I.2*. In the *Collection VII*, Pappus records that Apollonius spent a long time with the disciples of Euclid in Alexandria. Apollonius himself referred to Euclid's achievement with the statement that his method surpasses Euclid's *Conics*. Proclus Diadochus (411 CE–485 CE) and Joannes Stobabeing (the fifth century CE) recorded the oft-told (but not trustworthy) anecdotes about Euclid. The sure thing is that he is the author of the definitive mathematical treatise *Elements* (Στοιχεῖα),[79] which is a crystallization of the plane and space geometry known in his day (ca. 300 BCE) and was so predominant that all earlier texts were driven out. It is solely to the *Elements*, one of the most influential books in long history, which Euclid owes his abiding fame in spite of his shadowy profile.

The *Elements*, consisting of thirteen books, starts with the construction of equilateral triangles and ends up with the classification of the regular solids; *tetrahedron*, *cube*, *octahedron*, *dodecahedron*, and *icosahedron* (Book XIII, Prop. 465). This suggests the influence of Pythagorean mysticism that put great emphasis on aesthetic issues, which, nowadays, are described in terms of symmetry (συμμετρία).[80] Moreover, it has an orderly organization consisting of 23 fundamental definitions, 5 postulates, 5 axioms, and 465 propositions. The "reductio ad absurdum" is made

[78]F. K. Schweikart, a professor of law and an uncle of Taurinus, had a germinal idea of non-Euclidean geometry ("astral geometry" in his term), and asked for Gauss, through Gerling, to comment on the idea in 1818. In response to the request (March 16, 1819), Gauss communicated, with compliments, that he concurred in Schweikart's observation (*Werke*, VIII, p. 181). However Schweikart neither elaborated his idea nor published any result.

[79]Besides the *Elements*, at least five works of Euclid have survived to today; *Data, On Divisions of Figures, Catoptrics, Phaenomena, Optics*. There are a few other works that are attributed to Euclid but have been lost: *Conics, Porisms, Pseudaria, Surface Loci, Mechanics.*

[80]"Symmetry" is explicated by *group actions*, with which both spatial homogeneity and isotropy are described. Klein's Erlangen Program is also explained in this framework.

Aetius (first- or second-century CE) says that Pythagoras discovered the five regular solids (*De Placitis*), while Proclus argues that it was the historical Theaetetus (ca. 417 BCE–ca. 369 BCE) who theoretically constructed the five solids. Plato's *Timaeus* (53b5–6) provides the earliest known description of these solids as a group. Nowadays, regular polyhedra are discussed often in relation to finite subgroups (*polyhedral groups*) of SO(3) or O(3). For example, the *icosahedral group* is the rotational symmetry group of the icosahedron (see Remark 21.1).

full use of as a powerful gambit (Prop. 6 in Book I is the first one proved by reductio ad absurdum).

As often said, only a few theorems are thought of having been discovered by Euclid himself. The theorem of angle-sum (Book I, Prop. 32), the Pythagorean Theorem (Book I, Prop. 47), and perhaps the *triangle inequality* (Book I, Prop. 20) are attributed to the Pythagoreans. The theory of proportions (Book V) and the method of exhaustion (Book X, Prop. 1, Book XII) are, as we mentioned before, greatly indebted to Eudoxus.

We shall talk about a checkered history of the *Elements*, originally written on fragile papyrus scrolls.[81] The book survived in Alexandria and Byzantium under the control of the Roman Empire. In contrast, Roman people—the potentates of the world during this period—by and large expressed little interest in pure sciences as testified by Marcus Tullius Cicero (BCE 106–BCE 43), who confesses, "Among the Greeks, nothing was more glorious than mathematics. We, however, have limited the usability of this art to measuring and calculating" (*Tusculanae Disputationes*, ca. 45 BCE). Luckily, Alexandria is far away from the central government of the Empire, disintegrating in scandal and corruption. Situated as it was and thanks to the Hellenistic tradition, pure science flourished there. The people involved were Heron (ca. 10 CE–ca. 70 CE), Menelaus (ca. 70 CE–ca. 140 CE), Ptolemy, Diophantus (ca. 215 CE–ca. 285 CE), Pappus, Theon Alexandricus (ca. 335 CE–ca. 405 CE), and Hypatia (ca. 350 CE–415 CE).[82]

But even in the glorious city, scientific attitude declined with the lapse of time, as the academic priority had been given, if anything, to annotations on predecessor's work, and eventually fell into a state of decadence. This was paralleled by the steady decay of the city itself that all too often suffered overwhelming natural and manmade disasters. After a long tumultuous period, the great city was finally turned over to Muslim hands (641 CE).

In the Dark Ages, Europeans could no longer understand the *Elements*. In this circumstance, Anicius Manlius Boethius (ca. 480 CE–524 CE) and Flavius Magnus Aurelius Cassiodorus Senator (ca. 485 CE–ca. 585 CE) are very rare typical scholars in this period who were largely influential during the Middle Ages. Meanwhile, the *Elements* was brought from Byzantium to the Islamic world in ca. 760 CE, and

[81]In the tiny fragment "Papyrus Oxyrhynchus 29" housed in the library of the University of Pennsylvania, one can see a diagram related to Prop. 5 of Book II, which can be construed in modern terms as $ab + \left(\frac{a+b}{2} - b\right)^2 = \left(\frac{a+b}{2}\right)^2$.

[82]Diophantus stands out as unique because he applied himself to some algebraic problems in the time when geometry was still in the saddle. Theon is known for editing the *Elements* with commentaries. His version was the only Greek text known, until a hand-written copy of the *Elements* in the tenth century was discovered in the Vatican library (1808). Hypatia, a daughter of Theon, is the earliest female mathematician in history, and is more known for her awful death; she was dragged to her death by fanatical Christian mobs.

was taught in Bagdad, Cordova, and Toledo where there were large-scale libraries comparable with the Great Museum in Alexandria.[83]

After approximately 400 years of blank period in Europe, Adelard of Bath (Remark 2.1 (2)), who undertook a journey to Cordova around 1120 to delve into Arabic texts of Greek classics, obtained a copy of the book and translated it into Latin. Later on, Gerard (Footnote 5) translated it into Latin from another Arabic version procured in Toledo. His translation is considered being close to the Greek original text. Then, in the mid-renaissance, only 27 years later from the type printing of the *Bible* by Johannes Gutenberg in Mainz, the first printed edition—based on the Latin version from the Arabic by Campanus of Novara who probably had access to Adelard's translation—was published in Venice. Thereafter, versions of the *Elements* in various languages had been printed. The first English-language edition was printed in 1570 by H. Billingsley. Then in 1607, Matteo Ricci and Xu Guangqi translated into Chinese the first six volumes of the Latin version published in 1574 by Christopher Clavius.

One of the first multicolored books *The First Six Books of the Elements of Euclid* by O. Byrne was printed in 1847 in London. A. De Morgan was very critical of its non-traditional style embellished by pictorial proofs, and judged this inventive attempt to be nonsense (*A Budget of Paradoxes*, 1872). At that time, there was considerable debate about how to teach geometry; say, the pros and cons of whether to adopt a new approach to Euclidean geometry in teaching. C. Dodgson in Oxford, known as Lewis Carroll, assumed a critical attitude to his colleague Playfair who tried to simplify the Euclid's proofs by introducing algebraic notations. In the little book *Euclid and his Modern Rivals* printed in 1879, he argues, "no sufficient reasons have yet been shown for abandoning [Euclid's *Elements*] in favour of any one of the modern Manuals which have been offered as substitutes." G. B. Halsted says, in the translator's introduction to J. Bolyai's *Scientiam spatii*, "Even today (1895), in the vast system of examinations carried out by the British Government, by Oxford, and by Cambridge, no proof of a theorem in geometry will be accepted which infringes Euclid's sequence of propositions." As a matter of course, it is now seldom to make direct use of Euclid's propositions for university examinations, but the book went through more than 2000 editions to date, and has been (and still is) the encouragement and guide of scientific thought. Furthermore, modern mathematics inherits much of its style from the *Elements*; in particular, many mathematical theories, even if not all of them, begin with axiomatic systems (Sect. 6.18).

What about the original Greek text? It was Heiberg (Footnote 11) who tracked down all extant manuscripts—the aforementioned Greek Vatican manuscript is one of them—to produce a definitive Greek text together with its Latin translation and

[83]Worthy to mention is the *House of Wisdom* (Bayt al-Hikma) in Bagdad. It was founded by Caliph Harun al-Rashid (ca. 763 CE–809 CE) as an academic institute devoted to translations, research, and education, and culminated under his son Caliph al-Mamoon. Al-Khwārizmī was a scholar at this institute.

prolegomena (*Euclidis Opera Omnia*, 8 vols, 1883–1916). This is a trailblazing achievement that has become the basis of later researches on Euclid.

6.10 The Fifth Postulate

Euclid's greatest contribution is his daring selection of a few postulates as major premises (if not completely adequate from today's view), of which Greek geometers thought as self-evident truth requiring no proof. He built up a deductive system in which every theorem is derived from these postulates. Among his five postulates, the last one, called the *Fifth Postulate* (FP henceforward), constitutes the core of our story on non-Euclidean geometry. It says:

> If a straight line falls on two straight lines in such a manner that the interior angles on the same side are together less than two right angles, then the straight lines, if produced indefinitely, meet on that side on which are the angles less than the two right angles [7, Book I].

The theorem of angle-sum and the Pythagorean Theorem rely on this postulate. For example, what is required for the former is the fact that "if a straight line fall upon two parallel straight lines it will make the alternate angles equal to one another" (Prop. 29 in Book I), which is a consequence of the FP.[84]

The first four postulates have fairly simple forms; say, the first one declares, in today's terms, that given two points, there exists a straight line through these points, whereas the necessity of the last postulate is by no means overt, owing not only to the intricacy of its formulation, but also to the fact that the converse of Prop. 29 is proved without an appeal to the last postulate (Prop. 27). Actually the last one had been believed to be redundant and hence to be a "proposition" all through modern times until the second decade of the nineteenth century [2].

An elaborate and stalwart attempt to vindicate Euclidean geometry by "proving" the FP was done by Girolamo Saccheri (1667–1733), an Italian Jesuit priest, who explored the consequences under the negation of the postulate, hoping to reach a contradiction. For this sake, he made use of quadrilaterals of which two opposite sides are equal to each other and perpendicular to the base, and set up the following three hypotheses (Fig. 6.8)[85];

(i) *the hypothesis of the right angle*: $\angle C = \angle D = \angle R$,
(ii) *the hypothesis of the obtuse angle*: $\angle C = \angle D > \angle R$, and
(iii) *the hypothesis of the acute angle*: $\angle C = \angle D < \angle R$,

[84] In contrast to these two theorems, the triangle inequality is proved without invoking the FP (Book I, Prop. 20).

[85] Omar Khayyam (1048–1131) dealt with the FP in his *Explanation of the Difficulties in the Postulates of Euclid*. He is considered a predecessor of Saccheri.

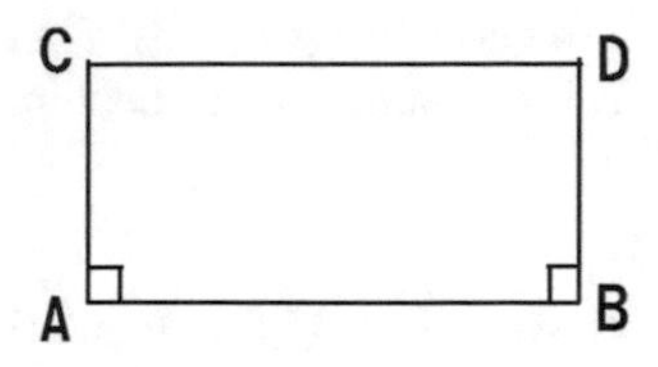

Fig. 6.8 Saccheri's quadrilateral

where the first hypothesis is equivalent to the FP. Examining these hypotheses meticulously, he concluded that hypothesis (ii) contradicts the infinite extent of plane, and is false, but could not exclude hypothesis (iii). At the final stage of his argument, Saccheri appeals to intuition about our space, and says, "*Hypothesis anguli acuti est absolute falsa; quia repugnans naturae liniae rectae*" ["the hypothesis of the acute angle is absolutely false; because it is repugnant to the nature of the straight line"] [2]. In summary, he virtually proved that, under the negation of the FP, the angle-sum of any triangle is less than π (thus the existence of a triangle with angle-sum of π implies the FP).[86]

Johann Heinrich Lambert (1728–1777)—conceivably familiar with Saccheri's work because, in his *Theorie der Parallellinien* (1766), he quoted the work of G. S. Klügel who listed nearly 30 attempts to prove the FP including Saccheri's work[87]—investigated the FP in alignment with an idea resembling that of Saccheri [2, p. 44].[88] His work, published after the author's death, bristles with far-sighted observations and highly speculative remarks (see [26]). One of his revelations is that, under the negation of the FP, the *angular defect* $\pi - (\angle A + \angle B + \angle C)$ for a triangle $\triangle ABC$ is proportional to the area of the triangle. This is inferred from the fact that if we define $m(P)$ for a polygon P with n sides by setting $m(P) = (n-2)\pi - $(the sum of inner angles of P), then $m(\cdot)$ and the area functional Area$(\cdot)$ share the "additive property"; i.e., $m(P) = m(P_1) + m(P_2)$ for a polygon P composed of two polygons P_1, P_2. With this observation in hand, he reasoned that non-Euclidean plane-geometry (if it exists) has a close resemblance to *spherical geometry*.

Let us briefly touch on spherical geometry, which is nearly as old as Euclidean geometry, and was developed in connection with geography and astronomy. In his *Sphaerica*, extant only in an Arabic translation, Menelaus introduced the notion of *spherical triangle*, a figure formed on a sphere by three great circular arcs intersecting pairwise in three vertices. He observed that the angle-sum of a spherical triangle is greater than π [28]. Greek spherical geometry gave birth to *spherical trigonometry* that deals with the relations between trigonometric functions of the

[86]Saccheri, *Euclides ab omni naevo vindicatus sive conatus geometricus quo stabiliuntur prima ipsa geometriae principia* (1733). See [32].

[87]*Conatuum praecipuorum theoriam parallelarum demonstrandi recensio* (1763).

[88]Lambert proved that the transfer time from $\boldsymbol{x}_1$ to $\boldsymbol{x}_2$ along a Keplerian orbit is independent of the orbit's eccentricity and depends only on $\|\boldsymbol{x}_1\| + \|\boldsymbol{x}_2\|$ and $\|\boldsymbol{x}_1 - \boldsymbol{x}_2\|$ (*Über die Eigenschaften der Kometenbewegung*, 1761). He was the first to express Newton's second law of motion in the notation of differential calculus (*Vis Viva*, 1783).

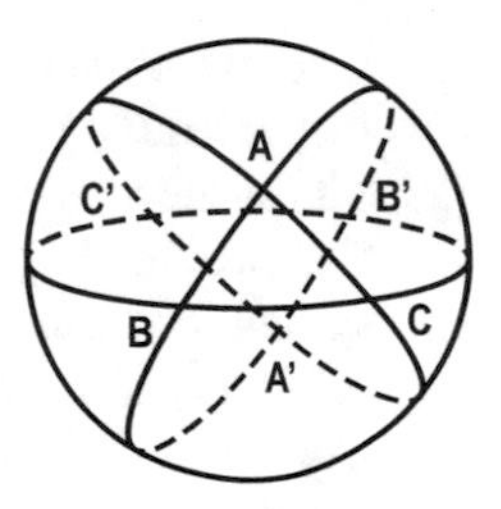

Fig. 6.9 A spherical triangle

sides and angles of spherical triangles. At a later time, it was treated in Islamic mathematics and then by Viète, Napier, and others. A culmination is the following fundamental formulas given by Euler.[89]

$$\cos \angle A = -\cos \angle B \cos \angle C + \sin \angle B \sin \angle C \cos \frac{a}{R}, \tag{6.10.2}$$

$$\cos \frac{a}{R} = \cos \frac{b}{R} \cos \frac{c}{R} + \sin \frac{b}{R} \sin \frac{c}{R} \cos \angle A, \tag{6.10.3}$$

where $\triangle ABC$ is a spherical triangle on the sphere of radius R, and a, b, c are the arc-length of the edges corresponding to A, B, C, respectively (Fig. 6.9).

Now, what Lambert paid attention to is the formula $(\angle A + \angle B + \angle C) - \pi = R^{-2}\text{Area}(\triangle ABC)$, a refinement of Menelaus' observation on spherical triangles discovered by Harriot (1603) and Girard (*Invention nouvelle en algebra*, 1629), and probably known by Regiomontanus.[90] Replacing the radius R by $\sqrt{-1}R$ in this formula, we obtain $\pi - (\angle A + \angle B + \angle C) = R^{-2}\text{Area}(\triangle ABC)$, which fits in with his outcome on the angular defect, and brought him to the speculation that an "imaginary sphere" (under a suitable justification) provides a model of "non-Euclidean plane," but he did not pursue this idea any further.

Remark 10.1

(1) Here is a list (by no means exhaustive) of people who tried to deduce the FP from other propositions: Posidonius of Apameia (ca. 135 BCE–ca. 51 BCE), Geminus of Rhodes (the first century BCE), Ptolemy, Proclus, Nasīr al-Dīn Tūsī (1201–1274), C. Clavius (1538–1612), P. A. Cataldi (1548–1626), G. A. Borelli (1608–1679), G. Vitale (1633–1711), Leibniz, Wallis, J. S. König (1712–1757), Adrien-Marie Legendre (1752–1833), B. F. Thibaut (1775–1832), and F. Bolyai. Gauss, J. Bolyai, and Lobachevsky were, for that matter, no exception at the beginning of their studies of parallels. What is in common to their arguments

[89] *Trigonometria sphaerica universa, ex primis principiis breviter et dilucide derivata*, Acta Academiae Scientarum Imperialis Petropolitinae, **3** (1782), 72–86 (see [25]). The spherical cosine law (6.10.3) is seen in al-Battānī's work *Kitāb az-Zīj* (Footnote 20).

[90] The Latin name of Johannes Müller von Königsberg (1436–1476), the most important astronomer of the fifteenth century who may have arrived at the heliocentricism.

(except for the great trio) is that they beg the question; that is, they assume what they intend to prove.
(2) The system of postulates in the *Elements* is not complete. An airtight system was provided by Hilbert in the text *Grundlagen der Geometrie* (1899). His system consists of five main axioms; I. *Axioms of Incidence*, II. *Axioms of Order*, III. *Axioms of Congruence*, IV. *Axiom of Parallels*, V. *Axioms of Continuity*.[91] The Axiom of Parallels says, "Given a line and a point outside it, there is exactly one line through the given point which lies in the plane of the given line and point so that the two lines do not meet." This statement, clearly given by Proclus in his commentary, was adopted by Playfair in his edition of the *Elements* (1795) as an alternative for the FP. Hilbert's Axioms of Continuity consists of the axiom of completeness related to Dedekind cuts (Remark 4.2 (4)) and the Axiom of Archimedes. □

6.11 Discovery of Non-Euclidean Geometry

We now go back to Gauss's discovery of non-Euclidean geometry. In his letter to Taurinus in 1824, Gauss mentions, "The assumption that angle-sum [of triangle] is less than 180° leads to a curious geometry, quite different from ours [the euclidean] but thoroughly consistent, which I have developed to my entire satisfaction. The theorems of this geometry appear to be paradoxical, and, to the uninitiated, absurd, but calm, steady reflection reveals that they contain nothing at all impossible" (see [1]).

Now, when did Gauss become aware of the new geometry? A few tantalizing pieces of evidence are in Gauss's *Mathematisches Tagebuch*, a record of his discoveries from 1796 to 1814 (*Werke*, X).[92] The 72nd item written on July 28, 1797 says, "Plani possibilitatem demostravi" ["I have demonstrated the *possibility* of a plane"]. This seems to indicate Gauss's interest in the foundation of Euclidean geometry, of which he never lost sight from that time onward. Following that, in the 99th entry dated September 1799, he wrote, "In principiis Geometriae egregios progressus fecimus" ["On the foundation of geometry, I could make a remarkable progress"]. It is difficult to imagine what this oblique statement means, but it could have something to do with non-Euclidean geometry.[93] In the same year, Gauss

[91] Moritz Pasch (1843–1930), Giuseppe Peano (1858–1932), and Mario Pieri (1860–1913) contributed to the axiomatic foundation of Euclidean geometry.

[92] Gauss's diary—which contains 146 entries written down from time to time, and most of which consist of brief and somewhat cryptical statements—was kept by his bereaved until 1899. The 80th entry (October, 1797) announces the discovery of the proof of the Fundamental Theorem of Algebra.

[93] In a letter to F. Bolyai, March 6, 1832 (*Werke* VIII, p. 224), Gauss alluded that Kant was wrong in asserting that space is only the form of our intuition.

confided to F. Bolyai to the effect that he had doubt about the truth of geometry.[94] However, his letter to F. Bolyai in 1804 witnessed that he was still suspicious of the new geometry (*Werke*, VIII, p. 160). All of the available witnesses point to the possibility that it is in 1817 at the latest that Gauss disengaged himself from his preconceptions, and began to deal seriously with the new geometry. In a letter to his friend H. Olbers (April 28, 1817), he elusively says, "I am becoming more and more convinced that the necessity of *our geometry* cannot be proved" (*Werke*, VIII, p. 177).

In all likelihood, Gauss had a project to bring out an exposition of the "Nicht-Euklidische Geometrie". His unpublished notes about the new theory of parallels might be part of this exposition (*Werke*, VIII, pp. 202–209). A letter addressed to Schumacher on May 17, 1831 says, "In the last few weeks, I have begun to write up a few of my own meditations, which in part are already about 40 years old."[95] His project was, however, suspended upon the receipt of a copy of J. Bolyai's work. Accordingly, the detailed study of the new geometry was put in the hands of Bolyai and Lobachevsky, who established, under the negation of the FP, an analogue of trigonometry, with which they could be confident of the consistency of the new geometry [17, 24].

The fundamental formulas in non-Euclidean trigonometry are given by

$$\cos\angle A = -\cos\angle B\cos\angle C + \sin\angle B\sin\angle C\cosh\frac{a}{R}, \tag{6.11.4}$$

$$\cosh\frac{a}{R} = \cosh\frac{b}{R}\cosh\frac{c}{R} - \sinh\frac{b}{R}\sinh\frac{c}{R}\cos\angle A. \tag{6.11.5}$$

These are obtained from the trigonometric formulas established by Bolyai and Lobachevsky.[96] Notice here that, thanks to Euler's formula, we get (6.11.4) and (6.11.5) by transforming R into $\sqrt{-1}R$ in (6.10.2) and (6.10.3), thereby justifying Lambert's anticipation of a new geometry on the imaginary sphere.

We should underline here that non-Euclidean geometry involves a positive parameter R just like spherical geometry involving the radius of a sphere. As Lambert took notice of it, this implies that, once R is selected, the *absolute unit* of length is determined.[97] Actually, the non-existence of an absolute unit of length is

[94] In the letter, he also said, "If one could prove the existence of a triangle of arbitrarily great area, all Euclidean geometry would be validated" (see [4]).

[95] In a letter to Schumacher of July 12, 1831, Gauss said that the circumference of a non-Euclidean circle of radius r is $\pi k(e^{r/k} - e^{-r/k})$ where k is a positive constant (*Werke*, VIII, p. 215), which turns out to coincide with the radius R of the imaginary sphere.

[96] The non-Euclidean formulas were found by Taurinus (*Geometriae Prima Elementa*, 1826), but the outcome did not wholly convince him of the possibility of a new geometry.

[97] In the ordinary geometry, only a relative meaning is attached to the choice of a particular unit in the measurement. The concept of "similarity" arises exactly from this fact.

equivalent to the FP. Failing to notice this fact, Legendre believed that he succeeded in proving the FP.[98]

Thus the issue of the FP, which had perplexed many able scholars for more than 2000 years, was almost settled (the reason why we say "almost" will become clear later). More than that, mathematicians grew up to realize that postulates in general are *not* statements which are regarded as being established, accepted, or *self-evidently true*, but a sort of predefined rules.

On reflection, Euclid's choice of the FP shows his keen insight and deserves great veneration. Thomas Little Heath (1861–1940), a leading authority of Greek mathematics, remarked that "when we consider the countless successive attempts made through more than twenty centuries to prove the Postulate, many of them by geometers of ability, we cannot but admire the genius of the man who concluded that such a hypothesis, which he found necessary to the validity of his whole system of geometry, was really indemonstrable" [7, p. 202].

6.12 Geodesics

In Sect. 6.7, we stressed the role of shortest curves in the intrinsic measurement in a surface. In order to outline another work of Gauss related to non-Euclidean geometry, we shall take a look at such curves from a general point of view.

The length of a curve $c(t) = (x(t), y(t), z(t))$, $a \leq t \leq b$, on a surface S is given by $\ell(c) = \int_a^b \sqrt{\dot{x}(t)^2 + \dot{y}(t)^2 + \dot{z}(t)^2}\, dt$. Thus our issue boils down to finding a curve c that minimizes $\ell(c)$ among all curves on S satisfying the boundary condition $c(a) = p$, $c(b) = q$. Johann Bernoulli raised the problem of the shortest distance between two points on a convex surface (1697). Following his senescent teacher Johann, Euler obtained a differential equation for shortest curves on a surface given by an equation of the form $F(x, y, z) = 0$ (1732).

Finding shortest curves is a quintessential problem in the *calculus of variations*, a central plank of mathematical analysis originated by Euler, which deals with maximizing or minimizing a general *functional* (i.e., a function whose "variable" is functions or maps). A bud of this newly-born field is glimpsed in Fermat's principle of least time, but the derivation of Snell's law is a straightforward application of the usual extreme value problem. The starting point in effect was the *brachistochrone curve problem* raised by Johann Bernoulli (1696);[99] the problem of finding the shape of the curve down which a sliding particle, starting at rest from $P_1 = (x_1, y_1)$ and

[98] *Léflexions sur différentes manières de démontrer la théorie des paralléles ou le théorème sur la somme des trois angles du triangle*, 1833. In his futile attempt to prove the FP, Legendre rediscovered Saccheri's result. The error in his "proof" was pointed out by Gauss in a letter to Gerling (1816; *Werke*, VIII, pp. 168, 175). (By 1808, Gauss realized that the new geometry must have an absolute unit of length; *Werke* VIII, p. 165.)

[99] "Brachistochrone" is from the Greek words βράχιστος (shortest) and χρόνος (time). The first to consider the brachistochrone problem is Galileo (*Two New Sciences*).

accelerated by gravity, will slip (without friction) to a given point $P_2 = (x_2, y_2)$ in the least time. To be exact, the problem is to find a function $y = y(x)$ which minimizes the integral $\int_{x_1}^{x_2} \sqrt{\frac{1+y'^2}{2gy}}\,dx$.

In the paper *Elementa Calculi Variationum* read to the Berlin Academy on September 16, 1756, Euler unified various variational problems. At about the same time, Joseph-Louis Lagrange (1736–1813) simplified Euler's earlier analysis and derived what we now call the *Euler-Lagrange equation* (E-L equation). Especially, his letter to Euler dated August 12, 1755 (when he was only 19 years old!) contains the technique used today. Here in general (however restricted to the case of one-variable), the E-L equation associated with the functional of the form

$$\mathcal{F}[\boldsymbol{q}] = \int_a^b L\big(t, \boldsymbol{q}(t), \dot{\boldsymbol{q}}(t)\big)\,dt, \quad \boldsymbol{q}(t) = (q_1(t), \ldots, q_n(t)) \tag{6.12.6}$$

is given by

$$\frac{\partial L}{\partial q_i} - \frac{d}{dt}\frac{\partial L}{\partial \dot{q}_i} = 0 \qquad (i = 1, \ldots, n). \tag{6.12.7}$$

What Lagrange proved is that $\boldsymbol{q}(t)$ minimizing $\mathcal{F}$ is a solution of (6.12.7).

Applying the E-L equation to length-minimizing curves on a surface, we obtain an ordinary differential equation for $x(t), y(t), z(t)$. What should be aware of is that a solution of this equation is only a *locally* length-minimizing curve; not necessarily shortest. Such a curve is called a *geodesic*, the term (coming from geodesy) coined by Joseph Liouville (1809–1882).[100]

A geodesic may be defined as a trajectory of *force-free motion*. To explain this, we regard a curve $c(t)$ in a surface S as the trajectory of a particle in motion. The velocity vector $\dot{c}(t)$ is a vector tangent to S at $c(t)$, but the acceleration vector $\ddot{c}(t)$ is not necessarily tangent to S. If the *constraint force* confining the particle to S is the only force, then $\ddot{c}(t)$ must be perpendicular to the tangent plane of S at $c(t)$ because the constraint force always acts vertically on S. Such a curve c turns out to be a geodesic. Hence, if we denote by $\frac{D}{dt}\frac{dc}{dt}$ the tangential part of $\ddot{c}$, then c is a geodesic if and only if $\frac{D}{dt}\frac{dc}{dt} = 0$. Since this is a second-order ordinary differential equation (see (6.17.18) and Sect. 6.20), we have a unique geodesic c defined on an interval containing 0 such that $c(0) = p$, $\dot{c}(0) = \boldsymbol{\xi}$ for given a point p and a vector $\boldsymbol{\xi}$ tangent to S at p.

There is an alternative derivation of the equation $\frac{D}{dt}\frac{dc}{dt} = 0$, based on the *principle of least action* that was enunciated by Pierre Louis Moreau de Maupertuis

[100] *De la ligne géodésique sur un ellipsoide quelconque*, J. Math. Pures Appl., **9** (1844), 401–408. Strictly speaking, as a parameter of a geodesic, we adopt the one proportional to arc-length. We should note that $\ell(c)$ is left invariant under a parameter change.

(1698–1759) and occupies its position in the physicist's pantheon,[101] according to which a particle travels by the *action*-minimizing path between two points. The action he defined is the "sum of mvs" along the entire trajectory of a particle with mass m, where s is the length of a tiny segment on which the velocity of the particle is v.[102]

The obscurity of Maupertuis' formulation was removed by Euler (1744) and Lagrange (1760). Specifically the *action integral* $\mathcal{S}$ applied to a particle in motion under a given *potential energy* V (the concept conceived by Lagrange in 1773) is: $\mathcal{S}=\int_a^b \left(\frac{1}{2}m\|\dot{c}(t)\|^2 - u(c(t))\right)dt$. The E-L equation for $\mathcal{S}$ reduces to the Newtonian equation: $\ddot{c} = -\mathrm{grad}\, u$. (For example, the gravitational potential energy generated by the mass distribution μ is $u(\boldsymbol{x})=-G\int \|\boldsymbol{x}-\boldsymbol{y}\|^{-1}\, d\mu(\boldsymbol{y})$.) This suggests us to define the action integral for a curve c in S by $E(c)=\int_a^b \|\dot{c}(t)\|^2 dt$. Indeed, the E-L equation for this functional coincide with $\frac{D}{dt}\frac{dc}{dt} = 0$.

The minimum (maximum) problems come up in various scenarios in geometry. For instance, the (classical) *isoperimetric theorem* whose origin goes back to ancient Greece states, "Among all planar regions with a given perimeter, the circle encloses the greatest area." Needless to say, minimal surfaces are solutions of a minimum problem. In particular, finding a surface of the smallest area with a fixed boundary curve is known as *Plateau's problem*, the problem raised by Lagrange in 1760 and investigated in depth in the twentieth century onward.[103] Minimum/maximum principles hold emphatic importance even in different fields; say, in *potential theory* (Remark 15.1 (1)), in *statistical mechanics*, in *information science*, in the design of crystal structures [38], and much else besides. Even more surprisingly, so too does in the characterization of the "shape" of the universe as will be explained later.

Remark 12.1

(1) Hamilton reformulated the E-L equation in the context of mechanics as follows (1835). For simplicity, suppose that the integrand L in (6.12.6) (called a *Lagrangian*) does not involve the time variable t. Making the change of variables $(q_1,\ldots,q_n,\, \dot{q}_1,\,\ldots,\dot{q}_n) \mapsto (q_1,\ldots,q_n,\, p_1,\ldots,\, p_n)$ where $p_i = \partial L/\partial \dot{q}_i$, and defining the function H (the *Hamiltonian*) with the new variables

[101] *Accord de différentes loix de la nature qui avoient jusqu'ici paru incompatibles*, read to the French Academy on April 15, 1744. His paper *Loi du repos des corps*, Histoire Acad. R. Sci. Paris (1740), 170–176, is its forerunner.

[102] On reflection, the principle of least action seems to tell us that present events are dependent on later events in a certain manner; so, though a bit exaggerated, a natural occurrence seems to be founded on an *intention* directed to a certain end. For such a teleological and purposive flavor (dating back to Aristotle), Maupertuis (and Euler alike) thought that his cardinal principle proves a rational order in nature, and buttresses the existence of a rational God.

[103] Plateau conducted experiments with soap films to study their configurations (1849).

$(\boldsymbol{p}, \boldsymbol{q})$ by $H(\boldsymbol{p}, \boldsymbol{q}) = \sum_i p_i \dot{q}_i - L(\boldsymbol{q}, \dot{\boldsymbol{q}})$, we may transform (6.12.7) into what we call the *Hamiltonian equation*:

$$\frac{dp_i}{dt} = -\frac{\partial H}{\partial q_i}, \quad \frac{dq_i}{dt} = \frac{\partial H}{\partial p_i}, \quad (i = 1, \dots, n).$$

A general form of the *law of the conservation of energy* is then embodied by the fact that $H(\boldsymbol{p}(t), \boldsymbol{q}(t))$ is constant for any solution $(\boldsymbol{q}(t), \boldsymbol{p}(t))$. In the case where $L = \frac{m}{2}\sum_{i=1}^{n} \dot{q}_i{}^2 - V(\boldsymbol{q})$, we have $\boldsymbol{p} = m\dot{\boldsymbol{q}}$ and $H = \frac{1}{2m}\sum_{i=1}^{n} p_i{}^2 + V(\boldsymbol{q})$ $(= \frac{m}{2}\sum_{i=1}^{n} \dot{q}_i{}^2 + V(\boldsymbol{q}))$. Hence $\boldsymbol{p}$ in this case is the *momentum*, and H is the *total mechanical energy*; that is, the sum of the *kinetic energy* and the potential energy. Hamilton's formulation gained further prominence with the advent of statistical mechanics and quantum mechanics.

(2) To outline a maximum principle in statistical mechanics, consider a simple quantum mechanical system with possible microscopic states $1, \dots, n$ whose energies are $E_1, \dots, E_n$.[104] A (macroscopic) *state* of the system is described by a probability distribution $P = \{p_i\}_{i=1}^{n}$, where p_i is the probability that the microscopic state i occurs during the system's fluctuations. In this setting, the *internal energy* $U(P)$ and the *entropy* $S(P)$ are defined by $U(P) = \sum_{i=1}^{n} E_i p_i$, $S(P) = -k_B \sum_{i=1} p_i \log p_i$, respectively, where k_B is *Boltzmann's constant*. Among all states P with $U = U(P)$, there is a unique $P^0 = \{p_i^0\}_{i=1}^{n}$ with the maximum entropy; i.e. $S(P) \leq S(P^0)$. It is given by the *Gibbs distribution*[105] representing the equilibrium state $p_i^0 = \exp(-E_i/k_B T)/Z(1/k_B T)$; $Z(\beta) := \sum_{i=1}^{n} \exp(-\beta E_i)$, where T is the *temperature*, determined by the equation $U = -d/d\beta|_{\beta=1/k_B T} \log Z(\beta)$.

The *second law of thermodynamics* (formulated by Clausius) says that entropy of an isolated system invariably increases because the system evolves towards equilibrium. Therefore the total entropy of the universe is continually increasing. □

6.13 Curvature and Non-Euclidean Geometry

In a nutshell, what Gauss discovered during his study of the new geometry is that homogeneity and isotropy are not enough to characterize Euclidean space; namely, his discovery suggests that there is another "space" with these properties, which is to be called "non-Euclidean space". However, we have to stress that the way Gauss

---

[104] A quantum mechanical system is described by a self-adjoint operator $\widehat{H}$ acting on a Hilbert space. A unit eigenvector of $\widehat{H}$ with eigenvalue E represents a state with energy E.

[105] The term "entropy," from Greek εν ("in") + τροπή ("a turning"), was coined in the early 1850s by R. J. E. Clausius. In the 1870s, L. Boltzmann gave its statistical definition.

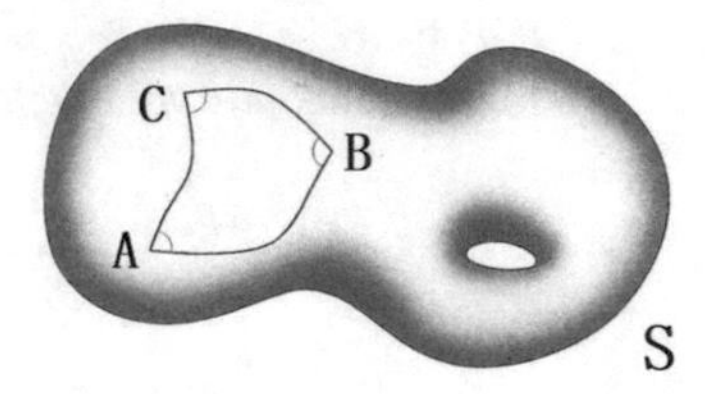

Fig. 6.10 A triangle on a surface

and others searched for their new geometry was synthetic. At this stage, therefore, the "new space" has not yet been explicitly described.

Gauss did not refer to the non-Euclidean space, but perhaps he was aware that the "new space" may yield an alternative model of the universe. In truth, the moment when he perceived this is by far the most significant turning point in cosmology.[106] What is more, Gauss seemed to know an intimate relation between his intrinsic theory of surfaces and non-Euclidean (plane) geometry because he obtained the arresting formula (*Disquisitiones generales*, Art. 20):

$$\iint_{\triangle ABC} K_S d\sigma = (\angle A + \angle B + \angle C) - \pi, \tag{6.13.8}$$

where $\triangle ABC$ is a *geodesic triangle* (a triangle on S formed by the arcs of three geodesics segments; Fig. 6.10), and $d\sigma$ is the *surface element* of S (Sect. 6.15). Note that this is a generalization of Harriot-Girard's formula.

From (6.13.8), a "plane-like surface" of constant negative curvature (if exists) is expected to be a non-Euclidean plane-model, and hence finding such a surface became an urgent issue. In a note dated about 1827, Gauss jotted down his study about the surface of revolution generated by the *tractrix* (Fig. 6.11),[107] which he called "das Gegenstück der Kugel" (the opposite of the sphere).[108] This is an example of a surface with constant negative curvature, but he did not mention clearly that it has to do with the new geometry (*Werke*, VIII, p. 265).[109] The reason might

[106]In the *Disquisitiones generales* (Art. 28), Gauss refers to the numerical data for the triangle formed by the three mountain peaks Brocken, Hohenhagen, and Inselsberg. Some say that with these data Gauss wished to determine whether the universe is Euclidean. What we can say at the very least is that he compared a geodesic triangle on the earth surface with a plane rectilinear triangle. Meanwhile, Lobachevsky attempted to measure the angle sum of a triangle by analyzing data on the parallax of stars (*Exposition succincte*; see [19]).

[107]The tractrix is the graph of the function $y = R\log((R + \sqrt{R^2 - x^2})/x) + \sqrt{R^2 - x^2}$, which shows up as a locus of an object obtained from dragging it along a line with an inextensible string. The architect C. Perrault proposed a problem related to the tractrix (1670) for the first time. The appellation "tractrix" was coined by Huygens in 1692.

[108]This was to be called the *pseudosphere* by Eugenio Beltrami (1835–1900); see Footnote 126.

[109]F. Minding found that the trigonometry for geodesic triangles on this surface enjoys analogy to spherical trigonometry (1840). However he did not notice that his finding is equivalent to the non-Euclidean trigonometric relations.

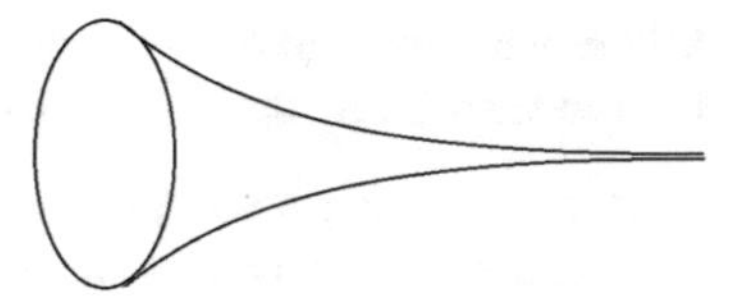

Fig. 6.11 Das Gegenstück der Kugel

be because he could not construct such a surface with a planar shape. The truth is that the non-Euclidean plane cannot be realized as an ordinary surface in space, as Hilbert showed in 1901. In order to apprehend its true identity, we need the notion of manifold (Sect. 6.16).

Gauss may have known (6.13.8) as early as 1816, so it may have served as a forerunner to the Theorema Egregium. In his *Disquisitiones generales*, Art. 20, he says, "This formula, if I am not wrong, should be counted among the most elegant ones of the theory of surfaces."

Behind (6.13.8), there seems to be his commitment to electromagnetism[110]; in 1813, he proved a special case of what we now call *Gauss's flux law* that correlates the distribution of electric charge with the resulting electric field. The law is expressed by a formula connecting an integral inside a closed surface with a surface integral over the boundary (*divergence theorem*; *Werke*, V, pp. 5–7).

Remark 13.1 The divergence theorem asserts that

$$\iiint_D \operatorname{div} X \, d\boldsymbol{x} = \iint_S \langle \boldsymbol{n}, X \rangle \, d\sigma$$

for a vector field X defined on a domain $D \subset \mathbb{R}^3$ with boundary S, where $\boldsymbol{n}$ is the outer unit normal of the surface S. A special case was treated by Lagrange (1762), and by Gauss (1813). It was M. V. Ostrogradsky who gave the first proof for the general case (1826). The 2D version is known as Green's theorem (1828):

$$\iint_D \left(\frac{\partial g}{\partial x} - \frac{\partial f}{\partial y} \right) dxdy = \int_C f dx + g dy, \qquad (6.13.9)$$

where D is the domain surrounded by a piecewise smooth, simple closed counter-clockwise oriented curve C. Another important formula is *Stokes' formula* for a bordered surface S with a normal unit vector field $\boldsymbol{n}$:

$$\iint_S \langle \boldsymbol{n}, \operatorname{rot} X \rangle \, d\sigma = \int_C \langle \dot{c}(s), X(c(s)) \rangle \, ds,$$

[110]The rapid progress of electromagnetism originated with H. C. Oersted who verified by experiment that electricity and magnetism are not independent (1820).

where we represent the boundary curve C of S by an arc-length parameterization $c(s)$ in such a way that $\dot{c}(s) \times \boldsymbol{n}(c(s))$ points towards the interior of S.[111] □

A formula that deserves to be mentioned is derived from (6.13.8). For a closed surface S with a division by a collection of geodesic triangles such that each triangle side is entirely shared by two adjacent triangles, we have

$$\int_S K_S d\sigma = 2\pi(v - e + f), \tag{6.13.10}$$

where v, e, f are the number of vertices, edges, and faces, respectively. This formula, due to Walther Franz Anton von Dyck (1856–1934), is called the *Gauss-Bonnet formula*, though they never referred to (6.13.10) in their work.[112] It tells us that $\chi(S) := v-e+f$ (the *Euler characteristic* of S) is independent of the choice of a triangulation. In particular, applying to the unit sphere S^2, we get the *polyhedron formula* $\chi(S^2) = 2$ discovered by Euler in 1750 (Sect. 6.21).

The proof of (6.13.10) goes as follows. The sum of the interior angles gathering at each vertex is 2π, so that

$$\int_S K_S d\sigma = \sum_{\triangle ABC} \iint_{\triangle ABC} K_S d\sigma = \sum_{\triangle ABC} \{(\angle A + \angle B + \angle C) - \pi\} = 2\pi v - f\pi,$$

while $3f = 2e$ because each edge is counted twice when we count edges of all triangles. Consequently, the left-hand side of (6.13.10) is equal to $\pi(2v - f) = \pi(2v - 2e + 2f) = 2\pi(v - e + f)$, as desired.

The Gauss-Bonnet formula opened the doorway to *global analysis* that knits together three fields; differential geometry, topology, and analysis (Sect. 6.21).

After Gauss's monumental feat, cosmology began to depart from religion and even philosophy, and to snuggle up to higher mathematics which is unconstrained by perceptual experience; thus God walked away from the center stage. However the outside of space still remains as a backdrop. It was Riemann, the next protagonist in our story, who at last erased the outside.

6.14 The Dimension of Space

Deferring the account of Riemann's achievement to the next section, we shall pause for a moment to discuss *higher-dimensional spaces*.

[111]The statement of Stokes' formula appeared as a postscript to a letter (July 2, 1850) from William Thomson (Lord Kelvin, 1824–1907) to G. Stokes. Stokes set the theorem as a question on the 1854 Smith's Prize Examination at Cambridge.

[112]von Dyck, *Beiträge zur Analysis situs I. Aufsatz. Ein- und zweidimensionale Mannigfaltigkeiten*, Math. Ann., **32** (1888), 457–512. Pierre Ossian Bonnet (1819–1892) generalized (6.13.8) to triangles with non-geodesic sides (1848).

Dimension is, in a naive sense, a measurable extent of a particular kind, such as length, breadth and depth, or the maximum number of independent directions in a refined sense. Physically it is the number of degrees of freedom available for movement in a space. Aristotle was among the earliest who contemplated dimension; he says, "magnitude which is continuous in one dimension is length; in two breadth, in three depth, and there is no magnitude besides these" (*Metaphysics*, V, 11–14). Ptolemy likewise asserted in his book *On Distance* that one cannot consider "space of dimension more than three," and even gave a proof as testified by Simplicius of Cilicia (ca. 490 CE–ca. 560 CE).

In Sect. 6.7, we speculated what the inhabitants in a surface would do when they want to know the shape of their universe. Surfaces are 2D, whilst our universe is 3D. Why are we in the 3D space, not in another?[113] However difficult it may have been to furnish a definite answer to this question, there is no barrier to thinking of higher-dimensional models of space. Actually we have a clue in analytic geometry to formulate coordinate spaces of general dimension.[114]

It was in the eighteenth century that mathematicians began to contemplate the possibility of the use of $\mathbb{R}^d$ as a gadget to develop mechanics. As Lagrange and D'Alembert suggested, it is natural to add the time parameter as the fourth dimension. Moreover, it is technically convenient to describe a mass system of N particles located at $(x_1, y_1, z_1), (x_2, y_2, z_2), \ldots, (x_N, y_N, z_N)$ as the single point $(x_1, y_1, z_1, x_2, y_2, z_2, \ldots, x_N, y_N, z_N)$ in $\mathbb{R}^{3N}$. Then, around the 1830s, the theory of n-ple integrals was evolved. For instance, Carl Gustav Jacob Jacobi (1804–1851) computed the volume of a higher-dimensional sphere in connection with quadratic forms though he deliberately avoided geometrical language (1834).

Certain it is that, by the early nineteenth century, the time was ripe for developing higher-dimensional geometry. In 1843, Arthur Cayley published a small memoir on $\mathbb{R}^n$, however the paper is characteristically algebraic. In the same year, Hamilton conceived the 4D number system (Sect. 6.7). A year later, *Die Lineale Ausdehnungslehre, ein neuer Zweig der Mathematik*, a pioneer work on vector spaces of general dimension, was published by Hermann Grassmann (1809–1877).[115]

A point at issue is how to imagine $\mathbb{R}^d$ as a higher-dimensional analogue of plane and space. A lighthearted way is to execute a geometry in $\mathbb{R}^d$ by mimicking geometry in $\mathbb{R}^2$ and $\mathbb{R}^3$ (thus geometric terminology may be of psychological

---

[113] Closely related to this question is the *anthropic principle* (Brandon Carter, 1973), the philosophical deliberation that the universe is inseparable from mankind's very existence.

[114] Analytic geometry acquired its whole range in the eighteenth century with A. Parent, Clairaut, Euler, G. Cramer, Lagrange and many others. La Hire'swork *Nouveaux élémens des sections coniques, les lieu géométriques* (1679) summarized the progress of analytic geometry during a few decades since Descartes. His work contains some ideas leading to the extension of space to more than three dimensions.

[115] In this book, he created what we now call the *exterior algebra* (or *Grassmann algebra*), which, denigrated at the time, was to exert a profound influence on the formation of algebra and geometry; see Sect. 6.20. In this sense, he was quite ahead of his time.

assistance when we imagine $\mathbb{R}^d$). It is to the credit of Ludwig Schläfli (1814–1895) to have inaugurated the study of the geometry of $\mathbb{R}^d$ along this line. In his magnum opus, *Theorie der vielfachen Kontinuität*, worked out from 1850 to 1852,[116] Schläfli defined the distance between $A = (x_1, \ldots, x_d)$ and $B = (y_1, \ldots, y_d)$ in $\mathbb{R}^d$ by $d(A, B) := \sqrt{(x_1 - y_1)^2 + \cdots + (x_d - y_d)^2}$.

As expected, many results in $\mathbb{R}^d$ are just direct generalizations of the facts that hold in plane and space, but there are several results depending heavily on dimension. Schläfli's classification of higher-dimensional convex regular polytopes is such an example. Specifically, there are six regular polytopes in $\mathbb{R}^4$, while there are only three regular polytopes in $\mathbb{R}^d$, $d > 4$ (the analogues of the tetrahedron, the cube and the octahedron). Another dimension-dependent example—rather recent one originating with Newton's curiosity regarding celestial bodies—is the maximum possible kissing number $k(d)$ for $\mathbb{R}^d$, where the *kissing number* is the number of non-overlapping unit spheres $S^d = \{(x_1, x_2, \ldots, x_{d+1}) | (x_1 - a_1)^2 + \cdots + (x_{d+1} - a_{d+1})^2 = 1\}$ that can be arranged in such a way that each of them touches another given unit sphere. For example, $k(2) = 6$, $k(3) = 12$ and $k(4) = 24$, but curiously $k(d)$ is unknown for $d > 4$ except for $k(8) = 240$ and $k(24) = 196,560$.[117] Such phenomena enrich geometry; thereby, as opposed to Kant's dictum, higher-dimensional geometry turns out to be non-empty and fruitful.

Remark 14.1 As mentioned in Sect. 6.7, there exists no 3D number system that inherits, in part, the arithmetic properties of real numbers, while a 4D system exists. Hence a dimension-dependence phenomenon occurs in number systems. In 1843, J. T. Graves discovered the *octonions*, a non-associative 8D number system, inspired by Hamilton's discovery of quaternions.[118] We thus have four number systems. These systems share the following properties; writing $\boldsymbol{a} \cdot \boldsymbol{b}$ for multiplication of $\boldsymbol{a} = (a_1, \ldots, a_d), \boldsymbol{b} = (b_1, \ldots, b_d)$ $(d = 1, 2, 4, 8)$, we have $(\boldsymbol{a} + \boldsymbol{b}) \cdot \boldsymbol{c} = \boldsymbol{a} \cdot \boldsymbol{c} + \boldsymbol{b} \cdot \boldsymbol{c}$, $\boldsymbol{a} \cdot (\boldsymbol{b} + \boldsymbol{c}) = \boldsymbol{a} \cdot \boldsymbol{b} + \boldsymbol{a} \cdot \boldsymbol{c}$, $k(\boldsymbol{a} \cdot \boldsymbol{b}) = (k\boldsymbol{a}) \cdot \boldsymbol{b} = \boldsymbol{a} \cdot (k\boldsymbol{b})$ $(k \in \mathbb{R})$, and $\|\boldsymbol{a} \cdot \boldsymbol{b}\| = \|\boldsymbol{a}\| \|\boldsymbol{b}\|$. Here $\|\boldsymbol{a}\|^2 = {a_1}^2 + \cdots + {a_d}^2$. Interestingly, if $\mathbb{R}^d$ has multiplication having these properties, then d must be 1,2,4, or 8 as A. Hurwitz showed in 1898.

Quaternions and octonions—the products of pure thought— have applications in modern physics; thus being more than mathematical fabrications. □

[116] It was in 1901, after his death, that the entire manuscript was published in Denkschriften Der Schweizerischen Naturforschenden Gesellschaft, **38** (1901), 1–237.

[117] The fact $k(3) = 12$ was conjectured by Newton, while D. Gregory, a nephew of James Gregory, insisted that $k(3) = 13$ in a discussion taking place at Cambridge in 1694. A proof in favor of Newton was given by K. Schütte and van der Waerden (1953).

[118] Slightly before the publication of his paper, Cayley discovered the octonions. Octonions are often referred to as *Cayley numbers*.

6.15 Riemann: The Universe as a Manifold

Gauss brought about a revolution in geometry as he envisaged. Indeed, his study of surfaces turned out to be the germ of a "science of generalized spaces" that was essentially embodied by Georg Friedrich Bernhard Riemann (1826–1866) in his Habilitationsschrift *Über die Hypothesen welche der Geometrie zu Grunde liegen* presented to the Council of Göttingen University.[119]

It was Gauss—only 1 year before the death at the age of 78—who asked his brilliant disciple to prepare a thesis on the foundations of geometry, out of the three possible topics that Riemann proposed. The two other topics were on electrical theory, and unlike the one chosen by his mentor, he was well-prepared for them. So Riemann had to work out the topic Gauss recommended, and in a matter of months he could bring it to completion without qualm.

The aged Gauss, albeit being sick at that time, attended Riemann's probationary lecture delivered on 10 June 1854 in anticipation of hearing something new related to his old work. "Among Riemann's audience, only Gauss was able to appreciate the depth of Riemann's thoughts. ... The lecture exceeded all his expectations and greatly surprised him. Returning to the faculty meeting, he spoke with the greatest praise and rare enthusiasm to W. Weber about the depth of the thoughts that Riemann had presented" [19].

In his lecture aimed to a largely non-mathematical audience, Riemann posed profound questions about how geometry is connected to the world we live in. The thrust being as such, the content was much more philosophical than mathematical though he humbly denied this (Part I).

At the outset, he says that the traditional geometry assumes the notion of space and some premises which are merely nominal, while the relation of these assumptions remains in darkness because the general notion of *multiply extended magnitude* remained entirely unworked. He thus sets himself the task of constructing the notion of *manifold* of general dimension.

There is something of note here. In his setup, Riemann does not set a limit to the dimension of space. Remember that, when he inaugurated his study, higher-dimensional geometry as an abstract theory was just formulated. This might be a precipitating factor for his commitment to higher-dimensional curved spaces. His general treatment of space was, however, not a mere idea or pipe dream, and had truly a decisive influence on modern cosmology that employs manifolds of dimension more than four (in superstring theory, space-time is 10D).

Riemann continues his deliberation on manifolds, which is bolstered by his immense insight into the principles behind the proper understanding of our space; but to go straight to the heart of his scheme, let us take a shortcut at the expense of his true motive.

[119]*Habilitationsschrift* is a postdoctoral thesis required for qualification as a lecturer. This is his second stage of the procedure (see Footnote 158 for his thesis at the first stage). The title of Riemann's thesis is an implicit criticism of Kant.

To single out a reasonable model of the universe, we must take into account the fact that our universe looks Euclidean in so far as we examine our vicinity. In the light of this observation, we define a d-dimensional manifold to be a "generalized space" on which one can set up a curvilinear coordinate system $(x_1, \ldots, x_d)$ around each point p_0, so that we may indicate each point p near p_0 by its coordinates $(x_1, \ldots, x_d)$. In addition, the higher-dimensional Pythagorean Theorem is supposed to hold in the infinitesimal sense. To explore what this means, we recall that the distance Δs between $(y_1, \ldots, y_d)$ and its displacement $(y_1 + \Delta y_1, \ldots, y_d + \Delta y_d)$ in $\mathbb{R}^d$ is given by $(\Delta s)^2 = (\Delta y_1)^2 + \cdots + (\Delta y_d)^2$. When a skew coordinate system $(x_1, \ldots, x_d)$ is taken instead of the Cartesian one, this expression is modified as $(\Delta s)^2 = \sum_{i,j=1}^{d} g_{ij} \Delta x_i \Delta x_j$. Here if the coordinate transformation between $(y_1, \ldots, y_d)$ and $(x_1, \ldots, x_d)$ is given by $y_i = \sum_{j=1}^{d} r_{ij} x_j$, then $\Delta y_i = \sum_{j=1}^{d} r_{ij} \Delta x_j$, and hence $g_{ij} = \sum_{h=1}^{d} r_{hi} r_{hj}$. The square matrix (g_{ij}) is a positive symmetric matrix. Since in the infinitesimally small scale, the curvilinear coordinate system is regarded as a skew coordinate system, it is natural to express the "infinitesimal" Pythagorean Theorem as

$$ds^2 = \sum_{i,j=1}^{d} g_{ij} dx_i dx_j, \tag{6.15.11}$$

where the "line element" ds stands for the "distance between infinitesimally nearby points" $(x_1, \ldots, x_d)$ and $(x_1 + dx_1, \ldots, x_d + dx_d)$, and (g_{ij}) is, in turn, a function of the variables $x_1, \ldots, x_d$ with values in positive symmetric matrices. The right-hand side is called *the first fundamental form* (or *Riemannian metric* in today's term; Sect. 6.20), which is literally fundamental in Riemann's theory of manifolds. With it, one can calculate the length of a curve $c(t) = (u_1(t), \ldots, u_d(t))$ $(a \leq t \leq b)$ by the integral

$$\int_a^b \left\{ \sum_{i,j=1}^{d} g_{ij} \frac{du_i}{dt} \frac{du_j}{dt} \right\}^{1/2} dt. \tag{6.15.12}$$

In imitation of the case of surfaces, we say that c is a geodesic if it is locally length-minimizing. The distance of two points p, q is defined to be the infimum of the length of curves joining p, q.

The next task—a highlight of his thesis—is to define "curvature" in his setting. To this end, Riemann takes a union of geodesics passing through a given point p to form a surface S, and then defines what we now call the "sectional curvature" to be $K_S(p)$ (Part II, §3; see Sect. 6.20 for a modern definition).

The manifold is said to have constant curvature if $K_S(p)$ does not depend on the choice of p and S. He observes that a manifold with constant curvature α has a coordinate system $(x_1, \ldots, x_d)$ around each point such that

$$ds^2 = \sum_{i=1}^{d} dx_i{}^2 \Big/ \Big(1 + \frac{\alpha}{4} \sum_{i=1}^{d} x_i{}^2\Big)^2. \tag{6.15.13}$$

In particular, the curvature of a manifold covered by a single coordinate system $(x_1, \ldots, x_d)$ with the fundamental form $\sum_{i=1}^{d} dx_i{}^2$ vanishes everywhere. Obviously this manifold, for which he used the term "flat," is the d-dimensional coordinate space $\mathbb{R}^d$ with the distance introduced by Schläfli.

Related to the curvature is a question about *unboundedness* (Unbegrenzheit) and *infinite extent* (Unendlichkeit) of the universe (Part III, §2). He said that this kind of inquiry is possible even when our empirical determination is beyond the limit of observation. For example, *if the independence of bodies from positions could be assumed*, then the universe is of constant curvature, and hence must be *finite* provided that the curvature is positive (see Sect. 6.18 for the exact meaning of finiteness). He justifies his argument by saying that a manifold with positive constant curvature has a sphere-like shape.[120]

Riemann's consideration is not limited to finite dimensions; he suggests the possibility to contrive the theory of infinite-dimensional manifolds which may allow to handle spaces of functions or mappings *en bloc* from a view similar to the finite-dimensional case (Part I, §3). Thus, at this point of time, he might already have had an inspiration that variational problems are regarded as extreme value problems on infinite-dimensional manifolds, which a little later led to his use of what he called *Dirichlet's principle* in complex analysis (1857).[121]

At the end of his lecture, Riemann ponders on the question of the validity of the hypotheses of geometry in the infinitely small realm. He says that the question of the validity of the hypotheses of geometry in the infinitely small is bound up with the question of the ground of the metric relations of space, and concluded the lecture with the following confident statement.

> Researches starting from general notions, like the investigation we have just made, can only be useful in preventing this work from being hampered by too narrow views, and progress in knowledge of the interdependence of things from being checked by traditional prejudices. This leads us into the domain of another science, of physics, into which the object of this work does not allow us to go to-day (translated by W. K. Clifford, *Nature*, Vol. VIII, 1873; see also [34]).

---

[120] Riemann's claim should be phrased as "a d-dimensional manifold with positive constant curvature is locally a portion of the sphere S^d." A typical example besides S^d is the *projective space* $P^d(\mathbb{R})$ obtained from S^d by identifying $(x_1, \ldots, x_{d+1})$ with $(-x_1, \ldots, -x_{d+1})$, on which projective geometry is developed (see Sect. 6.16). In the 3D case, the classification of such manifolds is related to polyhedral groups. We will come across an example in Sect. 6.21.

[121] *Theorie der Abel'schen Funktionen*, J. für die Reine und Angew. Math., **54**, 115–155.

Remark 15.1

(1) Dirichlet's principle was already conceived by Gauss in 1839 to solve a boundary value problem for harmonic functions by means of a minimum principle.[122] Specifically, it says that a solution u of the *Laplace equation* $\Delta u = 0$ on a domain D of $\mathbb{R}^d$ with boundary condition $u = g$ on ∂D can be obtained as the minimizer of the functional $E(u) = \int_D \frac{1}{2}\|\mathrm{grad}\, u\|^2\, dx$. Yet, as pointed out by Karl Theodor Wilhelm Weierstrass (1815–1897) in 1870, the existence of the minimizer is not obvious. It was in 1899 that Riemann's use of the principle was warranted (Hilbert).

(2) An "infinitesimal" that shows up as the symbol dx_i in (6.15.11) is paradoxically thought of as an object which is smaller than any feasible measurement. The symbol dx (called the *differential* and exploited by Leibniz for the first time) is convenient when expressing tidily formulas that involve differentiation and integration. For example, the expression $\int_c f\, dx + g\, dy$ (see (6.13.9)) for a line integral along a curve $c(t) = (x(t), y(t))$ $(a \leq t \leq b)$ is originally $\int_a^b \left(f(x(t), y(t))\frac{dx}{dt} + g(x(t), y(t))\frac{dy}{dt}\right)dt$. Because of such a background, calculus had been referred to as the *infinitesimal analysis*. Typical are the *Analyse des infiniment petits* (1696) by Guillaume François Antoine Marquis de L'Hôpital (1661–1704), which contributed greatly to the dissemination of calculus in the Continent, and Euler's *Introductio in Analysin Infinitorum*, the first systematic exposition of calculus of several variables.

The paradoxical nature of infinitesimal was criticized as incorrect by Berkeley who described "0/0" in differentiation as the "ghost of departed quantities,"[123] and provoked controversy among the successors of Leibniz. For instance, for Euler, infinitely small quantities are actually equal to zero, and differential calculus is simply a heuristic procedure for finding the value of the expression 0/0" (*Institutiones Calculi Differentialis* [8]. Yet, with the hindsight of modern geometry, infinitesimals are not altogether absurd and could survive today as "duals" of certain "operations" (Sect. 6.20).

It is a significant progress in the history, however inconspicuous, that differentiation was recognized as an operation that can be manipulated independently of functions to which they are applied. L. F. A. Arbogast was one of the first to make such a conceptual leap (*Calcul des dérivations*, 1800). It goes without saying that the Leibnizian notations played a key part in this comprehension. □

To appreciate Riemann's work, we shall step back into Gauss's work, wherein we see the elements of continuity between them as forthrightly expressed in Riemann's

[122]Kelvin used the principle in 1847 in his study of electric fields. The most complete formulation was given by Johann Peter Gustav Lejeune Dirichlet (1805–1859) around 1847 in his lectures on potential theory at Berlin.

[123]*The analyst, or a discourse addressed to an infidel mathematicians* (1734). Berkeley denied the material existence of the world; he says, "Esse est percipi (To be is to be perceived)."

words, "In the comprehension of the geometry of surfaces, the data on how a surface sits in space is incorporated with the intrinsic measure-relation in which only the length of curves on the surface is considered" (Part II, §3).

We let $(u, v) \mapsto \boldsymbol{S}(u, v)$ be a *local parametric representation* of a surface S in space; that is, $\boldsymbol{S}$ is a smooth one-to-one map from a domain U of $\mathbb{R}^2$ into S such that $\boldsymbol{S}_u(u, v)$ and $\boldsymbol{S}_v(u, v)$ are linearly independent for each $(u, v) \in U$ (thus span the tangent plane of S at $\boldsymbol{S}(u, v)$). Note that the inverse map of $\boldsymbol{S}$ yields a curvilinear coordinate system (u, v) of S.

Consider a curve $c(t)$ ($t \in [a, b]$) on S whose coordinates are $\big(u(t), v(t)\big)$, and differentiate both sides of $c(t) = \boldsymbol{S}\big(u(t), v(t)\big)$ to obtain

$$\left\|\frac{dc}{dt}\right\|^2 = \left\|\frac{du}{dt}\boldsymbol{S}_u + \frac{dv}{dt}\boldsymbol{S}_v\right\|^2 = \langle \boldsymbol{S}_u, \boldsymbol{S}_u\rangle\left(\frac{du}{dt}\right)^2 + 2\langle \boldsymbol{S}_u, \boldsymbol{S}_v\rangle\frac{du}{dt}\frac{dv}{dt} + \langle \boldsymbol{S}_v, \boldsymbol{S}_v\rangle\left(\frac{dv}{dt}\right)^2.$$

so the length of c is given by

$$\int_a^b \left\|\frac{dc}{dt}\right\| dt = \int_a^b \left\{\langle \boldsymbol{S}_u, \boldsymbol{S}_u\rangle\left(\frac{du}{dt}\right)^2 + 2\langle \boldsymbol{S}_u, \boldsymbol{S}_v\rangle\frac{du}{dt}\frac{dv}{dt} + \langle \boldsymbol{S}_v, \boldsymbol{S}_v\rangle\left(\frac{dv}{dt}\right)^2\right\}^{1/2} dt.$$

Putting $E = \langle \boldsymbol{S}_u, \boldsymbol{S}_u\rangle$, $F = \langle \boldsymbol{S}_u, \boldsymbol{S}_v\rangle$, $G = \langle \boldsymbol{S}_v, \boldsymbol{S}_v\rangle$ and comparing this formula with (6.15.12), we see that the first fundamental form is $\mathrm{I} := Edu^2 + 2Fdudv + Gdv^2$.

Besides, Gauss introduced the coefficients of what we now call *the second fundamental form* $\mathrm{II} := Ldu^2 + 2Mdudv + Ndv^2$ by setting $L = \langle \boldsymbol{S}_{uu}, \boldsymbol{n}\rangle$, $M = \langle \boldsymbol{S}_{uv}, \boldsymbol{n}\rangle$, $N = \langle \boldsymbol{S}_{vv}, \boldsymbol{n}\rangle$, where $\boldsymbol{n}$ is the unit normal defined by $\boldsymbol{n} = (\boldsymbol{S}_u \times \boldsymbol{S}_v)/\|\boldsymbol{S}_u \times \boldsymbol{S}_v\|$. Then $K_S(p) = (LN - M^2)/(EG - F^2)$ (*Disquisitiones generales*, Art. 10). To show this, we employ the coordinate system (x, y, z) and the function $f(x, y)$ introduced in Sect. 6.7. We may assume that $\boldsymbol{S}(0, 0) = (0, 0, 0)$ and the direction of the z-axis coincides with that of $\boldsymbol{n}$. Let $(x(u, v), y(u, v))$ be the xy-coordinate of the orthogonal projection of $\boldsymbol{S}(u, v)$ onto the tangent plane, so that $\boldsymbol{S}(u, v) = \big(x(u, v), y(u, v), f(x(u, v), y(u, v))\big)$.[124] We then have

$$\boldsymbol{S}_{uu} = \big(x_{uu}, y_{uu}, a{x_u}^2 + 2bx_u y_u + c{y_u}^2\big), \quad \boldsymbol{S}_{vv} = \big(x_{vv}, y_{vv}, a{x_v}^2 + 2bx_v y_v + c{y_v}^2\big),$$
$$\boldsymbol{S}_{uv} = \big(x_{uv}, y_{uv}, ax_u x_v + b(x_u y_v + y_u x_v) + cy_u y_v\big) \quad ((u, v) = (0, 0)).$$

From the assumption on the vector $\boldsymbol{n}$, it follows that the coefficients L, M, N are the z-coordinates of $\boldsymbol{S}_{uu}, \boldsymbol{S}_{uv}, \boldsymbol{S}_{vv}$, respectively, so that $L = a{x_u}^2 + 2bx_u y_u + c{y_u}^2$, $N =$

[124] For the local parametric representation of S given by the map $(x, y) \mapsto (x, y, f(x, y))$, the second fundamental form at $(0, 0)$ is given by $adx^2 + 2bdxdy + cdy^2$.

$ax_v{}^2 + 2bx_v y_v + cy_v{}^2$, $M = ax_u x_v + b(x_u y_v + y_u x_v) + cy_u y_v$. In matrix form, these formulas together may be written as

$$\begin{pmatrix} L & M \\ M & N \end{pmatrix} = \begin{pmatrix} x_u & y_u \\ x_v & y_v \end{pmatrix} \begin{pmatrix} a & b \\ b & c \end{pmatrix} \begin{pmatrix} x_u & x_v \\ y_u & y_v \end{pmatrix}.$$

Therefore, taking the determinant of both sides, we have $LN - M^2 = (x_u y_v - y_u x_v)^2(ac - b^2) = (x_u y_v - y_u x_v)^2 K_S$, while, noting that $\boldsymbol{S}_u = (x_u, y_u, 0)$ and $\boldsymbol{S}_v = (x_v, y_v, 0)$, we find that $EG - F^2 = \langle \boldsymbol{S}_u, \boldsymbol{S}_u\rangle\langle \boldsymbol{S}_v, \boldsymbol{S}_v\rangle - \langle \boldsymbol{S}_u, \boldsymbol{S}_v\rangle^2 = (x_u y_v - y_u x_v)^2$, from which the claim follows.

We add a few words about the surface element $d\sigma$ (Sect. 6.13). For a point $p = \boldsymbol{S}(u, v)$, define $\boldsymbol{S}_0 : \mathbb{R}^2 \to E$ by setting $\boldsymbol{S}_0(x, y) = p + x\boldsymbol{S}_u + y\boldsymbol{S}_v$, which is a parametric representation of the tangent plane at p. Consider the small parallelogram $P = \boldsymbol{S}_0\big(\{(x, y) | 0 \le x \le \Delta u,\ 0 \le y \le \Delta v\}\big)$. This approximates $\boldsymbol{S}\big(\{(x, y) | 0 \le x \le \Delta u,\ 0 \le y \le \Delta v\}\big)$. Hence, the surface area of $\boldsymbol{S}\big(\{(x, y) | 0 \le x \le \Delta u,\ 0 \le y \le \Delta v\}\big)$ is approximated by $\mathrm{Area}(P) = \big\{\langle \boldsymbol{S}_u, \boldsymbol{S}_u\rangle\langle \boldsymbol{S}_v, \boldsymbol{S}_v\rangle - \langle \boldsymbol{S}_u, \boldsymbol{S}_v\rangle^2\big\}^{1/2}\Delta u\Delta v = \sqrt{EG - F^2}\Delta u \Delta v$, so the surface area of $\boldsymbol{S}(U)$ is given by $\iint_U \sqrt{EG - F^2}du dv$, and $d\sigma = \sqrt{EG - F^2}du dv$ (*Disquisitiones generales*, Art. 17).

The reader might suspect that Riemann's approach is more or less a direct generalization of Gauss's one. That is not quite correct since Riemann's formulation is entirely intrinsic; therefore, the universe can exist *without the outside* (in particular, we may consider surfaces not realized in space, thereby providing a possibility of constructing a model of the non-Euclidean plane). This should be the point that Gauss spoke of with the greatest praise and rare enthusiasm.

6.16 Hyperbolic and Projective Spaces

Soon after Riemann passed away at the age of 39 on June 28, 1866, his recondite thesis was published by his friend Dedekind,[125] and was taken up by Beltrami to put an end to the argument on non-Euclidean geometry. In fact, most people at that time were reluctant to accept the unaccustomed geometry because it might very well be possible to fall in with inconsistency after a more penetrating investigation. So as to convince incredulous people, a *model* of non-Euclidean geometry needed to be constructed, resting on well-established geometric ingredients, with which inconsistency can be kept at bay.

[125]Abhandlungen der Königlichen Gesellschaft der Wissenschaften zu Göttingen, **13** (1868), 133–152.

In 1868, the Italian journal *Giornale di Mathematiche* carried Beltrami's article *Teoria fondamentale degli spazii de curvatura constante*,[126] in which it was shown that the non-Euclidean space is realized as the 3D open ball $D = \{(x_1, x_2, x_3) | x_1{}^2 + x_2{}^2 + x_3{}^2 < 4\}$ with $ds^2 = (dx_1{}^2 + dx_2{}^2 + dx_3{}^2)/\bigl(1 - (x_1{}^2 + x_2{}^2 + x_3{}^2)/4\bigr)^2$, where geodesics are circular arcs in space crossing perpendicularly $\partial D = \{(x_1, x_2, x_3) | \ x_1{}^2 + x_2{}^2 + x_3{}^2 = 4\}$. The apparent boundary ∂D is at infinity, so that the universe modeled by D has infinite extent.[127] In the same vein, the non-Euclidean plane is represented by the 2D open disk $\{(x_1, x_2) | \ x_1{}^2 + x_2{}^2 < 4\}$ with $ds^2 = (dx_1{}^2 + dx_2{}^2)/\bigl(1 - (x_1{}^2 + x_2{}^2)/4\bigr)^2$.[128]

Here, a short comment is in order. Regarding geodesics as straight lines, one may logically build up a theory that satisfies all premises of non-Euclidean geometry. What should be emphasized is that straight lines are not necessarily the ones that we visualize on a piece of paper. Actually, as postulated by Hilbert (Remark 10.1 (2)), "point" and "straight line" are *undefined objects*, and only the *relations* between them such as "a point p *is on* a straight line ℓ" and "there is a unique line *passing through* given two points" are a vital necessity. Of great significance in this comprehension is that all the non-Euclidean objects are described in the Euclidean terminology. Thus the logical consistency of non-Euclidean geometry reduces to that of Euclidean geometry, and then to that of the real number system because $\mathbb{R}^3$ is a model of Euclidean space.[129]

In Sect. 6.5, we touched on projective geometry as a branch of classical geometry. For the pedagogical sake, we shall present an alternative perspective that was developed in the same period as Poncelet and Jakob Steiner (1796–1863) embarked on it in view of synthetic geometry. We owe the idea in the main to August Ferdinand Möbius (1790–1868).[130] It provides another exemplar of the transition to a "geometry of space itself."

A "plane" that one assumes a priori in projective geometry is called today the *projective plane*. Plane projective geometry is, if we distill its content down to a few

---

[126] This is his second paper on a non-Euclidean model. He did not know Riemann's work when writing the first paper *Saggio di Interpretazione della Geometrica Non-euclidea* (1868), wherein the pseudosphere (Footnote 108) is exploited in a tricky manner.

[127] In the *Teoria fondamentale*, it is indicated that $H^d = \{(x_1, \ldots, x_d) \in \mathbb{R}^d | \ x_d > 0\}$ with $ds^2 = (\sum_{i=1}^d dx_i{}^2)/x_d{}^2$ has constant negative curvature. This manifold, together with other ones, was later to be called the *hyperbolic space* (F. Klein, 1871); see [35].

[128] Thinking back over Riemann's thesis, we wonder why he did not refer to the non-Euclidean geometry since (6.15.13) for $\alpha < 0$ is connotative of Beltrami's model. In a letter to J. Hoüel dated April 4, 1868, Beltrami says, "What amazes me is that for all the time I talked with Riemann (during the 2 years he spent in Pisa, shortly before his sad end), he never mentioned these ideas to me, though they must have occupied him for quite a long time, for a fine draft cannot be the work of a single day, even for such a brilliant genius" (see [11]).

[129] At the ICM 1900, Hilbert proposed to prove the consistency of the real number system within his *formalistic* framework. Yet, this turns out to be not feasible as Gödel showed (1931).

[130] Möbius, *Der barycentrische Calcül*, 1827. A similar approach was taken by J. Plücker. Meanwhile, Steiner stuck to the synthetic method.

Fig. 6.12 Projective plane

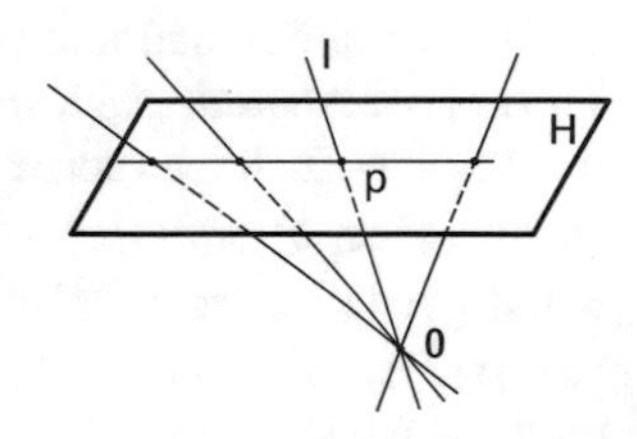

essential premises, built on the following simple system of axioms that are stated in terms of relations between points and lines:

(i) Any two distinct points are contained in one and only one line.
(ii) Any two distinct lines have one and only one point in common.
(iii) there exists four points, no three of which are contained in one line.

Although this might cause a sense of discomfort at first, a concrete model of the projective plane with "points" and "lines", symbolically indicated by $\mathbb{P}$, is given by the totality of lines in $\mathbb{R}^3$ passing through the origin O. Here a "point" is just an element of $\mathbb{P}$. A "line" is defined to be an ordinary plane in space passing through O. We say that a "point" ℓ is contained in a "line" H if the line ℓ is contained in the plane H in the usual sense. The axioms (i)–(iii) above are easy consequences of the facts that hold for lines and planes.

In Sect. 6.5, we said that, in plane projective geometry, points at infinity are adjoined to the Euclidean plane. To confirm this in our setting, select a plane $\mathbb{H}$ not containing O, and to take the line ℓ_p containing both $p \in \mathbb{H}$ and O. We let $\mathbb{H}(\mathbb{P})$ be the collection of lines $\ell \in \mathbb{P}$ not parallel to $\mathbb{H}$, and let $\mathbb{H}_\infty(\mathbb{P})$ be the plane passing through O and parallel to $\mathbb{H}$. Then $p \mapsto \ell_p$ is a one-to-one correspondence between $\mathbb{H}$ and $\mathbb{H}(\mathbb{P})$, which allows us to identify $\mathbb{H}(\mathbb{P})$ with $\mathbb{H}$. Figure 6.12 illustrates that, as a point in $\mathbb{H}$ tends to infinity, the line ℓ_p converges to a line parallel to $\mathbb{H}$. Accordingly, one may regard $\mathbb{H}_\infty(\mathbb{P})$ as the line at infinity.

Now, $\mathbb{P}$ is identified with $P^2(\mathbb{R})$ in a natural manner. We call $(x, y, z) \in \mathbb{R}^3\backslash\{\mathbf{0}\}$ the *homogeneous coordinate* of the "point" $\ell \in \mathbb{P}$ when ℓ passes through (x, y, z), and write $\ell = [x, y, z]$ by convention. This concept is immediately generalized to $P^d(\mathbb{R})$, and also to the realm of complex numbers.[131] The simplest case is the *complex projective line* $P^1(\mathbb{C}) = \{[z, w] | \ (z, w) \neq (0, 0), \ z, w \in \mathbb{C}\}$, which coincides with the *Riemann sphere*, the complex plane plus a point at infinity, introduced by Riemann (Footnote 121). A projective transformation in this case is represented as $T(z) = (az + b)/(cz + d)$ $(a, b, c, d \in \mathbb{C}, \ ad - bc \neq 0)$.

Henri Poincaré (1854–1912) observed a conflation of $P^1(\mathbb{C})$ and non-Euclidean geometry in his study of *Fuchsian functions* (a sort of "periodic" functions of one complex variable). Indeed, the above T with $a, b, c, d \in \mathbb{R}$ satisfying $ad - bc > 0$

[131] One can consider the projective planes over quaternions and octonions as well. It is interesting to point out that Desargues' theorem holds for quaternions, but does not for octonions, and that Pappus' hexagon theorem does not hold for quaternions.

preserves not only $H=\{z=x+yi \in \mathbb{C} | y>0\}$, but also $ds^2 = |dz|^2/y^2$ ($|dz|^2 = dx^2+dy^2$). Since (H, ds^2) is a model of the non-Euclidean plane (Footnote 126), T is a non-Euclidean congruent transformation.[132] More conspicuously, this idea led up to an entirely new phase of *analytic number theory* (initiated by Euler and definitively established by Dirichlet) through the notion of *automorphic form*, a generalization of Fuchsian functions.

Another non-Euclidean model that is cross-fertilized by projective geometry was given by Beltrami in his *Saggio* and independently by F. Klein (1871). This model, called the *Klein model* and first constructed in 1859 by A. Cayley in the context of real projective geometry without reference to non-Euclidean geometry (1859), is realized as the open unit disk $D = \{(x, y) \in \mathbb{R}^2 | x^2 + y^2 < 1\}$, in which the "line segments" are chords in D. Worthy of mention is that the cross-ratio pops up in the distance between two points. To be exact, the distance between two points P and Q in D can be expressed as $d(P, Q) = \frac{1}{2}\log\left|[A, P, Q, B]\right|$, where A, B are the points of intersection on the circle ∂D with the line connecting P, Q (see Remark 17.2 (5)). This distance is the one associated with $ds^2 = \{(1-(x^2+y^2))(dx^2+dy^2)+(xdx+ydy)^2\}/(1-(x^2+y^2))^2$.

6.17 Absolute Differential Calculus

Riemann's work was evolved further by Elwin Bruno Christoffel (1829–1900) and Rudolf Lipschitz (1832–1903). The significance of Christoffel's work lies in his introduction of the concept later called *covariant differentiation* and of the *curvature tensor* which crystallizes the notion of Riemann's curvature (1869).[133] After a while, Gregorio Ricci-Curbastro (1853–1925) and Tullio Levi-Civita (1873–1941) initiated "absolute differential calculus"[134]—calculus, in line with Christoffel's work, dealing with "invariant" geometric quantities (called *tensor fields*). Levi-Civita further brought in *parallel transports* as a partial generalization of parallel translations.[135] Before long, absolute differential calculus was renamed "tensor calculus" and used effectively by Einstein for his general relativity.[136] In this

[132]Poincaré, *Sur les Fonctions Fuchsiennes*, Acta Math., **1** (1882), (1882), 1–62. For this reason, (H, ds^2) is called the *Poincaré half plane*. The line element ds was previously obtained by Liouville through transformation of the line element on the pseudosphere (1850).

[133]*Über die Transformation der homogenen Differentialausdrücke zweiten Grades*, J. für die Reine und Angew. Math., **70** (1869), 46–70; *Über ein die Transformation homogener Differentialausdrücke zweiten Grades betreffendes Theorem*, ibid., 241–245.

[134]*Méthodes de calcul différentiel absolu et leurs applications*, Math. Ann., **54** (1900), 125–201.

[135]*Nozione di parallelismo in una varietàqualunque e consequente specificazione geometrica della curvatura Riemanniana*, Rend. Circ. Mat. Palermo, **42** (1917), 73–205.

[136]*Die Feldgleichungen der Gravitation*. Sitzungsberichte der Preußischen Akademie der Wissenschaften zu Berlin: 844–847 (November 25, 1915).

groundbreaking theory, a slight modification of Riemann's theory is required; namely, the matrix (g_{ij}) associated with the first fundamental form is a four-by-four symmetric matrix with three positive and one negative eigenvalue. A manifold with such a fundamental form is called a curved space-time, or a *Lorentz manifold* after Hendrik Anton Lorentz (1853–1928).

We shall now plunge ourselves into tensor calculus. Henthforth the matrix (g_{ij}) is supposed to be invertible, so that our discussion includes the case of Lorentz manifolds. A tensor field (in the classical sense) is roughly a system of multi-indexed functions depending on the choice of a coordinate system and satisfying a certain transformation rule under coordinate transformations so as to yield a *coordinate-free* quantity. Tensor fields thus defined fit in with the *principle of relativity* which Einstein set up to claim that the laws of physics should have the same form in all admissible coordinate systems of reference.

An example is the coefficients $\{g_{ij}\}$ of the first fundamental form (called the *metric tensor*). Here we should note that $ds^2 = \sum_{i,j=1}^{d} g_{ij}dx_i dx_j$ is a coordinate-free quantity. If we take another coordinate system $(y_1, \ldots, y_d)$, we have a different expression $ds^2 = \sum_{h,k=1}^{d} \overline{g}_{hk} dy_h dy_k$. The chain rule for differentiation applied to the coordinate transformation $x_i = x_i(y_1, \ldots, y_d)$ $(i = 1, \ldots, d)$ is written as $dx_i = \sum_{h=1}^{d} \frac{\partial x_i}{\partial y_h} dy_h$. So substituting this for (6.15.11), we obtain $ds^2 = \sum_{h,k=1}^{d} \sum_{i.j=1}^{d} g_{ij} \frac{\partial x_i}{\partial y_h} \frac{\partial x_j}{\partial y_k} dy_h dy_k$. Comparing this with $ds^2 = \sum_{h,k=1}^{d} \overline{g}_{hk} dy_h dy_k$, we have the transformation rule $\overline{g}_{hk} = \sum_{i.j=1}^{d} g_{ij} \frac{\partial x_i}{\partial y_h} \frac{\partial x_j}{\partial y_k}$.

Another example is a *contravariant vector* ξ^i $(i = 1, \ldots, d)$ obeying the rule $\overline{\xi}^i = \sum_{j=1}^{d} \xi^j \frac{\partial y_i}{\partial x_j}$. For instance, $\xi^i = \frac{dx_i}{dt}$ for a curve $c(t) = \big(x_1(t), \ldots, x_d(t)\big)$ is a contravariant vector because $\frac{dy_i}{dt} = \sum_{j=1}^{d} \frac{dx_j}{dt} \frac{\partial y_i}{\partial x_j}$ $\big(y_i(t) = y_i\big(x_1(t), \ldots, x_d(t)\big)\big)$. Hence, a contravariant vector is reckoned as a vector tangent to the manifold.

A general tensor field (of type (h, k)) is a system of multi-indexed functions $\xi^{i_1 \cdots i_h}_{j_1 \cdots j_k}$ $(i_1, \ldots, i_h, j_1, \ldots, j_k = 1, 2, \ldots, d)$ satisfying the transformation formula

$$\overline{\xi}^{a_1 \cdots a_h}_{b_1 \cdots b_k} = \sum_{i_1, \ldots, i_h = 1}^{d} \sum_{j_1, \ldots, j_k = 1}^{d} \xi^{i_1 \cdots i_h}_{j_1 \cdots j_k} \frac{\partial y_{a_1}}{\partial x_{i_1}} \cdots \frac{\partial y_{a_h}}{\partial x_{i_h}} \frac{\partial x_{j_1}}{\partial y_{b_1}} \cdots \frac{\partial x_{j_k}}{\partial y_{j_k}}.$$

Hence g_{ij} is a tensor field of type $(0, 2)$. For the inverse matrix $(g^{ij}) = (g_{ij})^{-1}$, we have $\overline{g}^{hk} = \sum_{i.j=1}^{d} g^{ij} \frac{\partial y_i}{\partial x_h} \frac{\partial y_j}{\partial x_k}$, so that g^{ij} is a tensor field of type $(2, 0)$.

The *covariant derivative* of $\xi^{i_1 \cdots i_h}_{j_1 \cdots j_k}$ is a tensor field of type $(h, k + 1)$ given by

$$\nabla_i \xi^{i_1 \cdots i_h}_{j_1 \cdots j_k} := \frac{\partial}{\partial x_i} \xi^{i_1 \cdots i_h}_{j_1 \cdots j_k} + \sum_{a=1}^{d} \left\{ \begin{matrix} i_1 \\ ai \end{matrix} \right\} \xi^{a i_2 \cdots i_h}_{j_1 \cdots j_k} + \cdots + \sum_{a=1}^{d} \left\{ \begin{matrix} i_h \\ ai \end{matrix} \right\} \xi^{i_1 \cdots i_{h-1} a}_{j_1 \cdots j_k}$$

$$- \sum_{a=1}^{d} \left\{ \begin{matrix} a \\ j_1 i \end{matrix} \right\} \xi^{i_1 \cdots i_h}_{a j_2 \cdots j_k} - \cdots - \sum_{a=1}^{d} \left\{ \begin{matrix} a \\ j_k i \end{matrix} \right\} \xi^{i_1 \cdots i_h}_{j_1 \cdots j_{k-1} a},$$

where

$$\left\{ \begin{matrix} h \\ ij \end{matrix} \right\} = \sum_{a=1}^{d} \frac{1}{2} g^{ha} \left(\frac{\partial g_{ia}}{\partial x^j} + \frac{\partial g_{ja}}{\partial x^i} - \frac{\partial g_{ji}}{\partial x^a} \right) \tag{6.17.14}$$

is what we call the *Christoffel symbol*.[137] The term involving the Christoffel symbols is thought of as making the derivation intrinsic. The curvature tensor R^i_{jkl} is then defined to be a tensor that measures non-commutativity of covariant differentiation; namely, $\nabla_l \nabla_k \xi^j - \nabla_k \nabla_l \xi^j = \sum_{i=1}^{d} R^j_{ilk} \xi^i$, where

$$R^i_{jkl} = \frac{\partial}{\partial x_k} \left\{ \begin{matrix} i \\ lj \end{matrix} \right\} - \frac{\partial}{\partial x_l} \left\{ \begin{matrix} i \\ kj \end{matrix} \right\} + \sum_{a=1}^{d} \left\{ \begin{matrix} i \\ ka \end{matrix} \right\} \left\{ \begin{matrix} a \\ lj \end{matrix} \right\} - \sum_{a=1}^{d} \left\{ \begin{matrix} i \\ la \end{matrix} \right\} \left\{ \begin{matrix} a \\ kj \end{matrix} \right\}.$$

Although implicit, the curvature tensor turns up in Riemann's paper submitted to the Académie des Sciences on July 1, 1861 (remained unknown until 1876).[138] In the second half of this paper, which was his answer to the prize question on heat distribution posed by the Academy in 1858, he gave a condition in order that $ds^2 = \sum_{i,j=1}^{d} g_{ij} dx_i dx_j$ be flat, or equivalently that there exists a curvilinear coordinate system $(y_1, \ldots, y_d)$ such that $ds^2 = dy_1{}^2 + \cdots + dy_d{}^2$. His condition can be transliterated into $R^i_{jkl} \equiv 0$ (see [9] for the detail).

We defer the details of how the curvature tensor is linked with Riemann's sectional curvature (see Sect. 6.20), and only cite the fact that the Gaussian curvature K_S of a surface S is expressed as $K_S = R^1_{221}/(g_{11}g_{22} - g_{12}{}^2)$, so that K_S is a rational function (not depending on S) in E, F, G and their derivatives up to second order as shown by Gauss with "bare hands" (*Disquisitiones generales*, Arts. 9–11). More specifically (in the present notation),

$$K_S = \frac{1}{\sqrt{EG - F^2}} \Big[\frac{\partial}{\partial u} \Big(\frac{\sqrt{EG - F^2}}{G} \left\{ \begin{matrix} 1 \\ 22 \end{matrix} \right\} \Big) - \frac{\partial}{\partial v} \Big(\frac{\sqrt{EG - F^2}}{G} \left\{ \begin{matrix} 1 \\ 12 \end{matrix} \right\} \Big) \Big], \tag{6.17.15}$$

$$\left\{ \begin{matrix} 1 \\ 22 \end{matrix} \right\} = \frac{2GF_v - GG_u - FG_v}{2(EG - F^2)}, \quad \left\{ \begin{matrix} 1 \\ 12 \end{matrix} \right\} = \frac{GE_v - FG_u}{2(EG - F^2)}.$$

Now take this for granted. For an isometry $\Phi : S \to \overline{S}$, the composition $\overline{\boldsymbol{S}}(u, v) = \Phi(\boldsymbol{S}(u, v))$ gives a local parametric representation of $\overline{S}$ such that $\overline{E}(u, v) = E(u, v)$, $\overline{F}(u, v) = F(u, v)$, $\overline{G}(u, v) = G(u, v)$ because Φ preserves

---

[137] The symbol Christoffel used is $\left\{ \begin{matrix} ij \\ h \end{matrix} \right\}$. R. Lipschitz (reputable for "Lipschitz continuity") brought in a similar symbol in his *Untersuchungen in Betreff der ganzen homogenen Functionen von n Differentialen*, J. für die Reine und Angew. Math., **70** (1869), 71–102.

[138] *Commentatio Mathematica, qua respondere tentatur quaestioni ab* Illma, Academia Parisiensi propositae.

ds^2. Therefore the derivatives of $\overline{E}, \overline{F}, \overline{G}$ of any order coincide with the corresponding derivatives of E, F, G, from which Gauss's Theorema Egregium follows.

Remark 17.1 The following formulas are due to Gauss, K. M. Peterson, G. Mainardi, and D. Codazzi:

$$L_v - M_u = \left\{ {1 \atop 12} \right\} L + \left(\left\{ {1 \atop 11} \right\} - \left\{ {2 \atop 12} \right\} \right) M + \left\{ {2 \atop 11} \right\} N, \quad (6.17.16)$$

$$N_u - M_v = \left\{ {1 \atop 22} \right\} L + \left(\left\{ {2 \atop 22} \right\} - \left\{ {1 \atop 12} \right\} \right) M - \left\{ {2 \atop 12} \right\} N. \quad (6.17.17)$$

In 1867, Bonnet proved the converse (the *fundamental theorem of surface theory*): if functions E, F, G and L, M, N in variables u and v satisfy (6.17.16), (6.17.17), (6.17.15), while $EG - F^2 > 0$, then there exists a unique surface (up to motion) in space which admits E, F, G and L, M, N as the coefficients of the first and second fundamental forms. □

In terms of the Christoffel symbol, the equation of a geodesic $c(t) = \big(x_1(t), \ldots, x_d(t)\big)$ is expressed as

$$\frac{d^2x_i}{dt^2} + \sum_{j,k=1}^{d} \left\{ {i \atop jk} \right\} \frac{dx_j}{dt}\frac{dx_k}{dt} = 0 \qquad (i = 1, \ldots, d). \quad (6.17.18)$$

This expression in a curved space-time—in alliance with imaginary experiments carried out within a closed chamber in interstellar space to confirm the equivalence of gravity and acceleration[139]—is the starting point of Einstein's theory; that is, we think of (6.17.18) as an analogue of the Newtonian equation of motion $d^2x_i/dt^2 = -\partial u/\partial x_i$ under gravitational potential u (i.e., the equation for "free fall").[140] Pursuing this analogy, he concluded that the first fundamental form may be loosely regarded as a generalization of the gravitational potential,[141] and that the fundamental interaction of gravitation as a result of *space-time being curved by matter and energy* is manifested by the *field equations*:

$$R_{ij} - \frac{1}{2}Rg_{ij} + \Lambda g_{ij} = \frac{8\pi G}{c^4}T_{ij}, \quad (6.17.19)$$

[139]The freely floating chamber produces a gravity-free state inside, while the accelerated chamber produces a force indistinguishable from gravitation.

[140]*Grundlage der allgemeinen Relativitätstheorie*, Ann. der Physik, **49** (1916), 769–822.

[141]In his thesis (Part III, §3), Riemann prophetically says, "We must seek the ground of its metric relations outside it, in *binding forces* which act upon it."

where $R_{ij} = \sum_{k=1}^{d} R_{ikj}^{k}$ is the *Ricci tensor*, $R = \sum_{i,j=1}^{d} g^{ij} R_{ij}$ is the *scalar curvature* (note that $R = 2K_S$ for a surface S), Λ is the *cosmological constant*,[142] c is the speed of light in vacuum, and T_{ij} is the *stress–energy tensor* (an attribute of matter, radiation, and non-gravitational force fields).

The left-hand side of (6.17.19) as an operator acting on g_{ij} is a non-linear generalization of the *d'Alembertian* $-\partial^2/\partial t^2 + c^2\Delta$ showing up in the description of wave propagation with speed c. This observation brought Einstein to the prediction of *gravitational waves*, which was successfully detected by the Laser Interferometer Gravitational-Wave Observatory (11th February, 2016).

What is remarkable is, as observed by Hilbert (1915), that the vacuum field equations ($R_{ij} = 0$) are the E-L equation associated with the action integral: $S = \int R\sqrt{-\det(g_{ij})}dx_1dx_2dx_3dx_4$.

Remark 17.2

(1) D'Alembert showed that solutions of the wave equation $\frac{\partial^2 y}{\partial t^2} - a^2\frac{\partial^2 y}{\partial x^2} = 0$ are of the form $f(x+at) + g(x-at)$, where f and g are *any* functions (1747). On the other hand, Daniel Bernoulli (1700–1782), a son of Johann Bernoulli and a friend of Euler, came up with *trigonometric series* for the first time, when he tried to solve the wave equation (1747). The two ways to express solutions stirred up controversy on the nature of "arbitrary" functions (Remark 19.1).
(2) The Laplacian Δ named after Laplace appears in the theory of heat conduction by Jean Baptiste Joseph Fourier (1768–1830). While he was Governor of Grenoble (appointed by Napoleon I), he carried out experiments on the propagation of heat along a metal bar, and used the trigonometric series to solve the heat equation $\frac{\partial y}{\partial t} = a^2\frac{\partial^2 y}{\partial x^2}$ that describes the distribution of heat over a period of time (1807 and 1822); see Remark 19.1. An interesting fact is that a special solution is given by the function $(1/\sqrt{4\pi a^2 t})\exp(-x^2/4a^2 t)$, which shows up as the *normal distribution* in mathematical statistics that A. de Moivre, Gauss, and Laplace pioneered.
(3) (6.17.19) is also regarded as an analogue of *Poisson's equation* $\Delta u = 4\pi G\rho$, where u is the *gravitational potential* associated with a mass density ρ.[143] As noticed by Green (1828) and Gauss (1839; *Werke* V, 196–242), this equation embodies the law of universal gravitation (and hence $\Delta u = 0$ in the vacuum case). Indeed, $u(\boldsymbol{x}) = -G\int \|\boldsymbol{x}-\boldsymbol{y}\|^{-1}\rho(\boldsymbol{y})d\boldsymbol{y}$ is a solution of Poisson's equation.

A heuristic (and intrepid) way to derive (6.17.19) in the vacuum case is to take a look at the tendency of free-falling objects to approach or recede from one another, which is described by the "variation vector field" $J(t) := \partial\boldsymbol{x}/\partial s|_{s=0}$, where $\boldsymbol{x}(t,s) = (x_i(t,s))$ $(-\epsilon < s < \epsilon)$ is a family of solutions of $d^2x_i/dt^2 = -\partial u/\partial x_i$. Clearly $J(t)$ satisfies $d^2J/dt^2 + H(J) = 0$, where H is the linear

[142]This constant is used to explain the observed acceleration of the expansion of the universe. (In 1929, E. Hubble discovered that the universe is expanding.)

[143]The density ρ for a mass distribution μ is defined by the relation $\rho(\boldsymbol{x})d\boldsymbol{x} = d\mu(\boldsymbol{x})$.

map defined by $H(z) = \sum_{j=1}^{d} \frac{\partial^2 u}{\partial x_i \partial x_j} z_j$. Notice that $\mathrm{tr}(H) = \Delta u = 0$. In turn, the variation vector field $J(t)$ for the geodesic equation (6.17.18) satisfies the apparently similar equation $\frac{D}{dt}\left(\frac{D}{dt}J\right) + K(J) = 0$ (called the *Jacobi equation*), where $K(z) = \sum_{j=1}^{d} R^i_{hjk}\dot{x}_h\dot{x}_k z_j$, and $\frac{DX}{dt} = \frac{dX_i}{dt} + \sum_{h,k=1}^{d} \left\{ {i \atop hk} \right\} \frac{dx_h}{dt} X_k$, the covariant derivative of a vector field X along the curve $\boldsymbol{x}(\cdot, 0)$ (Sect. 6.20). Since $\mathrm{tr}(K) = \sum_{h,k=1}^{d} R_{hk}\dot{x}_h\dot{x}_k$, we may regard the equation $\sum_{h,k=1}^{d} R_{hk}\dot{x}_h\dot{x}_k = 0$ (or $R_{hk} \equiv 0$ since $\dot{\boldsymbol{x}}$ is arbitrary) as an analogue of $\Delta u = 0$.

(4) The theory of *special relativity* created by Einstein in 1905 is literally a special case of general relativity, which has radically changed our comprehension of time and space.[144] Roughly speaking, it is a theory of space-time under absence of gravitation. Einstein assumed that (i) the speed of light c ($= 2.99792458 \times 10^8$ m/s) in a vacuum is the same for all observers, regardless of the motion of the light source, and (ii) the physical laws are the same for all non-accelerating observers (this is a special case of the principle of relativity). His setup squares with the theory of electromagnetic waves due to James Clerk Maxwell (1831–1879); see Sect. 6.22. The special relativity was mathematically formulated by Hermann Minkowski (1864–1909).[145] His model (called the *Minkowski space-time*) is a 4D *affine space* with a coordinate system (x_1, x_2, x_3, t) for which the line element is given by $ds^2 = dx_1{}^2 + dx_2{}^2 + dx_3{}^2 - c^2dt^2$. Such a coordinate system (stemming from the invariance of the speed of light) is called an *inertial system*, which an observer employs to describe events in space-time. If (y_1, y_2, y_3, s) is another inertial system moving in uniform velocity relative to (x_1, x_2, x_3, t), then the relation between them is given by an inhomogeneous *Lorentz transformation*:

$$\begin{pmatrix} \boldsymbol{y} \\ s \end{pmatrix} = \begin{pmatrix} A & -A\boldsymbol{v} \\ \mp c^{-2}\left(1 - \frac{\|\boldsymbol{v}\|^2}{c^2}\right)^{-1/2} {}^t\boldsymbol{v} & \pm\left(1 - \frac{\|\boldsymbol{v}\|^2}{c^2}\right)^{-1/2} \end{pmatrix} \begin{pmatrix} \boldsymbol{x} \\ t \end{pmatrix} + \begin{pmatrix} \boldsymbol{b} \\ t_0 \end{pmatrix} \tag{6.17.20}$$

($\boldsymbol{v}, \boldsymbol{b} \in \mathbb{R}^3$, and $\boldsymbol{x} = {}^t(x_1, x_2, x_3)$, $\boldsymbol{y} = {}^t(y_1, y_2, y_3)$); here A is a 3×3 matrix satisfying

$${}^tAA = I_3 + \frac{\|\boldsymbol{v}\|^2}{c^2 - \|\boldsymbol{v}\|^2} P_{\boldsymbol{v}}, \quad |\det A| = \left(1 - \frac{\|\boldsymbol{v}\|^2}{c^2}\right)^{-1/2},$$

where $P_{\boldsymbol{v}}$ is the orthogonal projection of $\mathbb{R}^3$ onto the line $\mathbb{R}\boldsymbol{v}$. Since $\boldsymbol{y} = A(\boldsymbol{x} - t\boldsymbol{v}) + \boldsymbol{b}$, the vector $\boldsymbol{v}$ is the *relative velocity* of two systems. Equation (6.17.20) tells us that the relative speed $\|\boldsymbol{v}\|$ cannot exceed c. Notice that, formally

[144] *Zur Elektrodynamik bewegter Körper*, Ann. der Physik, **17** (1905), 891–921.

[145] Minkowski, *Raum und Zeit*, Physikalische Zeitschrift **10** (1907), 75–88.

making c tend to infinity, we obtain a *Galilei transformation* compatible with Newtonian mechanics:

$$\begin{pmatrix} \boldsymbol{y} \\ s \end{pmatrix} = \begin{pmatrix} A & -A\boldsymbol{v} \\ 0 & \pm 1 \end{pmatrix} \begin{pmatrix} \boldsymbol{x} \\ t \end{pmatrix} + \begin{pmatrix} \boldsymbol{b} \\ t_0 \end{pmatrix}, \qquad A \in O(3).$$

With (6.17.20), one may interpret the *Lorentz-FitzGerald contraction* of moving objects. Before special relativity came out, Lorentz and G. F. FitzGerald posited the idea of contraction to rescue the existence of *aether* (or *ether*) from a paradox caused by the outcome of Michelson-Morley experiment (1887).[146]

(5) The linear part of the transformation given in (6.17.20) with $c = 1$ belongs to $O(3, 1)$, where $O(n, 1)$ is the matrix group (called the *Lorentz group*) that preserves the quadratic form $Q(x_1, \ldots, x_n, x_{n+1}) = x_1{}^2 + \cdots + x_n{}^2 - x_{n+1}{}^2$. Interestingly, the subgroup $O^+(2, 1)$ of $O(2, 1)$ that preserves the sign of the last coordinate comes up as the congruence group for the Klein model D (note that the set $P = \{[x_1, x_2, x_3] \in P^2(\mathbb{R}) | x_1{}^2 + x_2{}^2 - x_3{}^2 < 0\}$ is identified with D, and that the projective transformation associated with a matrix in $O^+(2, 1)$ leaves P (and D) invariant and preserves the distance $d(z, w)$ on D because of the projective invariance of cross-ratios). □

6.18 Cantor's Transfinite Set Theory

Mathematical terminology to formulate what Riemann wanted to convey had not yet matured enough in his time. Here is an excerpt from Riemann's paper referred to above, which indicates how he took pains to impart the idea of manifolds with immatured terms.

> If in the case of a notion whose specializations form a continuous manifold, one passes from a certain specialization in a definite way to another, the specializations passed over form a simply extended manifold, whose true character is that in it a continuous progress from a point is possible only in two directions, forwards or backwards. If one now supposes that this manifold in its turn passes over into another entirely different, and again in a definite way, namely so that each point passes over into a definite point of the other, then all the specializations so obtained from a doubly extended manifold. In a similar manner one

[146]Aether, whose existence was discarded by Einstein, had been thought to permit the determination of our absolute motion and also to allow electromagnetic waves to pass through it as elastic waves. Going back into history, ancient thinkers (e.g. Parmenides, Empedocles (ca. 490 BCE–ca. 430 BCE), Plato, Aristotle) assumed that aether (Αἰθήρ) filled the celestial regions. Descartes insisted that the force acting between two bodies not touching each other is transmitted through it (*Traité du monde et de la lumière*, 1629–1633). Newton suggested its existence in a paper read to the Royal Society (1675). Riemann assumed that both "gravitational and electrostatic effects" and "optical and magnetic effects" are caused by aether (1853), while Kelvin, holding Newton in reverence, said that he could not be satisfied with Maxwell's work until a mechanical model of the aether could be constructed (1884).

obtains a triply extended manifold, if one imagines a doubly extended one passing over in a definite way to another entirely different; and it is easy to see how this construction may be continued. If one regards the variable object instead of the determinable notion of it, this construction may be described as a composition of a variability of $n + 1$ dimensions out of a variability of n dimensions and a variability of one dimension (English translation by Clifford).

Poincaré attempted to give a definition of manifold for his own motivation. In the ground-breaking work *Analysis Situs* (J. École Polytech, **1** (1895) 1–121), he offered two definitions, both of which rely on "constructive procedures." The first is to describe a manifold as the zero set $f^{-1}(0)$ of a smooth function $f : U \longrightarrow \mathbb{R}^k$, where U is an open set of $\mathbb{R}^{d+k}$, and the Jacobian of f is of maximal rank everywhere. In the second, a manifold is produced by a "patchwork" of a family of open sets of $\mathbb{R}^d$. Although his definitions are considered a precursor to the modern formulation of manifolds, it was not yet satisfactory enough ([33] and [23]). To remove the ambiguity in the early formulation and to study a global aspect of manifolds, it was necessary to establish the notion of *topological space*, an ultimate concept of generalized space allowing us to talk about "finite and infinite" in an entirely intrinsic manner (thus one can say that topological spaces are "stark naked" models of the universe).

What should be accentuated in this development, is the invention of (*transfinite*) *set theory* by Georg Cantor (1845–1918). His *Grundlagen einer allgemeinen Mannigfaltigkeitslehre* (1883) was originally motivated by the work of Dirichlet and Riemann concerning trigonometric series (Remark 19.1),[147] and made a major paradigmatic shift in the sense that since then mathematicians have accepted publicly "the actual infinity", as opposed to the traditional attitude towards infinity.[148] To grasp the atmosphere before set theory came in, we shall quote Gauss's words. In a letter addressed to Schumacher (July 12, 1831), he says, "I protest against the use of infinite magnitude as something completed, which is never permissible in mathematics. Infinity is merely a way of speaking, the true meaning being a limit which certain ratios approach indefinitely close, while others are permitted to increase without restriction" (*Werke*, VIII, p. 216). What he claimed is not altogether inapposite as far as "differentiation" is concerned, but cannot apply to infinity embodied by set theory.

A set (Mengenlehre) is, as Cantor's declares, a collection of definite, well-distinguishable objects of our intuition or of our thought to be discerned as a whole.[149] To tell how set theory is linked to the issue of infinity, let us take a look at natural numbers. We usually recognize natural numbers $1, 2, 3, \ldots$ as potential infinity; that is, we identify them by a non-terminating process such as adding 1

[147] In *Über die Ausdehnung eines Satzes aus der Theorie der trigonometrischen Reihen*, Math. Ann., **5** (1872), 123–132, he defined real numbers, using Cauchy sequences of rationals.

[148] Bernhard Bolzano (1781–1848) conceived the notion of set from a philosophical view (*Wissenschaftslehre*, 1837).

[149] *Beiträge zur Begründung der transfiniten Mengenlehre*, Math. Ann., **46** (1895), 481–512. This is Cantor's last major publication.

to the previous number, while Cantor thinks that one may capture the "set" of all natural numbers as a completed totality (which we shall denote by $\mathbb{N}$). Once the notion of set as the actual infinity is accepted, one can exploit one-to-one correspondences (OTOC) to compare two infinite sets as Archimedes, Bradwardine, and Oresme did (Sect. 6.2). In this context, infiniteness of a set is characterized by the property that it has a OTOC with a proper subset. This characterization (due to Dedekind) resolves Galileo's paradox mentioned in Sect. 6.2, simply by saying that the set of natural numbers is infinite.

The most significant discovery by Cantor is the existence of different sizes of infinite sets; for instance, the real numbers are more numerous than the natural numbers. Thus $\mathbb{R}$, the set of real numbers, cannot be captured as potential infinity whatsoever. This was mentioned in a letter[150] to Dedekind dated December 7, 1873 and published in 1874. [151] Furthermore, he introduced *cardinal numbers* as a generalization of natural numbers; that is, a cardinal number is defined to be a name assigned to an arbitrary set, where two sets have the same name if and only if there is a OTOC between them.

At any rate, the notion of set *per se* is so simple that it seem not necessary to take it up expressly. Indeed, "set" is nearly an everyday term with synonyms such as "collection," "family," "class," and "aggregate." In truth, people before Cantor more or less adapted themselves to "sets" without noticing that the notion cannot be discussed separately from actual infinity.

Greek thinkers were timid with infinity since it often leads to falsity (remember the erroneous argument by Antiphon and Bryson in "squaring the circle"). Cantor, too, struck a snag when he put forward his theory, if not the same snag the Greeks struck. In fact, the simplicity of the notion of set is deceptive. Unless the concept of set is suitably introduced, we run into a contradiction like Russell's paradox discovered in 1901, which had disquieted, for some time, mathematicians of the day who were favorably disposed towards set theory.[152]

Surely, Cantor's innovation was unorthodox and daring at that time; he met with belligerent resistance from a few conservative people. For instance, Poincaré poignantly said that there is no actual infinity since mathematics accepting it has fallen into contradiction (1906), and L. J. J. Wittgenstein bluntly dismissed set theory as "utter nonsense," and "wrong."[153]

---

[150] *Briefwechsel Cantor-Dedekind*, eds. E. Noether and J. Cavailles, Paris: Hermann.

[151] His *diagonal argument* was given in *Über eine elementare Frage der Mannigfaltigkeitslehre*. Jahresbericht der Deutschen Mathematiker-Vereinigung, **1** (1890–1891), 75–78.

[152] His paradox results from admitting that $X = \{x \mid x \notin x\}$ is a set. In fact, if $X \notin X$, then $X \in X$ by definition, and if $X \in X$, then $X \notin X$ again by definition.

[153] See "Wittgenstein's Philosophy of Mathematics" in the Stanford Encyclopedia of Philosophy. The frustration and despair caused by the disinterest and cold rejection directed Cantor to a theological explanation of his theory, which, he believed, would open up a whole new landscape in Christian theology (a letter to C. Hermite, dated January 22, 1894 [18].

Fortunately, various paradoxes derived from a vague understanding were resolved by axiomatizing set theory,[154] and most mathematicians responded favorably to his theory. Hilbert, who stands squarely against the outcry of conservative people, exclaims, "No one shall expel us from the Paradise that Cantor has created" (*Über das Unendliche*, Math. Ann., **95** (1926), 161–190).[155]

6.19 Topological Spaces

The swift change caused by set theory delimited the history of mathematics. The old thought that mathematics is the science of quantity, or of space and number, has largely disappeared. From then on, mathematics has been built eventually on set theory, indissolubly combined with the concept of *mathematical structure*—in short, any set of objects along with certain relations among those objects—which was put forward in a definitive form by Bourbaki in the campaign for modernization of mathematics. An instructive example is $\mathbb{N}$ whose structure is characterized by *Peano's axioms*, the setting that epitomizes the transition from potential infinity to actual infinity (1889):

"The set $\mathbb{N}$ has an element 1 and an injective map $\varphi : \mathbb{N} \longrightarrow \mathbb{N}$ such that (i) $1 \notin \varphi(\mathbb{N})$, (ii) if S is a subset of $\mathbb{N}$ with $1 \in S$ and $\varphi(S) \subset S$, then $S = \mathbb{N}$."

Algebraic systems such as *groups*, *rings*, and *vector spaces* are sets with structures as well. Topological spaces are another exemplary, which has, in no small part, evolved out of a long process of understanding the meaning of "limit" and "continuity" of functions.

Remark 19.1 A primitive form of functions is glimpsed in the *Almagest*, Book 1, Chap. 11. He made a table of chords of a circle (dating back to Hipparchus), which, from a modern view, can be thought of as associating the elements of one set of numbers with the elements of another set. After Oresme (1350), Galileo (1638), and Descartes (1637) adumbrated germinal ideas, the great duo of Johann Bernoulli and Leibniz adopted the word "functio" for "a quantity formed from indeterminate and constant quantities" (1694). A more lucid formulation was given by Euler, who introduced the notation $f(x)$ (1734) and defined a function to be an *analytic expression* composed in any way whatsoever of the variable quantity and numbers or constant quantities" (*Introductio*; [13]). But, after a while, he altered this definition as "When certain quantities depend on others in such a way that they undergo a change when the latter change, then the first are called functions of the second" [8].

---

[154] Axiomatic set theory, established by E. Zermelo in 1908 and A. Fraenkel in 1922, is formulated with a formal logic.

[155] In the late 1920s, Hilbert preferred to compare the use of actual infinity to the addition of points at infinity in projective geometry.

Interest in the true nature of functions was renewed when Fourier claimed that an *arbitrary* function $f(x)$ on $[-\pi, \pi]$ can be expressed as a trigonometric series

$$f(x) = \frac{1}{2}b_0 + (a_1 \sin x + b_1 \cos x) + (a_2 \sin 2x + b_2 \cos 2x) + \cdots, \quad (6.19.21)$$

$$a_n = \frac{1}{\pi}\int_{-\pi}^{\pi} f(x) \sin nx\, dx, \quad b_n = \frac{1}{\pi}\int_{-\pi}^{\pi} f(x) \cos nx\, dx.$$

However, he did not turn his eye to the convergence of the series, let alone the meaning of the equality, the unavoidable issues as pointed out by Cauchy (1820).[156] It was in this context that Cauchy gave a faultless definition of continuous function, using the notion of "limit" for the first time.[157] Following Cauchy's idea, Weierstrass popularized the ϵ-δ argument in the 1870s, which made it possible to discuss diverse aspects of convergence. In the meantime, Dirichlet made Fourier's work rigorous on a clearer understanding of (dis)continuity. With some prodding from Dirichlet, Riemann made further progress by giving a precise meaning to integrability of function.[158]

The rigor of calculus whose absence was deplored by Abel was now in place (Remark 4.2 (4)). Analysis that developed on it (and was accompanied with Cantor's set theory) contributed to the advent of the concept of topological space. □

A *topology* on a set X is a family of subsets (called *neighborhoods*) which makes it possible to express the idea of getting closer and closer to a point and also to introduce the notion of continuous maps as a generalization of continuous functions. In his magnum opus *Grundzüge der Mengenlehre* (1914), Felix Hausdorff (1869–1942) sets the four axioms in terms of neighborhoods, the last being known as the *Hausdorff separation axiom*; thus his topological spaces are what we now call *Hausdorff spaces*, a proper generalization of *metric spaces*.[159]

Now, a manifold with the first fundamental form is a metric space, so we are tempted to discuss *finiteness* of the universe in the topological framework. Finiteness of a figure in $\mathbb{R}^N$ is interpreted by the terms *closed* and *bounded*. Thus the issue boils down to the unequivocal question of how we define such terms in an intrinsic

[156]The bud of the issue is seen in the debate among D'Alembert, Euler, and Daniel Bernoulli about solutions of the wave equation (Remark 17.2 (1)).

[157]Euler thought of "$\lim_{n\to\infty}$" (resp. "$\lim_{x\to 0}$") as "the value for n infinitely large (resp. the value for x infinitely small"). The modern definition was furnished by Bolzano in his *Der binomische Lehrsatz*, 1816; however, it was not noticed at that time.

[158]Riemann, *Über die Darstellbarkeit einer Function durch eine trigonometrische Reihe*, Abhandlungen der Königlichen Gesellschaft der Wissenschaften zu Göttingen, **13** (1868), 87–132. This was presented in December of 1853 as his Habilitationsschrift at the first stage.

[159]The earlier results on topology—ascribed to F. Riesz, M. R. Fréchet, and others—fitted naturally into the framework set up by Hausdorff. Currently, there are several ways to define topological spaces; e.g. by the axioms on the closure operation (1922) due to K. Kuratowski, and by the axioms on open sets (1934) due to W. F. Sierpiński. The notion of metric space was first suggested by Fréchet in his dissertation paper (1906).

manner, without referring to "outside." The answer is afforded by the notion of *compactness* (Fréchet, 1906), which expresses "closed and bounded in itself" so to speak, whose origin is the Bolzano-Weierstrass Bolzano-Weierstrass theorem and the Heine-Borel theorem in analysis.

Armed with the impregnable language of finiteness, one may call into question, without assessing relevance to metaphysical or theological meaning, whether our universe is finite. Yet, this overwhelming question is beyond the scope of this essay and belongs entirely to astronomy. The recent observation (by the *Planck*, the space observatory operated by the European Space Agency) lends support to an "almost" flat model of the universe in the large. However this does not entail its infiniteness since "flatness" does not imply "non-compactness." In turn, if there are only finitely many particles in the whole universe, then the universe should be finite (here it is assumed that the mass would be uniformly distributed in the large).[160] This seems not to be a foolhardy hypothesis altogether. The *cosmological principle* in modern cosmology claims that the distribution of matter in the universe is homogeneous and isotropic when viewed on a large enough scale. Einstein says, "I must not fail to mention that a theoretical argument can be adduced in favor of the hypothesis of a finite universe."[161] Truth be told, however, nobody could definitively pronounce on the issue at the moment. On this point, our intellectual adventure cannot finish without frustration since the issue of finiteness is the most fundamental in cosmology.

6.20 Towards Modern Differential Geometry

After the notion of topological space took hold, mathematicians got ready to move forward. What had to be done initially was to link the concept of curvilinear coordinate system with topological spaces. An incipient attempt was made by Hilbert, who, using a system of neighborhoods, tried to characterize the plane.[162] Subsequently, Hilbert's former student Hermann Klaus Hugo Weyl (1885–1955) published *Die Idee der Riemannschen Fläche* (1913), a classical treatise that lays a solid foundation for complex analysis initiated by Riemann (Footnote 121). His perspicuous exposition, though restricted to the 2D case, opened up the modern synoptic view of geometry and analysis on manifolds. Stimulated with this, O. Veblen and his student J. H. C. Whitehead gave the first general definition of manifolds in their book *The foundations of differential geometry*, 1933. At around the same time, Hassler Whitney (1907–1989) and others clarified the foundational aspects of manifolds during the 1930s. One of significant outcomes is that any

[160]In the *Letter to Herodotus*, Epicurus (341 BCE–270 BCE), an adherent of atomism and advocate of "billiard ball universe," states that the number of atoms is infinite.

[161]*Geometry and Experience*, an address to the Berlin Academy on January 27th, 1921.

[162]Appendix IV (1902) to *Grundlagen der Geometrie*.

smooth manifold (with an additional mild property) can be *embedded* in a higher-dimensional Euclidean space (1936), thus giving a justification to one of Poincaré's definitions of manifold (Sect. 6.18).

Meanwhile, differential geometry of surfaces was expanded by Jean-Gaston Darboux (1842–1917). He published, between 1887 and 1896, four enormous volumes entitled *Leçons sur la théorie générale des surfaces et les applications géométriques du calcul infinitésimal*, including most of his earlier work and touching on global aspects of surfaces to some extent. Darboux's spirit was then inherited by Élie Cartan (1869–1951). Most influential is his meticulous study of *symmetric spaces* (1926), a class of manifolds with an analogue of point symmetry at each point, including $\mathbb{R}^d$, H^d, S^d, and $P^d(\mathbb{R})$; thus unifying various geometries in view of symmetry and also following the Greek tradition in seeking symmetric shapes. What is remarkable is that Cartan's list includes symmetric spaces described in terms of quaternions and octonions. Furthermore Cartan enlarged Klein's "Erlangen Program" (Footnote 42) so as to encompass general geometries.

A decade from 1930 was the period when geometry began to be intensively studied from a global point of view, in concord with a new field launched by Riemann and Poincaré (see Sect. 6.21). What is more, the fundamental concepts have been made more transparent by means of a *coordinate-free setup*, which, in a sense, conforms to the principle of relativity perfectly.[163]

In what follows, just to make our story complete, we shall quickly review, at the cost of some repetition, how some of the ingredients introduced by the forerunners are reformulated in modern terms. The first is the definition of smooth manifolds as a *terminus ad quem* of our journey to "generalized curved spaces."

Definition

(1) A Hausdorff space M is said to be a *d-dimensional topological manifold* (or *d-manifold*) if each point of M has an open neighborhood homeomorphic to an open set in $\mathbb{R}^d$; therefore we have an *atlas of local charts*; that is, a family of curvilinear coordinate systems that covers M.
(2) A topological manifold M is called smooth if there is an atlas such that every coordinate transformation is smooth (such an atlas is said to be smooth).

Using a smooth atlas, we may discuss "smoothness" of various objects attached to smooth manifolds; say, *smooth functions* and *smooth maps*. What comes next to mind is the question whether it is possible to define *tangent space* as the generalization of tangent planes of a surface. The idea to conceptualize tangent space without reference to a coordinate system and ambient spaces dates back to the work of C. Bourlet who, paying special attention to the product rule, gave an algebraic characterization of differentiation.[164]

[163]Grassmann was the first to realize the importance of the coordinate-free concepts.

[164]*Sur les opérations en général et les équations différentielles linéaires d'ordre infini*, Ann. Ec. Normale, **14** (1897), 133–190.

To explain the idea in depth, we shall first observe that a tangent plane of a surface can be intrinsically defined. Let T_pS be the space of vectors tangent to a surface S at p, and let $C_p^\infty(S)$ be the set of smooth functions defined on a neighborhood of p. Define the action of $\boldsymbol{\xi} \in T_pS$ on $C_p^\infty(S)$ by setting $\boldsymbol{\xi}(f) = \frac{d}{dt} f\big(c(t)\big)\big|_{t=0}$, where $c : (-\epsilon, \epsilon) \longrightarrow S$ is a curve with $c(0) = p$ and $\dot{c}(0) = \boldsymbol{\xi}$. The correspondence $f \in C_p^\infty(S) \mapsto \boldsymbol{\xi}(f) \in \mathbb{R}$ has the following properties:

1. (Linearity) $\boldsymbol{\xi}(af + bg) = a\boldsymbol{\xi}(f) + b\boldsymbol{\xi}(g) \quad a, b \in \mathbb{R},\ f, g \in C_p^\infty(S)$.
2. (Product rule) $\boldsymbol{\xi}(fg) = f(p)\boldsymbol{\xi}(g) + g(p)\boldsymbol{\xi}(f)$.
3. $(a\boldsymbol{\xi} + b\boldsymbol{\eta})(f) = a\boldsymbol{\xi}(f) + b\boldsymbol{\eta}(f) \quad a, b \in \mathbb{R},\ \boldsymbol{\xi}, \boldsymbol{\eta} \in T_pS,\ f \in C_p^\infty(S)$.
4. If $\boldsymbol{\xi}(f) = 0$ for any $f \in C_p^\infty(S)$, then $\boldsymbol{\xi} = \mathbf{0}$.

We let τ_pS be the vector space consisting of maps $\boldsymbol{\omega} : C_p^\infty(S) \to \mathbb{R}$ with the properties $\boldsymbol{\omega}(af + bg) = a\boldsymbol{\omega}(f) + b\boldsymbol{\omega}(g)$ and $\boldsymbol{\omega}(fg) = f(p)\boldsymbol{\omega}(g) + g(p)\boldsymbol{\omega}(f)$. In view of (1), (2), $\boldsymbol{\xi}$ as a map of $C_p^\infty(S)$ into $\mathbb{R}$ belongs to τ_pS. Hence we obtain a map ι of T_pS into τ_pS, which turns out to be a linear isomorphism. Identifying T_pS with τ_pS via ι, we have an intrinsic description of T_pS.

The foregoing discussion suggests how to define the tangent spaces of a smooth manifold M. Let $C_p^\infty(M)$ be the set of smooth functions defined on neighborhoods of $p \in M$, and define T_pM to be the vector space consisting of operations $\omega : C_p^\infty(M) \to \mathbb{R}$ satisfying linearity and the product rule.[165] For a curvilinear coordinate system $(x_1, \ldots, x_d)$ around p, partial differentiations $\partial_i|_p := \partial/\partial x_i\big|_p$ $(i = 1, \ldots, d)$ form a basis of T_pM. Moreover, the *velocity vector* $\dot{c}(t) \in T_{c(t)}M$ for a curve c in M is defined by putting $\dot{c}(t)f = \frac{d}{dt} f(c(t))$.

The notion of tangent space allows us to import various concepts in calculus into the theory of manifolds. An example is the *differential* $\Phi_{*p} : T_pM \longrightarrow T_{\Phi(p)}N$ of a smooth map $\Phi : M \longrightarrow N$ defined by $\Phi_{*p}(\boldsymbol{\xi})(f) = \boldsymbol{\xi}(f \circ \Phi),\ \boldsymbol{\xi} \in T_pM,\ f \in C_{\Phi(p)}^\infty(N)$, which is an intrinsic version of the *Jacobian matrix*.

A *Riemannian manifold* is a manifold with a (*Riemannian*) *metric*, a smooth family of inner products $\langle \cdot, \cdot \rangle_p$ on T_pM $(p \in M)$. The relation between the metric and the first fundamental form is given by $g_{ij}(p) = \langle \partial_i|_p, \partial_j|_p \rangle_p$. A smooth map $\Phi : M \longrightarrow N$ between Riemannian manifolds is called an *isometry* if $\langle \Phi_{*p}(\boldsymbol{\xi}), \Phi_{*p}(\boldsymbol{\eta}) \rangle_{\Phi(p)} = \langle \boldsymbol{\xi}, \boldsymbol{\eta} \rangle_p \quad (\boldsymbol{\xi}, \boldsymbol{\eta} \in T_pM)$. The differential Φ_{*p} of an isometry Φ is obviously injective for every p. Conversely, for a smooth map Φ of M into a Riemannian manifold N such that Φ_{*p} is injective for every p, one can equip a metric on M (called the *induced metric*) which makes Φ an isometry. The first fundamental form on a surface S is nothing but the metric induced by the inclusion map of S into $\mathbb{R}^3$.[166] Furthermore, the metric (6.15.13) given by Riemann coincides essentially with the standard metric on the sphere $S^d(R) = \{(x_1, \ldots, x_{d+1}) | {x_1}^2 + \cdots + {x_{d+1}}^2 = R^2\}$ that is read in the curvilinear coordinate

[165] This definition is given in Claude Chevalley's *Theory of Lie Groups* (1946).

[166] In 1956, J. F. Nash showed that every Riemannian manifold has an isometric inclusion into some Euclidean space $\mathbb{R}^N$.

system derived by the stereographic projection $\varphi : S^d(R)\setminus\{(0,\ldots,0,R)\} \longrightarrow \mathbb{R}^d$: $\varphi(x_1,\ldots,x_{d+1}) = \big(Rx_1/(R-x_{d+1}),\ldots,Rx_d/(R-x_{d+1})\big)$.

No less important than tangent spaces is the dual space T_p^*M of T_pM, which is distinguished from T_pM though they are isomorphic as vector spaces.[167] We denote by $\{dx_1|_p,\ldots,dx_d|_p\}$ the *dual basis* of $\{\partial_1|_p,\ldots,\partial_d|_p\}$; i.e., $dx_i|_p(\partial_j|_p) = \delta_{ij}$. The symbol $dx_i|_p$ defined in this way justifies the notion of differential (Remark 15.1 (2)). A *tensor of type* (h,k) is then a multilinear functional $T : \underbrace{T_p^*M\times\cdots\times T_p^*M}_{h}\times\underbrace{T_pM\times\cdots\times T_pM}_{k} \longrightarrow \mathbb{R}$. which is related to a classical tensor by $T^{i_1\cdots i_h}_{j_1\cdots j_k}(p) := T(dx_{i_1}|_p,\ldots,dx_{i_h}|_p,\partial_{j_1}|_p,\ldots,\partial_{j_k}|_p)$. A *tensor field of type* (h,k) is a smooth assignment to each $p\in M$ of a tensor $T(p)$ of type (h,k). A *vector field* is a tensor field of type $(1,0)$. The vector space consisting of all vector fields on M is denoted by $\mathfrak{X}(M)$.

Essential to global analysis are *differential forms* and their *exterior derivative* introduced by Poincaré (1886) and Cartan (1899). To give a current definition of differential forms, we denote by $\wedge^k T_p^*M$ the totality of tensors ω of type $(0,k)$ satisfying $\omega(\boldsymbol{\xi}_{\sigma(1)},\ldots,\boldsymbol{\xi}_{\sigma(k)}) = \mathrm{sgn}(\sigma)\omega(\boldsymbol{\xi}_1,\ldots,\boldsymbol{\xi}_k)$ $(\boldsymbol{\xi}_i\in T_pM)$ for every permutation σ of the set $\{1,\ldots,k\}$. A smooth assignment to each $p\in M$ of $\omega(p)\in\wedge^k T_p^*M$ is called a (*differential*) *k-form*, all of which constitute a vector space, denoted by $A^k(M)$. A k-form is informally expressed as

$$\omega = \sum_{1\le i_1<\cdots<i_k\le d} f_{i_1\cdots i_k}(x_1,\ldots,x_d)dx_{i_1}\wedge\cdots\wedge dx_{i_k},$$

where the "wedge product" $\wedge$ is supposed to satisfy $(dx_h\wedge dx_i)\wedge dx_j = dx_h\wedge(dx_i\wedge dx_j)$, $dx_i\wedge dx_j = -dx_j\wedge dx_i$.[168] With this notation, the *exterior derivation* $d_k : A^k(M)\longrightarrow A^{k+1}(M)$ is defined by

$$d_k\omega(=d\omega) := \sum_{i=1}^{d}\sum_{1\le i_1<\cdots<i_k\le d}\frac{\partial f_{i_1\cdots i_k}}{\partial x_i}dx_i\wedge dx_{i_1}\wedge\cdots\wedge dx_{i_k}.$$

Note here that $df = f_{x_1}dx_1+\cdots+f_{x_d}dx_d$ for a smooth function $f\in A^0(M)$. A significant fact derived from the equality $f_{xy} = f_{yx}$ is the identity $d_k\circ d_{k-1} = 0$. Thus we obtain the vector space $H^k_{\mathrm{dR}}(M) := \mathrm{Ker}d_k/\mathrm{Image}d_{k-1}$, which is called the *de Rham cohomology group*; Remark 21.2 (3).

---

[167]The distinction between a vector space and its dual was definitively established following the work of S. Banach and his school, though it had been already reflected for a long time by the distinction between *covariant* and *contravariant*, and also between *cogredient* and *contragredient*, the concepts in group-representation theory.

[168]The totality of $\wedge^k T_p^*M$ $(k=0,1,\ldots,d)$, equipped with the wedge product, constitutes the *exterior algebra* introduced by Grassmann (Sect. 6.14).

Differential forms give an incentive to justify the formal expression $dx_1 \cdots dx_d$ in multiple integrals, because the basic rules for the wedge product make the appearance of *Jacobian determinants* automatic: $dy_1 \wedge \cdots \wedge dy_d = \frac{\partial(y_1,\ldots,y_d)}{\partial(x_1,\ldots,x_d)} dx_1 \wedge \cdots \wedge dx_d$, where $y_i = y_i(x_1, \ldots, y_d)$ is a coordinate transformation. This is in conformity with a change of variables in multiple integrals:[169] $dy_1 \cdots dy_d = \left|\frac{\partial(y_1,\ldots,y_d)}{\partial(x_1,\ldots,x_d)}\right| dx_1 \cdots dx_d$. Thus we are motivated to make the following definition: "M is said to be *orientable* if it has a smooth atlas such that the Jacobian determinant of every coordinate transformation is positive" (Poincaré, *Analysis Situs*, 1895; see Sect. 6.22). Then, what springs to mind is to define the integral over an orientable d-manifold M of a d-form $\omega = f(x_1, \ldots, x_d)dx_1 \wedge \cdots \wedge dx_d$ by setting $\int_M \omega = \int f(x_1, \ldots, x_d)dx_1 \cdots dx_d$.

Now, given $\omega \in A^k(N)$ and a map Φ of an orientable k-manifold M (possibly with boundary) into N, we define the integral of ω along Φ by $\int_\Phi \omega := \int_M \Phi^*\omega$, where $(\Phi^*\omega)(p) \in \wedge^k T_p^*M$ is defined by setting $(\Phi^*\omega)(\boldsymbol{\xi}_1, \ldots, \boldsymbol{\xi}_k) = \omega(\Phi_*(\boldsymbol{\xi}_1), \ldots, \Phi_*(\boldsymbol{\xi}_k))$. This is a generalization of line integrals, with which we may unify Stokes' formula, Gauss's divergence theorem, and Green's theorem, as the *generalized Stokes' formula*:

$$\int_M d\omega = \int_{\partial M} \omega \left(= \int_\iota \omega \right),$$

where ω is a $(d-1)$-form on an orientable d-manifold M with boundary ∂M (appropriately oriented), and $\iota : \partial M \longrightarrow M$ is the inclusion map.

The notion of vector field is quite old. In his *Principes généraux du mouvement des fluides* (1755), Euler used it to represent a fluid's velocity. Related to vector fields is the work *Theorie der Transformationsgruppen* (1888–1893) by Sophus Lie (1842–1899). Under the influence of Klein's Erlangen Program, he made the study of *infinitesimal group actions*, which eventually evolved into the theory of *Lie groups*. As he observed, giving such an action is equivalent to giving a subspace $\mathfrak{g}$ of $\mathfrak{X}(M)$ satisfying "$X, Y \in \mathfrak{g} \Rightarrow [X, Y] \in \mathfrak{g}$," where $[X, Y](p)f = X(p)(Yf) - Y(p)(Xf)$. Here the binary operation $[X, Y]$ (the *Lie bracket*) satisfies the *Jacobi identity* $[[X, Y], Z]+[[Y, Z], X]+[[Z, X], Y] = 0$. An algebraic system with such an operation was to be called a *Lie algebra*.[170]

A *covariant differentiation* acting on tensor fields is an intrinsic generalization of directional differentiation acting on vector-valued functions, which is also designated a *connection*, a term introduced by H. Weyl (1918). In the case of vector fields, it is a bilinear map $\nabla : (\boldsymbol{\xi}, X) \in T_pM \times \mathfrak{X}(M) \mapsto \nabla_{\boldsymbol{\xi}} X \in T_pM$ satisfying

[169] This formula was first proposed by Euler for double integrals (1769), then generalized to triple ones by Lagrange (1773). Ostrogradski extended it to general multiple integrals (1836).

[170] Lie algebra ("infinitesimal group" in Lie's term) was independently invented by W. Killing in the 1880s with quite a different purpose.

$\nabla_{\xi}(fX) = \xi(f)X + f\nabla_{\xi}X$ $(a, b \in \mathbb{R},\ f \in C^{\infty}(M))$ (J.-L. Koszul). Define the functions $\left\{ {k \atop ij} \right\}$ by the relation $\nabla_{\partial_i}(\partial_j) = \sum_{k=1}^{d} \left\{ {k \atop ij} \right\} \partial_k$. Then

$$\nabla_{\xi}X = \sum_{i=1}^{d} \xi_i \Big(\frac{\partial f_k}{\partial x_i} + \sum_{j=1}^{d} \left\{ {k \atop ij} \right\} f_j \Big) \partial_k \quad (\xi = \sum_{i=1}^{d} \xi_i \partial_i,\ X = \sum_{j=1}^{d} f_j \partial_j).$$

The *Levi-Civita connection* is a unique connection satisfying

$$\nabla_X Y - \nabla_Y X = [X, Y],\ \ \xi\langle X, Y\rangle = \langle \nabla_{\xi}X, Y\rangle + \langle X, \nabla_{\xi}Y\rangle. \tag{6.20.22}$$

The first equality is equivalent to $\left\{ {k \atop ij} \right\} = \left\{ {k \atop ji} \right\}$ $(i, j, k = 1, \ldots, d)$, while the second one tells us that $\left\{ {k \atop ij} \right\}$ coincides with the Christoffel symbol (6.17.14).

Let $c(t) = (c_1(t), \ldots, c_d(t))$ $(a \leq t \leq b)$ be a smooth curve in M. If we write $f_i(t)$ for $f_i(c(t))$ for brevity, then

$$\nabla_{\dot{c}(t)}X = \sum_{i,k=1}^{d} \frac{dc_j}{dt} \Big(\frac{\partial f_k}{\partial x_i} + \sum_{j=1}^{d} \left\{ {k \atop ij} \right\} f_j \Big) \partial_k = \sum_{i,k=1}^{d} \Big(\frac{df_k}{dt} + \sum_{j=1}^{d} \left\{ {k \atop ij} \right\} \frac{dc_i}{dt} f_j \Big) \partial_k.$$

With this formula in mind, we define the covariant derivative of a vector field $X(t) = f_1(t)\partial_1|_{c(t)} + \cdots + f_d(t)\partial_d|_{c(t)}$ along the curve c by setting

$$\frac{D}{dt}X = \sum_{k=1}^{d} \Big(\frac{df_k}{dt} + \sum_{i,j=1}^{d} \left\{ {k \atop ij} \right\} \frac{dc_i}{dt} f_j \Big) \partial_k.$$

If $DX/dt \equiv 0$, then X is said to be *parallel*. Since $DX/dt \equiv 0$ is a system of linear equations of first order, there exists a parallel vector field X along c satisfying $X(c(a)) = \xi \in T_{c(a)}M$. The *parallel transport* $P_c : T_{c(a)}M \longrightarrow T_{c(b)}M$ is then defined by $P_c(\xi) = X(c(b))$; thus a connection literally yields a "bridge" between two tangent spaces, originally unrelated to each other. For the Levi-Civita connection, we have $\langle P_c(\xi), P_c(\eta)\rangle_{c(b)} = \langle \xi, \eta\rangle_{c(a)}$ in view of (6.20.22). Therefore P_c preserves the angle between two tangent vectors.

A geodesic is defined to be a curve c whose velocity vector $\dot{c}(t)$ is parallel along c, i.e., $\frac{D}{dt}\frac{dc}{dt} = 0$, which, in the case of Levi-Civita connection, turns out to be the E-L equation associated with the functional $E(c) = \int_a^b \|\dot{c}(t)\|^2 dt$.

Finally we shall give a modern formulation of curvature. A multilinear map $R(\cdot, \cdot)\cdot : T_pM \times T_pM \times T_pM \longrightarrow T_pM$ may be defined so as to satisfy

$$R(X, Y)Z = \nabla_X \nabla_Y Z - \nabla_Y \nabla_X Z - \nabla_{[X,Y]}Z \ \ (X, Y, Z \in \mathfrak{X}(M)).$$

Fig. 6.13 Parallel transport

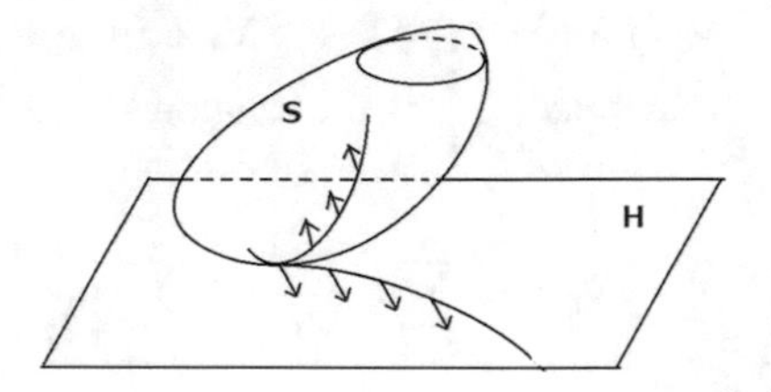

Putting $R(\omega, \boldsymbol{\xi}, \boldsymbol{\eta}, \boldsymbol{\zeta}) := \omega(R(\boldsymbol{\xi}, \boldsymbol{\eta})\boldsymbol{\zeta})$ ($\omega \in T_p^*M,\ \boldsymbol{\xi}, \boldsymbol{\eta}, \boldsymbol{\zeta} \in T_pM$), we obtain the tensor field R of type $(1, 3)$ such that $R(dx_l, \partial_i, \partial_j, \partial_k) = R^l_{kij}$. Moreover, taking the subspace H of T_pM spanned by linearly independent vectors $\boldsymbol{\xi}, \boldsymbol{\eta} \in T_pM$, we obtain the sectional curvature $K(H) = \langle R(\boldsymbol{\xi}, \boldsymbol{\eta})\boldsymbol{\eta}, \boldsymbol{\xi}\rangle / \{\|\boldsymbol{\xi}\|^2\|\boldsymbol{\eta}\|^2 - \langle \boldsymbol{\xi}, \boldsymbol{\eta}\rangle^2\}^{1/2}$. The *Ricci curvature* is the quadratic form $R(\boldsymbol{\xi}, \boldsymbol{\eta}) := \mathrm{tr}(\boldsymbol{\zeta} \mapsto R(\boldsymbol{\zeta}, \boldsymbol{\eta})\boldsymbol{\xi})$ on T_pM, to which the Ricci tensor is linked by $R_{ij} = R(\partial_i, \partial_j)$.

Remark 20.1

(1) Parallel transport on a surface S is related to classical parallelism. Let X be a vector field along a curve $c : [0, a] \to S$, and let H be the plane tangent to S at $c(0)$. Imagine that c and X are freshly painted and still wet. Roll S on H in such a way that $c(t)$ $(0 \leq t \leq a)$ is a point where the rolled surface at time t is tangent to H (*Cartan rolling*; Fig. 6.13). Then the curve c is transferred to a curve c' in H, $c(t)$ to a point $c'(t)$, and $X(t)$ to a vector $X'(t)$ at $c'(t)$. With a little nudge, we can show that X is parallel if and only if X' along c' is parallel in the classical sense.

Pushing further this idea, we may construct a one-to-one correspondence between smooth curves c in a manifold M with $c(0) = p$ and smooth curves c_0 in T_pM with $c_0(0) = \mathbf{0}$. Moreover this correspondence extends to the one for continuous curves by means of *stochastic integrals*, with which we may define *Brownian motion* on M,[171] the starting point of "stochastic differential geometry" initiated by K. Ito and P. Malliavin.

(2) The parallel transport around closed loops informs us how a manifold is curved. We shall see this by looking at the parallel transport $P_{\triangle ABC} : T_AS \longrightarrow T_AS$ along the perimeter $A \to B \to C \to A$ of a geodesic triangle $\triangle ABC$ on a surface S. It coincides with the rotation of vectors in T_AS through the angle $\theta_{\triangle ABC} := |\pi - (\angle A + \angle B + \angle C)|$ since the parallel transport preserves the angle of two tangent vectors. By virtue of Gauss's formula (6.13.8), we find $\theta_{\triangle ABC} = \left|\iint_{\triangle ABC} K_S\, d\sigma\right|$.[172]

[171] "Brownian motion" originally means the random motion of small particles suspended in fluids. It was named for the botanist R. Brown, the first to study such phenomena (1827). In 1905, Einstein made its statistical analysis, and observed that the probabilistic behavior of particles is described by the heat equation (Remark 17.2 (2)).

[172] The idea explained here is generalized in terms of *holonomy groups*, the concept introduced by É. Cartan (1926).

(3) The idea of connection makes good sense for a *vector bundle*, a family of vector spaces parameterized by a manifold which locally looks like a direct product, and are used to formulate *gauge theory* in modern physics that provides a unified framework to describe the fundamental forces of nature. □

6.21 Topology of the Universe

It cannot be completely denied that the universe may have a complicated structure. Hence it is irresistible to ponder the question "how can we describe the complexity of the universe?" In pursuing this, it will be appropriate to use topological terms as in the case of the question about finiteness.

In passing, the term "topology" used to describe a structure on an abstract set also indicates the branch of geometry which principally put a premium on the properties preserved under homeomorphisms. What, in addition, is important is the concept of *homotopy*, coined by Dehn and P. Heegaard in 1907, and employed by L. E. J. Brouwer with the current meaning of the word (1911).

A distinct advantage of topology compared with classical geometry is that we can employ lenient operations such as "gluing (pasting)" figures together. For instance, we obtain a sphere by gluing boundaries of two disks (Fig. 6.14).

Likewise one can glue boundaries (spheres) of two balls to get the 3-sphere $S^3 = \{(x, y, z, w) \in \mathbb{R}^4 |\ x^2 + y^2 + z^2 + w^2 = 1\}$ where one of the two balls is the mirror image of another. But this operation cannot be illustrated with drawings, so we shall proceed the other way around; that is, we start with the resulting figure S^3, and put $S^3_+ = \{(x, y, z, w) \in S^3 |\ w \geq 0\}$, $S^3_- = \{(x, y, z, w) \in S^3 |\ w \leq 0\}$. Clearly $S^3 = S^3_+ \cup S^3_-$ and $S^3_+ \cap S^3_- = \{(x, y, z) |\ x^2 + y^2 + z^2 = 1\} = S^2$. Furthermore, $(x, y, z, w) \mapsto (x, y, z)$ yields homeomorphisms of $S^3_\pm$ onto the ball $D = \{(x, y, z) \in \mathbb{R}^3 |\ x^2 + y^2 + z^2 \leq 1\}$. This signifies that S^3 is obtained by gluing two balls along their boundaries.

This construction of S^3 reminds us of an allegorical vision of Christian afterlife in Dante's poem *Divina Commedia*. In Canto XXVIII of *Paradiso* (line 4–9), he "constructs" the empyrean as a *mirror image* of the Aristotelian universe. Thus, if it were taken at face value (though not a little farfetched), Dante would have obtained S^3 as a model of the "universe" [27].

Apart from the Dante's fantastic imagination, the 3-sphere is surely the easiest conceivable model of a (homogeneous and isotropic) finite universe, if not possible

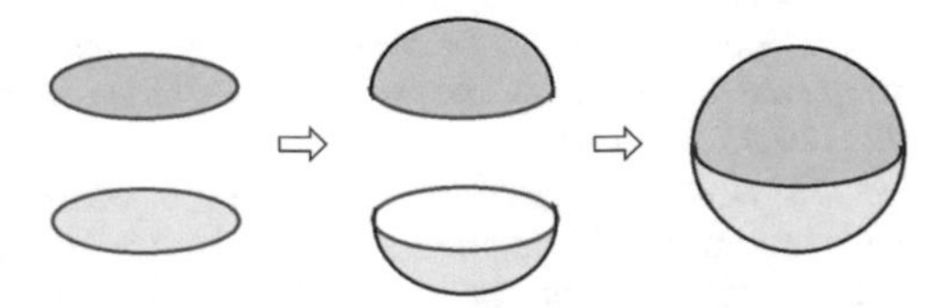

Fig. 6.14 From two disks to a sphere

to view it all at once like S^2. Einstein once said that the universe (at any instant of time) can be viewed globally as S^3 (see Sect. 6.18). He wrote, "Now this is the place where the reader's imagination boggles. 'Nobody can imagine this thing,' he cries indignantly. 'It can be said, but cannot be thought. I can imagine a spherical surface well enough, but nothing analogous to it in three dimensions.' We must try to surmount this barrier in mind, and the patient reader will see that it is by no means a particularly difficult task" [6].

J.-P. Luminet and his colleagues put forward an alternative model of a finite universe with constant positive curvature, which, called the *Poincaré homology sphere*, is the quotient of S^3 by the *binary icosahedral group*, a finite subgroup of the spin group Spin(3).[173] Their model—pertinent to the Poincaré conjecture on which we shall comment in due course—explains an apparent periodicity in the *cosmic microwave background*, electromagnetic radiation left over from an early stage of the universe in Big Bang cosmology.

Remark 21.1 The binary icosahedral group is the preimage of the icosahedral group under the homomorphism of Spin(3) onto SO(3) (Sect. 6.7), and is explicitly given as the union of 24 quaternions $\{\pm 1, \pm i, \pm j, \pm k, \frac{1}{2}(\pm 1 \pm i \pm j \pm k)\}$ with 96 quaternions obtained from $\frac{1}{2}(0 \pm i \pm \phi^{-1} j \pm \phi k)$ by an even permutation of all the four coordinates $0, 1, \phi^{-1}, \phi$, and with all possible sign combinations (ϕ being the golden ratio). □

Solely for the convenience of the reader, let us provide a shorthand historical account of topology. The etymology of "topology" is the German word "Topologie," coined by J. B. Listing, in his treatise "Vorstudien zur Topologie" (1847) as a synonym for the "geometry of position". He learned this discipline from Gauss, and launched the study of several geometric figures like screws, knots, and links from the topological perspective. It was Leibniz, if traced back to the provenance of topology, who offered a first rung on the ladder. In a letter to Huygens dated September 8, 1679, Leibniz communicated that he was not satisfied with the algebraic methods in geometry and felt for a different type of calculation leading to a new geometry to be called *analysis situs*. Leibniz himself did not put his plan into practice, but his intuition was not leading him astray. It was embodied by Euler in the solution of the celebrated problem "Seven Bridges of Königsberg" (1736), and also in his polyhedron formula $v - e + f = 2$ for a convex polyhedron with v vertices, e edges, and f faces, which, stated in a letter to his friend C. Goldbach (November 14, 1750), is recognized as giving a topological characteristic of the sphere (Sect. 6.13),[174] and contains some seeds of *combinatorial topology* pioneered by Poincaré that

[173] J.-P. Luminet, et al., *Dodecahedral space topology as an explanation for weak wide-angle temperature wide-angle temperature correlations in the cosmic microwave background*, Nature, **425** (2003), 593–595.

[174] A copy of Descartes' work around 1630, which was taken by Leibniz on one of his trips to Paris, reveals that he obtained an expression for the sum of the angles of all faces of a polyhedron, from which Euler's polyhedron formula can be deduced.

deals with geometric figures based on their decomposition into combinations of elementary ones.

Gauss took part in the formation of topology, though he opted not to publish any work on topology as usually happened with him. On January 22, 1833 (still in his prime), he noted down a summary of his consideration over the past few months,[175] which was to usher us into the theory of *knots* and *links*.

> Of the geometria situs, which was foreseen by Leibniz, and into which only a pair of geometers (Euler and Vandermonde were granted a bare glimpse, we know and have, after a century and a half, little more than nothing.
>
> A principal problem at the interface of geometria situs and geometria of magnitudinis will be to count the intertwining of two closed or endless curves.
>
> Let x, y, z be the coordinates of an undetermined point on the first curve; x', y', z' those of a point on the second and let
>
> $$\iint \frac{(x'-x)(dydz'-dzdy')+(y'-y)(dzdx'-dxdz')+(z'-z)(dxdy'-dydx')}{\left((x'-x)^2+(y'-y)^2+(z'-z)^2\right)^{3/2}} = V \tag{6.21.23}$$
>
> then this integral taken along both curves is $4m\pi$, m being the number of intertwinings [called the linking number today]. The value is reciprocal, i.e., it remains the same if the curves are interchanged.

His pretty observation has something to do with electromagnetism. Indeed, (6.21.23) reminds us of the *Biot–Savart law* (1820), an equation describing the magnetic field $\boldsymbol{B}$ generated by an electric current $\boldsymbol{i}$, which is named after J.-B. Biot and F. Savart. It is stated as

$$\boldsymbol{B}(\boldsymbol{x}) = \frac{\mu_0}{4\pi}\int_{\mathbb{R}^3} \frac{\boldsymbol{i}(\boldsymbol{y}) \times (\boldsymbol{x}-\boldsymbol{y})}{\|\boldsymbol{x}-\boldsymbol{y}\|^3} d\boldsymbol{y}, \tag{6.21.24}$$

where μ_0 is the magnetic permeability of vacuum (Footnote 189).

Another (more pedagogical) example in topology to which Gauss slightly contributed is the notion of *winding number*. It is defined analytically by

$$W(c, p_0) = \frac{1}{2\pi}\int_c \frac{-(y-y_0)dx + (x-x_0)dy}{(x-x_0)^2+(y-y_0)^2}, \tag{6.21.25}$$

where c is a closed (not necessarily simple) plane curve that does not pass through $p_0 = (x_0, y_0)$. This is an integer because, if we parameterize the curve c as $c(t) = (x_0 + r(t)\cos\theta(t), y_0 + r(t)\sin\theta(t))$ $(0 \le t \le 1)$ with continuous functions $r(t)$ and $\theta(t)$, then, without costing much effort, we see that the integral in (6.21.25) is transformed into $\frac{1}{2\pi}\int_0^1 d\theta(t) = \frac{1}{2\pi}[\theta(1) - \theta(0)]$, which, because $c(0) = c(1)$, represents the total (net) number of times that c travels around p_0.

From the analytic expression (6.21.25) and the fact that an integral-valued continuous function is constant, it follows that, if c_1 and c_2 are homotopic as maps of

[175] *Nachlass zur Electrodynamik* in Gauss *Werke*, V, translated by Ricca and Nipoti [29].

the circle S^1 into $\mathbb{R}^2\backslash\{p_0\}$, then $W(c_1, p_0) = W(c_2, p_0)$. Moreover, $W(c, p)$ does not depend on p so far as p is in the connected component D_0 of $\mathbb{R}^2\backslash c$ containing p_0; hence one may put $W(c, D_0) := W(c, p_0)$.

The homotopy invariance of winding numbers stands us in good stead in proving the fundamental theorem of algebra (cf. Gauss *Werke* III, 31–56). Given a polynomial $f(z) = z^n + a_1 z^{n-1} + \cdots + a_{n-1}z + a_n$ $(a_n \neq 0)$, one can take $R > 0$ such that $c_s(t) := f_s(Re^{2\pi\sqrt{-1}t}) \neq 0$ $(0 \leq s, t \leq 1)$, where $f_s(z) = z^n + s(a_1 z^{n-1} + \cdots + a_{n-1}z + a_n)$, so that $W(c_1, 0) = W(c_0, 0) = n(\neq 0)$. On the other hand, if $f(z) \neq 0$ for any $z \in \mathbb{C}$, then $c^r(t) := f(re^{2\pi\sqrt{-1}t}) \neq 0$ $(0 \leq r \leq R,\ 0 \leq t \leq 1)$, so that $W(c_1, 0) = W(c^R, 0) = W(c^0, 0) = 0$; thereby a contradiction, and hence $f(z) = 0$ must have a solution.

What deserves a good deal of attention is that the integral (6.21.25) is neatly expressed as the complex line integral: $\frac{1}{2\pi\sqrt{-1}} \int_c \frac{1}{z-z_0}\, dz$ $(z_0 = x_0 + y_0\sqrt{-1})$. Thus as Gauss briefly noticed in the letter to Bessel in 1811 (Sect. 6.7), the winding number reveals the nature of the complex logarithm (see Remark 8.1).

Winding numbers appeared in the study of "signed areas" by Albrecht Ludwig Friedrich Meister (1724–1788).[176] Here, the signed area *surrounded* by a general closed curve c is defined to be $\frac{1}{2}\int_c(-ydx + xdy)$. If c is simple and counterclockwise, this is the "genuine" area Area(D) of the domain D surrounded by c by virtue of Green's theorem (6.13.9). For a general c and the bounded connected components $D_1, \ldots, D_N$ of $\mathbb{R}^2\backslash c$, we have

$$\frac{1}{2}\int_c (-ydx + xdy) = \sum_{k=1}^{N} W(c, D_k)\mathrm{Area}(D_k).$$

In fact, this is a consequence of the formula $\int_c \omega = \sum_{k=1}^N W(c, D_k) \int_{\partial D_k} \omega$ holding for any 1-form ω on $\mathbb{R}^2$, where the boundary ∂D_k is assumed to have counterclockwise orientation. This suggests a formal framework allowing us to write "$c = \sum_{k=1}^N W(c, D_k)\partial D_k$". Indeed, this identity can be justified by the notion of "chain" in *homology theory* (Remark 21.2 (2)).

Moving away from the small treasure trove of Gauss's "toys", we shall see how topology evolved thereafter.

The period from the mid-nineteenth century to the early twentieth century was the infancy of topology. It coincides with the era when mathematics began to develop autonomously on a rigid base. Prompted by such a state of affairs, geometers created a wealth of novel ideas. One of the marked achievements in this period is the discovery of *one-sided* surfaces that differ from ordinary surfaces in the way of their disposition in the space. Remember that a smooth surface S is two-sided if and only if it admits a global continuous (unit) normal vector field, or equivalently S is orientable. (This equivalence is deduced from the fact that, if

[176] *Generalia de genesifigurarum planarum, et inde pendentibus earum affectionibus*, Novi Commentarii Soc. Reg. Scient. Gott., **1** (1769/1770), 144–180.

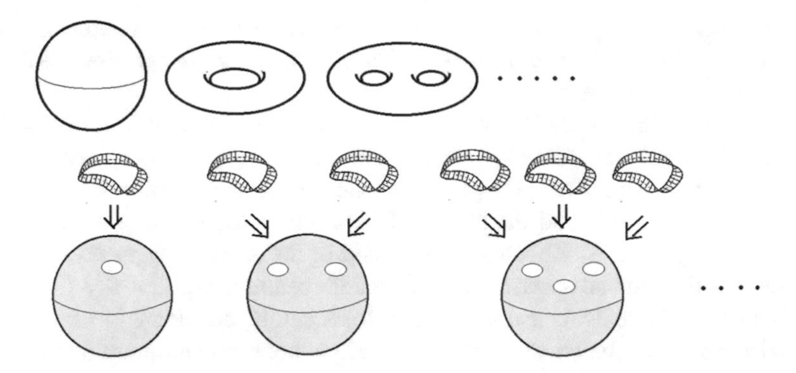

Fig. 6.15 Classification of closed surfaces

$S_1 : U_1 \longrightarrow S$ and $S_2 : U_2 \longrightarrow S$ are two local parametric representations of a surface S with $S_1(U_1) \cap S_2(U_2) \neq \emptyset$, then $(S_1)_u \times (S_1)_v = \frac{\partial(z,w)}{\partial(u,v)}(S_2)_z \times (S_2)_w$, where ${S_2}^{-1} \circ S_1(u, v) = (z(u, v), w(u, v))$ on $U_1 \cap {S_1}^{-1}(S_2(U_2))$.) The foremost examples of one-side surfaces are the *Möbius band* and the *Klein bottle*. The former was discovered by Möbius (1865) and Listing (1862). The latter was constructed by Klein (1882). The projective plane $P^2(\mathbb{R})$ is another example.

The classification of closed surfaces is the problem attacked in the late nineteenth century. The case of two-sided surfaces was independently treated by Möbius (1863) and M. E. C. Jordan (1866). One-sided closed surfaces were classified by von Dyck (1888). As depicted in Fig. 6.15, every closed one-sided surface can be constructed from S^2 by attaching a finite number of Möbius bands (the leftmost is the projective plane, and the next is the Klein bottle). Thus the "shapes" of 2D models of the *finite* universe are completely comprehended.

Now, one may ask, "what about the higher-dimensional case?" If restricted to 3- or 4-manifolds, this is a problem concerning a possible structure of our universe, and hence has a much higher profile. More specifically, one may pose the question: Does our universe have a structure like a surface with many holes? Here, the meaning of "hole" is well elucidated in terms of *algebraic topology* which borrows tools from abstract algebra crystallized in the early twentieth century.

Riemann had a clear perception of what we now call the *Betti numbers*, significant topological invariants related to the Euler characteristic and defined today as the rank of the *homology group* (see the remark below). His grandiose program—showing again that he was far ahead of the times—was published posthumously as the *Fragment aus der Analysis Situs* [31]. Later, Enrico Betti (1823–1892), an Italian friend of Riemann, made clearer what Riemann thought.[177] Finally, their idea came to fruition as homology theory by Poincaré (1899), who also introduced

[177] *Sopra gli spazi di un numero qualunque di dimensioni*, Ann. Mat. Pura Appl., **4** (1871), 140–158. A letter dated October 6, 1863 from Betti to his colleague tells us that he got an accurate conception about the connectivity of spaces through a conversation with Riemann.

the *fundamental group*, another algebraic tool and a modern version of "Greek geometric algebra" in a sense because the algebraic system in question are directly constructed from geometric figures.

In the meantime, G. de Rham related homology to differential forms, proving what we now call *de Rham's theorem*, one of the earliest outcomes in global analysis (1931). This line—coupled with Chern's generalization of the Gauss-Bonnet formula (1944) and the theory of *harmonic integrals* developed by Weyl, W. V. D. Hodge and K. Kodaira—culminated in the *Atiyah-Singer theory*, one of the most exhilarating adventures of the twentieth century. Additionally, the work of M. Gromov on large-scale aspects of manifolds can be considered as far-reaching generalizations of a batch of results on relations between curvature and topology obtained by geometers on and after the 1950s.

The success of the taxonomy of closed surfaces impelled topologists to attack the case of closed 3-manifolds, but the matters turned out to be more complicated than expected. In 1982, W. P. Thurston offered the *geometrization conjecture* as an initial template, which daringly says that all 3-manifolds admit a certain kind of decomposition involving the *eight geometries* (one of them is hyperbolic geometry). This spectacular supposition was proved by G. Perelman in 2003. His arguments include the proof of the long-standing *Poincaré conjecture* which claims, "a simply connected closed 3-manifold is homeomorphic to the 3-sphere."[178] A conspicuity of his proof is that a non-linear evolutional equation involving Ricci curvature is exploited in an ingenious way; thereby displaying a miraculous trinity of differential geometry, topology, and analysis.

The situation gets out of hand when trying to classify closed manifolds M with $\dim M \geq 4$. This is because an arbitrary *finitely presented group* (a group defined by a finite number of generators, and a finite number of defining relations) can be the fundamental group of a closed 4-manifold, and there is no algorithm to decide whether two finitely presented groups are isomorphic.[179]

We close this section with an instructive example in *differential topology* which has been intensively studied in the latter half of the last century.

We shall say that a smooth curve c in $\mathbb{R}^2$ is *regular* if $\dot{c}(t) \neq \mathbf{0}$ for every t. Given a closed regular curve c, we define the *rotation number* $R(c)$ by setting $R(c) = W(\dot{c}, \mathbf{0})$. In plain language, $R(c)$ is the total number of times that a person walking once around the curve turns counterclockwise.[180] This notion came up in Gauss's

[178]Poincaré, *Cinquième complément à l'analysis situs*, Rend. Circ. Mat. Palermo, **18** (1904), 45–110. He claimed, at first, that homology is sufficient to tell if a closed 3-manifold is homeomorphic to S^3 (1900), but 4 years later, he found that what we now call the *Poincaré homology sphere* gives a counterexample.

[179]A. Turing and A. Church rigorously mathematized the concept of algorithm by analyzing the meaning of computation. They independently treated the *decision problem* challenged by Hilbert in 1928, and showed that a general solution is impossible (1936).

[180]As known, at least informally, by Meister (*Generalia*, 1770), the rotation number is defined for a closed oriented polygonal curve. It is the sum of all (signed) exterior angles divided by 2π (the right of Fig. 6.16). In the *Geometria Speculativa* (ca. 1320), Bradwardine (Sect. 6.2) studied a class

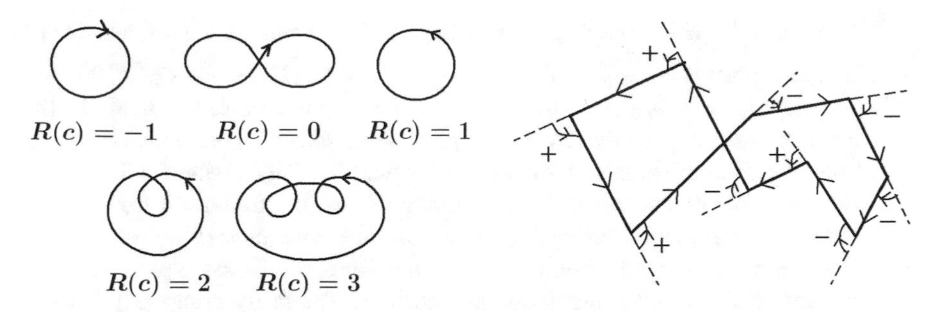

Fig. 6.16 Rotation number and exterior angles

work of experimental nature; he observed, among others (*Werke*, VIII, 271–281), that, if we denote by $n(c)$ the number of self-intersection points of c, then we have the sharp inequality: $n(c) \geq ||R(c)| - 1|$.

In 1937, H. Whitney proved that a closed regular curve can be deformed to another one through a "smooth" family of closed regular curves (this being the case, two curves are said to be *regularly homotopic*) if and only if two curves have the same rotation number. In particular, a clockwise oriented circle ($R = -1$) is not regularly homotopic to a counterclockwise oriented circle ($R = 1$). This fact is rephrased as "one cannot turn a circle *inside-out*," because giving an orientation to a closed regular curve c is equivalent to choosing one of two normal unit vector fields along c, and the inside and outside of a circle correspond to the inner normal and outer normal vector fields along it, respectively. Of particular note is the unexpected result by S. Smale that one can turn the sphere S^2 inside-out, quite by contrast to the case of the circle (1959).

Remark 21.2

(1) The rotation number is linked with the signed curvature (Remark 7.1) via the formula $\int_c \kappa(c(s))ds = 2\pi R(c)$. Indeed,

$$2\pi R(c) = \int_c \frac{\ddot{y}(s)\dot{x}(s) - \ddot{x}(s)\dot{y}(s)}{\dot{x}(s)^2 + \dot{y}(s)^2} ds = \int_c [\ddot{y}(s)\dot{x}(s) - \ddot{x}(s)\dot{y}(s)]ds = \int_c \kappa(c(s))ds.$$

We thus have a fusion of topology and differential geometry at a very basic level.

(2) In the *Analysis situs*, Poincaré defined "homology classes" in a somewhat fuzzy way. The group structure on homology classes, which he did not explicitly indicate, was studied by E. Noether and others in the period 1925–1928. Later on, S. Eilenberg developed the *singular homology theory* that allows to define the homology group for a general topological space (1944).

---

of general polygonal curves, which seems to have a relevance with rotation number (Jeff Erickson, *Generic and Regular Curves*, PDF, 2013).

A central role in homology theory is played by a *chain complex*, a series of homomorphisms $\cdots \xrightarrow{\partial_{k+1}} C_k(X, \mathbb{Z}) \xrightarrow{\partial_k} \cdots \xrightarrow{\partial_2} C_1(X.\mathbb{Z}) \xrightarrow{\partial_1} C_0(X, \mathbb{Z}) \xrightarrow{\partial_0} 0$ satisfying $\partial_k \circ \partial_{k+1} = 0$, where, in the singular case, $C_k(X, \mathbb{Z})$ is the free abelian group generated by "singular k-simplices" (continuous maps σ_k from the k-simplex into a topological space X). The k-simplex here is a generalization of a segment ($k=1$), a triangle ($k=2$), and a tetrahedron ($k=3$). The homomorphism ∂_k brings σ_k to the sum of the singular $(k-1)$-simplices represented by the restriction of σ_k to the faces of the k-simplex, with an alternating sign to take orientation into account. The factor group $H_k(X, \mathbb{Z}) = \operatorname{Ker} \partial_k / \operatorname{Image} \partial_{k+1}$ is what we call the k-th homology group of X. The k-th Betti number $b_k(X)$ is the rank of $H_k(X, \mathbb{Z})$ (if finite). When $b_k(X) = 0$ for $k > d$, the Euler characteristic of X is defined to be $\chi(X) = \sum_{k=0}^{d} (-1)^k b_k(X)$, which, for a surface, agrees with the definition relying on a triangulation (Sect. 6.13).

(3) For a k-form ω on a convex domain $D \subset \mathbb{R}^d$ satisfying $d\omega = 0$, there is a $(k-1)$-form η with $d\eta = \omega$; that is, $H^k_{\mathrm{dR}}(D) = \{0\}$. This (the *Poincaré lemma*),[181] in tandem with the generalized Stokes' formula, is crucial in the proof of de Rham's theorem asserting that $b_k(M) = \dim H^k_{\mathrm{dR}}(M)$ for a closed manifold M.

(4) Define the d-form Ω on a closed even-dimensional orientable manifold M by

$$\Omega = \frac{(-1)^n}{2^{2n}\pi^n n!} \sum_{\sigma=(i_1,\ldots,i_d)} \operatorname{sgn}(\sigma) g^{i_1 i_2} g^{i_3 i_4} \cdots g^{i_{2n-1} i_{2n}} \Omega_{i_1 i_2} \wedge \Omega_{i_3 i_4} \wedge \cdots \wedge \Omega_{i_{2n-1} i_{2n}},$$

$$\Omega_{ij} = \sum_{m,k,l=1}^{d} g_{im} R^m_{jkl} dx_k \wedge dx_l \quad (\dim M = d = 2n).$$

Then Chern's generalization of the Gauss-Bonnet formula is expressed as

$$\chi(M) = \int_M \Omega.$$

(5) Hodge's theorem says that the space of *harmonic forms* $\mathcal{H}^k(M) = \{\omega \in A^k(M) | d_k\omega = d^*_{k-1}\omega = 0\}$ is isomorphic to $H^k_{\mathrm{dR}}(M)$, where d^*_{k-1} is the adjoint of d_{k-1} (with respect to natural inner products on $A^*(M)$). Using this fact, we obtain

$$\dim \operatorname{Ker} D - \dim \operatorname{Coker} D = \chi(M), \tag{6.21.26}$$

[181] Poincaré, *Les Méthodes nouvelles da la Mécanique céleste*, vol. 3, Gauthier-Villars, Paris, 1899, pp. 9–15. This fact, which, in the case of 1-forms, dates back to Euler (1724/1725) and Clairaut, is essentially ascribed to Vito Volterra (1860–1940).

where $D = d + d^*$, an operator from the space of even forms to the space of odd forms. A special feature of D is that it is *elliptic*; i.e., its *principal symbol* is invertible.[182]

The left-hand side of (6.21.26) is called the *analytic index* of D. The Atiyah-Singer theorem asserts that the analytic index ($\text{Ind}\, P$) of an elliptic operator P equals the *topological index* defined in terms of topological data, and that there is a d-form Ω involving both curvature tensor and principal symbol of P such that $\text{Ind}\, P = \int_M \Omega$. □

6.22 Right and Left in the Universe

An intriguing question about our universe is whether it has "one side" or "two sides." In discussing sidedness, it seems necessary, if looking back to the case of surfaces (Sect. 6.21), to assume the existence of "outer side" of the universe. The fact is that it is genuinely intrinsic in character as in the case of finiteness. Why is it so? This time, we bring in the idea of "right and left," the terms used not only in everyday life, but also in mathematics, which are usually designated by the human hands. No less obvious perhaps to our eyes, but no less essential, is the recognition that the human hands are represented by *frames* (ordered basis), the notion obtained by whittling away the extraneous details of hands; see Fig. 6.17 illustrating the *right-handed frame* $(\boldsymbol{a}, \boldsymbol{b}, \boldsymbol{c})$.[183]

To clarify the nature of right and left, we shall refer to the notorious *mirror paradox* that is best stated as a question: Why do flat mirrors reverse left and right, but not top and bottom? *Prima facie*, this sounds puzzling, but the mirror paradox is simply a paradox of "a red herring" or a word play to confuse or startle people by making them believe that the wording "left and right" is the same sort of "top and bottom." Indeed, the latter is not described by a frame, but by a single vector; say, the vector represented by the directed segment joining bottom and top.[184]

Now, the claim "the universe has two sides" is rephrased as "the right-handed frame and the left-handed frame are distinct wherever they are" in the sense that these two frames cannot be superposed no matter how one is moved to another in the universe. Otherwise expressed, if the universe has "only one side," then the

[182] A differential operator P of order m on M (acting on vector-bundle valued functions) is locally expressed as $P = \sum_{|\alpha| \leq m} A_\alpha(\boldsymbol{x}) D^\alpha$, where $\alpha = (\alpha_1, \ldots, \alpha_d)$ denotes a d-tuple of non-negative integers, $|\alpha| = \alpha_1 + \cdots + \alpha_d$, $D^\alpha = (\partial^{\alpha_1}/\partial x_1^{\alpha_1}) \cdots (\partial^{\alpha_d}/\partial x_d^{\alpha_d})$, and $\{A_\alpha(\boldsymbol{x})\}$ are matrix-valued functions. The principal symbol of P is defined to be $\sigma_P(\boldsymbol{x}, \boldsymbol{\xi}) := \sum_{|\alpha|=m} A_\alpha(\boldsymbol{x}) \xi_1^{\alpha_1} \cdots \xi_d^{\alpha_d}$ ($\boldsymbol{\xi} \neq \boldsymbol{0} \in T^*M$).

[183] The fact that there are just two kinds of frames in $\mathbb{R}^d$ is equivalent to that the general linear group $\text{GL}_d(\mathbb{R})$ has two connected components.

[184] Shin-itiro Tomonaga argues that this paradox cannot be resolved by neither geometric optics nor mathematics (1963), while Martin Gardner says that the mirror does not reverse right and left, but does reverse front and back [10].

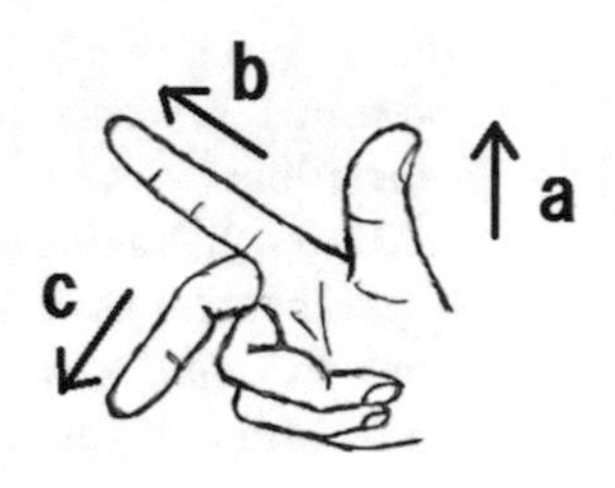

Fig. 6.17 The frame corresponding to the gun-shaped right hand

distinction between the right-handed frame and the left-handed frame does not make sense; namely there is only one kind of frame. What is more, as in the case of surfaces, two-sidedness and orientability of the universe are equivalent when we regard the universe as a smooth manifold. Thus the distinguishability between right and left depends on the global intrinsic structure of the universe.

Even if our universe is Euclidean, there remains a subtle problem on "right and left" that dates back to the dispute between Newton and Leibniz and have fascinated philosophers and cosmologists since then [12]. Newton insisted that there is a 'left' and 'right' in the universe, while Leibniz opposed this view and argued that left and right are in no way different from each other.[185] Puzzled by *enantiomorphs* for decades, Kant mused about this issue in connection with Leibniz's claim that all spatial facts should be reduced to facts about relative distances between material bodies (*Prolegomena*, 1738). He surmised in his prolix discussion that the difference between similar but not congruent things cannot be made intelligible by understanding about any concept, and are known only through sensuous intuition. He even set up an extreme thought-experiment to adduce conclusive evidence of his assertion.[186] According to him, this suggests, after all, that space is not independent of the mind that perceives it.

The purport of what Newton, Leibniz, and Kant thought up is clarified if we replace Kant's claim by the statement, "one cannot furnish any mathematical characterization of right-handedness of a frame." This might perplex the reader because, in several places of this essay, we used *vector product*, whose (casual) definition relies on the right-handed frame; thus seemingly incomplete as a mathematical concept. As a matter of fact, we may select any frame as a reference frame; if we choose another kind of frame, then the resulting vector has the opposite direction, and there is no trouble caused at all.

[185]Leibniz's critique of Newton is unfolded in his third letter to S. Clarke (Footnote 54).

[186]*Von dem ersten Grunde des Unterschieds der Gegenden im Raume*, 1768. In it, he considers a marble hand broken off a statue that is supposed to be the only object in the universe, and asks whether it makes sense to say that it is still either a right hand or a left.

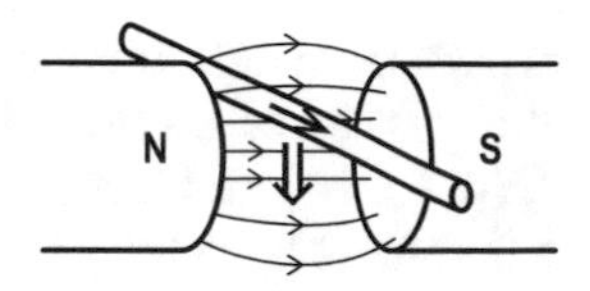

Fig. 6.18 Fleming's left-hand rule

How about the issue of right and left in physics? Ernst Mach (1838–1916)[187] cogently argued that only *asymmetric laws* can distinguish left and right (1886). Here, a law is said to be asymmetric (or *to violate parity*) if it is not invariant under transformations interchanging left and right (specifically, under the *parity transformation* $(x, y, z) \mapsto (-x, -y, -z)$). In this connection, the neophyte of physics might recollect *Fleming's left-hand rule* about the direction of force acting on a current-flowing wire under an external magnetic field (Fig. 6.18). As the name suggests, the rule is described by using human hands, and hence seem to imply that the electromagnetic phenomenon is asymmetric. But that is not actually the case because of the nature of magnetic fields. In order to put it clearly, let us recall the *Lorentz force* $e\,[\boldsymbol{E} + (\boldsymbol{v} \times \boldsymbol{B})]$ experienced by a particle of charge e moving with velocity $\boldsymbol{v}$ in the presence of an electric field $\boldsymbol{E}$ and a magnetic field $\boldsymbol{B}$. Here $\boldsymbol{B}$ is not an ordinary (*polar*) vector, but an *axial* vector gaining an additional sign flip under a reflection.[188]

The nature of magnetic fields is better explained by identifying $\boldsymbol{B} = (B_1, B_2, B_3)$ with the differential 2-form $\omega = B_3 dx_1 \wedge dx_2 + B_1 dx_2 \wedge dx_3 + B_2 dx_3 \wedge dx_1$ since ω is "handedness-invariant." This identification is not altogether artificial, though, at first sight, differential forms might look an elaborate mathematical fabrication, having nothing to do with the real world. To convince ourselves of this, let us employ differential forms to express Maxwell's equations.

The original form of Maxwell's equations in a vacuum is:

$$\begin{cases} \epsilon_0 \operatorname{div} \boldsymbol{E}(t, \mathbf{x}) = \rho(t, \boldsymbol{x}) \ \text{(Gauss's flux law)}, \\ \mu_0^{-1} \operatorname{rot} \boldsymbol{B}(t, \mathbf{x}) - \epsilon_0 \dfrac{\partial \boldsymbol{E}}{\partial t} = \boldsymbol{i}(t, \boldsymbol{x}) \ \text{(Ampère-Maxwell law)}, \\ \operatorname{div} \boldsymbol{B}(t, \boldsymbol{x}) = 0 \ \text{(Absence of magnetic monopoles)}, \\ \operatorname{rot} \boldsymbol{E}(t, \boldsymbol{x}) + \dfrac{\partial \boldsymbol{B}}{\partial t} = 0 \ \text{(Faraday's law)}, \end{cases} \tag{6.22.27}$$

[187]Mach is best-known for the sensation-based theory of reality. In *Die Mechanik in ihre Entwicklung historisch-kritisch dargestellt* (1883), he criticized Newton's conclusion on absolute motion based on the bucket experiment (Footnote 52), saying that centrifugal force may act on water in the stationary bucket if the universe rotates. According to him, the inertia and acceleration of a body are determined by all of the matter of the universe. This view was called the *Mach principle* by Einstein (1918).

[188]The distinction between "polar" and "axial" vectors was made by W. Voigt in 1896. Note that $polar \times polar = axial$ and $polar \times axial = polar$.

where ϵ_0 stands for the dielectric constant of a vacuum,[189] and ρ (resp. $\boldsymbol{i}$) is the density of electric charge (resp. the electric current density). Writing $\boldsymbol{E} = (E_1, E_2, E_3)$ and lumping together the magnetic field and the electric field, define a 2-form Ω by setting $\Omega = B_3 dx_1 \wedge dx_2 + B_1 dx_2 \wedge dx_3 + B_2 dx_3 \wedge dx_1 + E_1 dx_1 \wedge dt + E_2 dx_2 \wedge dt + E_3 dx_3 \wedge dt$. Writing $\boldsymbol{i} = (i_1, i_2, i_3)$ likewise, we put $\eta = \mu_0(i_1 dx_1 + i_2 dx_2 + i_3 dx_3) - \epsilon_0^{-1}\rho dt$. Then (6.22.27) are rewritten as $d\Omega = 0,\ d^*\Omega = \eta$, which is manifestly invariant under Lorentz transformations. Here d^* is the *adjoint* of the exterior derivation d with respect to the Minkowski metric; specifically,

$$d^*\omega = \Big(\frac{\partial f_3}{\partial x_2} - \frac{\partial f_2}{\partial x_3} - c^{-2}\frac{\partial g_1}{\partial t}\Big)dx_1 + \Big(\frac{\partial f_1}{\partial x_3} - \frac{\partial f_3}{\partial x_1} - c^{-2}\frac{\partial g_2}{\partial t}\Big)dx_2$$
$$+ \Big(\frac{\partial f_2}{\partial x_1} - \frac{\partial f_1}{\partial x_2} - c^{-2}\frac{\partial g_3}{\partial t}\Big)dx_3 - \Big(\frac{\partial g_1}{\partial x_1} + \frac{\partial g_2}{\partial x_2} + \frac{\partial g_3}{\partial x_3}\Big)dt$$

for $\omega = f_3 dx_1 \wedge dx_2 + f_1 dx_2 \wedge dx_3 + f_2 dx_3 \wedge dx_1 + g_1 dx_1 \wedge dt + g_2 dx_2 \wedge dt + g_3 dx_3 \wedge dt$.

It was unthinkable that anyone should question the validity of symmetry under a mirror reflection until a genuine asymmetric law was predicted by T.-D. Lee and C.-N. Yang in 1956 as the parity violation of the *weak interaction* (a force that governs all matter in the universe). Right after their announcement, the mind-boggling prediction was confirmed by C.-S. Wu and her collaborators through an experiment monitoring the beta decay of cobalt-60 atoms. Expressed in words, this implies that the nature at a very fundamental level can tell the characteristic difference between left- and right-handed (see [10]).

In addition to the weak interaction, there are the three other fundamental interactions in nature; say, the electromagnetic force, the strong interaction, and gravitation. The strong interaction is, loosely put, the mechanism responsible for the strong nuclear force and is known to be "P-symmetric", i.e., invariant under the parity transformation like the electromagnetic force. In this connection, a few more words about symmetry in quantum physics will not be out of place; besides the P-symmetry, we have the "C-symmetry" under the *charge conjugation* that reverses the electric charge and all the internal quantum numbers, and the "T-symmetry" under the *time reversal* replacing t by $-t$. Each of these symmetries can be violated individually. Theoretically, however, there exists no physical phenomenon that violates the "CPT-symmetry," the combination of all three symmetries. This fact is called the "CPT theorem."

---

[189] $\epsilon_0 = 8.854187817\ldots \times 10^{-12}$ F/m, $\mu_0 = 1.2566370614\cdots \times 10^{-6}$ H/m. Strictly speaking, these equations are Heaviside's version (1888/1889). Maxwell's original equations (consisting of 20 equations in 20 variables) are quite complicated. It follows from (6.22.27) that both $\boldsymbol{E}$ and $\boldsymbol{B}$ satisfy the wave equation $\mu_0\epsilon_0 \partial^2 \boldsymbol{f}/\partial t^2 - \Delta \boldsymbol{f} = \boldsymbol{0}$ under the absence of ρ and $\boldsymbol{i}$. Hence electromagnetic waves propagate at the speed $1/\sqrt{\mu_0\epsilon_0} = 0.299792458 \times 10^8$ m/s, which coincides with the speed of light. Maxwell thus predicted that light is an electromagnetic wave (1865). This was confirmed experimentally by H. Hertz in 1887.

6.23 Conclusion

In retrospect, the theorem of angle-sum and the Pythagorean Theorem in classical antiquity were the fresh impetus in the far-reaching feats by Gauss and Riemann. The triangle inequality as well was a fundamental source of the theory of topological spaces, with which we can get a better understanding of intrinsicness of curved spaces. Moreover, the effort to refine the ancient approach to "infinitesimals" came to fruition as calculus by Newton and Leibniz, which, combined with the idea of coordinate system (no little indebted to Descartes' method), provided us with a powerful tool to investigate our space. What we ought to remember in particular is Cantor's set theory that not only encompasses all the necessary stuff for mathematizing cosmology, but also spurred us on to give a probing interpretation to infinity whose nature had been a perennial controversial issue passed from antiquity.

Sullivan [36] says felicitously in a general context, "A history of mathematics is largely a history of discoveries which no longer exist as separate items, but are merged into some more modern generalization, these discoveries have not been forgotten or made valueless. They are not dead, but transmuted."

References

1. L. M. Blumenthal, *A Modern View of Geometry*, Dover, 1980.
2. R. Bonola, *Non-Euclidean Geometry: A Critical and Historical Study of its Development*, Dover, 2010.
3. M. Bossi and S. Poggi, ed. *Romanticism in Science: Science in Europe, 1790–1840*, Springer; 1993 edition.
4. H. S. M. Coxeter, Gauss as a geometer, *Historia Mathematica*, **4** (1977), 279–396.
5. Descartes, *The Geometry of Rene Descartes with a facsimile of the first edition*, translated by D. E. Smith and M. L. Latham, Dover, 1954.
6. A. Einstein, *Ideas and Opinions*, Broadway Books, 1995.
7. Euclid, *The Thirteen Books of Euclid's Elements*, translated by T. Heath, Dover, 1956 (1908).
8. L. Euler, *Foundations of Differential Calculus*, Springer, 2000.
9. R. Farwell and C. Knee, The missing link: Riemann's "Commentatio," differential geometry and tensor analysis, *Historia Math.* **17** (1990), 223–255.
10. M. Gardner, *The New Ambidextrous Universe: Symmetry and Asymmetry from Mirror Reflections to Superstrings*, Third Revised Edition, Dover, 2005.
11. L. Giaccardi, Scientific research and teaching problems in Beltrami's letters to Hoüel, in *"Using History to Teach Mathematics"*, edited by V. J. Katz, The Mathematical Association of America, (2000), 213–223.
12. C. G. Gross and M. H. Bornstein, *Left and right in science and art*, Leonard Vol. II (1978), 29–38.
13. H. N. Janke, Algebraic Analysis in the 18th Century, in *"A History of Analysis"* edited by Janke, London Math. Soc., (2003), 105–136.
14. V. J. Katz, *A History of Mathematics: An Introduction*, Addison-Wesley, 2009.
15. H. Koch, *Introduction to Classical Mathematics I: From the quadratic reciprocity law to the uniformization theorem*, Springer, 1991.
16. A. Koyré, *From the Closed World to the Infinite Universe*, The John Hopkins Press, 1957.

17. N. I. Lobachevsky, *Pangeometry*, edited and translated by A. Papadopoulos, European Math. Soc., 2010.
18. H. Meschkowski, Aus den Briefbüchern Georg Cantors, *Arch. Hist. Exact Sci.*, **2** (1965), 503–519.
19. M. Monastyrsky, *Riemann, Topology and Physics*, Boston-Basel, 1987.
20. S. Negrepontis, Plato on geometry and the geometers, in *Geometry in history* (S. G. Dani, A. Papadopoulos, ed.), Springer Verlag, 2018
21. R. Netz, K. Saito, and N. Tchernettska, A new reading of Method Proposition 14, preliminary evidence from the Archimedes Palimpsest (Part 1 and Part 2), *SCIAMVS*, **2**, (2001), 109–129.
22. O. Neugebauer, Zur geometrischen Algebra (Studien zur Geschichte der ntiken Algebra III), *Quellen und Studien zur Geschichte der Mathematik, Astronomie und Physik*, Abteilung, B. Bd. III (1936), 245–259.
23. K. Ohshika, The origin of the notion of manifold: from Riemann's Habilitationsvortrag onward, in *"From Riemann to Differential Geometry and Relativity"* (ed. L. Ji, A. Papadopoulos, S. Yamada), Springer, 2017, 295–309.
24. A. Papadopoulos, On Lobachevsky's trigonometric formulae, *Gaṇita Bhārātī (Bull. Indian Soc. Hist. Math.)*, **34** (2012), No.1–2, 203–224.
25. A. Papadopoulos, On the works of Euler and his followers on spherical geometry, *Gaṇita Bhārātī (Bull. Indian Soc. Hist. Math.)*, **36** (2014) No. 1–2, 237–292.
26. A. Papadopoulos and G. Théret, *La théorie des lignes parallèles de Johann Heinrich Lambert*, Sciences dans l'histoire, 2014.
27. M. Peterson, Dante and the 3-sphere, *Amer. J. Phys.*, **47** (1979), 1031.
28. R. Rashed and A. Papadopoulos, *Menelaus' 'Spherics': Early Translation and al-Māhānī/ al-Harawī's Version*, Walter de Gruyter 2017.
29. R. L. Ricca and B. Nipoti, Gauss' linking number revisited, *J. of Knot Theory and Its Ramifications*, **20** (2011), 1325–1343.
30. J. P. Richter, *The Literary Works of Leonardo da Vinci*, 1979.
31. B. Riemann, *Gesammelte Mathematische Werke und Wissenschaftlicher Nachlass*, edited with the support of R. Dedekind and H. Weber, Leipzig: Teubner (1876); 3rd ed. by R. Narasimhan, Springer (1990).
32. Saccheri, *Euclid Vindicated from Every Blemish*, edited and annotated by Vincenzo De Risi, translated by G. B. Halsted and L. Allegri, Birkhäser, 2014.
33. E. Scholz, The concept of manifold, 1850–1950, in *History of Topology*, edited by I. M. James, North-Holland, 1999, 25–64.
34. M. Spivak, *A Comprehensive Introduction to Differential Geometry*, Volume 2, Publish or Perish, 1970.
35. J. Stillwell, Sources of Hyperbolic Geometry, *Amer. Math. Soc.*, 1996.
36. J. W. N. Sullivan, *The History of Mathematics in Europe from the Fall of Greek Science to the Rise of the conception of Mathematical Rigour*, Oxford University Press, 1924.
37. T. Sunada, The Story of Area and Volume from Everyday Notions to Mathematical Concepts, Chap. 2 in *A Mathematical Gift, III: The Interplay Between Topology, Functions, Geometry, and Algebra*, translated by Eiko Nakayama Tyler, Amer. Math. Soc., 2005.
38. T. Sunada, *Topological Crystallography —With a View Towards Discrete Geometric Analysis—*, Springer, 2013.

Chapter 7
A Path in History, from Curvature to Convexity

Annette A'Campo-Neuen and Athanase Papadopoulos

> *God decided that quantity should exist before all other things so that there should be a mean for comparing a curved with a straight line.*
> (Johannes Kepler, *Mysterium cosmographicum*, 1596, [75, p. 93])

Abstract We describe a path in the history of curvature, starting from Greek antiquity, in the works of Euclid, Apollonius, Archimedes and a few others, passing through the works of Huygens, Euler, and Monge and his students, and ending in the twentieth century at the works of Bonnesen, Fenchel, Busemann, Feller and Alexandrov. Our goal is not to review the whole history of curvature, but to show how the approaches to curves, surfaces and curvature evolved from the synthetic point of view of the Greeks to the methods of analytic geometry founded by Fermat, Descartes, Newton and Leibniz, and eventually, in the twentieth century, experienced a return to the synthetic methods of the Greeks.

AMS Classification: 01-02, 01A20, 34-02, 34-03, 54-03, 92B99

7.1 Introduction

Our aim in this essay is to trace out a path between the early study of curves in Greek antiquity and the modern notion of curvature applied to convex (non-necessarily differentiable) surfaces that appears in the writings of Busemann–Feller and A.D. Alexandrov. The path traverses the work of Christiaan Huygens, which is

A. A'Campo-Neuen
Departement Mathematik und Informatik, Universität Basel, Basel, Switzerland
e-mail: annette.acampo@unibas.ch

A. Papadopoulos (✉)
Université de Strasbourg and CNRS, Strasbourg Cedex, France
e-mail: athanase.papadopoulos@math.unistra.fr

S. G. Dani, A. Papadopoulos (eds.), *Geometry in History*,
https://doi.org/10.1007/978-3-030-13609-3_7

in the direct lineage of those of the Greeks, then the work of Euler who developed the theory of curvature of differentiable surfaces, making extensive use of the newly introduced methods of analytic geometry and differential calculus, and then the works of the eighteenth and nineteenth-century French school of differential geometry founded by Monge, including his followers Meusnier, Dupin, Rodrigues and others. Thus, we made a choice in the topics that we survey. In particular, the works of Fermat and Descartes on the analytic geometry of curves, and those of Newton and Leibniz on differential calculus will hardly be mentioned, and we shall not talk about Gauss and Riemann.

Good references for the history of curvature are the article *Outline of a history of differential geometry* [108] by Struik and the historical notes in his book *Lectures on classical differential geometry* [109]. A rich source of information on differential geometry, which lies at the junction of the historical, mathematical and popularization literature, is Marcel Berger's book *Geometry revealed: A Jacob's ladder to modern higher geometry* [12].

7.2 The Precursors

There is no mention, nor indication, of any definition of curvature in the mathematical literature of Greek antiquity that survives. But the ground was ready there for the development of such a concept. First of all, a great variety of curves and surfaces were discovered and studied, and several qualitative features of their curvature were highlighted. Among these curves, the conics (intersections of planes with the surface of a cone of revolution) were studied extensively, but many other curves were also investigated, especially curves obtained by intersecting planes with surfaces (generally, boundaries of three-dimensional bodies). For example, a spiric is a curve of degree four obtained as the intersection of a spiric surface—a surface generated by a circle revolving about a straight line lying in its plane and disjoint from its center—with a plane parallel to its axis. As in the case of conics, there are three kinds of spiric curves; see the expositions by Heath [41, Vol. I, p. 163] and Tannery [110]. Among the other curves, we mention Aristotle's wheel, also called the "paradoxical curve", defined as the locus of a point situated on a small circle attached in the interior of a circle having the same center and a larger diameter, while the larger circle rolls on a straight line. This curve is called paradoxical because at first casual sight, the point situated on the smaller circle covers the same distance as a point situated on the bigger circle (say, on the same ray starting from the center of the wheel), but the two wheels do not have equal diameters! The fact is that the point on the smaller circle (like the one on the greater circle) does not describe a straight line, but, precisely, a curve called Aristotle's wheel. This curve is discussed in Aristotle's *Mechanica* [11]. It is more general than the cycloid, a curve that plays an important role in our exposition below (the cycloid is a special case of an Aristotle wheel where the small and the large circle coincide). Aristotle's wheel (and the cycloid) is an example of a "mechanical curve", a class of curves highlighted in

Greek antiquity, named so because they are constructed using a mechanical device. They form a collection of curves that is distinct from those defined by equations or by intersections of planes with boundaries of three-dimensional bodies. Other examples of mechanical curves are the conchoid and quadratrix, known to be useful for trisecting an angle, and the cissoid, useful for duplicating a cube. Archimedes' spiral is a curve that may be used for trisecting an angle and rectifying the circle. We recall though that in the mathematical construction problems of Greek antiquity, only the compass and ruler were allowed, therefore many mechanical curves were not permitted in the solution of these problems.

Besides curves, a great variety of surfaces were studied by the ancient Greek geometers. The spiric surfaces which we already mentioned are examples. A particular case of a spiric surface is our familiar two-dimensional torus (this is the case where the revolving circle does not intersect itself during its revolution). Cylinders were considered as limits of cones of revolution, and their intersections with planes are also conics (including degenerate conics). Archimedes (3rd c. BCE) wrote a book titled *Conoids and spheroids*, a far-reaching treatise including subtle geometric considerations on lengths of curves obtained by intersections of planes with surfaces, and on areas of pieces of surfaces bounded by such intersections. Conoids, in Archimedes' terminology, include the hyperboloids and the paraboloids of revolution. A spheroid is an ellipsoid of revolution (a surface obtained by rotating an ellipse around one of its axes). Cylindroids and other surfaces and curves obtained by intersecting these surfaces with planes were investigated by several Greek geometers, and many of them are described by Heron (1st c. CE) in his *Metrica* [68], by Pappus (4th c. CE) in his *Collection* [94] (especially Book IV), and by Proclus (5th c. CE) in his *Commentary on the first book of Euclid's Elements* [99]. Proclus was a great historian of mathematics and a great mathematician as well.

Let us return to Aristotle. In several treatises, he talks about the distinction between a straight line and a curved line. In the *Metaphysics*,[1] he gives a list of ten "opposites" or "contraries" which, he says, are fundamental in Pythagorean thought. The latter considered that opposites are the principles of things. The list includes the opposition between a straight line and a curved line.[2] In the treatise *On the heavens*,[3] Aristotle writes that there are two sorts of lines: the simple ones, namely, the straight line and the circle, and the others, which, he says, are a combination of these two. In the treatise *On the soul*,[4] he gives a dynamical definition of a curve, as an object produced by the motion of a point, a definition close to our intuition of a curve drawn by moving the point of a pencil on a sheet of paper. It may seem too

[1] *Metaphysics* [8] 985b23; cf. also 1016b25-27.

[2] The other nine oppositions are: limited-unlimited, odd-even, unity-plurality, right-left, male-female, rest-motion, light-darkness, good-bad, square-oblong.

[3] *On the heavens* [9] 268b17.

[4] *On the soul* [10] 409a.

much extrapolation to infer from these passages combined that Aristotle conceived a planar curve as a moving point together with a rolling circle of variable size representing the curvature at that point. But given the degree of geometric intuition and the sophisticated mathematical knowledge that the Philosopher had, we do not find this conjecture too unsound.

An important notion related to the curvature of a plane curve is that of normal line. Indeed, (in the modern terminology) the normal at a point contains the center of the osculating circle, and the curvature is the inverse of the radius of the osculating circle. Normals to conics were studied by Apollonius of Perga (3rd c. BCE). In Book V of the *Conics*, Apollonius investigates in detail lines that (locally) maximize or minimize the distance from a given point to the conic. He shows that these lines are normal to the conic, that is, perpendicular to the tangent to the conic at the point where they intersect it. As a matter of fact, the study of these maxima and minima is the main object of Book V. This is stated explicitly by Apollonius himself, in the first book where he gives a summary of all of the *Conics*; cf. [100, p. 252]. The work is all the more remarkable in that it does not make use of any differential calculus, analytic geometry or algebraic formalism. The methods are ingenious, purely synthetic.

Book III of Euclid's *Elements* is entirely dedicated to the circle and its properties, and Euclid studies there the tangents and normals to that curve. For instance, Proposition 18 says that if a straight line is tangent to a circle, and another straight line joins the center to the point of contact, then the two lines are perpendicular to each other. Apollonius' work on this topic in Book V of the *Conics* may be considered as an extension of Euclid's propositions on the circle to the case of ellipses and more general conics.

Apollonius was also aware of the idea of evolute. In particular, in Book V of the *Conics*, he studies a notion that is equivalent to the one of the evolute of the family of normals to the three types of conics. The importance of this notion in his works has been highlighted by many authors. Chasles, in his *Aperçu historique sur l'origine et le développement des méthodes en géométrie* [27] (1837), writes:

> The fifth book is the most valuable monument of Apollonius' genius. This is where the questions of *maxima* and *minima* appeared for the first time. We find there all the things that todays' analytic methods teach us on this subject, and we recognize there the germ of the beautiful theory of *evolutes*. Indeed, Apollonius proves that there exists, on each side of the axis of a conic, a set of points from where we can draw a unique normal onto the opposite part of the curve. He gives the construction of these points, and he observes that their continuity separates two spaces that present a remarkable difference, namely, that from any point of the first one we can draw two normals to the curve, and from any other point of the other one we cannot draw any normal. Therefore, here are perfectly determined the *osculation centers* and the *evolute* of a conic.[5]

[5]Le cinquième livre est le monument le plus précieux du génie d'Apollonius. C'est là qu'ont paru pour la première fois les questions de *maxima* et de *minima*. On y retrouve tout ce que les méthodes analytiques d'aujourd'hui nous apprennent sur ce sujet; et l'on y reconnaît le germe de la belle théorie des *développées*. En effet, Apollonius prouve qu'il existe, de chaque côté de l'axe d'une conique, une suite de points d'où l'on ne peut mener à la partie opposée de la courbe qu'une normale; il donne la construction de ces points, et observe que leur continuité sépare deux espaces

We recall in this context that the envelope of the family of normals to a plane curve is its evolute, and that the evolute of a curve is the locus of its centers of curvature. On the same subject, Dieudonné writes in his *History of algebraic geometry* [35]:

> In the geometry of the Greeks, algebraic curves are also introduced as "loci" with respect to problems of purely geometric origin. The most beautiful example is undoubtedly the most profound part of Apollonius' work on conics, the study of normals to conics, in which the evolutes of conics are completely characterized and studied. Apollonius' theorems translate immediately in our notation into the equation of the evolute that only the undeveloped state of Greek algebra prevents him from writing.

Ver Eecke, the author of the French translation of the *Conics*, also expressed the fact that Apollonius was close to discovering the notion of curvature: "Most of the propositions of this book are, in a striking way, close to the modern theories of normals, sub-normals and radii of curvature, and we already find there the germ of the theory of evolutes."[6] [5, p. xix].

Two propositions from Book V of the *Conics* (Propositions 51 and 52) are called by Heath "the great propositions" and are commented on by him in a section of his *History of Greek Mathematics* under the title *Propositions leading immediately to the determination of the evolute of a conic* [67, Vol. 2, pp. 168–179].

Even though the idea of evolute is contained in Apollonius' work, for a precise definition one had to wait for the seventeenth century, and more precisely the work of Christiaan Huygens, who thoroughly studied this notion and found a spectacular application of it. We survey his work in the next section of the present paper.

In Book VII of the *Conics*, Apollonius introduces the notion of conjugate diameters of a conic. Roughly speaking, for an ellipse, a diameter (line passing through its center) is conjugate to a second diameter if it is parallel to the tangents of the ellipse at the endpoints of the first diameter. The theory works for any kind of conic. It was generalized by Dupin in the eighteenth century to a theory of conjugate curves on a surface; we shall review this in Sect. 7.6 below.

Several commentators consider that the Greek theory of conic sections has a practical origin. The mathematician Philippe de La Hire, who continued the works of Desargues and Pascal on conics, published in 1682 a treatise titled *Gnomonique ou l'art de tracer des cadrans ou horloges solaires sur toutes les surfaces, par différentes pratiques, avec les démonstrations géométriques de toutes les opérations* (Gnomonics, or the art of tracing sundials over all kind of surfaces by different methods, with geometrical proofs of all the operations) [70], in which he claims that conic sections originate in the art of sundials, which can be traced back in

---

qui présentent cette différence remarquable, savoir: que de chaque point de l'un on peut mener deux normales à la courbe, et que d'aucun point de l'autre on n'en peut mener aucune. Voilà donc les *centres d'osculation*, et la *développée* d'une conique parfaitement déterminés. [In this paper, the translations from the French and German are ours].

[6] La plupart des propositions de ce livre se rapprochent d'ailleurs d'une manière frappante des théories modernes sur les normales, les sous-normales et les rayons de courbure, et l'on y trouve déjà le germe de la théorie des développées.

Greek antiquity to the fourth century BCE. In the Foreword to his book, he writes (p. 4):

> I could easily demonstrate that we owe to the solar clocks the discovery of these wonderful curved lines of which we find great usage in all parts of mathematics; because one cannot consider the shadow of the extremity of some sharp body on a surface without noticing at the same time the curvature marking the path of the sun, which is very similar to the one of the section of the right cone having as basis a circle parallel to the equator, where we can assume that the sun moves while making this shadow, and whose tip is the extremity of the body that makes the shadow.[7]

Neugebauer, in his paper *The Astronomical Origin of the Theory of Conic Sections* [92], made the same conjecture, namely, that the study of conic sections originates in the mathematical study of sundials. This brings us to a class of curves studied by the Greeks which are more general than the conics, namely, those traced by the shadows of paths of the sun on sundials, at the various stages of the year. Such curves were investigated by Ptolemy, Pappus and others. We note that the sundials were not all planar: some of them were convex surfaces engraved in stone blocks. We refer the interested reader to the survey articles by Rinner [103] and by Jones [74]. In the latter, the author notes (the trivial but nonetheless remarkable fact) that a surface of arbitrary shape could be used as a sundial, provided that one does not require the day curves to be circles, straight lines, or conic sections. The article contains pictures of sundials of various forms. From here, we reach the theory of curves drawn on an arbitrary surface, that is, space curves. We shall talk about (the modern theory of) space curves in Sect. 7.5 below.

In 1888, the English architect, archaeologist and astronomer Francis Penrose, who was a specialist of Greek monuments, published a book titled *An investigation of the principles of Athenian architecture; or, the results of a survey conducted chiefly with reference to the optical refinements exhibited in the construction of the ancient buildings at Athens* [95]. In this book, Penrose makes a thorough investigation of Greek architecture of the fourth century BCE, and he shows that a large number of curves (conics are just a small class of them) were used in the various parts of the monuments of that epoch (Figs. 7.1 and 7.2). He also claims that the Greeks of that period were acquainted with the notion of osculating circle and radius of curvature. See in particular p. 118ff of [95].

[7]Je pourrais facilement démontrer que nous sommes redevables aux horloges solaires de la découverte de ces admirables lignes courbes dont nous trouvons de très grands usages dans toutes les parties des mathématiques; car l'on ne peut considérer l'ombre de l'extrémité de quelque corps pointu sur une surface, sans s'apercevoir en même temps de la courbure que marque le chemin du soleil, qui est très semblable à celle de la section du cône droit qui aurait pour base un cercle parallèle à l'équateur, dans lequel on peut supposer que le soleil marche lorsqu'il fait cette ombre, et dont le sommet est l'extrémité du corps qui fait l'ombre.

Fig. 7.1 Space curves drawn on a capital. From Penrose's *Investigation of the principles of Athenian architecture; or, the results of a survey conducted chiefly with reference to the optical refinements exhibited in the construction of the ancient buildings at Athens*

Coolidge, in his paper *The unsatisfactory story of curvature* [33], writes that the first author to have talked explicitly of curvature and given a hint towards the modern definition of this notion is Nicolas Oresme (fourteenth century). The latter used the name *curvitas* and stated that the curvature of a circle is inversely proportional to its radius. Oresme also expressed the property that if two curves touch the same line at the same point, then the smaller curve has the greater curvature. There is a discussion of this work of Oresme in the paper [107] by Serrano and Suceava (following Coolidge's work), in which these authors describe the content of his treatise *De configurationibus*. They note that in Chapter LXXI of this book, Oresme, for what concerns curves, refers to Aristotle.

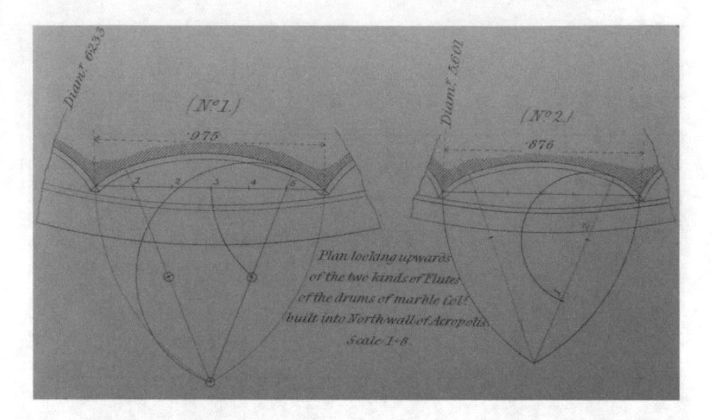

Fig. 7.2 Another picture from Penrose's treatise

7.3 Christiaan Huygens

Christiaan Huygens is the mathematician *par excellence* who made the link between the mathematics of Greek antiquity and that of the modern period. He knew very well the works of the Ancients, with whom he shared a purely geometric point of view, in particular, on curves. At the same time, he was knowledgeable in the analytic geometry newly developed by Fermat and Descartes, and he became acquainted with the emerging differential calculus through his relations with several mathematicians including Leibniz, Newton, the brothers Johann and Jacob Bernoulli, and the Marquis de l'Hôpital. The important edition of Huygens' complete works (*Œuvres complètes* [72]), published in 22 volumes, including his papers, books, notes and extensive correspondence, is an invaluable source of information on his life and work, and more generally on scientists and science in the seventeenth-century. It is apparent from these writings that Huygens disliked the mechanical and algebraic methods of differential and integral calculus. We shall say more about this below, after spending a few more words on Huygens' background.

Christiaan Huygens acquired most of his teaching from his father, Constantijn Huygens, who belonged to the higher Dutch aristocracy. Constantijn was a diplomat, poet and composer, and above all, a Renaissance humanist, infused of classical Greek culture and science. In Huygens' complete works edition [72], we find a list of mathematical books that the mathematician Jan Stampioen, who was his private teacher at the age of fifteen, asked him to read.[8] The list includes, besides works of contemporary authors like Kepler, Copernicus, Descartes, Viète and a few others, Ptomely's *Almagest*, Diophantus' *Arithmetic* and Apollonius' *Conics*.

At the University of Leiden, where he enrolled at the age of sixteen, in a law curriculum, Christiaan Huygens continued to be nurtured by the writings of the Greek geometers. J.A. Vollgraff, the editor of Volume XXII of his complete works,

[8] Huygens [72, Vol. I, No. 5].

writes, on p. 421 of that volume: "The study of law for Christiaan Huygens was an imposed task; but that of geometry and its applications, and also of practical mechanics (at least during his youth) a pure delight. It was indeed, as far as geometry is concerned, Archimedes' work that attracted him. He could not be satisfied with the modern authors."[9] Because of Christiaan Huygens' expertise in Greek geometry, and in particular in the methods of Archimedes, his father used to call him *Mon petit Archimède* (My little Archimedes). The surname was used for several years by several friends and correspondents of his father, including Mersenne.[10]

One of the professors at Leiden University, the mathematician and orientalist Jacob Golius, who had been Descartes' teacher, was fluent in Arabic. He stayed in Aleppo, and then in Constantinople where he was the advisor of the representative of the Netherlands to the Sublime Porte. He was also a collector of Arabic manuscripts and he translated some of them into Latin. In 1653, he published an Arabic-Latin dictionary. We mention these details because Golius was a friend of Constantijn Huygens, and had a private copy of an Arabic translation of Books V, VI and VII of Apollonius' *Conics*, three books that do not survive in Greek. (We already mentioned that the theory of normals is contained in Book V.) Vollgraff, in his comments in Volume XXII of Huygens' complete works, reports that Golius obtained that copy from an uncle of Christiaan Huygens who used to travel in the Middle East and brought it from there.[11] Golius made a draft of a first translation of the three books of Apollonius, and he communicated it to Descartes. It is possible that Christiaan Huygens had access to Golius' translation, and it is also conceivable that he made his own translation of them. Indeed, like his father, Christiaan Huygens was talented in languages, and both men were fluent in several European as well as old and more modern Middle-Eastern languages. It is also possible that Christiaan Huygens had access to a translation of the Arabic books of Apollonius made by the Italians. Indeed, in a letter written to Christiaan Huygens in 1661,[12] prince Leopoldo de' Medici informed him about a translation of the *Conics* from the Arabic which he ordered. The Huygens family had good relations with the Medicis, like with other princes in Europe. The translation was eventually made by the Italian mathematician Giovanni Alfonso Borelli together with Abraham Ecchellensis, a Syrian linguist who lived in Rome (his original name was Ibrahim Al-Haqilani) [6]. Christiaan Huygens received a copy of that translation the following year.[13] There is a correspondence between Christiaan Huygens and the French astronomer

[9] L'étude du droit fut pour Christiaan une tâche imposée; mais celle de la géométrie et de ses applications, et aussi de la mécanique pratique (du moins dans sa jeunesse) un pur plaisir. C'étaient bien, en matière de géométrie, les travaux d'Archimède qui l'intéressaient. Il ne pouvait se contenter des auteurs modernes.

[10] See e.g. letters 25, 48, 200, 210 of Volume I of the *Complete works*, and there are others.

[11] Huygens [72, Vol. XXII, p. 405].

[12] Huygens [72, Vol. XXII, letter No. 37].

[13] Huygens [72, Vol. IV, letter No. 1029].

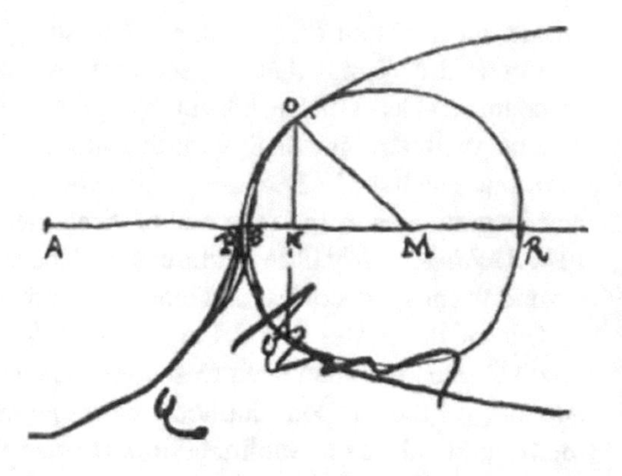

Fig. 7.3 A drawing showing an osculating circle and a radius of curvature, from Huygens' Volume XXII of his complete works (p. 211)

Ismaël Boulliau in which they discuss the Arabic manuscripts of Apollonius and Archimedes that belong to Leopoldo de' Medici.[14]

Toomer's edition of Book V of the *Conics* [4] contains a translation of a memoir by Huygens on the solution of the following problem of Pappus: *Given a parabola and a point, to draw from it a straight line to meet the parabola at right angles* (from Proposition 30 of Book IV of Pappus' *Collection* [94]). The original was published in Huygens' *Oeuvres complètes* [72], Vol. I, No. 365. pp. 533–534. In his edition of the *Conics*, Heath notes that this problem of finding the normal to a parabola was not treated by Apollonius.

Besides the conics, Christiaan Huygens investigated a great variety of curves. One of his first discoveries, which he made the year he enrolled at Leiden's university, concerns the hanging chain, also called the catenoid. Huygens showed that this curve cannot be a parabola, thus correcting a statement made by Galileo Galilei. His result is discussed in a letter he wrote to Mersenne, dated October 28, 1646.[15] Huygens returned to the catenoid a few years before the end of his life. In an article written in 1690, he computes its radius of curvature at the vertex.[16] He uses in this article the words "radius curvitatis" and "centro curvitatis." Vollgraff, commenting on a drawing by Huygens reproduced here (Fig. 7.3, extracted from p. 211 of Volume XXII of the complete works), in which he indicated the osculating circle of the curve of equation $a^2x = y^3$, writes: "I cannot understand how Huygens obtained this result [...] One can see that before 1658, he was already interested in what he will call later the radius of curvature."[17]

The notion of radius of curvature also often appears in Huygens' writings on Dioptrics, in the context of radii of curvature of lenses; cf. [72], Vol. XIII, XXI and others.

[14]See e.g. the letter No. 536 in volume II of [72], dated October 18, 1658 and the letter No. 547 in the same volume, dated November 8, 1658.

[15]Huygens [72, Vol. I, No. 14].

[16]Huygens [72, Vol. IX, No. 2625].

[17]Nous ne voyons pas comment Huygens a obtenu ce résultat [...] On voit qu'avant 1658 il s'intéressait déjà à ce qu'il appellera plus tard le rayon de courbure. [...].

Besides the catenary and the other curves we mentioned (conics, conchoid, cycloid, Archimedean spiral, etc.) we find many other curves in the works of Huygens, some of them with exotic names: the epicycloid, the hypercycloid, the hypocycloid, the cassinoid, the cardioid, the Beaune curve, the Gutschoven curve, the Tschirnhaus curve, the Descartes folium, the Dinostratus quadratrix, the sail curve, the lemniscate, the spring curve, the paracentric isochronous curve, etc. Huygens also studied caustics, which he observed during his experiments on optics. In his *Traité de la lumière*, written in 1678, he made the relation between caustics and the theory of evolutes and involutes which he developed in his 1673 *Horologium oscillatorium* and of which we shall talk later in this paper. Huygens showed that the caustic is the evolute of the wave front. Most of all, he liked to study the curves that appear in nature and that have some practical use. He expressed this on several occasions, and in particular in a letter to Leibniz, dated September 1st, 1691[18]:

> I have often considered that the curved lines which the nature presents frequently to our view, and which it describes, so to say, itself, all comprise very remarkable properties. Such are the circle which we encounter everywhere, the parabola, which the water jets describe, the ellipse and the hyperbola, which the shadow of the edge of the stylet makes and which we also encounter elsewhere, the cycloid which a nail on the circumference of a wheel describes, and finally, our hanging chain that was noticed by so many centuries, without analyzing it. [...] But to build new ones, only in order to exert one's geometry, without predicting any other use, is *difficiles agitare nugas*,[19] and I have the same opinion on all the problems that concern numbers.[20]

Huygens studied the locus of the center of a rolling hexagon (Fig 7.4). One may add here that at the time of Huygens, the conics, which had been studied in a purely theoretical manner by the Greeks, have found their place in nature: Kepler saw that planets move in ellipses and Galileo showed that objects projected obliquely upward move in parabolas.

The drawing in Fig. 7.5 is by Huygens. It is extracted from the paper [18] by Bos, and it originates in a manuscript Huygens wrote on the tractrix in 1692. It represents an instrument that may be used for drawing such a curve. At the top left, Huygens wrote: "A little cart or a boat would serve for squaring the hyperbola."[21]

[18]Huygens wrote two versions of that letter, a preliminary one and the one he sent. Both versions are reproduced in the complete works (Huygens [72] Vol. X, No. 2693) and we quote from both.

[19]The Latin expression *agitare nugas* means to deal with bagatelles, or child's plays. Thus, the expression *difficiles agitare nugas* means to occupy oneself with questions that are difficult but with no importance, which, in this context means questions that have no practical relevance.

[20]J'ai souvent considéré que les lignes courbes que la nature présente souvent a notre vue, et qu'elle décrit, pour ainsi dire, elle-même, renferment toutes des propriétés fort remarquables. Telles sont le cercle que l'on rencontre partout, la parabole, que décrivent les jets d'eau, l'ellipse et l'hyperbole, que l'ombre du bout du stile parcourt et qu'on rencontre aussi ailleurs, la cycloïde qu'un clou qui est dans la circonférence d'une roue décrit, et enfin notre chaînette qu'on a remarquée par tant de siècles sans l'examiner. [...] Mais d'en forger de nouvelles, seulement pour y exercer sa géométrie, sans y prévoir d'autre utilité, il me semble que c'est *difficiles agitare nugas*, et j'ai la même opinion de tous les problèmes touchant les nombres.

[21]Une charrette, ou un bateau servira à quarrer l'hyperbole.

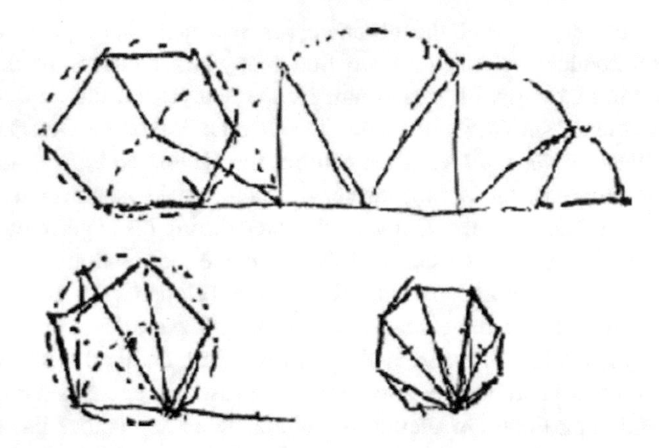

Fig. 7.4 A curve studied by Huygens: the locus of the center of a hexagon rolling along a line. Vol. XVIII p. 403

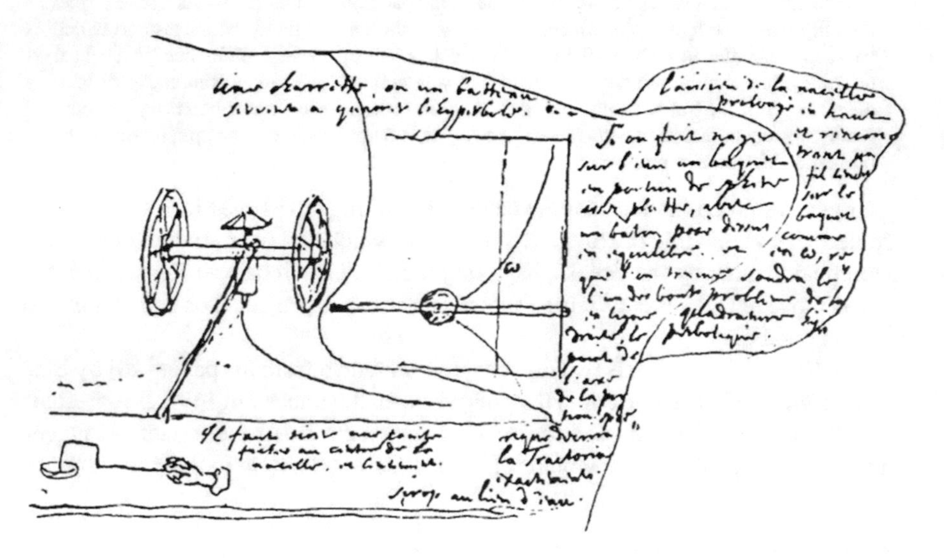

Fig. 7.5 An instrument imagined by Huygens, to construct a tractrix (Figure extracted from the article [18] by Bos)

Huygens was the first to introduce a word for *evolute* (he used the French word "développée"), and he studied it extensively. No progress on this notion was made between the work of Apollonius and that of Huygens. The latter used both properties of this curve, being the locus of centers of curvature of the original curve (the involute), and the envelope of the normals of this involute. Furthermore, he gave a beautiful application of this notion, in his work on the pendulum clock (*Horologium oscillatorium*), published in several versions. Namely, he found that the cycloid is a shape on which a pendulum may roll so that it is isochronous, that is, it has the same

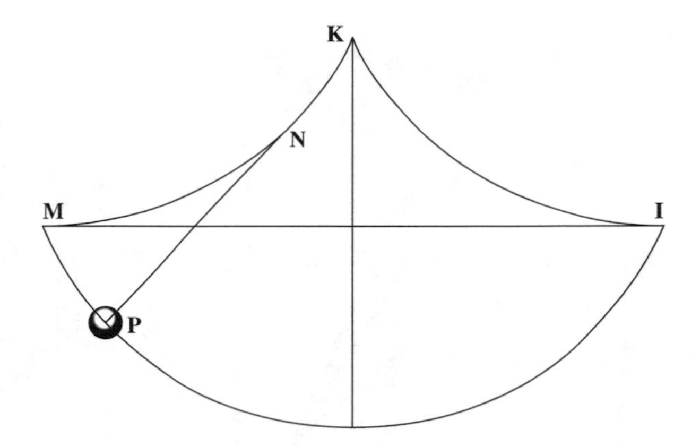

Fig. 7.6 From Huygens' *Horologium oscillatorium*. The pendulum swings along two cycloids, KM and KI, and the curve MPI that the bob makes is also a cycloid (1673)

period, independently of the amplitude. In Fig. 7.6, extracted from his complete works and representing a swinging pendulum, the curves KM and KI are cycloids. Huygens also proved that the curve made by the bob of the pendulum, (the curve MPI in Fig. 7.6), which is the involute of the cycloid, is also a cycloid.

Talking about the cycloid in the introduction of the 1673 version of his *Horologium oscillatorium*, Huygens writes[22]: "The geometers, in our times, called it cycloid and examined it carefully because of its various properties, whereas we considered it because of that faculty we mentioned, namely, that of measuring time."[23]

In Fig. 7.7, we have reproduced two drawings of Huygens, extracted from a manuscript titled *De linearum curvarum evolutione et dimensione* (on the evolute and the dimension of curved lines) which is a part of his 1673 *Horologium oscillatorium* (see [72] Vol. XVII, p. 142 ff). In these drawings, the upper curve, a cycloid, is the involute of the lower curve, its evolute. Huygens' goal in these drawings is to prove that the evolute, which is obtained here as the locus of intersection points of infinitesimally close normals to the curve (this is the definition of the center of curvature), is also a cycloid.

The cycloid was already studied before Huygens, in particular by Galileo Galilei, who, as is well known, also worked on the pendulum, but without noticing the tautochronous property of this curve. We also noted that the cycloid appears in Aristotle's works, as a particular case of Aristotle's wheel. Finally, it is worth noting that Descartes studied the cycloid among other mechanical curves before writing his *Géometrie* (1637), and that he used in his study a classical approach to tangents (without the new infinitesimal apparatus), considering the rolling circle as a limit of polygons with a number of sides growing to infinity; cf. the article [101] by Rashed

[22]Huygens [72, Vol. XVIII, p. 86].

[23]Les géomètres de notre temps l'ont appelée cycloïde et l'ont examinée avec soin à cause de ses diverses autres propriétés. Quant à nous, nous l'avons considérée à cause de cette faculté dont nous parlions, savoir celle de mesurer le temps.

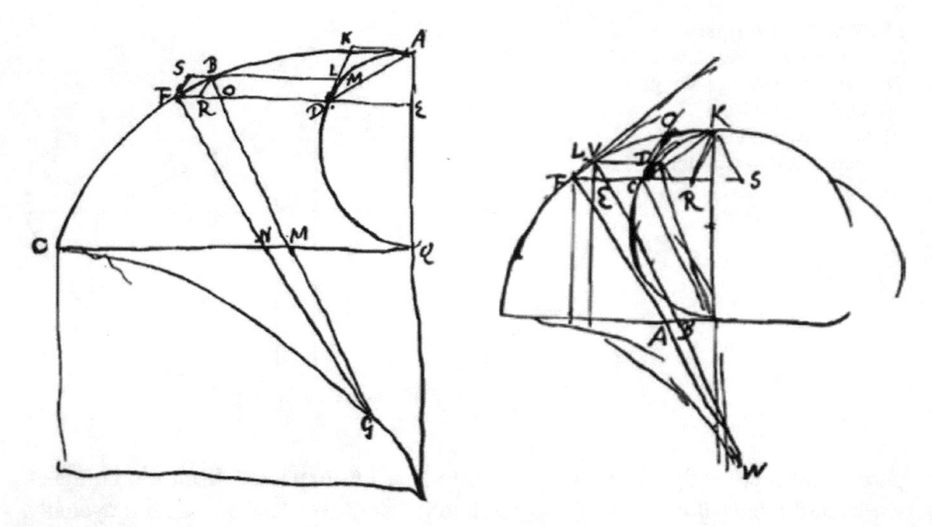

Fig. 7.7 Finding the centers of curvature of the cycloid. Figures extracted from Huygens' manuscript *De linearum curvarum evolutione et dimensione* (1673) [72] Vol. XVII, pp. 144–145

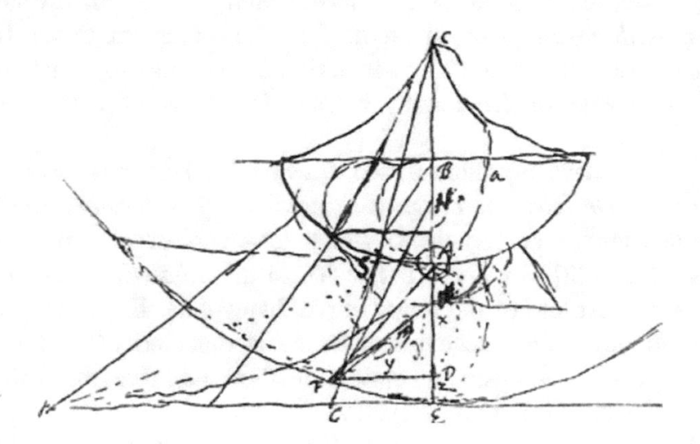

Fig. 7.8 A study of the cycloid, [72, Vol. XVIII, p. 391]

on *Descartes and the infinitely small.* The reader may also recall here Huygens' drawing of a rolling polygon reproduced in Fig. 7.4 above.

Figure 7.8 is extracted from another study of Huygens on the cycloid [72, Vol. XVIII, p. 391].

There is a set of letters exchanged between Leibniz and Huygens where these two mathematicians discuss evolutes and curvature. For instance, in a letter to Huygens dated January 8, 1692, Leibniz writes[24]:

[24]Huygens [72, Vol. X, letter 2727].

I really think that you saw clearly the circle which appears when the point evolves on the curve and whose radius is the shortest segment that can be drawn from that point to the given curve. But maybe you had not thought of considering it primarily as a measure of the curvature. And myself, when I first thought of considering the greatest circle that touches the curve internally as the measure of the curvature or the contact angle, I did not have the idea of considering evolutes.[25]

After developing the theory of the evolute of the cycloid, Huygens worked on the evolutes of the parabola, ellipse and other curves. He addressed the question of what are the curves, other than the cycloid, whose evolute is a similar curve.[26] He showed in particular that the evolute of the epicycloid is also an epicycloid.[27]

Huygens also worked on the problem of curvature of sails of ships under the pressure of the wind.[28] This is an extremely difficult problem which was posed on several occasions by Johann Bernoulli[29] and which Newton had addressed in Book II of his *Principia*. In 1694, one year before his death, Huygens corresponded with Leibniz on this subject.[30] This topic was also studied later by Euler. We shall say more about this later.

By the end of his life, Christiaan Huygens became interested in the new methods developed by Leibniz, Johann Bernoulli and the Marquis de l'Hôpital to study curves, and, more generally, in the new differential calculus. In a letter to Leibniz, dated September 1st, 1691, he writes[31]:

Apart from the reduction to the construction of the squaring of the hyperbola or to logarithms, I can see the foundations of what you and Mr. Bernoulli have more than me. But this reduction, for which I have high esteem, I don't see up to now how you obtained it, and I would be very happy if you could teach it to me. I want to believe that [your new calculus] is useful for noticing more easily the various properties of the lines that we examine, because I see that yourself as well as Mr. Bernoulli discovered things regarding this catenary, which I did not propose to myself to investigate, because I thought they were too far off. But to you, it seems that they reveal themselves.[32]

---

[25]Je crois bien que vous avez vu le cercle qui se décrit du point de la courbe évolue, et dont le rayon est la moindre droite qu'on peut mener de ce point à la courbe décrite; mais peut-être n'aviez vous pas songé d'abord à le considerer comme la mesure de la courbure, et moi lorsque j'avais considéré le plus grand cercle qui touche la courbe intérieurement comme la mesure de la courbure ou de l'angle de contact, je ne m'étais pas avisé de songer aux évolutions.

[26]Huygens [72, Vol. XVIII, p. 104].

[27]Huygens [72, Vol. XVIII], appendix III to Pars Tertia of the *Horologium Osillatorium*, p. 399 ff.

[28]*Calculs et considérations sur les résistances éprouvées par différentes surfaces appartenant à des corps animés d'un mouvement uniforme à travers l'air ou l'eau et sur les vitesses que le vent peut donner à des voiliers à une seule voile supposée plane* (1691), published in Vol. XXII. of [72].

[29]Johann Bernoulli, *Solution du problème de la courbure que fait une voile enflée par le vent*, 1692 and *Essai d'une nouvelle théorie de la manoeuvre des vaisseaux*, 1714.

[30]Huygens [72, Vol. XX, p. 552].

[31]Huygens [72, Vol. X, letter No. 2693].

[32]Hormis la réduction de la construction à la quadrature de l'hyperbole, ou aux logarithmes, je vois les fondements de tout ce que vous et M. Bernoulli avez de plus que moi; mais cette réduction, que j'estime fort, je ne vois pas jusqu'ici comment vous y êtes parvenus, et vous me ferez plaisir de me l'apprendre. [...] je veux croire que [votre nouveau calcul] sert à faire remarquer plus facilement

In a letter sent in 1692,[33] three years before his death, to the Swiss mathematician and astronomer Fatio de Duillier who lived in London and who was, like Huygens, a member of the Royal Society, Huygens enquires about a treatise that Newton promised to publish on curves, expressing the will to learn more about the latter's method of fluxions. He writes:

> [Newton's] treatise on curved lines, according to what my brother sent me (he learned it from you, Sir), should soon be born, and I am waiting for this impatiently, hoping to find there all these beautiful things which you mention in your last [letter], and which, the more I conceive the difficulty, the more I respect. [...] Newton has general rules [...] Also like when the equation of the tangent is given, whether he knows if it belongs to some curve.[34] Besides, I don't understand what the fluxion of the fluxion means, it seems that this would mean the tangent of a curved line on which depends the curve of the first fluxion, but then I don't see why the difficulty is greater. Please Sir, implore Mr. Newton to publish this treatise which will have marvelous usefulness and will do him great honour.[35]

Huygens eventually started to use the new differential calculus. During the last years of his life, he spent a large part of his time trying to solve problems using the methods of Leibniz. In one of his last memoirs, concerning a curve whose tangents make a constant angle with the parts of the axis (1693), he writes that without these methods he would not have succeeded in solving the problem addressed [32, p. 67].

Newton had a great admiration for Huygens. He used to call him *Summus Hugenius*[36] (Huygens, the greatest). Henry Pemberton, the editor of the third edition of Newton's *Principia*, writes in the preface of his book *A View of Sir Isaac Newton's Philosophy* [96]: "Sir Isaac Newton several times has particularly recommended to me Huygens' style and manner. He thought him to be the most elegant of any mathematical writer of modern times, and the most just imitator of the Ancients." In his article on Huygens, published in Volume 21 of the *Biographie universelle* [83], J.-F.-Th. Maurice writes: "The high esteem that Newton carried for the truly geometrical style of Huygens is most probably the reason for the method

---

les diverses propriétés des lignes qu'on examine, parce que je vois que M. Bernoulli aussi bien que vous a découvert des choses touchant cette chaînette, que je ne me suis pas proposées à chercher, parce que je les croyais trop éloignées; mais à vous et lui il semble qu'elles se soient offertes.

[33]Huygens [72, Vol. XXII, letter LXXVIII].

[34]We have tried to keep the translation close to Huygens' French, at the expense of making some passages, like the present one, look awkward.

[35]Le traité [de Newton] des lignes courbes, à ce que mon frère me mande, (qui le tenait de vous Monsieur) devait bientôt voir le jour, ce que j'attends avec impatience, espérant d'y apprendre toutes ces belles choses dont vous faites mention dans votre dernière, et que j'estime d'autant plus que j'en conçois la difficulté. [...] Newton a des règles générales [...] Comme aussi lorsque l'équation de la tangente est donnée s'il peut connaître qu'elle appartient a quelque ligne courbe. Au reste je n'entends pas ce que signifie la fluxion de la fluxion; il semble que cela veuille dire la tangente d'une ligne courbe dont dépend la courbe de la premiere fluxion mais je ne vois pas alors en quoi la difficulté devient plus grande. Je vous prie de solliciter auprès de Monsr. Newton la publication de ce traité qui sera d'une utilité merveilleuse, et lui fera grand honneur.

[36]See e.g. his letter to Leibniz, dated Octobre 16, 1693, in Turnbull (ed.), Correspondence of Isaac Newton, Vol. 3, p. 285.

of exposition that he carried himself in his great work of the *Principia*, where he made use of only synthetic proofs and constructions, hiding the thread that guided him."[37]

In his *Meditatio nova de natura anguli contactus et osculi: horumque usu in practica mathesi, ad figuras faciliores succedaneas difficilioribus substituendas* (New meditation on the nature of the angles of tangency and osculation and their mathematical application in order to successfully replace complex figures with simpler ones) [80] (1686), Leibniz defined the notion of osculating circle for a point on a plane curve, as a limit of circles determined by three points on that curve. In the same work, he addressed the question of the existence of an osculating sphere for a point on a surface and of the definition of the curvature of the surface at that point using this osculating sphere. Vollgraff discusses this notion on p. 42 of Volume XVIII of Huygens' complete works [72], recalling that Leibniz was mistaken in his definition of the osculating circle. He also notes that Jacob Bernoulli pointed out this mistake a few years later in an article he published in the Acta Eruditorum, *J.B. Additamentum ad solutionem curvae causticae fratris Jo. Bernoulli, una cum meditatione de natura evolutarum, & varus osculationum generibus* (Addition to the solution of caustic curves of my brother Jo. Bernoulli together with a meditation on the nature of evolutes and other more general osculations) (1692). Leibniz acknowledged his mistake and published a correction in his article *Generalia de natura linearum, anguloque contactus et osculi, provolutionibus, aliisque cognatis, et eorum usibus nonnullis* (General observations on the nature of lines, angle of tangency and osculation, other related matters and several of their applications [81]). In these last two works, the name of Huygens is mentioned several times. The reader is referred to Vollgraff's comments in [72] (p. 42 ff) for the intricate history of the notions of contact angle, osculating circle, center of curvature and radius of curvature in that period.

7.4 Euler

After Huygens, we pass to Euler, who developed a new theory of curves and surfaces.

Since his youth, Euler was interested in the differential geometry of curves and surfaces. He wrote a large number of essays and a book dedicated to the subject. His memoir, *Recherches sur la courbure des surfaces* (Researches on the curvature of surfaces), published in 1767 and which we shall survey in some detail, constitutes

[37]La haute estime que faisait Newton du style vraiment géométrique d'Huygens est la cause très probable de la méthode d'exposition qu'il a suivie lui-même dans son grand ouvrage des *Principes*, où il n'a guère fait usage que de démonstrations et de constructions synthétiques, en déguisant le fil qui l'avait guidé.

a big step in the theory of curvature of differentiable surfaces. Before doing so, we briefly review a few other writings of Euler on curves and surfaces.

First of all, we mention three of his memoirs that make the link with Huygens' work. They concern isochronous curves (also called tautochronous; Euler uses the two terminologies). The titles are: *Constructio linearum isochronarum in medio quocunque resistente* (Construction of isochronous curves in a resistant medium) [42] (1926), *De innumerabilibus curvis tautochronis in vacuo* (On the innumerable tautochronous curves in a vacuum) [45] (1935) and *Curva tautochrona in fluido resistentiam faciente secundum quadrata celeritatum* (Tautochronous curves in a fluid making a resistance proportional to the square of the speed) [46] (1935). The memoir [42] is the first that Euler ever published (he was nineteen). In this and in the memoir [46], Euler considers isochronous curves in a resisting medium. In the classical case, a curve is submitted to the force of gravity, whose directions at any two points are parallel (the force is directed towards the center of the earth, considered at an infinite distance). This is the situation which Huygens considered in his *Horologium*. Newton, in his *Principia* (Book II, Proposition 16) studied the question of isochronous curves in the case where the forces of gravity are not parallel but are directed towards a point at finite distance, and he also addressed the same question in a resisting medium with resistance proportional to speed. He proved that in these cases, the cycloid is also tautochronous. Euler worked on tautochronous curves in a setting which is more general than Newton's, and he found other curves. In his memoir [45], he considered the question of isochronous curves in the vacuum, and he found many curves satisfying this property, besides the cycloid. In his solution, he made use of Huygens' theory of evolutes. Euler studied evolutes in two other memoirs, *Investigatio curvarum quae evolutae sui similes producunt* (Investigation of curves which produce evolutes similar to themselves) (1750) [54] and *Investigatio curvarum quae similes sint suis evolutis vel primis vel secundis vel tertiis vel adeo ordinis cuiuscunque* (Investigation of curves that are similar to their evolutes of first or second or third or whatever order) (1787) [60]. He addressed there the natural question of finding curves whose evolutes are similar to themselves, a question on which Huygens had already worked,[38] and he found, like Huygens did, that, besides the cycloid, the epicycloid satisfies this property.[39] Euler also studied the problem of finding the curves that are similar to their higher evolutes (evolutes of the evolute, etc.) In his memoir *De linea celerrimi descensus in medio quocunque resistente* (On the curve of fastest descent in whatever resistant medium) [48], he studied the brachistochrone in a resistant medium.

Another memoir of Euler on plane curves is his *De curvis triangularibus* (On triangular curves) [57], in which we find the origin of curves of constant breadth. These are convex curves in the plane having the property that the distance between any two distinct parallel supporting lines (that is, lines intersecting the curves but

[38]Huygens [72] Vol. XVIII, appendix III to Pars Tertia of the *Horologium Osillatorium*, p. 399 ff.

[39]Cf. p. 104 of volume XVIII of Huygens' works [72], and the editor's note on p. 40 of the same volume.

not the open convex region it bounds) is constant. The property is satisfied by the circle, and it is an interesting question to find what other curves satisfy it. The title of the memoir, "triangular curves", stems from the fact that Euler finds these curves as involutes of curves having three singularities (cusps, or non-smooth points), which he calls triangular curves. Euler calls the curves of constant breadth orbiforms (*orbiformis*), a name that reminds us of a modern notion. The notion of breadth became later a fundamental one in convexity theory and there is a large literature on the notion of breadth of convex bodies, see e.g. the papers [15] by Blaschke and Hessenberg, [102] by Reidemeister, [28] by Chebotarev and the survey on convex bodies of constant breadth in §16 of the book by Bonnesen and Fenchel [17]. Euler's memoir *De curvis triangularibus* also contains results on developable surfaces. In fact, these surfaces, introduced by Euler, were used later as a tool to deform space curves. We refer the reader to the exposition in the book by Hilbert and Cohn-Vossen [69], in particular §§30 and 31. Euler had already introduced these surfaces in his memoir *De solidis quorum superficiem in planum explicare licet* (On solids whose entire surface can be unfolded onto a plane) [56]. He obtained a partial differential equation that describes such a surface and concluded that this surface must be either a plane, a cone, a cylinder or a surface obtained as the set of tangents to a curve in 3-space. Euler's work on developable surfaces was continued and developed to a high degree by Monge; we shall talk about this below.

Among the other published works of Euler on surfaces, we mention the early memoir *De linea brevissima in superficie quacunque duo quaelibet puncta jungente* (Concerning the shortest line on any surface by which any two points can be joined together) [44] in which he gives the differential equation satisfied by a geodesic joining two points on a differentiable convex surface. He applies his methods to cylinders, cones and surfaces of revolution. Several other memoirs by Euler concern geodesics on surfaces. In particular, in his major treatise *Mechanica*, published in 1736 [47], he proves that a point-mass moving freely (that is, without any force exerted on it) on a surface describes a geodesic, that is, a curve which is locally the shortest path between its two endpoints. As a matter of fact, the main subject of Volume 2 of the *Mechanica* is the motion of a point-mass on a given curve or surface. Euler derives in this book the differential equations of geodesics.

Among the many other works of Euler on surfaces, we mention his study of minimal surfaces. These surfaces first appear in his treatise on the calculus of variations, a field of which he was the founder. Minimal surfaces were one of the first applications he gave of that theory. (Cf. Chapter V, Sect. 47 of [50]).

In his work on curves and surfaces, Euler relied heavily on differential calculus and especially differential equations, a field in which he was a master. Curves and surfaces in his work are almost always defined by equations: mostly algebraic, but also transcendental, by which Euler meant that they are defined by functions that cannot be expressed by quotients of polynomials. He introduced the notion of transcendence in several memoirs (see [49] and [59]), and he expanded on it thoroughly in the treatise *Introductio in analysin infinitorum* (Introduction to the analysis of the infinite) [51, 52] (1748) of which we talk now.

The *Introductio* is a complete two-volume treatise on the differential geometry of space curves and surfaces intended for students, and it is written in the modern language of differential geometry. Curves are defined by parametric equations $x = x(t), y = y(t), z = z(t)$. In Volume I, osculating circles and radii of curvature of plane curves are given in coordinates, and convexity and concavity of curves are discussed in relation with the signs of the radii of curvature. Algebraic curves are classified, and transcendental curves are also considered. Volume II is concerned with curves in 3-space, and it contains a long appendix (about 100 pages) on surfaces. Surfaces are represented by parametric equations $x = x(u, v), y = y(u, v), z = z(u, v)$. The appendix includes a section on the intersection of a surface with an arbitrary plane, a study of plane sections of cylinders, cones and spheres, a classification of second-degree surfaces and a chapter on the intersection of two arbitrary surfaces.

Euler published his memoir *Recherches sur la courbure des surfaces* [55] in 1767, about 20 years after the publication of his *Introductio*. With this memoir, he opened a new era in the differential geometry of surfaces. We shall highlight below some of the works done by several preeminent geometers, based on Euler's ideas.

In the memoir [55], Euler introduced a concept of curvature at a point of a differentiable surface, based on the curvature of curves that pass through that point. He started by pointing out that instead of studying the curvature of the totality of curves passing through a point on a surface, it suffices to consider *normal curves*, that is, curves that are intersections of the surface with planes containing the normal vector to the surface at the given point, i.e., the vector normal to the tangent plane. Such a plane is called a *normal plane*, or a *normal section*.[40] In the same appendix, Euler studied the osculating circles and the radii of curvature of normal curves. The curvature of a curve obtained in this way is called a *normal curvature* of the surface. It turns out that the collection of these normal curvatures contains all the information on the curvature of the surface at that point. Euler proved that at any given point of the surface, the *maximal* and *minimal* curvatures, that is, the maximum and minimum of the normal curvatures taken over all the normal sections passing through that point, determine all the other normal curvatures. More precisely, he showed that at any point on a surface where the normal curvatures are not all equal, the directions of the planes that realize the extremal curvatures are orthogonal to each other, and he proved the following (we use Euler's notation): If at the given point, f and g are the maximal and minimal radii of curvature for the normal sections through it, then for any normal section containing a tangent vector making an angle φ with the tangent vector contained in the normal section

[40]The fact that the curvature of the normal sections determines the curvature of all the other sections is nicely formulated by a theorem of Meusnier which we recall below.

corresponding to the greatest osculating circle, its radius of curvature r is given by the equation

$$r = \frac{2fg}{(f+g) - (f-g)\cos(2\varphi)}.$$

The normal sections that correspond to the greatest or smallest curvature at a point on a surface (provided these two quantities are distinct) are called the *principal sections*. They play an important role, as we shall see below.

The following is a convenient way of re-writing Euler's equation (the form is due to Dupin, from his *Développements de géométrie* [37, p. 109], which we shall discuss below):

$$\frac{1}{r} = \frac{\cos^2\varphi}{f} + \frac{\sin^2\varphi}{g}.$$

Note that f and g may take any real value.

Let us quote the conclusion of Euler's memoir [55] (Réflexion VI, p. 143):

> Thus, the judgement of the curvature of surfaces, however complicated it may seem at the beginning, is reduced for each element to the knowledge of two osculating radii, one being the largest and the other the smallest at that element. These two objects determine entirely the nature of the curvature, displaying for us the curvature of all the possible sections that are perpendicular to the proposed element.[41]

Late in his life, Euler wrote a memoir on curves in 3-space, *Methodus facilis omnia symptomata linearum curvarum non in eodem plano sitarum investigandi* (An easy method to investigate all properties of curves that do not lie in a plane) [58]. The memoir was published in 1782, one year before his death. In this memoir, to study the curvature at one point, he introduced a sphere centered at that point and made use of spherical trigonometry, which was one of his favorite subjects. In a sense this inaugurated the spherical map that Gauss introduced more formally 40 years later.

We shall review the history of curves in 3-space in the next section.

Euler, like Huygens before him, worked extensively on the theory of lenses, and he wrote several articles on catoptrics, where the notion of center of curvature is central.

Finally, we mention another question that Euler addressed and which involves curvature, namely the question of sails of ships. He first investigated this question in 1727, at the age of twenty, in a memoir he sent to the French Academy of Sciences, as a solution to a problem which the Academy proposed as a contest. The memoir is

---

[41] Ainsi le jugement sur la courbure des surface, quelque compliqué qu'il ait paru au commencement, se réduit pour chaque élément à la conaissance de deux rayons osculateurs, dont l'un est le plus grand et l'autre le plus petit dans cet élément; ces deux choses déterminent entièrement la nature de la courbure en nous découvrant la courbure de toutes les sections possibles qui sont perpendiculaires sur l'élément proposé.

titled *Meditationes super problemate nautico, quod illustrissima regia Parisiensis Academia scientiarum proposuit* (Thoughts on a nautical problem, proposed by the illustrious Royal Academy of Sciences in Paris) [43]. Johann Bernoulli,[42] who was Euler's teacher in Basel and who encouraged him to work on that problem, had already published a memoir on the same subject, in 1714, which he called *Théorie de la manœuvre des vaisseaux* (Theory of maneuver of ships). In the memoir submitted to the Academy, Euler extended ideas of Archimedes from his treatise *On floating bodies*, introducing the techniques of differential calculus, in particular partial differential equations. Euler remained interested in these questions for the rest of his life. Several years later, he completed his major opus on ship building, the *Scientia navalis* (Naval science) [53], a two-volume treatise which appeared in 1749 in Saint Petersburg [53].

Talking about Euler, one may also mention Lagrange, his younger colleague, sometimes competitor, for whom Euler had a real admiration, and who is considered to be the co-founder of the calculus of variations. At the same time as Euler, he introduced the methods of the calculus of variations in the study of curves and surfaces. Lagrange's *Théorie des fonctions analytiques* [77], which is a written version of the lectures he gave at the École Polytechnique during the years 1795 and 1796, is, like Euler's *Introductio*, an important monument of our mathematical literature. The treatise has two parts, Part I containing the principles of differential and integral calculus freed from the notion of *infinitely small*, in a language close to our modern one. Part II, titled *Application de la théorie à la géométrie et à la mécanique*, starts with a paragraph referring to the "Ancient geometers," with a review of how they conceived tangent lines to curves, and a comparison with the new point of view using techniques of fluxions and infinitesimals. Lagrange declares there that the modern techniques do not provide the rigor and the evidence of the ancient proofs, and that the methods he develops in his treatise allow him to deal with the problems of tangents and related problems according to the principles of the Ancients. This part constitutes a treatise on the differential geometry of curves and surfaces, with a study of curvature of space curves, osculating planes, tangent planes to surfaces, order of contact of surfaces, developable surfaces, and the use of the methods of the calculus of variations in various problems of maxima and minima related to surfaces.

7.5 Curves in 3-Space

In this section, we briefly review works on space curves by Alexis Clairaut (1713–1765), a contemporary of Euler with whom he held an important correspondence, of Gaspard Monge (1746–1818), one of the main founders of the famous French school of geometry that flourished in the last quarter of the eighteenth century

[42]This is Johann I Bernoulli the father (1667–1748).

and continued through the nineteenth and part of the twentieth centuries, and of Michel-Ange Lancret (1774–1807), who was a student of Monge at the École Polytechnique.

At the beginning of Chapter I of the second volume of the *Introductio*, Euler mentions the work of Clairaut on curves in space, writing: "Curves of this kind have two kinds of curvature, which has been beautifully discussed by the brilliant geometer Clairaut. Since this material is closely connected with the nature of surfaces, which we will now discuss, We have decided not to give a separate treatment, but to explain both simultaneously."[43]

Clairaut wrote a memoir titled *Recherches des courbes à double courbure* [30] in 1729, at the age of 16. The work was published 2 years later. The reason commonly accepted for this delay is that Clairaut worked so hard on his paper that he became ill, and it took him 2 years to recover. In his memoir, Clairaut studies curves that are not contained in a plane, which he calls "courbes à double courbure."[44] A surface is always in the background of such a curve. In fact, at the beginning of his memoir, Clairaut says that the curves he considers can be traced on the surfaces of solids, like, for instance, the one we get by making a compass turn on a cylinder or on another arbitrary surface. He considers himself as the first to investigate such curves, even if he notes that Descartes planned such a study, and he adds that he learned from Descartes' work that to investigate such a curve, one has to project it onto two perpendicular planes, and transform it into curves contained in these planes.

Like Euler, Clairaut works with a fixed coordinate system (the two perpendicular planes he refers to). His space curves are generally given by two polynomial equations in three variables (he calls such curves "geometric", but he notes in the introduction to his memoir that the same methods apply to transcendental curves). The surface upon which the curves are drawn is given by an equation in three variables. Clairaut assumes that the curve he studies is contained in a right solid angle, and he projects it onto the three planes that bound this solid angle. He develops two approaches to the study of such a curve, the first one by considering its projection onto the coordinate planes and the second one by considering it as a curve drawn on a curved surface. The fact that a curve is defined by two

[43] Vol. II, p. 378 of [52].

[44] The name "courbes à double courbure" was already used by Henri Pitot (1695–1771) in a memoir titled *Sur la quadrature de la moitié d'une courbe qui est la compagne des arcs, appelée la compagne de la cycloïde* (On the quadrature of half of a curve which is the companion of arcs called the companion of the cycloid), presented to the Académie royale in 1724 and published 2 years later [97]. Talking about a spiral on a cylinder, Pitot writes: "The Ancients called this curve spiral or helix, because its construction on the cylinder follows the same analogy as the construction of the ordinary spiral on a plane, but it is very different from the ordinary spiral, being one of these curves with double curvature [*à double courbure*] or a line which one can conceive as traced out on the curved surface of a solid. [Les anciens ont nommé cette courbe spirale ou hélice, parce que sa formation sur le cylindre suit la même analogie que la formation de la spirale ordinaire sur un plan, mais elle est bien différente de la spirale ordinaire, étant une des courbes à double courbure ou une des lignes qu'on conçoit tracée sur la surface courbe des solides.] Pitot, like many other French mathematicians of that epoch, was an engineer.

equations in three variables makes it naturally the intersection of two surfaces. When Clairaut talks about a normal to a space curve, he means a normal to a surface that contains the curve. He exploits his two approaches in the description of tangents and tangent planes and he calculates path integrals in many explicit examples. In his development, Clairaut uses only first derivatives. He studies objects such as loci of intersections with the coordinate planes of lines that are tangent or normal to the given curve. Such loci were thoroughly investigated later by Monge and his students.

Clairaut does not introduce any explicit notion of curvature, nor of osculating plane for a space curve. Such notions were introduced by Monge and his students, in the work we review below. Clairaut notes in his memoir that a theory of surfaces in three-space needs to be developed, and that he once had the intention to write a treatise on that subject. He mentions that the only known fact about such surfaces is that they can be expressed by a three-variable equation, a fact which, he says, is contained in an article by Bernoulli published in the Leipzig Acta.

There is a solid which, in the eighteenth century literature, is called the *Clairaut spheroid*. This is our familiar earth. The name was coined after Clairaut's treatise, *Théorie de la figure de la terre, tirée des principes de l'hydrostatique* (Theory of the shape of the earth, drawn from the principles of hydrostatics) [31], published in 1743, in which he develops a theory that supports the fact that the earth is not spherical but has the form of an ellipsoid of revolution. Clairaut's arguments came from physics, and at the same time they were based on measurements made during a famous expedition in Lapland done in 1736–1737. The expedition, whose aim was to settle a question that was hanging on since several decades concerning the shape of the earth, lasted 16 months. It was supported by the French Academy of sciences, it was headed by Pierre-Louis de Maupertuis, and it included several French scientists as well as foreign ones like the Swedish astronomer Anders Celsius.

The question of the spheroidal shape of the earth started with Newton who stated in 1687 that the earth is flattened at the poles, and, in fact, predicted a flattening whose magnitude is of the order of 1/230 (*Principia*, Book III, Proposition XIX). This led to a dispute between the English scientists and a large part of the French, led by the astronomer Jacques Cassini. The latter believed, on the contrary, that the earth was elongated at the poles. Johann Bernoulli, with his so-called vortex theory, was on the side of Cassini, whereas Huygens, Maupertuis and Clairaut were on the side of the English. Clairaut, in his *Théorie de la figure de la terre*, confirmed Newton's conclusions, even though he showed that the latter had made a mistake in his computations.

We now pass to the work of Monge on space curves.

In 1769, Gaspard Monge, who was twenty-three, sent a letter to the editors of the *Journal encyclopédique*, concerning the curvature of space curves. The letter was published in that journal and it was reprinted with a commentary by René Taton in [113]. It is 2 pages long, and it constitutes the first published work of Monge. It carries the title *Sur les développées des courbes à double courbure* (On the evolutes of curves with double curvature). The letter is an announcement of results that concern the evolutes of space curves. These results were expanded in a paper Monge presented at the Academy of sciences 2 years later, titled *Mémoire sur les développées des courbes à double courbure et leurs inflexions*.

At the beginning of the letter, Monge points out to the editors of the *Journal* that before he started working on the subject, nobody had developed any theory of evolutes of non-planar curves, and that even in the planar case, people worked out only the planar evolutes of these curves. In his theory, a plane curve has, like a space one, infinitely many evolutes (except in one single instance of a plane curve). The evolutes of a curve are space curves, and Monge showed that they lie on a developable surface, of which he gave a differential equation. After recalling how to find the center of curvature of a plane curve at a point (or, as Monge says, of an infinitely small arc), as the intersection of two infinitely close perpendiculars, he considers the case of space curves, and of plane curves but considered in 3-space. In this case, the perpendiculars are not unique, and at a given point, they form a plane. The set of intersection points of two infinitely close such planes constitutes a surface which Monge calls the "surface of centers of curvature", and he shows that this surface is also a developable surface. In the case where the curve we started with is planar, the surface is a cylinder whose basis is the usual evolute of the plane curve. Monge says in his announcement that he can write the equations of such a surface, and that this surface is the locus of all the evolutes of the given curve. He gives a way of constructing such an evolute using a rod which is tangent to the surface, in a way that generalizes the classical construction of the evolute of a plane curve. The paper ends with statements on the inflection points of space curves.

Monge's *Mémoire sur les développées des courbes à double courbure et leurs inflexions* was published in 1785, and it was also reprinted in his treatise *Application de l'analyse à la géométrie* [88]. In this memoir, Monge gives the expressions of the normal plane, of the center of radius of the first curvature, of the evolutes, of the polar lines, of the osculating sphere, of the simple inflection points (where the second curvature vanishes) and of the double inflection points (where the first curvature vanishes).

We shall say more on Monge's work in the next section which concerns surfaces. Before that, we need to review the work of Monge's student Lancret on curves.

Michel-Ange Lancret, who studied at the École Polytechnique, where Monge was teaching, is known for his participation (like Monge, Fourier, Berthollet, Ampère, Geoffroy Saint-Hilaire and many other scholars) in Napoleon's expedition to Egypt. In his short life, during which he was very active as an explorer and civil engineer, he wrote two important memoirs on space curves, the *Mémoire sur les courbes à double courbure* (Memoir on curves with double curvature) [78] (published in 1806) and the *Mémoire sur les développoïdes des courbes planes, des courbes à double courbure, et des surfaces développables* (memoir on evolutoids of plane curves, of curves with double curvature and of developable surfaces) [79] (published after his death, in 1811).

Lancret's work was based on those of Monge and Fourier. His work involves an interplay between curvature properties of space curves and surfaces. In the first memoir, he defines the concepts of curvature and torsion of a space curve in terms of the infinitesimal rotation angles of normal planes and osculating planes. He uses for these concepts the terms *first flexion* and *second flexion*, which correspond respectively to our notions of contingence angle (the angle made by two infinitesimally close normal planes) and torsion (the angle between two infinitesimally close osculating planes). In considering space curves, Lancret, following Monge, studies their evolutes. In his first memoir, he starts by giving a proof of an unpublished theorem he learned from Fourier, saying that the first flexion of the involute of a curve is equal to the second flexion of its plane evolute, and that the converse is true: the first flexion of the plane evolute of a curve is equal to the second flexion of its involute. Lancret then introduces a developable surface which is the envelope of the planes that pass through the points of a curve that are perpendicular to the principal normals. He establishes relations between this surface, the osculating surface and the envelope of the normal planes. He makes several connections between the two flexions of a space curve and those of its evolutes.

In his second memoir [79], Lancret develops a new theory, namely, the theory of the evolutoid of a plane or a space curve. This is a space curve whose tangent lines cut a given space curve at a constant angle different from a right angle.

The theory of space curves and the interplay between that theory and that of developable surfaces continued to be a subject of intense investigation by French geometers. We mention the *Mémoire sur la théorie des courbes à double courbure* (1850) [14] by Joseph Bertrand, in which the author studies the principal normals of a space curve and in particular conditions under which two curves have the same principal normals. Bertrand also investigated conditions under which the generatrices of a ruled surface are the lines carrying the radii of curvature of a curve traced on some surface, and conditions under which a ruled surface is the locus of the principal normals of a curve.

In the second half of the nineteenth century, the study of space curves was the subject of several doctoral dissertations in France. We mention as examples the dissertation of Jean-Frédéric Frenet, *Sur les courbes à double courbure*, defended in 1847, in which the author introduces the so-called Frenet-Serret formulas found independently by Serret in his paper *Sur quelques formules relatives à la théorie des courbes a double courbure* published in 1851. We also mention the dissertation of Paul Appell, *Sur les cubiques gauches et le mouvement hélicoïdal d'un corps solide*, defended in 1876, and that of Émile Picard, *Application de la théorie des complexes linéaires à l'étude des surfaces et des courbes gauches*, defended in 1877.

7.6 Curvature of Surfaces: Monge and His School

We now return to surfaces, reviewing works of Monge, Meusnier, Dupin and Olinde Rodrigues.

Monge was in several ways the continuator of Euler's work on curvature of surfaces. For instance, he highlighted the significance of the two orthogonal line fields that are tangent to Euler's minimal and maximal directions of radii of

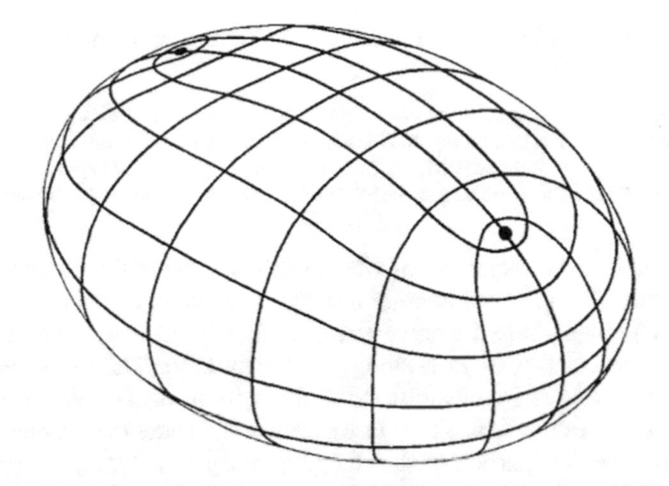

Fig. 7.9 The pair of singular foliations of an ellipsoid by lines of curvature, discovered by Monge (Extracted from Hilbert and Cohn Vossen's *Geometry and the imagination*)

curvature. These line fields define two orthogonal foliations on the surface, and this constitutes one of the first explicit appearances in the mathematical literature of the notion of foliation. Monge called these foliations *lines of curvature* [87]. Figure 7.9 shows two orthogonal foliations by lines of curvature of an ellipsoid. Dupin, in his *Essai historique sur les services et les travaux scientifiques de Gaspard Monge*, reports that when Monge presented his construction of lines of curvature of an ellipsoid at a lecture he gave at the École polytechnique, Lagrange, who was present, was delighted and said to Monge: "I would have liked to be the author!"[45] Dupin adds that Monge was most flattered by Lagrange's reaction and that during the rest of his lifetime, he liked to repeat this episode to his friends [39, p. 120].

Monge coined the expression *umbilical point*, to denote the points where the two principal curvatures have the same value. These points, which Euler mentioned without further discussion in his study of principal directions, are the singular points of the foliations by the lines of curvature.

Monge continued Euler's work on developable surfaces, in his *Mémoire sur les propriétés de plusieurs genres de surfaces courbes et particulièrement sur celles des surfaces développables avec une application à la théorie générale des ombres et des pénombres* (Memoir on the properties of several species of curved surfaces and particularly those of developable surfaces with an application to the theory of shadow and twilight) [84] and his *Mémoire sur les développées, les rayons de courbure et les différents genres d'inflexions des courbes à double courbure* (Memoir on evolutes, radii of curvature, and the various species of inflexions of curves with double curvature) [86]. He expressed several times his debt to Euler,

[45] Je voudrais en être l'auteur!

whose works have motivated his investigations. In the first memoir just mentioned, he writes:

> Having resumed this matter, at the occasion of a memoir that Mr. Euler gave in the 1771 volume of the Petersburg Academy on developable surfaces, in which this famous geometer gives formulae to determine whether a given surface may or may not be applied onto a plane, I reached results on the same subject which seem to me much simpler, and whose usage is much easier.[46]

Monge published a memoir [85] whose subject is the optimal transport of two given volumes of sand (or of another material constituted of infinitesimally small particles). The memoir is titled *Mémoire sur la théorie des déblais et remblais* (Memoir on the theory of excavating and backfilling). This is an engineering problem which acted as a motivation for his theory of lines of curvature of a surface and for his theory of developable surfaces. He worked under the hypothesis that the trajectories of the sand particles, when they are transported during the operations of excavation and backfill, are straight lines, and he showed that these lines are normal to a unique surface. He collected these trajectories into groups of developable surfaces whose geometry is related to the curvature of the unique surface. This topic was further developed by Dupin and by other prominent mathematicians; we shall talk about this below.

In 1807, Monge published a treatise, *Applications de l'analyse à la géométrie* [88], which became one of the most influential mathematical textbooks in nineteenth-century mathematics. It contains the lectures he gave at the *École polytechnique*, edited in the form of a collection of papers.[47] In this treatise, Monge systematically translates every geometric question concerning curves and surfaces into the language of partial differential equations. A comprehensive reference on Monge is the book [112] by René Taton.

Another important complement to Euler's theory of curvature of surfaces was provided by Jean-Baptiste Meusnier (1754–1793), who was a student of Monge at the military school of Mézières. It is interesting to read a report that Monge left on his first meeting with the young Meusnier; it shows clearly the lineage with Euler[48]:

> I saw the young Meusnier for the first time when he entered as a student the *École du génie* of Mézières, where I was a professor. He was eighteen [...] Right on the day of his arrival at Mézières, he came to see me in the evening and expressed his wish that I propose to him a question which would allow me to be able to appreciate his degree of knowledge and to be

[46]Ayant repris cette matière, à l'occasion d'un mémoire que M. Euler a donné dans le volume 1771 de l'Académie de Pétersbourg, sur les surfaces développables, et dans lequel cet illustre géomètre donne des formules pour reconnaître si une surface courbe proposée jouit ou non de la propriété de pouvoir être appliquée sur un plan, je suis parvenu à des résultats qui me semblent beaucoup plus simples, et d'un usage bien plus facile pour le même sujet.

[47]The fifth edition of Monge's *Applications* is the most well known, because it contains extended notes by Liouville.

[48]Extracted from the biographical article by Taschereau [111] where the author reproduces handwritten notes by Gayvernon, vice-director of the École Polytechnique, where Monge was also teaching.

able to judge his capabilities. In order to satisfy him, I talked with him about Euler's theory of maximal and minimal radii of curvature of curved surfaces. I explained to him the main results and I suggested to him to look for the proof. The next morning, in the classrooms, he gave me a small piece of paper containing that proof. But what was the most remarkable is that the considerations he made were much more direct, and his approach much faster than the one Euler used. The elegance of that solution and the very short time it costed him gave me an idea of his wisdom and of that exquisite feeling for the nature of things of which he has given multiple proofs in all the works he did after that. I then showed him the volume of the Berlin Academy which contains Euler's memoir on this subject. He soon recognised that the methods he used were more direct than those of his model. It turned out they were also more productive, and he reached results that Euler had missed. He wrote a memoir which I sent on his behalf to the Academy of Sciences and which was published among those of the associate members.[49]

On the 14th and 21st of February 1776, Meusnier, who was twenty-one, presented his first discoveries on the curvature of surfaces at the Académie des Sciences. We learn from a biography of Meusnier written by Darboux [34] that d'Alembert, who was present at the Academy, declared: "Meusnier starts while I finish." The same year, and despite his very young age, Meusnier was made corresponding member of the Academy. He wrote only one mathematical paper, the *Mémoire sur la courbure des surfaces* [82], published in 1785. He became a general in the French army and pursued a military career.

In his paper, Meusnier gave a formula that complements Euler's formula on the curvature of curves obtained by normal sections, namely, he gave, for a given point on the surface, the curvature of the curves obtained by intersecting the surface by oblique planes, relating them to the curvature at that point of the corresponding normal sections, that is, the intersection of the surface with the normal sections having the same tangent vector. The precise formula says that at a point on the surface, the curvature κ_ϕ of a curve obtained by intersecting the surface by a plane

[49]La première fois que j'eus l'occasion de voir le jeune Meusnier, ce fut lorsqu'il vint en qualité d'élève à l'École du génie de Mézières où j'étais professeur, il avait dix-huit ans. [...] Le jour même de son arrivée à Mézières, il vint me voir le soir, et il me témoigna le désir qu'il avait que je lui proposasse une question qui me mît à portée et de connaître le degré de son instruction et de juger ses dispositions. Pour le satisfaire je l'entretins de la théorie d'Euler sur les rayons de courbure *maxima* et *minima* des surfaces courbes; je lui en exposai les principaux résultats, et lui proposai d'en chercher la démonstration. Le lendemain matin, dans les salles, il me remit un petit papier qui contenait cette démonstration ; mais ce qu'il y avait de remarquable, c'est que les considérations qu'il avait employées étaient beaucoup plus directes, et la marche qu'il avait suivie était beaucoup plus rapide que celles dont Euler avait fait usage. L'élégance de cette solution, et le peu de temps qu'elle lui avait coûté, me donnèrent une idée de la sagacité et de ce sentiment exquis de la nature des choses dont il a donné des preuves multipliées dans tous les travaux qu'il a entrepris depuis. Je lui indiquai alors le volume de l'Académie de Berlin dans lequel était le mémoire d'Euler sur cet objet; il reconnut bientôt que les moyens qu'il avait employés étaient plus directs que ceux de son modèle; ils devaient être aussi plus féconds, et il parvint à des résultats qui avaient échappé à Euler. Il en composa un mémoire que j'adressai de sa part à l'Académie des Sciences et qui fut imprimé parmi ceux des savants étrangers.

containing a tangent vector making an angle ϕ with the normal plane containing the same tangent vector satisfies the relation

$$\kappa_\phi \cos \phi = \kappa,$$

where κ is the curvature of the curve cut out by the normal section. He proved at the same time that at any point on a surface and for any tangent curve at that point, the osculating circles to a curve obtained by the various oblique sections having the given tangent vector lie on the same sphere.

In the same paper, Meusnier introduced a notion of osculating torus for a surface at a point, generalizing the notion of osculating circle for a plane curve. In fact, he defined, at a given point of a surface, two osculating tori at each point, each one obtained by rotating the osculating circle of one principal section along the axis of the osculating circle of the second principal section. Each such torus has first and second derivatives equal to those of the surface at the given point. His proof of Euler's theorem and of the so-called Meusnier formula which we mentioned above are based on these tori. Meusnier also used these tori to give a characterization of minimal surfaces. He also gave a characterization of the surfaces for which the two extremal radii of curvature are always equal, a condition satisfied at the so-called umbilical points, and he solved a problem on developable surfaces. In his paper, Meusnier also showed that the helicoid obtained by rotating and displacing a straight line around an axis is a minimal surface of average curvature zero.

Charles Dupin is another student of Monge who made significant advances in the theory of curvature, establishing relations between differential geometry and the theory of conics and, more generally, projective geometry. Needless to say, as a student of Monge, Dupin was knowledgeable in projective geometry.

As a first-year student at the École Polytechnique, Dupin became known for a solution of a problem concerning spheres that are tangent to three given spheres. He found a property of the envelope of that family of spheres, namely, that its lines of curvature are two systems of circles. Such a surface is now called a Dupin cyclide. In his article on Monge, Dupin writes that the latter already considered surfaces that are envelopes of families of surfaces, and he says that in his last two memoirs, written amidst the dangers and the works of the Egyptian expedition, Monge applied the theory of curvature of surfaces to the study of the equations and the properties of a surface that is the envelope of a family of spheres of varying radii distributed along an arbitrary curve [39, p. 113 and p. 122]. The problem of spheres tangent to three given spheres is a generalization of a famous set of problems of Apollonius, asking for a circle tangent to three circles, where the latter may have radius zero or infinity, that is, they may be a point or a line. Apollonius wrote two books on the subject, which he titled *On contacts*. The books are lost, but later authors refer to them (see e.g. Propositions 9, 10, 14 and others in Book IV of Pappus' *Collection* [94]).

Dupin, like his teacher, expressed at several places his admiration for Euler. In his treatise *Développements de géométrie* [37, p. 107], he talks about the "wonderful theorem of Euler who was, for modern geometry, the source of the greatest progress:

At any point of a surface, the two directions of greatest and smallest curvature are always at right angles."[50]

Dupin considered himself (and he indeed was) a true disciple of Monge. His treatise *Développements de géométrie* [37] is dedicated to his master. The treatise is a collection of several memoirs, divided into two sections. The first section contains three memoirs, and the second one contains two. The topics studied in this treatise include the osculation of surfaces by ellipsoids, the theory of conjugate tangents, the so-called Dupin indicatrix and the theory of orthogonal trajectories applied to the determination of lines of curvature. Let us briefly review some of these topics.

An osculating ellipsoid, now called *Dupin ellipsoid*, is introduced in the first memoir. The osculating circles of all the normal sections of the surface are also the osculating circles of the normal sections of this ellipsoid. The set of curvatures of the surface at the given point is the set of curvatures of the ellipsoid at its vertex (the contact point with the surface). One of the axes of this ellipsoid coincides with the normal at that point. This says in particular that the directions of smallest and largest curvature of the surface at that point are perpendicular, as for the ellipsoid. One may deduce, from the existence of this ellipsoid with its properties, Euler's theorem saying that the curvatures of all the normal sections may be deduced from the two principal curvatures.

The *Dupin indicatrix* is an assignment, at every point of the surface, of a conic in the tangent space, centered at the origin, which Dupin called *curvature indicatrix* ("indicatrice de courbure"; it indicates the directions of the two principal curvatures at that point). The system of *conjugate diameters* of the conic represents the system of conjugate tangents. In the case where the indicatrix is an ellipse (parabolic point), its axes represent the directions of largest and smallest curvature of the surface at the given point, and the normal radius of curvature in any direction at a point in the surface is equal to the square of the distance from the corresponding point of the ellipse to its center. The Dupin indicatrix is reduced to a line when one of the principal curvatures vanishes, while the other is infinite. If the indicatrix is an ellipse (respectively a hyperbola), then the sum (respectively the difference) of the two radii of curvature of the sections which correspond to two conjugate tangents is constant, and it is equal to the sum (respectively the difference) of the two principal radii. If the indicatrix is a parabola, one of these radii becomes infinite, and the curvature vanishes. This happens for instance at any point of a developable surface.

Dupin gave a construction of the indicatrix. In the case of a parabolic point, the first approximation of the indicatrix is obtained by intersecting the given surface with a plane parallel to the tangent plane at that point and infinitesimally close to that tangent plane, then projecting this approximation onto the tangent plane. In the case of a hyperbolic point, one has to take two infinitesimally close planes, one from each side of the surface. Dupin also developed a theory of *asymptotic directions* of

[50][...] admirable théorème d'Euler, qui a été pour la géométrie moderne la source des plus grands progrès: *En chaque point d'une surface, les deux directions de plus grande et de moindre courbure sont constamment à angles droit.*

the surface (defined by the asymptotic directions of the indicatrix) and of *conjugate tangents* at each point. The latter are obtained by constructing a developable surface touching our surface along a certain curve. Then, at the given point, the tangent to the curve and the line of the developable surface at this point, form two conjugate tangents.

Dupin's theory of conjugate directions led him to define a notion of conjugate families of curves on the surface. This is a generalization of Apollonius' theory of conjugate diameters, to the setting of differentiable surfaces. The two lines of curvature of a surface are examples of conjugate lines. Dupin gave a relation between the curvature in some direction of the surface and the curvature in the conjugate direction. He proved that the sum of the radii of curvature in conjugate directions is constant.

In the third memoir of his *Développements de géométrie*, Dupin determines points where the indicatrix is a circle (that is, all the curvatures of normal sections are equal), making relations with Monge's umbilics. Dupin also introduced an object that he called a "system of triply orthogonal coordinates in space", that is, a system of three families of surfaces that meet everywhere orthogonally. He showed that any surface in any such system meets the other surfaces in its lines of curvature.

Dupin, like Monge and others, who were engineers, published memoirs and books on the application of the theory of curves and surfaces to practical problems. We mention in particular his memoir *Applications de la géométrie à la mécanique* [40], in which he studied the stability of floating bodies, that reminds us of Archimedes. One may also mention his memoir on earthwork [36] and his *Exercices* [38], concerned with the optimal transport of a volume of particles of sand, continuing the work of Monge on the subject. Dupin had a more general approach compared to his master, namely he allowed the paths taken by the particles to be curved. He writes in [38]: "The only way to add something to the results of a famous predecessor is to consider the case where the paths cannot all be rectilinear, but depend on the form and the slope of the ground on which they must be traced."[51] This theory of optimal transport was further developed by Paul Appell in his booklet *Le problème géométrique des déblais et des remblais* (The geometrical problem of cut and fill) [7], in which the author also reviews the works of Monge and Dupin on this subject.[52]

We also mention Dupin's memoir on optics, *Mémoire sur les routes de la lumière, dans les phénomènes de la réflection et de la réfraction* (Memoir on the routes of light in the phenomena of reflexion and refraction), in which he gives, among other things, a complete theory of *cyclides*, that is, surfaces whose lines of curvature are circles. In his article on Monge, he writes that the questions of transport of particles

[51]Le seul moyen d'ajouter quelque chose aux recherches d'un illustre devancier est de considérer le cas où les routes ne sauraient être toutes rectilignes, mais dépendent de la forme et de la pente du terrain sur lequel elles doivent être tracées. (quoted by Appell in [7]).

[52]There seems to be a "strange case" of plagiarism in this booklet by Appell, [7] which contains several pages copied verbatim from a memoir by Albert de Saint-Germain published in 1886; cf. the paper [106] by Roitman and Le Ferrand where this issue is discussed.

of sand have led, by extension, to general theorems on mathematical optics [39, p. 131].

Dupin later became a naval engineer, and a friend and a confident of Napoleon. He was a lover of modern Greece. He spent several years in the island of Corfu and he founded there, in 1808, together with Antoine-Marie Augoyat, a fellow student at the École polytechnique, an academy which became known as the Ionian Academy. He became a senator under Napoleon III.

To conclude with the work on Dupin, let us note that Poincaré, in 1847, published a paper titled *Démonstration nouvelle des propriétés de l'indicatrice d'une surface* (A new proof of properties, of the indicatrix of a surface) [98] in which he gives new proofs and complements to results of Dupin. Poincaré starts by giving a proof of the following result: If at each normal section at a given point on a surface we take, at the intersection with the tangent plane, a segment whose length is the square root of the radius of curvature of that section, then the result obtained is a conic. He then gives a new proof of Meusnier's theorem as well as a generalization, and a series of properties of the Dupin indicatrix, including results on conjugate tangents. Poincaré's proofs do not involve any calculation or equation, but only the synthetic theory of conics.

Olinde Rodrigues (1795–1851), who belonged to the circle of Monge,[53] wrote in 1815 and 1816 two memoirs on lines of curvature, *Sur quelques propriétés des intégrales doubles et des rayons de courbure des surfaces* (On some problems of double integrals and radii of curvature of surfaces) [104], and *Recherches sur la théorie des lignes et des rayons de courbure des surfaces, et sur la transformation d'une classe d'intégrales doubles qui ont un rapport direct avec les formules de cette théorie* (Researches on the theory of lines and of radii of curvature of surfaces, and on the transformation of a class of double integrals which have a direct relation with the formulas of this theory) [105]. Rodrigues is known for a formula on this subject carrying his name, and for developing the idea of a representation of a differentiable surface on a sphere of radius one which amounts to the Gauss map. He used such a representation to study the ratio of the area of (a piece of) the original surface to the area of its image on the sphere, arriving at a measure of curvature which in fact is the Gaussian curvature at a point of a surface.

7.7 Twentieth Century: Return to Euclid

The work of Busemann and Feller on which we shall comment below is strongly influenced by Hilbert's vision on mathematics. Both Busemann and Feller obtained their PhD in Göttingen, and although their thesis advisor was Courant, they were

[53]Strictly speaking, Rodrigues was not a student of Monge, but according to Taton [112], Monge considered him as his student. Rodrigues became a follower of Saint-Simon (1760–1825), the famous utopian socialist whom he met in 1823, 2 years before Saint-Simon's death, and he became the main promoter of his ideas.

shaped by Hilbert's view on the foundations of geometry. Feller obtained his PhD in 1927, and Busemann in 1931. Before talking about their work, let us say a few words on some work of Hilbert related to our topic.

In 1932, together with Stephan Cohn-Vossen,[54] Hilbert published his remarkable book *Geometry and the Imagination*[55] [69] which is based on Hilbert's lectures held in 1920/21. Let us quote Hilbert from the preface, about the aim of this book:

> Here we want to look at geometry in its present condition from the point of view of the imagination. With imagination, we can approach the manifold geometric facts and questions, and moreover in many cases we can give an idea in imaginative form of the methods of investigation and proof leading to the discovery of the facts, without having to give all the details of the conceptual theories and the calculations.[56]

The point of view is indeed purely geometric, unlike the one of Euler or Monge and his students, which is based on analytic geometry. In a sense, it is a return to the methods of the Greeks. Let us quote from the chapter on differential geometry [69, p. 155] their description of the osculating circle at a point P_1 on a plane curve, a notion which we encountered several times in this essay. It says the following:

> We draw a circle through P_1 and two neighbouring points on the curve. If we let the two neighbouring points approach P_1, the corresponding circle converges to a limit position [...] This limit circle is called the osculating circle at P_1. Because of the construction described, it is usual to say that the curvature circle has three points running together in common with the curve. Similarly, one says that the tangent and the curve have two points running together in common.[57]

The osculating circle of a curve at a point P typically changes sides of the curve at the touching point P. If we consider all circles through P touching the curve at P, then in general those circles of radius greater than the radius of curvature r lie on one side of the curve whereas those of radius less than r lie on the other side of the curve. This is represented in Fig. 7.10, which is extracted from Cohn-Vossen's book (see [69, Figure 186]).

These ideas directly translate to space curves. To construct the osculating plane of a space curve at a point P, we can consider the limiting position of the

[54]For a comprehensive review on Cohn-Vossen's work, we refer the reader to the article [1] by Alexandrov.

[55]The original German title is: Anschauliche Geometrie.

[56]Wir wollen hier die Geometrie in ihrem gegenwärtigen Zustand von der Seite des Anschaulichen aus betrachten. An Hand der Anschauung können wir uns die mannigfachen geometrischen Tatsachen und Fragestellungen nahebringen, und darüber hinaus lassen sich in vielen Fällen auch die Untersuchungs- und Beweismethoden, die zur Erkenntnis der Tatsachen führen, in anschaulicher Form andeuten, ohne daß wir auf die Einzelheiten der begrifflichen Theorien und der Rechnung einzugehen brauchen.

[57]Wir legen einen Kreis durch P_1 und zwei benachbarte Punkte auf der Kurve. Wenn wir dann die beiden Nachbarpunkte auf der Kurve gegen P_1 rücken lassen, nähert sich der Kreis einer Grenzlage. [...] Man nennt diesen Kreis den Krümmungskreis der Kurve in P_1. [...] Der angegebenen Konstruktion wegen pflegt man zu sagen, daß der Krümmungskreis drei zusammenfallende Punkte mit der Kurve gemein hat. Ebenso sagt man, die Tangente hat mit der Kurve zwei zusammenfallende Punkte gemein.

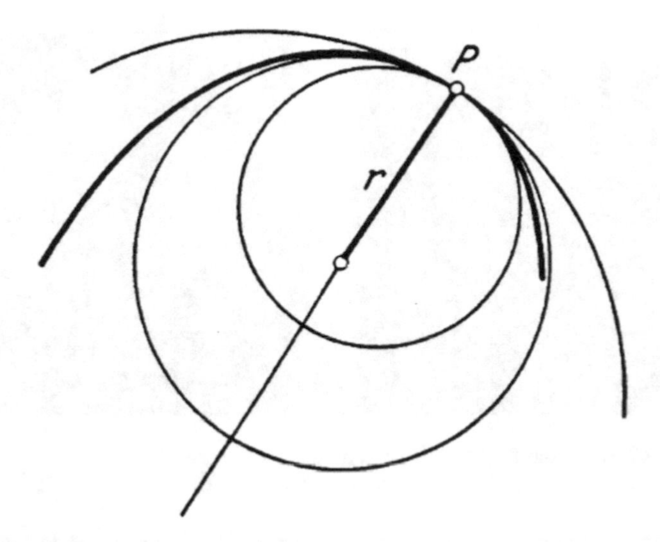

Fig. 7.10 The osculating circle at P separates two kinds of circles tangent to the curve at P: those that lie on one side of the curve and those that lie on the other side of the curve (in the neighbourhood of the point P). Figure from [69]

plane through the tangent at P and a neighbouring point that approaches P [69, p. 158]: "In the same sense as above, the osculating plane has three points that are running together in common with the curve."[58] Let us note incidentally that the term "osculating plane" can be traced back to Johann Bernoulli, specifically to his memoir *Problema: in superficie curva ducere lineam inter duo puncta brevissimam* (Problem: to draw on a curved surface a shortest line between two points) [13].[59] Bernoulli used this notion in the context of his study of geodesics on surfaces. In fact, he characterized geodesics as the curves on the surface whose osculating plane is always perpendicular to the tangent plane of the surface.

In [69], §32, Hilbert and Cohn-Vossen discuss eleven properties of the sphere, some of which characterize it and others do not. For example the third property says that the sphere is a surface of constant breadth and all the projections have constant circumference. The property of having constant breadth means that the distance between any two parallel tangent planes is constant. It is a generalization for surfaces of the property with the same name that we encountered in Euler's memoir on triangular curves [57]. As Hilbert and Cohn-Vossen explain, there are many other convex surfaces sharing this property. Figure 7.11 is extracted from Hilbert and Cohn-Vossen's book and reproduces pictures of such surfaces. In the

[58]Die Schmiegungsebene hat im früher erläuterten Sinn drei zusammenfallende Punkte mit der Kurve gemein.

[59]Bernoulli writes: Voco autem *planum osculans* quod transit per tria curvae quaesitae puncta infinite sibi invicem propinqua. (I also call an *osculating plane* one that passes through three points on the given curve that are infinitely close to each other).

Fig. 7.11 Surfaces of constant breadth, from [69]

picture on the left hand side, one can see a surface having as a section a curve with three cusps (a triangular curve in the sense of Euler). The fourth property in §32 says that all points on a sphere are umbilical, which in fact characterizes the sphere. The fifth property concerns the focal points of a surface. Let us explain this notion. The set of centers of curvature of normal sections at a point of a surface is in general a segment contained in the normal section at that point, whose extremities are the centers of curvature of the principal directions. These extremities are called *focal points*, and we already saw that these two points coincide if and only if the given point is umbilical. The fifth property then states that the focal points of the sphere do not form a surface. Indeed the sphere is the only closed surface that has a unique focal point. But there are other surfaces where the set of focal points degenerates to a curve. These are the surfaces we mentioned, discovered by Dupin, called Dupin cyclides.

After Hilbert and Cohn-Vossen, and before going over to the work of Busemann and Feller, we briefly review the work of Bonnesen and Fenchel on convexity.

It is interesting to note that there was a strong interaction between the whole group of four mathematicians: Fenchel, Feller, Bonnesen and Busemann. Werner Fenchel obtained his PhD in 1928 in Berlin under Bieberbach. The title of his dissertation was *Über Krümmung und Windung geschlossener Raumkurven* (On curvature and torsion of closed space curves) [62]. The topic was differentiable curves in 3-space. The main results concern the image of the curve obtained by taking the tangent vectors of a given curve parametrized by arc length. This curve of tangents is considered as a curve lying on the unit sphere. Fenchel proves that (1) the total curvature of a space curve is bounded below by 2π, with equality if and only if the curve is planar and convex; (2) if the curve of tangents has a double point, then the torsion changes sign unless it is constant. The first of these two results is known as "Fenchel's theorem", and it was generalized in several ways by various people. The proofs of the results in Fenchel's thesis are very geometric in spirit. After his PhD, Fenchel moved to Göttingen, where he became Landau's assistant.

Busemann and Feller were also working in Göttingen, as young mathematicians. Fenchel remained there until 1933, when Hitler was appointed chancellor, with two interruptions between 1930 and 1932 during which he spent some time in Copenhagen, where Tommy Bonnesen was working. In 1933, Busemann, Feller and Fenchel fled to Denmark. Fenchel spent essentially the rest of his career there, whereas Busemann and Feller eventually moved to the US. The reader interested in this turbulent period of the history of Europe may refer to Feller and Busemann's biographies contained in their Selected Works editions, [61, Vol. 1] and [25, Vol. 1]. We also refer to Busemann's biography in [93].

The book [17] by Bonnesen and Fenchel, published in 1934, a year after Fenchel's move to Denmark, belongs to the Göttingen tradition of Hilbert's foundations of geometry. It is one of the founding books of modern convexity theory and, at the same time, of the synthetic point of view on curvature. §17 of the book starts with the following statement: "Far-reaching assertions can be made about the curvature relations of convex curves without making any assumption of differentiability" [17, p. 153]. Indeed, in that section and the next one, the authors attempt to develop a geometric theory of curvature of convex curves and surfaces without differentiability conditions.

Convex curves in the plane (respectively convex surfaces in 3-space) have support lines (respectively support planes) which sometimes can play the role of tangent lines (respectively planes). We recall that for any point on a convex hypersurface (that is, the boundary of a proper open convex set of maximal dimension in some Euclidean space), a support hyperplane is a hyperplane that contains that point but no point from the open convex subset it bounds. Any point on a convex surface is contained in a (not necessarily unique) support hyperplane. If the surface is differentiable at the given point, then the support hyperplane is unique and coincides with the tangent plane at that point.

Bonnesen and Fenchel call an *element* of a convex curve C in the plane a pair (P, s) consisting of a point P and a support line s at P. A *normal line* to this element is a line passing through P and perpendicular to s. An *osculating circle* at (P, s) is defined as a geometric limit of circles each of which touches s at P and which passes through a sequence of points on the curve that converges to P. The radius of such a circle is then a *radius of curvature* of the curve C at P. The values 0 and ∞ are allowed; in such cases the osculating circle is a point or a line respectively. To each element (P, s) is associated a set of osculating radii and of radii of curvature, each being an interval, the former contained in the latter. A *circle of curvature* at (P, s) is a limit of circles through P whose centers are the intersections of the normal of s at P with the normals of a sequence of elements that converge to (P, s). One also defines the *center of curvature* of C at P as the limit (if this limit exists) of the intersection of the normal line at p with the normal lines of a sequence of points on C that converge to P. The center of curvature, when it exists, may or may not coincide with the center of the osculating circle of C at P. For a given element (P, s), there may be several osculating circles and several circles of curvature. An osculating circle is a circle of curvature but the converse is not necessarily true. In the case where the osculating circle is unique, there exists also a unique circle of curvature, and both circles coincide. This occurs for almost all points of the convex curve, with respect to the Lebesgue measure of arc length on the curve; cf. Jessen [73]. In the cases where the limits do not exist (at a set of points of measure zero), one may define, by taking the smallest upper bound and the greatest lower bound, the upper and lower centers of curvature on the normal line n and then define the *upper* and *lower* curvature of the curve at the point P.

Bonnesen and Fenchel refer to the papers by Hjelmslev [71] and by Jessen [73] for these notions and their basic properties. They relate this notion of greatest lower bound of curvature radii to the question of the largest circle that can roll unhindered on a convex curve. This kind of questions remind us of questions that Aristotle addressed, and which we recalled in Sect. 7.2 of this paper. Bonnesen and Fenchel refer to papers by Bohr and Jessen [16], Mukhopadhyaya [90] and Brunn [20]. Then they discuss osculating conics to convex curves, referring to papers by Hayashi [65], Mukhopadhyaya [89] and others. In his paper *Circles incident on an oval of undefined curvature* [91] published in 1931, Mukhopadhyaya studied contact and osculation of circles relative to a convex continuous curve in the plane under the assumption that a definite tangent exists at each point but with no assumption on the existence of an osculating circle.

All this concerns curves. Regarding surfaces, Bonnesen and Fenchel state that "systematic investigations of the curvature properties of convex surfaces without differentiability assumptions are not available" [26, p. 154]. They mention preliminary tools and methods of attack due to Bouligand [19], Hadamard [64], Cohn-Vossen [29] and a few others.

In their article [26], Busemann and Feller investigate curvature properties of convex surfaces in Euclidean space. They describe their goal in the introduction thus:

> The aim of this paper is to develop the elements of differential geometry of convex surfaces without making the usual regularity assumptions. Hereby necessarily, purely geometric considerations often replace the usual analytic arguments.[60]

Consequently, they start from purely geometric definitions of tangent, tangent plane, osculating circle, radius of curvature etc. in the spirit of the book of Hilbert and Cohn-Vossen [69]. Their approach allows them to consider topological surfaces in 3-space with mild extra assumptions, and in particular convex surfaces, without further regularity conditions. Their starting point is a definition of tangent line, approximating circle, curvature and osculating plane, generalized from the setting of curves in space to that of sequences of points in space converging to a given point (Section 1 of [26]). At this point, the reader may stop and think about the meaning of these words, in the simplest situation, and without any differentiability assumption. Here is how Busemann and Feller proceed.

We start with a sequence of points $\{P_n\}$ in 3-space converging to a point P. If the lines joining P_n to P converge to a line t, this limit is called the *tangent line* of

[60] Cf. the English translation of the whole article in [25, Vol. 1, p. 235].

$\{P_n\}$. As in the treatise by Bonnesen and Fenchel, the pair (P, t) is said to be a *line element*. If the sequence $\{P_n\}$ has a tangent line t, then the circle in the plane (t, P_n) through P_n that is tangent to t is called the *approximating circle* of P_n with respect to the line element (P, t). If ρ_n denotes the radius of this circle and if the limit ρ of ρ_n exists, then this limit is called the *radius of curvature*, and its inverse $\frac{1}{\rho}$ the *curvature* of the sequence of points $\{P_n\}$. If the planes through P_n and t converge to a plane E, then this plane E is called the *osculating plane* of $\{P_n\}$.

Other metric notions associated with a sequence of points $\{P_n\}$ are introduced in the same paper, such as the *foot of the perpendicular*, *normal segment* and *tangent segment* from P_n to P. Formulae for curvature radii, etc. are given in terms of limits of normal segments, tangent segments, etc. generalizing the classical formulae given in terms of derivatives.

From this, Busemann and Feller introduce as follows the notions of tangent, curvature and osculating plane for a space curve:

Let κ be a continuous space curve and P a point on κ. If for any sequence P_n of points on κ converging to P the tangent and curvature exist and are independent of the choice of P_n, then these tangent and curvature are defined to be the *tangent* and *curvature of* κ at P. The osculating plane of κ at P is defined similarly.

Several other metric definitions in the same spirit are given. In particular, if the tangent and the osculating plane exist but the curvature depends on the chosen sequence, then the upper limit of all the possible curvatures is called the *upper curvature of* κ. Choosing the sequence P_n to be on the same side of P on the curve, one obtains notions of right-sided and left-sided tangent and curvature.

Busemann and Feller compare these definitions with the classical ones by taking derivatives in the special case of plane curves and they show that their concept is broader. Namely, if a curve κ is given as the graph of a continuous function f of one variable x and if at some point x_0 of the interval of definition the right second derivative of f exists, then the usual notion of right curvature of κ at x_0 coincides with the one they introduce. They show that the converse is not true: there exists a C^1-function f with right curvature at x_0 (in the sense of the authors) admitting no right second derivative at x_0.

The case of convex plane curves is of particular interest. Here, at each point of such a curve, a left and a right tangent exists. Because the slope of these one-sided tangents is monotone with respect to the point, the curve has a two-sided tangent almost everywhere (up to countably many exceptions). This monotonicity property also implies that up to a set of measure zero, f has also almost everywhere a second derivative.

In the same trend, the authors give a geometric definition of a strict tangent plane at a point of a surface [26, § 2]: given a point P on a surface Φ in the Euclidean space, a plane Π is called *strict tangent plane* of Φ at P if for every sequence of secants $Q_n P_n$ of Φ converging to P, all the limits of the lines joining Q_n and P_n are contained in Π. Again, the case of a convex surface is a typical example: any tangent plane is automatically strict. Busemann and Feller indicate that this notion of strict tangent plane coincides with that of *paratangent* introduced by Bouligand [19].

Using these definitions, Busemann and Feller present generalizations of classical results on the differential geometry of surfaces. In particular, Euler's theorem on normal curvatures which we surveyed in Sect. 7.4 and Meusnier's formula which we mentioned in Sect. 7.6 serve as testing ground to demonstrate that the general setting is reasonable enough, that is, non-trivial results may be proven in it. More precisely, the general form of Meusnier's theorem, which holds also for non-convex surfaces, is the following: If at a point with a strict tangent plane some oblique plane section through a tangent t has curvature, then all plane sections through t have curvature as well, and the corresponding osculating circles lie on a common sphere. In particular, for a convex surface, Meusnier's formula holds almost everywhere.

For convex surfaces in 3-space, Busemann and Feller show that Euler's theorem on normal curvatures holds almost everywhere. They formulate their result in terms of Dupin's indicatrix. We already recalled in Sect. 7.6 that Dupin observed that Euler's equation relating to the normal curvatures at a point of an analytic surface define a conic (called Dupin indicatrix) in the tangent plane at the given point. In the case of a convex surface, the indicatrix can either be an ellipse or a pair of parallel lines unless all normal curvatures at the point vanish.

Busemann and Feller use the following construction as a definition of the indicatrix: Consider the tangent plane T at a point P of a convex surface where such a plane exists, and choose coordinates such that this tangent plane is not vertical and P is its origin. Take the plane parallel to T and lying above T, at distance ϵ. It intersects the surface in a curve, call it c_ϵ. Then, for almost every P, the curve c_ϵ rescaled by the factor $\frac{1}{\epsilon}$ converges, as $\epsilon \to 0$, to a curve, which the authors call the indicatrix at that point. Moreover, they show that it is either an ellipse or a pair of parallel lines. This means that for a convex surface, at almost all points normal curvatures exist and satisfy Euler's theorem. They then show that Rodrigues' formula for principal curvature directions holds in this general setting. They also obtain an umbilical theorem. We recall that in its classical form, the theorem says that an analytic surface in 3-space in which all points are umbilics (i.e. the maximal and minimal normal curvatures coincide) is either a piece of a sphere or a piece of a plane. In particular, this property characterizes the sphere among all convex closed analytic surfaces in 3-space (see also [69], §32, Property 4). Busemann and Feller generalize this result by proving that a differentiable convex surface where almost all points are umbilics and the upper normal curvatures are bounded is either a piece of a sphere or a plane. They also show, through a counterexample, that the extra condition of boundedness of the upper normal curvatures is essential. More precisely, they construct an example of a closed differentiable convex surface where almost all points are umbilics with the same curvature c and that is neither a piece of a sphere nor a plane. This example is obtained by a subtle algorithm starting with the unit ball and stepwise replacing parts by caps of curvature c and smoothing out the edges with canal surfaces of curvature greater than c.

In the last section of their paper, Busemann and Feller study geodesics on convex surfaces. They prove that a geodesic passing through (or emanating from) any point at which a tangent plane exists is in fact differentiable. Moreover, such a geodesic has a tangent line (or tangent ray) at that point, and an osculating plane which is perpendicular to the tangent plane. This generalizes a result of Johann Bernoulli which we already mentioned, from his article published in his 1728, *Problema: in superficie curva ducere lineam inter duo puncta brevissimam* [13] in which he used the term "osculating plane" for the first time.

For the proofs of all these theorems, Busemann and Feller acknowledge the strong influence of Bonnesen and Fenchel's *Theorie der konvexen Körper* [17] (1934), and this influence is also stressed in Busemann's monograph *Convex surfaces* [24] published in 1958. In the first chapter of that monograph, Busemann gives more detailed proofs of all these theorems. In the introduction, he writes that the first two chapters form a complement to Bonnesen and Fenchel's book, dealing in particular with subjects conjectured or suggested there. Zalgaller, in his paper [114], writes that Busemann took part in the preparation of the book by Bonnesen and Fenchel.

In the introduction of his monograph *Convex surfaces*, Busemann declares that his purpose is to present a subject, convex surfaces, which "during the past 25 years has experienced a striking and beautiful development, principally in Russia, but has remained largely unknown, at least in the USA." Indeed, the book is a tribute to the work of A. D. Alexandrov. As a matter of fact, this is the way Busemann presented his book project to the publisher.[61] We are not going to attempt any description of the results from Alexandrov or the school he founded, because this would lead us too far, but since our main subject matter in this article is curvature, we shall review his notion of nonpositive curvature.

We start with the definition of the length of a curve γ in a metric space (X, d), that is, of a continuous path $\gamma : [0, 1] \to X$. This is defined as

$$L(\gamma) = \sup \sum_i d(\gamma(t_i), \gamma(t_{i-1}))$$

where the supremum is taken over the set of subdivisions $0 = t_0 \leq t_1 \leq \ldots \leq t_n = 1$ of the interval $[0, 1]$. The definition is reminiscent of the notion of length of a curve as defined by Archimedes in his *On the sphere and the cylinder* [66]. If $\gamma(0) = A$ and $\gamma(1) = B$, we say that the curve joins A and B. We also say that the curve starts at A.

A metric on a surface (2-dimensional manifold) M is said to be intrinsic if the distance between two points is equal to the infimum of the lengths of curves in that surface that joins them.

A geodesic in M joining two points is a curve joining them which is the isometric image of a Euclidean interval.

[61] A letter written on July 23, 1956, from the editor-in-chief of Interscience Publishers Inc. (with copies to Courant, Stocker and Bers) starts with: "It is with very special pleasure indeed that we learn from Dr. Stocker of your willingness to preset in our Tracts series an account of important recent development of the work of certain Russian geometers." (A copy of the letter is kept with Courant's correspondence at the Elmer Holmes Bobst Library in New York University).

A triangle Δ in M is the union of three points in M (called the vertices of Δ) together with three minimizing geodesics (called the sides of Δ) joining them.

To each triangle Δ in M, one can associate a triangle Δ_E in the Euclidean plane E, together with a natural marking between the sides of Δ and the sides of Δ_E. The triangle Δ_E is called a comparison triangle of Δ. Now consider two curves γ_1 and γ_2 in S starting at a common point p. We wish to define the angle these curves make at p. We take a sequence of points x_n and y_n on the images of γ_1 and γ_2 respectively, both converging to p.

We then define the upper angle between γ_1 and γ_2 at p as the upper limit of the sequence of angles of the comparison triangles of p, x_n, y_n, all the angles being taken at the image of the point p in the Euclidean plane.

With this definition, each triangle in the metric surface M has a well-defined upper angle at each vertex. Let S be the sum of these three upper angles. The *angle excess* of such a triangle is then equal to $S - \pi$.

The surface M with an intrinsic metric is said to have (local) non-positive curvature in the sense of Alexandrov if (locally) each triangle has non-positive angle excess.

There are straightforward variations on this definition that lead to that of surfaces of bounded curvature (from below and from above), where, instead of taking comparison triangles in the Euclidean plane, one takes comparison triangles in one of the constant curvature geometries. Alexandrov thoroughly investigated this notion in several books and articles, see e.g. [2] and [3].

In his monograph [23] (and in other books and papers), Busemann introduced other (close but different) metric notions of curvature, and we recall the definition. He starts with the notion of a metric space *midpoint convex*, that is, for any two points x and y in that space, there exists a point z whose distances from x and from y are both equal to half the distance between x and y. Such a point z is called a *midpoint* of x and y (it is not unique). In a midpoint convex geodesic metric space which is complete, any two points may be joined by a geodesic.

Busemann defines a midpoint convex metric space (M, d) to be *non-positively curved* if for any three points a, b, c in M, if a' is a midpoint of a and c and b' a midpoint of b and c, then

$$d(a', b') \leq \frac{1}{2} d(a, b).$$

There are variations of this definition, namely, the space is called *negatively curved, non-negatively curved, or positively curved* depending on whether the above inequality is replaced by a strict inequality, or the reverse inequality or strict reverse inequality respectively. If the inequality is replaced by an equality, the space is said to be *zero curved*. These notions are used today under the name *Busemann non-positive (resp. non-negative, positive, negative, zero) curvature*.

In the same monograph, Busemann studied other notions of metric curvature, and we would like to mention three of them. The first one uses the notion of a *peakless function*. This is a continuous function f defined on an interval I of the real line such that for all $t_1 < t_2 < t_3$ in I we have

$$f(t_2) \leq \max(f(t_1), f(t_3))$$

with equality if and only if $f(t_1) = f(t_2)$. Using this, Busemann introduced the notion of *space whose distance function is peakless*, another generalization of nonpositive curvature.

Another notion of curvature, introduced by Busemann, uses the notion of *capsules* in metric spaces: If T is a segment in a metric space (M, d) and ρ a nonnegative number, a subset of the form

$$\{x \in M \mid d(x, T) \leq \rho\}$$

is said to be a *capsule* of axis T and radius ρ.

Busemann proves a local-implies-global result for convexity of capsules [23, Theorem 36.20], and he proves that a space which has Busemann non-positive (respectively negative) curvature has convex (respectively strictly convex) capsules. [23, Theorem 36.20]

In the same book, Busemann introduced a notion of curvature in terms of angle excess of triangles in spaces he calls G-spaces (G-stands for "geodesic"). These are metric spaces that are finitely compact (every bounded infinite sequence contains a convergent subsequence) and that satisfy a property of local uniqueness of extension of geodesics.

The notion of G-space, like many of the other metrical notions introduced by Busemann, is already contained in his early book *Metric Methods in Finsler Spaces and in the Foundations of Geometry* [21] (1942) and in his paper *On spaces in which points determine a geodesic* [22] (1943).

The metric notions of curvature introduced by Alexandrov and by Busemann turned out to be the bases of whole new fields in geometry, called after their names[62] and there is a multitude of theorems and research problems that rely on these definitions.

Like Busemann, Fenchel and some other twentieth-century geometers we mentioned, Alexandrov's interest abides in the most basic notions of geometry, putting isoperimetry, isoepiphany and other problems of classical mathematics at the forefront of research, with a dislike for a certain Riemannian geometry based on linear algebra and tensor calculus. We refer the reader to Alexandrov's book *Intrinsic geometry of convex surfaces* [2], in which he provided proofs of the theorems of Euler, Meusnier and Rodrigues in the general setting of convex surfaces, attributing these generalizations to Busemann and Feller. He proved in the same book [2] that any convex two-dimensional manifold is non-negatively curved in the sense he introduced it and which we recalled above, and that every non-negatively

[62]There are AMS Mathematics Subject Classification codes that carry the names of Alexandrov and Busemann geometries.

curved two-dimensional manifold homeomorphic to the sphere is isometric to some convex surface.

As another link between the group of geometers we mentioned, let us note that in his treatise *Convex surfaces*, Busemann gives a proof of a theorem in convexity theory that was independently obtained by himself and Fenchel and Jessen [63] stating that if two convex bodies in Euclidean n-space have equal p-th order area measure (we shall not go into the definition here) for some p satisfying $1 \leq p \leq n-1$, then the two bodies are the same up to a translation.

In a tribute in memory of Alexandrov, S.S. Kutateladze, who belongs to Alexandrov's school, writes that Alexandrov's slogan in mathematics was "Retreat to Euclid" and that he used to declare that "the pathos of contemporary mathematics is the return to Ancient Greece" [76].

A geometry freed from the dominion of differential calculus kept growing since the last quarter of the twentieth century, reinvigorated in the works of Thurston whose methods constitute a return to the synthetic methods of non-Euclidean geometry, and of Gromov, closer in spirit to the Busemann and Alexandrov approaches.

References

1. A. D. Alexandrov, On the works of S. E. Cohn-Vossen, *Uspehi Matem. Nauk (N.S.)* 2, (1947), no. 3 (19), pp. 107–14. English translation by A. Iacob.
2. A. D. Alexandrov, *Intrinsic geometry of convex surfaces*, ed. S. S. Kutateladze, Y. G. Reshetnyak, transl. S. S. Kutateladze, CRC Press, 2005.
3. A. D. Alexandrov and V. A. Zalgaller, *Intrinsic geometry of surfaces*, Academy Sc. USSR, Moscow, 1962. English translation by J. M. Danskin, *AMS Translations of Mathematical Monographs*, American Mathematical Society, Providence, 1967.
4. Apollonius of Perga, *Conics, Books V to VII*, ed. G. J. Toomer, Springer-Verlag, 1990.
5. Apollonius of Perga, *Les coniques d'Apollonius de Perge* (P. Ver Eecke, ed.), Librairie Albert Blanchard, Paris, 1963.
6. Apollonius of Perga: Apollonii Pergaei Conicorum lib. V. VI. VII. paraphraste abalphato asphahanensi nunc primum editi. Additus in calce Archimedis assumptorum, ex codocibus arabicis manuscriptis. Serenissimi Magni Ducis Etruriae Abrahamus Ecchellensis Maronita. In alma Vrbe Linguar. Orient. Professor Latinos reddidit. Io. Alfonsus Borellus In Pisana Academia Matheseos Professor curam in Geometricis versioni contulit, & notas vberiores in vniuersum opus adiecit, Firenze, 1661.
7. P. Appell, Le problème géométrique des déblais et des remblais, Mémorial des sciences mathématiques, fascicule 27, Gauthier-Villars, Paris, 1928.
8. Aristotle, Metaphysics, transl. W. D. Ross, In: *The works of Aristotle translated into English*, ed. W. D. Ross and J. A. Smith, Clarendon Press, Oxford, 1908, 2nd. ed. 1928.
9. Aristotle, On the Heavens, transl. J. L. Stocks, In: *The works of Aristotle translated into English*, ed. W. D. Ross and J. A. Smith, Clarendon Press, Oxford, 1922.
10. Aristotle, On the Soul, transl. J. A. Smith, In: *The works of Aristotle translated into English*, ed. W. D. Ross and J. A. Smith, Clarendon Press, Oxford, 1927.
11. Aristotle, *Minor Works*, transl. W. S. Hett, Harvard University Press, Cambridge/London: William Heinemann, London, 1936.
12. M. Berger, *Geometry revealed: A Jacob's ladder to modern higher geometry*, Springer, Berlin, 2010.

13. J. Bernoulli, Problema: in superficie curva ducere lineam inter duo puncta brevissimam, Opera Omnia IV (reed. Georg Olms 1968), (1728), pp. 108–128.
14. J. Bertrand, Mémoire sur la théorie des courbes à double courbure, *Journal de mathématiques pures et appliquées 1re série*, tome 15 (1850), pp. 332–350.
15. W. Blaschke and G. Hessenberg, Lehrsätze über konvexe Körper. *Jber. Deutsch. Math-Vereinig*. Vol. 26 (1917), pp. 215–220.
16. H. Bohr and B. Jessen, Om Sandsynlighedsfordelinger ved Addition af konvexe Kurver. *Danske. Vid. Selsk. Skr.* (8) Vol. 12 (1929), pp. 326–405.
17. T. Bonnesen and W. Fenchel, *Theorie der konvexen Körper*, Springer, Berlin, 1934, English translation: Theory of convex bodies, BCS associates, Moscow, Idaho, 1987.
18. H J. M. Bos, Tractional motion and the legitimation of transcendental curves, *Centaurus* 31(1) (1988), pp. 9–62.
19. G. Bouligand, *Introduction à la géométrie infinitésimale directe (avec une préface d'E. Cartan)*, Vuibert, Paris, 1932.
20. H. Brunn, Über Ovale und Eiflächen, Inaugural Dissertation, München, 1887, 42 p.
21. H. Busemann, *Metric Methods in Finsler Spaces and in the Foundations of Geometry.* Annals of Mathematics Studies, no. 8. Princeton University Press, Princeton, N. J., 1942.
22. H. Busemann, On spaces in which points determine a geodesic. *Trans. Am. Math. Soc.* 54 (1943), 171–184.
23. H. Busemann, *The geometry of geodesics*, Academic Press (1955), reprinted by Dover in 2005.
24. H. Busemann, *Convex surfaces*, Interscience tracts in pure and applied mathematics, Inter science Publishers, New York and London, 1958.
25. H. Busemann, *Selected Works*, 2 volumes, ed. A. Papadopoulos, Springer International, 2018.
26. H. Busemann and W. Feller, Krümmungseigenschaften konvexer Flächen, *Acta Math.* 66 (1936), pp. 1–47. English translation by A. A'Campo Neuen: Curvature Properties of Convex Surfaces, In: Herbert Busemann, Selected works, Vol. II, Springer International (2018), pp. 235–270.
27. M. Chasles, Aperçu historique sur l'origine et le développement des méthodes en géométrie. Mémoires couronnés par l'Académie royale des sciences et belles-lettres de Bruxelles, tome XI, Hayez, Bruxelles, 1837.
28. N. G. Chebotarev, On the breadth of contours and bodies (Russian), *Ber. Wiss. Forschgsinst. Odessa* Vol. 1 (1924), pp. 29–41.
29. S. Cohn-Vossen, Singularitäten konvexer Flächen. *Math. Ann.* Vol. 97 (1927), pp. 377–386.
30. A.-C. Clairaut, *Recherches sur les courbes à double courbure*, Nyon, Didot et Quillau, Paris, 1731.
31. A. Clairaut, *Théorie de la figure de la terre, tirée des principes de l'hydrostatique*, David Fils, Paris, 1743.
32. N. de Condorcet, Éloge d'Huyghens, In : Œuvres de Condorcet, publiées par A. Condorcet O'Connor et M. F. Arago, tome II, Firmin Didot Frères, Paris (1847–1849), pp. 54–72.
33. J. L. Coolidge, The unsatisfactory story of curvature, *American Mathematical Monthly* 59 (1952), pp. 375–379.
34. G. Darboux, Notice historique sur le général Meusnier, membre de l'Ancienne Académie des Sciences, lue dans la séance publique annuelle du 20 décembre 1909, Paris, Gauthier-Villars, 1910, and Mém. Acad. Sciences, Vol. 51, 2e série (1910), pp. 1–40.
35. J. Dieudonné, *History of algebraic geometry*, Wadsworth Inc., Monterey, California, 1985.
36. Ch. Dupin, Le problème géométrique des déblais et remblais, Correspondance de l'École polytechnique, VIIe cahier (1807), pp. 218–225.
37. Ch. Dupin, Développements de géométrie, avec des applications à la stabilité des vaisseaux, aux déblais et remblais, au défilement, à l'optique, etc. Paris, Coursier, 1813.
38. Ch. Dupin, Exercices de géométrie et d'algèbre, 1818.
39. Ch. Dupin, *Essai historique sur les services et les travaux scientifiques de Gaspard Monge*, Bachelier, Paris, 1819.
40. Ch. Dupin, *Applications de la géométrie à la mécanique*, Bachelier, Paris, 1822.

41. Euclid (Th. L. Heath, ed.) *The thirteen books of Euclid's Elements*, 3 volumes, Cambridge University Press, 1908, Reprint, Dover.
42. L. Euler, Constructio linearum isochronarum in medio quocunque resistente. *Acta Eruditorum* (1726), pp. 361–363, Opera Omnia: Series 2, Volume 6, pp. 1–3. English translation by I. Bruce available on the internet.
43. L. Euler, Meditationes super problemate nautico, quod illustrissima regia Parisiensis Academia scientarum proposuit, Pièce qui a remporté le prix de l'Academie Royale des sciences, 1727 (1728), pp. 1–48, Opera Omnia, Series 2, Volume 20, pp. 1–35.
44. E. Euler, De linea brevissima in superficie quacunque duo quaelibet puncta jungente, *Commentarii academiae scientiarum Petropolitanae* 3 (1732), 110–124, Opera Omnia, Series 1, Volume 25, pp. 1–12.
45. L. Euler, De innumerabilibus curvis tautochronis in vacuo, *Commentarii academiae scientiarum Petropolitanae* 4 (1735), pp. 49–67, Opera Omnia: Series 2, Volume 6, pp. 15–31. English translation by I. Bruce available on the internet.
46. L. Euler, Curva tautochrona in fluido resistentiam faciente secundum quadrata celeritatum, *Commentarii academiae scientiarum Petropolitanae* 4 (1735), pp. 67–89, Opera Omnia: Series 2, Volume 6, pp. 32–50. English translation by I. Bruce available on the internet.
47. L. Euler, *Mechanica*, 2 volumes, published as a book in 1736, Opera Omnia: Series 2, Volumes 1 and 2.
48. L. Euler, De linea celerrimi descensus in medio quocunque resistente, *Commentarii academiae scientiarum Petropolitanae* 7 (1740), pp. 135–149, Opera Omnia: Series 1, Volume 25, pp. 41–53
49. L. Euler, De fractionibus continuis dissertatio Commentarii academiae scientiarum Petropolitanae 9 (1744), pp. 98–137, Opera Omnia: Series 1, Volume 14, pp. 187–216.
50. L. Euler, Methodus inveniendi lineas curvas maximi minimive proprietate gaudentes, sive solutio problematis isoperimetrici lattissimo sensu accepti, Lausanne and Geneva, 1744, Opera Omnia, Series 1, vol. 24
51. L. Euler, Introductio in analysin infinitorum, First edition: Lausannae: Apud Marcum-Michaelem Bousquet & socios., 1748. Opera omnia, Series 1, Vol. VIII. English translation by J. T. Blanton, 2 vol., Springer-Verlag, New York, 1988, 1990.
52. L. Euler, *Introduction to Analysis of the Infinite*, English translation of [51] by Blanton, 2 vol., Springer-Verlag, New York, 1988.
53. L. Euler, *Scientia navalis seu tractatus de construendis ac dirigendis navibus*, 2 volumes, Saint Petersburg, 1749. Opera Omnia, Series 2, Vol. 18 and 19, Zürich and Basel, 1967 and 1972.
54. L. Euler, Investigatio curvarum quae evolutae sui similes producunt, *Commentarii academiae scientiarum Petropolitanae* 12 (1750), pp. 3–52, Opera Omnia: Series 1, Volume 27, pp. 130–180.
55. L. Euler, Recherches sur la courbure des surfaces, *Mémoires de l'académie des sciences de Berlin* 16 (1767), pp. 119–143, Opera Omnia, Series 1, Volume 28, pp. 1–22.
56. L. Euler, De solidis quorum superficiem in planum explicare licet, *Novi Commentarii academiae scientiarum Petropolitanae* 16 (1772), pp. 3–34. Opera Omnia Series 1, Volume 28, pp. 161–186.
57. L. Euler, De curvis triangularibus, *Acta Academiae Scientarum Imperialis Petropolitinae* (1778), pp. 3–30, Opera Omnia, Series 1, Volume 28, pp. 298–321.
58. L. Euler, Methodus facilis omnia symptomata linearum curvarum non in eodem plano sitarum investigandi, *Acta Academiae Scientarum Imperialis Petropolitinae* (1782), pp. 19–57, Opera Omnia, Series 1, Volume 28, pp. 348–381.
59. L. Euler, De plurimis quantitatibus transcendentibus, quas nullo modo per formulas integrales exprimere licet, *Acta Academiae Scientarum Imperialis Petropolitinae* 4, 1784, pp. 31–37, Opera Omnia: Series 1, Volume 15, pp. 522–527.
60. L. Euler, Investigatio curvarum quae similes sint suis evolutis vel primis vel secundis vel tertiis vel adeo ordinis cuiuscunque, *Nova Acta Academiae Scientarum Imperialis Petropolitinae* 1 (1787), pp. 75–116 Opera Omnia: Series 1, Volume 29, pp. 72–111.

61. W. Feller, *Selected Works*, 2 volumes, ed. R. L. Schilling Z. Vondraček and W. A. Woyczyński, Springer Verlag International, 2015.
62. W. Fenchel, Über Krümmung und Windung geschlossener Raumkurven, *Math. Ann.*, 101 (1929), pp. 238–252.
63. W. Fenchel and B. Jessen, Mengenfunktionen und konvexe Körper, *Danske Vid. Selskab Mat.-Fys.Medd.* 16, 3 (1938), pp. 1–31.
64. J. Hadamard, Sur certaines propriétés des trajectoires en dynamique. *J. Math. Pures et Appl.* 5 (1897), pp. 331–387.
65. T. Hayashi, The least number of the sextatic points of a central oval, *Sci. Rep. Tohoku Uni.* Vol. 12 (1924) pp. 393–395.
66. L. Heath (ed.) *The works of Archimedes*, Cambridge University Press, 1897, Reprint, Dover, 2002.
67. T. L. Heath, *A history of Greek Mathematics*, Cambridge University Press, Cambridge, 1921, reprint, Dover.
68. Heron of Alexandria, Metrica. In: *Codex Constantinopolitanus Palatii Veteris*, ed. E. M. Bruins, Brill, 1964.
69. D. Hilbert and S. Cohn-Vossen, *Anschauliche Geometrie*, Springer, Berlin and Heidelberg, 1932.
70. Ph. de La Hire, Gnomonique ou l'art de tracer des cadrans ou horloges solaires sur toutes les surfaces, par différentes pratiques, avec les démonstrations géométriques de toutes les opérations, Paris Estienne Michallet, 1682.
71. J. Hjelmslev, Über die Grundlagen der kinematischen Geometrie, *Acta Math.* 47 (1926), pp. 143–188.
72. C. Huygens, *Œuvres complètes*, 22 volumes, Société Hollandaise des Sciences, La Haye, 1888–1950.
73. B. Jessen, Om konvexe Kurvers Krumning. *Mat. Tidssr. B* (1929) pp. 50–62.
74. A. R. Jones, The roofed spherical sundial and the Greek geometry of curves, In: *Studies in the ancient exact sciences in honor of Lis Brack-Bernsen*, ed. J. Steele and M. Ossendrijver, Berlin studies of the ancient world, Topoi, Berlin, 2017.
75. J. Kepler, *Mysterium cosmographicum* (1596). English translation: The secret of the universe, transl. A. M. Duncan, Abaris, New York, 1981.
76. S. S. Kutateladze, Alexandrov of Ancient Hellas, *Siberian Electronic Mathematical Reports*, 9 (2012), pp. A6-A11.
77. J.-L. de Lagrange, Théorie des fonctions analytiques contenant les principes du calcul différentiel dégagés de toute considération d'infiniment petits ou d'évanouissans de limites ou de fluxions, Paris, Imprimerie de la République, 1797.
78. M.-A. Lancret, Mémoire sur les courbes à double courbure, Académie des Sciences de Paris, Mémoire des savants étrangers, Vol. I (1806), pp. 416–454.
79. M.-A. Lancret, Mémoire sur les développoïdes des courbes planes, des courbes à double courbure, et des surfaces développables, Mémoire des savants étrangers, Vol. I (1811), pp. 1–79.
80. G. W. Leibniz, Meditatio nova de natura anguli contactus et osculi: horumque usu in practica mathesi ad figuras faciliores succedaneas difficilioribus substituendas, Acta Eruditorum, 1686, apud J. Grossium & J. F. Gletitschium, 1687.
81. G. W. Leibniz, Generalia de natura linearum, anguloque contactus et osculi, provolutionibus, aliisque cognatis, et eorum usibus nunnullis, Acta Eruditorum, 1692.
82. J.-B. Meusnier, Mémoire sur la courbure des surfaces, Mém. div. sav. Paris, Vol. X (1785), pp. 477–510.
83. J.-F.-Th. Maurice, article sur Christiaan Huygens, Biographie Universelle, Paris, Michaud, 1818.
84. G. Monge, Mémoire sur les propriétés de plusieurs genres de surfaces courbes et particulièrement sur celles des surfaces développables avec une application à la théorie générale des ombres et des pénombres, Mém. Div. savants 9 (1780), pp. 382–440.
85. G. Monge, Mémoire sur la théorie des déblais et remblais, Mém. Div. sav., 1781.

86. G. Monge, Mémoire sur les développées, les rayons de courbure et les différents genres d'inflexions des courbes à double courbure, Imprimerie Royale, Paris, 1785.
87. G. Monge, Sur les lignes de courbure de la surface de l'ellipsoïde, *J. Éc. Polytechnique*, 2e cahier (1796), pp. 145–165.
88. G. Monge, Applications de l'analyse à la géométrie, Paris, 1807.
89. S. Mukhopadhyaya, Note on T. Hayashi's paper on the osculating ellipse of a plane curve, *Rend. Circ. palermo* vol. 51 (1927), pp. 394.
90. S. Mukhopadhyaya, Extended minimum number theorems of cyclic and sextatic points on a plane oval, *Math. Z.* 33 (1931) pp. 648–662.
91. S. Mukhopadhyaya, Circles incident on an oval of undefined curvature, *Tohoku vo.* 34 (1931) pp. 115–129.
92. O. Neugebauer, The astronomical origin of the theory of conic sections, *Proc. Am. Phil. Soc.* 92, No. 3 (1948), pp. 136–138.
93. A. Papadopoulos, Herbert Busemann, *Notices of the American Mathematical Society*, 2018, Vol. 65, No. 3, March 2018, pp. 341–343.
94. Pappus, the Collection, French translation by P. Ver Eecke, Fondation universitaire de Belgique, 1932. Reprint, Blanchard, Paris, 1982.
95. F. C. Penrose, *An Investigation of the Principles of Athenian Architecture; or, The Results of a Survey Conducted Chiefly with Reference to the Optical Refinements Exhibited in the Construction of the Ancient Buildings at Athens*, Macmillan, London, 1888.
96. H. Pemberton, *A View of Sir Isaac Newton's Philosophy*, printed by S. Palmer, London, 1728.
97. H. Pitot, Sur la quadrature de la moitié d'une courbe qui est la compagne des arcs, appelée la compagne de la cycloïde, *Histoire de l'Académie des sciences* (1976), pp. 107–113.
98. H. Poincaré, Démonstration nouvelle des propriétés de l'indicatrice d'une surface. *Nouvelles annales de mathématiques*, 2e série, Vol. 13 (1874) pp. 449–456.
99. Proclus, *A commentary on the first book of Euclid's Elements*, Princeton University Press, 1970.
100. R. Rashed, Apollonius : Les Coniques, tome 1.1 : Livre I, commentaire historique et mathématique, édition et traduction du texte arabe, de Gruyter, 2008.
101. R. Rashed, Descartes et l'infiniment petit, *Boll. Stor. Sci. Mat.* 33, No. 1, 151–169 (2013).
102. K. Reidemeister, Über Körper konstanten Durchmessers, *Math. Z.* Vol. 10 (1921), pp. 214–216.
103. E. Rinner, Ancient Greek sundials and the theory of conics reconsidered, In: *Studies in the ancient exact sciences in honor of Lis Brack-Bernsen*, ed. J. Steele and M. Ossendrijver, Berlin studies of the ancient world, Topoi, Berlin, 2017.
104. O. Rodrigues, Sur quelques propriétés des intégrales doubles et des rayons de courbure des surfaces, *Bulletin scientifique de la société philomatique de Paris* (1815), pp. 34–36.
105. O. Rodrigues, Recherches sur la théorie des lignes et des rayons de courbure des surfaces, et sur la transformation d'une classe d'intégrales doubles qui ont un rapport direct avec les formules de cette théorie, *Correspondance de l'Ecole Polytechnique*, 3 (1815), pp. 162–182.
106. P. Roitman and H. Le Ferrand, The strange case of Paul Appell's last memoir on Monge's problem: "sur les déblais et remblais", *Historia Mathematica*, Vol. 43, Issue 3, August 2016, pp. 288–309.
107. I. M. Serrano and B. D. Suceava, A Medieval Mystery: Nicole Oresme's Concept of Curvitas. *Notices of the AMS*, 62 (October 2015), pp. 1030–1034.
108. D. J. Struik, *Outline of a history of differential geometry*, Saint Catherine Press Limited, 1933.
109. D. J. Struik, *Lectures on classical differential geometry*, Addison Wesley, Reading, Mass. 1961.
110. P. Tannery, Pour l'histoire des lignes et des surfaces courbes dans l'antiquité, *Bulletin des Mathématiques et astronomiques*, 8 (1884), 25–27.
111. J.-A. Taschereau, Notice sur le général Meusnier d'après, des notes biographiques de Monge et d'autres documents également inédits, Revue rétrospective, ou Bibliothèque historique, contenant des mémoires et documents authentiques (2) 4 (1835), pp. 82–91.

112. R. Taton, *L'œuvre scientifique de Monge*, Presses Universitaires de France, Paris, 1951.
113. R. Taton, La première note mathématique de Gaspard Monge (juin 1769), *Revue d'histoire des sciences et de leurs applications*, tome 19, no. 2 (1966), pp. 143–149.
114. V. A. Zalgaller, Memoirs of A. D. Alexandrov and his Leningrad geometry seminar, *Siberian Electronic Mathematical Reports*, 9 (2012), pp. A26-A80.

Chapter 8
The Axiomatic Destiny of the Theorems of Pappus and Desargues

Victor Pambuccian and Celia Schacht

Non c'è un unico tempo: ci sono molti nastri
che paralleli slittano
spesso in senso contrario e raramente
s'intersecano.

Eugenio Montale, *Tempo e tempi*

Abstract We present the largely twentieth century history of the discovery of the significance of Pappus and Desargues for the axiomatics of geometry. Their significance is followed in projective, affine, and orthogonality contexts. There is an extensive bibliography, that should allow the interested reader to take a comprehensive look at the research literature on these axioms.

8.1 Introduction

There are mathematical results that are significantly ahead of their time. Having materialized during a time which had no use for them, they did not find their rightful place in the tissue of mathematical results of their own time, having to wait many centuries to reveal their consequential worth. Results born long before their *kairos*, their right time. One tends to think of Apollonius's conics, the findings of which were of rather limited use until more than 1800 years later, when Kepler uncovered their centrality for understanding planetary motion.

The most remarkable of these early results are Pappus's and Desargues's theorems, whose peripheral nature is even more surprising than that of the conic sections due to the fact that the context that endows them with meaning did not become fully actualized for centuries. In the meantime, their hibernation traversed mathematically active centuries during which they remained unused and, to a great

---

V. Pambuccian (✉) · C. Schacht
School of Mathematical and Natural Sciences, Arizona State University, Phoenix, AZ, USA
e-mail: pamb@asu.edu

S. G. Dani, A. Papadopoulos (eds.), *Geometry in History*,
https://doi.org/10.1007/978-3-030-13609-3_8

extent, forgotten. In times preceding the axiomatic turn, the value of a theorem resided in the range of its consequences or in the effect it had in organizing a whole domain of investigation. None of the two appeared to have any influence on matters of geometric, algebraic, or analytic concern.

Pappus's Theorem can be found in Book VII of Pappus's *Collection*, where it is spread over Lemmas XII, XIII, XV, and XVII in the part of Book VII containing lemmas to the first book of Euclid's *Porisms* (see [189, pp. 270–273] and [188]). It thus goes back to the first half of the fourth century A.D.

Desargues's Theorem appeared in *Manière universelle de M. Desargues pour practiquer la perspective*, a practical book on the use of perspective, published in 1648 by Desargues's student Abraham Bosse (see [62] for a French edition of Desargues's works and [74] for an English translation).

Although Pappus's theorem was generalized by Blaise Pascal—in a note written in 1639, when he was 16 years old, and published the following year as the broadside "Essay pour les coniques. Par B. P." (see [191])—the deeper meaning of these results came to be understood only beginning with the last decade of the nineteenth century.

In fact, it took all of the twentieth century and beyond to reveal the wealth of meaning encoded by these results. The subject of this chapter is not simply the story of the late recognition that befell these results, but the adventure through which their true importance became revealed; the significance of Pappus's and Desargues's theorems not only became relevant, but spurred the twentieth century discoveries of other like-minded results that are either special cases of the two theorems or different configuration theorems that turn out to be logically equivalent to some form of one of Pappus's and Desargues's theorems. At one point, we will also elucidate the role played by another older and perhaps even more enigmatic result, that of Menelaus's Theorem, which goes back to Menelaus's *Sphaerica*, a work which was probably written in the beginning of the second century A.D.

The story of the gradual understanding of these theorems is a chapter in the history of the axiomatic foundation of geometry. A significant aspect of the developments we will survey is the surprisingly close relationship between geometry with important configuration theorems and algebra. This relationship was emphasized very early in the history of projective geometry, perhaps most emphatically by Gaspard Monge in his *Géométrie descriptive*:

> There is no construction belonging to descriptive geometry that could not be translated into analysis; and whenever the matter involved does not require more than three unknowns, each analytic operation can be seen as the script of a spectacle in geometry.[1]

Our narrative is structured as follows: In a first section we present the history of the role of configuration theorems deriving from Pappus and Desargues inside geometries with incidence and sometimes order as their only primitive notions. In a

[1] Il n'y a aucune construction de Géométrie descriptive qui ne puisse être traduite en Analyse; et lorsque les questions ne comportent pas plus de trois inconnues, chaque opération analytique peut être regardée comme l'écriture d'un spectacle en Géométrie. (All translations are by V.P.).

second section we group all results in which the above configuration theorems play a major role inside geometries which have, in addition to incidence and perhaps order, congruence, orthogonality, and line reflections as primitive notions. We will exclude from this survey the results that belong to finite geometries, with the exception of very few, in which the only additional assumption is finiteness. The story of finite geometries constitutes a separate subplot, which has been summarized in several monographs [28, 61, 106–108, 115, 151, 235]. Both historical bias against explicitly finite worlds, and a bias towards first-order logic, warrants an attitude of agnosticism with regard to the cardinality of the models of the geometries under consideration.

The pre-history, as it were, of our story is that of the attempts to present projective geometry in a *more geometrico* manner, to separate it as far as possible from the bond with the real numbers. The major figures of this pre-history are J. Steiner, K. G. C. von Staudt, and M. Pieri. Their pioneering contributions have been surveyed in [154, 172, 266] and [13].

8.2 Desargues and Pappus in the Projective Setting

8.2.1 Configuration Theorems

Both theorems of Pappus and of Desargues belong to the class of *universal configuration theorems*. In a language with *points* (uppercase Latin letters) and *lines* (lowercase Latin letters) as individual variables, and the binary *incidence* relation $\in$ (for which we will use words such as "lying on" or "passing through", as well as "collinear" for three points incident with the same line), these can be defined, following [163] and [197, p. 26] (see [146–149] and [171] for variants, and [82, 141, 202], and [129] for comprehensive treatments of configurations), as universal statements, in which the antecedent is a conjunction of some of the following: (i) incidences, (ii) negated incidences, (iii) negated equalities between point variables, and (iv) negated equalities between line variables, and the consequent is a conjunction of the same kind.

In its projective form, the Desargues axiom can be stated as: "If two triangles are in perspective centrally then they are in perspective axially." Two triangles $A_1A_2A_3$ and $B_1B_2B_3$ with pairwise different vertices are said to be *in perspective centrally* if the lines A_1B_1, A_2B_2, and A_3B_3 are concurrent (have a point C in common), and they are said to be *in perspective axially* if the three intersection points C_{ij} of the corresponding sides A_iA_j with B_iB_j (with $1 \leq i < j \leq 3$) of the two triangles are collinear. Formally, it can be stated as (Fig. 8.1)

^{P}**Des** *If* $C, A_i, B_i \in c_i$, $C_{ik}, A_i, A_k \in a_{ik}$, $C_{ik}, B_i, B_k \in b_{ik}$, C, A_i, *and* B_i *are pairwise different points,* c_1, c_2, *and* c_3 *are three pairwise different lines,* $a_{12} \neq a_{13}$, $b_{12} \neq b_{13}$, *and* $C_{12}, C_{13} \in c$ *(for all* i *and* k *in* $\{1, 2, 3\}$ *with* $i < k$*), then* $C_{23} \in c$.

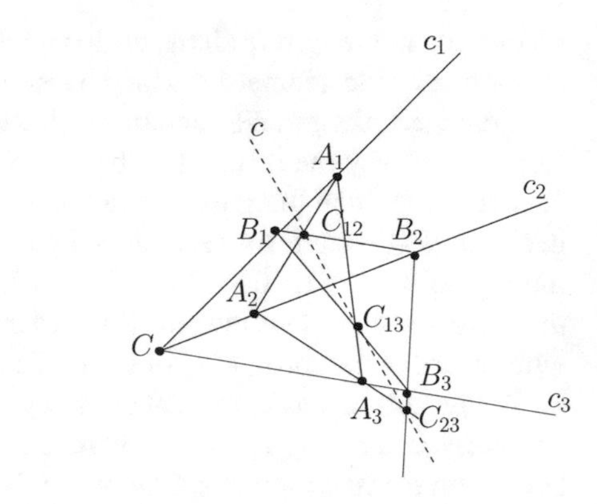

Fig. 8.1 The projective form of the Desargues axiom

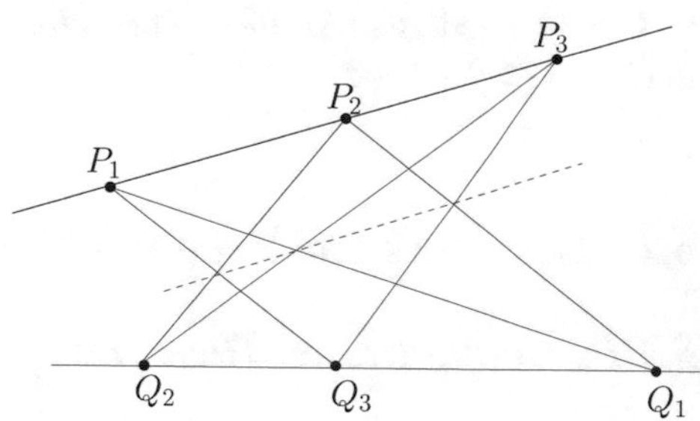

Fig. 8.2 The projective form of the Pappus axiom

In its projective form, the Pappus axiom states that (Fig. 8.2):

^{p}**Papp** *If the three points* P_1, P_2, *and* P_3 *are on a line* a, *the three points* Q_1, Q_2, *and* Q_3 *are on a line* b *but not on* a, *then the points of intersection of the lines: (i)* P_1Q_1 *and* Q_2P_3, *(ii)* P_2Q_2 *and* Q_3P_1, *(iii)* P_3Q_3 *and* Q_1P_2, *are collinear.*

They will be considered in this section as statements inside plane projective geometry, whose other axioms can be stated in the following duality-emphasizing manner (as in [131]):

P 1 *There is a line passing through any two given points.*

P 2 *There is a point incident with any two given lines.*

P 3 *If two points A and B are incident with the lines a and b, then* $a = b$ *or* $A = B$.

P 4 *There are four points* A, B, C, D *and four lines* a, b, c, d, *such that A is incident with a and b, but with neither c nor d, B is incident with d and a, but with neither b nor c, C is incident with c and d, but with neither a nor b, and D is incident with b and c, but with neither a nor d.*

This axiom system for plane projective geometry without any configuration theorem will be denoted by $\mathscr{P} = \{$P1–P4$\}$. If the points A and B are distinct, then the unique line (by P1 and P3) joining them will be denoted by AB.

In [105, p. 85], configurations are defined in a much more restrictive manner, as a "system of p points and g lines, which lie in a plane in such a manner, that every point of the system is incident with the same number γ of lines of the system and at the same time every line of the system is incident with the same number π of points of the system. Such a configuration is denoted by the symbol $(p_\gamma g_\pi)$." When $p = g$ and $\gamma = \pi$, we will denote $(p_\gamma g_\pi)$ by just (p_γ). Both ${}^p Des$ and ${}^p Papp$ fit this narrower definition, as (10_3) and (9_3) respectively, but the axiom of the fourth harmonic point from Sect. 8.2.6, although a universal configuration theorem in the sense of Pickert, does not fit the above definition.

8.2.2 The Fundamental Theorem of Projective Geometry

Our story begins with the result announced in 1890, without proof, by H. Wiener in [270], which states that one can prove the fundamental theorem of projective geometry from the axioms P1–P4, ${}^p Papp$, and ${}^p Des$. The novelty lied in the fact that for the first time, ever since von Staudt had proved it in 1847, all need for some kind of continuity had disappeared. That need had been present, without being explicitly mentioned among the assumptions, in von Staudt's work (see [266]). The claim was first proved in 1899 by Schur [232].

To understand what the fundamental theorem of projective geometry states, we first define *perspectivities*. Given two lines l and m and a point P on neither line, the mapping between the points on l and those on m assigning to a point X incident with l, the point X' on m that is the intersection of the line PX with m is referred to as the *perspectivity* with center P. The composition of a finite number of perspectivities is called a *projectivity*.

Following [24], we denote by (P_n) the statement that a projectivity of a line l to itself with n fixed points is the identity mapping of l.

With this notation, the fundamental theorem of projective geometry is (P_3). The statement (P_3) as such is not a first-order sentence, given that a projectivity is the composition of an unspecified number of perspectivities. As determined by Hessenberg, in 1905 [98], who raised precisely this question, asking for the minimal number of perspectivities needed to obtain a given projectivity, if ${}^p Des$ holds, then every projectivity of a line onto itself must be the composition of *at most three perspectivities* (see also [220]). However, in the absence of any configuration theorem, there is no a priori bound on the number of perspectivities involved (see [25] for more on perspectivities). If we denote, as in [199], by $(P_{n,m})$ the statement (P_n) makes, in which the projectivity is a composition of at most m perspectivities, then we do, indeed obtain a result in first-order logic, namely that, for all $m \geq 2$, in the presence of ${}^p Papp$ and ${}^p Des$, a projectivity which is the composition of at most m perspectivities of a line l to itself which fixes 3 points is the identity mapping of l, or formally (here and in the sequel, "$\Sigma, \alpha, \beta \vdash \gamma$" should be read as "$\gamma$ can be deduced from α, β, and the statements in Σ")

$$\mathscr{P},\ {}^p Papp,\ {}^p Des \vdash P_{3,m}. \tag{8.1}$$

Such concerns were certainly not in the minds of Wiener or Schur, for the concept of first-order logic had only emerged in the early 1920s (see [73] and [65]).

The significance of (8.1) lies in the fact that von Staudt had already shown in 1856 that, by means of his *Wurfrechnung*, one can introduce the operations of addition and multiplication in a projective plane which satisfies the fundamental theorem (P_3). This led Lüroth, in 1875, to provide a streamlined development of analytic projective geometry based on the *Wurfrechnung*. Thus, at the latest, by the time Hilbert published the first edition of his *Grundlagen der Geometrie* in 1899, in which Chapter V is devoted entirely to the significance of Desargues's theorem, it had become apparent that any projective plane which satisfies ^{p}Des is a projective coordinate plane over a skew field (Hilbert referred to skew fields as "Desarguessche Zahlensysteme" as the actual term had not yet been coined). This effect encompassed by ^{p}Des, that it allows the "introduction of number" into geometry, which "in any exact science" is "a most highly prized aim,"[2] has great epistemological significance for Hilbert, which he expressed in his 1989–1999 lectures on the foundations of geometry [104, p. 222], expressing the thought:

> However, if science is not to fall prey to a sterile formalism, then it must reflect on itself in some later stage of its development, and should at least check the foundations that led to the introduction of number.[3]

Although Hilbert's axiom system included the order axioms, these were not used in the coordinatization itself. One also reads off the same book, whose Chapter VI is devoted to Pappus's theorem (referred to by Hilbert as *Pascal*'s theorem), that in the presence of both ^{p}Des and $^{p}Papp$, the coordinatizing skew field must be commutative, thus a field (see [217] for a history of the influence of Hessenberg's work on Desargues and Pappus mentioned below had on the various editions of the *Grundlagen der Geometrie*). Anne Lucy Bosworth, Hilbert's first female doctoral student, provided in her doctoral dissertation [33] a coordinatization, by means of a segment calculus, of ordered incidence planes satisfying ^{p}Des and $^{p}Papp$ (in which certain lines need to intersect, if the additive and the multiplicative inverses are to exist), and showed that segment addition and multiplication satisfy the usual associativity, commutativity, and distributivity properties. A different, purely projective coordinatization was proposed by Hessenberg in 1904 [96], based solely on ^{p}Des (an alternative coordinatization was put forward by Schwan in 1919 [234]; see also [9, 10]). (8.1) thus meant that *synthetic and analytic projective geometry are identical* in the presence of ^{p}Des or of ^{p}Des and $^{p}Papp$. This is notable, for while synthetic geometry is expressed in terms of the purest geometrical form of intuition, that of a point lying on a line, analytic projective geometry is expressed

[2] Nun ist in der That in jeder exakten Wissenschaft die Einführung der Zahl ein vornehmstes Ziel.

[3] Aber, wenn die Wissenschaft nicht einem unfruchtbaren Formalismus anheimfallen soll, so wird sie auf einem späteren Stadium der Entwicklung sich wieder auf sich selbst besinnen müssen und mindestens die Grundlagen prüfen, auf denen sie zur Einführung der Zahl gekommen ist.

in terms of two operations satisfying a variety of axioms, none of which stem from our understanding of space.[4]

The question regarding the need for both ${}^{p}Des$ and ${}^{p}Papp$ in (8.1) was soon settled by Hessenberg in 1905. He showed in [98] that ${}^{p}Des$ can be derived from ${}^{p}Papp$ ("Pappus implies Desargues"), or formally

$$\mathscr{P},\ {}^{p}Papp \vdash {}^{p}Des. \tag{8.2}$$

The proof was later found to be incomplete, as it did not consider all cases in which some coincidences might happen. Several complete proofs have since been published. One fixing the original proof can be found in [64], while other proofs were presented in [58, 118, 197, 236]. All existing proofs use the Pappus axiom three times to prove Desargues. It was shown in [186], using Guggenheimer's [83] axiom system for Desarguesian affine planes in terms of the axiom of Menelaus (see Sect. 8.3.1 below), that there is no proof of ${}^{p}Des$ that uses ${}^{p}Papp$ less than three times. That the number of uses of an axiom in a proof can be made precise in a certain axiomatic set-up has been pointed out in [181]. One can look at axiom use in a proof as resource use, and resources may be depleted after the first use, or after a certain finite number of uses, or repeated uses may simply turn the proof into a more expensive affair if one thinks that each accessing of an axiom comes with a price tag, in which case repeated use of an axiom—in particular repeated use of *expensive* axioms, such as the very strong statement ${}^{p}Papp$—renders the proof exceedingly expensive (see [182] for more on the history of this concern for the number of uses of an assumption in a proof).[5]

Equation (8.2) itself led Hilbert and Cohn-Vossen [105, p. 117] to state that "the theorem of Pascal [which is the name they use for ${}^{p}Papp$] is the only essential configuration theorem of the plane", that it "represents the most important figure of plane geometry."[6]

Given the existence of ordered proper skew fields, the first one being constructed in [103] (see also [51]), (8.2) with ${}^{p}Papp$ and ${}^{p}Des$ switched, does not hold even in the presence of order axioms. However, given that there are no *finite* skew fields, as shown by J. Wedderburn and L. E. Dickson in 1905 (see [190] for the history of this discovery and [8] for a short proof), we do have $\mathscr{P}, \exists^{\leq n},\ {}^{p}Des \vdash {}^{p}Papp$, for all natural numbers n (or, if we want to have non-vacuous statements before the $\vdash$ sign, for all $n \geq 7$, since there are no projective planes with less than 7 elements), where $\exists^{\leq n}$ denotes the statement that there are at most n points.

[4]What's more, in the higher-dimensional case, there is not even a need for any configuration theorem to hold, for any at least 3-dimensional projective space must satisfy ${}^{p}Des$ and thus surprisingly gives rise to the intricate algebraic structure of a skew field in terms of trivial geometrical statements.

[5]All known proofs that theorems π equivalent to Pappus (in its projective or in its affine form) imply theorems δ equivalent to Desargues resort to three uses of π (see, for example, [127]).

[6]["Wir] können [...] sagen, daß der Satz von PASCAL der einzig wesentliche Schnittpunktsatz der Ebenen ist, daß die Konfiguration $(9_3)_1$ die wichtigste Figur der ebenen Geometrie darstellt."

Following Beniamino Segre [235], geometers have tried to find a geometric proof of this fact; the key idea in this is not to use the fact that projective planes satisfying ${}^{P}Des$ can be coordinatized by skew fields. The most successful of these, by Bamberg and Penttila in 2015 [21], is significantly more convincing than the previous attempts in [137] and [257], by not giving the impression of an algebraic proof clad in geometric garb.

That neither ${}^{P}Des$ nor ${}^{P}Papp$ follow from plain $\mathscr{P}$ was settled by Veblen and Maclagan-Wedderburn in 1907 [265] by constructing a finite model of projective geometry, in which ${}^{P}Des$ (and thus *a fortiori* ${}^{P}Papp$) does not hold. Although the concept of *freeness* had been around for groups since Walther von Dyck's 1882 paper, it was only in 1943 that free projective planes were defined by Hall [85], and thus a very direct procedure for constructing projective planes that do not satisfy any non-trivial configuration theorem. The procedure is very simple. One starts with four distinct points and builds a structure in steps. The initial set of four points forms the structure at step 0. At odd steps, one adds to the existing points and lines a new line for each pair of distinct points that do not yet have a joining line (all the lines thus introduced are distinct). At even steps, one adds an intersection point for any two lines that do not have an intersection point (all the newly added points are distinct). The union of all these structures is the free plane constructed starting from the original four points.

Returning now to (P_3), the fundamental theorem of projective geometry, by Hessenberg's theorem [98] and the observation that

$$\mathscr{P}, P_{3,3} \vdash {}^{P}Papp \tag{8.3}$$

one gets that (P_3), ($P_{3,3}$), and ${}^{P}Papp$ are equivalent statements with respect to $\mathscr{P}$. After Barlotti proved in [24] that free projective planes (introduced by Hall in [85])—which do not satisfy any non-trivial configuration theorem—satisfy (P_6), the question whether ${}^{P}Papp$ follows from (and thus is equivalent with) some (P_n) with $3 < n < 6$ led Schleiermacher [225] (a different proof being provided in [78]) to prove that ${}^{P}Papp$ already follows from (P_5), which is the strongest result one could expect. In fact, his proof shows that ($P_{5,10}$) suffices. Later, Pickert [199] showed, by modifying the proof in [78], that ($P_{5,5}$) suffices to prove ${}^{P}Papp$. Schleiermacher also showed in [226] that, for any $n \geq 6$, ${}^{P}Papp$ follows from ${}^{P}Des$ and ($P_{n,4}$). Moreover, in case the projective plane is assumed to be finite, ${}^{P}Papp$ already follows from (P_6). Schleiermacher's result was improved upon in [63], where it is shown that ${}^{P}Papp$ follows from ${}^{P}Des$ and ($P_{n,3}$).

In the intuitionist setting—in which the underlying logic is intuitionistic, which means that it is compatible with a mathematics practiced in the manner proposed by L. E. J. Brouwer, avoiding any use of the Law of Excluded Middle and of its corollaries—one can prove some of the main theorems, but one needs somewhat stronger assumptions, such as that ${}^{P}Des$ and P_3 implies ${}^{P}Papp$ (rather than just P_3 itself implies ${}^{P}Papp$). A detailed analysis of the difficulties encountered in this case can be found in [153].

8.2.3 The Missing Link Between Desargues and Pappus

Since ^{P}Des is a strictly weaker statement than $^{P}Papp$ (given that there are skew fields that are not commutative, as was known since the discovery of quaternions by Hamilton in 1843; that proper skew fields exist in the presence of order as well was first proved by Hilbert, by constructing an ordered skew field, in 1903, in the second edition of the *Grundlagen der Geometrie*), the question regarding a missing link between ^{P}Des and $^{P}Papp$ was raised by Bottema in 1935 [34].

A *missing link* between two statements α and β, for which we know that α is strictly stronger than β (that is, $\alpha \rightarrow \beta$ ("α implies β") but $\beta \not\rightarrow \alpha$ ("β does not imply α") holds), is a statement γ such that $\alpha \wedge \gamma \leftrightarrow \beta$ ("the conjunction of α and γ is equivalent to β"), but such that γ alone does not imply β (all implications are considered with respect to some background theory, in our case $\mathscr{P}$). That missing statement was the restriction of $^{P}Papp$ in which one fixes the points P_1, P_2, Q_1, Q_2 (formally speaking, the statement $^{P}Papp$ had all variables universally quantified; the restricted statement makes the same statement about the variables involved, but its prefix starts with four existential quantifiers binding the variables P_1, P_2, Q_1, Q_2, followed by universal quantifiers binding all other variables).

8.2.4 Local Forms of the Desargues and Pappus Theorems

Hessenberg [95] was the first to have noticed that a configuration need not be assumed valid for all its point and line variables. He showed that it is enough to assume that all the open configurations of ^{P}Des close for a fixed pair of incident C_{23} and c, or else assume them for a configuration of a certain set of two lines and two points fixed. By this we mean that all the sentences formed by having C_{23} and c are quantified existentially at the beginning of the sentence, all other variables are quantified universally, $C_{23} \in c$ is part of the hypothesis, as are all the negated equalities, as well as all but one of the incidences between the universally quantified variables, and the remaining incidence is the conclusion. From this finite set of configuration theorems with fixed elements one can derive ^{P}Des.

We have mentioned that Pascal proved a generalization of Pappus's theorem. It states that if the two lines a and b in the statement of Pappus's theorem are replaced by a non-degenerate conic section, the theorem remains true. Pappus's theorem is simply a degeneration of Pascal's, in which the conic section reduces to two lines. The main problem with using Pascal's theorem as an axiom is that plane projective geometry, as introduced by us and as considered generally, does not have individual variables for "conic sections" in its language, creating major difficulties in simply expressing Pascal's theorem. This obstacle was, however, overcome in the formulation of Buekenhout's theorem [38] to which we turn.

First, let us mention that it is surprising and remarkable that the general form of both Pappus's and Pascal's theorems follows from a very restricted form of their general statements. As shown by Pickert in 1959 [198] (and reproved, with different methods in [42]), ${}^{p}Papp$ follows from its specialization in which a and b are two *fixed* lines.

In the same vein, F. Buekenhout in 1966 [38] showed that if Pascal's theorem is valid whenever the six vertices of the hexagon lie on a fixed *oval*, then ${}^{p}Papp$ holds and the oval is a conic section. An *oval* stands here for a non-empty set $\mathscr{S}$ of points in the projective plane with the property that no line intersects S in more than two points, and for every point P in $\mathscr{S}$ there exists precisely one line which intersects $\mathscr{S}$ only in P. In axiomatic parlance, the theorem states that if we enlarge the language of plane projective geometry with a unary predicate ω (whose intended interpretation is "if A is a point variable, then $\omega(A)$ means that A lies on the oval ω defines," and which can be defined not to hold for line variables), and we add to $\mathscr{P}$ the axioms (the defining axioms for an "oval")

O 1 *There exists A such that* $\omega(A)$.

O 2 *If* A, B, *and* C *are collinear points such that* $\omega(A)$, $\omega(B)$, *and* $\omega(C)$, *then* $A = B$ *or* $B = C$ *or* $C = A$.

O 3 *If* $\omega(A)$, *then there is exactly one line a through A such that no point* $P \neq A$ *on a satisfies* $\omega(P)$.

then (with $Papp_\omega$ standing for ${}^{p}Papp$ in which we add $\omega(P_i)$ and $\omega(Q_i)$ for $i \in \{1, 2, 3\}$ to the hypothesis)

$$\mathscr{P}, O2, O3, Papp_\omega \vdash {}^{p}Papp.$$

Buekenhout's theorem, together with variants and more general forms, has been reproved with different means in [12, 39] (using results from [11]), [22, 52, 67, 68, 111, 123, 124, 221].

8.2.5 What Is so Special About Desargues and Pappus?

8.2.5.1 Dehn's Question

A plethora of configuration theorems have sprung up during the twentieth century that are equivalent to ${}^{p}Des$ and ${}^{p}Papp$, or such that their conjunction is equivalent to one of the two. Many of these can be found in G. Pickert's comprehensive monograph [197] and the literature cited therein, and more have come to the fore since its publication, such as [75, 92, 93, 200, 201, 204, 223, 224, 255]. Faced with such an abundance of statements, all equivalent to ${}^{p}Des$ or ${}^{p}Papp$, one wonders if there is anything really distinguishing about these two configurations, or if they are simply artifacts of history. Both were found to be expressible as self-dual statements [131, 160], both can be seen as "generalized equivalences" (i.e. can be written as

$\bigwedge_{1\leq j\leq n}\left(\bigwedge_{1\leq i\leq n, i\neq j}\varphi_i \rightarrow \varphi_j\right)$), as pointed out in [156], but these properties are not uniquely applied to these two configuration theorems. However, early on, two authors raised different questions meant to showcase the central importance of these particular configuration theorems.

The first one was Max Dehn, a student of Hilbert, who in 1922 [59] (see also [60]) asked whether there was any configuration theorem that was strictly between ^{p}Des and $^{p}Papp$, in the sense that it was true in all Desarguesian projective planes, but not in all Pappian projective planes, without being equivalent to ^{p}Des. It was known, for quite some time, that such a configuration theorem could not exist in ordered Desarguesian projective planes of a special kind (a survey of how order can be introduced in a projective plane can be found in [183]), a result proved by Wagner in 1937 [267]. That such a configuration theorem cannot exist in any ordered Desarguesian projective plane was proved only in 1966 by Amitsur [2] (see also [27]). His proof shows more, namely that such configuration theorems are impossible in any Desarguesian geometry whose skew field satisfies an additional condition that is of a purely algebraic nature, unrelated to a relevant geometric statement. In this sense, one may say that, in ordered projective planes, all configuration theorems strong enough to imply ^{p}Des are equivalent to one of ^{p}Des or $^{p}Papp$. Dehn's question and Wagner's [267] pioneering treatment of that question gave rise to the field of rings with polynomial identities, now an autonomous field within algebra (see [3] or [222]).

However, this result still leaves open the questions whether the actual formal statements of these two configuration theorems are distinguished by more than historical accident from their many equivalent formulations.

8.2.5.2 Why ^{p}Des and $^{p}Papp$ Stand Out Among All (n_3) Configurations

Rachevsky's Answer

Before moving on to Pyotr Konstantinovich Rachevsky, the second author who attempted to distinguish the two historical theorems from their competitors—and whose work seems to have never been mentioned (not even in the otherwise encyclopaedic [197] or [82])—it is worth mentioning that all ordered projective planes satisfy an axiom introduced by Fano in 1891 [69]: "The three diagonal points of a complete quadrangle are non-collinear":

Fano *If A_1, A_2, A_3, A_4 are such that no three among them are collinear, then the points of intersection of (i) A_1A_2 and A_3A_4, (ii) A_1A_3 and A_2A_4, (iii) A_1A_4 and A_2A_3 are not collinear.*

Let us also notice that all three statements, ^{p}Des, $^{p}Papp$, and *Anti-Fano* (which, given the same hypothesis as *Fano*, concludes that the three diagonal points of a complete quadrangle are collinear, are (n_3) configurations. In case there are only finitely many points, *Anti-Fano* implies ^{p}Des, as shown by Gleason [81]—but not in the general, infinite case (see [197]). An (n_3) configuration consists of a set **L** of n lines and of a set **P** of n points, on every line from **L** there are precisely three points from **P** and through every point from **P** go precisely three lines from **L**. ${}^{p}Des$ is a (10_3), ${}^{p}Papp$ a (9_3), and *Anti-Fano* a (7_3).

We are now ready to state the 1940 result of Rachevsky [210]. Assume, for a fixed natural number $n \geq 7$, the following axioms: P1–P3, P5, (P_{cofin}), *Fano*, C_n, and all the purely existential statements that state the existence of every constructible configuration (see [163], where the concept was introduced and [197, p. 29]) of $< n$ points and $\leq n$ lines (or $< n$ lines and $\leq n$ points) without any additional incidences or coincides. Here P5 stands for

P 5 *Through every point there pass at least three distinct lines.*

By (P_{cofin}) we understand the following set of statements $\varphi_{m,n}$, for every pair of natural numbers n and m, with $m \geq 2$: a projectivity composed of at most m perspectivities that fixes all but n points of a line is the identity mapping of that line. Given Barlotti's result on free planes satisfying (P_6), this is a very weak requirement. By C_n, we mean the configuration theorem that states that from all but one of the incidences of a specific (n_3) configuration one can derive the missing one. A constructible configuration is one in which one starts with a point or a line and introduces, in stages, either points or lines incident with the existing ones, lines joining two points, or intersection points of existing lines. Still, one cannot stipulate that additional ("surprising") collinearities or incidences will result. Thus, for a given n, there are finitely many constructible configurations with $< n$ points and $\leq n$ lines (or vice-versa), and we add an axiom stating their existence without any surprises in our projective plane.

Under these modest assumptions, Rachevsky proves that C_n must be ${}^{p}Des$ or ${}^{p}Papp$. He could not show that the *Fano* assumption was needed, and the problem has not been revisited since.

Kocay's Answer

Another characterization of ${}^{p}Des$ and ${}^{p}Papp$ inside the class of all (n_3) configurations was established by Kocay in 2016 [130]. It is based on the impossibility of obtaining these configurations by a certain 1-point extension. This impossibility is a very rare one, being shared only with two other configurations.

The idea of extending (n_3) configurations to $((n+1)_3)$ configurations goes back to Martinetti [155]. A new extension procedure that characterizes ${}^{p}Des$ or ${}^{p}Papp$, *Anti-Fano*, and an anti-Fano-type configuration has been presented in [130].

Given an (n_3) configuration (Σ, Π), where Σ is the set of points and lines and Π is the set of incidences of the configuration, we pick three distinct points A_1, A_2, A_3 and three distinct lines l_1, l_2, l_3 in Σ, such that $A_1 = l_1 \cap l_2$, $A_2 = l_2 \cap l_3$, $A_3 \in l_3$, $A_3 \notin l_1$, and such that, if l' is the third line containing A_1, then $l' \cap l_3 \neq \emptyset$ implies $l' \cap l_3 = A_3$. We construct a new configuration (Σ', Π'), the 1-point extension of (Σ, Π), which turns out to be an $((n+1)_3)$ configuration, by setting $\Sigma' = \Sigma \cup \{A_0, l_0\}$, where a_0 is a new point and l_0 is a new line, and $\Pi' = (\Pi \setminus \{(A_1, l_1), (A_2, l_2), (A_3, l_3)\}) \cup \{(A_1, l_3), (A_2, l_0), (A_3, l_0), (A_0, l_0), (A_0, l_1), (A_0, l_2)\}$.

Kocay [130] shows that the only $((n+1)_3)$ configurations that cannot be obtained by a 1-point extension from an (n_3) configuration are the Anti-Fano configuration, the Pappus configuration, the Desargues configuration, and a specific Anti-Fano-type configuration.

Desargues and Pappus as Geometries in Their Own Right
Inspired by Bachmann's approach to the axiomatic foundation of plane absolute geometry in [19], Struve [250] provided in 1984 a remarkable characterization of the two configurations as geometries in their own right. In it the two configurations sit, once more, particularly close to one another.

The setting is that of a group G with trivial center (i.e., such that the only element x, for which $xg = gx$ for all $g \in G$, is the identity of G), together with two subsets S and P (the elements of S, which stand for *line reflections*, will be denoted by lower-case letters, those of P, which stand for *point reflections*, by upper-case letters), such that all elements of S and of P are involutory, i.e., elements x different from the identity 1, but such that $x^2 = 1$, and such that the sets S and P are invariant under conjugation—i.e., for any $a \in S$ and for any $A \in P$, $g^{-1}ag \in S$ and $g^{-1}Ag \in P$ for all $g \in G$—and generate together G (i.e., every element of G is a product of elements from S and P). In this setting, we can speak in a recognizably geometric language, by considering elements of P to be *points*, those of S to be *lines*, and introducing an *incidence relation* $\mid$ between points and lines by defining $A \mid b$ to mean that Ab is involutory, i.e., that $A \neq b$ and $Ab = bA$, as well as a notion of parallelism $\parallel$ between lines, defined so that $a \parallel b$ holds if and only if either $a = b$ or there is no point C such that $C \mid a$ and $C \mid b$. To any triple (G, S, P) of a group with two distinguished subsets S and P, one can thus associate an incidence structure, the *group plane* $(S, P, \mid)$. Two additional axioms, one of which is Playfair's form of the Euclidean parallel postulate, the other of which is asking the geometry to be an (n_3) configuration, turn out to be sufficient to determine ${}^{P}Des$ and ${}^{P}Papp$. These axioms can be expressed as

PA *For all* $A \in P$ *and* $g \in S$ *with* $A \neq g$ *there is exactly one* h *with* $A \mid h$ *and* $h \parallel g$.

N$_3$ *For all* $d \in S$ *there are exactly three elements* A, B, *and* C *in* P *such that* $A, B, C \mid d$, *and for all* $D \in P$ *there are exactly three elements* a, b, *and* c *in* S *with* $D \mid a, b, c$. *There are* A *and* b *such that* $A \neq b$ *and* $A \nmid b$.

The surprising result in [250] is that the group plane of a triple (G, S, P) in the above setting that satisfies PA and N$_3$ is either such that $S = P$, in which case it is the projective Desargues configuration, or $S \cap P = \emptyset$, in which case it is the projective Pappus configuration.

8.2.5.3 Desargues and Pappus as Cayley–Klein Geometries

The configurations of Desargues and Pappus can also be conceived as finite plane Cayley–Klein geometries. Plane Cayley–Klein geometries have been axiomatized in Bachmann's reflection-geometric style in [254]. As pointed out in [253, p. 182–183], the configuration described by $^p Des$ is an elliptic Cayley–Klein geometry over the field with five elements, whereas the configuration described by $^p Papp$ is a Galileian geometry over the field with three elements.

The groups of automorphisms of the Desargues and Pappus configurations have been the subject of a series of papers by Coxeter [53–55], a line of thought that has been pursued in [252] as well.

8.2.5.4 Strambach and Conic Sections

Conic sections have received two definitions, by J. Steiner and K. G. C. von Staudt, during the nineteenth century, and another one, by Krüger [132] (see also [90]) during the twentieth century. One would expect that each of these produces ovals (non-empty sets $\mathscr{S}$ of points in the projective plane with the property that no line intersects S in more than two points, and for every point P in $\mathscr{S}$ there exists precisely one line which intersects $\mathscr{S}$ only in P), and that the objects they define coincide. However, it was noticed by Ostrom in 1981 [177] that this is not the case in projective planes that do not satisfy $^p Papp$. The question regarding the equivalence of the three definitions led Strambach [249] to determine in 1995 that objects defined according to some of these definitions coincide with others or are ovals only if $^p Papp$ or, in one case, $^p Des$ holds in that projective plane.

In a projective plane, let $\mathfrak{P}(A)$ stand for the pencil of lines going through A. For two distinct points A_1 and A_2, let σ denote a projectivity mapping the lines of $\mathfrak{P}(A_1)$ onto those of $\mathfrak{P}(A_2)$, that does not fix the line A_1A_2. A *Steiner conic section* $K(A_1, A_2, \sigma)$ is the set $\{l \cap \sigma(l) \mid l \in \mathfrak{P}(A_1)\}$.

Strambach [249, Satz 1] proved that a Steiner conic section is an oval (i.e., satisfies O2 and O3) if and only if $^p Papp$ holds in that projective plane.

To define a Krüger conic section, we denote by $[l, \mathfrak{P}(A)]$, for $A \not\in l$, the perspectivity mapping the points of l onto the lines in $\mathfrak{P}(A)$, and similarly $[\mathfrak{P}(A), l]$ the perspectivity mapping the lines of $\mathfrak{P}(A)$ onto the points of l. For any four points A, B, C, and D, with no three collinear, the notion of a *Krüger conic section* $\mathfrak{K}(A, B, C; D)$ is the set $\{l \cap \sigma(l) \mid l \in \mathfrak{P}(A)\}$, where $\sigma = [\mathfrak{P}(A), BC][BC, \mathfrak{P}(D)][\mathfrak{P}(D), AC][AC, \mathfrak{P}(B)]$ (that is, σ is the projectivity obtained by composing the four perpsectivities $[\mathfrak{P}(A), BC]$, $[BC, \mathfrak{P}(D)]$, $[\mathfrak{P}(D), AC]$, and $[AC, \mathfrak{P}(B)]$).

Every Steiner conic section is a Krüger conic section precisely if $^p Des$ holds in that projective plane [249, Satz 2].

To define *strict von Staudt conic sections*, we first introduce a few notions of projective geometry. A *correlation* of a projective plane is a one-one mapping of the points onto the lines and lines onto the points which preserves incidence. A *polarity* is a correlation of order 2; it maps a point into its *polar* and a line into its *pole*. An *absolute* point (line) of a polarity is one which is incident with its polar (pole). An *orthogonal polarity* is a polarity whose set of absolute points is non-empty and forms an oval. A *strict von Staudt conic section* is precisely the above-mentioned set of absolute points of an orthogonal polarity.

Steiner conic sections and strict von Staudt conic sections coincide if and only if ^{P}*Papp* and *Fano* hold in that projective plane [249, Satz 6].

8.2.6 Moufang Planes

In 1843, spurred by the spectacular discovery of the quaternions that same year by William Rowan Hamilton, John T. Graves and, independently, Arthur Cayley discovered *octonions*, a *normed division algebra* that is eight-dimensional over the reals. The octonions can be also seen as a *non-associative* division algebra that uses real scalars, obtained by "doubling" the quaternions.[7] While quaternions might be said to have a loose connection with the foundations of geometry, as projective planes satisfying ^{P}*Des* but not ^{P}*Papp* are coordinatized by skew fields and the quaternions present one example of a proper skew field, it was realized very early on, with Hilbert's model of an ordered skew field in 1903, that quaternions are in no way emblematic of proper skew fields. In fact, it was rather an accident of history that they were the first non-commutative field to be discovered. Octonions (not necessarily built over the reals, for no elementary (first-order) axiomatization can capture a specific infinite field such as the reals), on the other hand, did turn out to be precisely the algebraic structure that corresponds to a natural class of projective planes.

It all started in the early 1930s when Ruth Moufang, a student of Max Dehn, provided a geometric home for these once obscure numerical abstractions in a series of papers [163–169] devoted to a specialization of ^{P}*Des*, in which one adds $C \in c$ to the hypothesis, nowadays referred to as the *projective minor form of the Desargues axiom*, to be denoted by ^{P}*des*. Projective planes that satisfy ^{P}*des* are called *Moufang planes*, following [197]. Moufang [165, 168] (and later, by more geometrical means, [157]) showed that ^{P}*des* is equivalent to a geometrically more illuminating requirement: the *axiom of the fourth harmonic point*.

To understand the statement of that axiom, consider three distinct collinear points A, B, and C, and let M be any point not on AB. Let c be any line through C distinct from AB and not passing through M, intersecting point MA in L, and MB in E. Let AE intersect BL in R and let MR intersect AB in D. Whenever A, B, C, and D are related by such a diagram, A, B, C, and D are said to be a *harmonic set of four collinear points*. The axiom of the fourth harmonic point states that, given three distinct collinear points A, B, and C, there is exactly one point D, such that A, B, C, and D form a harmonic set, and that D is different from C. This can be phrased as a configuration theorem, stating that for any collinear and different triple

[7] The word *associative* was coined by Hamilton in 1848 precisely to describe the fact that the octonions are *non*-associative.

of points (A, B, C), and any two pairs (M, c) and $(M'c')$, satisfying the conditions stated above for the construction of the fourth harmonic point, the construction leading to D starting with (M, c) also leads to D when one starts with (M', c').

Having found a wide class of configuration theorems that follow from ${}^{p}des$, Moufang showed (and a different proof was provided in [157]) that the Moufang planes can be coordinatized by algebraic structures that had just been defined by Artin and Zorn [277, 278] as *alternative algebras*. In these, the multiplication is not only not commutative, but it is not associative either. To be precise, these are structures $\mathfrak{A}$ with an addition and multiplication operation, such that (i) $\langle \mathfrak{A}, + \rangle$ is an Abelian group, with neutral element 0, (ii) multiplication has a two-sided neutral element 1, (iii) for all a and b different from 0, the equations $a \cdot x = b$ and $y \cdot a = b$ have unique solutions x and y; and for all a, b, and c we have (iv) $a \cdot (a \cdot b) = (a \cdot a) \cdot b$ and $a \cdot (b \cdot b) = (a \cdot b) \cdot b$, (v) $a \cdot (b + c) = a \cdot b + a \cdot c$, and (vi) $(b + c) \cdot a = b \cdot a + c \cdot a$. That there are Moufang planes that do not satisfy ${}^{p}Des$, i.e., that we are in the presence of the discovery of some new structures, was pointed out by Moufang in 1933 [168]. For at the time of their discovery, it was apparent that the octonions, not only when built upon the field of the real numbers as constructed by their discoverers Graves and Cayley, but when built up in the same manner over an arbitrary field K (a structure to be denoted by $\mathbb{O}(K)$), are an example of a non-associative alternative algebra. Despite some partial results characterizing alternative algebras in [278], the problem was solved in its entirety only in 1950–1951, by Skornyakov [240–243] (after preliminary work in [239]) and Bruck and Kleinfeld [37], where it was shown that any non-associative alternative algebra of characteristic different from 2 (corresponding to a Moufang plane which does not satisfy ${}^{p}Des$ and in which *Fano* holds) is an octonion algebra $\mathbb{O}(K)$ for some field K of characteristic $\neq 2$. The case of characteristic 2 was solved in [125]. These proofs were simplified in [41] and presented in the monograph [197, 6. Kapitel & Anhang 4] and in the textbook [71].

That certain local forms of ${}^{p}des$ imply ${}^{p}des$ was shown in [238].

Given that octonion algebras are not orderable, in the presence of order, a Moufang plane must satisfy ${}^{p}Des$. Also, every finite Moufang plane must satisfy ${}^{p}Papp$ (see [197, p. 301]). In the spirit of Schleiermacher's results, Pickert [199] proved the following stronger version of a result from [226], that ${}^{p}Papp$ follows already in Moufang planes from the assumption that a composition of at most 6 perspectivities that fixes 4 points must be the identity (Schleiermacher had shown that ${}^{p}Papp$ follows in Desarguesian planes from that same assumption), or formally

$$\mathscr{P},\ {}^{p}des,\ P_{6,4} \vdash {}^{p}Papp.$$

8.2.6.1 The Missing Link Between Moufang and Pappus in Fanoian Planes

We have mentioned earlier that O. Bottema determined in 1935 the missing link between ${}^{p}Des$ and ${}^{p}Papp$. The question regarding a missing link between ${}^{p}des$ and ${}^{p}Papp$ has been answered in 1981 by Weiß [269], but only in the presence of *Fano*. The missing link is the projective version ${}^{p}NG$ of the following Newton–Gauss theorem: *The three midpoints of the diagonals of a complete quadrilateral are collinear* (if x, y, z, and w denote the sides of a quadrilateral, A the intersection of x and z, A' that of y and w, B that of z and y, B' that of x and w, C that of y and x, and C' of z and w, then the midpoints of the diagonals AA, BB, CC are collinear). Thus, in the presence of the Fano axiom, ${}^{p}Papp$ can be replaced by Moufang's ${}^{p}des$ and the Newton–Gauss theorem ${}^{p}NG$, or formally

$$\mathcal{P},\ Fano \vdash {}^{p}Papp \leftrightarrow {}^{p}des \wedge {}^{p}NG.$$

8.2.7 The Lenz–Barlotti Classification

A specialization of ${}^{p}Des$ lies at the heart of a classification of all projective planes that satisfy configuration theorems that are not stronger than ${}^{p}Des$. The specialization, commonly referred to as the (C, c)-*Desargues* differs from the same statement of ${}^{p}Des$ just by fixing in advance C and c, and allowing all other variables to be free. Thus a projective plane may satisfy the (C, c)-Desargues for some pairs of (C, c) but not for others. Lenz [138] provided in 1954 a first classification of projective planes based on the sets of pairs (C, c), with $c \in C$, for which the plane satisfies the (C, c)-Desargues. Six non-empty classes of projective planes took shape from that analysis. By removing the requirement that $c \in C$, Barlotti [23] obtained in 1957 a refinement of Lenz's classification, referred to as the *Lenz–Barlotti classification*. The classes that interest us here are the Lenz classes IV, V, VII, corresponding to *translation planes*, *Moufang planes*, and *Desarguesian planes*.

The Lenz–Barlotti classification can also be understood in terms of a certain notion of *transitivity* that the group of collineations could act on the lines of a projective plane. A bijection α of the point and line sets of a projective plane onto itself is called a *collineation* if it preserves the incidence relation, i.e., if $P \in g$ implies $\alpha(P) \in \alpha(g)$. A collineation α is called a *perspective collineation* with *center* C and *axis* c if $\alpha(g) = g$ for all lines g with $C \in g$ and $\alpha(P) = P$ for all $P \in g$. For a fixed projective plane $\mathfrak{P}$, let $\Gamma_{C,c}$ stand for the group of all perspective collineations with center C and axis c. For a point C and a line c, we say that $\mathfrak{P}$ is (C, c)-transitive if for any two points P and Q, none of which lies on c, that are collinear with C, but none of which is C, there exists a collineation α in $\Gamma_{C,c}$, such that $\alpha(P) = Q$. It turns out that (C, c)-transitivity amounts to the transitive action of the group $\Gamma_{C,c}$ on the points different from C and the intersection of c with g of a single line $g \neq c$ passing through C. The (C, c)-transitivity of $\mathfrak{P}$ is equivalent to the validity of the (C, c)-Desargues in $\mathfrak{P}$ (see [197, p. 76]). Desarguesian planes are (C, c)-transitive for any pair (C, c), Moufang planes are (C, c)-transitive for any pair (C, c) with $C \in c$, whereas translation planes are (C, c)-transitive for all C lying on a fixed line c.

We will now turn to a definition of translation planes in the context of affine planes.

8.3 The Affine Setting

Given the particularly simple and direct connection between affine and projective planes, any affine plane being extendable to a projective plane by adding a new point to every line as well as a new line incident with all new points, and any projective plane becoming an affine plane by deleting one of its lines together with all points incident with it, the story narrated here could have been told in the projective setting as well. However, given that certain statements have been discovered in the affine setting and tend to have a more appealing statement in the language of affine geometry, we have decided to devote a special section to the affine forms of Pappus and Desargues.

Plane affine geometry can be expressed in the same language as projective geometry, but it is helpful to use the defined notion of *parallelism* (defined by $g \parallel h$ if and only if either $g = h$ or else there is no point P incident with both g and h) when expressing configuration theorems. It is axiomatized by $\mathscr{A} = \{A1 - A4\}$, where

A 1 *There is precisely one line passing through two different points A and B (which will be denoted by AB).*

A 2 *There are at least two points on every line.*

A 3 *There is exactly one parallel through P to l for any point P and any line l.*

A 4 *There are three non-collinear points.*

The affine version of ${}^{P}Des$ is obtained by letting c be the "line at infinity", which turns it into (Fig. 8.3)

^{a}**Des** *If* $C, A_i, B_i \in c_i$, $A_i, A_k \in a_{ik}$, $B_i, B_k \in b_{ik}$, C, A_i, *and* B_i *are pairwise different points (for all i and k in* $\{1, 2, 3\}$ *with* $i < k$*),* c_1, c_2, *and* c_3 *are three*

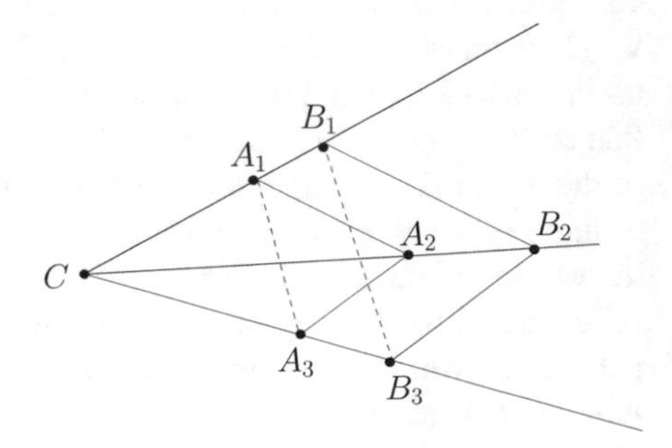

Fig. 8.3 The affine form of the Desargues axiom

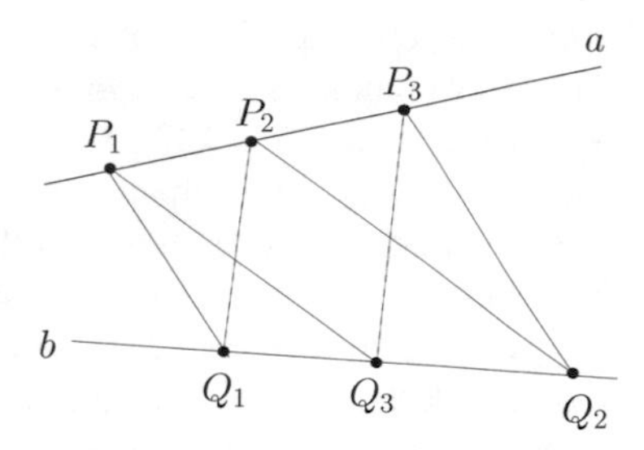

Fig. 8.4 The affine form of the Pappus axiom

pairwise different lines, $a_{12} \neq a_{13}$*,* $b_{12} \neq b_{13}$*,* $A_1A_2 \parallel B_1B_2$*, and* $A_2A_3 \parallel B_2B_3$*, then* $A_1A_3 \parallel B_1B_3$*.*

The affine version of ^{p}*Papp*, to be denoted by ^{a}*Papp*, is obtained by adding (i) and (ii), in which the respective pairs of lines are parallel, to the hypothesis, and asking for the pair in (iii) to be parallel as conclusion (Fig. 8.4).

^{a}**Papp** *If the three points* P_1*,* P_2*, and* P_3 *are on a line a, the three points* Q_1*,* Q_2*, and* Q_3 *are on a line b but not on a,* $P_1Q_1 \parallel Q_2P_3$*, and* $P_2Q_2 \parallel Q_3P_1$*, then* $P_3Q_3 \parallel Q_1P_2$*.*

If in ^{a}*Papp*, the lines a and b are required to be parallel, then we get the *minor form of the affine Pappus axiom*, ^{a}*Papp*, and if in ^{a}*Des*, C is a "point at infinity", i.e. all lines c_i are parallel, then we get *minor form of the affine Desargues axiom*, ^{a}*des*. Affine planes satisfying ^{a}*des* are called *translation planes* and came to prominence after André [4, 5] showed that there are translation planes that do not satisfy ^{a}*Des*. These can, however, be coordinatized by *quasifields*, an algebraic structure that had shown up earlier, in [265]. These are structures $\mathfrak{A}$ endowed with an addition and a multiplication, both with neutral elements, 0 and 1 respectively, satisfying the conditions (i), (ii), (v) mentioned earlier for alternative algebras, as well as (vii) $0 \cdot x = x \cdot 0 = 0$ for all x; (viii) for all a, b, and c with $a \neq b$ there exists y such that $a \cdot y - b \cdot y = c$.

So far, our story, which is one of interactions between geometry and algebra, has brought algebraists into the narrative fold. We now meet a major figure of differential and convex geometry, Wilhelm Blaschke. Educated in a variety of centers of geometry, he developed a particular interest for the *geometry of webs*, in a style born out of multiple influences, from Klein, Hilbert, Engel, Study, and Bianchi, expounding upon the early works of Lie and Poincaré.

At the 1928 International Congress of Mathematicians in Bologna, Blaschke posed the question of how these webs might be considered in an axiomatic setting. Reidemeister[218, 219], Thomsen [260], and Bol [261] took it upon themselves to answer this call and produced a great number of configuration theorems, all of which, although can be understood in projective planes, are best stated in the language of affine geometry (see [122] and [161] for a survey of these configuration theorems). It was eventually shown that all these configuration theorems were equivalent to one of ^{a}*Papp*, ^{a}*Des* or ^{a}*des*. Wilhelm Klingenberg—at the time influenced by Friedrich Bachmann, himself influenced by Kurt Reidemeister (to whom Bachmann's *opus magnum* [19] is dedicated)—showed many of these equivalences in [126, 127], the results of which surprised Moufang, the leading expert in configuration theorems at that time. In fact, of particular interest was the startlingly unexpected strength of the Reidemeister configuration theorem. A *major Reidemeister configuration* is a configuration consisting of 9 points and 12 lines, and can be best thought of as the drawing of rectangle $ABCD$ illuminated from a source of light O outside its plane, in which the rays of light are shown emanating from O and projecting the vertices of $ABCD$ onto another rectangle $A'B'C'D'$ on a plane parallel to that of $ABCD$. The surprise resided in the fact that the major Reidemeister configuration theorem is strong enough to entail ${}^{a}Des$.[8]

In the meantime, many more affine configurations emerged, such as those in [32, 79, 80, 134, 135, 175, 176, 272]. The textbook [256] (as well as [144]) presents synthetic proofs of the dependencies (here ${}^{a}Papp_1$ stands for ${}^{a}Papp$ in which the lines a and b have an intersection point; it is also shown there that $\mathscr{A} \vdash {}^{a}Papp_1 \leftrightarrow {}^{a}Papp$)

$$\mathscr{A} \vdash {}^{a}Papp_1 \rightarrow {}^{a}Des \rightarrow {}^{a}des \rightarrow {}^{a}papp. \tag{8.4}$$

It should come as no surprise that ${}^{a}Des$ leads to skew fields as coordinates and ${}^{a}Papp$ to commutative fields, since the projective planes in which ${}^{p}Des$ and ${}^{p}Papp$ are satisfied are planes over skew fields and commutative fields, respectively. That ${}^{p}Des$ (${}^{p}Papp$) holds in affine planes satisfying ${}^{a}Des$ (resp. ${}^{a}Papp$) is apparent if one thinks in terms of the coordinatization of these structures. For the Desarguesian case, a direct geometric proof can be found in [203]. The first two implications in (8.4) cannot be reversed (given that there are quasifields that are not skew fields and skew fields that are not commutative), but it is still not known whether the last implication is reversible, i.e. if ${}^{a}papp \rightarrow {}^{a}des$ does or does not hold.

As pointed out by Reinhold Baer in 1944 [14] (see also [197, p. 211–212]), translation planes of characteristic $\neq 2$, i.e., satisfying the affine version of *Fano*, stating that the diagonals of a parallelogram do intersect, can also be axiomatized in a language enlarged with a binary operation μ on points, whose intended interpretation is that of a *midpoint operation*, by the axioms of $\mathscr{A}$, together with

[8] The connection between Blaschke—who would later author two books on webs ([31] (with G. Bol) and [30]) and a textbook on projective geometry [29]—and the foundations of affine and projective geometry was not only one giving an impetus to the discovery of so many configuration theorem central to the foundational approach, but also one of a personal friendship with Reidemeister, for whom he intervened in June 1933, after Reidemeister was dismissed from his position in Königsberg—which he had held since 1925, and where he had worked among others with Ruth Moufang—for having been a longtime critic of the National Socialists, by organizing a petition to persuade the government that forcing Reidemeister to retire at 40 was not in the interest of the teaching of mathematics and of mathematical research in Germany. The petition was successful and Reidemeister was appointed to Hensel's chair at Marburg.

the following axioms (Mid 4 states that midpoints are invariant under parallel projection):

Mid 1 *For $X \neq Y$, we have $\mu(X, Y) \in XY$; $\mu(X, X) = X$.*

Mid 2 *For all $X \neq Z$, there is exactly one Y such that $\mu(X, Y) = Z$.*

Mid 3 $\mu(X, Y) = \mu(Y, X)$.

Mid 4 *For $X \neq Y$, $X' \notin XY$, $XX' \parallel YY'$ or $Y = Y'$, we have $\mu(X, Y)\mu(X', Y') \parallel XX'$.*

This characterization of translation planes of characteristic $\neq 2$ led in [179] to the simplest possible axiom system in a language with points and parallelism as the only primitive notions, as each axiom is a statement about at most 5 points.

In a similar vein, Lüneburg [150], has shown in 1967 that translation planes can be characterized as affine planes admitting the notion of a *ratio*—i.e., a binary operation μ with points as variables and values, with $\mu(A, B)$ to be interpreted as the point on the line AB dividing it in a fixed ratio. The axiom system consists of the axioms of $\mathscr{A}$ together with Mid1 and the axiom requesting that for all X there exists a point Y such that $\mu(X, Y) \neq X$ and $\mu(X, Y) \neq Y$.

Affine Moufang planes (affine planes obtained from a projective Moufang plane by removing a line and all of it points) of characteristic $\neq 2$ can be axiomatized (as shown in [134]) by the axioms of $\mathscr{A}$, the affine form of the Fano axiom, stating that the diagonals of a parallelogram intersect, and the *Trapezoid-Desargues*, which has the same conclusion as ${}^a\mathit{Des}$, but to whose hypothesis $A_1A_2 \parallel A_3B_3$ has been added. An alternative axiomatization was proposed by Zimmer in 1958 [275] (and a different proof was offered in [134], the author being apparently unaware of the existence of [275]), where it was shown that the *Trapezoid-Desargues* can be replaced by one of its specializations, called *Parallelogram-Desargues*, in which one adds to the hypothesis of the Trapezoid-Desargues axiom the statement that $A_2A_3 \parallel A_1B_1$, provided that one replaces the affine Fano axiom by a statement on the existence and uniqueness of midpoints. It states that, for any two distinct points P and Q there is a point M on PQ, such that, for any S and R, not on PQ, and such that $PQ \parallel QS$ and $QR \parallel PS$, we have $M \in RS$ (it states the existence of the midpoint M for any pair of distinct points P and Q, as the intersection point of the diagonals of a parallelogram $PRQS$, as well as its independence from the particular choice of a parallelogram with PQ as one of its diagonals).

It is also worth mentioning that André [6] also introduced *central translation structures*. These structures are more general than translation planes in the sense that non-parallel lines do not need to intersect (parallelism being a primitive notion in those structures, that is not definable by means of the point-line incidence relation). They can be thought of as incidence structures—in which every line is incident with at least two points, and which contains three non-collinear points—endowed with a parallelism relation that satisfies not only the Euclidean parallel postulate (existence and uniqueness of a parallel through a given point to a given line), but also all the universal statements valid in arbitrary affine planes, such that all of its translations (fixed point-free collineations which map lines into parallel lines, and the identity) are *central* (if τ is a proper translation and P and Q are two points, then the lines $P\tau(P)$ and $Q\tau(Q)$ are parallel), form a group, and act transitively on the set of points. Central translation structures were provided with a different, simple axiom system in [180].

A reflection-geometric axiomatization of all major classes of translation planes (Moufang, Desarguesian, and Pappian), in the style of Bachmann's [19] was proposed in [207] and [208].

The constructive (in the sense of Errett Bishop) content of the results needed to introduce coordinates in Desarguesian affine planes was analyzed in [152].

8.3.1 The Role of the Theorem of Menelaus

Desargues's original proof for his theorem could be of purely historical interest, as perhaps revealing something of the inspirational background leading to its discovery, or it could be of methodological interest, raising the question whether the assumptions used in that proof could be turned into a set of axioms to provide a foundation for some Desarguesian geometry. For his theorem, Desargues utilized both Menelaus's theorem and its converse. Menelaus's theorem, going back to Menelaus of Alexandria's *Sphaerica* [158], states that, given a triangle ABC, and a line g that goes through none of the vertices of this triangle and which intersects the lines AB, BC, and CA in the points P, Q, and R, respectively, then (ratios being considered *oriented*, i.e., $\frac{AP}{BP} = -\frac{PA}{BP}$)

$$\frac{AP}{PB} \cdot \frac{BQ}{QC} \cdot \frac{CR}{RA} = -1.$$

The converse states that if three points P, Q, and R, lie on the sides AB, BC, and CA of a triangle ABC, different from its vertices, such that the above equality holds, then P, Q, and R are collinear. Since ratios of segments lying on the same line are notions of affine geometry, from an axiomatic point of view the question is one regarding the possible role of Menelaus's theorem taken as an axiom in the foundations of affine planes satisfying ${}^{a}\mathit{Des}$. This question was both raised and solved by Guggenheimer in 1974 [83], the explanation of which is delineated below.

The axiomatic set-up is inside a two-sorted first-order language. One sort of variable is for *elements of a multiplicative group*, to be denoted by lowercase Greek letters, the other for *points*, to be denoted by uppercase Latin letters. There is a unary operation $^{-1}$ standing for the inverse element operation, binary operation $\cdot$, whose arguments and values are elements of the group, two individual constants, 1 (the neutral element) and ω (an element of the multiplicative group), a ternary operation f, with the first two arguments points and the third one an element of the group, taking point values, a ternary function λ, with all arguments points, taking elements of the group as values.

The intended interpretation of $f(A, B, \alpha)$ is the point X on the 'line' determined by A and B, for which $\overrightarrow{AX} = \alpha \cdot \overrightarrow{XB}$ (by abuse of language we also write $\overrightarrow{AX} : \overrightarrow{XB} = \alpha$). The intended interpretation of $\lambda(X, A, B)$ is the ratio $\overrightarrow{AX} : \overrightarrow{XB}$ in case $A \neq B$ and X is a point on the line determined by A and B, different from A and B.

The axioms state that: (M1) $\cdot$ is associative, has a neutral element, and every element has an inverse; (M2) ω is an involutory element of the group that belongs to its center; (M3–M6) describe basic properties of the operations f and μ; (M7) Menelaus' theorem; (M8) its converse.

M 1 $x \cdot (y \cdot z) = (x \cdot y) \cdot z \wedge 1 \cdot x = x \wedge x^{-1} \cdot x = 1.$

M 2 $\omega \cdot x = x \cdot \omega \wedge \omega \neq 1 \wedge \omega^2 = 1.$

M 3 $A \neq B \wedge \alpha \neq \omega \wedge \beta \neq \omega \wedge f(A, B, \alpha) = f(A, B, \beta) \to \alpha = \beta.$

M 4 $A \neq B \wedge \alpha \neq \omega \to \lambda(f(A, B, \alpha), A, B) = \alpha.$

M 5 $(\forall AB\alpha)(\exists \beta\gamma)\, A \neq B \wedge \alpha \neq \omega \wedge X = f(A, B, \alpha)$
$\to A = f(B, X, \beta) \wedge \beta \neq \omega \wedge B = f(X, A, \gamma) \wedge \gamma \neq \omega$

M 6 $A \neq B \wedge \alpha \neq \omega \to \lambda(f(A, B, \alpha), B, A) = \alpha^{-1}.$

M 7 $P \neq Q \wedge f(P, Q, \eta) = R \wedge \eta \neq \omega \wedge A \neq B \wedge f(A, B, \alpha) = P \wedge \alpha \neq \omega$
$\wedge f(B, C, \beta) = Q \wedge \beta \neq \omega \wedge C \neq A \wedge f(C, A, \gamma) = R \wedge \gamma \neq \omega$
$\to \lambda(P, A, B) \cdot (\lambda(Q, B, C) \cdot \lambda(R, C, A)) = \omega$

M 8 $(\forall PQRABC\alpha\beta\gamma)(\exists \eta)\, P \neq Q \wedge A \neq B \wedge f(A, B, \alpha) = P \wedge \alpha \neq \omega$
$\wedge f(B, C, \beta) = Q \wedge \beta \neq \omega \wedge C \neq A \wedge f(C, A, \gamma) = R \wedge \gamma \neq \omega$
$\wedge \lambda(P, A, B) \cdot (\lambda(Q, B, C) \cdot \lambda(R, C, A)) = \omega \to f(P, Q, \eta) = R \wedge \eta \neq \omega$

One can show that one *needs* three applications of the direct form (M7) and one of the converse of Menelaus's theorem (M8) to prove ${}^{a}Des$, precisely the number of times Desargues himself used Menelaus' theorem (see [186]). It is remarkable that, without having postulated that there is a skew field involved at all, since there is no addition operation in this language, Guggenheimer was able to show that these structures naturally *emerge*, that the lines are coordinatizable by skew fields, and that the multiplicative group in our axiom system is the multiplicative group of that skew field.

If we add to the above axiom system an axiom stating that $\cdot$ is commutative, then we can prove, by a fivefold application of M7 and by one of M8, as in [56, p. 68], ${}^{p}Papp$.

8.3.2 Area

In [103, p. 80], Hilbert points out that some basic theorems regarding area in Euclidean geometry are dependent upon and can be used to prove ${}^{a}Papp$. This remark has been made precise by Hotje in 1979 [113], by showing that the existence of an oriented area function in an arbitrary affine plane satisfying certain common sense conditions implies that the affine plane satisfies $^{a}Papp$ (see also [112] and [84]). Another set of area related conditions, this time on *equiaffinities* (affine transformations that preserve the area), entail, in the case of Desarguesian affine planes of characteristic $\neq 2$, the validity of $^{a}Papp$, which is given in [128, pp. 126–131]. Less stringent conditions on the oriented area function than those stipulated by Hotje allowed Lesieur [139] to enlarge the class of affine planes endowed with an area function to affine Moufang planes. Likewise, Petit [193–195] found an even wider class of affine planes that includes certain translation planes that are not Moufang planes, to which an area function can be associated.

Although no general area theory can be developed to allow the comparison of any two triangles in ordered translation planes without additional properties, in at least a manner that would satisfy the very modest requirements of J.-C. Petit, a particular problem involving area comparison of triangles sharing a side can be expressed in a perfectly meaningful manner and proved inside the theory of ordered translation planes, as shown in [116]. The problem is an elementary one, proposed by H. Debrunner in print, but going back to P. Erdős and E. Trost. It states that of the four triangles formed by three points A', B', and C' on the sides BC, CA, and AB, one of the corner triangles, $AB'C'$, $BC'A'$, and $CA'B'$, has the smallest area. This is an indication that ordered translation planes are not just abstract structures that result from specializing Desargues's theorem, but occur naturally as just the right structures in which certain elementary geometry statements can be expressed and proved.

8.4 The Effect of the Archimedean Axiom

The Archimedean axiom, although not expressible in first-order logic (see [183] for the logics extending first-order logic in which it is expressible), has played a central role in the foundations of geometry ever since its discovery by Eudoxus and extensive use by Archimedes. It is thus worth looking at its effect on the configuration theorems of our story. The aim of the Archimedean axiom is to exclude the possibility of the existence of "infinitely small" or "infinitely large" magnitudes, i.e., to ensure that, given $0 < x < y$, there exists a natural number n such that $nx > y$. The first to notice its surprising effect was Hilbert in [103, §32, pp. 72–73]. What he noticed, was that the commutative law for multiplication follows in skew fields from the presence of the Archimedean axiom. With *Arch* denoting the Archimedean axiom, this means that (here $\mathcal{OA}$ stands for the axioms for ordered affine planes, such as those in [256])

$$\mathcal{O}A,\ {}^{a}Des,\ Arch \vdash {}^{a}Papp. \tag{8.5}$$

It is easily noticed that Hilbert's proof, which operates entirely on the algebraic level, goes through in alternative algebras, since proper alternative algebras cannot be ordered at all. The question thus arises what the lowest level in the Lenz–Barlotti classification is, in which Hilbert's result can be extended, i.e., in which the addition of the Archimedean axiom implies ^pPapp. As shown in [205, p. 224], using results from [57], the answer to the above question, i.e. the algebraic structure corresponding to the lowest level of the Lenz–Barlotti classification for which the addition of Archimedeanity forces it to turn into an Archimedean ordered commutative field, is that of a quasifield, the coordinatizing structures for translation planes. Thus (8.5) remains valid even with ^aDes replaced by ^ades, and $^a\ des$ is the weakest (C, c)-Desargues configuration that can replace ^aDes in (8.5). A purely geometric proof of the derivability of ^pPapp from *Arch*, $\mathscr{P}$ and some configuration theorems that were shown by Moufang to follow from ^pdes, has been provided in [26]. Later on, a purely geometric proof of (8.5) was provided in [211].

Rather surprisingly, (8.5) was turned into first-order logic by Rautenberg (in [214], after having done a similar feat in an algebraic setting in [212]), who has introduced a certain first-order version Σ_{ar} of *Arch*, which can replace the latter in (8.5). This version is much weaker than all the statements true in first-order logic in all Archimedean ordered affine planes with ^aDes, for that theory is not even recursively axiomatizable, as shown in [213]. With $Z(ABC)$ standing for 'B lies strictly between A and C' and $AB \cong CD$ standing for $\mu(A, D) = \mu(B, C)$, where μ denotes the midpoint operation, it states that

$$\Sigma_{ar}\quad (\forall PQRST)(\exists UV)\, P \neq Q \wedge \varphi(P) \wedge \varphi(Q) \wedge [PQ \cong RS \rightarrow (\varphi(R) \leftrightarrow \varphi(S))] \\ \rightarrow [T \in PQ \rightarrow \varphi(U) \wedge \varphi(V) \wedge Z(UTV)],$$

where $\varphi(\cdot)$ is a formula containing none of the variables P, Q, R, S, T, U, V, and PQ stands for the line determined by P and Q.

This axiom schema states that, if $\mathfrak{A}$, a subset of the point set of a model $\mathfrak{M}$ of plane affine geometry, is a definable set of points (defined by φ, which means that $\mathbf{X} \in \mathfrak{A}$ if and only if $\varphi(\mathbf{X})$ holds in $\mathfrak{M}$), containing distinct points $\mathbf{P}$ and $\mathbf{Q}$, and such that, whenever $\mathbf{PQ} \equiv \mathbf{RS}$, $\mathbf{R}$ is in $\mathfrak{A}$ if and only if $\mathbf{S}$ is in $\mathfrak{A}$, then every point $\mathbf{T}$ on the line $\mathbf{PQ}$ is between two points of $\mathfrak{A}$.

8.5 Desargues's Axiom as Indicator of a Projective Plane's Embeddability in a Projective Space

After asking, in the lectures [104, p. 318] (see also [7] and [185]) preceding the publication of the *Grundlagen der Geometrie*, whether ^aDes is a *sufficient* condition for a plane to be a part of three-dimensional space (that it is a *necessary* condition was known, given that one can easily prove ^pDes from considerations involving intersections of planes), Hilbert answered in [103] this question in the affine case. What he actually did show in Theorem 35 of [103] was that Desarguesian ordered affine planes can be embedded in ordered affine spaces, but he did this in an algebraic manner. After coordinatizing the affine plane by a skew field, he added a third coordinate to each point (x, y) of the plane, which becomes $(x, y, 0)$ in the affine space in which the plane is embedded.

A *more geometrico* proof of Hilbert's Theorem 35 was provided in [228], while purely geometric embeddings of projective planes satisfying ${}^{p}Des$ into 3-dimensional projective spaces (without any order relation) were provided in [15, 20, 72, 76, 77, 89, 94, 99, 142], and [146, Theorem 9.2.6].

If we now drop the requirement that the plane be affine or projective, and ask the necessity version of Hilbert's question for *ordered planes*, as defined by the plane axioms of the first two groups of axioms of Hilbert's [103], then even the fact that any plane in an ordered space (i.e., in a model of Hilbert's incidence and order axioms) has to satisfy ${}^{p}Des$, is no longer a simple matter. That this is indeed the case was first shown in [192, pp. 46–55] (see also [1]). That the validity of the betweenness axioms is essential for this has been pointed out in [88]. To understand why ordered planes pose a problem and why one cannot just transfer the relevant results from projective geometry, one should visualize an ordered plane as possibly the interior of any convex set in, say, an affine ordered plane. Thus, if the domain of discourse is, say, the interior of a triangle, and one knows that a configuration theorem τ is valid in it, that means that we *only* know that τ holds when all its points belong to that domain. One must then find a way to transfer the local information to an entire projective plane that is needed to construct an extension of the ordered plane.

The first step toward answering the sufficiency question in the case of ordered planes was taken by Owens in 1910 [178], where it was shown that an ordered plane, in which certain strong forms of the Desargues axiom and its converse hold, can be embedded in a projective ordered plane satisfying ${}^{p}Des$. Athough Owens' work was reviewed by Dehn for the *Jahrbuch für die Fortschritte der Mathematik*, it is not referred to by Moufang [164] (see also [244]), who takes the next step by showing that an ordered plane that satisfies ${}^{p}des$ (whenever all points involved belong to the ordered plane) can be embedded in a projective ordered plane, which in turn satisfies ${}^{p}des$. In 1938 Emanuel Sperner [247] completely answered Hilbert's question in the case of ordered planes by showing that ordered planes satisfying ${}^{p}Des$ can be embedded in projective ordered planes satisfying ${}^{p}Des$. It will turn out later, with the results of Skornyakov and Bruck and Kleinfeld referred to earlier, which imply that all ordered Moufang planes satisfy ${}^{p}Des$, that Moufang's 1931 result actually implies Sperner's 1938 result. However, since *no synthetic geometric* deduction of ${}^{p}Des$ from the axioms of ordered projective planes and ${}^{p}des$ is known, Sperner's proof is the only purely geometric one we have. These results thus imply that any ordered plane satisfying ${}^{p}Des$ is indeed part of an ordered three-dimensional space as defined in [103].

In its most general version, the problem of embedding a part of a projective plane that satisfies some form of the Desargues axiom into a projective plane satisfying ${}^{p}Des$ has been solved in [66] (and later, by a different method in [121]). The set-up is a very general one. There are points, lines, and incidence in the language, as usual, but there is also a unary predicate π which applies to points only (i.e., for all lines g, we have $\neg\pi(g)$), with $\pi(A)$ to be interpreted as "point A is a *proper* point", in contradistinction from points A with $\neg\pi(A)$, which will be referred to as *ideal* points. One can visualize this again in terms of the points in the interior $\mathfrak{J}$ of some convex set in an ordered projective plane ϵ being proper points, whereas all other points are ideal, and the set of lines consists only of those lines of ϵ that intersect $\mathfrak{J}$. However, we should bear in mind that the class of possible models for our axiom is *much wider* and that there is actually no notion of order, so the above model is for visualization purposes only. The axiom system consists of a certain form of ^{P}Des, in which the relevant points are proper, and:

ES 1 *There are two different lines. Any two different lines have precisely one point in common.*

ES 2 *Every line is incident with at least three proper points.*

ES 3 *If A is a proper point and B is any point different from A, then there is a line incident with both A and B.*

What Ellers and Sperner show in [66] is that these few requirements suffice for any model of the above axiom system to be embedded in a projective plane satisfying ^{P}Des. Obviously, this is a much stronger statement than all those for ordered planes previously mentioned.

Whether a similar theorem is true for $^{P}Papp$ is not known, at least in the ordered setting. That is, whether an ordered plane satisfying $^{P}Papp$ can be embedded in an ordered space satisfying $^{P}Papp$ remains open. L. W. Szczerba told the first author in 2003 that he had once proved this to be true, but did not write the proof down anywhere. If this were true, this would simplify Skala's [237] incidence-based axiomatization of plane hyperbolic geometry, which uses $^{P}Papp$, ^{P}Des, as well as Pascal's theorem for all hexagons with vertices 'ends', which lie on the absolute conic, the inside of which constitutes the hyperbolic plane.

While ^{P}Des follows from the axioms of a projective space of more than two dimensions, this is not the case for $^{P}Papp$, as one can construct projective spaces of any dimension as left vector fields over skew fields. There is, however, a statement with a special higher-dimensional flavor, which is equivalent to $^{P}Papp$, given the usual axioms of three-dimensional projective geometry. It is sometimes referred to as *Dandelin's theorem*, or as the *Sixteen Points Proposition*. It states that (see [100, p. 97, Th. 4.2.1]): "If $a_0, a_1, a_2, a_3, b_0, b_l, b_2, b_3$ are different lines, not in one plane, and if fifteen of the intersections $a_i \cap b_k$ exist, then the sixteenth exists as well." A higher-dimensional version of Dandelin's theorem was considered in [91], and it was shown that it is equivalent to $^{P}Papp$.

8.6 One-Dimensional Characterizations

Our story so far has been one of an astonishing dialogue between configuration theorems in projective or affine geometry and algebra. Since the proud two-dimensional configuration theorems of Pappus and Desargues become, in this dialogue, properties of algebraic structures constructed on lines living in projective planes, one wonders why one would need a two-dimensional statement to tell a one-dimensional story. Couldn't these algebraic structures themselves be axiomatized in purely geometric terms? By *geometric* we mean here some story that allows for transformations that leave the structure invariant. The algebra itself is definitely not "geometric" in this sense. If we look at the real number field, for example, we notice that the only transformation (bijection) that preserves addition and multiplication is the identity transformation.

The first steps in this direction were taken by Hans Zassenhaus, who approached in 1935 [273, 274] the problem in its finite case. He started from where our story began, namely with (P_3), with the simple observation that the group of Möbius transformations of a projective line $\mathbb{P}^1(K)$ over a field K—which can be thought of as consisting of points as one-dimensional subspaces of K^2, those of the form $K\cdot(1, a)$ to be identified with $a \in K$, and those of the form $K\cdot(0, 1)$ with ∞, so that one can think of the projective line over K as $K \cup \{\infty\}$—which are transformations defined for all a, b, c, and d in K with $ad - bc \neq 0$ by $x \mapsto \frac{ax+b}{cx+d}$, for $x \in K$, with the understanding that, if $c = 0$, then $\infty \mapsto \infty$, and if $c \neq 0$, $-\frac{d}{c} \mapsto \infty$ and $\infty \mapsto \frac{a}{c}$, acts *sharply* 3-*transitively* on $\mathbb{P}^1(K)$. To understand the statement, we need to define *group actions*. If G is a multiplicatively written group and X is a set, then a group action α of G on X is a function $\alpha : G \times X \to X$ satisfying: (i) $\alpha(e, x) = x$ for all $x \in X$, where e denotes the neutral element of G, and (ii) $\alpha(g \cdot h, x) = \alpha(g, \alpha(h, x))$ for all $g, h \in G$ and all $x \in X$. The action α is sharply 3-transitive if, for any two triples (x_1, x_2, x_3) and (x'_1, x'_2, x'_3), with $x_i \neq x_j$ and $x'_i \neq x'_j$ for all $1 \leq i < j \leq 3$, there is exactly one $g \in G$ such that $g(x_i) = x'_i$. In fact, Zassenhaus could determine the structure of all sharply 3-transitive group actions on a finite set X. The characterization involved an algebraic structure generalizing fields, called a *nearfield*, and amounted to the fact that these group actions are much like both the group of Möbius transformations and that of affine linear transformations, i.e. the $x \mapsto ax + b$. One would have thought that, based on Zassenhaus' work, someone would find a group-theoretic characterization of the projective line. In the finite realm, this would have been an easy task, but no one knew whether the nearfield connection was valid for sharply 3-transitive groups acting on infinite sets. In fact, it was only in 2016 that Katrin Tent [258] proved that this is not the case, that there are sharply 3-transitive groups that do not arise from a nearfield. So the path from Zassenhaus's papers to solving the general case was indeed, unbeknown to anyone at the time, barred. Yet history decided to take another route altogether. The 19 year old Jacques Tits, originally unaware of Zassenhaus's work, started from scratch and provided in 1949 [262–264] several simple characterizations of the projective line over a field. The most memorable, perhaps, is that for any sharply 3-transitive group G acting on a set X that satisfies the additional property that there are no "pseudo-involutions", i.e., that

$$\text{If } x \neq y, \alpha(g, x) = y, \alpha(g, y) = x, \text{ then } g \cdot g = e$$

one can endow X with the structure of a projective line over a field, and G must be isomorphic to the Möbius group of X.

Tits also provided like-minded axiomatization of the affine line over a field. A completely different axiomatization, inspired by the calculus of ends in plane hyperbolic geometry, of the projective line over a field was found later by Bachmann [18], and axiomatizations of affine and projective lines over skew fields were provided in [40, 114], and [68]. Purely geometric axiomatizations of Desarguesian and Moufang lines, in terms of a sexternary relation for "quadrangle sections," were presented in [70].

8.7 Pappus and Desargues in Geometries with Notions of Congruence, Orthogonality, and Reflection

8.7.1 *Hilbert's Questions*

Returning to Hilbert's questions in [104] (see also [185]), we find that perhaps the most pertinent questions, those which opened unexpected vistas, were ones asking:

> Is ^{P}*Des* provable with the help of the congruence axioms alone? ([104, p. 172])

> Prove ^{P}*Papp* (and ^{P}*Des*) in the plane based only on axioms in the groups I, II, and III (i.e., on the basis of the axioms for absolute geometry, without using the Parallel Postulate) [104, pp. 284, 392]; in another form: ^{P}*Papp* "arises by the elimination of the congruence axioms, indeed [^{P}*Papp*] is the sufficient condition that ensures that a definition of congruence is possible." [104, p. 261]

On the one hand, Hilbert was interested in these question, in "introducing number" into geometry in a purely geometrical manner—for which the theorems of Desargues and Pappus are instrumental—to justify geometry as an autonomous discipline. On the other hand, there are deeper metamathematical reasons for wanting "numbers" to be part of geometry, for the arithmetization of geometry is a means of transferring the consistency problem of geometry to that of a "number system" (a certain field (or a certain class of fields) in today's language).

These questions can be viewed in either the absolute setting, i.e., in the absence of any assumption regarding the existence or uniqueness of parallels, or in an affine setting. We first consider the absolute setting.

8.7.2 *The Absolute Setting*

The first contribution on the effect the congruence axioms (in the absence of any assumption on parallels) have on ^{P}*Papp* and ^{P}*Des*, came in the ground-breaking paper by Hjelmslev [109]. Max Dehn, in his 1926 afterword to the second edition of Pasch's [192], refers to the significance of Hjelmslev's paper in the following terms:

> Hjelmslev's result represents the highest point reached by modern mathematics going beyond Euclid in setting up elementary geometry: in the plane, with assumptions only regarding a limited part of the plane, without continuity, one lays the foundations for analytic geometry.[9]

Although Hjelmslev's name is unfamiliar to the vast majority of mathematicians living today, the methods he pioneered having remained hidden from what has become the mathematical mainstream, there is no exaggeration in these words. Hjelmslev's achievement was to realize that line reflections have certain properties that are independent of any assumption regarding parallels, and thus *absolute*. Line reflections, and in particular the central *three reflections theorem*, stating that the composition of three reflections in lines that have a common perpendicular or a common point must be a line reflection, were not entirely new. They had been treated earlier in [97, 102, 232, 271], yet in these works line reflections were not treated independently of the particular geometry in which they were defined (Euclidean, hyperbolic, or elliptic), as they were by Hjelmslev. The ideas espoused in 1907 [109] were developed further in a series of papers [110] that did away with several of the assumptions present in [109]. Many more geometers—their contributions are chronicled in [122]—have helped in the understanding of geometry in terms of line reflection as primitive notions. They helped remove assumptions regarding both the order of the plane and the free mobility of the plane (i. e. the possibility of transporting segments on any given line). The final touch in carving a particularly austere axiom system came from Friedrich Bachmann in 1951 [17] (whose contributions to the subject go back to 1936 [16]), who showed that two axioms proposed by Hilbert's student Arnold Schmidt [227] for that theory are superfluous.

In its final version, Bachmann's axiom system for *metric planes*, about which a monograph [19] and more than a hundred additional papers were written, states so very few facts about line reflections, that it is surprising that those assumptions suffice to give rise to the "analytic geometry" to which Dehn refers. In terms of points, lines, and incidence, using line reflections as a 'figure of speech',—defined as bijections of the collection of all points and lines, which preserve incidence and orthogonality, are involutory transformations, different from the identity, and fix all the points of a line—the axioms state that:

MP 1 *There are at least two points.*

MP 2 *For every two different points there is exactly one line incident with those points.*

[9]Mit dem *Hjelmslev*schen Resultat haben wir den höchsten Punkt bezeichnet, den die moderne Mathematik über *Euklid* hinausgehend in der Begründung der Elementargeometrie erreicht hat: in der Ebene, mit Voraussetzungen nur über einen beschränkten Teil der Ebene, ohne Stetigkeit ist die analytische Geometrie zu begründen.

MP 3 *If a is orthogonal to b, then b is orthogonal to a.*

MP 4 *Orthogonal lines intersect.*

MP 5 *Through every point P there is to every line l a perpendicular, which is unique if P is incident with l.*

MP 6 *To every line there is at least one reflection in that line.*

MP 7 *The composition of reflections in three lines a, b, and c which have a point or a perpendicular in common is a reflection in a line d.*

These structures, referred to as *metric planes*, can be axiomatized in terms of orthogonality alone (see [184]), and, in the non-elliptic case, in terms of incidence and segment congruence (or just in terms of the latter alone)—in the manner required by Hilbert's formulation of the question—as shown in [246].[10]

The axioms for metric planes are strong enough to imply both ^{p}*Papp* and ^{p}*Des*, for metric planes are embeddable in projective planes coordinatized by fields of characteristic $\neq 2$ and endowed with an orthogonality relation extending that of the metric plane, as shown in [19]. The difficulty of that embedding process has been likely responsible for its absence from the collective consciousness of present-day mathematicians. One of the important steps involves proving Brianchon's theorem, which is just a dual form of ^{p}*Papp*. In metric planes that satisfy additional conditions and correspond to the three classical geometries, Euclidean, hyperbolic, and elliptic, the proof of ^{p}*Papp* can be somewhat simplified. These simplified proofs are all presented with great care in [19]. For example, one can find six different proofs from the literature that ^{a}*Papp* holds in metric planes in which there is a rectangle. Formal proofs of ^{a}*Papp* from Tarski's axioms for plane Euclidean geometry, and of an absolute version of Pappus's axiom from Tarski's axioms for plane absolute geometry, to be found in [233] (which is also the geometry axiomatized by the plane axioms of groups I, II, and III of [103]), were carried out using Coq, a second-order formal proof management system, and took some ten thousand lines, as reported in [35].

In fact, an even weaker axiom system, for structures that we will refer to as *Sperner planes*, in which the three reflections theorem is weakened, implies both ^{p}*Des* and ^{p}*Papp*. That axiom system was developed by Sperner in [248] with the declared aim of being just strong enough to prove ^{p}*Des*. However, Sperner's Ph.D. student Kannenberg showed in [117] that, if Sperner planes can be embedded in projective planes satisfying ^{p}*Des* and *Fano*, then one can deduce the commutativity of the underlying skew field, and line reflections have their usual form in terms of a quadratic form. That these results hold even in characteristic 2 (i.e., that *Fano*

[10] Something that Hilbert apparently considered impossible in 1898–1899: "Der umgekehrte Weg, die Kongruenzaxiome und -sätze mit Hülfe des Bewegungsbegriffs zu beweisen ist falsch, da sich die Bewegung ohne den Kongruenzbegriff gar nicht definieren lässt." [104, p. 335] ("The converse, proving the congruence axiom and theorems in terms of the concept of rigid motion is wrong, for rigid motions cannot be defined at all in the absence of the congruence notion.").

is not needed to deduce the validity of ^{p}*Papp* in Sperner planes, should these be embeddable in projective planes with ^{p}*Des*) was shown in [119, 120]. According to [122, p. 182], around 1958, Harms, a student of Sperner, proved that all Sperner planes can be embedded in Desarguesian projective planes. Her proof was then simplified and became [66]. Sperner's axiom system and Sperner planes were studied in great depth in [143] and a monograph [145] was devoted to them, one that can be seen as the counterpart of [19] for Sperner planes.

A weakening of the axiom system for metric planes in a different direction, to become an axiom system for all Cayley–Klein geometries with the exception of the doubly-hyerbolic, was presented by Rolf Struve in 2016 [254]. Its axioms also imply ^{p}*Papp*.

Thus, it appears that there is a little understood obstruction preventing absolute axiom system more general than that of metric planes for plane reflection geometry to imply ^{p}*Des* but not ^{p}*Papp*.

That one does need somewhat strong congruence axioms to prove ^{a}*Des* has been pointed out by models of independence that verify all of Hilbert's plane axioms except the triangle congruence axiom ("Side-Angle-Side"), ever since the first edition of the *Grundlagen der Geometrie*. The independence in question is implicit in the work of Beltrami going back to 1865 and explicit in a paper of Peano of 1894. Non-Desarguesian geometries with several other properties were a subject of several other papers (see the survey [50]), the one by Moulton from 1902 [170] being adopted in all later editions of Hilbert's *Grundlagen der Geometrie* [103].

In a different direction, Smid [245] has shown that ^{p}*Papp* and ^{p}*Des* can be proved even if one weakens, inside the axiom system for absolute geometry, Hilbert's segment transport axiom, in such a manner that it is still valid in bounded regions of an absolute plane. One cannot define line reflections in these planes, so Hjelmslev's results cannot be applied directly.

8.7.3 The Affine Plane with Orthogonality Setting

Unlike the absolute case, in the affine case there are several axiom systems ([133, 173, 174, 206, 209, 215, 229, 230, 256, 276] (it is also shown in [136] that one cannot obtain a non-Pappian three-dimensional geometry in the same manner)) that provide in the Euclidean case a precise answer to Hilbert's question, i.e., provide orthogonality or congruence axioms that are strong enough to imply ^{a}*Des* but not ^{a}*Papp*. On the other hand, as shown in [216] even in the affine setting, if one postulates the existence, in a translation plane, of an orthogonality relation on lines, such that (i) if $a \perp b$, $a \parallel a'$, and $b \parallel b'$, then $a' \perp b'$; (ii) if $a \perp b$ and $a \perp c$, then $b \parallel c$, (iii) if two of the altitudes of a triangle exist and intersect, then the third altitude exists and passes through that intersection point, (iv) there are four lines a_1, a_2, a_3, a_4, such that $a_1 \perp a_2$, $a_1' \perp a_2'$, such that none of a_1 and a_2 is parallel to any of a_1' and a_2', then ^{a}*Papp* holds. Whether that statement remains true even without the assumption that the affine plane is a translation plane is not known, according to [216]. Other conditions leading to ${}^{a}Papp$ can be found in [251].

Moreover, by weakening some orthogonality axioms from [229], Schütte succeeded in [231] to construct affine planes with a symmetric orthogonality relation over certain alternative fields that need not be skew fields (the algebraic description of possible models can be found in [36]). In [140], it is shown that a symmetric orthogonality relation can be introduced in an even larger class of affine parallel structures.

8.8 The Desargues Property in Busemann's G-Spaces

8.8.1 Hilbert's Fourth Problem

Desargues theorem also appears as an axiom in a rather different context. Although motivated by a fundamental question in the axiomatic foundation of geometry, *Hilbert's Fourth Problem*, as it became known by being the fourth problem on Hilbert's list of twenty-three problems—one actually read at the 1900 International Congress of Mathematicians in Paris—it very soon became apparent that the methodology required for its solution was of a non-elementary nature. It concerns the construction of all the metrics in which the ordinary lines, that is (pieces of) lines in n-dimensional real projective space $P^n(\mathbb{R})$, are the shortest curves or *geodesics*. It became clear very soon, already with the dissertation of Hilbert's student Hamel [86], parts of a revised version of which were published in 1903 [87], that one needs to assume the Desargues axiom to obtain a result resembling a classification. While Hamel assumed the differentiability of a certain length function that shows up, Busemann [43] introduced a set-up in which all assumptions of differentiability are banned from the axioms. What remains is still non-elementary, as it assumes a metric with real number values as well as topological notions, but the spirit is very close to that of the axiomatic foundation of geometry that was the subject of the previous sections.

Hilbert's Fourth Problem, as well as Busemann's contributions were surveyed in depth in [187], so we will mention only the main aspects of the axiomatic set-up and the main results that involve what Busemann calls "the Desargues property."

8.8.2 G-Spaces

We first take care of notation. In any metric space (X, d), one can define a notion of *metric betweenness* β_d in the manner of Menger [159] (see also [268]), by stipulating that $\beta_d(x, y, z)$ holds if and only if x, y, z are distinct and $d(x, y) + d(y, z) = d(x, z)$. Here $\beta_d(x, y, z)$ is read as "y lies between x and z." The open ball with center p and radius ϱ, denoted by $S(p, \varrho)$ is the set of all x in X with $d(p, x) < \varrho$.

A *G-space* is a metric space (X, d) satisfying the following axioms: (1) (Finite compactness) Every bounded infinite set has an accumulation point; (2) (Menger convexity) For any x and z in X, there exists a point y which is between x and z (that is, for which $\beta_d(x, y, z)$ holds); (3) (Local extendability) To every point p of X there corresponds a positive ϱ_p, such that, for two distinct points x and y in $S(p, \varrho_p)$ there is a point z in $S(p, \varrho_p)$ with $\beta_d(x, y, z)$; (4) (Uniqueness of extension) If $\beta_d(x, y, z_1)$ and $\beta_d(x, y, z_2)$ and if $d(y, z_1) = d(y, z_2)$ then $z_1 = z_2$.

A *geodesic* is a map $x : \mathbb{R} \to X$, such that, for any real number τ_0 there is a positive number $\epsilon(\tau_0)$ such that for all τ_1, τ_2 with $|\tau_1 - \tau_0| < \epsilon(\tau_0)$ and $|\tau_2 - \tau_0| < \epsilon(\tau_0)$, we have $d(x(\tau_1), x(\tau_2)) = |\tau_1 - \tau_2|$ (that is, if x is a locally isometric map). A geodesic is called a *straight line* if the equation $d(x(\tau_1), x(\tau_2)) = |\tau_1 - \tau_2|$ holds for all τ_1 and τ_2.

Under the *Desargues property* Busemann understands the conjunction of a modified version of the statement ${}^pDes$ makes and of its converse (by which we mean that from the collinearity of C_{12}, C_{23}, and C_{31} one deduces the concurrence of A_1B_1, A_2B_2, and A_3B_3). The modified version of ${}^pDes$ states, with geodesics instead of lines, that if A_1B_1, A_2B_2, and A_3B_3 have the point C in common, if A_1A_2 and B_1B_2 have C_{12} in common, if A_2A_3 and B_2B_3 have C_{23} in common, and if two of the three intersections $C_{12}C_{23} \cap A_3A_1$, $C_{12}C_{23} \cap B_3B_1$, and $A_3A_1 \cap B_3B_1$ exists, then they coincide. The version of the converse of ${}^pDes$ states that, if the intersections of A_1A_2 and B_1B_2, A_2A_3 and B_2B_3, and A_1A_3 and B_1B_3 exist and are collinear (that is, lie on a geodesic), and if two of the three intersections if $A_1B_1 \cap A_2B_2$, $A_2B_2 \cap A_3B_3$, and $A_3B_3 \cap A_1B_1$ exist, then they coincide.

A G-space in which the geodesic through two points is unique and in which the Desargues Property holds is called a *Desarguesian space*. Using the Menger-Urysohn notion of topological dimension, one may speak of n-dimensional Desarguesian spaces.

In a similar vein to the results reported in Sect. 8.5, Busemann proved in [44, (14.6), (14.8)] (and dealt with an unsolved case in [46, p. 32] to complete the proof) that

> For a given n-dimensional Desarguesian G-space R there is an $(n + 1)$-dimensional Desarguesian space R^* such that R is a hyperplane in R^* and the restriction of the metric of R^* to R is the given metric in R.

Extending to G-spaces a result of Hamel [87], Busemann also proved the following characterization of two-dimensional Desarguesian spaces [44, Theorems 13.1] (a similar characterization is valid for the n-dimensional case, even if the metric is not assumed to be symmetric, as shown in [44, Theorem 14.1] and in [46, p. 37]);

> Let R be a two dimensional Desarguesian space. Then
> Either all geodesics are great circles of the same length and R can be mapped topologically on the projective plane $P^2(\mathbb{R})$ in such a way that each great circle in R goes into a line in $P^2(\mathbb{R})$
> Or all geodesics are straight lines and R can be mapped topologicallv on an open convex subset C of the affine plane $A^2(\mathbb{R})$ in such a way that each straight line in R goes into the intersection of C with a line in $A^2(\mathbb{R})$.

Desarguesian spaces in which the metric is not necessarily symmetric, which are noncompact and for which the closed balls, $\overline{S}(p, \varrho)$ (with the balls defined in either way, as the set of all x in X with $d(p, x) < \varrho$ or as the set of all x in X with $d(x, p) < \varrho$), also allow for a simple joint characterization of the Minkowski and Hilbert geometries (the former being a generalization of the Euclidean plane in which the unit circle is replaced by an arbitrary convex curve (and it shows up for the first time in [162] (see [259] for a comprehensive treatment of these geometries) and the latter being a generalization of the Beltrami-Cayley–Klein model of hyperbolic geometry, in which the circle is being replaced by an arbitrary convex curve (and which shows up for the first time in a letter of Hilbert to Klein, published in [101] and as Appendix I in [103]) by the property that an isometry of one geodesic on another or itself is a projectivity.

A G-space R is called *straight* if $\varrho(p) = \infty$ for all $p \in R$, where ϱ is the function introduced in axiom (3) for G-spaces. A straight two-dimensional G-space is *quasihyperbolic* if it possesses all translations along two geodesics G and H, where G is an asymptote to H, but not parallel to H (here the terms "asymptote" and 'parallel" have rather involved definitions, see [44]). As shown in [45], hyperbolic geometry is the only Desarguesian quasihyperbolic geometry.

The last surprising result of the theory of G-spaces, with an elementary flavor, that we will state out of the vast œuvre of Busemann [49] and of his students of results emphasizing the importance of the Desargues axiom, is due to Busemann's student Phadke [196], who has shown that a straight two-dimensional G-space which satisfies the following property (P), stipulating the existence of a ray having the property of the Euclidean angle bisector,

Inside a nonstraight convex angle with legs N_1, N_2 and vertex v, there is a ray M with origin v, such that any segment with endpoints a_1 and a_2, with $a_1 \in N_1$ and $a_2 \in N_2$, with $a_1 \neq v$, $a_2 \neq v$, intersects M in a point $b = b(a_1, a_2)$, for which $\frac{d(v,a_1)}{d(v,a_2)} = \frac{d(b,a_1)}{d(b,a_2)}$.

must be Desarguesian (that is, it must satisfy ${}^{a}Des$). Given the results in [47] and [48], where the problem of characterizing two-dimensional G-spaces satisfying property (P) was raised, it follows that a straight two-dimensional G-space which satisfies (P) must be a Minkowski plane.

Acknowledgements Thanks are due to Hans Havlicek, Horst and Rolf Struve, and the referee for their very valuable suggestions and corrections. The second author was nominally supported by an NCUIRE Fellowship of the New College of Interdisciplinary Arts and Sciences at the West campus of Arizona State University.

References

1. P. Abellanas, Estructura analitica del segmento abierto definido por los postulados de incidencia y orden de Hilbert. *Rev. Mat. Hisp.-Amer. (4)* 6 (1946), 101–126.
2. S. A. Amitsur, Rational identities and applications to algebra and geometry. *J. Algebra* 3 (1966), 304–359.

3. S. A. Amitsur, Polynomial identities. *Israel J. Math.* 19 (1974), 183–199.
4. J. André, Über nicht-Desarguessche Ebenen mit transitiver Translationsgruppe. *Math. Z.* 60 (1954), 156–186.
5. J. André, Projektive Ebenen über Fastkörpern. *Math. Z.* 62 (1955), 137–160.
6. J. André, Über Parallelstrukturen. II: Translationsstrukturen. *Math. Z.* 76 (1961), 155–163.
7. A. Arana, P. Mancosu, On the relationship between plane and solid geometry. *Rev. Symb. Log.* 5 (2012), 294–353.
8. E. Artin, Über einen Satz von Herrn J. H. Maclagan Wedderburn. *Abh. Math. Semin. Hamb. Univ.* 5 (1927), 245–250.
9. E. Artin, Coordinates in affine geometry. *Rep. Math. Colloquium (Second Series)* 2 (1940). 15–20.
10. E. Artin, *Geometric algebra*. Interscience, New York, 1957.
11. R. Artzy, Non-Euclidean incidence planes. *Israel J. Math.* 4 (1966), 43–53.
12. R. Artzy, Pascal's theorem on an oval. *Amer. Math. Monthly* 75 (1968), 143–146.
13. M. Avellone, A. Brigaglia, C. Zappulla, The foundations of projective geometry in Italy from De Paolis to Pieri. *Arch. Hist. Exact Sci.* 56 (2002), 363–425.
14. R. Baer, The fundamental theorems of elementary geometry. An axiomatic analysis. *Trans. Amer. Math. Soc.* 56 (1944), 94–129.
15. R. Baer, *Linear algebra and projective geometry*. Academic Press, New York, 1952.
16. F. Bachmann, Eine Begründung der absoluten Geometrie in der Ebene. *Math. Ann.* 113 (1936), 424–451.
17. F. Bachmann, Zur Begründung der Geometrie aus dem Spiegelungsbegriff. *Math. Ann.* 123 (1951), 341–344.
18. F. Bachmann, Eine Kennzeichnung der Gruppe der gebrochen-linearen Transformationen. *Math. Ann.* 126 (1953), 79–92.
19. F. Bachmann, *Aufbau der Geometrie aus dem Spiegelungsbegriff*. Zweite ergänzte Auflage. Springer-Verlag, Berlin, 1973.
20. J. T. Baldwin, Formalization, primitive concepts, and purity, *Rev. Symb. Log.* 6 (2013), 87–128.
21. J. Bamberg, T. Penttila, Completing Segre's proof of Wedderburn's little theorem. *Bull. Lond. Math. Soc.* 47 (2015), 483–492.
22. J. Bamberg, T. Harris, T. Penttila, On abstract ovals with Pascalian secant lines. *J. Group Theory* 21 (2018), 1051–1064.
23. A. Barlotti, Le possibili configurazioni del sistema delle coppie punto-retta (A, a) per cui un piano grafico risulta (A, a)-transitivo. *Boll. Un. Mat. Ital. (3)* 12 (1957), 212–226.
24. A. Barlotti, Sul gruppo delle proiettività di una retta in sè nei piani liberi e nei piani aperti. *Rend. Sem. Mat. Univ. Padova* 34 (1964), 135–159.
25. A. Barlotti, K. Strambach, Remarks on projectivities. *Aequationes Math.* 28 (1985), 212–228.
26. F. Bennhold, Zur synthetischen Begründung der projektiven Geometrie der Ebene. *Math. Ann.* 129 (1955), 213–229 .
27. G. M. Bergman, Skew fields of noncommutative rational functions, after Amitsur (preliminary version). Séminaire M. P. Schützenberger, A. Lentin et M. Nivat, 1969/70: Problèmes Mathématiques de la Théorie des Automates, Exp. 16, 1–18, Secrétariat mathématique, Paris, 1970.
28. M. Biliotti, V. Jha, N. L. Johnson, *Foundations of translation planes*. Marcel Dekker, New York, 2001.
29. W. Blaschke, *Projektive Geometrie*. 3. verbesserte Auflage. Birkhäuser Verlag, Basel, 1954.
30. W. Blaschke, *Geometrie der Waben*. Birkhäuser Verlag, Basel, Stuttgart, 1955.
31. W. Blaschke, G. Bol, *Geometrie der Gewebe. Topologische Fragen der Differentialgeometrie*. Julius Springer, Berlin, 1938.
32. R. Borges, Eine mit dem Satz von Pappos äquivalente Version des Scherensatzes. *Math. Semesterber.* 34 (1987), 246–249.
33. A. L. Bosworth, *Begründung einer vom Parallelenaxiome unabhängigen Streckenrechnung*. Dietrich, Göttingen, 1900.

34. O. Bottema, Eine Bemerkung über den Desarguesschen und den Pascalschen Satz. *Math. Ann.* 111 (1935), 68–70.
35. G. Braun, J. Narboux, A synthetic proof of Pappus' theorem in Tarski's geometry. *J. Automat. Reason.* 58 (2017), 209–230.
36. E. Brožíková, Existence of Schütte semiautomorphisms. *Časopis Pěst. Mat.* 107 (1982), 143–158, 188.
37. Bruck, R. H., Kleinfeld, E.: The structure of alternative division rings. *Proc. Amer. Math. Soc.* 2, 878–890 (1951).
38. F. Buekenhout, Plans projectifs à ovoides pascaliens. *Arch. Math. (Basel)* 17 (1966), 89–93.
39. F. Buekenhout, B Études intrinsèque des ovales. *Rend. Mat. e Appl.* (5) 25 (1966), 333–393.
40. F. Buekenhout, X. Hubaut, Groupes affins et projectifs. *Acad. Roy. Belg. Bull. Cl. Sci. (5)* 52 (1966), 368–381.
41. E. Burger, Über die Einzigkeit der Cayley-Zahlen. Bemerkung zu einer Arbeit von L. A. Skorniakov. *Arch. Math. (Basel)* 3 (1952), 298–302.
42. R. P. Burn, Bol quasi-fields and Pappus' theorem. *Math. Z.* 105 (1968), 351–364.
43. H. Busemann, Paschsches Axiom und Zweidimensionalität. *Math. Ann.* 107 (1932), 324–328.
44. H. Busemann, *The geometry of geodesics.* Academic Press, New York, 1955.
45. H. Busemann, Quasihyperbolic geometry. *Rend. Circ. Mat. Palermo, II. Ser.* 4 (1955), 256–269.
46. H. Busemann, *Recent synthetic differential geometry*. Springer-Verlag, New York-Berlin, 1970.
47. H. Busemann, Planes with analogues to Euclidean angular bisectors. *Math. Scand.* 36 (1975), 5–11.
48. H. Busemann, Remark on 'Planes with analogues to Euclidean angular bisectors'. *Math. Scand.* 38 (1976), 81–82.
49. H. Busemann, *Selected works I & II.* Edited by A. Papadopoulos. Springer, Cham, 2018.
50. C. Cerroni, Non-Desarguian geometries and the foundations of geometry from David Hilbert to Ruth Moufang. *Hist. Math.* 31 (2004), 320–336.
51. P. M. Cohn, *Skew fields. Theory of general division rings.* Cambridge University Press, Cambridge, 1995.
52. G. Conti, Piani proiettivi dotati di un ovale pascaliano. *Boll. Unione Mat. Ital., IV. Ser. /* 11 (1975), 330–338.
53. H. S. M. Coxeter, Desargues configurations and their collineation groups. *Math. Proc. Cambridge Philos. Soc.* 78 (1975), 227–246.
54. H. S. M. Coxeter, The Pappus configuration and its groups. *Nederl. Akad. Wet. Verslag Afd. Natuurk.* 85 (1976), 44–46.
55. H. S. M. Coxeter, The Pappus configuration and the self-inscribed octagon. I, II, III. *Nederl. Akad. Wet. Proc. Ser. A* 80 (1977), 256–269, 270–284, 285–300.
56. H. S. M. Coxeter, S. L. Greitzer, *Geometry revisited.* The Mathematical Association of America, Washington, 1967.
57. S. Crampe, Schließungssätze in projektiven Ebenen und dichten Teilebenen. *Arch. Math. (Basel)* 11 (1960), 136–145.
58. A. Cronheim, A proof of Hessenberg's theorem. *Proc. Amer. Math. Soc.* 4 (1953), 219–221.
59. M. Dehn, Über die Grundlagen der projektiven Geometrie und allgemeine Zahlsysteme. *Math. Ann.* 85 (1922), 184–194.
60. M. Dehn, Über einige neue Forschungen in den Grundlagen der Geometrie. *Mat. Tidsskr. B* 3/4 (1931), 63–83.
61. P. Dembowski, *Finite geometries.* Springer-Verlag, Berlin-New York 1968.
62. G. Desargues, *L'œuvre mathématique de G. Desargues.* Textes publiés et commentés avec une introduction biographique et historique par René Taton. Presses Universitaires de France, Paris, 1951.
63. G. Donati, Pappus' configuration in non commutative projective geometry with application to a theorem of A. Schleiermacher. *Rend. Circ. Mat. Palermo (2)* 50 (2001), 325–328.

64. M. L. Dubreil-Jacotin, L. Lesieur, R. Croisot, *Leçons sur la théorie des treillis des structures algébriques ordonnées et des treillis géométriques*. Gauthier-Villars, Paris, 1953.
65. M. Eklund, On how logic became first-order. *Nordic J. Philos. Logic* 1 (1996), 147–167.
66. E. Ellers, E. Sperner, Einbettung eines desarguesschen Ebenenkeimes in eine projektive Ebene. *Abh. Math. Semin. Univ. Hamburg* 25 (1962), 206–230.
67. G. Faina, Una estensione agli ovali astratti del teorema di Buekenhout sugli ovali pascaliani. *Boll. Un. Mat. Ital. Suppl.*, 2 (1980), 355–364.
68. G. Faina, G. Korchmáros, Una caratterizzazione del gruppo lineare $PGL(2, K)$ e delle coniche astratte nel senso di Buekenhout. *Boll. Un. Mat. Ital. Suppl.* 2 (1980), 195–208.
69. G. Fano, Sui postulati fondamentali della geometria projettiva in uno spazio lineare a un numero qualunque di dimensioni. *Giornale di Matematiche* 30 (1892), 106–132.
70. J. R. Faulkner, Lines with quadrangle section. *J. Geom.* 22 (1984), 31–46.
71. J. R. Faulkner, *The role of nonassociative algebra in projective geometry*. American Mathematical Society, Providence, 2014.
72. C.-A. Faure, A. Frölicher, *Modern projective geometry*. Kluwer, Dordrecht, 2000.
73. J. Ferreirós, The road to modern logic—an interpretation. *Bull. Symbolic Logic* 7 (2001), 441–484.
74. J. V. Field, J. J. Gray, *The geometrical work of Girard Desargues*. Springer-Verlag, New York, 1987.
75. W. K. Forrest, The fundamental configurations of linear projective geometry. *Z. Math. Logik Grundlagen Math.* 32 (1986), 289–306.
76. H. Freudenthal, Een niet-Desargues'se meetkunde en de invoeging van een vlakke Desargues'se meetkunde in de ruimtelijke projectieve meetkunde. *Chr. Huygens* 13 (1935), 247–256.
77. R. Fritsch, Synthetische Einbettung Desarguesscher Ebenen in Räume, *Math. -Phys. Semesterber.* 21 (1974), 237–249.
78. R. Fritsch, Ein "affiner" Beweis des Satzes von v. Staudt-Schleiermacher. *Monatsh. Math.* 86 (1978/79), 177–184.
79. M. Funk, Octagonality conditions in projective and affine planes. *Geom. Dedicata* 28 (1988), 53–75.
80. M. Funk, On the existence of little Desargues configurations in planes satisfying certain weak configurational conditions. *Geom. Dedicata* 31 (1989), 207–235.
81. A. M. Gleason, Finite Fano planes. *Amer. J. Math.* 78 (1956), 797–807.
82. B. Grünbaum, *Configurations of points and lines*. American Mathematical Society, Providence, 2009.
83. H. Guggenheimer, The theorem of Menelaus in axiomatic geometry. *Geom. Dedicata* 3 (1974), 257–261.
84. H. Hähl, B. Weber, Area functions on affine planes. *J. Geom.* 107 (2016), 483–507.
85. M. Hall, Projective planes. *Trans. Amer. Math. Soc.* 54 (1943), 229–277, correction 65 (1949), 473–474.
86. G. Hamel, *Über die Geometrieen, in denen die Geraden die Kürzesten sind*. Dissertation. Universität Göttigen, 1901.
87. G. Hamel, Über die Geometrieen, in denen die Geraden die Kürzesten sind. *Math. Ann.* 57 (1903), 231–264.
88. Z. R. Hartvigson, A non-Desarguesian space geometry. *Fund. Math.* 86 (1974), 143–147.
89. H. Havlicek, Einbettung projektiver Desargues-Räume. *Abh. Math. Semin. Univ. Hamb.* 52 (1982), 228–231.
90. H. Havlicek, Applications of results on generalized polynomial identities in Desarguesian projective spaces. R. Kaya, P. Plaumann and K. Strambach (eds.), *Rings and geometry* (Istanbul, 1984), pp. 39–77, Reidel, Dordrecht, 1985.
91. A. Herzer, Eine Verallgemeinerung des Satzes von Dandelin. *Elem. Math.* 27 (1972), 52–56.
92. A. Herzer, Dualitäten mit zwei Geraden aus absoluten Punkten in projektiven Ebenen. *Math. Z.* 129 (1972), 235–257.

93. A. Herzer, Ableitung zweier selbstdualer, zum Satz von Pappos äquivalenter Schließungssätze aus der Konstruktion von Dualitäten. *Geom. Dedicata* 2 (1973), 283–310.
94. A. Herzer, Neue Konstruktion einer Erweiterung von projektiven Geometrien. *Geom. Dedicata* 4 (1975), 199–213.
95. G. Hessenberg, Desarguesscher Satz und Zentralkollineation. *Arch. der Math. u. Phys. (3)* 6 (1903), 123–127.
96. G. Hessenberg, Über einen geometrischen Calcül (Verknüpfungs-Calcül). *Acta Math.* 29 (1904), 1–24.
97. G. Hessenberg, Neue Begründung der Sphärik, *S.-Ber. Berlin. Math. Ges.* 4 (1905), 69–77.
98. G. Hessenberg, Beweis des *Desargues*schen Satzes aus dem *Pascal*schen. *Math. Ann.* 61 (1905), 161–172.
99. G. Hessenberg, J. Diller, *Grundlagen der Geometrie*. 2. Aufl. Walter de Gruyter, Berlin 1967.
100. A. Heyting, *Axiomatic projective geometry*. Second edition. V. Wolters-Noordhoff, Groningen; North-Holland, Amsterdam, 1980.
101. D. Hilbert, Über die gerade Linie als kürzestes Verbindung zweier Punkte. *Math. Ann.* 44(1895), 91–96.
102. D. Hilbert, Neue Begründung der *Bolyai-Lobatschefsky*schen Geometrie, *Math. Ann.* 57 (1903), 137–150.
103. D. Hilbert, *Grundlagen der Geometrie*. Leipzig, Teubner (1899) – translated by L. Unger, Open Court, La Salle, Ill., under the title: *Foundations of Geometry*, 1971, 14th German edition 1999.
104. D. Hilbert, *David Hilbert's lectures on the foundations of geometry, 1891–1902*. Edited by M. Hallett and U. Majer. David Hilbert's Foundational Lectures, 1. Springer-Verlag, Berlin, 2004.
105. D. Hilbert, S. Cohn-Vossen, *Anschauliche Geometrie*. J. Springer, Berlin, 1932.
106. J. W. P. Hirschfeld, *Projective geometries over finite fields*. Oxford University Press, New York, 1979.
107. J. W. P. Hirschfeld, *Finite projective spaces of three dimensions*. Oxford University Press, New York, 1985.
108. J. W. P. Hirschfeld, J. A. Thas, *General Galois geometries*. Springer, London, 2016.
109. J. Hjelmslev, Neue Begründung der ebenen Geometrie. *Math. Ann.* 64 (1907), 449–474.
110. J. Hjelmslev, Einleitung in die allgemeine Kongruenzlehre. I, II., *Mat.-Fys. Medd. K. Dan. Vidensk. Selsk.* 8 (1929), Nr. 11, 1–36; 10 (1929), Nr. 1, 1–28.
111. C. E. Hofmann, III, Specializations of Pascal's theorem on an oval. *J. Geom.* 1 (1971), 143–153.
112. H. Hotje, Geometrische Bewertungen in affinen Räumen. *Abh. Math. Sem. Univ. Hamburg* 45 (1976), 82–90.
113. H. Hotje, Zur Definition des Flächeninhalts in affinen Ebenen. *Elem. Math.* 34 (1979), 25–31.
114. X. Hubaut, Groupes affins et groupes projectifs. *Acad. Roy. Belg. Bull. Cl. Sci. (5)* 53 (1967) 1266–1275.
115. N. L. Johnson, V. Jha, M. Biliotti, *Handbook of finite translation planes*. Chapman & Hall/CRC, Boca Raton, 2007.
116. F. Kalhoff, V. Pambuccian, An axiomatic look at the Erdős-Trost problem. *J. Geom.* 107 (2016), 379–385.
117. K.-R. Kannenberg, *Grundgedanken einer Theorie der Gebilde zweiter Ordnung in Schiefkörpergeometrien*. Dissertation, Universität Bonn, 1954.
118. J. Karamata, Eine elementare Herleitung des Desarguesschen Satzes aus dem Satze von Pappos-Pascal. *Elem. Math.* 5 (1950), 10–11.
119. H. Karzel, Ein Axiomensystem der absoluten Geometrie. *Arch. Math. (Basel)* 6 (1954), 66–76.
120. H. Karzel, Verallgemeinerte absolute Geometrien und Lotkerngeometrien. *Arch. Math. (Basel)* 6 (1955), 284–295.
121. H. Karzel, H.-J. Kroll, Zur projektiven Einbettung von Inzidenzräumen mit Eigentlichkeitsbereich. *Abh. Math. Semin. Univ. Hamburg* 49 (1979), 82–94.

122. H. Karzel, H.-J. Kroll, *Geschichte der Geometrie seit Hilbert*. Wissenschaftliche Buchgesellschaft, Darmstadt, 1988.
123. H. Karzel, K. Sörensen, Projektive Ebenen mit einem pascalschen Oval. *Abh. Math. Semin. Univ. Hamburg* 36 (1971), 123–125.
124. H. Karzel, K. Sörensen, Die lokalen Sätze von Pappus und Pascal. *Mitt. Math. Ges. Hamburg* 10 (1971), 28–55.
125. E. Kleinfeld, Alternative division rings of characteristic 2. *Proc. Nat. Acad. Sci. U. S. A.* 37 (1951), 818–820.
126. W. Klingenberg, Beziehungen zwischen einigen affinen Schließungssätzen. *Abh. Math. Sem. Univ. Hamburg* 18 (1952), 120–143.
127. W. Klingenberg, Beweis des Desarguesschen Satzes aus der Reidemeisterfigur und verwandte Sätze. *Abh. Math. Sem. Univ. Hamburg* 19 (1955), 158–175.
128. B. Klotzek, Ebene äquiaffine Spiegelungsgeometrie. *Math. Nachr.* 55 (1973), 89–131.
129. N. Knarr, B. Stroppel, M. J. Stroppel, Desargues configurations: minors and ambient automorphisms. *J. Geom.* 107 (2016), 357–378.
130. W. L. Kocay, One-point extensions in n_3 configurations. *Ars Math. Contemp.* 10 (2016), 291–322.
131. M. Kordos, Bisorted projective geometry. *Bull. Acad. Polon. Sci. Sér. Sci. Math.* 30 (1982), 429–432.
132. W. Krüger, Kegelschnitte in Moufangebenen. *Math. Z.* 120 (1971), 41–60.
133. E. Kusak, Desarguesian Euclidean planes and their axiom system. *Bull. Polish Acad. Sci. Math.* 35 (1987), 87–92.
134. E. Kusak, On some configurational axioms characteristic of affine Moufang planes. *Bull. Polish Acad. Sci. Math.* 36 (1988), 583–592.
135. E. Kusak, K. Prażmowski, On affine reducts of Desargues axiom. *Bull. Polish Acad. Sci. Math.* 35 (1987), 77–86.
136. E. Kusak, K. Prażmowski, Non-Pappian Euclidean space does not exist. *Bull. Polish Acad. Sci. Math.* 36 (1988), 567–569.
137. F. Luci, On the theorem of Pappus-Pascal. *J. Geom.* 19 (1982), 154–165.
138. H. Lenz, Kleiner Desarguesscher Satz und Dualität in projektiven Ebenen. *Jahresber. Deutsch. Math. Verein.* 57 (1954), 20–31.
139. L. Lesieur, Sur la mesure des triangles en géométrie affine plane. *Math. Z.* 93 (1966), 334–344.
140. J. Lettrich, The orthogonality in affine parallel structure. *Math. Slovaca* 43 (1993), 45–68.
141. F. Levi, *Geometrische Konfigurationen*. S. Hirzel, Leipzig, 1929.
142. F. W. Levi, On a fundamental theorem of geometry. *J. Indian Math. Soc. (2)* 3 (1939), 182–192
143. R. Lingenberg, Über Gruppen mit einem invarianten System involutorischer Erzeugender, in dem der allgemeine Satz von den drei Spiegelungen gilt. I, II, III, IV. *Math. Ann.* 137 (1959), 26–41, 83–106, 142 (1961), 184–224, 158 (1965), 297–325.
144. R. Lingenberg, *Grundlagen der Geometrie*. 3. Auflage. Bibliographisches Institut, Mannheim, 1978.
145. R. Lingenberg, *Metric planes and metric vector spaces*. Wiley, New York, 1979.
146. L. Lombardo-Radice, Sull'immersione de un piano grafico in uno spazio grafico a tre dimensioni. *Atti Accad. Naz. Lincei. Rend. Cl. Sci. Fis. Mat. Nat.* (8), 10 (1951), 203–205.
147. L. Lombardo-Radice, Sur la définition de proposition configurationnelle et sur certaines questions algébro-géométriques dans les plans projectifs. Colloque d'algèbre supérieure (Bruxelles, 1956), 217–230. Centre Belge de Recherches Mathématiques Établissements Ceuterick, Louvain; Gauthier-Villars, Paris, 1957.
148. L. Lombardo-Radice, Sur la classification des théorèmes en géométrie projective plane abstraite. Hommage au Professeur Lucien Godeaux, 169–179. Librairie Universitaire, Louvain, 1968.

149. L. Lombardo-Radice, Sulla classificazione combinatoria dei teoremi proiettivi. Colloquio Internazionale sulle Teorie Combinatorie (Roma, 1973), Tomo I, 371–380. Atti dei Convegni Lincei, No. 17, Accad. Naz. Lincei, Roma, 1976.
150. H. Lüneburg, An axiomatic treatment of ratios in an affine plane. *Arch. Math. (Basel)* 18 (1967), 444–448.
151. H. Lüneburg, *Translation planes*. Springer-Verlag, Berlin-New York, 1980.
152. M. Mandelkern, Constructive coordinatization of Desarguesian planes. *Beitr. Algebra Geom.* 48 (2007), 547–589.
153. M. Mandelkern, A constructive real projective plane. *J. Geom.* 107 (2016), 19–60.
154. E. A. Marchisotto, The projective geometry of Mario Pieri: a legacy of Georg Karl Christian von Staudt. *Historia Math.* 33 (2006), 277–314.
155. V. Martinetti, Sulle configurazioni piane μ_3. *Ann. Mat. Pura Appl.* 15 (1887), 1–26.
156. T. A. McKee, Generalized equivalence and the phraseology of configuration theorems. *Notre Dame J. Formal Logic* 21 (1980), 141–147.
157. N. S. Mendelsohn, Non-Desarguesian projective plane geometries which satisfy the harmonic point axiom. *Canad. J. Math.* 8 (1956), 532–562.
158. Menelaus of Alexandria, *Spherics*. Early translation and al-Mahanı / al-Harawı's version by Roshdi Rashed and Athanase Papadopoulos. De Gruyter, Berlin, 2017.
159. K. Menger, Untersuchungen uber allgemeine Metrik. *Math. Ann.* 100 (1928), 75–163.
160. K. Menger, Independent self-dual postulates in projective geometry. *Rep. Math. Colloq.* (2), 8 (1948), 81–87.
161. M. Menghini, On configurational propositions. *Pure Math. Appl. Ser. A*, 2 (1991), 87–126.
162. H. Minkowski, *Geometrie der Zahlen*. B. G. Teubner, Leipzig, 1896.
163. R. Moufang, Zur Struktur der projektiven Geometrie der Ebene. *Math. Ann.* 105 (1931), 536–601.
164. R. Moufang, Die Einführung der idealen Elemente in die ebene Geometrie mit Hilfe des Satzes vom vollständigen Vierseit. *Math. Ann.* 105 (1931), 759–778.
165. R. Moufang, Die Schnittpunktsätze des projektiven speziellen Fünfecksnetzes in ihrer Abhängigkeit voneinander. (Das A-Netz). *Math. Ann.* 106 (1932), 755–795.
166. R. Moufang, Ein Satz über die Schnittpunktsätze des allgemeinen Fünfecksnetzes. (Das (A, B)-Netz.). *Math. Ann.* 107 (1932), 124–139.
167. R. Moufang, Die Desarguesschen Sätze vom Rang 10. (Eine Ergänzung zu "Ein Satz über die Schnittpunktsätze des allgemeinen Fünfecksnetzes. (Das (A, B)-Netz)"). *Math. Ann.* 108 (1933), 296–310.
168. R. Moufang, Alternativkörper und der Satz vom vollständigen Vierseit (D_9). *Abh. Math. Sem. Hamb. Univ.* 9 (1933), 207–222.
169. R. Moufang, Zur Struktur von Alternativkörpern. *Math. Ann.* 110 (1934), 416–430. .
170. F. R. Moulton, A simple non-Desarguesian plane geometry. *Trans. Amer. Math. Soc.* 3 (1902), 192–195.
171. H. J. Munkholm, On the classification of incidence theorems in plane projective geometry. *Math. Z.* 90 (1965), 215–230.
172. P. Nabonnand, La théorie des "Würfe" de von Staudt—une irruption de l'algèbre dans la géométrie pure. *Arch. Hist. Exact Sci.* 62 (2008), 201–242.
173. H. Naumann, Eine affine Rechtwinkelgeometrie. *Math. Ann.* 131 (1956), 17–27.
174. H. Naumann, K. Reidemeister, Über Schließungssätze der Rechtwinkelgeometrie. *Abh. Math. Sem. Univ. Hamburg* 21 (1957), 1–12.
175. M. Neumann, L. Stanciu, Über einige Rechenregeln und Äquivalenzrelationen in einem Alternativkörper. *Inst. Politehn. Traian Vuia Timişoara Lucrăr. Sem. Mat. Fiz.* (1982), 41–44.
176. M. Neumann, L. Stanciu, Über eine Begründung der Moufang-Ebene. *Inst. Politehn. Traian Vuia Timişoara Lucrăr. Sem. Mat. Fiz.* No. 2 (1984), 37–40.
177. T. G. Ostrom, Conicoids: conic-like figures in non-Pappian planes. P. Plaumann and K. Strambach (eds), *Geometry—von Staudt's point of view* (Bad Windsheim, 1980), pp. 175–196, D. Reidel, Dordrecht 1981.

178. F. W. Owens, The introduction of ideal elements and a new definition of projective n-space. *Trans. Amer. Math. Soc.* 11 (1910), 141–171.
179. V. Pambuccian, Simple axiom systems for affine planes. *Zeszyty Nauk. Geom.* 21 (1995), 59–74.
180. V. Pambuccian, Two notes on the axiomatics of structures with parallelism. *Note Mat.* 20 (2000/2001), 93–104.
181. V. Pambuccian, Fragments of Euclidean and hyperbolic geometry. *Sci. Math. Jpn.* 53 (2001), 361–400.
182. V. Pambuccian, Early examples of resource-consciousness. *Studia Logica* 77 (2004), 81–86.
183. V. Pambuccian, The axiomatics of ordered geometry I. Ordered incidence spaces. *Expo. Math.* 29 (2011), 24–66.
184. V. Pambuccian, Orthogonality as single primitive notion for metric planes. With an appendix by H. and R. Struve. *Beitr. Algebra Geom.* 48 (2007), 399–409.
185. V. Pambuccian, Review of: M. Hallett and U. Majer (eds.), *David Hilbert's lectures on the foundations of geometry, 1891–1902*. Springer-Verlag, Berlin, 2004, *Philos. Math. (III)* 21 (2013), 255–277.
186. V. Pambuccian, Prolegomena to any theory of proof simplicity. *Philos. Trans. Royal Soc. A* 377 (2019), 20180035.
187. A. Papadopoulos, Hilbert's fourth problem. A. Papadopoulos and M. Troyanov (eds.), *Handbook of Hilbert geometry*, pp. 391–431, European Mathematical Society, Zürich, 2014.
188. Pappus d'Alexandrie, *La collection mathématique*. Œuvre traduite pour la première fois du grec au français avec une intriduction et des notes par Paul Ver Eecke. Desclée de Brouwer, Paris-Bruges, 1933.
189. Pappus of Alexandria, *Book 7 of the Collection*. Edited, with translation and commentary by Alexander Jones. Springer-Verlag, New York 1986.
190. K. H. Parshall, In pursuit of the finite division algebra theorem and beyond: Joseph H. M. Wedderburn, Leonard E. Dickson, and Oswald Veblen. *Arch. Internat. Hist. Sci.* 33 (1983), 274–299.
191. B. Pascal, *Œuvres complètes*. Texte établi, présenté et annoté par Jacques Chevalier. Bibliothèque de la Pléiade, Gallimard, Paris, 1976.
192. M. Pasch, *Vorlesungen über neuere Geometrie*. 1. Aufl. (1882), 2. Aufl. (Mit einem Anhang von M. Dehn: *Die Grundlegung der Geometrie in historischer Entwicklung*) J. Springer, Berlin, 1926.
193. J.-C. Petit, Caractérisation algébrique des plans affines munis d'une mesure des triangles. *Algèbre Theorie Nombres, Sem. P. Dubreil, M.-L. Dubreil-Jacotin, L. Lesieur et C. Pisot* 19 (1965/66), No. 22.
194. J.-C. Petit, Mesure des triangles et mesure des vecteurs à supports parallèles dans une géométrie plane affine. *Math. Z.*, **94**, 271–306 (1966), Zbl 156.19303.
195. J.-C. Petit, Mesure des vecteurs sur un groupe G et ternaires G-mesurables. *Math. Z.* 110 (1969), 223–256.
196. B. B. Phadke, The theorem of Desargues in planes with analogues to Euclidean angular bisectors. *Math. Scand.* 39 (1976), 191–194.
197. G. Pickert, *Projektive Ebenen*. Springer-Verlag, Berlin, 1955 (1st ed), 1975 (2nd ed).
198. G. Pickert, Der Satz von Pappos mit Festelementen. *Arch. Math. (Basel)* 10 (1959), 56–61.
199. G. Pickert, Projectivities in projective planes. P. Plaumann and K. Strambach (eds), *Geometry—von Staudt's point of view* (Bad Windsheim, 1980), pp. 1–49, D. Reidel, Dordrecht, 1981.
200. G. Pickert, Bemerkungen zu einem Inzidenzsatz von W. Kroll. *Math. Semesterber.* 40 (1993), 55–62.
201. G. Pickert, Ceva-Transitivität. *Geom. Dedicata* 74 (1999), 73–78.
202. B. Polster, *A geometrical picture book*. Springer, New York, 1998.
203. M. Prażmowska, A proof of the projective Desargues axiom in the Desarguesian affine plane. *Demonstratio Math.* 37 (2004), 921–924.

204. K. Prażmowski, On affine localisations of minor Desargues axiom and related affine axioms. *Bull. Pol. Acad. Sci. Math.* 36 (1988), 571–581.
205. S. Prieß-Crampe, *Angeordnete Strukturen. Gruppen, Körper, projektive Ebenen*. Springer-Verlag, Berlin, 1983.
206. E. Quaisser, Metrische Relationen in affinen Ebenen. *Math. Nachr.* 48 (1971), 1–31.
207. E. Quaisser, Zum Aufbau affiner Ebenen aus dem Spiegelungsbegriff. *Z. Math. Logik Grundlag. Math.* 27 (1981), 131–140.
208. E. Quaisser, Zum Stufenaufbau von Translationsebenen in spiegelungsgeometrischer Darstellung. *Z. Math. Logik Grundlag. Math.* 34 (1988), 89–95.
209. E. Quaisser, Zu einer Orthogonalitätsrelation in Desarguesschen Ebenen (der Charakteristik $\neq 2$). *Beitr. Algebra Geom.* 4 (1975), 71–84.
210. P. Rachevsky, Sur l'unicité de la géométrie projective dans le plan. *Rec. Math. Moscou (Nouv. Sér.)* 8 (1940), 107–119.
211. W. Rautenberg, Ein Beweis des Satzes von Pappus-Pascal in der affinen Geometrie. *Math.-Phys. Semesterber.* 12 (1965), 197–210.
212. W. Rautenberg, Beweis des Kommutativgesetzes in elementar-archimedisch geordneten Gruppen. *Z. Math. Logik Grundlag. Math.* 11 (1965), 1–4.
213. W. Rautenberg, Elementare Schemata nichtelementarer Axiome. *Z. Math. Logik Grundlag. Math.* 13 (1967), 329–366.
214. W. Rautenberg, The elementary Archimedean schema in geometry and some of its applications. *Bull. Acad. Polon. Sci. Sér. Sci. Math. Astronom. Phys.* 16 (1968), 837–843.
215. W. Rautenberg, Euklidische und Minkowskische Orthogonalitätsrelationen. *Fund. Math.* 64 (1969), 189–196.
216. W. Rautenberg, E. Quaisser, Orthogonalitätsrelationen in der affinen Geometrie. *Z. Math. Logik Grundlag. Math.* 15 (1969), 19–24.
217. K. Reich, Der Desarguessche und der Pascalsche Satz: Hessenbergs Beitrag zu Hilberts Grundlagen der Geometrie. H. Hecht et al.(eds.), *Kosmos und Zahl. Beiträge zur Mathematik- und Astronomiegeschichte, zu Alexander von Humboldt und Leibniz*, pp. 377–393, Franz Steiner Verlag, Stuttgart, 2008.
218. K. Reidemeister, Topologische Fragen der Differentialgeometrie V. Gewebe und Gruppen. *Math. Z.* 29 (1928), 427–435.
219. K. Reidemeister, *Vorlesungen über Grundlagen der Geometrie*. J. Springer, Berlin, 1930.
220. J. F. Rigby, Collineations, correlations, polarities, and conics. *Canad. J. Math.* 19 (1967) 1027–1041.
221. J. F. Rigby, Pascal ovals in projective planes. *Canad. J. Math.* 21 (1969), 1462–1476.
222. L. H. Rowen, *Polynomial identities in ring theory*. Academic Press, New York-London, 1980.
223. A. Saam, Ein neuer Schließungssatz für projektive Ebenen. *J. Geom.* 29 (1987), 36–42.
224. A. Saam, Schließungssätze als Eigenschaften von Projektivitäten. *J. Geom.* 32 (1988), 86–130.
225. A. Schleiermacher, Bemerkungen zum Fundamentalsatz der projektiven Geometrie. *Math. Z.* 99 (1967) 299–304.
226. A. Schleiermacher, Über projektive Ebenen, in denen jede Projektivität mit sechs Fixpunkten die Identität ist. *Math. Z.* 123 (1971), 325–339.
227. A. Schmidt, Die Dualität von Inzidenz und Senkrechtstehen in der absoluten Geometrie, *Math. Ann.* 118 (1943), 609–625.
228. D. Schor, Neuer Beweis eines Satzes aus den „Grundlagen der Geometrie" von Hilbert. *Math. Ann.* 58 (1904), 427–433.
229. K. Schütte, Ein Schließungssatz für Inzidenz und Orthogonalität. *Math. Ann.* 129 (1955), 424–430.
230. K. Schütte, Die Winkelmetrik in der affin-orthogonalen Ebene. *Math. Ann.* 130 (1955), 183–195.
231. K. Schütte, Schließungssätze für orthogonale Abbildungen euklidischer Ebenen. *Math. Ann.* 132 (1956), 106–120.

232. F. Schur, Über den Fundamentalsatz der projectiven Geometrie, *Math. Ann.* 51 (1899), 401–409.
233. W. Schwabhäuser. W. Szmielew, A. Tarski, *Metamathematische Methoden in der Geometrie*. Springer-Verlag, Berlin, 1983 (re-issued by ISHI Press, New York, 2011).
234. W. Schwan, Streckenrechnung und Gruppentheorie. *Math. Z.* 3 (1919), 11–28.
235. B. Segre, *Lectures on modern geometry*. With an appendix by Lucio Lombardo-Radice. Edizioni Cremonese, Roma 1961.
236. A. Seidenberg, Pappus implies Desargues. *Amer. Math. Monthly* 83 (1976), 190–192.
237. H. L. Skala, Projective-type axioms for the hyperbolic plane, *Geom. Dedicata* 44 (1992), 255–272.
238. L. A. Skornyakov, Right-alternative fields. (Russian) *Izvestiya Akad. Nauk SSSR. Ser. Mat.* 15, (1951). 177–184.
239. L. A. Skornyakov, On the theory of alternative fields. (Russian) *Uspehi Matem. Nauk (N.S.)* 5 (1950), no. 5(39), 160–162.
240. L. A. Skornyakov, Alternative fields. (Russian) *Ukrain. Mat. Žurnal* 2 (1950), no. 1, 70–85.
241. L. A. Skornyakov, Alternative fields of characteristic 2 and 3. (Russian) *Ukrain. Mat. Žurnal* 2 (1950), no. 3, 94–99.
242. L. A. Skornyakov, The configuration D_9. (Russian) *Mat. Sbornik (N.S.)* 30(72) (1952), 73–78.
243. L. A. Skornyakov, Concerning the note "On the theory of alternative fields." (Russian) *Uspehi Matem. Nauk (N.S.)* 9 (1954), no. 2(60), 185–188.
244. L. J. Smid, Über die Einführung der idealen Elemente in die ebene Geometrie mit Hilfe des Satzes vom vollständigen Vierseit. *Math. Ann.* 111 (1935), 285–288.
245. L. J. Smid, Eine absolute Axiomatik der Geometrie in einer begrenzten Ebene. *Math. Ann.* 112 (1935), 125–138.
246. K. Sörensen, Ebenen mit Kongruenz. *J. Geom.* 22 (1984), 15–30.
247. E. Sperner, Zur Begründung der Geometrie im begrenzten Ebenenstück. *Schr. Königsberger Gel. Ges. Naturw. Kl.* 14 (1938), Nr. 6, 121–143 = Max Niemeyer, Halle a. S., 1938.
248. E. Sperner, Ein gruppentheoretischer Beweis des Satzes von Desargues in der absoluten Axiomatik. *Arch. Math. (Basel)* 5 (1954), 458–468.
249. K. Strambach, Projektive Ebenen und Kegelschnitte. *Arch. Math. (Basel)* 64 (1995), 170–184.
250. H. Struve, Eine spiegelungsgeometrische Kennzeichnung der Desargues- und der Pappus-Konfiguration. *Arch. Math. (Basel)* 43 (1984), 89–96.
251. H. Struve, Affine Ebenen mit Orthogonalitätsrelation. *Z. Math. Logik Grundlag. Math.* 30 (1984), 223–231.
252. H. Struve, R. Struve, The Desargues configuration and inversive geometry. *Beitr. Algebra Geom.* 23 (1986), 15–25.
253. H. Struve, R. Struve, Endliche Cayley–Kleinsche Geometrien. *Arch. Math. (Basel)* 48 (1987), 178–184.
254. R. Struve, An axiomatic foundation of Cayley–Klein geometries. *J. Geom.* 107 (2016), 225–248.
255. Z. Szilasi, Notes on the Cevian nest property and the Newton property of projective planes. *Elem. Math.* 66 (2011), 1–10.
256. W. Szmielew, *From affine to Euclidean geometry. An axiomatic approach*. D. Reidel, Dordrecht; PWN-Polish Scientific Publishers, Warszawa, 1983.
257. H. Tecklenburg, A proof of the theorem of Pappus in finite Desarguesian affine planes. *J. Geom.* 30 (1987), 172–181.
258. K. Tent, Sharply 3-transitive groups. *Adv. Math.* 286 (2016), 722–728.
259. A. C. Thompson, *Minkowski geometry*. Cambridge University Press. Cambridge, 1996.
260. G. Thomsen, Un teorema topologico sulle schiere di curve e una caratterizzazione geometrica delle superficie isotermo-asintotiche. *Boll. Un. Mat. Ital.* 6 (1927), 80–85.
261. G. Thomsen, Topologische Fragen der Differentialgeometrie. XII: Schnittpunktssätze in ebenen Geweben. *Abh. Math. Semin. Hamb. Univ.* 7 (1929), 99–106.
262. J. Tits, Généralisations des groupes projectifs. *Acad. Roy. Belgique. Bull. Cl. Sci. (5)* 35 (1949), 197–208.

263. J. Tits, Généralisations des groupes projectifs. II. *Acad. Roy. Belgique. Bull. Cl. Sci. (5)* 35 (1949), 224–233.
264. J. Tits, Généralisations des groupes projectifs basées sur leurs propriétés de transitivité. *Acad. Roy. Belgique. Cl. Sci. Mém. Coll. in 8°* 27 (1952). no. 2.
265. O. Veblen, J. H. Maclagan-Wedderburn, Non-Desarguesian and non-Pascalian geometries. *Trans. Amer. Math. Soc.* 8 (1907), 379–388.
266. J.-D. Voelke, Le théorème fondamental de la géométrie projective: évolution de sa preuve entre 1847 et 1900. *Arch. Hist. Exact Sci.* 62 (2008), 243–296 .
267. W. Wagner, Über die Grundlagen der projektiven Geometrie und allgemeine Zahlensysteme. *Math. Ann.* 113 (1937), 528–567.
268. A. Wald, Axiomatik des Zwischenbegriffes in metrischen Räumen, *Math. Ann.* 104 (1931), 476–484.
269. G. Weiß, Fano'sche Moufang-Ebenen mit Gauss-Figur sind Pappos'sch. *J. Geom.* 17 (1981), 46–49.
270. H. Wiener, Über Grundlagen und Aufbau der Geometrie. *Jahresber. Dtsch. Math.-Ver.* 1 (1890–1891), 45–48.
271. H. Wiener, *Sechs Abhandlungen über das Rechnen mit Spiegelungen, nebst Anwendungen auf die Geometrie der Bewegungen und auf die projective Geometrie*. Breitkopf & Härtel, Leipzig, 1893.
272. L. Włodarski, On the Pappus, Desargues and related theorems of affine geometry. *Bull. Acad. Polon. Sci. Sér. Sci. Math.* 27 (1979), 869–873.
273. H. Zassenhaus, Kennzeichnung endlicher linearer Gruppen als Permutationsgruppen. *Abh. Math. Semin. Hamb. Univ.* 11 (1935), 17–40.
274. H. Zassenhaus, Über endliche Fastkörper. *Abh. Math. Semin. Hamb. Univ.* 11 (1935), 187–220.
275. H.-G. Zimmer, Bemerkungen zum Trapezdesargues. *Math. Ann.* 135 (1958), 274–278.
276. H.-G. Zimmer, Über Quadrate der affinen Rechtwinkelgeometrie. *Math. Ann.* 135 (1958), 340–351.
277. M. Zorn, Theorie der alternativen Ringe. *Abh. Math. Semin. Hamb. Univ.* 8 (1930), 123–147.
278. M. Zorn, Alternativkörper und quadratische Systeme. *Abh. Math. Semin. Hamb. Univ.* 9 (1933), 395–402.

Chapter 9
Projective Configuration Theorems: Old Wine into New Wineskins

Serge Tabachnikov

Abstract We survey some recent results concerning projective configuration theorems in the spirit of the classical theorems of Pappus, Desargues, Pascal, ... We hope that this modern take on the old theorems makes this evergreen topic fresh again. We connect configuration theorems to completely integrable systems, identities in Lie algebras of motion, modular group, and other subject of contemporary interest.

9.1 Introduction: Classical Configuration Theorems

Projective configuration theorems are among the oldest and best known mathematical results. The next figures depict the famous theorems of Pappus, Desargues, Pascal, Brianchon, and Poncelet (Figs. 9.1, 9.2, 9.3, 9.4).

The literature on configuration theorems is vast; the reader interested in a panoramic view of the subject is recommended [6, 36].

Configuration theorems continue to be an area of active research. To a great extent, this is due to the advent of computer as a tool of experimental research in mathematics. In particular, interactive geometry software is a convenient tool for the study of geometric configurations. The illustrations in this article are made using such a software, Cinderella 2 [52].

Another reason for the popularity of configuration theorems is that they play an important role in the emerging field of discrete differential geometry and the theory of completely integrable systems [7].

The goal of this survey is to present some recent results motivated and inspired by the classical configuration theorems; these results make the old theorems fresh again. The selection of topics reflects this author's taste; no attempt was made to present a comprehensive description of the area. In the cases when proofs are discussed, they are only outlined; the reader interested in details is referred to the original papers.

S. Tabachnikov (✉)
Department of Mathematics, Penn State University, University Park, PA, USA
e-mail: tabachni@math.psu.edu

S. G. Dani, A. Papadopoulos (eds.), *Geometry in History*,
https://doi.org/10.1007/978-3-030-13609-3_9

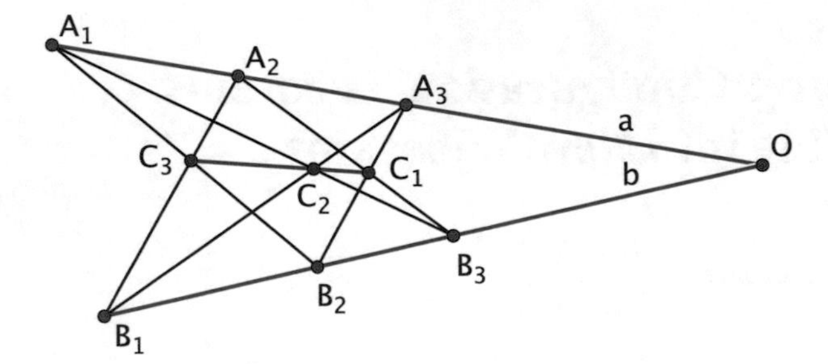

Fig. 9.1 The Pappus theorem: if A_1, A_2, A_3 and B_1, B_2, B_3 are two collinear triples of points, then C_1, C_2, C_3 is also a collinear triple

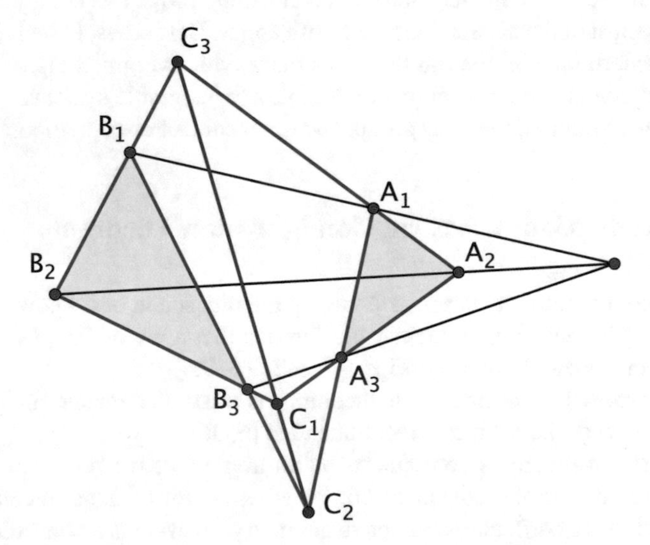

Fig. 9.2 The Desargues theorem: if the lines A_1B_1, A_2B_2 and A_3B_3 are concurrent, then the points C_1, C_2, C_3 are collinear

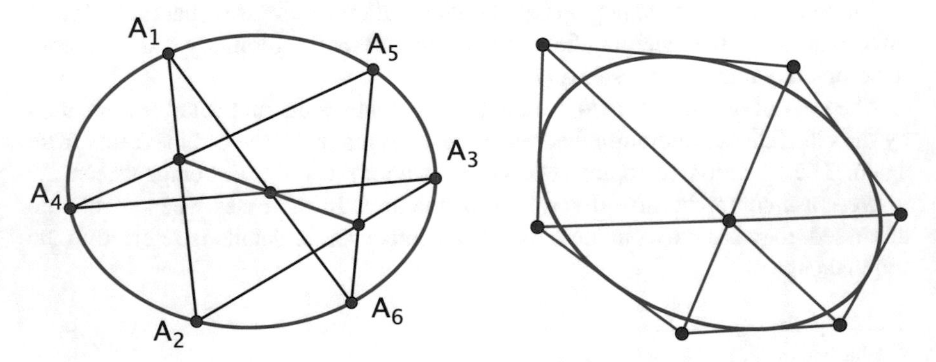

Fig. 9.3 The Pascal theorem, a generalization of the Pappus theorem: the points $A_1, \ldots, A_6$ lie on a conic, rather than the union of two lines. The Brianchon theorem is projectively dual to Pascal's

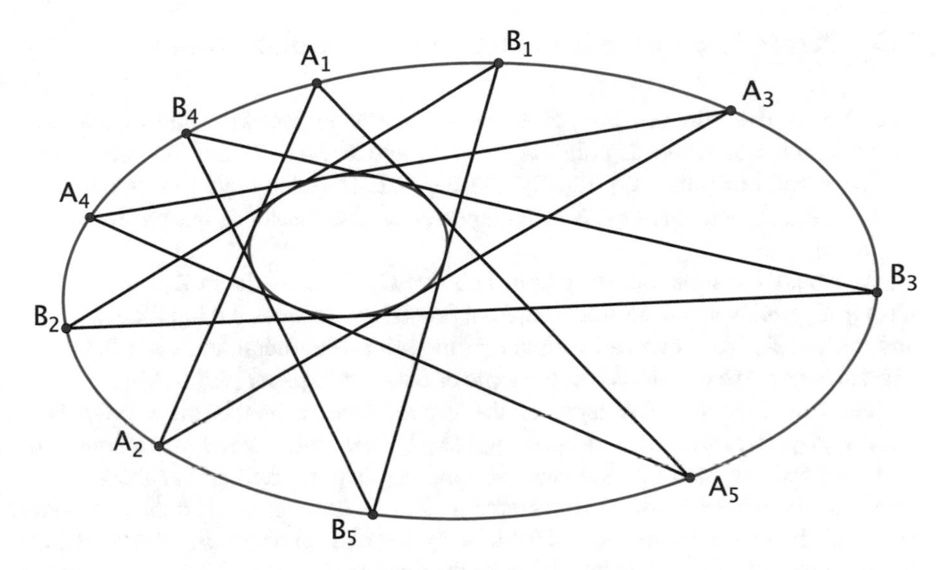

Fig. 9.4 The Poncelet Porism, case $n = 5$: if the polygonal line $A_1A_2A_3A_4A_5$, inscribed into a conic and circumscribed about a conic, closes up after five steps, then so does any other polygonal line $B_1B_2B_3B_4B_5$

We assume that the reader is familiar with the basics of projective, Euclidean, spherical, and hyperbolic geometries. One of the standard references is [5], and [22] is as indispensable as ever.

Now let us specify what we mean by configuration theorems in this article. The point of view is dynamic, well adapted for using interactive geometry software.

An initial data for a configuration theorem is a collection of labelled points A_i and lines b_j in the projective plane, such that, for some pairs of indices (i, j), one has the incidence $A_i \in b_j$. If, in addition, a polarity is given, then one can associate the dual line to a point, and the dual point to a line. In presence of polarity, the initial data includes information that, for some pairs of indices (k, l), the point A_k is polar dual to the line b_l.

One also has an ordered list of instructions consisting of two operations: draw a line through a pair of points, or intersect a pair of lines at a point. These new lines and points also receive labels. If polarity is involved, one also has the operation of taking the polar dual object, point to line, or line to point.

The statement of a configuration theorem is that, among so constructed points and lines, certain incidence relations hold, that is, certain points lie on certain lines.

It is assumed that the conclusion of a configuration theorem holds for almost every initial set of points and lines satisfying the initial conditions, that is, holds for a Zariski open set of such initial configurations. This is different from what is meant by a configuration of points and lines in chapter 3 of [22] or in [20]: the focus there is on whether a combinatorial incidence is realizable by points and lines in the projective plane.

9.2 Iterated Pappus Theorem and the Modular Group

The Pappus theorem can be viewed as a construction in $\mathbb{RP}^2$ that inputs two ordered triples of collinear points A_1, A_2, A_3 and B_1, B_2, B_3, and outputs a new collinear triple of points C_1, C_2, C_3, see Fig. 9.1. One is tempted to iterate: say, take A_1, A_2, A_3 and C_1, C_2, C_3 as an input. Alas, this takes one back to the triple B_1, B_2, B_3.

To remedy the situation, swap points C_1 and C_3. Then the input A_1, A_2, A_3 and C_1, C_2, C_3 yields a new collinear triple of points, and so does the input C_1, C_2, C_3 and B_1, B_2, B_3. And one can continue in the same way indefinitely, see Fig. 9.5. The study of these iterations was the topic of Schwartz's paper [39].

Return to Fig. 9.1. The input of the Pappus construction is the *marked box* $(A_1, A_3, B_3, B_1; A_2, B_2)$, a quadrilateral $A_1A_3B_3B_1$ with the top distinguished point A_2 and the bottom distinguished point B_2. The marked box is assumed to satisfy the *convexity condition*: the points A_1 and A_3 are separated by the points A_2 and O on the projective line a, and likewise for the pairs of points B_1, B_3 and B_2, O on the line b. Marked boxes that differ by the involution

$$(A_1, A_3, B_3, B_1; A_2, B_2) \leftrightarrow (A_3, A_1, B_1, B_3; A_2, B_2)$$

are considered to be the same.

A convex set in $\mathbb{RP}^2$ is a set that is disjoint from some line and that is convex in the complement to this line, the affine plane. Two points in $\mathbb{RP}^2$ can be connected by a segment in two ways. The four points A_1, A_3, B_3, B_1, in this cyclic order, define 16 closed polygonal lines, but only one of them is the boundary of a convex quadrilateral, called the interior of the convex marked box.

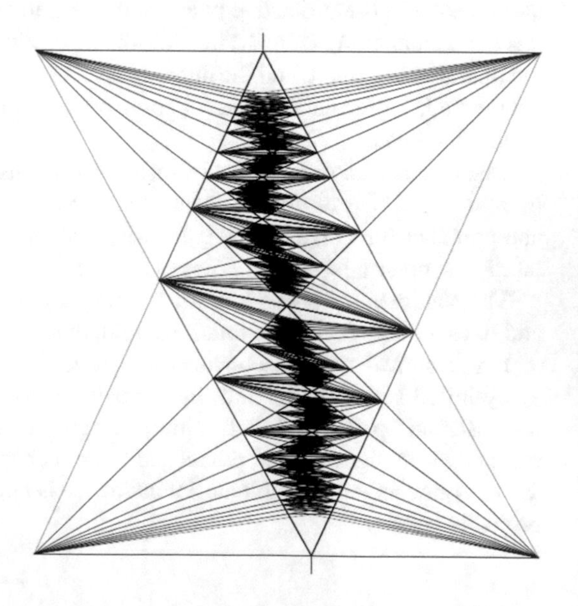

Fig. 9.5 Iterations of the Pappus construction produced by Schwartz's applet [53]

Recall that the points of the dual projective plane are the lines of the initial plane. Let $\Theta = (A_1, A_3, B_3, B_1; A_2, B_2)$ be a convex marked box in $\mathbb{RP}^2$. Its dual, Θ^*, is a marked box in the dual projective plane whose points are the lines

$$(A_2B_1, A_2B_3, A_1B_2, A_3B_2; a, b).$$

The dual marked box is also convex.

The moduli space of projective equivalence classes of marked boxes is 2-dimensional. One can send the points A_1, A_3, B_3, B_1 to the vertices of a unit square; then the projective equivalence class of a convex marked box is determined by the positions of the points A_2 and B_2 on the horizontal sides of the square. Namely, let $x = |A_1A_2|$, $y = |B_1B_2|$. Then the equivalence class given by

$$(x, y) \sim (1 - x, 1 - y), \tag{9.1}$$

where $0 < x, y < 1$, determines the projective equivalence class of a convex marked box. We denote this equivalence class by $[x, y]$.

The Pappus construction defines two operations on convex marked boxes, see Fig. 9.6:

$$\tau_1 : (A_1, A_3, B_3, B_1; A_2, B_2) \mapsto (A_1, A_3, C_3, C_1; A_2, C_2),$$
$$\tau_2 : (A_1, A_3, B_3, B_1; A_2, B_2) \mapsto (C_1, C_3, B_3, B_1; C_2, B_2).$$

Add to it a third operation

$$i : (A_1, A_3, B_3, B_1; A_2, B_2) \mapsto (B_1, B_3, A_1, A_3; B_2, A_2),$$

also shown in Fig. 9.6.

The three operations form a semigroup G. The operations satisfy the following identities, proved by inspection.

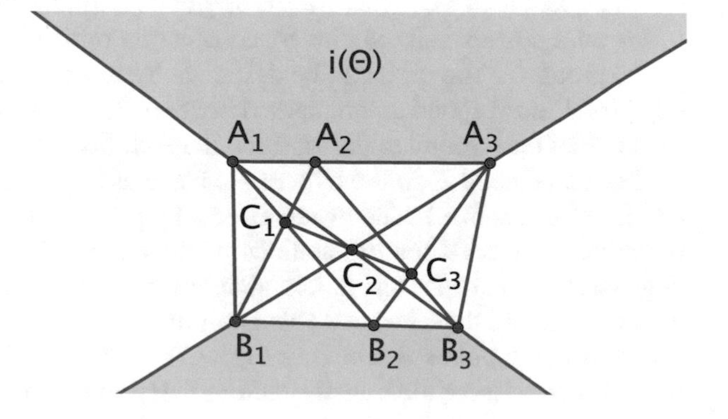

Fig. 9.6 The interior of the convex marked box $i(\Theta)$ is bounded by the segments A_1A_3, A_3B_1, B_1B_3 and B_3A_1. Two of these segments cross the line at infinity

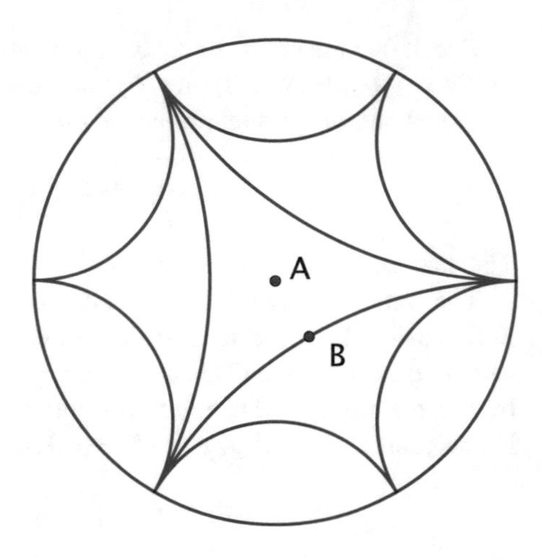

Fig. 9.7 A tiling of the hyperbolic plane, in the Poincaré disk model, by ideal triangles

Lemma 2.1 *One has:*

$$i^2 = 1, \; \tau_1 i \tau_2 = \tau_2 i \tau_1 = i, \; \tau_1 i \tau_1 = \tau_2, \; \tau_2 i \tau_2 = \tau_1.$$

As a consequence, G is a group; for example, $\tau_1^{-1} = i\tau_2 i$.

Recall that the modular group M is the group of fractional-linear transformations with integral coefficients and determinant one, that is, the group $\mathrm{PSL}(2, \mathbb{Z})$. Since $\mathrm{PGL}(2, \mathbb{R})$ is the group of orientation preserving isometries of the hyperbolic plane, the modular group M is realized as a group of isometries of H^2.

Consider the tiling of H^2 by ideal triangles obtained from one such triangle by consecutive reflections in the sides, see Fig. 9.7 for the beginning of this construction. The modular group is generated by two symmetries of the tiling: the order three rotation about point A and the order two rotation (central symmetry) about point B. Algebraically, M is a free product of $\mathbb{Z}_3$ and $\mathbb{Z}_2$.

Return to the group G. It is generated by the elements $\alpha = i\tau_1$ and $\beta = i$. Lemma 2.1 implies that $\alpha^3 = \beta^2 = 1$. One can prove that there are no other relations, and hence $G = \mathbb{Z}_3 * \mathbb{Z}_2$ is identified with the modular group.

Given a convex marked box Θ, consider its orbit $\Omega = G(\Theta)$ under the action of the group G. The orbit can be described by its oriented incidence graph Γ. The edges of Γ correspond to the marked boxes of Ω, oriented from top to bottom, and the vertices correspond to the tops and the bottoms of the boxes.

One can embed Γ in the hyperbolic plane as in Fig. 9.7 (the orientations of the edges are not shown). The group G acts by permutations of the edges of Γ. The operation i reverses the orientations of the edges. The operation τ_1 rotates each edge counterclockwise one 'click' about its tail, and τ_2 rotates the edges one 'click' clockwise about their heads. (This is a different action from the one generated by rotations about points A and B in Fig. 9.7). Denote by G' the index two subgroup of G that consists of the transformations that preserve the orientations of the edges.

The orbit Ω of a convex marked box Θ has a large group of projective symmetries, namely, an index two subgroup M' of the modular group M. This is one of the main results of [39]. Specifically, one has

Proposition 2.2 *Given a convex marked box* Θ, *there is an order three projective transformation with the cycle*

$$i(\Theta) \mapsto \tau_1(\Theta) \mapsto \tau_2(\Theta).$$

In addition, there exists a polarity that identifies $i(\Theta)$ *with the dual box* Θ^*.

Proof For the proof of the first statement, one can realize the box Θ in such a way that the three-fold rotational symmetry is manifestly present, see Fig. 9.8. Namely,

$$\Theta = (B_3, B_1, A_1, A_3; B_2, A_2),\ i(\Theta) = (A_1, A_3, B_1, B_3; A_2, B_2),$$
$$\tau_1(\Theta) = (B_1, B_3, C_1, C_3; B_2, C_2),\ \tau_2(\Theta) = (C_1, C_3, A_1, A_3; C_2, A_2).$$

In terms of the marked box coordinates (x, y), described in (9.1), the three operations, i, τ_1, and τ_2, act in the same way: $[x, y] \mapsto [1 - y, x]$.

For the second statement, consider another realization depicted in Fig. 9.9. The marked points A_2 and B_2 are at infinity, and $|OA_1||OA_3| = |OB_1||OB_3| = 1$. Then the polarity with respect to the unit circle centered at point O acts as follows:

$$A_1 \mapsto A_3B_2,\ A_3 \mapsto A_1B_2,\ B_1 \mapsto B_3A_2,\ B_3 \mapsto B_1A_2,$$

providing the desired projective equivalence. □

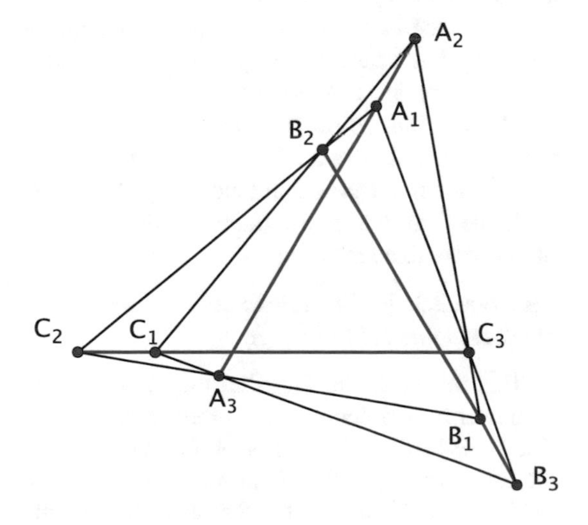

Fig. 9.8 A symmetric realization of the marked boxes $i(\Theta)$, $\tau_1(\Theta)$, $\tau_2(\Theta)$

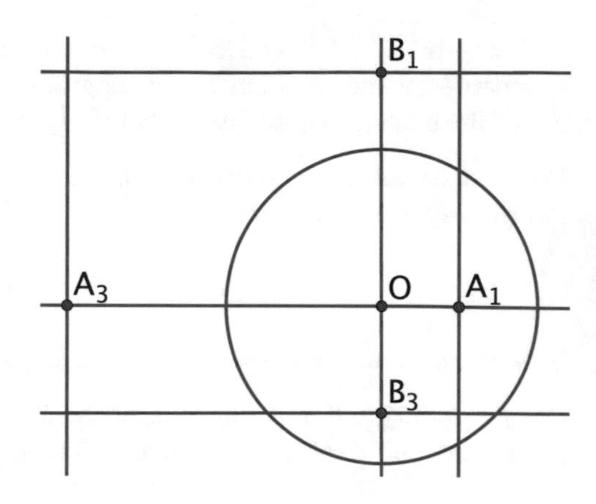

Fig. 9.9 Projective equivalence of $i(\Theta)$ and Θ^*

If one identifies the projective plane with its dual by a polarity, then the above discussion describes a faithful representation of the modular group M as the group of projective symmetries of the G-orbit Ω of a convex marked box.

A marked box Θ determines a natural map f of the set of vertices of the graph Γ to the set of the marked points of the orbit Ω. The map f conjugates the actions of the group G' on the graph Γ and the group M' of projective symmetries of the orbit Ω. The set of vertices of Γ is dense on the circle at infinity of the hyperbolic plane S^1, see Fig. 9.7. Using the nested properties of the interiors of the boxes in Ω and estimates on their sizes (in the elliptic plane metric), Schwartz proves the following theorem.

Theorem 2.1 *The map f extends to a homeomorphism of S^1 to its image.*

The image $\Lambda = f(S^1)$ is called the *Pappus curve*; see Fig. 9.5 that provides an approximation of this curve.

The above discussion shows that the Pappus curve is projectively self-similar. In the exceptional case of $x = y = 1/2$, the curve Λ is a straight line. Otherwise, it is not an algebraic curve, see [21].

The tops and bottoms of the marked boxes form a countable collection of lines that also extends to a continuous family, a curve L in the dual projective plane.

Define a transverse line field along Λ as a continuous family of lines such that each line from the family intersects the curve at exactly one point and every point of Λ is contained in some line.

Theorem 2.2 *If the Pappus curve Λ is not a straight line, then L is a unique transverse line field along Λ.*

This theorem, the fact that the Pappus curve is projectively self-similar, and computer experiments suggest that Λ is a true fractal (unless it is a straight line). The thesis [26] contains some preliminary numerical results on the box dimension of the Pappus curve and its dependence on the coordinates $[x, y]$ of the initial convex marked box. According to these experiments, the maximal possible box dimension of Λ is about 1.25.

Finding the fractal dimensions of the Pappus curves as a function of $[x, y]$ or, at least, proving that this dimension is greater than one in all non-exceptional cases $[x, y] \neq [1/2, 1/2]$, is an outstanding open problem.

9.3 Steiner Theorem and the Twisted Cubic

This section is based on another recent ramification of the Pappus theorem, the work of Rigby [37] and Hooper [23].

Let us start with the dual Pappus theorem, see Fig. 9.10 where the objects dual to the ones in Fig. 9.1 are denoted by the same letters, with the upper and lower cases swapped (the Pappus theorem is equivalent to its dual). As an aside, let us mention that the dual Pappus theorem has an interpretation in the theory of webs: the 3-web, made of three pencils of lines, is flat, see [16], lecture 18.

Now consider Pascal's theorem, Fig. 9.3. The six permutations of the points on the conic yield 60 Pascal lines. These lines and their intersection points, connected by further lines, form a intricate configuration of 95 points and 95 lines, the *hexagrammum mysticum*. There is a number of theorems describing this configuration, due to Steiner, Plücker, Kirkman, Cayley, and Salmon. See [9, 10] for a contemporary account of this subject.

The Pappus theorem is a particular case of Pascal's theorem, and in this case, the number of lines that result from permuting the initial points (say, points B_1, B_2, B_3 in Fig. 9.1) reduces to six, shown in Fig. 9.11.

Let us introduce notations. Consider Fig. 9.1 and denote the triples of points:

$$\mathcal{A} = (A_1, A_2, A_3), \ \mathcal{B} = (B_1, B_2, B_3).$$

The Pappus theorem produces a new triple, $\mathcal{C} = (C_1, C_2, C_3)$. The lines containing these triples are denoted by a, b, c, respectively. We write: $c = \ell(\mathcal{A}, \mathcal{B})$.

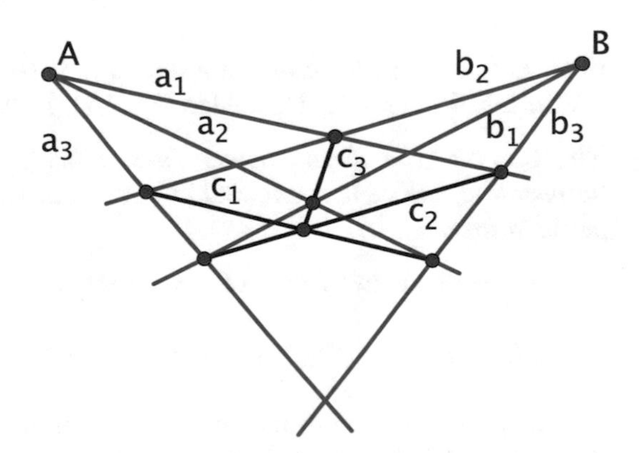

Fig. 9.10 Dual Pappus theorem

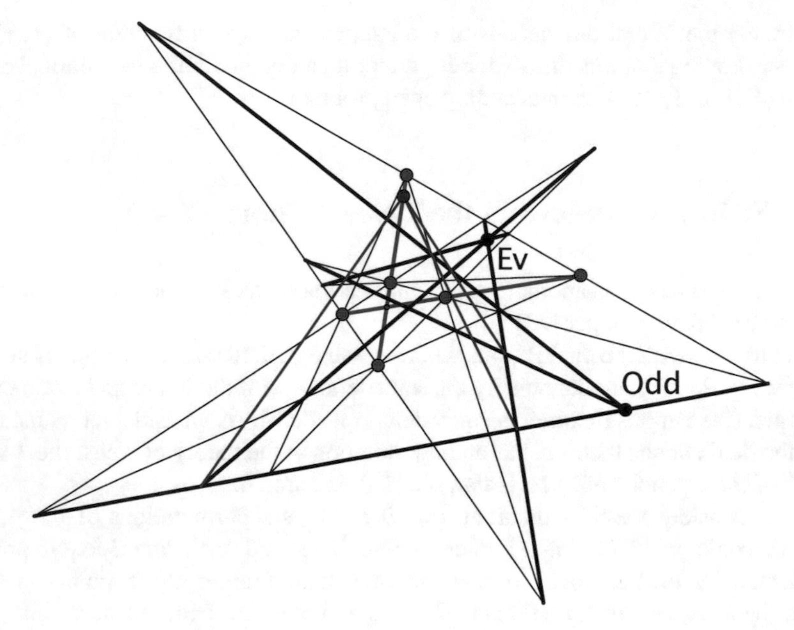

Fig. 9.11 Two Steiner points, corresponding to even and odd permutations, are labelled. One of the points $\ell^*(\varphi, s(\psi))$ is shown

We use a similar notation for the dual Pappus theorem: if

$$\alpha = (a_1, a_2, a_3), \ \beta = (b_1, b_2, b_3)$$

are two triples of concurrent lines, then $\ell^*(\alpha, \beta)$ is the point of intersection of the triple of lines (c_1, c_2, c_3), see Fig. 9.10.

The permutation group S_3 acts on triples by the formula

$$s(\mathcal{B}) = (B_{s^{-1}(1)}, B_{s^{-1}(2)}, B_{s^{-1}(3)}).$$

Let $\sigma \in S_3$ be a cyclic permutation, and $\tau \in S_3$ be a transposition of two elements.

The following result, depicted in Fig. 9.11, is due to Steiner.

Theorem 3.1 *The three Pappus lines $\ell(\mathcal{A}, s(\mathcal{B}))$ where $s \in S_3$ is an even permutation, are concurrent, and so are the three lines corresponding to the odd permutations.*

Thus we obtain two triples of concurrent lines; denote them by

$$\varphi = (\ell(\mathcal{A}, \mathcal{B}), \ell(\mathcal{A}, \sigma(\mathcal{B}), \ell(\mathcal{A}, \sigma^2(\mathcal{B})), \psi = (\ell(\mathcal{A}, \tau(\mathcal{B})), \ell(\mathcal{A}, \tau\sigma(\mathcal{B}), \ell(\mathcal{A}, \tau\sigma^2(\mathcal{B})).$$

Apply the dual Pappus theorem to the permutations of these triples of lines. By the dual Steiner theorem, the six points $\ell^*(\varphi, s(\psi)), \ s \in S_3$, are collinear in threes.

More is true. The next two theorems are due to Rigby [37].

Theorem 3.2 *The points $\ell^*(\varphi, s(\psi))$ lie on line a when s is an even permutation, and on line b when s is odd.*

Let $\mathcal{B}'$ be another collinear triple of points such that the line b' still passes through point $O = a \cap b$. Applying the above constructions to $\mathcal{A}$, $\mathcal{B}'$, we obtain new triples of lines φ', ψ', and a new triple of points $\ell^*(\varphi', s(\psi'))$ on line a where s is an even permutation.

Theorem 3.3 *The new triple of points coincides with the old one: for even permutations s, one has $\ell^*(\varphi', s(\psi')) = \ell^*(\varphi, s(\psi))$.*

Theorems 3.2 and 3.3 are stated by Rigby without proof; to quote,

> The theorems in this section have been verified in a long and tedious manner using coordinates. There seems little point in publishing the calculations; it is to be hoped that shorter and more elegant proofs will be found.

Conceptual proofs are given in [23]; the reader is referred to this paper and is encouraged to find an alternative approach to these results.

The above theorems make it possible to define the *Steiner map*

$$S_O : (A_1, A_2, A_3) \mapsto (\ell^*(\varphi, \psi), \ell^*(\varphi, \sigma^2(\psi)), \ell^*(\varphi, \sigma(\psi))).$$

This map depends on the point O, but not on the choice of the triple $\mathcal{B}$.

The Steiner map commutes with permutations of the points involved, and hence it induces a map of the space of unordered triples of points of the projective line. Abusing notation, we denote this induced map by the same symbol. Hooper [23] gives a complete description of the Steiner map.

Assume that the ground field is the field of complex numbers. The space of unordered triples of points of $\mathbb{CP}^1$, that is, the symmetric cube $S^3(\mathbb{CP}^1)$, is identified with $\mathbb{CP}^3$. This is a particular case of the Fundamental Theorem of Algebra, one of whose formulations is that $S^n(\mathbb{CP}^1) = \mathbb{CP}^n$ (given by projectivizing the Vieta formulas that relate the coefficients of a polynomial to its roots). Thus S_O is a self-map of $\mathbb{CP}^3$.

The set of cubic polynomials with a triple root corresponds to a curve $\Gamma \subset \mathbb{CP}^3$, the twisted cubic (the moment curve). The secant variety of the twisted cubic, that is, the union of its tangent and secant lines, covers $\mathbb{CP}^3$, and the lines are pairwise disjoint, except at the points of Γ.

The set of cubic polynomials with a zero root corresponds to a plane in $\mathbb{CP}^3$. Denote this plane by Π. The Steiner map $S_O : \mathbb{CP}^3 \to \mathbb{CP}^3$ is described in the next theorem.

Theorem 3.4

(i) The map S_O preserves the secants of the twisted cubic Γ that do not pass through the origin (the image of the cubic polynomial z^3).

(ii) *One can choose projective coordinates on these secant lines so that the map is given by the formula* $x \mapsto x^2$.

(iii) *The choice of coordinates is as follows: the two points of intersection of the secant line with* Γ *have coordinates* 0 *and* ∞, *and the intersection point of the secant with the plane* Π *has coordinate* -1.

In homogeneous coordinates of $\mathbb{CP}^3$, the map S_O is polynomial of degree 6; see [23] for an explicit formula for a particular choice $O = (0 : 1)$.

In the real case, the secant lines are identified with the circle $\mathbb{R}/\mathbb{Z}$, and the Steiner map becomes the doubling map $t \mapsto 2t$ mod 1, a well-known measure-preserving ergodic transformation.

9.4 Pentagram-Like Maps on Inscribed Polygons

This section, based on [43], concerns eight configuration theorems of projective geometry that were discovered in the study of the pentagram map.

The pentagram map, whose study was put forward by Schwartz [38], is a transformation of the moduli space of projective equivalence classes of polygons in the projective plane depicted in Fig. 9.12. The pentagram map has become a popular object of study: it is a discrete completely integrable system, closely related with the theory of cluster algebras. See [17–19, 33, 34, 46] for a sampler of the current literature on this subject.

To formulate the results, let us introduce some notations.

By a polygon in the projective plane we mean a cyclically ordered collection of its vertices (that also determines the cyclically ordered collection of lines, the sides of the polygon).

Let $\mathcal{C}_n$ and $\mathcal{C}_n^*$ be the spaces of n-gons in the projective plane $\mathbb{RP}^2$ and its dual $(\mathbb{RP}^2)^*$. Define the k-diagonal map $T_k : \mathcal{C}_n \to \mathcal{C}_n^*$: for $P = \{p_1, \ldots, p_n\}$,

$$T_k(P) = \{(p_1 p_{k+1}), (p_2 p_{k+2}), \ldots, (p_n p_{k+n})\}.$$

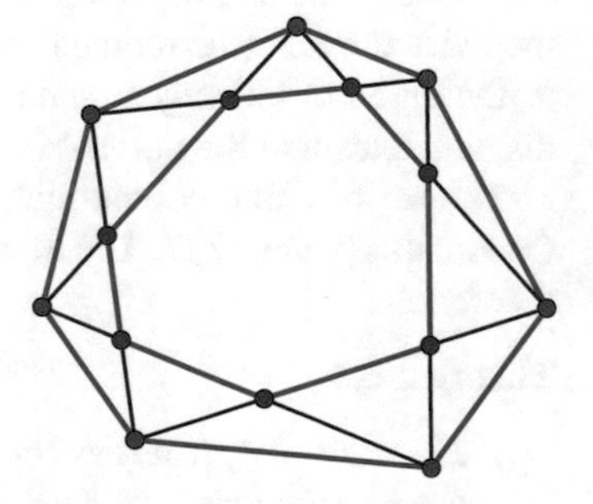

Fig. 9.12 The pentagram map takes an n-gon P to the polygon made by the intersection points of the short (skip one) diagonals of P. Here $n = 7$

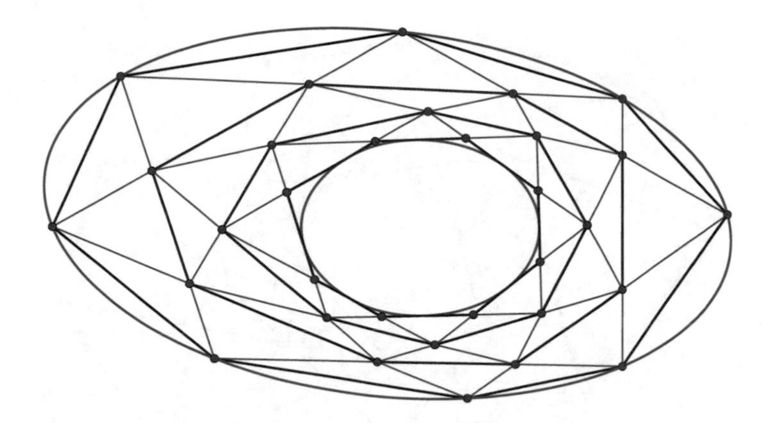

Fig. 9.13 The third iteration of the pentagram map on an inscribed octagon yields a projectively dual octagon

Each map T_k is an involution; the map T_1 is the projective duality that sends a polygon to the cyclically ordered collection of its sides.

Extend the notation to muti-indices: $T_{ab} = T_a \circ T_b$, $T_{abc} = T_a \circ T_b \circ T_c$, etc. For example, the pentagram map is T_{12}. If P is a polygon in $\mathbb{RP}^2$ and Q a polygon in $(\mathbb{RP}^2)^*$, and there exists a projective transformation $\mathbb{RP}^2 \to (\mathbb{RP}^2)^*$ that takes P to Q, we write: $P \sim Q$.

Now we are ready to formulate our results; they concern polygons inscribed into a conic or circumscribed about a conic.

Theorem 4.1

(i) *If P is an inscribed* 6*-gon, then* $P \sim T_2(P)$.
(ii) *If P is an inscribed* 7*-gon, then* $P \sim T_{212}(P)$.
(iii) *If P is an inscribed* 8*-gon, then* $P \sim T_{21212}(P)$.

Surprisingly, this sequence does not continue! Theorem 4.1 (iii) is depicted in Fig. 9.13. See also Schwartz's applet [54] for illustrations of this and other results of this section.

Theorem 4.2 *If P is a circumscribed* 9*-gon, then* $P \sim T_{313}(P)$.

See Fig. 9.14.

Theorem 4.3 *If P is an inscribed* 12*-gon, then* $P \sim T_{3434343}(P)$.

The next results have a somewhat different flavor: one does not claim anymore that the final polygon is projectively related to the initial one.

Theorem 4.4

(i) *If P is an inscribed* 8*-gon, then* $T_3(P)$ *is circumscribed.*
(ii) *If P is an inscribed* 10*-gon, then* $T_{313}(P)$ *is circumscribed.*
(iii) *If P is an inscribed* 12*-gon, then* $T_{31313}(P)$ *is circumscribed.*

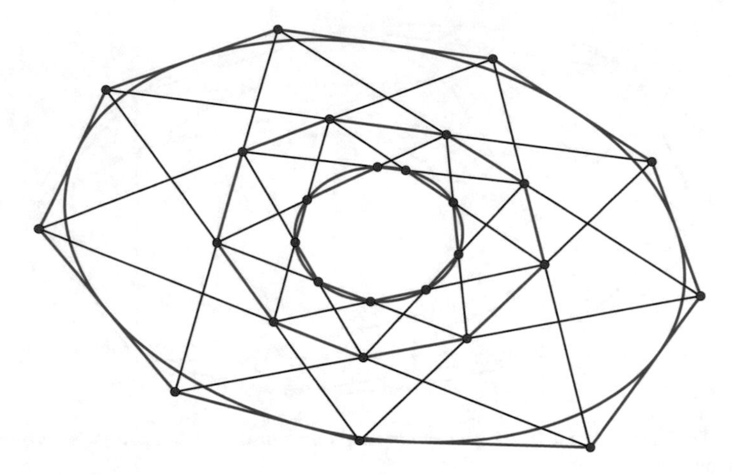

Fig. 9.14 Theorem 4.2

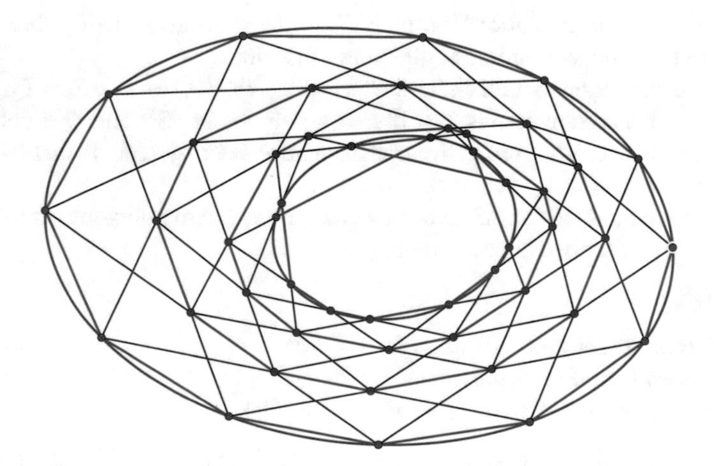

Fig. 9.15 Theorem 4.4 (iii)

Again, contrary to one's expectation, this sequence does not continue. Theorem 4.4 (iii) is illustrated in Fig. 9.15.

Now about the discovery of these results and their proofs. Theorems 4.1 (i) and (ii) were discovered in our study of the pentagram map. Then Valentin Zakharevich, a participant of the 2009 Penn State REU (Research Experience for Undergraduates) program, discovered Theorem 4.2. Inspired by this discovery, we did an extensive computer search for this kind of configuration theorems; the results are the above eight theorems. We think that the list above is exhaustive, but this remains a conjecture.

Note that one may cyclically relabel the vertices of a polygon to deduce seemingly new theorems. Let us illustrate this by an example. Rephrase the statement of

Theorem 4.4 (iii) as follows: *If P is an inscribed dodecagon then $T_{131313}(P)$ is also inscribed.* Now relabel the vertices by $\sigma(i) = 5i \bmod 12$. The map T_3 is conjugated by σ as follows:

$$i \mapsto 5i \mapsto 5i + 3 \mapsto 5(5i + 3) = i + 3 \mod 12,$$

that is, it is the map is T_3 again, and the map T_1 becomes

$$i \mapsto 5i \mapsto 5i + 1 \mapsto 5(5i + 1) = i + 5 \mod 12,$$

that is, the map is T_5. One arrives at the statement: *If P is an inscribed dodecagon then $T_{535353}(P)$ is also inscribed.*

We proved all of the above theorems, except Theorem 4.4 (iii), by uninspiring computer calculations (the symbolic manipulation required for a proof of Theorem 4.4 (iii) was beyond what we could manage in Mathematica).

Of course, one wishes for elegant geometric proofs. Stephen Wang found proofs of Theorems 4.4 (i) and (ii) which are presented below, and Maria Nastasescu, a 2010 Penn State REU participant, found algebraic geometry proofs of the same two theorems. Fedor Nilov proved Theorem 4.4 (iii) using a planar projection of hyperboloid of one sheet. Unfortunately, none of these proofs were published.

Here is Wang's proof of Theorem 4.4 (i).

Consider Fig. 9.16. We need to prove that the points $B_1, \ldots, B_8$ lie on a conic. The hexagon $A_6A_1A_4A_7A_2A_5$ is inscribed so, by Pascal's theorem, the points B_1, B_6 and C are collinear. That is, the intersection points of the opposite sides of the hexagon $B_1B_2B_3B_4B_5B_6$ are collinear. By the converse Pascal theorem, this hexagon is inscribed.

A similar argument shows that the hexagon $B_2B_3B_4B_5B_6B_7$ is inscribed. But the two hexagons share five vertices, hence they are inscribed in the same conic. Likewise, B_8 lies on this conic as well.

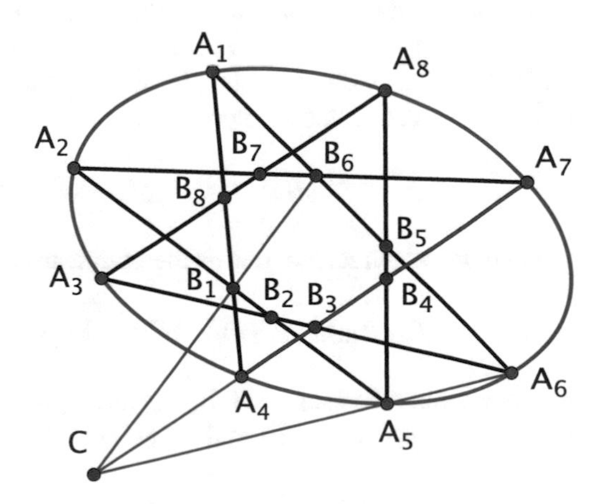

Fig. 9.16 Proof of Theorem 4.4 (i)

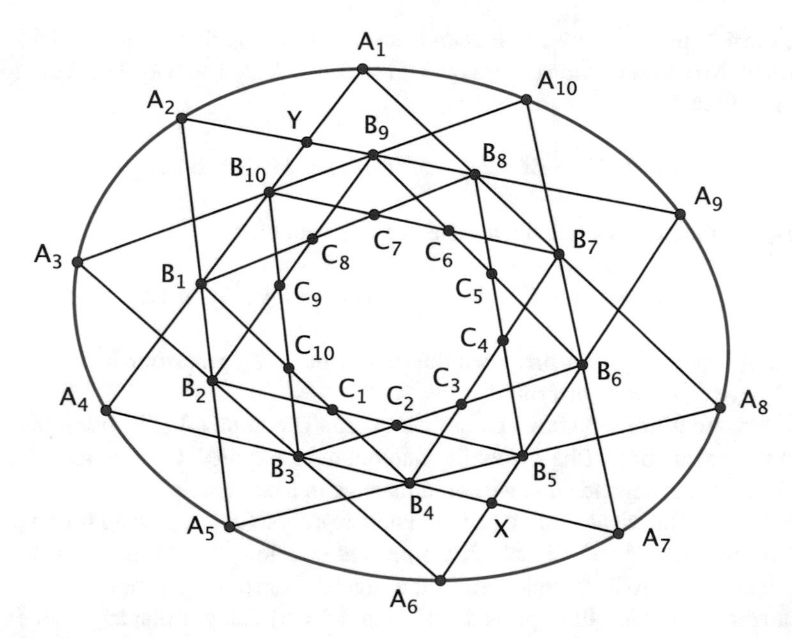

Fig. 9.17 Proof of Theorem 4.4 (ii)

Now, to the proof of Theorem 4.4 (ii), see Fig. 9.17.

Consider the inscribed hexagon $A_3A_6A_9A_{10}A_7A_4$. By Pascal's theorem, the points

$$(A_3A_6) \cap (A_{10}A_7),\ (A_6A_9) \cap (A_7A_4),\ (A_9A_{10}) \cap (A_4A_3)$$

are collinear. Hence the triangles $A_3B_3A_4$ and $A_{10}A_9B_6$ are perspective. By the Desargues theorem, the points A_4, A_9, $(B_3B_6) \cap (A_3A_{10})$ are collinear.

It follows that the triangles $B_9B_{10}Y$ and B_3B_6X are perspective. By the Desargues theorem, the points X, Y, $(B_6B_9) \cap (B_3B_{10})$ are collinear. The same argument, with all indices shifted by five, implies that the points X, Y, $(B_1B_4) \cap (B_8B_5)$ are collinear as well. Hence the points

$$(B_6B_9) \cap (B_3B_{10}),\ (B_1B_4) \cap (B_8B_5),\ \text{and } X$$

are collinear. Reinterpret this as the collinearity of

$$(C_{10}B_{10}) \cap (C_5B_9),\ (C_{10}B_4) \cap (C_5B_5),\ (B_3B_4) \cap (B_5B_6).$$

It follows that the triangles $B_3B_4C_{10}$ and $B_5B_6C_5$ are perspective. By the Desargues theorem, the points B_4, B_5 and $(C_{10}C_5) \cap (C_2C_3)$ are collinear. That is, the points

$$(C_{10}C_5) \cap (C_2C_3),\ (C_{10}C_1) \cap (C_3C_4),\ (C_1C_2) \cap (C_4C_5)$$

are collinear, and by the converse Pascal theorem, the points

$$C_{10}, C_1, C_2, C_3, C_4, C_5$$

lie on a conic. The rest of the argument is the same as in the previous proof. □

One can add to Theorems 4.1–4.4 a statement about pentagons. Consider the following facts:

(1) every pentagon is inscribed in a conic and circumscribed about a conic;
(2) every pentagon is projectively equivalent to its dual;
(3) the pentagram map sends every pentagon to a projectively-equivalent one.

Therefore one may add the following theorem to our list: *for a pentagon P, one has $P \sim T_2(P)$.*

The following result of Schwartz [18, 40, 42] also has a similar flavor.

Theorem 4.5 *If P is a $4n$-gon inscribed into a degenerate conic (that is, a pair of lines) then*

$$(T_1 T_2 T_1 T_2 \dots T_1)(P) \qquad (4n - 3 \text{ terms})$$

is also inscribed into a degenerate conic.

One wonders whether there is a unifying theme here. A possibly relevant reference is [19].

9.5 Poncelet Grid, String Construction, and Billiards in Ellipses

A Poncelet polygon is a polygon that is inscribed into an ellipse Γ and circumscribed about an ellipse γ. Let $L_1, \dots, L_n$ be the lines containing the sides of a Poncelet n-gon, enumerated in such a way that their tangency points with γ are in the cyclic order. The *Poncelet grid* is the collection of $n(n+1)/2$ points $L_i \cap L_j$, where $L_i \cap L_i$ is the tangency point of the line L_i with γ. To simplify the exposition, assume that n is odd (for even n, the formulations are slightly different).

One can partition the Poncelet grid in two ways. Define the sets

$$P_k = \cup_{i-j=k} \ell_i \cap \ell_j, \quad Q_k = \cup_{i+j=k} \ell_i \cap \ell_j,$$

where the indices are understood mod n. There are $(n+1)/2$ sets P_k, each containing n points, and n sets Q_k, each containing $(n+1)/2$ points. The sets P_k are called concentric, and the sets Q_k are called radial, see Fig. 9.18.

The following theorem is proved in [41].

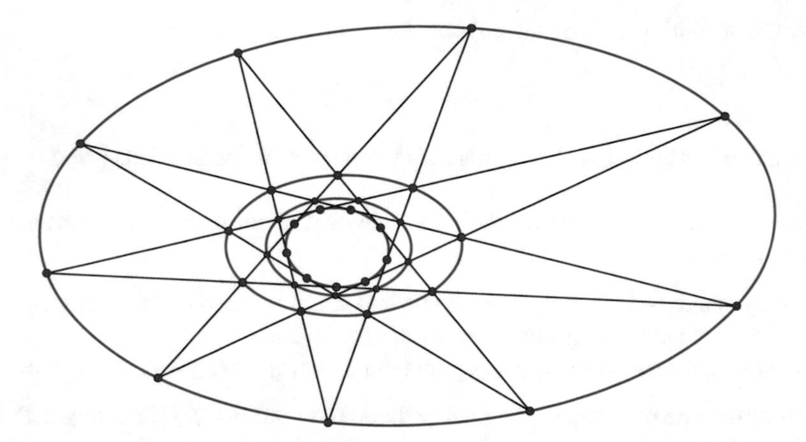

Fig. 9.18 Poncelet grid, $n = 9$: shown are the concentric sets P_0, P_2, P_3, and P_4 that lie on four ellipses

Theorem 5.1

(i) The concentric sets lie on nested ellipses, and the radial sets lie on disjoint hyperbolas.
(ii) The complexified versions of these conics have four common tangent lines.
(iii) All the concentric sets are projectively equivalent to each other, and so are all the radial sets.[1]

In this section, following [28], we prove this projective theorem using Euclidean geometry, namely, the billiard properties of conics. As a by-product of this approach, we establish the Poncelet theorem and prove the theorem of Reye and Chasles on inscribed circles. See [13, 15] for general information about the Poncelet theorem, and [27, 47, 48] for the theory of billiards.

The reduction to billiards goes as follows. Any pair of nested ellipses $\gamma \subset \Gamma$ can be taken to a pair of confocal ellipses by a suitable projective transformation. This transformation takes a Poncelet polygon to a periodic billiard trajectory in Γ.

The billiard inside a convex domain with smooth boundary is a transformation of the space of oriented lines (rays of light) that intersect the domain: an incoming billiard trajectory hits the boundary (a mirror) and optically reflects so that the angle of incidence equals the angle of reflection.

The space of oriented lines has an area form, preserved by the optical reflections (independently of the shape of the mirror). Choose an origin, and introduce coordinates (α, p) on the space of rays: α is the direction of the ray, and p is its signed distance to the origin. Then the invariant area form is $\omega = d\alpha \wedge dp$.

[1] See also a recent paper [50] for an extension of Kasner's theorem from pentagons to Poncelet polygons.

Fig. 9.19 String construction

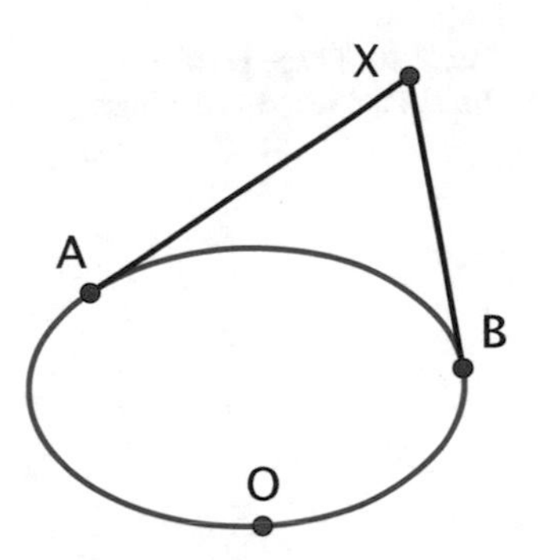

A *caustic* of a billiard is a curve γ with the property that if a segment of a billiard trajectory is tangent to γ, then so is each reflected segment.

There is no general method of describing caustics of a given billiard curve,[2] but the converse problem, to reconstruct a billiard table Γ from its caustic γ, has a simple solution given by the following *string construction*: wrap a non-stretchable closed string around γ, pull it tight, and move the farthest point around γ; the trajectory of this point is the billiard curve Γ. This construction yields a 1-parameter family of billiard tables sharing the caustic γ: the parameter is the length of the string.

The reason is as follows, see Fig. 9.19. For a point X outside of the oval γ, consider two functions:

$$f(X) = |XA| + |\breve{AO}|, \ g(X) = |XB| + |\breve{BO}| \, .$$

The gradients of these functions are the unit vectors along the lines AX and BX, respectively. It follows that these two lines make equal angles with the level curves of the functions $f+g$ and $f-g$, and that these level curves are orthogonal to each other. In particular, the level curves of $f+g$ are the billiard tables for which γ is a caustic.

Note that the function $f+g$ does not depend on the choice of the auxiliary point O, whereas the function $f-g$ is defined up to an additive constant, so its level curves are well defined.

Here is a summary of the billiard properties of conics. The interior of an ellipse is foliated by confocal ellipses: these are the caustics of the billiard inside an ellipse. Thus one has Graves's theorem: *wrapping a closed non-stretchable string around an ellipse yields a confocal ellipse*.

The space of rays A that intersect an ellipse is topologically a cylinder, and the billiard system inside the ellipse is an area-preserving transformation $T : A \to A$.

---

[2]The existence of caustics for strictly convex and sufficiently smooth billiard curves is proved in the framework of the KAM theory.

Fig. 9.20 Phase portrait of the billiard map in an ellipse

The cylinder is foliated by the invariant curves of the map T consisting of the rays tangent to confocal conics, see Fig. 9.20.

The curves that go around the cylinder correspond to the rays that are tangent to confocal ellipses, and the curves that form 'the eyes' to the rays that are tangent to confocal hyperbolas. A singular curve consists of the rays through the foci, and the two dots to the 2-periodic back and forth orbit along the minor axis of the ellipse.

One can choose a cyclic parameter, say, x modulo 1, on each invariant circle, such that the map T becomes a shift $x \mapsto x + c$, where the constant c depends on the invariant curve. Here is this construction (a particular case of the Arnold-Liouville theorem in the theory of integrable systems).

Choose a function H whose level curves are the invariant curves that foliate A, and consider its Hamiltonian vector field sgrad H with respect to the area form ω. This vector field is tangent to the invariant curves, and the desired coordinate x on these curves is the one in which sgrad H is a constant vector field d/dx. Changing H scales the coordinate x on each invariant curve and, normalizing the 'length' of the invariant curves to 1, fixes x uniquely up to an additive constant. In other words, the 1-form dx is well defined on each invariant curve.

The billiard map T preserves the area form and the invariant curves, therefore its restriction to each curve preserves the measure dx, hence, is a shift $x \mapsto x + c$.

An immediate consequence is the Poncelet Porism: if a billiard trajectory in an ellipse closes up after n reflections, then $nc \equiv 0$ mod 1, and hence all trajectories with the same caustic close up after n reflections.[3]

Note that the invariant measure dx on the invariant curves does not depend on the choice of the billiard ellipse from a confocal family: the confocal ellipses share their caustics. This implies that the billiard transformations with respect to two confocal ellipses commute: restricted to a common caustic, both are shifts in the same coordinate system. This statement can be considered as a configuration theorem; see Fig. 9.21.

[3] See [44] for a curious property of the centroids of Poncelet polygons.

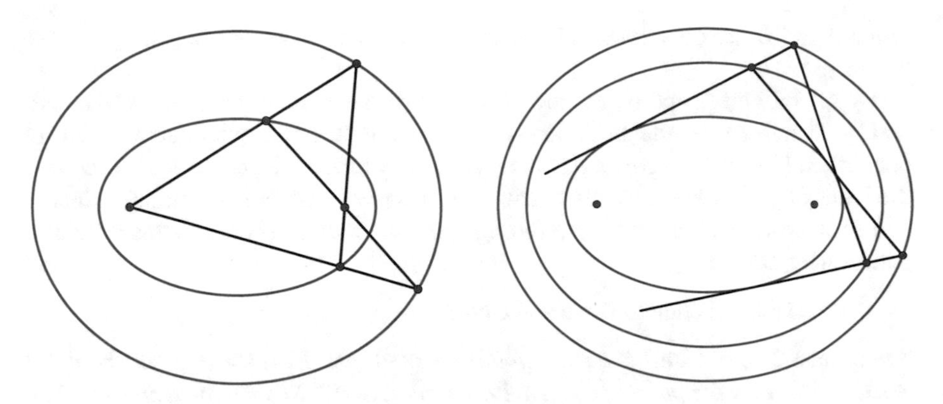

Fig. 9.21 Left: the billiard reflections of the rays from a focus in two confocal ellipses commute. Right: the general case

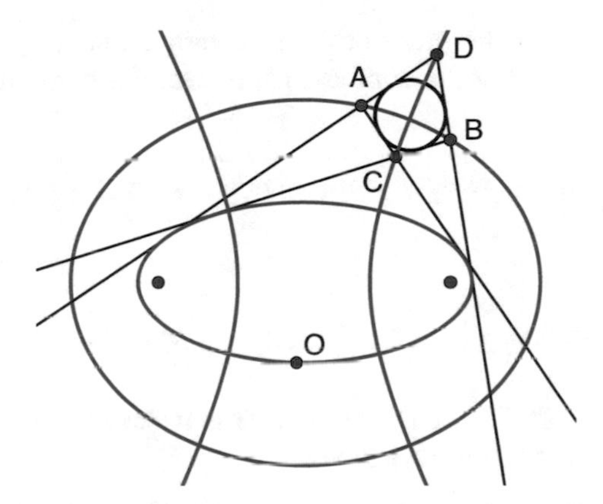

Fig. 9.22 Two pairs of tangents from an ellipse to a confocal ellipse

To summarize, an ellipse is a billiard caustic for the confocal family of ellipses. It carries a coordinate x, defined up to an additive constant, in which the billiard reflection in confocal ellipses is given by $x \mapsto x + c$. We refer to the coordinate x as the canonical coordinate.

Consider an ellipse γ, and let x be the canonical coordinate on it. Define coordinates in the exterior of the ellipse: the coordinates of a point X outside of γ are the coordinates x_1 and x_2 of the tangency points of the tangent lines from X to γ. Let us call (x_1, x_2) the string coordinates of point X. The confocal ellipses are given by the equations $x_2 - x_1 = \text{const}$.

Lemma 5.1 *The confocal hyperbolas have the equations* $x_2 + x_1 = const.$

Proof Consider Fig. 9.22. Let the canonical coordinates of the tangency points on the inner ellipse, from left to right, be x_1, x_2, x_3, x_4, so that the string coordinates are as follows:

$$A(x_1, x_3),\ B(x_2, x_4),\ C(x_2, x_3),\ D(x_1, x_4).$$

Since A and B are on a confocal ellipse, $x_4 - x_2 = x_3 - x_1$, and hence $x_2 + x_3 = x_1 + x_4$.

By the billiard property, the arc of an ellipse AB bisects the angles CAD and CBD. Therefore, in the limit $A \to B$, the infinitesimal quadrilateral $ABCD$ becomes a kite: the diagonal AB is its axis of symmetry. Hence $AB \perp CD$, and the locus of points given by the equation $x_1 + x_4 = \text{const}$ and containing points C and D is orthogonal to the ellipse through points A and B. Therefore this locus is a confocal hyperbola. □

The next result is due to Reye and Chasles.

Theorem 5.2 *Let A and B be two points on an ellipse. Consider the quadrilateral ABCD, made by the pairs of tangent lines from A and B to a confocal ellipse. Then its other vertices, C and D, lie on a confocal hyperbola, and the quadrilateral is circumscribed about a circle, see Fig. 9.22.*

Proof In the notation of the proof of the preceding lemma, $x_2 + x_3 = x_1 + x_4$, hence points C and D lie on a confocal hyperbola. Furthermore, in terms of the string construction,

$$f(A) + g(A) = f(B) + g(B), \quad f(C) - g(C) = f(D) - g(D),$$

hence

$$f(D) - f(A) - g(A) + g(C) + f(B) - f(C) - g(D) + g(B) = 0,$$

or $|AD| - |AC| + |BC| - |BD| = 0$. This is necessary and sufficient for the quadrilateral $ABCD$ to be circumscribed. □

Now, consider a Poncelet n-gon, an n-periodic billiard trajectory in the ellipse Γ. One can choose the canonical coordinates of the tangency points of the sides of the polygon with the confocal ellipse γ to be

$$0, \ \frac{1}{n}, \ \frac{2}{n}, \ \ldots, \ \frac{n-1}{n}.$$

Then the string coordinates of the points of the concentric set P_k are

$$\left(0, \frac{k}{n}\right), \left(\frac{1}{n}, \frac{k+1}{n}\right), \left(\frac{2}{n}, \frac{k+2}{n}\right), \ldots,$$

that is, their difference equals k/n, a constant. It follows that P_k lies on a confocal ellipse. Likewise for the radial sets Q_k, proving the first claim of Theorem 5.1.

Theorem 5.2 implies that each quadrilateral of the Poncelet grid is circumscribed, see Fig. 9.23. We refer to [3] for circle patterns related to conics.

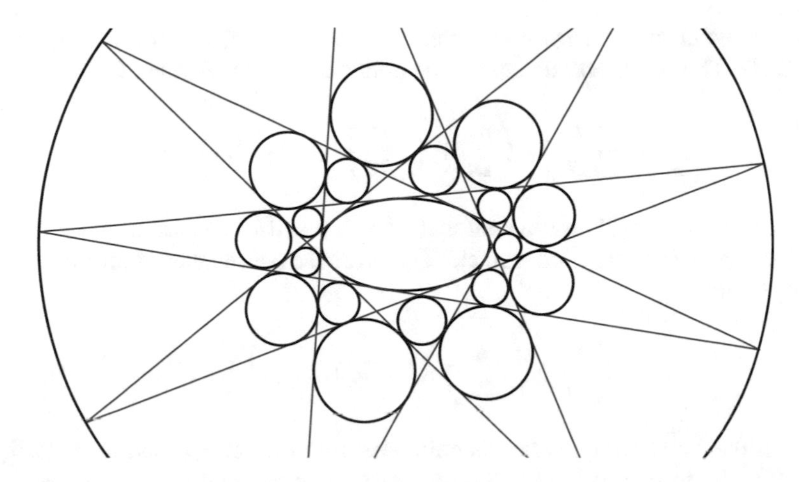

Fig. 9.23 Poncelet grid of circles

Next, we prove the second claim of Theorem 5.1. The confocal family of conics is given by the equation

$$\frac{x_1^2}{a_1^2+\lambda}+\frac{x_2^2}{a_2^2+\lambda}=1,$$

where λ is a parameter. Its dual family is the pencil

$$(a_1^2+\lambda)x_1^2+(a_2^2+\lambda)x_2^2=1$$

that consists of the conics that share four points, possibly complex. Hence the confocal family consists of the conics that share four tangent lines, also possibly complex.

To prove the last claim of Theorem 5.1, we need the following classical result. Let γ and Γ be confocal ellipses, centered at the origin and symmetric with respect to the coordinate axes, and let A be the diagonal matrix with positive entries that takes γ to Γ.

Lemma 5.2 (Ivory) *For every point $P \in \gamma$, the points P and $A(P)$ lie on a confocal hyperbola.*

Let us show that the linear map A takes P_k to P_m or to its centrally symmetric set; the argument for the radial sets is similar.

It is convenient to change the string coordinates (x, y) to $u = (x + y)/2, v = (y - x)/2$. The (u, v)-coordinates of the points of the sets P_k and P_m are

$$\left(\frac{k}{2n} + \frac{j}{n}, \frac{k}{2n}\right), \left(\frac{m}{2n} + \frac{j}{n}, \frac{m}{2n}\right) \quad (j = 0, 1, \ldots, n-1).$$

We know that P_k and P_m lie on confocal ellipses γ and Γ. According to Lemma 5.2, the map A preserves the u-coordinate. Therefore the coordinates of the points of the set $A(P_k)$ are

$$\left(\frac{k}{2n} + \frac{j}{n}, \frac{m}{2n}\right) \quad (j = 0, 1, \ldots, n-1).$$

If m has the same parity as k, this coincides with the set P_m, and if the parity of m is opposite to that of k, then this set is centrally symmetric to the set P_m. This completes the proof.[4]

9.6 Identities in the Lie Algebras of Motions

It is well known that the altitudes of a Euclidean triangle are concurrent. It is a lesser known fact that an analogous theorem holds in the spherical and hyperbolic geometries.

In this section, we describe Arnold's observation [4] that these results have interpretations as the Jacobi identity in the Lie algebras of motions of the respective geometries of constant positive or negative curvatures; see also [24, 45]. Following [2, 51], we shall discuss the relation of other classical configuration theorems with identities in these Lie algebras.

In spherical geometry, one has the duality between points and lines that assigns the pole to an equator. There are two poles of a great circle; one can make the choice of the pole unique by considering oriented great circle, or by factorizing by the antipodal involution, that is, by replacing the sphere by the elliptic plane.

This spherical duality can be expressed in terms of the cross-product: if A and B are two vectors in $\mathbb{R}^3$ representing points in the elliptic plane, then the vector $A \times B$ represents the point dual to the line AB. In the following argument, we do not distinguish between points and their dual lines.

Given a spherical triangle ABC, the altitude dropped from C to AB is the great circle connecting the pole of the great circle AB and point C. Using the identification of points and lines, and cross-product, this altitude is represented by the vector $(A \times B) \times C$, see Fig. 9.24.

[4] See also [25] for another theorem of Ivory and its relation to billiards in conics.

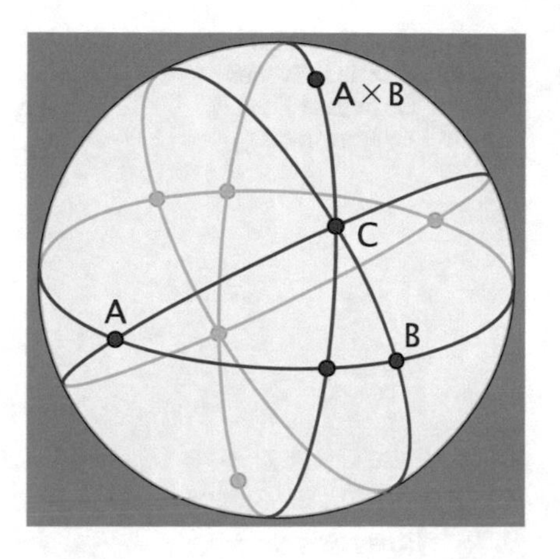

Fig. 9.24 An altitude of a spherical triangle

Two other altitudes are given by similar cross-products, and the statement that the three great circles are concurrent is equivalent to linear dependence of the these three cross-products. But

$$(A \times B) \times C + (B \times C) \times A + (C \times A) \times B = 0,$$

the Jacobi identity for cross-product, hence the three altitudes are concurrent.

Note that the Lie algebra $(\mathbb{R}^3, \times)$ is $so(3)$, the algebra of motions of the unit sphere. Thus the Jacobi identity in $so(3)$ implies the the existence of the spherical orthocenter.

A similar, albeit somewhat more involved, argument works in the hyperbolic plane, with the Lie algebra of motions $sl(2, \mathbb{R})$ replacing $so(3)$. Note that these algebras are real forms of the complex Lie algebra $sl(2, \mathbb{C})$.

Interestingly, the Euclidean theorem on concurrence of the three altitudes of a triangle does not seem to admit an interpretation as the Jacobi identity of the Lie algebra of motions of the plane.

Developing these ideas, Tomihisa [51] discovered the following identity.

Theorem 6.1 *For every quintuple of elements of the Lie algebra $sl(2)$ (with real or complex coefficients), one has*

$$[F_1, [[F_2, F_3], [F_4, F_5]]] + [F_3, [[F_2, F_5], [F_4, F_1]]] + [F_5, [[F_2, F_1], [F_4, F_3]]] = 0.$$

Note that the indices 1, 3, 5 permute cyclically, while 2 and 4 are frozen.

As above, the Tomihisa identity can be interpreted as a configuration theorem: the Lie bracket corresponds to one of the two basic operations: connecting a pair of points by a line or intersecting a pair of lines at a point. See Fig. 9.25 for such an interpretation.

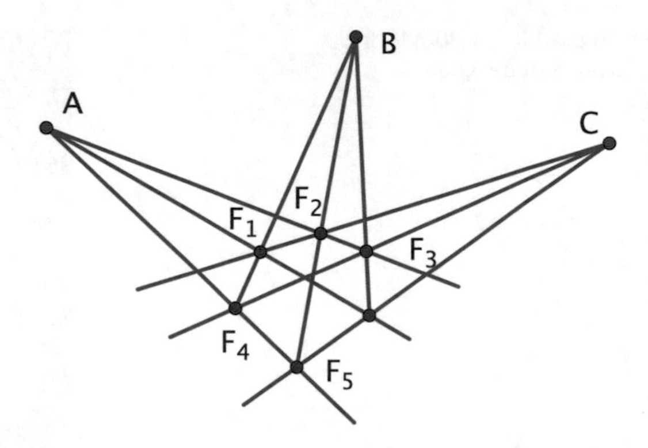

Fig. 9.25 The Tomihisa identity as the dual Pappus theorem: the lines AF_1, BF_3, and CF_5 are concurrent

9.7 Skewers

This section is based upon the recent paper [49]. The main idea is that planar projective configuration theorems have space analogs where points and lines in the projective plane are replaced by lines in space, and the two operations, connecting two points by a line and intersecting two lines at a point, are replaced by taking the common perpendicular of two lines.

The *skewer* of two lines in 3-dimensional space is their common perpendicular. We denote the skewer of lines a and b by $S(a, b)$. In Euclidean and hyperbolic spaces, a generic pair of lines has a unique skewer; in the spherical geometry, a generic pair of lines (great circles) has two skewers, similarly to a great circle on S^2 having two poles. We always assume that the lines involved in the formulations of the theorems are in general position.

Here is the 'skewer translation' of the Pappus theorem, as depicted in Fig. 9.1:

Theorem 7.1 *Let a_1, a_2, a_3 be a triple of lines with a common skewer, and let b_1, b_2, b_3 be another triple of lines with a common skewer. Then the lines*

$$S(S(a_1, b_2), S(a_2, b_1)), \ S(S(a_1, b_3), S(a_3, b_1)), \text{ and } S(S(a_2, b_3), S(a_3, b_2))$$

share a skewer.

This theorem, as well as in the following ones, holds in $\mathbb{R}^3$, S^3 and H^3.

And here is the skewer version of the Desargues theorem, as depicted in Fig. 9.2:

Theorem 7.2 *Let a_1, a_2, a_3 and b_1, b_2, b_3 be two triples of lines such that the lines $S(a_1, b_1)$, $S(a_2, b_2)$ and $S(a_3, b_3)$ share a skewer. Then the lines*

$$S(S(a_1, a_2), S(b_1, b_2)), \ S(S(a_1, a_3), S(b_1, b_3)), \text{ and } S(S(a_2, a_3), S(b_2, b_3))$$

also share a skewer.

The 'rules of translation' should be clear from these examples.

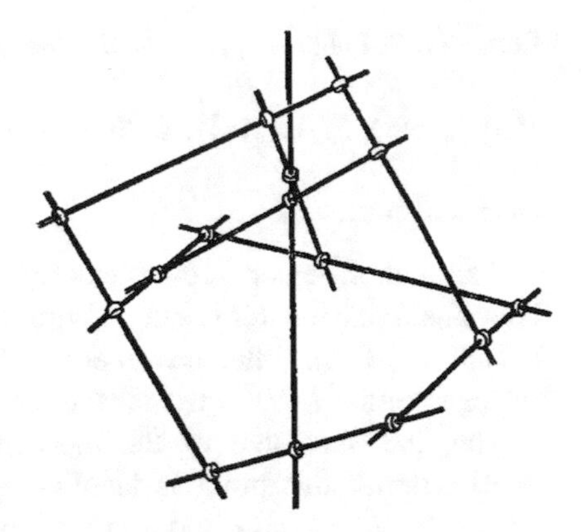

Fig. 9.26 Petersen-Morley configuration of ten lines

As a third example, consider a configuration theorem that involves polarity, namely, the theorem that the three altitudes of a triangle are concurrent that was discussed in Sect. 9.6. In its skewer version, one does not distinguish between polar dual objects, such as a great circle and its pole. This yields

Theorem 7.3 *Given three lines a, b, c, the lines*

$$S(S(a, b), c), \; S(S(b, c), a), \; \text{ and } \; S(S(c, a), b)$$

share a skewer.

This is the Petersen-Morley, also known as Hjelmslev-Morley, theorem [30]. An equivalent formulation: *the common normals of the opposite sides of a rectangular hexagon have a common normal.* See Fig. 9.26, borrowed from [32].

Denote the 2-parameter family of lines that meet a given line ℓ at right angle by $\mathcal{N}_\ell$. The sets $\mathcal{N}_\ell$ play the role of lines in the skewer versions of configuration theorems. Two-parameter families of lines in 3-space are called congruences.

Next we describe line analogs of circles. Let ℓ be an oriented line in 3-space (elliptic, Euclidean, or hyperbolic). Let G_ℓ be the 2-dimensional subgroup of the group of orientation-preserving isometries that preserve ℓ. The orbit $G_\ell(m)$ of an oriented line m is called the *axial congruence* with ℓ as axis (an analog of the center of a circle).

In $\mathbb{R}^3$, the lines of an axial congruence with axis ℓ are at equal distances from ℓ and make equal angles with it. In the hyperbolic space, one can define the complex distance between oriented lines, see [29]. The complex distance between the lines of an axial congruence and its axis is constant.

Axial congruences share the basic properties of circles: if two generic axial congruences share a line, then they share a unique other line; and three generic oriented lines belong to a unique axial congruence.

The next result is a skewer analog of the Pascal theorem, see Fig. 9.3, in the particular case when the conic is a circle.

Theorem 7.4 *Let $A_1, \ldots, A_6$ be lines from an axial congruence. Then*

$$S(S(A_1, A_2), S(A_4, A_5)),\ S(S(A_2, A_3), S(A_5, A_6)),\ \text{and}\ S(S(A_3, A_4), S(A_6, A_1))$$

share a skewer.

As another, lesser known, example, consider the Clifford's Chain of Circles. This chain of theorems starts with a collection of concurrent circles labelled $1, 2, 3, \ldots, n$. The intersection point of the circles i and j is labelled ij. The circle through points ij, jk and ki is labelled ijk.

The first statement of the theorem is that the circles ijk, jkl, kli and lij share a point; this point is labelled $ijkl$. The next statement is that the points $ijkl, jklm, klmi, lmij$ and $mijk$ are concyclic; this circle is labelled $ijklm$. And so on, with the alternating claims of being concurrent and concyclic; see [11, 32], and Fig. 9.27 where the initial circles are represented by lines (circles of infinite radius sharing a point at infinity).

The next theorem, in the case of $\mathbb{R}^3$, is due to Richmond [37].

Theorem 7.5

1) Consider axial congruences $\mathcal{C}_i$, $i = 1, 2, 3, 4$, sharing a line. For each pair of indices $i, j \in \{1, 2, 3, 4\}$, denote by ℓ_{ij} the line shared by $\mathcal{C}_i$ and $\mathcal{C}_j$. For each triple of indices $i, j, k \in \{1, 2, 3, 4\}$, denote by $\mathcal{C}_{ijk}$ the axial congruence

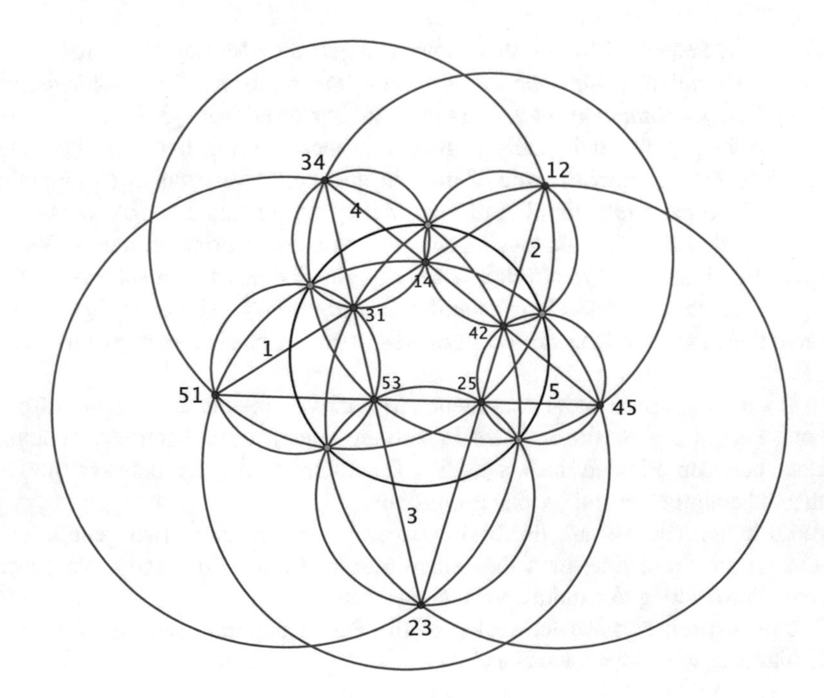

Fig. 9.27 Clifford's Chain of Circles ($n = 5$)

containing the lines $\ell_{ij}, \ell_{jk}, \ell_{ki}$. *Then the congruences* $\mathcal{C}_{123}, \mathcal{C}_{234}, \mathcal{C}_{341}$ *and* $\mathcal{C}_{412}$ *share a line.*

2) *Consider axial congruences* $\mathcal{C}_i$, $i = 1, 2, 3, 4, 5$, *sharing a line. Each four of the indices determine a line, as described in the previous statement of the theorem. One obtains five lines, and they all belong to an axial congruence.*
3) *Consider axial congruences* $\mathcal{C}_i$, $i = 1, 2, 3, 4, 5, 6$, *sharing a line. Each five of them determine an axial congruence, as described in the previous statement of the theorem. One obtains six axial congruences, and they all share a line. And so on...*

Next one would like to define line analogs of conics. A first step in this direction is taken in [49], but much more work is needed. In particular, one would like to have skewer analogs of various configuration theorems involving conics, including the Pascal theorem and the whole hexagrammum mysticum, the Poncelet Porism, and the theorems described in Sect. 9.4. As of now, this is an open problem.

Now we outline two approaches to proofs of the above theorems and the skewer versions of other planar configuration theorems. The first approach is by way of the spherical geometry, and the second via the hyperbolic geometry. Either approach implies the results in all three classical geometries by 'analytic continuation'. This analytic continuation principle is well known in geometry; see, e.g., [1, 35] where it is discussed in detail.

9.7.1 Elliptic Approach

The space of oriented great circles in S^3, or lines in the elliptic space $\mathbb{RP}^3$, is the Grassmannian $G(2, 4)$ of oriented 2-dimensional subspaces in $\mathbb{R}^4$. Below we collect pertinent facts concerning this Grassmannian.

To every oriented line ℓ in $\mathbb{RP}^3$ there corresponds its dual oriented line ℓ^*: the respective oriented planes in $\mathbb{R}^4$ are the orthogonal complements of each other. The dual lines are equidistant and they have infinitely many skewers.

The Grassmannian is a product of two spheres: $G(2, 4) = S^2_- \times S^2_+$. This provides an identification of an oriented line in $\mathbb{RP}^3$ with a pair of points of the unit sphere S^2: $\ell \leftrightarrow (\ell_-, \ell_+)$. The antipodal involutions of the spheres S^2_- and S^2_+ generate the action of the Klein group $\mathbb{Z}_2 \times \mathbb{Z}_2$ on the space of oriented lines generated by reversing the orientation of a line and by taking the dual line.

Two lines ℓ and m intersect at right angle if and only if $d(\ell_-, m_-) = d(\ell_+, m_+) = \pi/2$, where d denotes the spherical distance in S^2. A line n is a skewer of lines ℓ and m if and only if n_- is a pole of the great circle $\ell_- m_-$, and n_+ is a pole of the great circle $\ell_+ m_+$.

The set of lines that intersect ℓ at right angle coincides with the set of lines that intersect ℓ and ℓ^*. A generic pair of lines has exactly two skewers (four, if orientation is taken into account), and they are dual to each other.

It follows that a configuration involving lines in elliptic space and their skewers can be identified with a pair of configurations on the spheres S^2_- and S^2_+. Under this identification, the great circles of these spheres are not distinguished from their poles, just like in the proof described in Sect. 9.6. That is, the operation of taking the skewer of two lines is represented, on both spheres, by the cross-product.

In this way, a configuration of lines in space becomes the direct product of the corresponding planar configurations. For example, the Petersen-Morley Theorem 7.3 splits into two statements that the altitudes of triangles, on the spheres S^2_- and S^2_+, are concurrent.

9.7.2 *Hyperbolic Approach*

In a nutshell, a skewer configuration theorem in 3-dimensional hyperbolic space is a complexification of a configuration theorem in the hyperbolic plane. We follow the ideas of Morley [31, 32], Coxeter [12], and Arnold [4].

Consider the hyperbolic space in the upper halfspace model. The isometry group is $\mathrm{SL}(2, \mathbb{C})$, and the sphere at infinity (the celestial sphere of [31]) is the Riemann sphere $\mathbb{CP}^1$.

A line in H^3 intersects the sphere at infinity at two points, hence the space of (non-oriented) lines is the configuration space of unordered pairs of points. As we mentioned in Sect. 9.3, $S^2(\mathbb{CP}^1) = \mathbb{CP}^2$, namely, to a pair of points in the projective line one assigns the binary quadratic form having zeros at these points:

$$(a_1 : b_1, a_2 : b_2) \longmapsto (a_1 y - b_1 x)(a_2 y - b_2 x).$$

Thus a line in H^3 can be though of as a complex binary quadratic form, up to a factor.

The space of binary quadratic forms $ax^2 + 2bxy + cy^2$ has the discriminant quadratic form $\Delta = ac - b^2$ and the respective bilinear form. The equation $\Delta = 0$ defines the diagonal of $S^2(\mathbb{CP}^1)$; this is a conic in $\mathbb{CP}^2$ that does not correspond to lines in H^3.

The next result is contained in §52 of [32].

Lemma 7.1 *Two lines in H^3 intersect at right angle if and only if the respective binary quadratic forms $f_i = a_i x^2 + 2b_i xy + c_i y^2$, $i = 1, 2$, are orthogonal with respect to Δ:*

$$a_1 c_2 - 2b_1 b_2 + a_2 c_1 = 0. \tag{9.2}$$

If two lines correspond to binary quadratic forms $f_i = a_i x^2 + 2b_i xy + c_i y^2$, $i = 1, 2$, then their skewer corresponds to the Poisson bracket (the Jacobian)

$$\{f_1, f_2\} = (a_1 b_2 - a_2 b_1)x^2 + (a_1 c_2 - a_2 c_1)xy + (b_1 c_2 - b_2 c_1)y^2.$$

If $(a_1 : b_1 : c_1)$ and $(a_2 : b_2 : c_2)$ are homogeneous coordinates in the projective plane and the dual projective plane, then (9.2) describes the incidence relation between points and lines. In particular, the set of lines in H^3 that meet a fixed line at right angle corresponds to a line in $\mathbb{CP}^2$.

Suppose a configuration theorem involving polarity is given in $\mathbb{RP}^2$. The projective plane with a conic provides the projective model of the hyperbolic plane, so the configuration is realized in H^2. Consider the complexification, the respective configuration theorem in $\mathbb{CP}^2$ with the polarity induced by Δ. According to Lemma 7.1, this yields a configuration of lines in H^3 such that the pairs of incident points and lines correspond to pairs of lines intersecting at right angle.

Remark 7.2 (On Lie algebras) From the point of view of the identities in Lie algebras, discussed in Sect. 9.6, the relation between configuration theorems in the hyperbolic plane and the hyperbolic space is the relation between $sl(2,\mathbb{R})$ and $sl(2,\mathbb{C})$: an identity in the former implies the same identity in the latter.

As to the Lie algebras in space, in the elliptic case, the Lie algebra of motions is $so(4) = so(3) \oplus so(3)$, and in the hyperbolic case, it is $sl(2,\mathbb{C})$. Accordingly, an elliptic skewer configuration splits into two configurations in S^2, and a hyperbolic skewer configuration is obtained from a configuration in H^2 by complexification.

We finish the section by discussing two results concerning lines in 3-space that do not follow the above described general pattern. The first of them is the skewer version of the Sylvester Problem.

Given a finite set S of points in the plane, assume that the line through every pair of points in S contains at least one other point of S. Sylvester asked in 1893 whether S necessarily consists of collinear points. See [8] for the history of this problem and its generalizations.

In $\mathbb{RP}^2$, the Sylvester Problem, along with its dual, has an affirmative answer (the Sylvester-Galai theorem), but in $\mathbb{CP}^2$ one has a counter-example: the 9 inflection points of a cubic curve (of which at most three can be real, according to a theorem of Klein), connected by 12 lines.

The skewer version of the Sylvester Problem concerns a finite collection of pairwise skew lines in space such that the skewer of any pair intersects at least one other line at right angle. The question is whether a collection of lines with this skewer Sylvester property necessarily consists of the lines that intersect some line at right angle.

Theorem 7.6 *The skewer version of the Sylvester-Galai theorem holds in the elliptic and Euclidean geometries, but fails in the hyperbolic geometry.*

Proof In the elliptic case, we argue as in the above described elliptic proof. A collection of lines becomes two collections of points, in $\mathbb{RP}^2_-$ and in $\mathbb{RP}^2_+$, and the skewer Sylvester property implies that each of these sets has the property that the line through every pair of points contains another point, so one applies the Sylvester-Galai theorem on each sphere.

In the hyperbolic case, we argue as in the hyperbolic proof. Let $a_1, \dots, a_9$ be the nine inflection points of a cubic curve in $\mathbb{CP}^2$, and let $b_1, \dots, b_{12}$ be the respective lines. Let $b_1^*, \dots, b_{12}^*$ be the polar dual points. Then the points a_i correspond to nine lines in H^3, and the points b_j^* to their skewers. We obtain a collection of nine lines that has the skewer Sylvester property but does not possess a common skewer.

In the intermediate case of $\mathbb{R}^3$, the argument is due to Timorin (private communication).

Let us add to $\mathbb{R}^3$ the plane at infinity H; the points of H are the directions of lines in space. One has a polarity in H that assigns to a direction the set of the orthogonal directions, a line in H.

Therefore, if three lines in $\mathbb{R}^3$ share a skewer, then their intersections with the plane H are collinear. Let $L_1, \dots, L_n$ be a collection of lines with the skewer Sylvester property. Then, by the Sylvester-Galai theorem in H, the points $L_1 \cap H, \dots, L_n \cap H$ are collinear. This means that the lines $L_1, \dots, L_n$ lie in parallel planes, say, the horizontal ones.

Consider the vertical projection of these lines. We obtain a finite collection of non-parallel lines in the plane such that through the intersection point of any two there passes at least one other line. By the dual Sylvester-Galai theorem, all these lines are concurrent. Therefore the respective horizontal lines in $\mathbb{R}^3$ share a vertical skewer. □

The second result is a different skewer version of the Pappus theorem.

Theorem 7.7 *Let ℓ and m be a pair of skew lines. Choose a triple of points A_1, A_2, A_3 on ℓ and a triple of points B_1, B_2, B_3 on m. Then the lines*

$$S((A_1B_2), (A_2B_1)), \ S((A_2B_3), (A_3B_2)), \text{ and } S((A_3B_1), (A_1B_3))$$

share a skewer.

We proved this result, in the hyperbolic case, by a brute force calculation using the approach to hyperbolic geometry, developed in [14]; see [49] for details. It is not clear whether this theorem is a part of a general pattern.

Let us close with an invitation to the reader to mull over the skewer versions of other constructions of planar projective geometry. For example, one can define the skewer pentagram map that acts on cyclically ordered tuples of lines in space:

$$\{L_1, L_2, \dots\} \mapsto \{S(S(L_1, L_3), S(L_2, L_4)), S(S(L_2, L_4), S(L_3, L_5)), \dots\}$$

Is this map completely integrable?

Acknowledgements I am grateful to Schwartz for numerous stimulating discussions and to Hooper for an explanation of his work. I was supported by NSF grant DMS-1510055. This article was written during my stay at ICERM; it is a pleasure to thank the Institute for its inspiring and friendly atmosphere.

References

1. N. A'Campo, A. Papadopoulos. Transitional geometry. in *Sophus Lie and Felix Klein: The Erlangen Program and its Impact in Mathematics and Physics*, 217–235. European Math. Soc., Zürich, 2015.
2. F. Aicardi. Projective geometry from Poisson algebras. *J. Geom. Phys.* **61** (2011), 1574–1586.
3. A. Akopyan, A. Bobenko. Incircular nets and confocal conics. *Trans. Amer. Math. Soc.* **370** (2018), 2825–2854.
4. V. Arnold. Lobachevsky triangle altitude theorem as the Jacobi identity in the Lie algebra of quadratic forms on symplectic plane. *J. Geom. Phys.* **53** (2005), 421–427.
5. M. Berger. *Geometry. I. II.* Springer-Verlag, Berlin, 1987.
6. M. Berger. *Geometry revealed. A Jacob's ladder to modern higher geometry.* Springer, Heidelberg, 2010.
7. A. Bobenko, Yu. Suris. *Discrete differential geometry. Integrable structure.* Amer. Math. Soc., Providence, RI, 2008.
8. P. Borwein, W. O. J. Moser. A survey of Sylvester's problem and its generalizations. *Aequationes Math.* **40** (1990), 111–135.
9. J. Conway, A. Ryba. The Pascal mysticum demystified. *Math. Intelligencer* **34** (2012), no. 3, 4–8.
10. J. Conway, A. Ryba. Extending the Pascal mysticum. *Math. Intelligencer* **35** (2013), no. 2, 44–51.
11. J. L. Coolidge. *A treatise on the circle and the sphere.* Chelsea Publ. Co., Bronx, N.Y., 1971.
12. H. S. M. Coxeter. The inversive plane and hyperbolic space. *Abh. Math. Sem. Univ. Hamburg* **29** (1966), 217–242.
13. V. Dragović, M. Radnović. *Poncelet porisms and beyond. Integrable billiards, hyperelliptic Jacobians and pencils of quadrics.* Birkhäuser/Springer, Basel, 2011.
14. W. Fenchel. *Elementary geometry in hyperbolic space.* Walter de Gruyter, Berlin, 1989.
15. L. Flatto. *Poncelet's theorem.* Amer. Math. Soc., Providence, RI, 2009.
16. D. Fuchs, S. Tabachnikov. *Mathematical omnibus. Thirty lectures on classic mathematics.* Amer. Math. Soc., Providence, RI, 2007.
17. M. Gekhtman, M. Shapiro, S. Tabachnikov, and A. Vainshtein, *Integrable cluster dynamics of directed networks and pentagram maps*, with appendix by A. Izosimov, Adv. Math. **300** (2016), 390–450.
18. M. Glick. The pentagram map and Y-patterns. *Adv. Math.* **227** (2011), 1019–1045.
19. M. Glick, P. Pylyavskyy. Y-meshes and generalized pentagram maps. *Proc. Lond. Math. Soc.* **112** (2016), 753–797.
20. B. Grünbaum. *Configurations of points and lines.* Amer. Math. Soc., Providence, RI, 2009.
21. T. Hatase. *Algebraic Pappus Curves.* Ph.D. Thesis, Oregon State University, 2011.
22. D. Hilbert, S. Cohn-Vossen. *Geometry and the imagination.* Chelsea Publishing Co, New York, N. Y., 1952.
23. P. Hooper. From Pappus' theorem to the twisted cubic. *Geom. Dedicata* **110** (2005), 103–134.
24. N. Ivanov. Arnol'd, the Jacobi identity, and orthocenters. *Amer. Math. Monthly* **118** (2011), 41–65.
25. I. Izmestiev, S. Tabachnikov. Ivory's Theorem revisited. *J. Integrable Syst.* **2** (2017), no. 1, xyx006, 36 pp.
26. F. Kissler. *A family of representations for the modular group.* Master Thesis, Heidelberg, 2016.
27. V. Kozlov and D. Treshchev. *Billiards. A Genetic Introduction to the Dynamics of Systems with Impacts.* Amer. Math. Soc., Providence, RI, 1991.
28. M. Levi, S. Tabachnikov. The Poncelet grid and billiards in ellipses. *Amer. Math. Monthly* **114** (2007), 895–908.
29. A. Marden. *Outer circles. An introduction to hyperbolic 3-manifolds.* Cambridge Univ. Press, Cambridge, 2007.

30. F. Morley. On a regular rectangular configuration of ten lines. *Proc. London Math. Soc.* s1–29 (1897), 670–673.
31. F. Morley. The Celestial Sphere. *Amer. J. Math.* **54** (1932), 276–278.
32. F. Morley, F. V. Morley. *Inversive geometry*. G. Bell & Sons, London, 1933.
33. V. Ovsienko, R. Schwartz, S. Tabachnikov, The pentagram map: a discrete integrable system, *Comm. Math. Phys.* **299** (2010), 409–446.
34. V. Ovsienko, R. Schwartz, and S. Tabachnikov, Liouville-Arnold integrability of the pentagram map on closed polygons, *Duke Math. J.* **162** (2013), 2149–2196.
35. *Strasbourg master class on geometry.* A. Papadopoulos ed. European Math. Soc., Zürich, 2012.
36. J. Richter-Gebert. *Perspectives on projective geometry. A guided tour through real and complex geometry.* Springer, Heidelberg, 2011.
37. J. F. Rigby. Pappus lines and Leisenring lines. *J. Geom.* **21** (1983), 108–117.
38. R. Schwartz. The pentagram map. *Experiment. Math.* **1** (1992), 71–81.
39. R. Schwartz. Pappus' theorem and the modular group. *Inst. Hautes Études Sci. Publ. Math.* **78** (1993), 187–206 (1994).
40. R. Schwartz. Desargues theorem, dynamics, and hyperplane arrangements. *Geom. Dedicata* **87** (2001), 261–283.
41. R. Schwartz. The Poncelet grid. *Adv. Geom.* **7** (2007), 157–175.
42. R. Schwartz. Discrete monodromy, pentagrams, and the method of condensation. *J. Fixed Point Theory Appl.* **3** (2008), 379–409.
43. R. Schwartz, S. Tabachnikov. Elementary surprises in projective geometry. *Math. Intelligencer* **32** (2010), no. 3, 31–34.
44. R. Schwartz, S. Tabachnikov. Centers of Mass of Poncelet Polygons, 200 Years After. *Math. Intelligencer* **38** (2016), no. 2, 29–34.
45. M. Skopenkov. Theorem about the altitudes of a triangle and the Jacobi identity (in Russian). *Matem. Prosv., Ser.* 3 **11** (2007), 79–89.
46. F. Soloviev, Integrability of the pentagram map, *Duke Math. J.* **162** (2013), 2815–2853.
47. S. Tabachnikov. *Billiards.* Panor. Synth. No. 1, SMF, 1995.
48. S. Tabachnikov. *Geometry and billiards.* Amer. Math. Soc., Providence, RI, 2005.
49. S. Tabachnikov. Skewers. *Arnold Math. J.* **2** (2016), 171–193.
50. S. Tabachnikov. *Kasner meets Poncelet.* arXiv:1707.09267.
51. T. Tomihisa. Geometry of projective plane and Poisson structure. *J. Geom. Phys.* **59** (2009), 673–684.
52. http://www.cinderella.de/tiki-index.php.
53. https://www.math.brown.edu/~res/Java/App33/test1.html.
54. https://www.math.brown.edu/~res/Java/Special/Main.html.

Chapter 10
Poincaré's Geometric Worldview and Philosophy

Ken'ichi Ohshika

Abstract Poincaré is one of the pioneers in non-Euclidean geometry and topology. In this paper, we shall first review his work on non-Euclidean geometry and topology. Then we shall see how his researches in these fields are reflected in his philosophical work, especially his philosophical position often called "conventionalism".

AMS Subject Classification: 01A55, 01A60, 01A70, 57-03

10.1 Introduction

Poincaré's works extend to many fields of mathematics, such as analysis, algebra, arithmetic, geometry, topology, and celestial mechanics. Among these, geometry and topology, on which we focus in this paper, constitute very important parts. Poincaré interpreted non-Euclidean geometry, first discovered by Lobatchevsky and Bolyai, using a conformal model which was already mentioned by Riemann in his Habilitationsvorstrag [40] and is now called Poincaré model, and studied the isometry groups of hyperbolic spaces. This, along with his other papers on Fuchsian groups, is an important part of his work on geometry. As for topology, Poincaré's work started with a small piece entitled "Sur l'Analysis Situs", which appeared in 1892 [29], and culminated in a famous series of papers "Analysis Situs" and its five complements [30–35]. This series includes, among others, a formal definition of manifold, which is a concept invented by Riemann [40], the notion of Poincaré duality, the invention of fundamental groups, and the "Poincaré conjecture" which was solved more than 100 years later.

K. Ohshika (✉)
Department of Mathematics, Faculty of Science, Gakushuin University, Toshima-ku, Tokyo, Japan
e-mail: ohshika@math.gakushuin.ac.jp

S. G. Dani, A. Papadopoulos (eds.), *Geometry in History*,
https://doi.org/10.1007/978-3-030-13609-3_10

Poincaré wrote several papers whose topics would be labelled as philosophy or philosophy of science. Most of them were eventually included in his four books, *La science et l'hypothèse*, *La valeur de la science*, *Science et méthode*, and *Les dernières pensées* [36–39]. Reading these books, we can easily see that Poincaré's philosophy was motivated and influenced by his work on geometry and topology. In particular, it is evident that his work on non-Euclidean geometry was an important factor leading him to his geometric "conventionalism". His view on the relationship between the "real world" and mathematics was quite radical compared to that of his contemporaries, and was very far from the Kantian or neo-Kantian worldview which was influential in his time.

In this paper, we shall first review Poincaré's work on geometry and topology, particularly focusing on hyperbolic geometry and topology, "analysis situs" in his time. Then we shall look at his philosophical works, focusing on those concerning his view on space and time. We shall see how his work on and understanding of geometry and topology influenced his philosophical view.

Poincaré's philosophy would certainly be one of what Althusser called "spontaneous philosophy of scientists" [1]. Althusser's claim was that spontaneous philosophy of scientists, influenced by the ideology of their time, would always have idealistic elements, and that the philosophers' role should be to discern such elements and draw a "line of demarcation". Still, as far as we can see, Poincaré's epistemology was much freer from an idealistic worldview and more profound than most of the philosophers of his days. We shall explain this more in the last section.

The author would like to express his gratitude to Athanase Papadopoulos for inviting him to write this paper. He is also grateful to the anonymous referee for his/her valuable suggestions and drawing the author's attention to a magnificent book by Gray [12].

10.2 Hyperbolic Geometry

Non-Euclidean geometry usually means geometry for which Euclid's fifth postulate does not hold, and instead, its negation holds. The fifth postulate is equivalent to what is called Playfair's axiom, saying that for any straight line ℓ and a point P outside ℓ, there is a unique straight line ℓ' passing through P which is parallel to ℓ. Two kinds of non-Euclidean geometry can be considered: one is spherical geometry, where two "straight lines" always intersect, and the other is hyperbolic geometry, where there are more than one "straight lines" passing through P parallel to ℓ in the setting above.

Spherical geometry in dimension 2 is known to have been already studied in ancient Greek, and was extensively studied by Euler, but was regarded by the latter as being part of Euclidean space geometry rather than non-Euclidean geometry. (See Papadopoulos [24, 25].) In contrast, hyperbolic geometry in dimension 2, which was first found by Lobatchevsky [20] and Bolyai, was born from the beginning as a trial to construct a geometry without Euclid's fifth postulate. These two authors showed that they could prove non-Euclidean versions of many theorems in elementary geometry including trigonometry under the second type of the negation of the fifth postulate: the existence of more than one parallel lines as was described above. Still, at this stage, it was not clear that such a geometry is realisable without contradictions.

Riemann introduced the general notion of metric for spaces in all dimensions for the first time in his Habilitationsvorstrag [40]. Relying on Riemann's idea, Beltrami gave a concrete geometric object having a non-Euclidean geometry in all dimensions. This in particular settled the problem of realisability for hyperbolic geometry in dimension 2. Let us explain this in more detail.

Suppose that on an n-dimensional space $\mathbb{E}^n$, a Cartesian coordinate system $(x_1, \ldots, x_n)$ is given. By the Pythagorean law, the distance between two points $(x_1, \ldots, x_n)$ and $(y_1, \ldots, y_n)$ is expressed as $\sqrt{\sum_{j=1}^n (x_j - y_j)^2}$. Riemann in [40] regards its infinitesimal form: $ds = \sqrt{\sum_{j=1}^n dx_j^2}$ as a metric defining the Euclidean space. For a curve γ in $\mathbb{E}^n$, its length is defined to be $\int_\gamma ds$, and for two points in $\mathbb{E}^n$, the infimum of the lengths of arcs joining them coincides with their distance, which is realised by a straight segment joining them. Riemann noticed that by changing this infinitesimal form ds, we can get a different metric on the space. In particular, he mentioned that a form $ds = \dfrac{1}{1 + \frac{\alpha}{4} \sum_{j=1}^n x_j^2} \sqrt{\sum_{j=1}^n dx_j^2}$ gives a metric of "curvature" α. Although his definition of curvature is rather intuitive, what he meant was the sectional curvature in modern terminology.

Developing this idea of Riemann, Beltrami considered the n-dimensional space with curvature -1 in [4]. He showed that if we consider the upper half-space $\{(x_1, \ldots, x_n) \mid x_n > 0\}$ and equip it with a metric given by $ds = \dfrac{\sqrt{\sum_{j=1}^n dx_j^2}}{x_n}$, we get a model of n-dimensional hyperbolic geometry. Mapping the upper half-space to the open unit ball $\{(x_1, \ldots, x_n) \mid \sum_{j=1}^n x_j^2 < 1\}$ conformally, he also got the same expression as Riemann's for $\alpha = -1$ on the open unit ball. In Beltrami's upper-half space model of hyperbolic geometry, for any two distinct points, the shortest path connecting them is a part of either a line or a circle intersecting the hyperplane $\{(x_1, \ldots, x_n) \mid x_n = 0\}$ perpendicularly. In the open unit ball model, the shortest path is a part of a circle intersecting the unit sphere perpendicularly. A different model of hyperbolic geometry was given in another paper by Beltrami [6], and also by Klein [18] from the viewpoint of projective geometry. The base space of Klein's model is also the open unit ball, but the shortest path between two distinct points is a Euclidean segment connecting them.

Poincaré used the same model as Beltrami's to study hyperbolic geometry, although it is not certain whether Poincaré was aware of Beltrami's work. This choice of model was important for his argument regarding "conventionalism". In fact, in Chapitre 4 of [36], he considered a universe where the metric is deformed conformally by temperature. Poincaré determined the isometry group of the hyperbolic plane in [28], which turned out to be the group of linear fractional transformations with real coefficients. Furthermore, Poincaré studied discrete groups of linear fractional transformations, which he called Fuchsian groups, using fundamental domains in the hyperbolic plane. His work in particular shows that every closed orientable surface of genus greater than 1 has a hyperbolic metric. This is very important in the history of hyperbolic geometry: by proving that hyperbolic geometry is a natural geometry on closed surfaces of high genera, Poincaré showed that hyperbolic geometry has the same right as spherical geometry, which is a natural geometry for genus-0 surface, and as Euclidean geometry, a natural geometry for genus-1 surface. This anticipated the same kind of naturalness of hyperbolic geometry in dimension 3, which would be formulated by Thurston much later, in the 1980s (see [47]).

In Chapitre 3 in Livre I of [38], Poincaré described the process in which he found Fuchsian groups to be isometry groups of the hyperbolic plane. According to his description there, his invention of Fuchsian groups did not arise from his study of non-Euclidean geometry, but from his study of Fuchsian functions, i.e. automorphic functions. He called groups preserving Fuchsian functions Fuchsian groups. He found that Fuchsian groups are also isometry groups of the hyperbolic plane only afterward, while he was taking a "course géologique" planned by the École des Mines, forgetting his mathematics on the surface of consciousness, but probably thinking about it subconsciously.

10.3 Topology-Analysis Situs

Poincaré wrote nine papers on topology, two short papers with the same title "Sur l'Analysis Situs", published in 1892 and 1901 respectively, "Analysis Situs", published in 1895, a one-page paper entitled "Sur les nombres de Betti", published in 1899, "Sur la connection des surfaces algébriques", published in 1901, and five complements to the "Analysis Situs", published in 1899, 1900, two in 1902, and 1904 respectively. The short papers "Sur les nombres de Betti", the second "Sur l'Analysis Situs", and "Sur la connection des surfaces algébriques" are just announcements of the results whose details are contained in the first, the third and the fourth complements respectively. Therefore we have seven papers to discuss here. (There is one more paper entitled "Sur un théorème de géométrie" in the section on topology of Poincaré's collected works, which is an unfinished work and whose topics is not closely related to the subject of the present paper.) The fifth complement to the "Analysis Situs" contains a very famous conjecture, which is now called the Poincaré conjecture, saying that every simply connected closed 3-manifold should be homeomorphic to the 3-sphere. The conjecture was finally resolved by Perelman about 100 years after the conjecture was raised. The Poincaré conjecture has been one of the strongest driving forces in research of topology throughout this 100 years. (See Berevstokii's paper in this volume [5] for more on the Poincaré conjecture.) We fully admit the importance of this conjecture, but we must never forget other works of Poincaré on topology, which also contain many important topics. We refer the reader to Sarkaria's paper [43] for more detailed accounts about Poincaré's papers on topology.

In the first "Sur l'Analysis Situs", Poincaré considered a question asking whether the Betti numbers determine the homeomorphism type of manifolds. This is the first paper in history which deals with a long-standing question of topology, from which modern topology has developed and is still developing: " how do algebraic invariants determine homeomorphism types of manifolds?" In [6], Betti considered manifolds embedded in a Euclidean space, and introduced the notion called "l'ordine di connessione", which should be interpreted as the real dimension of the k-th homology plus one, i.e. one greater than what we call the k-th Betti number today. His definition does not involve simplices or cycles in the way we use to define homology groups in modern textbooks. He just considered k-dimensional submanifolds embedded in a given manifold, and counted how many independent k-dimensional submanifolds there are. There is a subtle problem in the way to define the independence of submanifolds, but anyway, this work of Betti gave birth to homology theory.

Poincaré asked in this paper whether two manifolds having the same Betti numbers can be deformed from one to the other continuously, and showed that the answer is no, by giving examples of 3-manifolds having the same Betti numbers without being homeomorphic. His examples are torus bundles over the circle. He claimed that a torus bundle over the circle with monodromy $A \in SL_2(\mathbb{Z})$ has first Betti number 3, which he called "quadruple connection" using Betti's term, if and only if $A = E$; 2, which he called "triple connection", if and only if $\mathrm{Tr}\, A = 2$; and 1, which he called "double connection", otherwise. Poincaré also claimed that two torus bundles, with monodromies A and B, are homeomorphic if and only if A and B are conjugate in $GL_2(\mathbb{Z})$. We note that it is not so simple to prove the last claim, even using modern techniques in low-dimensional topology. Poincaré discussed this example again in the following paper "Analysis Situs" and gave a "proof" of it. We are not sure whether a rigorous proof could be given using only the techniques known in Poincaré's time.

Poincaré's second paper on topology, "Analysis Situs", was published in the Journal de l'École Polytechnique in 1895. This paper, consisting of 121 pages, gave foundations for the theory of manifolds, first invented by Riemann in his Habilitationsvortrag [40], and for the homology theory of manifolds. Poincaré gave two definitions of manifold. In the first of them, a manifold is defined to be a set of points $(x_1, \dots, x_n)$ in the n-dimensional Euclidean space $\mathbb{E}^n$ satisfying a system of p (differentiable) equations $F_1, \dots, F_p$ and q inequalities such that the rank of the Jacobian matrix of $F_1, \dots, F_p$ has rank p. In other words, Poincaré's first definition of a manifold is that of what we call an $(n - p)$-dimensional submanifold in the n-dimensional Euclidean space defined by implicit functions. Local charts or local coordinate systems which we use in defining manifolds today did not appear in this definition. This is quite similar to the definition of "spazio" of dimension $(n - p)$ by Betti in [6] although in Betti's paper the condition for the independence of equations is quite obscure.

In the second definition, Poincaré considered m-dimensional sets in $\mathbb{E}^n$ parameterised by m variables in such a way that two parameterisations are transformed from one to the other by analytic functions at their intersection. In this definition, for the first time, a construction of a manifold by patching up local coordinate systems appeared, which would lead to a later definition of manifolds using local charts due to Hilbert, Weyl, Kneser, and Veblen-Whitehead. (See Scholz [44] and Ohshika [23].)

Poincaré then introduced the notion of homology, clarifying and refining Betti's work. He considered a q-dimensional manifold W with boundary in a p-dimensional manifold V, and he regarded the boundary components of W, which are $(q-1)$-dimensional submanifolds of V, as being related by a homology. This definition is more like that of cobordism in today's terminology. He then defined the $(q-1)$-th Betti number to be the maximal number of linearly independent $(q-1)$-dimensional submanifolds with regard to homologies. Here he took into account the possibility that more than one of the boundary components represent parallel copies of the same $(q-1)$-dimensional manifold. This makes his definition slightly different from Betti's original definition. This difference concerns the topic of the first complement. Poincaré also showed the duality of Betti numbers using intersection number: for a closed orientable manifold of dimension n, the p-th Betti number is equal to the $(n-p)$-th Betti number.

In the latter part of the paper, Poincaré studied 3-dimensional manifolds, systematically constructing examples from polyhedra. This can be regarded as a generalisation of his construction of torus bundles in the previous paper. Poincaré in particular observed the following from this construction.

- 3-manifolds can be obtained from properly discontinuous actions of groups on Euclidean space just like Fuchsian groups.
- There are distinct 3-manifolds with the same Betti numbers, as was already mentioned in "Sur l'Analysis Situs", such that the groups corresponding to these manifolds (fundamental groups) are different.
- There are 3-manifolds having finite fundamental groups.

Poincaré also posed the following essential problems:

(1) Etant donné un groupe G défini par un certain nombre d'équivalences fondamentales, peut-il donner naissance à une variété fermée à n dimensions ? Comment doit-on s'y prendre pour former cette variété ?
(Given a group G defined by a certain number of fundamental equivalences, can it give birth to a closed n-dimensional manifold? How should we proceed to form this manifold?)[1]

(2) Deux variétés d'un même nombre de dimensions, qui ont même groupe G, sont-elles toujours homéomorphes ?

[1]All translations put in parentheses in this paper are by the present author.

(Are two manifolds of the same dimension which have the same group G always homeomorphic?)

The condition for G to be defined by a certain number of fundamental equivalences should mean that it is finitely generated. We know today that for the first problem, we need to add the assumption that G is finitely presented, but with this condition, this has turned out to be true for $n \geq 4$, but false for $n = 3$. For the second problem, the answer is no for every dimension $n \geq 3$. What is amazing is that Poincaré was so far-sighted that he could consider such a problem which would decide the direction of research in topology for a long time to come and could be resolved only much later in the twentieth century.

In the first complement to the Analysis Situs, Poincaré clarified the duality of Betti numbers which he stated in his "Analysis Situs", responding to a criticism by Heegaard [14], and gave another proof, which is more rigorous than the one contained in the "Analysis Situs". It is interesting for us to see that in those days, results appearing in papers were sometimes incomplete, and could be rectified after discussion in other papers. The new proof is nearer to what we can find in a textbook on simplicial homology theory today. Poincaré defined Betti numbers using polyhedra and boundary operators. To prove the duality, he used the dual cell complex as is done in most textbooks today. In the second complement, using the same line of argument as he did in the first complement, Poincaré observed the existence of a torsion invariant, which corresponds to the torsion part of homology groups. At the end of the paper, Poincaré said:

> Tout polyhèdre qui a tous ses nombres de Betti égaux à 1 et tous ses tableaux T_q bilatères est simplement connexe, c'est-à -dire homéomorphe à l'hypersphère.
> (Every polyhedron all of whose Betti numbers are equal to 1 and all of whose charts T_q are bilateral is simply connected, i.e. is homeomorphic to the 3-sphere.)

In other words, Poincaré said that every combinatorial 3-manifold with trivial first homology group is homeomorphic to S^3. This is of course false, and indeed Poincaré himself realised that this is not the case as was to be shown in the fifth complement.

In the third complement, Poincaré studied specific 3-manifolds. First, he considered an algebraic surface in $\mathbb{C}^3$ expressed as $z = \sqrt{F(x, y)}$, where F is a polynomial. It was assumed that for every y except for finitely many singular points, $F(x, y) = 0$ has $2p+2$ simple roots as an equation in x. He then considered a loop on the complex y-plane going around singularities, and the set of corresponding points in the algebraic surface. This becomes a 3-manifold which is a closed surface bundle over the circle. Poincaré showed how to compute the fundamental groups of such 3-manifolds.

In the fourth complement, Poincaré studied the topology of algebraic surfaces defined by $f(x, y, z) = 0$ when f is a polynomial. The study of such surfaces was started by Picard in [26]. Poincaré calculated the Betti numbers of such surfaces. He also showed that there is a duality between the first and the third Betti numbers for 4-manifolds, which was later termed Poincaré duality.

The fifth complement is very famous for the fact that the Poincaré conjecture was formulated for the first time there. The paper also contains many other important results. In the first part of the paper, Poincaré introduced what is now called the Morse theory. We should have in mind that this paper was published in 1904 (dated on 3 November 1903), which is more than 20 years before Morse, to whom the invention of the "Morse theory" is usually attributed, started writing a series of paper on this topic. (See Morse [22] for his work.) Although Poincaré's setting is very special, it is interesting that this case is the same as the one which is shown as an example of an application of Morse theory in most textbooks. He considered an m-dimensional manifold V embedded in a higher-dimensional Euclidean space $\mathbb{E}^k$, and a function ϕ defined on $\mathbb{E}^k$. Then he considered to slice V at $\phi(x_1, \ldots, x_k) = t$, which gives a family of $m - 1$-dimensional manifolds $W(t)$ (allowing singularities for some t), and observed that the topological type of $W(t)$ changes only at points t where $W(t)$ has singularities. He also analysed how $W(t)$ changes before and after t passes a singular value, and how this affects the topology of V. As the first application of this theory, he gave an alternative proof of the topological classification of closed surfaces. The second application, which is the most important part of this paper, is for 3-manifolds. Poincaré showed that there exists a closed 3-manifold with trivial Betti numbers and trivial torsions which is not homeomorphic to the 3-sphere. In other words, he constructed a homology 3-sphere. In the case of dimension 3, slicing V into $W(t)$ corresponds to a Heegaard splitting of V, i.e. a decomposition of V into two handlebodies, which was first studied by Heegaard in [14]. Poincaré constructed a concrete Heegaard splitting, which gives a homology sphere, and calculated its fundamental group to show that the resulting 3-manifold is not homeomorphic to the 3-sphere. In the last page of the paper he posed the following question, which is the very famous Poincaré conjecture:

> Est-il possible que le group fondamental de V se réduise à la substitution identique et que pourtant V ne soit pas simplement connexe?
> (Is it possible that the fundamental group of V is reduced to the identical substitution and that nevertheless V is not simply connected?)

We should recall that Poincaré called a 3-manifold simply connected when it is homeomorphic to the 3-sphere. The paper is concluded with this sentence:

> Mais cette question nous entraînerait trop loin.
> (But this question would take us too far.)

Looking back at later development of topology, we see that the Poincaré conjecture was solved in higher dimensions by Smale [46] in a way extrapolated from Poincaré's own thinking. In fact, using Morse theory and handle decomposition to prove the h-cobordism theorem is a natural extension of Poincaré's idea of slicing a manifold to get a decomposition into a family of codimension-1 submanifolds. In dimension 4, Freedman [7] proved the conjecture in the topological category, using Casson handles instead of ordinary handle decomposition in higher dimensions. (See also Poénaru's paper in this volume [27] for related topics.) On the other hand, the original Poincaré conjecture, i.e. for dimension 3, was solved in a way which should have been quite unexpected to Poincaré. To get to the final solution by Perelman, it was necessary to pass through the geometric understanding of 3-manifolds by Thurston [47], and the deformation of Riemannian metrics by Ricci flows due to Hamilton (see Morgan-Tian [21]).

10.4 Epistemology of Space and Time

When we talk about epistemology in the nineteenth and the early twentieth centuries, we cannot ignore the strong impact of the work of Kant and his apriorism. From the ancient Greek period on, Euclidean geometry was considered to be a precise reflection of the reality. For empiricists, geometry should also be derived from experience. Kant's position is quite different from this. He maintained clearly that "space is not an empirical concept that has been drawn from outer experiences" in his major work "Kritik der reinen Vernunft" [17]. He regarded the notions of space and time as "synthetic a priori". This means that these notions are not what we construct based on our experience, but are foundations upon which all other recognitions should be built. This view of Kant was quite influential among (continental) European philosophers, in particular those who adhered to German idealism, and scientists, whether they agree on it or not. For instance, Herbart, an impact of whose philosophy can be also found in Riemann's Habilitaionsvorstrag, emphasised the aspect of the notion of space as a human construction, which can be regarded as a criticism of Kantian apriorism ([15], see also Banks [3]). Fries, while keeping the principle of apriorism by Kant, reexamined Kantian transcendental logic introducing psychologic basis of knowledge [10]. Still, we can say that these two philosophers are within a Kantian paradigm: their approaches are guided by problems posed by Kant, and their works are responses to the writings of Kant. (See §2 of Gray [12] for a more detailed account on these post-Kantian philosophers and their influences on mathematics.)

On the other hand, modern empiricists such as J-S. Mill developed ideas opposed to this Kantian apriorism. In his book entitled "A System of Logic, Ratiocinative and Inductive", which is famous for his emphasis on the importance of induction as a logic supporting science, he insisted that geometry is also a part of experimental science whose validity should be justified by induction. For him, even Euclidean geometry can be justified by experience.

Poincaré's view, which is often referred to under the name of (geometric) conventionalism, is quite different from both Kantianism and Mill's version of empiricism. By the discovery of non-Euclidean geometry, the apriorism concerning Euclidean geometry such as Kant proposed broke down. Still one could expect that it was possible to prove that Euclidean geometry is the right one in the universe where we live, through either experiments or observations. Poincaré did not think that this was the case. He further thought even properties of space common to both Euclidean and non-Euclidean geometry, such as the dimension of space, neither are given a priori nor can be proved by experiments or observations.

Poincaré published three books of a philosophical nature, *La science et l'hypothèse*, *La valeur de la science*, and *Science et méthode*. Another book was published posthumously under the title *Les dernières pensées*. Poincaré discussed epistemology of space and time throughout these four books. Here, we are going to examine what is expressed in the first two of his books.

The second part of his first book *La science et l'hypothèse* is entitled " l'espace" and contains three chapters "les géométries non-euclidiennes", " l'espace et la géométrie", and "l'expérience et la géométrie". In the first chapter of these three, Poincaré first explains how non-Euclidean geometries, spherical one and hyperbolic one, were born, through the work of Riemann, Lobatchevsky and Beltrami, and why these geometries are natural in the same way as Euclidean geometry. He then analyses what should be geometry in general. He emphasises the importance of the existence of a group acting on a space preserving the shapes of things lying there to make it possible to consider geometry. This view, which Poincaré attributes to S. Lie for its invention, is also a precursor of today's definition of geometric structures, which are also called (G, X)-structures. (The reader may refer to Goldman's paper [11] in this volume for more on (G, X)-structures.) We should recall here that for Poincaré, the definition of fundamental group is different from what we find today in textbooks: in modern terminology, his fundamental group is the covering translation group acting on the universal cover. Therefore, considering the group consisting of possible motions in each geometry should be very natural for him.

We should also note that Poincaré argues, contrary to Kant's view, that the axioms of geometry do not constitute synthetic a priori judgements. For Poincaré, the principle of induction is, for instance, a synthetic a priori judgement, for which he considered it was impossible to consider an alternative arithmetic. Since it is quite possible to consider the world in which non-Euclidean geometry holds instead of Euclidean geometry, the axioms of geometry could not be thrown into this category. On the other hand, Poincaré admits that Euclidean geometry reflects motions of solid bodies in the real world and that projective geometry is an abstraction of behaviours of light. Still he rejects the view that geometry is an experimental science, for it is not an object which is subject to revision. As a result, he concludes:

> Les axiomes géométriques ne sont donc ni des jugements synthétiques à priori ni des faits expérimentaux.
> (The geometric axioms are therefore neither synthetic a priori judgements nor experimental facts.)
> Ce sont des conventions; notre choix, parmi toutes les conventions possibles, est guidé par des faits expérimentaux: mais il reste libre et n'est limité que par la nécessité d'éviter toute contradiction.
> (They are conventions; our choice, among all possible conventions, is guided by the experimental facts, but it remains free and is limited only by the necessity to avoid any contradiction.)

He then insists that it is meaningless to ask if between two geometries, for example Euclidean and non-Euclidean geometries, one is truer than the other, and that what we can say is just one is more convenient than the other.

In the next chapter, "l'espace et la géométrie", Poincaré goes deeper into an epistemological aspect of our cognition of space and its relation to geometry. He characterises space as we understand it by the following five properties: (1) it is continuous; (2) it is infinite; (3) it has three dimensions; (4) it is homogenous; and (5) it is isotropic. He then analyses how these properties are derived from our "experience". He observes that our notion of space is derived from three different spaces which our senses directly perceive: visual space, tactile space, and motor space. He points out that the visual space has two dimensions, and observes that only by converging the views from two eyes and accommodating them, we get a sense of the third dimension: the distance. For the tactile and motor spaces, the process to get hold of three dimensional space is more complicated. Furthermore, Poincaré argues that our space and its geometry are not mere consequences of integrating these three kinds of spaces which we perceive. It is only through displacement of solid objects, causing changes in view or in touch, and their recoveries through our movement, we form our sense of geometry. This final point is very important in his epistemology of space. He says:

> Aucune de nos sensations, isolée, n'aurait pu nous conduire à l'idée de l'espace, nous y sommes amenés seulement en étudiant les lois suivant lesquelles ces sensations se succèdent.
> (None of our sensations, if they were isolated, would lead us to the idea of space. We are brought there only by studying the laws following which these sensations succeed to one another.)

Therefore, our space and time are products of our reasoning, not what was given a priori.

In the last of the three chapters, Poincaré discusses if geometry can be derived from experiments or observations. His answer is definitely no. For instance, he asks whether it is possible to determine if the space we are living in is either Euclidean or spherical or hyperbolic (Lobatchevskian in his words). Some may think that this is possible for instance by measuring parallaxes of stars. Poincaré says that this is not the case. To make this kind of idea acceptable, we must assume that light always proceeds along a geodesic. Since there is no way to prove this, Poincaré says, using parallaxes to measure the curvature is just shifting a problem to another one. More generally, Poincaré says,

> Aucune expérience ne sera jamais en contradiction avec le postulatum d'Euclide; en revanche aucune expérience ne sera jamais en contradiction avec postulatum de Lobatchevsky.
> (No experiment will ever lead to contradiction with the postulate of Euclid, on the other hand, no experiment will ever lead to contradiction with the postulate of Lobatchevsky.)

One may think that the existence of Euclidean solids would prove the "flatness" of space, but Poincaré points out that this is not true, for Euclidean solids can be realised also in a hyperbolic space as solids bounded by parts of horospherical surfaces. This observation is very shrewd, and reminds us of the fact that Lobatchevsky called his geometry "pangeometry" [20].

In the second book *La valeur de la science*, Poincaré takes up the epistemology of space again in two chapters, which are entitled "la notion d'espace" and "l'espace et ses trois dimensions". After reviewing what he showed in *La science et l'hypothèse*, Poincaré goes on to study the subject more deeply. Recall that both Euclidean geometry and (hyperbolic) non-Euclidean geometry presuppose three-dimensional space on which their metrics are defined. This space without any metric is called "amorphous continuum" by Poincaré. This continuum has some properties, which can be studied by "l'analysis situs", i.e. if we use modern terminology, topological properties. Poincaré poses the same questions for these topological properties as those which he posed for the geometry of space: whether they are given a priori or whether they can be verified by experiments and so forth. As one of the most important topological properties of space, he chooses its dimension and considers whether it can be determined a priori or by experiments. First, Poincaré asks how we can define the dimension of an amorphous continuum in which we live. In the real existing physical space, if two points are close enough, we cannot distinguish them. Therefore Poincaré thinks of the continuum not as a mere point-set, but as a set on which the relation of distinguishability is defined for any two points. We should note that at the time when Poincaré wrote this book, there was no notion of topological space. This notion was first introduced by Hausdorff in 1914, but Poincaré's idea is very similar to that of neighbourhood which appeared in Hausdorff's book [13] for the first time.

Poincaré considers what he called "coupures", i.e. cuts, for a continuum C. Cuts are a collection of either elements of C or continua contained in C. Suppose that one proceeds from one element A in C to another element B in C, by which we mean that there is a sequence of elements $E_1, \dots, E_n$ of C such that A is indistinguishable from E_1; E_j is indistinguishable from E_{j+1} for each j; B is indistinguishable from E_n; and two non-adjacent E_i and E_j are distinguishable. If there are cuts $\{e_k\}$ of C such that for any $E_1, \dots, E_n$ as above, some E_j is indistinguishable from some of e_k, then we say that the cuts $\{e_k\}$ divide C. Poincaré defines that C is one-dimensional if there are cuts consisting elements of C which divide C. Then inductively he defines C has n dimensions if there are cuts of C consisting of $(n-1)$-dimensional continua which divide C.

Having defined dimension for continua in this way, Poincaré returns to the original question. He first analyses how we can apply the above definition of dimension for the space in which we live. Then in the same way as in "l'espace et la géométrie" in the previous book, he analyses how we get the sense of three-dimensionality from our visual, tactile, and motor senses. Again, Poincaré says, we derive this from displacement of objects causing changes of senses and their recoveries by our movement. He concludes that our experience alone cannot prove that our space has three dimensions: in fact it may have more than three dimensions, but we choose three-dimensionality for its convenience. The experience has only guided this choice. He writes:

> Quel est alors le rôle de l'expérience ? C'est elle qui lui (l'esprit) donne les indications d'après lesquelles il fait son choix.

(What is then the role of experience? It is the experience that gives the mind the indications following which it makes choice.)

10.5 Spontaneous Philosophy: Conclusion

Althusser, arguably one of the most influential Marxist philosophers in the twentieth century, gave an inaugural lecture of a course on philosophy for scientists in 1967 at l'École Normale Supérieure. This lecture was later published as a book entitled *Philosophie et philosophie spontanée des savants* in 1974 [1]. Throughout the book, he described what he thought philosophy was and what role philosophy should play. What is relevant to the present article is his view on spontaneous philosophy of scientists, which is succinctly explained in his Thèse 25 as follows:

> Dans leur pratique scientifique, les spécialistes des différentes disciplines reconnaissent « spontanément » l'existence de la philosophie, et le rapport privilégié de la philosophie aux sciences. Cette reconnaissance est généralement inconsciente : elle peut devenir, en certaines circonstances, partiellement consciente. Mais elle reste alors enveloppée dans les formes propres de la reconnaissance inconsciente : ces formes constituent la « philosophie spontanée des scientifiques » (P.S.S.).
> (In the practice of science, the specialists of different disciplines recognise "spontaneously" the existence of philosophy, and the special relation of philosophy to science. This recognition is generally unconscious, and in some occasions it may become partially conscious. But then it remains covered in its own forms of unconscious recognition, and these forms constitute the "spontaneous philosophy of scientists" (P.S.S.).)

Althusser observed that P.S.S. contains often idealistic ideas coming from the outside of science, which he called "Élément 2", and that it constitutes an ideological aspect of P.S.S. For Althusser, the principal role of philosophers (as himself) with regard to P.S.S. is to draw a line of demarcation between science, which must be materialism, and ideology, which consists of idealism. We should have in mind that, as can been seen in [2], for Althusser a role model of such an intervention on the part of philosophers is Lenin's disparagement of empiriocriticism as one can find in [19].

It is evident that Poincaré's epistemology was born out of his work on geometry and topology. The progress in geometry and topology in Poincaré's day made it necessary to change the classical view on space and time as was formulated by Kant. In this sense, from Althusser's viewpoint, Poincaré's epistemology should be regarded as an example of P.S.S. (An analysis of Poincaré's philosophy in this line can be found in Rollet [41].) However, what we want to emphasise is that Poincaré's epistemology, which can be regarded as a precursor of empiriocriticism along with E. Mach, was quite ahead of his time, and that compared to naïve materialism such as the one propounded by Lenin in [19], it was based on far more profound insights into how our recognition of space and time is formed, as we saw in the previous section.

Putting aside a rather amateurish approach of Lenin, let us compare Poincaré's philosophy with his contemporary professional philosophers. Among them, we can in particular think of Frege, Husserl and Russell as those who studied seriously epistemological problem of space and time. All of them had an academic training in mathematics. Frege, who is often regarded as a precursor of logical positivism, studied the foundation of mathematics, as in [8], which inspired both Wittgenstein and the famous work of Russell-Whitehead, *Principia Mathematica*. Frege also had good knowledge on the work of Riemann. Still, his attitude toward the non-Euclidean geometries and epistemological problem of space can be regarded as staying within a Kantian framework: Frege seemed to believe that our space of intuition should be the Euclidean three-dimensional space, as can be seen in his correspondence with Hilbert and his review on Hilbert's book, both of which can be found in [9] (see also Shipley [45] for a more detailed account on Frege's epistemological view on geometry and space).

Husserl, who is now regarded as a precursor of phenomenology, also worked for philosophical aspects of mathematics. He gave a course on Riemann's theory of geometry at University of Halle in 1889 [16], where he dealt with Riemann's notion of metric critically. In particular, after criticising Riemann's definition of curvature, Husserl claimed that since Euclidean space is necessary to define Gaussian curvature, Euclidean space is a special entity which is given to us by pure intuition in Kantian sense. If we scrutinise his argument from the viewpoint of modern geometry, we notice that Husserl was confusing intrinsic and extrinsic natures of space. He did not seem to pay attention to the fact that spheres and Euclidean planes can be realised even in hyperbolic space. (Recall that this is why Lobachevsky coined the term "Pangeometry" as we noted in §4.)

Russell, who belongs to a younger generation than the other two, published in his twenties a book entitled *An essay on foundation of geometry* [42]. In this book, he criticised Kantian apriorism in geometry, and observed that there are two elements in geometry, what is a priori and what is empirical. Russell rejected the a-priority of Euclidean geometry, and claimed that experience can decide which of Euclidean or non-Euclidean geometry is valid. He further insisted that their common ground, which is projective geometry, should be a priori. This means that Russell embraced some part of conventionalism, but retained a weak version of Kantian apriorism. This position of Russell was harshly criticised by Poincaré in his review of this book.

Thus we have seen that even these three famous philosophers who were familiar with Riemann's work could not be as radical as Poincaré in abandoning Kantianism, and this seems to be caused by the fact that without experience of research on geometry and topology, they could not reach the level of Poincaré's understanding of space. In the latter half of the twentieth century, there appeared philosophers such as Merleau-Ponty, who analysed how our perception leads to our recognition of space, but in Poincaré's day no professional philosopher could approach the epistemology of space as deeply as Poincaré. According to Althusser, as we have explained above, the role of philosophers would be to draw a line of demarcation between science and ideology in spontaneous philosophy by scientists. However for this to be possible, philosophers need to have a deep understanding on what constitutes the core of spontaneous philosophy. In Poincaré's case, the core is nothing but his geometric conventionalism, which in turn depends on his research in geometry and topology. In his time, it was very difficult for other philosophers to reach this level for playing the role to draw such a line of demarcation in Poincaré's epistemology.

References

1. L. Althusser, *Philosophie et philosophie spontanée des savants*, François Maspero, Paris, 1974.
2. L. Althusser, *Lenine et la philosophie*, François Maspero, Paris, 1972.
3. E. C. Banks, Kant, Herbart and Riemann, Kant-Studien, *Philosophische Zitschrift*, 96 (2005) 208–234.
4. E. Beltrami, Teoria fondamentale degli spazii di curvatura costante, *Annali di Matematica Pura ed Applicata*, 2 (1868) 232–255.
5. V. N. Berestovskii, Poincaré conjecture and related statements, in *Geometry in History*, edited by S.G. Dani, A. Papadopoulos, Springer, Cham, 2019. https://doi.org/10.1007/978-3-030-13609-3_17
6. E. Betti, Sopra gli spazi de un numero qualunque di dimensioni, *Annali di Matematica* (2) 4, 140–158.
7. M. Freedman, The topology of four-dimensional manifolds, *J. Differential Geom.* 17 (1982), 357–453.
8. G. Frege, *Die Grundlagen der Arithmetik, eine logisch mathematische Untersuchung über den Begriff der Zahl*, Koebner, Breslau, 1884.
9. G. Frege, Kleine Schriften ; herausgegeben von Ignacio Angelell, Georg Olms, Hildesheim, 1967.
10. J. F. Fries, *Neue oder anthropologische Kritik der Vernunft*, Bande 1–3, 1828–1831, Winter, Heidelberg.
11. W. Goldman, Flat affine, projective and conformal structures on manifolds: a historical perspective, in *Geometry in History*, edited by S.G. Dani, A. Papadopoulos, Springer, Cham, 2019. https://doi.org/10.1007/978-3-030-13609-3_14
12. J. Gray, *Plato's ghost: the modernist transformation of mathematics*, Princeton University Press, Princeton NJ, 2008.
13. F. Hausdorff, *Grundzüge der Mengenlehre*, Veit, Leipzig, 1914.
14. P. Heegaard, Forstudier til en topologisk teori for de algebraiske Fladers Sammenhäng, Thesis; Traduction française: Sur l'Analysis Situs, *Bull. Soc. Math. France*, 44 (1916), 161–242.
15. J. F. Herbart, *Allgemeine Metaphysik, nebst den Anfängen der philosophischen Naturlehre* (2 Teile), A.W. Unzer, Königsberg, T.1 1828, T. 2, 1829.
16. E. Husserl, Husserliana : gesammelte Werke Bd. 21, Studien zur Arithmetik und Geometrie : Texte aus dem Nachlass (1886–1901), Nijhoff, Hague, 1983.
17. I. Kant, *Critik der reinen Vernunft*, Johann Friedrich Hartknoch, Riga 1781, 2 Aufl 1787.
18. F. Klein, Ueber die sogenannte Nicht-Euklidische Geometrie, *Math. Ann.*, 4, 573–625 (1871).
19. V. Lénine, *Matérialisme et empirio-criticisme*, Editions sociales Paris et Editions du progrès Moscou, (traduit du russe).
20. N. Lobatchevsky, *Pangeometry*, edited by A. Papadopoulos, EMS Heritage of European Mathematics, 4, 2010.
21. J. Morgan and G. Tian, *Ricci Flow and the Poincare Conjecture*, Clay Mathematics Monographs, 3. American Mathematical Society, Providence, RI; Clay Mathematics Institute, Cambridge, MA, 2007. xlii+521.
22. M. Morse, *Calculus of Variations in the Large*. Colloquium Publications Volume: 18; 1934; 368 pp; American Mathematical Society, Providence, R.I..

23. K. Ohshika, The origin of the notion of manifold: from Riemann's Habilitationsvortrag onward, in "*From Riemann to differential geometry and relativity*" edited by L. Ji, A. Papadopoulos and S. Yamada, Springer 2017.
24. A. Papadopoulos, On the work of Euler and his followers on spherical geometry, *Gaṇita Bhāratī* 36 (2014), no. 1, 53–108.
25. A. Papadopoulos, Euler, la géométrie sphérique et le calcul des variations, in "*Leonhard Euler : Mathématicien, physicien et théoricien de la musique*", CNRS Éditions, Paris, 349–392, 2015.
26. E. Picard, Sur une extension aux fonctions de deux variables du problème de Riemann relatif aux fonctions hypergéométriques. *Annales scientifiques de l'École Normale Supérieure*, Sér. 2, 10 (1881), pp. 305–322.
27. V. Poénaru, A glimpse into the problems of the fourth dimension, in this volume.
28. H. Poincaré, Théorie des groupes Fuchsiens, *Acta Math.*, 1, 1–62, (1882).
29. H. Poincaré, Sur l'Analysis Situs, *Comptes rendus hebdomadaires de l'Académie des sciences de Paris*, 115 (1892), 189–192.
30. H. Poincaré, Analysis situs, *Journal de l'École Polytechnique*, t. 1, pp.1–121 (1895).
31. H. Poincaré, Complément à l'Analysis situs, premier complément, *Rendiconti del Circolo Matematico di Palermo*, 13, 285–343 (1899).
32. H. Poincaré, Second complément à l'Analysis situs, *Proceedings of the London Mathematical Society*, 32, 277–308 (1900).
33. H. Poincaré, Sur certaines surfaces algébriques ; troisième complément à l'Analysis Situs, *Bulletin de la Société Mathématique de France*, 30, 49–70 (1902).
34. H. Poincaré, Sur les cycles des surfaces algébriques. Quatrième complément à l'Analysis Situs, *Journal de Mathématiques*, 8, 169–214 (1902).
35. H. Poincaré, Cinquième complément à l'Analysis Situs, *Rendiconti del Circolo matematico di Palermo*, 18, 45–110 (1904).
36. H. Poincaré, *La science et l'hypothèse*, Flammarion, Paris, 1902.
37. H. Poincaré, *La valeur de la science*, Flammarion, Paris,1905.
38. H. Poincaré, *Science et méthode*, Flammarion, Paris, 1908
39. H. Poincaré, *Dernières pensées*, Flammarion, Paris, 1913.
40. B. Riemann, Über die Hypothesen, welche der Geometrie zu Grunde liegen, *Abhandlungen der Königlichen Gesellschaft der Wissenschaften zu Göttingen*, 13, (1867).
41. L. Rollet, Henri Poincaré, des Mathématiques à la philosophie, Étude du parcours intellectuel, social et politique d' un mathématicien au début du siècle, Thèse, 1999, Université de Nancy.
42. B. Russell, *An essay on the foundations of geometry*, Cambridge University Press, Cambridge 1897.
43. K. S. Sarkaria, The topological work of Henri Poincaré, in "*History of Topology*", Edited by I.M. James, North-Holland, 123–167.
44. E. Scholtz, The Concept of Manifold, 1850–1950, in "*History of Topology*", Edited by I.M. James, North-Holland, 25–64.
45. J. Shipley, Frege on the foundation of geometry in intuition, *Journal for the History of Analytical Philosophy*, 3–6, (2015).
46. S. Smale, Generalized Poincaré's conjecture in dimensions greater than four, *Ann. of Math.* (2) 74 1961 391–406.
47. W. Thurston, Three-dimensional manifolds, Kleinian groups and hyperbolic geometry, *Bull. Amer. Math. Soc.* (N.S.) 6 (1982), no. 3, 357–381.

Chapter 11
Perturbing a Planar Rotation: Normal Hyperbolicity and Angular Twist

Alain Chenciner

Abstract In generic two-parameter families of local diffeomorphisms of the plane unfolding a local diffeomorphism with an elliptic fixed point, the tension between radial (hyperbolic) and tangential (elliptic) behaviour gives rise to phenomena where the whole wealth of the area preserving case is unfolded along some direction of the parameter space.

AMS Classification: 37E30, 37E40, 37D05, 34C23, 37G05, 37G15

11.1 Introduction

Perturbing the germ at the origin of a planar rotation $re^{2\pi i\theta} \mapsto re^{2\pi i(\theta+\omega)}$ leads to two celebrated results which describe geometrically the *dynamical* behaviour of the iterates of a diffeomorphism F obtained by perturbation, that is the structure of the *orbits* $\mathcal{O}(z) = \{z, F(z), F^2(z), \ldots, F^n(z), \ldots\}$: the *Andronov–Hopf–Neimark–Sacker bifurcation* of invariant curves under a generic radial hypothesis of weak attraction (or repulsion) and the *Moser invariant curve* theorem under an angular twist hypothesis in the area preserving case. The invariant curves whose existence is proved are normally hyperbolic with generic induced dynamics in the first case, with a dynamics smoothly conjugate to a diophantine rotation in the second one.

Statements and proofs illustrate the notion of *normal form*, introduced by Poincaré in his thesis in 1879. Closely related to the "averaging of perturbations" used by astronomers since the eighteenth century, it generalizes the Jordan normal form of a matrix to the nonlinear world. Namely, by introducing local coordinates which reveal an approximate geometry underlying the situation, it sets the scene for the application of refined analytic tools to the determination of which features of this geometry do really exist.

A. Chenciner (✉)
Observatoire de Paris, IMCCE (UMR 8028), ASD, Paris, France

Département de mathématique, Université Paris VII, Paris, France
e-mail: alain.chenciner@obspm.fr; https://perso.imcce.fr/alain-chenciner/

S. G. Dani, A. Papadopoulos (eds.), *Geometry in History*,
https://doi.org/10.1007/978-3-030-13609-3_11

After recalling these two classical contexts, say the one of nonlinear self-sustained oscillations (Lord Rayleigh, Van der Pol) and the one of the 3-body problem (Poincaré), I shall describe an old result of mine which in some sense makes the two worlds meet: in generic 2-parameter families of germs of diffeomorphisms of the plane near a fixed point, the tension between radial and angular (or hyperbolic and elliptic) behaviour leads to phenomena where the whole wealth of the area preserving situation is unfolded along some direction of the parameter space.

11.2 Elliptic Fixed Points

Let $F : (S, p) \to (S, p)$ be a local C^∞ (or analytic) diffeomorphism of a surface S defined in the neighborhood of a fixed point $p = F(p)$. The fixed point is said to be elliptic if the spectrum of the derivative $dF(p)$ is of the form $\{e^{2\pi i\omega}, e^{-2\pi i\omega}\}$ with $e^{2\pi i\omega} \neq \pm 1$. This is equivalent to the existence of a linear conjugation of $dF(p)$ with the rotation of angle $2\pi\omega$. Hence, after choosing good coordinates, one can suppose that $p = 0$ and that $F : (\mathbb{C}, 0) \to (\mathbb{C}, 0)$ is such that

$$F(\zeta) = \lambda\zeta + O(|\zeta|^2), \quad \text{with} \quad \lambda = e^{2\pi i\omega}.$$

In other words, F is a perturbation of a rotation.[1] Now, a rotation preserves each circle centered at the origin. This is a very strong property, very likely to be destroyed by the non-linear terms in the Taylor expansion of F. Nevertheless, reality is subtler and the study of the fate of these invariant circles is the starting point of two famous theories which correspond roughly to the dichotomy between *dissipative* and *conservative* dynamics:

1. *Andronov–Hopf–Neimark–Sacker bifurcation theory* which analyzes what happens when one considers a *generic*[2] diffeomorphism F with an elliptic fixed point at 0. The local behaviour of F itself is quite dull: indeed, the radial behaviour of the nonlinear terms turns the fixed point into an attractor or a repulsor and no other invariant object persists in its neighborhood. It is only when considering "generic" 1-parameter families F_μ of local diffeomorphisms stemming from $F_0 = F$ that the whole richness of the dynamics is regained (see [2, 3]): each small enough circle invariant under the rotation $dF(0)$ becomes a *normally hyperbolic*[3] closed curve invariant under some F_μ (Fig. 11.2).

[1] Beware that the notation $F(\zeta)$ does not mean that F is complex analytic, its expression depends on ζ and $\bar{\zeta}$.

[2] We shall not give a formal definition of this word; it means essentially that what is described is the general situation and that only special hypotheses could prevent the description from being correct.

[3] Roughly speaking this means that any attraction or repulsion normal to the curve under the iterates of F_μ dominates any attraction or repulsion inside the curve; this condition insures the robustness of the curve.

2. *Kolmogorov–Arnold–Moser (KAM) theory* which analyzes the case when F is area preserving, a hypothesis which is natural for diffeomorphisms with a mechanical origin, the paradigmatic example being *first return maps*[4] in the *restricted three body problem* first studied by Poincaré (see [8, 9]). In this case, it is the angular behaviour of the non-linear terms which plays the key part, the result being that "many" of the circles invariant under the rotation $dF(0)$ persist in the form of closed curves invariant under the action of F itself. Moreover the restriction of F to such an invariant closed curve is smoothly conjugated to a rotation whose angle is of the form $2\pi\alpha$ with α not rational and even "far from the rationals" in a precise sense.

11.3 Preparation: Poincaré's Theory of Normal Forms

The idea, which goes back to Poincaré's thesis in 1879, is the following: being a rotation, the derivative of F commutes with the whole group SO(2) of rotations. This is shown to imply that, provided some conditions on ω are satisfied, a high order approximation of F is locally invariant by an action of SO(2) close to the standard one. Equivalently, one proves the existence of local coordinates which reveal the approximate geometry of the map, in a spirit similar to the Jordan form of a matrix:

Theorem 1 *If $\lambda = e^{2\pi i\omega}$ is such that $\lambda^q \neq 1$ for all integers $q \in \mathbb{N}$ such that $q \leq 2n+2$, then there exists a local diffeomorphism*

$$H : (\mathbb{C}, 0) \to (\mathbb{C}, 0), \quad \zeta \mapsto z = H(\zeta) = \zeta + O(|\zeta|^2)$$

such that

$$H \circ F \circ H^{-1}(z) = N(z) + O(|z|^{2n+2}), \quad \text{where} \quad N(z) = z\left(1 + f(|z|^2)\right) e^{2\pi i(\omega + g(|z|^2))},$$

with f and g real polynomials of degree n such that $f(0) = g(0) = 0$. If moreover $\lambda^{2n+3} \neq 1$, one can achieve a rest which is $O(|z|^{2n+3})$.

The so-called *normal form* N, is characterized by the fact that it commutes with the whole group SO(2) of rotations:

$$\forall \alpha,\ N(e^{2\pi i\alpha} z) = e^{2\pi i\alpha} N(z).$$

[4]See section 1.4 of [7] for a brief introduction.

Proof Let us start with a local diffeomorphism of degree 2,

$$H_2 : (\mathbb{C}, 0) \to (\mathbb{C}, 0), \quad z = H_2(\zeta) = \zeta + \sum_{i+j=2} \gamma_{ij}\zeta^i\bar{\zeta}^j.$$

The direct computation of $H_2 \circ F \circ H_2^{-1}$ is illustrated in the following diagram.

$$\begin{array}{ccc}
\zeta & \xrightarrow{F} & \lambda\zeta + \sum_{i+j=2}\alpha_{ij}\zeta^i\bar{\zeta}^j + O\left(|\zeta|^3\right) \\
\downarrow H_2 & & \downarrow H_2 \\
\zeta + \sum_{i+j=2}\gamma_{ij}\zeta^i\bar{\zeta}^j & \xrightarrow{H\circ F\circ H^{-1}} & \lambda\zeta + \sum_{i+j=2}\alpha_{ij}\zeta^i\bar{\zeta}^j + \sum_{i+j=2}\gamma_{ij}\lambda^i\bar{\lambda}^j\zeta^i\bar{\zeta}^j + O\left(|\zeta|^3\right) \\
\| & & \| \\
z & & \lambda z + \sum_{i+j=2}\left(\alpha_{ij} + \gamma_{ij}\left(\lambda^i\bar{\lambda}^j - \lambda\right)\right) z^i\bar{z}^j + O\left(|z|^3\right)
\end{array}$$

Supposing that $F(\zeta) = \lambda\zeta + \sum_{i+j=2}\alpha_{ij}\zeta^i\bar{\zeta}^j + O(|\zeta|^3)$, we get

$$H_2 \circ F \circ H_2^{-1}(z) = \lambda z + \sum_{i+j=2}\left(\alpha_{ij} + (\lambda^i\bar{\lambda}^j - \lambda)\gamma_{ij}\right) z^i\bar{z}^j + O(|z|^3).$$

Hence, if no *resonance* relation of the form $\lambda^i\bar{\lambda}^j - \lambda = 0$ is satisfied with indices i, j such that $i + j = 2$, that is if $\lambda^3 \neq 1$ (otherwise $\bar{\lambda}^2 - \lambda = 0$), the choice of $\gamma_{ij} = -(\lambda^i\bar{\lambda}^j - \lambda)^{-1}\alpha_{ij}$ kills all degree 2 terms in the Taylor expansion of the transformed map $H_2 \circ F \circ H_2^{-1}$.

If one tries in the same way to simplify the terms of degree 3 in the Taylor expansion of $H_2 \circ F \circ H_2^{-1}$, one stumbles upon an *unavoidable resonance*

$$\lambda^2\bar{\lambda} - \lambda = 0$$

which merely reflects that $|\lambda| = 1$. Hence, if no other resonance of order 3 exists, which amounts to saying that $\lambda^4 \neq 1$ (otherwise $\bar{\lambda}^3 - \lambda = 0$), a local diffeomorphism H_3 of the form $H_3(z) = z + \sum_{i+j=3}\gamma_{ij}z^i\bar{z}^j$ can be found such that[5]

$$H_3 \circ H_2 \circ F \circ H_2^{-1} \circ H_3^{-1}(z) = \lambda z + c_1 z|z|^2 + O(|z|^4).$$

[5]In order to avoid too cumbersome notations we still call z the transformed coordinate $H_3(z)$.

Now, if $\lambda^q \neq 1$ for all $q \leq 2n+3$, one finds by induction a local diffeomorphism $H = H_{2n+2} \circ H_{2n+1} \circ H_3 \circ H_2$ tangent to Id at 0 such that

$$H \circ F \circ H^{-1}(z) = \lambda z + \sum_{k=1}^{n} c_k z|z|^{2k} + O(|z|^{2n+3}).$$

If $\lambda^{2n+3} = 1$, there is possibly a monomial $\gamma \bar{z}^{2n+2}$ which cannot be canceled. Finally, chosing polar coordinates, one writes $H \circ F \circ H^{-1}$ as in the conclusion of the theorem.

Remark Resonances of the form $\lambda^q = 1$ for $1 \leq q \leq 4$ are called *strong resonances*. They are characterized by the fact that the resonant monomial $\bar{z}^{q-1}$ is of smaller or comparable order to the first unvoidable resonant monomial $z|z|^2$ and hence could play a role in the geometry of the normal form N which could become invariant only by rotations by an angle multiple of $2\pi/q$. *In the sequel, the hypotheses always exclude strong resonances.*

Remark on Notations Theorem 1 allows us to suppose from the start that local coordinates z have been chosen so that F is in the form given, by Theorem 1. In other words, from now on we shall write $F(z)$ instead of $H \circ F \circ H^{-1}(z)$.

11.4 The Dissipative Case

11.4.1 Andronov–Hopf–Neimark–Sacker Bifurcation

The first two names are attached to the "continuous" case of a differential equation, the last two to the present "discrete" case of a map (see [2, 3, 13, 17]).

In general, the polynomial $f(s) = \sum_{k=1}^{n} a_k s^k$ is such that $a_1 \neq 0$. If $a_1 < 0$, one can scale the coordinates so that $a_1 = -1$ which, provided $\lambda^q \neq 1$ for all integers $1 \leq q \leq 4$, puts F into the form

$$F(z) = N(z) + O(|z|^4), \quad \text{where} \quad N(z) = z\left(1 - |z|^2\right) e^{2\pi i\left(\omega + b_1|z|^2\right)}.$$

As well as the rotation $dF(0)$, the normal form N still leaves invariant the *foliation* by circles centered at 0 but it sends the circle of radius r onto the circle of radius $r(1-r^2)$. This implies not only that $\lim_{m\to\infty} N^m(z) = 0$ but also that $\lim_{m\to\infty} F^m(z) = 0$ as soon as $|z|$ is small enough. Indeed, if $|z|$ is small enough, $|F(z)| < |z| \left|1 - \frac{1}{2}|z|^2\right|$.

One says that 0 is a *weak attractor* (Fig. 11.1), the adjective "weak" recalling that the attraction is due to a non-linear term.

Hence we completely understand the dynamics of F in some neighborhood $\mathcal{V}$ of the fixed point 0. Things become much more interesting if one perturbs F by

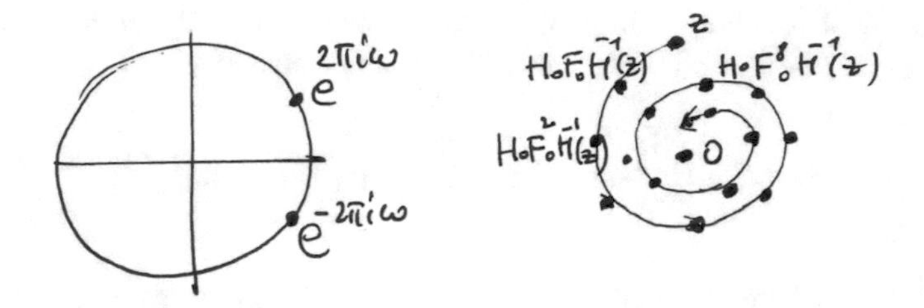

Fig. 11.1 Weak attraction

including it in a smooth one parameter family of local diffeomorphisms F_μ such that $F_0 = F$. A direct application of the implicit function theorem shows that, in the neighborhood of 0, the equation $F_\mu(z) - z = 0$ has a unique solution z_μ depending smoothly on μ and such that $z_0 = 0$. Hence, after a translation by z_μ of the coordinates, one can suppose that for all μ near 0, one has $F_\mu(0) = 0$.

For values of μ such that the spectrum of $dF_\mu(0)$ is not on the unit circle, there is no resonance and one could get a normal form which is linear up to any order. However, this would not be of much use: on the one hand the domain of definition of the conjugating diffeomorphism H_μ tends to 0 when the spectrum of $dF_\mu(0)$ tends to the unit circle and interesting phenomena occur outside of this domain, on the other hand, this would break the continuity with respect to μ of the coordinate change H_μ. In consequence, one chooses to eliminate in F_μ only the same terms as the ones we have eliminated in F_0, that is we mimic for H_μ the construction of H in Sect. 11.3. Doing so one gets a smooth family H_μ of local diffeomorphisms of $(\mathbb{C}, 0)$ defined in a fixed neighborhood of 0 which put F_μ into the form $F_\mu(z) = z(1 + f_\mu(|z|^2))e^{2\pi i(\omega + g_\mu(|z|^2))} + \cdots$ given by Theorem 1 except that $f_\mu(s) = \sum_{i=0}^n a_k s^k$ and $g_\mu(s) = \sum_{k=0}^n b_\mu(s) s^k$ now start with terms of degree 0. Finally, we shall suppose that $a_0(\mu)$ is monotone (say increasing) for μ close enough to zero. This is also a "generic" condition which amounts to saying that the spectrum of the derivative $dF_\mu(0)$ crosses transversally the unit circle when μ crosses the value 0. It allows us to change parameters and suppose that $a_0(\mu) = \mu$. At the end, we are reduced to study a family F_μ of local diffeomorphisms of the form

$$\begin{cases} F_\mu(z) = N_\mu(z) + O(|z|^4), \text{ where} \\ N_\mu(z) = z\left(1 + \mu + a_1(\mu)|z|^2\right) e^{2\pi i\left(b_0(\mu) + b_1(\mu)|z|^2\right)}, \text{ and} \\ a_1(\mu) = -1 + O(|\mu|), \; b_0(\mu) = \omega + O(|\mu|). \\ \text{The rest can be made } O(|z|^5) \text{ except if } \lambda^5 = 1, \text{ which can leave a term } \gamma\bar{z}^4. \end{cases}$$

Due to the commutation of N_μ with the group SO(2) of rotations, the study of its dynamics reduces to an elementary question in dimension 1. The results are summarized in Fig. 11.2: the origin, which is a strong (=linear) attractor when

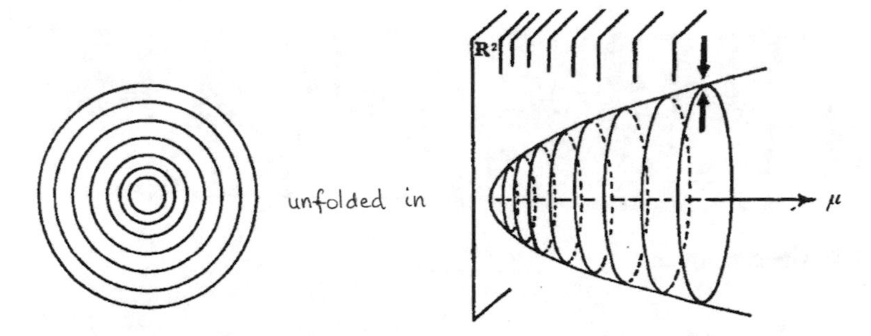

Fig. 11.2 Dynamics of the family of normal forms N_μ

$\mu < 0$, becomes a strong repellor when $\mu > 0$. But points far enough from the origin are still attracted and in between appears an invariant circle C_μ of radius r_μ the unique solution of the equation $\mu + a_1(\mu)r_\mu^2 = 0$.

Theorem 2 (Neimark 1959, Sacker 1964) *Under the above hypotheses, for each $\mu > 0$ small enough, F_μ possesses an invariant closed curve Γ_μ, close to C_μ, which attracts a uniform (that is independent of μ) neighborhood $\mathcal{V}$ of 0 (with 0 deleted). If the local diffeomorpisms F_μ are of class C^∞, these curves are of class C^k with k depending on μ and going to infinity when μ tends to 0.*

The proof proceeds in two steps:

1. One encloses the invariant circle C_μ in an annulus A_μ of width $O(|\mu|)$, say the one bounded by the circles whose radii $r_\mu^\pm$ are the two solutions of the equation $\mu + a_1(\mu)r^2 \pm r^3 = 0$. One checks that every point $z \neq 0$ in some uniform (i.e. independent of μ) neighborhood $\mathcal{V}$ of 0 is eventually sent inside A_μ under the iterates of F_μ (Fig. 11.3).
2. One shows that under the iterates of F_μ, every point inside the annulus tends asymptotically to some invariant curve Γ_μ close to the circle C_μ. For this, we choose coordinates in A_μ centered on C_μ of the form:

$$z = r_\mu(1 + \sqrt{\mu}\sigma)\, e^{2\pi i\theta}.$$

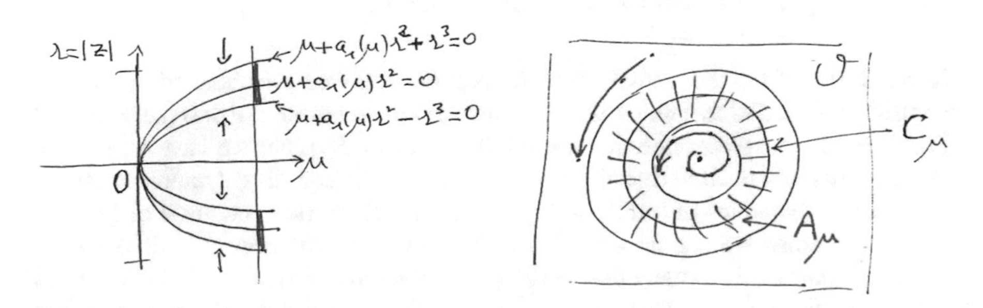

Fig. 11.3 The attracting annulus A_μ

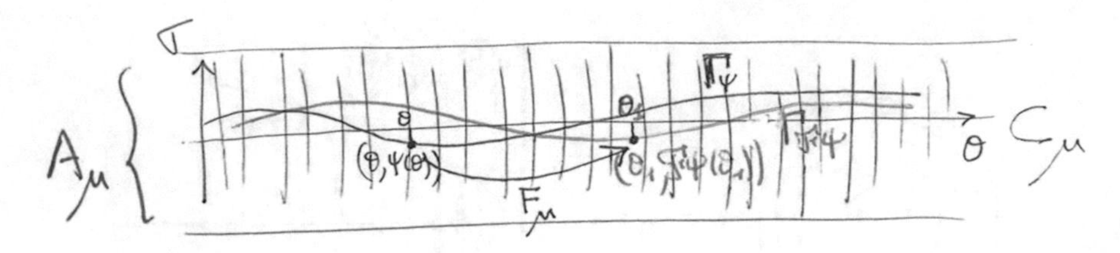

Fig. 11.4 Graph transform

If $\lambda^5 \neq 1$, the map F_μ becomes (we keep the same notation F_μ for convenience)

$$F_\mu(\sigma, \theta) = \Big((1 - 2\mu)\sigma + O(\mu^{3/2}), \theta + b\mu + O(\mu^{3/2})\Big).$$

(If $\lambda^5 = 1$ and the term $\gamma\bar{z}^4$ is present, a circle is not a good enough approximation of the invariant curve and a further change of variables is necessary to get to the above form, see [17]). Let $(\theta, \psi(\theta))$ be the graph of a function $\theta \mapsto \sigma = \psi(\theta)$ from the circle $\mathbb{R}/\mathbb{Z}$ to $\mathbb{R}$. If ψ is small enough, its graph Γ_ψ is contained in the annulus A_μ and the image by F_μ of its graph, also contained in A_μ, is the graph of a function $\mathcal{F}_\mu\psi$:

$$F_\mu(\Gamma_\psi) = \Gamma_{\mathcal{F}_\mu\psi}.$$

The map $\psi \mapsto \mathcal{F}_\mu\psi$ is called the *graph transform*. Thanks to the contracting factor $1 - 2\mu$ which dominates any contraction along the angular direction (a manifestation of the fact that the *normal hyperbolicity* of C_μ dominates the perturbation), one shows that $\mathcal{F}_\mu$ is a contraction in a well-chosen Banach space of C^k functions provided μ is close enough to 0 (a condition more and more stringent when k tends to $+\infty$). The attracting invariant curve $\Gamma_\mu \subset A_\mu$ we are looking for is the graph of the unique fixed point of this contraction (Fig. 11.4).

11.4.2 Dynamics on the Invariant Curves

In conclusion, from the "radial" hypothesis $a_1(0) < 0$ we have obtained a complete control on the radial dynamics of F_μ in a uniform neighborhood $\mathcal{V}$ of 0 (i.e. Fig. 11.2 is still pertinent to describe the normal dynamics of F_μ), but we have no control of the dynamics restricted to the invariant curves. Indeed, this dynamics may be a "generic" dynamics of a diffeomorphism of the circle (see section 4 of [7]). To be more precise we should add another "generic" assumption, this time on the "angular" part of F, namely that $b_1(0) \neq 0$, for example $b_1(0) > 0$. This implies that, for μ close enough to 0, the restriction of the normal form N_μ to its invariant circle C_μ is a rotation whose angle increases with μ. The two-parameter family $f_{\omega,\mu}$

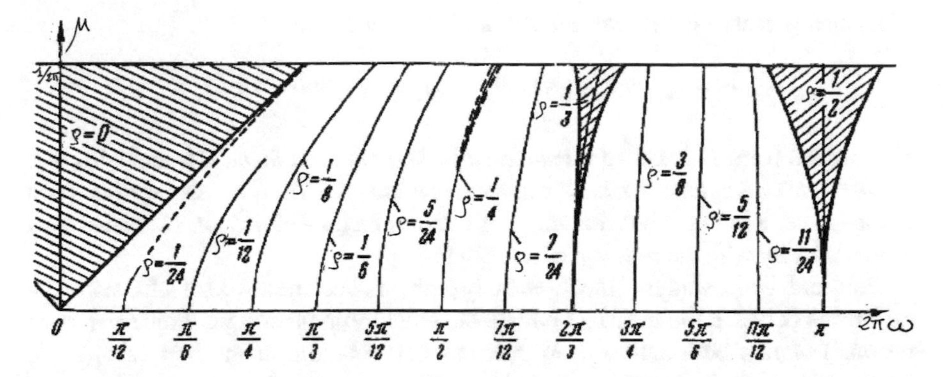

Fig. 11.5 Typical behavior of a 2-parameter family of circle diffeomorphisms (figure adapted from [4])

of diffeomorphisms of the circle defined by the restriction of F_μ to its invariant curve Γ_μ, the other parameter being ω, behaves in general as does Arnold's family $T_{\omega,\mu}$, described in [4]:

$$T_{\omega,\mu}(\theta) = \theta + \omega + \mu \cos 2\pi\theta.$$

In the interior of each of the so-called *Arnold's tongues*—values of the parameters for which the rotation number is rational—$f_{\omega,\mu}$ is in general a diffeomorphism of the circle with two periodic orbits of the same period q if the root of the tongue is the rotation of angle $(2\pi\omega = 2\pi\frac{p}{q})$. One orbit is attracting, the other repelling. Such periodic orbits cannot be destroyed by a small enough perturbation and hence persist over an interval of values of ω for each $\mu \neq 0$; the complement of the union of all these intervals is a Cantor set of values of ω for which $f_{\omega,\mu}$ is topologically (but not always smoothly) conjugated to a rotation. Moreover, for any $\mu \neq 0$, the set of ω for which the rotation number of $f_{\omega,\mu}$ is rational is in general big in the sense of topology, namely it is open and dense, but its complement is always big in the sense of measure, namely, its measure tends to 1 when $\mu \to 0$ (see [16]) (Fig. 11.5).

11.5 The Area Preserving Case

11.5.1 Moser's Invariant Curve Theorem [21]

We now suppose that, in addition to satisfying $\lambda^q \neq 1$ for all integers $1 \leq q \leq 4$, F is area preserving. It follows that the radial component f of the normal form N vanishes identically and one can show that it is possible to choose H area preserving. Hence, one is reduced to the study in the neighborhood of its elliptic fixed point 0 of an area-preserving diffeomorphism of $\mathbb{C}, 0$ of the form

$$F(z) = N(z) + O(|z|^4), \quad N(z) = ze^{2\pi i(\omega + b_1|z|^2)}.$$

The normal form N is called a *truncated Birkhoff normal form*. Dynamically, it is an *integrable monotone twist*: as well as the rotation $dF(0)$, it leaves invariant each circle C_r centered at 0 but the angle of rotation $2\pi(\omega + b_1 r^2)$ on C_r varies now monotonically with the radius r of this circle.

Poincaré, while studying the three body problem, became aware of a fundamental difference between the invariant circles on which N induces a periodic ($\omega + b_1 r^2$ rational) or non periodic ($\omega + b_1 r^2$ irrational) rotation: in the first case (angle $2\pi\omega = 2\pi p/q$) the invariant circle is simply the union of a continous family of q-periodic points z (i.e. of points z such that $N^q(z) = z$); in consequence, a small perturbation should in general break such a circle, with only a finite number of periodic points surviving the perturbation. On the other hand, if ω is irrational, the invariant circle being the closure $\overline{\cup_{n\geq 0} N^n(z)}$ of an orbit has a dynamical origin and hence has more chance to resist a perturbation. In the first volume of his famous book *The New Methods of Celestial Mechanics*, Poincaré even ventured to write that some arithmetic condition on ω could perhaps grant resistance to perturbations of such an invariant circle but that he considered such a possibility as quite improbable (Fig. 11.6).

Nevertheless, after the pioneering work of Kolmogorov in 1954, the so-called *KAM theory* (from the names of Kolmogorov, Arnold and Moser) showed that indeed, what Poincaré deemed improbable was in fact a dominant phenomenon. In the present case, the pertinent statement is the following

Theorem 3 (Moser 1962) *Given an area preserving diffeomorphism F as above, given $C > 0$ and $\beta > 0$, there exists $\epsilon(C, \beta) > 0$ such that each invariant circle C_{r_0} of the normal form N such that its rotation angle $2\pi\omega_{r_0} = 2\pi(\omega + b_1 r_0^2)$ satisfies the diophantine condition*

$$\forall \frac{p}{q} \in \mathbb{Q}, \left|\omega_{r_0} - \frac{p}{q}\right| \geq \frac{C|\omega_{r_0} - \omega|}{|q|^{2+\beta}} \quad \text{and} \quad |\omega_{r_0} - \omega| < \epsilon(C, \beta)$$

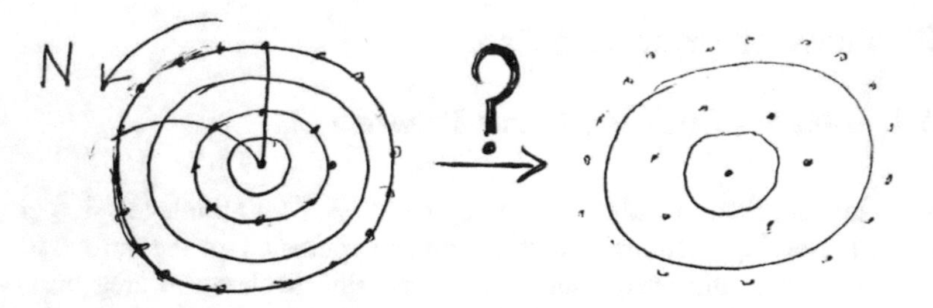

Fig. 11.6 Perturbation of a monotone twist ???

will give rise to a smooth (resp. analytic) closed curve Γ_{r_0} invariant under F and such that the restriction $F|_{\Gamma_{r_0}}$ of F is smoothly conjugate to the rotation of angle $2\pi\omega_{r_0}$.

The most transparent proof of theorem 2 is based on a version of the so-called "hard implicit function theorem" adapted to the problem of *small denominators* well known to astronomers since eighteenth century. The following consequence of area preservation, named *intersection property*, is crucial: the image $F(\Gamma)$ of a curve Γ surrounding the origin cannot be disjoint from Γ. Note that such a property is preserved even under changes of coordinates which do not preserve area. Fixing $r = r_0$ satisfying the hypotheses of the theorem, one chooses coordinates centered on C_{r_0} of the form:

$$z = r_0\sqrt{1+\sigma}\, e^{2\pi i\theta}.$$

The map F is now (as before we keep the same notation F)

$$F(\sigma,\theta) = \left(\sigma + O(r_0^4),\ \theta + \omega_{r_0} + b_1 r_0^2\sigma + O(r_0^4)\right).$$

As a further simplification, one replaces σ by $\rho = \sigma + O(r_0^2)$ so that the formula for F takes the form

$$F(\rho,\theta) = \left(\rho + \varphi(\rho,\theta),\ \theta + \omega_{r_0} + b_1 r_0^2\rho\right),$$

where the perturbation φ is $O(r_0^4)$. Following Rüssmann, it is enough to look for a curve of the form $\rho = \psi(\theta)$ which is sent by F to the *translated curve* $\rho = \psi(\theta)+\tau$ for some $\tau \in \mathbb{R}$. This is because the intersection property, still valid after the changes of coordinates, implies that τ must be equal to 0. This leads to the equation

$$\psi\big(g(\theta)\big) + \tau = \psi(\theta) + \varphi\big(\psi(\theta),\theta\big), \quad \text{where} \quad g(\theta) = \theta + \omega_{r_0} + b_1 r_0^2\psi(\theta).$$

Recall that in the dissipative case, the radial hypothesis $a_1(0) \neq 0$ implied the existence of a curve invariant under F_μ with a prescribed normal dynamics. Having now an angular hypothesis $b_1 \neq 0$, it is natural to look for invariant curves of F with a prescribed angular dynamics. It turns out that the right constraint to impose to the (translated) curve we are looking for is the existence of a diffeomorphism h of the circle $\mathbb{R}/\mathbb{Z}$ such that $g(\theta) = h^{-1}\circ R_{\omega_{r_0}} \circ h(\theta)$.

Finally, defining ψ by $\psi(\theta) = \frac{1}{b_1 r_0^2}\left[h^{-1}\circ R_{\omega_{r_0}}\circ h(\theta) - \theta - \omega_{r_0}\right]$, we must solve

$$\mathcal{F}(\varphi,\tau,h) := \psi(\theta) - \psi(h^{-1}\circ R_{\omega_{r_0}}\circ h(\theta)) - \tau + \varphi(\psi(\theta),\theta) = 0$$

in the neighborhood of the solution ($\varphi = 0$, $\tau = 0$, $h = Id$). This is typically a "hard implicit function problem" because even the best diophantine condition allows us only to invert the "derivative" of $\mathcal{F}$ in a weak sense (i.e. with loss of a finite number of derivatives on the target space of the inverse).

Warning Examples in [1] show that an area preserving C^∞-diffeomorphism of the disk D^2 with an elliptic fixed point such that ω is a *Liouville number*, too well approximated by rational numbers, may have a very wild dynamics, with dense orbits.

11.5.2 Periodic Orbits, Aubry-Mather Sets and Homoclinic Tangles

The curves Γ_{r_0} given by Theorem 3 form a Cantor family for which 0 is a density point (the relative measure of the Cantor set in smaller and smaller neighborhoods of 0 tends to 1). Nevertheless, this is far from being the whole story. The dynamics of such a generic area-preserving F in the complement of the invariant curves (the so-called *Birkhoff domains of instability*) is extremely complicated and, if the works of Birkhoff, Aubry, Mather, Herman, have shed considerable light on the way invariant circles of the normal form break (periodic points, invariant Cantor sets, see [14]), many questions remain open.

Some of the complexity of a generic area-preserving map of the disc is roughly suggested in Fig. 11.7. This figure, taken from [8], originates from [15]. It illustrates the dynamics of the monotone twist map of the annulus which arises when studying the *restricted three-body problem* at high values of the Jacobi constant (see [9] for explanations). To the periodic points are attached invariant stable (resp. unstable) manifolds along which the images of a point under the positive (resp. negative) iterates of F converge exponentially fast to the periodic orbit. The homoclinic tangles (see [8, 22]) created by the intersections of such invariant manifolds produce invariant Cantor sets on which the dynamics of F is the same as the one of throwing a dice (more technically, a Bernoulli shift, see [7, 18]) and hence possesses positive *topological entropy*. Also, orbits go from one boundary of a domain of instability to the other, but their *diffusion* is blocked by the invariant curves.

Remark One can check [9, 20] that Moser's invariant curve theorem applies to the Poincaré first return map on a surface of section of the planar circular restricted three body problem with any large enough energy in the rotating frame (i.e. *Jacobi constant*). This implies stability in a strong sense as the invariant tori corresponding to the invariant closed curves are of codimension 1 in the energy surface and hence serve as barriers confining the solutions. This is precisely because he lacked such a theorem that Poincaré tried to prove such a stability result using barriers made from invariant manifolds of periodic orbits, which lead to the famous error in the first version of his prize winning Memoir on the Three-body problem (see [8]).

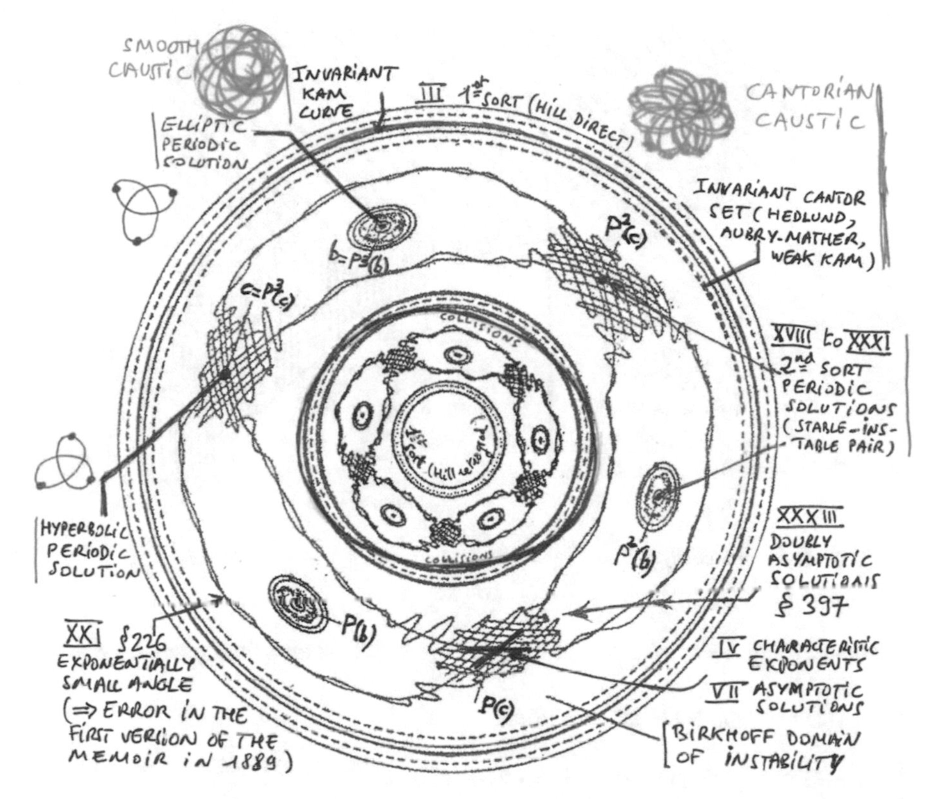

Fig. 11.7 The return map of the restricted 3-body problem at high Jacobi constant (figure reproduced (slightly modified) with the kind permission of *Encyclopædia Universalis*)

11.6 When Radial and Angular Behaviours Compete

Area-preserving maps form a subspace of infinite codimension within the set of all smooth maps and the same is true of rotations. If one views Neimark-Sacker bifurcation as an unfolding, due to the nonlinear terms, of the continuum of circles invariant by the rotation along the parameter μ (Fig. 11.2), the infinite codimension reflects the infinite number of events which happen for one and the same map while generically they happen for different values of μ. In a similar but subtler way one shows ([10], summarized in [5, 11, 12, 23]) that the whole complexity of the dynamics of an area-preserving map happens unfolded along some curve Γ of the parameter space in generic 2-parameter families $F_{\mu,a}$ of diffeomorphisms of the plane in the neighborhood of a degenerate elliptic fixed point which is a *very weak attractor*. Along the lines of Sect. 11.3, provided $F = F_{0,0}$ satisfies the non

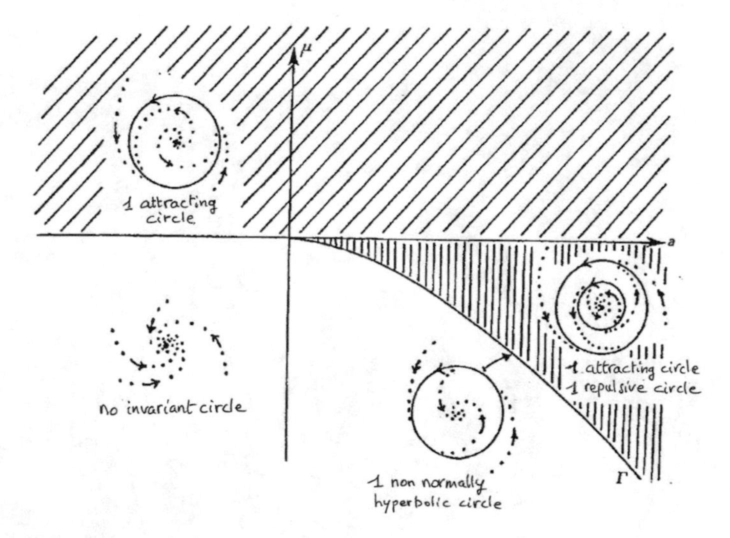

Fig. 11.8 Dynamics of $N_{\mu,a}$

resonance relations $\lambda^k \neq 1$ for all integers $1 \leq k \leq 6$, such a family can be written

$$\begin{cases} F_{\mu,a}(z) = N_{\mu,a} + O(|z|^6), \quad \text{where} \\ N_{\mu,a}(z) = z\left(1+\mu+a|z|^2+a_2(\mu,a)|z|^4\right) e^{2\pi i\left(b_0(\mu,a)+b_1(\mu,a)|z|^2+b_2(\mu,a)|z|^4\right)}, \\ a_2(0,0) = -1,\ b_1(0,0) \neq 0,\ b_1(0,0) + 2\dfrac{\partial b_0}{\partial a}(0,0) \neq 0. \end{cases}$$

Figure 11.8 shows the dynamics of $N_{\mu,a}$ in the different regions of the parameter plane around (0, 0). Along the curve Γ, $N_{\mu,a}$ possesses a non normally hyperbolic invariant curve, attracting from the outside and repelling from the inside. This is in some sense the closest dissipative approximation to an invariant curve of an area-preserving normal form.[6]

The complement of some cusp neighborhood of Γ belongs to the *hyperbolic domain*: here, the "normal" dynamics of $F_{\mu,a}$ is similar to the one of $N_{\mu,a}$ and the methods of proof are the ones of Sect. 11.4. On the contrary, in the cusp domain along Γ, the control is more on the angular dynamics of $F_{\mu,a}$ and the methods of proof are the ones of Sect. 11.5. This is a first approximation of the *elliptic domain* (Fig. 11.9).

More precisely, for a Cantor set of points (μ, a) near Γ the dynamics of $F_{\mu,a}$ is similar both in radial and angular directions to the one of $N_{\mu',a'}$ for some $(\mu', a') \in$

[6]All figures in this section are reproduced with the kind permission of *Publications mathématiques de l'IHÉS*.

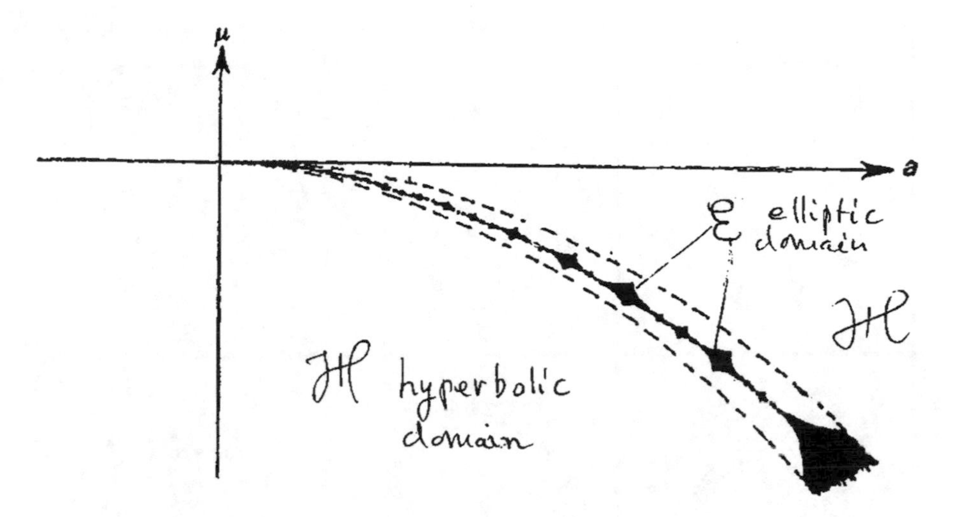

Fig. 11.9 Hyperbolic and elliptic domains

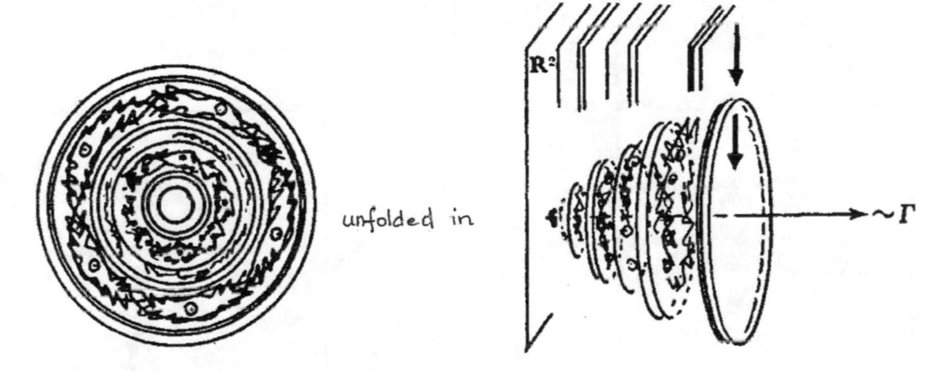

Fig. 11.10 "Unfolding" the dynamics of a monotone twist

Γ. Moreover, the hyperbolic domain extends to the complement of a countable number of bubbles having this Cantor set in their closure. The union of these bubbles is precisely the elliptic domain, the only place where complicated dynamics occurs. Figure 11.10, to be compared to Fig. 11.2, shows that one can describe heuristically the dynamics of $F_{\mu,a}$ along this elliptic domain as the unfolding of the dynamics of a generic area-preserving map as represented in Fig. 11.7.

Finally, in the neighborhood of an elliptic fixed point, generic one-parameter families of planar diffeomorphims displaying the elimination of a pair of invariant closed curves, one repelling and one attracting, may be thought of as being the dissipative analogues of the invariant subsets of a generic area-preserving diffeomorphism : in particular, to the Cantor set of KAM curves corresponds a Cantor set of families along which the elimination proceedes as simply as in the case of normal forms (or equivalently of time one maps of differential equations) with

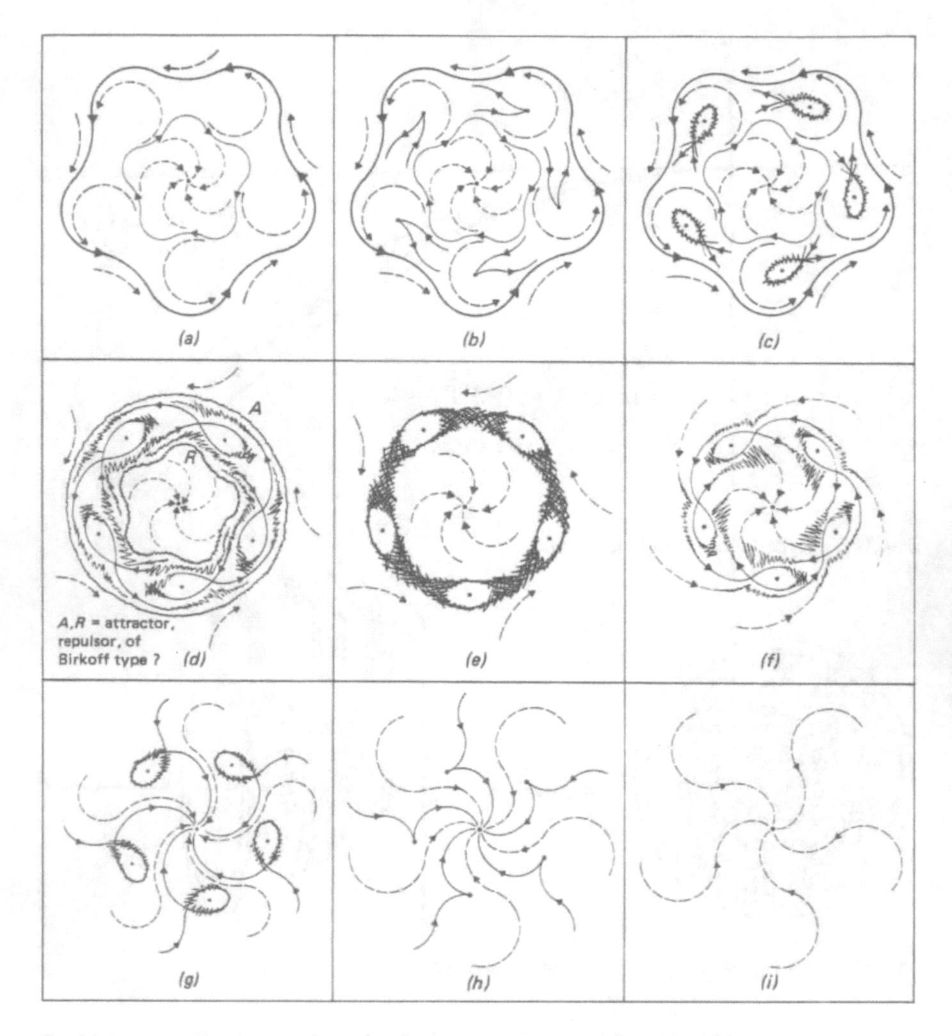

Fig. 11.11 Resonant elimination of a pair on invariant curves (from [10] III)

a single value of the parameter for which the diffeomorphism posesses a smooth invariant closed curve which is non normally hyperbolic and on which $F_{\mu,a}$ is smoothly conjugate to a diophantine rotation, while to the well ordered periodic orbits with rational rotation numbers p/q such that q is not too large with respect to the distance of the orbit to the fixed point 0, correspond one-parameter families along which the elimination process, much more complicated, is represented on Fig. 11.11.

The condition on p/q amounts to asking that in some annulus A containing the periodic points of rotation number p/q, the qth iterate $F^q_{\mu,a}$ of the map still be a small perturbation of the qth iterate $N^q_{\mu,a}$ of its normal form. The said periodic points are then interpreted as the trace left by a nearby resonant elliptic fixed point (compare Sect. 11.4.2) and resonant normal forms provide local coordinates (θ, y) in the annulus A which make the one-parameter subfamily $F_{\mu,a}$ depicted in Fig. 11.11 appear as a perturbation of the composition of the rotation $R_{p/q}$ (of angle $2\pi p/q$) with the time 1 map of a differential equation of the form

$$\frac{d\theta}{dt} = y, \quad \frac{dy}{dt} = \alpha + \gamma y^2 + \delta \cos 2\pi q\theta,$$

where $\gamma < 0$ and $\delta > 0$ are fixed and α is the parameter.

Finally, a surprizing consequence of this study is the strong organizing power of diophantine rotation numbers: if some $F_{\mu,a}$ possesses a closed invariant curve encircling 0 on which it induces a diffeomorphism with such a rotation number, it behaves like a normal form in a uniform (independent of (μ, a)) neighborhood of the origin, the sole possibly more complicated dynamics occuring in restriction to the second invariant closed curve when that curve exists.

Acknowledgements This text was written at the occasion of lectures given at IRMA (Université de Strasbourg), Southwest Jiaotong University (Chengdu), Tsinghua University and Capital Normal University (Beijing). The author warmly thanks the colleagues in these institutions who gave him the occasion to revisit these topics. Finally, thanks to both anonymous referees for a thorough reading which led to some clarifications and the elimination of typos.

References

1. D.V. Anosov and A.B. Katok, New examples in smooth ergodic theory. Ergodic diffeomorphisms, *Trans. Moscow Math. Soc.* Vol 23 (1970)
2. V.I. Arnold, *Geometrical Methods in the Theory of Ordinary Differential Equations*, Grundlehren vol. 250, 1983 Springer-Verlag
3. V.I. Arnold (editor) *Encyclopædia of Mathematical Sciences, Dynamical Systems V: Bifurcation theory*, Springer 1994
4. V.I. Arnold, *Small denominators. I: Mappings of the circumference onto itself*, AMS Translations, Ser. 2, 46 (1965), 213–284.
5. D.K. Arrowsmith & C.M. Place, *An introduction to Dynamical Systems*, Cambridge University Press 1990
6. C. Baesens and R. MacKay, Resonances for weak coupling of the unfolding of a saddle-node periodic orbit with an oscillator, *Nonlinearity* 20, 2007, 1283–1298
7. A. Chenciner, *Discrete dynamical systems*, Tsinghua, february–march 2017 https://www.imcce.fr/fr/presentation/equipes/ASD/person/chenciner/polys.html
8. A. Chenciner, Poincaré and the three-body problem, in "*Poincaré 1912–2012*", Birkhauser 2014 http://www.bourbaphy.fr/novembre2012.html
9. A. Chenciner, *The planar circular restricted three body problem in the lunar case*, minicourse at the Chern Institute of Mathematics, Nankai University, may 2014 https://www.imcce.fr/fr/presentation/equipes/ASD/person/chenciner/polys.html
10. A. Chenciner, Bifurcations de points fixes elliptiques I. Courbes invariantes, *Publications mathématiques de l'I.H.É.S.* tome 61, pp. 67–127 (1985) II. Orbites périodiques et ensembles de Cantor invariants, *Inventiones mathematicæ* 80, pp. 81–106 (1985) III. Orbites périodiques de "petites" périodes et élimination résonnante des couples de courbes invariantes, *Publications mathématiques de l'I.H.É.S.* tome 66, pp. 5–91 (1987)

11. A. Chenciner, Resonant Elimination of a Couple of Invariant Closed Curves in the Neighborhood of a Degenerate Hopf Bifurcation of Diffeomorphisms of $\mathbb{R}^2$, in *Dynamical Systems*, Proceedings of an IIASA Workshop on Mathematics of Dynamic Processes, Sopron 1985, Springer LN in Economics and Mathematical Systems 287
12. A. Chenciner, Hamiltonian-like phenomena in saddle-node bifurcations of invariant curves for plane diffeomorphisms, in *Singularities and dynamical systems*, S.N. Pneumaticos ed. North Holland 1985
13. A. Chenciner, Bifurcation de difféomorphismes de $\mathbb{R}^2$ au voisinage d'un point fixe elliptique, in *Chaotic Behaviour of Dynamical Systems* (G. Iooss, R.H.G. Helleman, R. Stora editors), Les Houches Session XXXVI (1981), North Holland 1983
14. A. Chenciner, *La dynamique au voisinage d'un point fixe elliptique : de Poincaré et Birkhoff à Aubry et Mather*, Séminaire N. Bourbaki, 1983–1984, exp. 622, 147–170
15. A. Chenciner, *Systèmes dynamiques différentiables*, Encyclopædia Universalis 1985
16. M. Herman, Mesure de Lebesgue et nombre de rotation, in *Geometry and Topology*, Lecture Notes in Mathematics 597, 271–293, Springer 1977
17. G. Iooss, *Bifurcation of Maps and Applications*, North-Holland 1979
18. A. Katok & B. Hasselblatt, *Introduction to the Modern theory of Dynamical Systems*, Cambridge University Press 1995
19. A. Keane and B. Krauskopf, Chenciner bubbles and torus break-up in a periodically forced delay differential equation, *Nonlinearity* 31, 2018, R165
20. M. Kummer, On the Stability of Hill's Solutions of the Plane Restricted Three Body Problem, *American Journal of Mathematics*, Vol. 101, No. 6 (Dec., 1979), 1333–1354
21. J. Moser, *Stable and random motions in dynamical systems*, Princeton University Press 1973
22. S. Smale, Differentiable Dynamical Systems, *Bull. Amer. Mat. Soc.* 1967
23. J.C. Yoccoz, *Bifurcations de points fixes elliptiques*, Séminaire N. Bourbaki, 1985–1986, exp. 668, 313–334

N.B. Study of the dynamics inside resonance bubbles has been further studied in [6], while the phenomena inside the bubbles were recently applied to a model of El Niño in [19] :

Chapter 12
René Thom and an Anticipated h-Principle

François Laudenbach

Abstract The first part of this chapter intends to present the role played by Thom in diffusing Smale's ideas about immersion theory, at a time (1957) where they sound counterintuitive: *it is **clearly** impossible to make the sphere inside out!* Around a decade later, M. Gromov transformed Smale's idea in what is now known as the *h-principle*. Here, the h stands for *homotopy*.

Shortly after the astonishing discovery by Smale, Thom gave a lecture in Lille (1959) announcing a theorem which would deserve to be named *a homological h principle*. The aim of our second part is to comment about this theorem which, at that time, was completely ignored by the topologists in Paris, but not in Leningrad. We explain Thom's statement and comment about it. The first idea is combinatorial. A beautiful subdivision of the standard simplex emerges from Thom's article. We connect it with the *jiggling* technique introduced by W. Thurston in his seminal work on foliations.

12.1 From Immersions Viewed by Smale to Gromov's h-Principle

12.1.1 Thom and Smale in 1956–1957

Important and reliable information[1] about Stephen Smale in these years is given by Hirsch [16, p. 36]:

[1]Curiously enough some biographies online give 1957 as the date of Smale's thesis even though the footnote in [28] is quite clear on this matter.

F. Laudenbach (✉)
Laboratoire de Mathématiques Jean Leray, UMR 6629 du CNRS, Université de Nantes, Nantes, France
e-mail: francois.laudenbach@univ-nantes.fr

S. G. Dani, A. Papadopoulos (eds.), *Geometry in History*,
https://doi.org/10.1007/978-3-030-13609-3_12

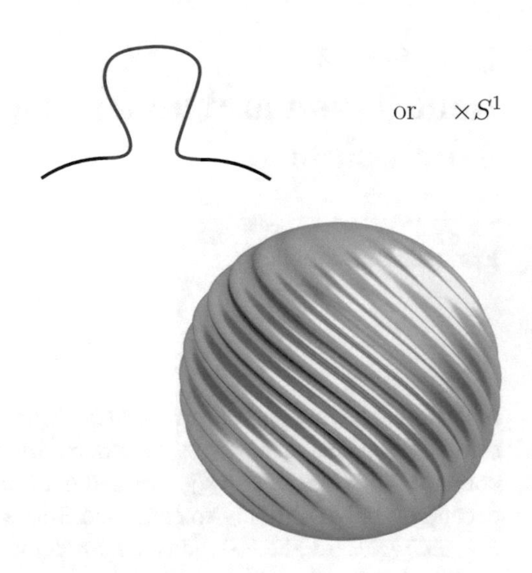

Fig. 12.1 Corrugation in dimension one and two

Fig. 12.2 First corrugating step for an isometric embedding of the unit sphere into the ball of radius 1/2. By courtesy of the Hevea Project

> I first learned of Smale's thesis at the 1956 Symposium on Algebraic Topology in Mexico City. I was a rather ignorant graduate student at the University of Chicago, Smale was a new PhD from Michigan ... I thought I could understand the deceptively simple geometric problem Smale addressed: *Classify immersed curves in a Riemannian manifold.*

René Thom gave an invited lecture at the same Symposium. Probably, it was the first occasion for Thom and Smale to meet. Let us continue reading Hirsch [16]: "In the Fall of 1956, Smale was appointed Instructor at the University of Chicago".

On January 2, 1957, Smale submitted an abstract to the Bulletin of the American Mathematical Society which was published in the issue of May 1957 [27, Abstract 380*t*]. This is a 14 lines piece[2] titled *A classification of immersions of the 2-sphere* where Smale announces[3] a complete classification of immersions of the 2-sphere valued in C^2 manifolds of dimension greater than two. He writes: "For example any two C^2 immersions of S^2 in E^3 are regularly homotopic".

In the Spring of 1957, Thom spent a semester as invited Professor at the University of Chicago. He spoke with Smale for hours until he had a full understanding of Smale's ideas on immersions. Back to France, Thom reported on Smale's work in a Bourbaki seminar of December 1957 [36] (or [38, p. 455–465]). It is remarkable that the written version of Thom's lecture contains the very first figure which has appeared in the theory of immersions. This is just a *bump*; according to [41], Thurston would have called this picture a *corrugation* (Fig. 12.1).

I should say that the theory of corrugations is still very lively for constructing concrete C^1 isometric embeddings (see Borrelli et al. [2] and Fig. 12.2).

In the rest of Sect. 12.1, I would like to present Smale's ideas, starting from the basics, and connect them with more recent ideas.

---

[2]This abstract is not included in [31].

[3]The complete paper following this announcement is [29].

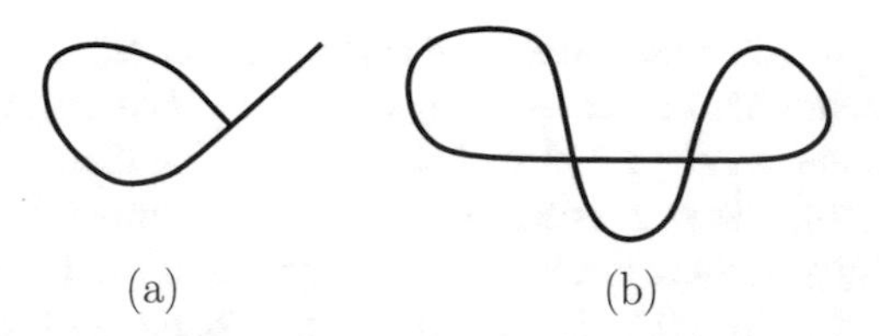

Fig. 12.3 (**a**) shows an embedding $(0, 1) \to \mathbb{R}^2$ whose image is not a submanifold. (**b**) shows an immersion $\mathbb{S}^1 \to \mathbb{R}^2$ which does not extend to an immersion of the 2-disc

12.1.2 Immersions

Given two smooth manifolds X and Y where the dimension of X is not greater than the dimension of Y, a C^1-map $f : X \to Y$ is said to be an immersion if its differential df is of maximal rank at every point of X. An immersion may have double points but no singular points like *folds*. Under a properness condition, an immersion with no double points is said to be an *embedding* and its image is a submanifold of Y; the condition is that f is a proper map[4] from X to some open subset of Y (Fig. 12.3a).

The space of immersions from X to Y, denoted by $Imm(X, Y)$, is an open set in $C^1(X, Y)$ if the space of C^1 maps is endowed with the so-called *fine Whitney* topology. When X is compact, there is no concern: a sequence (f_n) is convergent if and only if both sequences $(f_n(x))$ and $(df_n(x))$ converge uniformly. In what follows, we shall only consider immersions whose domain is compact. In that case, the set of immersions is locally contractible.

Two immersions $f_0, f_1 : X \to Y$ are said to be regularly homotopic if they are joined by a path in $Imm(X, Y)$ or equivalently if f_0 and f_1 belong to the same connected component of $Imm(X, Y)$.

12.1.3 Whitney-Graustein Theorem

The immersions from the circle to the plane have been classified by H. Whitney up to regular homotopy in the mid thirties [43]. The classification reduces to the degree of the Gauss map

$$\begin{aligned} G : \mathbb{S}^1 &\to \mathbb{S}^1 \\ x &\mapsto df_x(\partial_\theta)/\|df_x(\partial_\theta)\| , \end{aligned}$$

[4]Given two topological spaces A and B, a continuous map from A to B is said to be *proper* if the preimage of any compact set of B is a compact set of A.

where ∂_θ stands for the unit tangent vector to the circle $\mathbb{S}^1 := \mathbb{R}/2\pi\mathbb{Z}$. The reason why this theorem is named *Whitney-Graustein Theorem* is given by Whitney himself in a footnote on page 279 of his article: "This theorem, together with a straightforward proof, was suggested to me by W. C. Graustein".

It is worth noting that there is an interesting proof of the Whitney-Graustein Theorem given by S. Levy in [20, p. 33 - 37] following Thurston's idea of corrugation.

It would be wrong to think that this classification ends the story of the immersions of the circle to the plane. A much more difficult question is the following: *Which immersion extends to an immersion of the disc to the plane? For such an immersion, how many extensions are there?* An obvious necessary condition for a positive answer to the first question is that the degree of the Gauss map be equal to one. But this condition is not sufficient, as Fig. 12.3b shows. Actually, these questions have been solved by S. Blank in his unpublished thesis. Fortunately, V. Poénaru reported[5] on Blank's thesis in a Bourbaki seminar [26]. The analogous questions for immersions of the n-sphere into $\mathbb{R}^{n+1}$ can be raised and remain essentially open.

12.1.4 The Key Proposition in Smale's Thesis

Let f_0 denote the standard embedding of the 2-sphere $\mathbb{S}^2$ in $\mathbb{R}^3$. Choose an equator E on $\mathbb{S}^2$, a base point $p \in E$ and two hemispheres respectively named the northern and the southern hemisphere H_N and H_S. We consider the space of pointed immersions

$$Imm_p(\mathbb{S}^2, \mathbb{R}^3) := \left\{ f : \mathbb{S}^2 \looparrowright \mathbb{R}^3 \mid f(p) = f_0(p) \text{ and } df(p) = df_0(p) \right\}.$$

The spaces $Imm_p(H_S, \mathbb{R}^3)$ and $Imm_p(H_N, \mathbb{R}^3)$ are similarly defined. The space of immersions $H_N \looparrowright \mathbb{R}^3$ whose 1-jet $j^1 f(x) := \big(f(x), df(x)\big)$ coincides with $j^1 f_0(x)$ at every point $x \in E$ is denoted by $Imm_E(H_N, \mathbb{R}^3)$. Finally, $\widetilde{Imm}_p(E, \mathbb{R}^3)$ will denote the space of immersions $f : E \to \mathbb{R}^3$ enriched with a 2-framing along $f(E)$ which is fixed at p and whose span is tangent to $f(E)$.

The space of pointed immersions of the 2-disc into $\mathbb{R}^3$ is known to be contractible thanks to the Alexander's contraction which reads in the present setting:

$$(x, t) \mapsto p + \frac{1}{t}[f\big(p + t(x - p)\big) - f(p)]$$

[5]Note that Poénaru's report contains the drawing of the so-called Milnor example, that is, an immersion of the circle into the plane having two extensions to the disc which are not equivalent up to homeomorphism of $\mathbb{D}^2$.

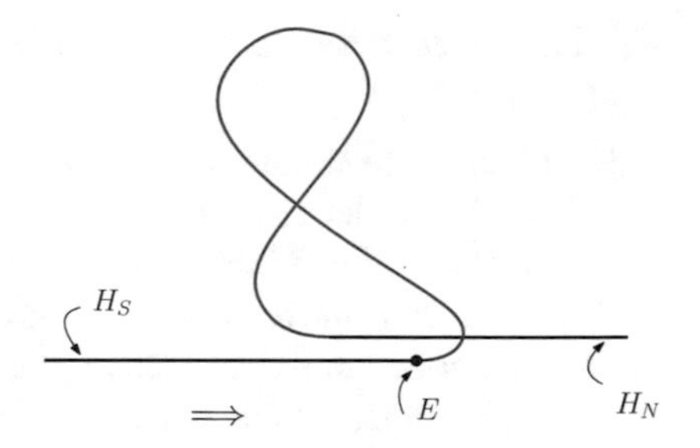

Fig. 12.4 The southern hemisphere moves to the right. Only a northern collar of the equator in H_N is deformed

where p lies in the boundary of $\mathbb{D}^2$ and $(x,t) \in \mathbb{D}^2 \times (0,1]$. When t approaches 0, the limit of the above expression, uniformly in x in the 2-disc, is the affine map $x \mapsto p + df(p)(x-p)$.

Proposition

(1) *The restriction map,* $Imm_p(\mathbb{S}^2, \mathbb{R}^3) \to Imm_p(H_S, \mathbb{R}^3)$, *is a Serre fibration. Its fibre over* f_0 *is homeomorphic to* $Imm_E(H_N, \mathbb{R}^3)$.
(2) *The 1-jet map along the equator,* $Imm_p(H_N, \mathbb{R}^3) \to \widetilde{Imm}_p(E, \mathbb{R}^3)$, *is a Serre fibration.*[6] *Its fibre over* $(j^1 f_0)|_E$ *is also homeomorphic to* $Imm_E(H_N, \mathbb{R}^3)$.

A map $\rho : X \to Y$ between two arcwise connected spaces is said to be a *Serre fibration* if it has the parametric *Covering Homotopy Property*. More precisely, for every $\gamma : [0,1] \to Y$ and every x_0 in X with $\rho(x_0) = \gamma(0)$, there exists a lift $\tilde{\gamma} : [0,1] \to X$ of γ starting from x_0; and similarly for families with parameters in the n-disc. In this case, there is a long exact sequence in homotopy.

It is worth noting that similar statements for a one-dimensional source were already in Smale's thesis (published in [28]).

The proof of the first item is sketched by a picture which shows the *flexibility* that the statement translates (Fig. 12.4).

Corollary *We have* $\pi_0\big(Imm_p(\mathbb{S}^2, \mathbb{R}^3)\big) = 0$, *that is, the space of pointed immersions of* $\mathbb{S}^2 \to \mathbb{R}^3$ *is arcwise connected.*

Proof Since the base of the first Serre fibration is contractible, the homotopy exact sequence yields $\pi_0\big(Imm_p(\mathbb{S}^2, \mathbb{R}^3)\big) \cong \pi_0\big(Imm_E(H_N, \mathbb{R}^3)\big)$. By the second Serre fibration whose total space is contractible, we have $\pi_0\big(Imm_E(H_N, \mathbb{R}^3)\big) \cong \pi_1\big(\widetilde{Imm}_p(E, \mathbb{R}^3)\big)$. Arguing similarly for the enriched immersions of E whose equator is a 0-sphere, we get

$$\pi_1\big(\widetilde{Imm}_p(E, \mathbb{R}^3)\big) \cong \pi_2\big(\widetilde{Imm}_p(S^0, \mathbb{R}^3)\big) \cong \pi_2(\{\text{2-frames in } \mathbb{R}^3\} \cong \pi_2(SO(3)) = \pi_2(\mathbb{S}^3) = 0.$$

□

[6] I am not saying that this 1-jet map is surjective!

12.1.5 Concrete Eversion of the Sphere

I do not intend to explain the history of this matter. I just give a list of references in chronological order and add a few comments: Phillips [22], Francis and Morin [7], Francis' book [6] and finally the text and video by Levy [20].

The first idea, due to Arnold Shapiro, is to pass through *Boy's surface*, here noted Σ, an immersion of the projective plane into the 3-space. Since the projective plane is non-orientable, a tubular neighborhood T of Σ is not a product. Therefore, T is bounded by an immersed sphere $\tilde{\Sigma}$. It turns out that $\tilde{\Sigma}$ is endowed with the involution which consists of intertwining the two end points in each fibre of T. This is realized by the regular homotopy

$$\tilde{\Sigma} \times [0, 1] \to T, \ (x, t) \mapsto x(1 - 2t),$$

where the product in the right hand side is associated to the affine structure of the fibre of x. If the two faces of $\tilde{\Sigma}$ are painted with different colors, this move has the effect of changing the color which faces Boy's surface. It remains to connect the standard embedding of $\mathbb{S}^2$ to Boy's surface by a regular homotopy in order to get an eversion of the sphere.

Remembering a walk with Nicolaas Kuiper during which he explained this construction to me, I had the feeling that he played a role in it. I did not know more until very recently, when Tony Phillips informed me about an article of Kuiper where his argument is written explicitly[7] [17, p. 88]. The video [20] does not follow the same idea: it goes the way of Thurston's corrugations and is not optimal for what regards the number of multiple points of multiplicity 3 or more.

12.1.6 Hirsch's Definitive Statement

The general statement in the *homotopy theory* of immersions is due to Hirsch [15]. He considers an arbitrary pair (X, Y) of smooth manifolds. For simplicity, assume X is connected. The main assumption is that $\dim X \leq \dim Y$, equality being allowed only when X is not closed (each connected component of X must have a non-empty boundary).

If $f : X \to Y$ is an immersion, we have a diagram

$$\begin{array}{ccc} TX & \xrightarrow{df} & TY \\ \downarrow & & \downarrow \\ X & \xrightarrow{f} & Y \end{array}$$

[7]The approaches by Shapiro and Kuiper were contemporary. As far as I know, nothing indicates some relationship between them.

where df is a fibre bundle map *over* f (between the total spaces of the respective tangent bundles) which is fibrewise linear and injective.

Even though this terminology has been in use only after Gromov's thesis, we are going to use it here. A *formal immersion* is a diagram

$$\begin{array}{ccc} TX & \xrightarrow{F} & TY \\ \downarrow & & \downarrow \\ X & \xrightarrow{f} & Y \end{array}$$

where f is only assumed to be continuous and F is a fibre bundle map which is fibrewise linear and injective. In the language of jet spaces, this is just a section of the 1-jet bundle $J^1(X, Y)$ over X valued in the open set of 1-jets whose linear part is of maximal rank.

With this vocabulary in hand, Hirsch's theorem states the following:

Theorem (Hirsch [15]) *The space $Imm(X, Y)$ of immersions from X to Y has the same homotopy type*[8] *as the space $Imm^{\text{formal}}(X, Y)$ of formal immersions.*

12.1.7 Phillips' Work on Submersions

When the dimension of X is greater than the dimension of Y, it is natural to consider submersions, that is, maps of maximal rank. When such maps exist they form a space that we denote by $Subm(X, Y)$. Using again the current terminology, a *formal submersion* is a section of the 1-jet bundle $J^1(X, Y)$ over X valued in the open set of 1-jets whose linear part is of maximal rank. Phillips' submersion Theorem sounds similar to Hirsch's immersion Theorem with, nevertheless, a fundamental difference: the domain needs to be an open manifold. Notice that the circle admits no submersion into the line even though there is a formal submersion; a similar claim holds for any parallelizable manifold like a compact Lie group.

Theorem (Phillips [23]) *If X is an open manifold, $Subm(X, Y)$ and $Subm^{\text{formal}}(X, Y)$ have the same homotopy type.*

Since a foliation is locally defined by a submersion into a local transversal, the next theorem can be viewed as an extension of the previous one. Let $\mathcal{F}$ be a smooth foliation of the manifold Y. Denote its normal bundle by $\nu(\mathcal{F})$; this is a vector bundle on Y whose rank is equal to the codimension of $\mathcal{F}$. Denote by $\pi : TY \to$

[8] In the literature on this topic, one generally speaks of the same *weak* homotopy type, meaning a map inducing an isomorphism between each respective homotopy group of the two considered spaces. Actually, Palais [21, Theorem 15] tells us that the two notions are equivalent for the topological spaces we are dealing with.

$\nu(\mathcal{F})$ the linear bundle morphism over Id_Y whose kernel is the sub-bundle of TY made of vectors which are tangent to the leaves of $\mathcal{F}$.

A smooth map $f : X \to Y$ is said to be transverse to $\mathcal{F}$ if the bundle morphism $\pi \circ df : TX \to \nu(\mathcal{F})$ over f is fibrewise surjective. In that case, the preimage of $f^{-1}(\mathcal{F})$ is a foliation of the same codimension as $\mathcal{F}$ and its normal bundle is the pull-back $f^*\big(\nu(\mathcal{F})\big)$. We denote by $C^{\pitchfork\mathcal{F}}(X, Y)$ the set of smooth maps transverse to Y.

Given a bundle morphism $F : TX \to TY$ over $f : X \to Y$, the pair (f, F) is said to be formally transverse to $\mathcal{F}$ if $\pi \circ F$ is fibrewise surjective. By abuse, one says also that f is formally transverse to $\mathcal{F}$.

Theorem (Phillips [24]) *The space $C^{\pitchfork\mathcal{F}}(X, Y)$ has the same homotopy type as the space of maps which are formally transverse to $\mathcal{F}$.*

Remark All the previous theorems reduce the understanding of immersions, submersions or maps transverse to foliations, from the homotopical point of view, to the understanding of the corresponding formal problems. And the latter reduce to classical homotopy theory: the matter is to find sections to some maps and thus it reduces to *well-known obstructions*. This does not mean that the homotopy type of the formal spaces in question is computable. In general it is not, as the homotopy groups of the spheres are not completely computed.

The aim of Gromov's approach which we are going to describe below is to consider all the previous problems as particular cases of a general *principle*.

12.1.8 Differential Relations After M. Gromov

The main reference here is Gromov's book [10]. A simplified approach is described in Eliashberg and Mishachev's book [5]; the new tool is their *holonomic approximation Theorem* which was first proved in [4].

The preface of [5] starts as follows: "A *partial differential relation* $\mathcal{R}$ is any condition imposed on the partial derivatives of an unknown function".

If the unknown function in question is a smooth map from X to Y, a simple definition consists of saying that $\mathcal{R}$ is a subset in a jet space[9] $J^r(X, Y)$ for some integer r. Recall that in coordinates an element of this jet space is just the data of a point $a \in X$, a point $b \in Y$ and a Taylor expansion of order r at a with constant term b.

The expression *h-principle* comes from the article of Gromov and Eliashberg [11]; they write: "The principle of weak homotopy equivalence for *etc.*" Later on,

[9]Since this is going to be forgotten, I recall that the concept of jet space is due to Charles Ehresmann.

this expression is abbreviated to *h-principle*.[10] With these authors we say that the *parametric h-principle* holds for $\mathcal{R}$ if the space

$$sec\,\mathcal{R} := \{\text{sections of } J^r(X,Y) \text{ valued in } \mathcal{R}\}$$

has the same homotopy type as

$$sol\,\mathcal{R} := \{f : X \to Y \mid j^r f \text{ is valued in } \mathcal{R}\}.$$

One can think of an element of $sec\,\mathcal{R}$ as a *formal* solution of the problem posed by $\mathcal{R}$. A section valued in $\mathcal{R}$ which is of the form $j^r f$ is said to be *holonomic* or *integrable*. The integrablity is prescribed by the vanishing on the section in question of a list of 1-forms (called a *Pfaff system*) which are naturally defined on the manifold $J^r(X,Y)$. For instance, when $r = 1$ and $Y = \mathbb{R}$, a section s is integrable if and only if its image is *Legendrian* for the canonical *contact form* α which reads $dz - \sum p_i dq^i$ in canonical coordinates, that is, if and only if $s^*\alpha = 0$.

Theorem (Gromov [8])) *If $\mathcal{R}$ is an open set in $J^r(X,Y)$ which is invariant by the natural right action of Diff(X) and if X is an* open *manifold (meaning that no connected component is closed), then the parametric h-principle holds for $\mathcal{R}$.*

The proof also goes through corrugations as in the case of Smale's theorem. Of course, the corrugations are not developed in the range; there is no room there for corrugating. They are developed in the domain. This is very clearly explained in Eliashberg-Mischaev's book [5].

Remark Another very important condition on a differential relation leads to an h-principle; it is when the relation is *ample*. In this case X does not have to be an open manifold. Here, the h-principle follows from the famous *convex integration* technique invented by Gromov in [9] (see Gromov's book [10]). A complete account on this is given in Spring's book [32]. The end of Eliashberg-Mishachev's book [5] focuses on the convex integration applied to the C^1 isometric embedding problem (Nash-Kuiper); Borrelli et al. [2] converted their theoretical result into an algorithm richly illustrated by pictures of C^1 *fractal objects*, as the authors say. In spite of the great interest of the subject, I do not intend to enter more deeply into it as it is less close to Thom's work than what follows.

Going back to the, say *open*, h-principle stated above, one sees that the previously mentioned results by Smale, Hirsch and Phillips are clearly covered by Gromov's theorem. One could be disappointed that only 1-jet spaces are involved. The simplest way for finding new examples with differential relations of higher order consists of

[10] I have already commented on the word *weak* in Footnote 8. Concerning the word *principle*, I feel uncomfortable with a principle which is not always true, and worse, whose domain of validity remains unknown. This means that the h-principle is not a gift from heaven.

the following construction, which naturally appears in Thom's singularity theory [35] as it is shown in the next subsection.

Consider a proper submanifold $\Sigma \subset J^{r-1}(X, Y)$ or a proper stratified set with nice singularities (for instance, with *conical singularities* in the sense of [18]); the important point is that transversality to any stratum $\Sigma_i \subset \Sigma$ implies transversality to all other strata in some neighbourhood of Σ_i in $J^{r-1}(X, Y)$. Assume that Σ is *natural*, that is, invariant by the action of $\mathrm{Diff}(X)$. The transversality to Σ is obviously a differential relation of order r. This differential relation which we denote by $\mathcal{R}_\Sigma$ is open and *natural*, that is, invariant by the action of $\mathrm{Diff}(X)$. Thus, if X is open, Gromov's theorem applies.

12.1.9 Examples from Singularity Theory

For a first concrete example, take $\dim X = \dim Y = 2$ and consider the stratified set Σ of 1-jets of rank less than 2. It is made of two strata: one stratum is the set of jets of rank 1; it has codimension 1 and is denoted by Σ^1 in the so-called Thom-Boardman notation [1]. The other stratum is made of 1-jets of rank 0; its codimension is 4 in our setting. Their union is a stratified set with conical singularities which is natural and proper. Thus $\mathcal{R}_\Sigma$ satisfies the open h-principle if X is open.

The next example leads to a differential relation of order 3. One starts with the first example and looks at a 2-jet $\xi \in \mathcal{R}_\Sigma$; say it is based in $a \in X$. On the one hand, as ξ is transverse to Σ, it does not project to the zero 1-jet for dimension/codimension reasons. Therefore, it determines the tangent space at a to the *fold locus* $L \subset X$ where the rank of any germ of map f realizing ξ is exactly 1. In our setting, L is one-dimensional. On the other hand, ξ determines the kernel K_a of the differential df_a. Thus, there is a natural stratification of $\mathcal{R}_\Sigma \subset J^2(X, Y)$: one stratum is $\Sigma^{1,0}$ which is made of 2-jets where K_a is transverse to T_aL; the second one, denoted by $\Sigma^{1,1}$, is made of 2-jets where K_a is tangent to T_aL. The stratum $\Sigma^{1,0}$ is an open set in $\mathcal{R}_\Sigma$ and $\Sigma^{1,1}$ has codimension 2 in $J^2(X, Y)$. Thus, if a 3-jet is transverse to $\mathcal{R}_\Sigma$ in a 2-jet in $\Sigma^{1,1}$, it is the jet of a germ having an isolated *cusp* from which emerge two branches of the fold locus (see Fig. 12.5).

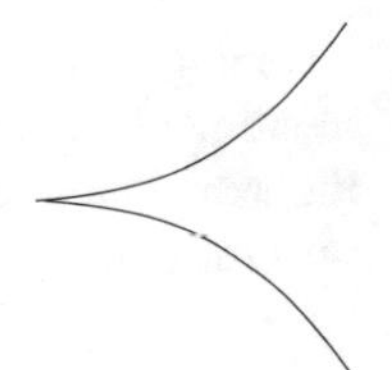

Fig. 12.5 Local image of a cusp in a two-dimensional manifold

12.1.10 Thom's Transversality Theorem in Jet Spaces

This was exactly the subject of Thom's lecture at the 1956 Symposium in Mexico City that I mentioned at the very beginning of this piece. Incidentally, this theorem will play a fundamental role in singularity theory, as the above discussion lets us foresee. The statement is the following:

Theorem (Thom [34])) *Let Σ be a submanifold in the total space an r-jet bundle $E^{(r)} \to X$ over a manifold. Then, generically,*[11] *an integrable section of $E^{(r)}$ is transverse to Σ.*

This theorem is remarkable in two ways:

(1) The usual transversality statement tells us that any section of $E^{(r)}$ can be approximated by a section transverse to Σ. But, the intregrability condition is a *closed constraint*[12] and even if we had started with an integrable section, the transverse approximation might be non-integrable.
(2) The same proof, by inserting the given map in a large family of maps which is transverse to Σ as a whole, works both for the usual transversality theorem and for the transversality theorem with constraints.

For a while, I tried to understand whether the statements of Thom and Gromov were somehow related. For instance, does the h-principle hold for the relations $\mathcal{R}_\Sigma$ from Sect. 12.1.9? In general, the answer is NO as it is shown in a note with Alain Chenciner [3]. Quoting from its abstract: "A section in the 2-jet space of Morse functions is not always homotopic to a holonomic section".

12.2 Integrability and Related Questions

12.2.1 Thom's Point of View in 1959

The title of the lecture given by René Thom at the 1959 conference organized by the CNRS in Lille (France) is striking when compared with the terminology that appears 10 years later: "Remarques sur les problèmes comportant des inéquations différentielles globales", which I translate by: *Remarks about problems involving global differential inequations.*

The setting is the same as in Gromov's theorem from Sect. 12.1.8 and, for consistency with what precedes, $\mathcal{R}$ still denotes an open set in the jet space

[11] A property is said to be generic in a given topological space F (here, it is the space of integrable sections with the C^1 or C^∞ topology, or the Whitney topology evoked in Sect. 12.1.2) if it is satisfied by every elements in a residual subset (that is, an intersection of countably many open dense subsets).

[12] The space of integrable sections is closed with empty interior in the space of all the sections.

$J^r(X, Y)$, except that now the openness of X is not assumed. There are two chain complexes naturally associated with $\mathcal{R}$:

(1) $C_*(\mathcal{R})$ is the complex of continuous[13] singular simplices.
(2) $C_*^{\text{int}}(\mathcal{R})$ is the sub-complex generated by the differentiable simplices valued in $\mathcal{R}$ which are integrable (or holonomic) in the sense that each 1-form from the integrability Pfaff system vanishes on them.

Here, a k-simplex is a map from the standard k-simplex $\Delta^k \subset \mathbb{R}^{k+1}$ into $\mathcal{R}$. In (2), thanks to the so-called *small simplices Lemma*, up to quasi-isomorphism, it is sufficient to consider holonomic smooth simplices of the form: $\sigma = j^r f \circ \underline{\sigma}$ where $\underline{\sigma}$ is a k-simplex of the base X and f is a smooth map defined near the image of $\underline{\sigma}$ with values in Y.

Theorem (Thom [37]) *The inclusion* $C_*^{\text{int}}(\mathcal{R}) \hookrightarrow C_*(\mathcal{R})$ *induces an isomorphism in homology for* $* < \dim X$ *and an epimorphism for* $* = \dim X$.

For instance, if X is closed and $s : X \to \mathcal{R}$ is a section, then the cycle $s(X)$ (at least with $\mathbb{Z}/2\mathbb{Z}$ coefficients when X is non-orientable) is homologous to a holonomic *zig-zag*, that is a cycle of the form $j^r f(X)$ where $f : X \to Y$ is multivalued.

12.2.2 What Happened Afterwards

The article [37] was actually only an announcement. The proof of the theorem was outlined in three pages, and remains still difficult to read even though some ideas were visibly emerging; an instance is the *sawtooth*, which is an antecedent to the *jiggling* intensively used by Thurston in the early seventies [39]. No complete proof has ever appeared. Unfortunately, the report by Smale in the Math. Reviews [30] was somewhat discouraging for anyone who would have tried to complete Thom's proof. Here is the final comment of this report: {The author has said to the reviewer that, although he believes his proof to be valid for $r = 1$, there seem to be further difficulties in case $r > 1$.}

Nevertheless, David Spring has known for a few years that Thom's statement holds (see his note [33]). His unpublished proof is based on the holonomic approximation Theorem of Eliashberg and Mishachev [4] when $* < \dim X$. In the remaining case, he also needs Poénaru's foldings theorem [25]. I should say that the holonomic approximation Theorem is in germ in Thom's announcement; his horizontal sawtooth is closely related to the construction made in [4].

[13] Replacing *continuous* with *smooth* changes the complex to a *quasi-isomorphic* sub-complex, meaning that the homology is unchanged.

When reading Thom's article for preparing the edition[14] of his collected mathematical works [38], I was not able to complete the proof in the way indicated by Thom. But, I discovered a beautiful construction in that article. I first translate the original few lines into English and then, in the next subsection, I shall state the lemma which I could extract from these lines.

> [The proof] mainly relies on the construction of a deformation (homotopy operator) from the complex of singular differentiable simplices to the sub-complex of integrable simplices. Such a deformation has to be *hereditary*, that is, compatible with the restriction to faces. Moreover, as the problem is local in nature, it will be sufficient to construct this deformation for an open set in $J'^r(\mathbb{R}^n, \mathbb{R}^p)$.
>
> ... ...
>
> Let b^k be a k-dimensional simplex, b^n an n-dimensional simplex, $n \geq k$; let b'^k be a subdivision of b^k and s a simplicial map from this subdivision b'^k of b^k to b^n. The finer the subdivision b' is, the more the map s has a "strong gradient" in the sense that the quotient $[s(x) - s(y)]/(x - y)$, for every pair of points $x, y \in b^k$ close enough, becomes larger and larger.

Here, the question is: why do such a subdivision and simplicial map exist?

12.2.3 The Thom Subdivision

Here is the statement that I cooked up for translating the preceding lines into a more precise language.

Lemma *There exists a sequence $(K_n, s_n)_n$, where K_n is a linear subdivision of Δ^n and $s_n : K_n \to \Delta^n$ is a simplicial map such that:*

(1) (Non-degeneracy) *For each n-simplex $\delta_n \subset K_n$, the restriction $s_n|_{\delta_n}$ is surjective;*
(2) (Heredity) *For each $(n-1)$-face F of Δ^n we have:*

$$\begin{cases} F \cap K_n \cong K_{n-1}\,, \\ s|_F \cong s_{n-1}\,. \end{cases}$$

Here, the symbol $\cong$ stands for a simplicial isomorphism. The non-degeneracy somehow translates Thom's strong gradient condition. After a time of thinking, I found a proof by induction on n in the way which is illustrated by passing from Figs. 12.6 to 12.7: put a small n-simplex δ^n *upside down* in the interior of Δ^n

[14] The team of editors of Thom's works was initiated by André Haefliger and is directed by Marc Chaperon.

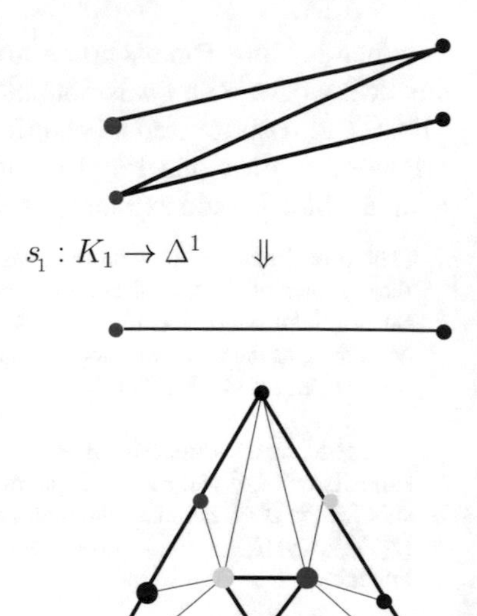

Fig. 12.6 The Thom subdivision in dimension 1 and its folding map

Fig. 12.7 The Thom subdivision in dimension two; s_2 is defined by the colour

and join each vertex of δ^n to the facet of Δ^n lying in front of it which is already subdivided by induction hypothesis.[15]

One can think of s_n as a folding map from Δ^n onto itself. Due to the heredity property, we have:

- Any polyhedron can be folded onto itself.
- The folding can be iterated r times:

$$\begin{aligned} K_n^{(r)} &= \left(s_n^{(r-1)}\right)^{-1} (K_n), \\ s_n^{(r)} &= s_n \circ s_n^{(r-1)}. \end{aligned}$$

Note that the folding map of any order is endowed with an hereditary *unfolding* homotopy to the *Identity* map.

12.2.4 Jiggling Formula

It is now easy to derive a natural *jiggling* formula, using the same terminology as Thurston's in [39], but without any measure consideration.

[15]Today, this subdivision is called the *standard chromatic subdivision* of the n-simplex. It appears in books of combinatorial topology or graph theory. It is also used by computer scientists [14]. I have never found any reference to Thom.

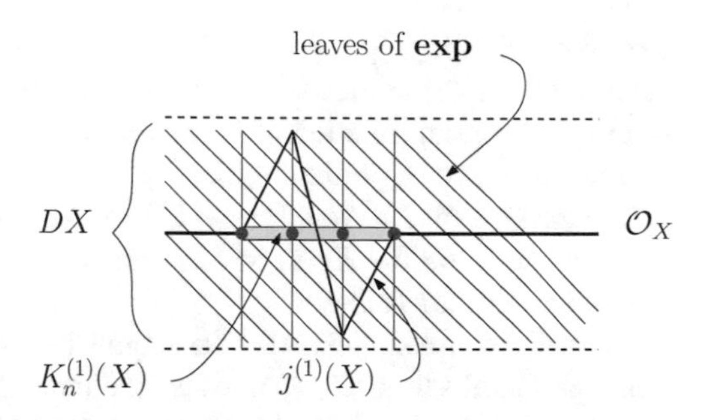

Fig. 12.8 The jiggling map of order $r = 1$. The vertical lines are fibres of TX

Equip X with a Riemannian metric. Let $DX \to X$ be a tangent disc bundle such that the *exponential* map $\exp : DX \to X$ is a submersion. Denote by $\mathcal{F}_{\exp}$ the *exponential foliation* of DX whose leaves are the fibres $\exp^{-1}(x)$, $x \in X$.

Choose a triangulation T of X finer than the open covering $\{\exp_x(D_x X) \mid x \in X\}$. Fix an integer r. The r-th jiggling map is the section of the tangent bundle defined by

$$j^{(r)} : X \to DX,$$
$$j^{(r)}(x) = \exp_x^{-1}\left(s_n^{(r)}(x)\right).$$

This map is piecewise smooth. Moreover, the larger r is, the more *vertical* the jiggling is (Fig. 12.8). As a consequence, for r large enough, $j^{(r)}(X)$ is *quasi-transverse* to the tangent space to the exponential foliation $\mathcal{F}_{\exp}$ in the sense that, for any simplex τ of the r-th Thom subdivision of T, the smooth image $j^{(r)}(\tau)$ shares no tangent vector with the tangent space to the leaves of $\mathcal{F}_{\exp}$; when the dimension of τ is n, quasi-transversality is equivalent to transversality.

Similarly, if X is compact and if we are given a compact family $\mathcal{P} = \{P_t\}_t$ of n-plane fields transverse to the fibres of TX, then taking r large enough makes $j^{(r)}(X)$ quasi-transverse to all the elements of $\mathcal{P}$ simultaneously.

12.2.5 Going Back to Immersions

This is contained in a joint work with Meigniez [19].

First, recall that one can reduce oneself to consider only immersions of codimension 0 whose domain has a non-empty boundary. Indeed, any formal immersion (f, F) from X to Y, when $\dim X < \dim Y$, has a normal bundle; this is the vector bundle over X which is the cokernel $\nu(f, F)$ of the monomorphism $TX \to f^*TY$ through which F factorizes. Thus, immersing X into Y is equivalent to immersing a disc bundle of $\nu(f, F)$ into Y and the latter is a codimension 0 immersion problem.

In what follows, we assume that X is compact with a non-empty boundary and has the same dimension as Y. For free, a formal immersion (f, F) from X to Y gives rise to a foliation $\mathcal{F}_X := F^{-1}(\mathcal{F}_{\exp_Y})$ which foliates a neighbourhood of the zero section O_X of TX. Indeed, since F maps fibres to fibres surjectively, F is transverse to the exponential foliation of Y (defined near the zero section O_Y of TY).

Such a (germ of) foliation like $\mathcal{F}_X$ is called a *tangential Haefliger structure* or a *Γ_n-structure* on X. We refer to [13] for more details on this important notion. As there is no reason for $\mathcal{F}_X$ to be transverse to O_X, the trace of $\mathcal{F}_X$ on $X = O_X$ is in general a *singular* foliation.

Actually, those singularities are responsible for the flexibility associated with that concept. Thanks to them, operations like induction (or pull-back) and homotopy (or concordance) are available for Γ_n-structures while they are not for foliations. Let us emphasize that a Γ_n-structure is mainly a Čech cocycle of degree one with values in the groupoid of germs of diffeomorphisms of $\mathbb{R}^n$. This allows one to induce such a structure on a polyhedron or a CW-complex without regarding the dimension of that space.

A concordance between two Γ_n-structures ξ_0 and ξ_1 on X is just a Γ_n-structure on $X \times [0, 1]$ which induces ξ_i on $X \times \{i\}$, $i = 0, 1$. There is a classifying space $B\Gamma_n$ in the following sense: the Γ_n-structures on X, up to concordance, are in 1-to-1 correspondence with the homotopy classes $[X, B\Gamma_n]$, as for vector bundles.

In our setting, the Haefliger structure in question is enriched with a *transverse geometric structure* invariant by holonomy: each transversal to $\mathcal{F}_X$ is endowed with a submersion into Y which is preserved when moving the transversal by isotopy along the leaves (this point being obvious since the leaves in question are contained in the inverse images of points in Y); such a Γ_n-structure will be named a Γ_n^Y-structure. In particular, if O_X were transverse to $\mathcal{F}_X$, then X would be endowed with a submersion into Y, that is an immersion into Y since $\dim X = \dim Y$. Therefore, the aim is to remove the singularities of the Γ_n^Y-structure, that is, to find a *regularizing concordance* of Γ_n^Y-structures from $\mathcal{F}_X$ to a Γ_n^Y-structure whose underlying foliation is transverse to the zero section O_X.

In the next subsection we give a brief review of the regularization problem. In the last subsection, a sketch for regularizing Γ_n^Y-structures is given in our setting of immersions in codimension 0 of compact manifolds with non-empty boundary and no closed connected components.

12.2.6 About the Regularization of Γ-Structures

Let ξ be a Γ_q-structure on an n-dimensional manifold X. In general, the underlying foliation $\mathcal{F}(\xi)$ is supported in a neighbourhood of the zero-section in a vector bundle $\nu(\xi)$ of rank q, called the normal bundle to ξ. This normal bundle remains unchanged along a concordance. If ξ is *regular*, that is, if $\mathcal{F}(\xi)$ is transverse to the 0-section of $\nu(\xi)$, then the trace of $\mathcal{F}(\xi)$ on X is a genuine foliation whose normal bundle is canonically isomorphic to $\nu(\xi)$. Therefore, a necessary condition for being regularizable is that:

(∗) $\nu(\xi)$ embeds into the tangent bundle TX[16]; in particular, $q \le n$.

André Haefliger was the first who proved that any Γ_q-structure on an *open* manifold X whose normal bundle embeds into TX is regularizable [12] (or [13, p. 148]). This follows from two things:

- First, the classifying property of the classifying space $B\Gamma_q$ which says that $B\Gamma_q$ is equipped with a *universal* Γ_q-structure in the sense that any Γ_q-structure on any CW-complex X is induced by pull-back from the universal structure through a map $X \to B\Gamma_q$.
- Second, Phillips' theorem about transversality to a foliation [24] when the domain is open.

Today, combining all the steps, this regularization theorem is frequently referred to as the Gromov-Haefliger-Phillips Theorem.

The next step was done by Thurston [39] who solved the case of closed manifolds. Namely, if $q > 1$ even when X is closed, any Γ_q structure satisfying the necessary condition (∗) is regularizable. The case $q = n$ is a toy case. The only technique is the famous *jiggling* lemma whose proof is quite tricky in terms of measure theory, even though Thurston considered it as an obvious statement. Exactly at this point, our jiggling, based on the Thom subdivision, is much simpler; in particular, it works in family, that is, with parameters.

The final step is the codimension-one case for closed manifolds, a piece of work indeed. Generally it is known in the following form:

Theorem (Thurston [40]) *Every hyperplane field is homotopic to an* integrable *hyperplane field, that is, a field tangent to some codimension-one foliation.*

Actually, the main part of this result is a regularization theorem for Γ_1-structures. In addition to the jiggling technique, there are many subtle points (simplicity of the group of diffeomorphisms, intricate constructions, *etc.*).

12.2.7 Regularization of Transversely Geometric Γ_n-Structures

In Sect. 12.2.5 we reduced the immersion problem to a problem of regularization of some Γ_n^Y-structure ξ on X associated with the given formal immersion and shown in Fig. 12.9. The exponent Y reminds us that we are considering a Γ_n-structure endowed with some transverse geometry which here consists of being a submersion into Y. The scheme shown in Fig. 12.10, and which I am going to comment on, summarizes an ordinary regularization (which would work even if X were closed).

[16]By abuse, we identify a vector bundle with its total space.

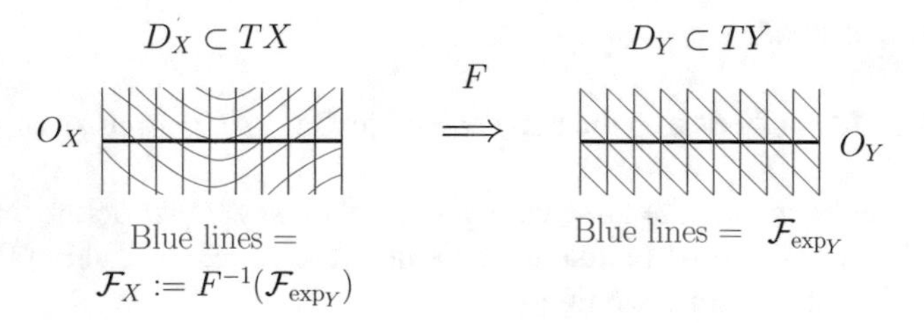

Fig. 12.9 The left part shows a tangential Haefliger structure

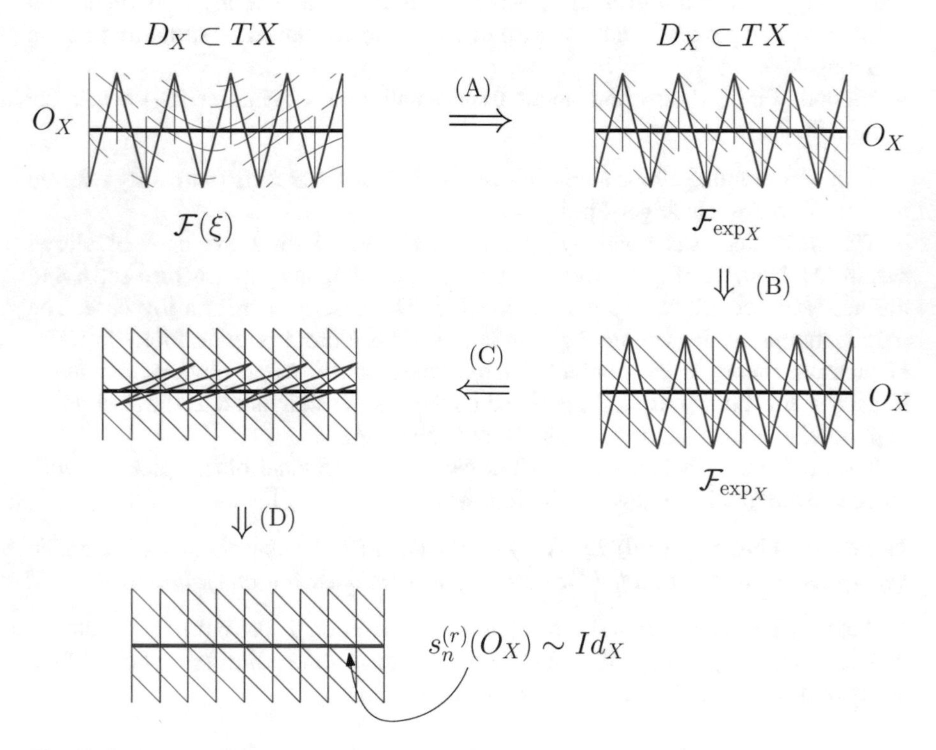

Fig. 12.10 The scheme of the regularization in four steps

It will appear in the end that this regularization is easily enriched with a transverse geometric structure when X is open. It is worth noting that the problem is the same whatever the transverse geometry is. In place of *submersion to Y* one could have a symplectic or contact structure, a complex structure or a codimension-one foliated structure *etc.* For any geometry,[17] the regularization is the same.

[17]We take the concept of geometry in the sense of Veblen and Whitehead [42] which could be rewritten in the more modern language of sheaves.

First, the jiggling is chosen, meaning that the order of the Thom subdivision r is fixed once and for all. This r is chosen so that $j^{(r)}(X)$ is quasi-transverse to the following codimension-n foliations or n-plane fields:

- the foliation $\mathcal{F}(\xi)$ underlying the given Γ_n-structure ξ (this foliation was denoted by $\mathcal{F}_X$ in the particular case of Fig. 12.9);
- the exponential foliation $\mathcal{F}_{\exp_X}$;
- every n-plane field which is a barycentric combination[18] of the two previous ones.

The homotopy from the zero-section to $j^{(r)}(X)$ gives rise to an obvious concordance which is not mentioned in the scheme of Fig. 12.10.

Step (A) is exactly Thurston's concordance in [39]. By using the above-mentioned barycentric combination, some generic $(n+1)$-plane field Π is chosen on $TX \times [0,1]$ quasi-transverse to $j^{(r)}(X) \times [0,1]$. Since the trace of Π on each simplex of the jiggling is 0- or 1-dimensional, such a trace is integrable. Thus, a C^0 approximation of Π is integrable in a neighbourhood of $j^{(r)}(X) \times [0,1]$. This gives the concordance (A) and explains the reason why some part of the tube D_X has been deleted from the initially foliated domain.

Step (B) is just the inclusion using the fact that the exponential foliation exists on the whole tube. Step (C) uses the interpolation $\exp^t$, $t \in [0,1]$, from Id_{DX} to $\exp : DX \to X$ given by:

$$\exp^t : (x,u) \longmapsto \left(\exp_x(tu), \left(D_{(x,tu)}\exp_x(tu)\right)(1-t)u\right).$$

It allows one to slide $j^{(r)}(X)$ along the leaves of the exponential foliation keeping the quasi-transversality to each simplex.[19] Observe that the vertical homothety does not have such a property. When $t = 1$, we finish with the folding map $s_n^{(r)}(X) \to X$. Step (D) is just the unfolding of $s_n^{(r)}$, that is, its hereditary homotopy to Id_X. Again, at each time of the homotopy, the image polyhedron (contained in the zero-section) is quasi-transverse to the exponential foliation. This finishes the regularization of ξ as a Γ_n-structure. In general, it is not possible to extend the transverse geometry of $\mathcal{F}(\xi)$ to the concordance. But, this is possible when X is an open manifold as we are going to explain[20] now.

[18]Recall that the space of n-planes tangent to the total space TX at (x,u), $x \in X$, $u \in T_xX$, and transverse to the *vertical tangent space* (that is, the kernel of $D_{(x,u)}\pi$ where $\pi : TX \to X$ denotes the projection) is an affine space.

[19]For the reader who does not like complicated formulas, I suggest a more topological approach of the previous interpolation. Let U be a *nice* tubular neighbourhood of the diagonal Δ in $X \times X$; here, nice means that the two projections π^v and π^h respectively given by $(x,y) \mapsto (x,x)$ and $(x,y) \mapsto (y,y)$ are n-disc bundle maps. If U is small enough, a Riemannian metric provides an identification of π^v with a tangent disc bundle to X. In that case, π^h is the corresponding exponential map. Hence, the above-mentioned interpolation is just a contraction of the fibres of π^h.

[20]Only the idea of the proof is given here. We refer to [19] for more details.

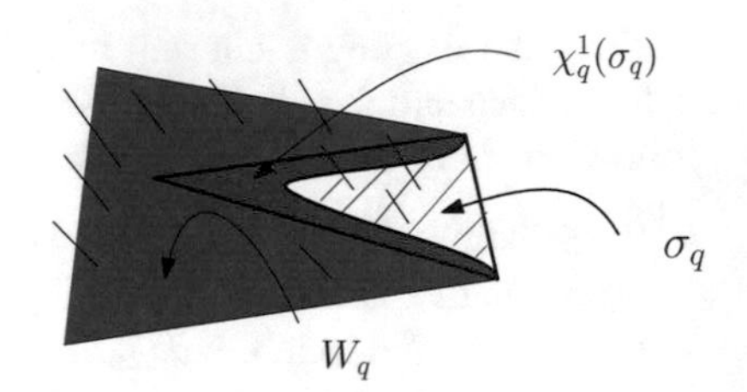

Fig. 12.11 A few leaves of $\mathcal{F}_W$ are drawn in black. The simplex σ_q is the union of the red full part and the red dashed part

If X is an n-dimensional manifold without closed connected component, endowed with a triangulation T, there exists a *spine* that is, a sub-complex K of dimension $n-1$ such that, for any neighbourhood $N(K)$, there is an isotopy of embeddings $\varphi_t : X \to X$ whose time one map sends X into $N(K)$ (see for instance [5, p. 40–41]).

Restricting ourselves to $K \subset X$, let us consider the concordance $(W, \mathcal{F}_W)$ of Γ_n-structures obtained by concatenation and time reparametrization of the four concordances described right above from $j^{(r)}(K)$ to $K \subset O_X$. Here, $W \subset TX \times [0, 1]$ is piecewise linear homeomorphic to $K \times [0, 1]$; and $\mathcal{F}_W$ is a codimension-n foliation defined near W and transverse to the fibres of $TX \times [0, 1] \to X \times [0, 1]$ which induces $\mathcal{F}(\xi)$ over $t = 0$ and $\mathcal{F}_{\exp_X}$ over $t = 1$. Moreover, $\mathcal{F}_W$ is quasi-transverse to every simplex of W. Therefore, since W is n-dimensional, every leaf meets each simplex of W in one point at most.[21]

By construction, W *collapses* onto its initial face $W_0 := j^{(r)}(K)$. We recall that a simplicial complex W collapses onto W_0 if there is a sequence of elementary collapses $W_{q+1} \searrow W_q$ starting with W and ending with W_0. An *elementary collapse* means that W_{q+1} is the union of W_q and a simplex σ_q so that $\sigma_q \cap W_q$ consists of the boundary of σ_q with an open facet removed. The elementary collapse $W_{q+1} \searrow W_q$ gives rise to an *elementary isotopy* χ_q^t pushing W_{q+1} into itself, keeping W_q fixed, and ending with $\chi_q^1(W_{q+1})$ as close to W_q as we want. Due to the quasi-transversality to $\mathcal{F}_W$, this isotopy extends to a neighbourhood of σ_q as a *foliated isotopy* $\tilde{\chi}_t$, meaning that each leaf is mapped to a leaf at each time (Fig. 12.11).

By induction on q, assume that the transverse geometric structure already exists on the foliation $\mathcal{F}_W|_{W_q}$. Then, by pulling back through $\tilde{\chi}_q^1$, this structure extends to the foliation $\mathcal{F}_W|_{W_{q+1}}$. Finally, the whole foliation $\mathcal{F}_W$ is enriched with the considered geometry, for instance a submersion into Y. And hence, $N(K) = N(K) \times \{1\}$ is endowed with a submersion into Y. As K is a spine of X, the submersion onto Y extends to X.

[21] Here, it is necessary to make a jiggling in the time direction. The cell decomposition of W is then prismatic (simplex×interval). Each prismatic cell has a Whitney triangulation (canonical up to the numbering of the vertices of X) [44, Appendix II].

12.2.8 Sphere Eversion Again

The main advantage of this proof based on the Thom subdivision and its associated jiggling is that it works for families (or with parameters). It is sufficient to choose the order r large enough so that a common jiggling is convenient for each member of the family.

For instance, if f_0 denotes the inclusion $\mathbb{S}^2 \hookrightarrow \mathbb{R}^3$ and $f_1 := -f_0$, these two immersions are formally homotopic[22] by:

$$(f_t, F_t) : (x, \vec{u}) \mapsto \big(tf_1(x) + (1-t)f_0(x), R^{\pi t}_{Ox}(\vec{u})\big).$$

Here, $t \in [0, 1]$ is the parameter of the homotopy, x is a point in $\mathbb{S}^2$ and $\vec{u}$ is a vector in $\vec{\mathbb{R}}^3$ tangent to $\mathbb{S}^2$ at x; finally, $R^{\pi t}_{Ox}$ stands for the Euclidian rotation of angle πt in $\mathbb{R}^3$ around the oriented axis directed by $\vec{x}$. When $t = 1$, we have indeed $F_1(x, -) = d_x f_1(-)$, the differential of f_1 at x.

By thickening, we have a one-parameter family F_t of formal submersions of $\mathbb{S}^2 \times (-\varepsilon, +\varepsilon)$ into $\mathbb{R}^3$. Thus, we have a one-parameter family of Γ_3-structures equipped with a transverse geometry (the local submersion to $\mathbb{R}^3$). The regularization by the Thom jiggling method – one jiggling working for all the foliations $F_t^{-1}(\mathcal{F}_{\exp_{\mathbb{R}^3}})$ – gives rise to a one-parameter family of submersions $\mathbb{S}^2 \times (-\varepsilon, +\varepsilon) \to \mathbb{R}^3$ joining the respective thickenings of f_0 and f_1. The restriction to $\mathbb{S}^2 \times \{0\}$ is a regular homotopy from f_0 to f_1. That is the desired sphere eversion.

12.2.9 Final Remark

Since f_0 and f_1 have the same image, we get that the space of *non-oriented* immersed 2-spheres in the 3-space is not simply-connected. It may be that those who are skeptical about the sphere eversion think the orientation should be preserved. Of course, if an orientation is chosen on the initial sphere, it propagates along any regular homotopy. But, as the image changes throughout such a homotopy, that f_0 and f_1 have the same image does not prevent us from a change of orientation. This is a phenomenon of *monodromy* well-known for detecting some non-simple connectedness.

Acknowledgements I am indebted to Anthony Phillips who gave me important information about the time of the sphere eversion. I thank Paolo Ghiggini, Peter Landweber and Athanase Papadopoulos who carefully read different versions of my article and made useful suggestions and corrections. I am also grateful to Michal Adamaszek who informed me about chromatic subdivisions and the content of Footnote 15.

[22] I learnt this very simple formula from Gaël Meigniez.

This work, supported by ERC Geodycon, is an expanded version of a lecture given in the conference: Hommage à René Thom, 1–3 Sept. 2016, IRMA, Strasbourg.

References

1. Boardman J.M., Singularities of differentiable maps, *Publ. Math. IHES* vol. 33, (1967), 21–57.
2. Borrelli V., Jabrane S., Lazarus F., Thibert B., Isometric embedding of the square flat torus in ambient space, *Ensaios Mat. 24, Soc. Bras. Mat.*, 2013.
3. Chenciner A. & Laudenbach F., Morse 2-jet space and h-principle, *Bull. Braz. Math. Soc.* 40(4) (2009), 455–463.
4. Eliashberg Y. & Mishachev N., Holonomic approximation and Gromov's h-principle, pp. 271–285 vol. I in: *Essays on geometry and related topics,* Monogr. Enseign. Math. 38, Geneva, 2001.
5. ———, *Introduction to the h-principle*, Graduate Studies in Math. 48, Amer. Math. Soc., 2002.
6. Francis G., *A Topological Picturebook*, Springer, 1987.
7. Francis G. & Morin B., Arnold Shapiro's eversion of the sphere, *Math. Intelligencer* 2 (1979), 200–203.
8. Gromov M., Stable mappings of foliations, *Izv. Akad. Nauk SSSR Ser. Mat.* 33 (1969), 707–734; English translation, *Math. Izv. USSR* 3 (1969), 671–694. MR0263103.
9. ———, Convex integration of partial differential relations I, *Izv. Akad. Nauk SSSR Ser. Mat.* 37 (1973), 329–343; English translation. *Math. Izv. USSR* 7 (1973), 329–343. MR0413206
10. ———, *Partial Differential Relations,* Ergeb. Math. vol. 9, Springer, 1986.
11. Gromov M. & Eliashberg Y., Removal of singularities of smooth mappings, *Izv. Akad. Nauk SSSR Esr. Mat.* 35 (1971), 600–627; English translation: *Math. USSR Izv.* 5 (1971), 615–639.
12. Haefliger A., Feuilletages sur les variétés ouvertes, *Topology* 9 (1970), 183–194.
13. ——, *Homotopy and integrability*, 133–175 in: Manifolds-Amsterdam 1970, Lect. Notes in Math. 197, Springer, 1971.
14. Herlihy M. & Shavit, N., A simple constructive computability theorem for wait free computation, *Proceedings of the Twenty-sixth Annual ACM Symposium on Theory of Computing (STOC 94)*, pp. 243–252, ACM, New York, ISBN:0-89791-663-8.
15. Hirsch M. W., Immersions of manifolds, *Trans. Amer. Math. Soc.* 93 (1959), 242–276.
16. ————, The work of Stephen Smale in differential topology, pp. 29–52 in: [31, vol. 1].
17. Kuiper N., Convex immersions in E^3, Non-orientable closed surfaces in E^3 with minimal total absolute Gauss-curvature, *Comment. Math. Helv.* 35 (1961), 85–92.
18. Laudenbach F., On the Thom-Smale complex, Appendix to: J.-M. Bismut-W. Zang, An extension of a Theorem by Cheeger and Müller, *Astérisque* 205 (1992), 219–233.
19. ——— & Meigniez G., Haefliger structures and symplectic/contact structures, *J. École polytechnique* 3 (2016), 1–29.
20. Levy S., *Making waves, A guide to the ideas behind Outside In*, (text and video), AK Peters, Wellesley (MA), 1995.
21. Palais R., Homotopy theory of infinite dimensional manifolds, *Topology* 5 (1966), 1–16.
22. Phillips A., Turning a surface inside out, *Scientific American* 214 (May 1966), 112–120.
23. ——, Submersions of open manifolds, *Topology* 6 (1967), 171–206. MR0208611.
24. ——, Smooth maps transverse to a foliation, *Bull. Amer. Math. Soc.* 76 (1970), 792–797. MR0263106.
25. Poénaru V., On regular homotopy in codimension 1, *Ann. of Math.* 83 (1966), 257–265.
26. ————, Extension des immersions en codimension 1 (d'après S. Blank), exposé n° 342, pp. 473–505 in: *Séminaire N. Bourbaki*, vol. 10, 1966–1968, Soc. Math. de France, 1995; (online) Eudml.

27. Smale S., A classification of immersions of the 2-sphere, Abstract 380*t*, *Bulletin Amer. Math. Soc.* 63 (3) (1957), pp. 196.
28. ———, Regular curves on Riemannian manifolds, *Trans Amer. Math. Soc.* 87 (1958), 492–512.
29. ———, A classification of immersions of the two-sphere, *Trans. Amer. Math. Soc.* 90 (1959), 281–290.
30. ———, *Review of [37],* MR0121807, Amer. Math. Soc. (1961).
31. ———, *Collected work of Stephen Smale*, Ed. F. Cucker & R. Wong, World Scientific Publishing, 2000.
32. Spring D., *Convex Integration Theory*, Monographs in Math. 92, Birkhäuser, 1998.
33. ———, Comments (to [37]), pp. 558–560 in: [38, vol. 1].
34. Thom R., Un lemme sur les applications différentiables, *Boletin de la Sociedad Matemática Mexicana* 1 (1956), 59–71.
35. ———, Les singularités des applications différentiables, *Ann. Institut Fourier* 6, (1955–1956), 43–87.
36. ———, La classification des immersions (d'après S. Smale), exposé n°157, pp. 279–289 in: *Séminaire N. Bourbaki*, vol. 4, 1956–1958, Soc. Math. de France, 1995; (online) Eudml.
37. ———, Remarques sur les problèmes comportant des inéquations différentielles globales, *Bull. Soc. Math. France* 87 (1959), 455–461; MR 22 #1253; or pp. 549–555 in [38].
38. ———, *Œuvres complètes de René Thom,* volume 1, Documents Mathématiques 15, Soc. Math. de France, 2017, ISBN 978-2-85629-816-9.
39. Thurston W., The theory of foliations of codimension greater than one, *Comment. Math. Helv.* 49 (1974), 214–231.
40. ———, Existence of codimension-one foliations, *Annals of Math.* 104 (1976), 249–268.
41. ———, Making waves: The theory of corrugations in: [20].
42. Veblen O. & Whitehead J. H. C., A set of axioms for differential geometry, *Proceedings of the National Academy of Sciences* 17, 10 (1931), 551–561.
43. Whitney H., On regular closed curves in the plane, *Compositio Math.* vol. 4 (1937), 276–284.
44. ———, *Geometric Integration Theory,* Princeton Univ. Press, 1957.

Chapter 13
Rigid and Flexible Facets of Symplectic Topology

Yakov Eliashberg

Abstract The article discusses in a historical perspective the notions of symplectic flexibility and rigidity and their role in the creation and the development of symplectic topology.

13.1 Rigid and Flexible Methods in Mathematics

In every area of Mathematics there always coexist two directions of explorations. The first is concerned with finding constraints and restrictions. I refer to this part as the *rigidity* side. On the other side, the *flexibility*, mathematicians are developing techniques for new constructions which test the boundaries of the mathematical world. The phenomena discovered are sometimes unexpected and counter-intuitive. Most famous examples of this kind were discovered in 1950s: these were the C^1-isometric embedding theorem by J. Nash [53], and S. Smale's 2-sphere eversion, which was developed to the immersion theory of Smale-Hirsch [37, 58]. Nash-Smale's ideas were greatly developed and generalized by M. Gromov, beginning from his 1968 PhD dissertation [31], and culminating in his book [34]. Since then the flexibility phenomena are usually referred as the h-principle.

Though rigid and flexible worlds coexist in most mathematical subjects, they come especially close to each other in symplectic topology.

13.2 Symplectic Basics

To set the stage we give here some basic definitions from symplectic and contact geometries.

Y. Eliashberg (✉)
Department of Mathematics, Stanford University, Stanford, CA, USA
e-mail: eliash-gt@math.stanford.edu

S. G. Dani, A. Papadopoulos (eds.), *Geometry in History*,
https://doi.org/10.1007/978-3-030-13609-3_13

Symplectic geometry was born as a geometric language of classical mechanics, and similarly contact geometry is a natural set-up for geometric optics and mechanics with non-holonomic constraints.

The cotangent bundle T^*M of any smooth n-dimensional manifold M carries a canonical *Liouville* 1-form λ, usually denoted pdq, which in any local coordinates $(q_1, \dots, q_n)$ on M and dual coordinates $(p_1, \dots, p_n)$ on cotangent fibers can be written as $\lambda = \sum\limits_1^n p_i dq_i$. The differential $\omega := d\lambda = \sum\limits_1^n dp_i \wedge dq_i$ is called the *canonical symplectic structure* on the cotangent bundle of M. Any diffeomorfism $f : M \to M$ lifts to a diffeomorphism $T^*f : T^*M \to T^*M$ which preserves the form pdq. The lift T^*f is given by the formula

$$T^*f(q, p) = \Big(f(q), (d_q f^*)^{-1}(p)\Big), \; q \in M, p \in T_q^*M.$$

In the Hamiltonian formalism of classical mechanics the cotangent bundle T^*M is viewed as the phase space of a mechanical system with configuration space M. The p-coordinates have a mechanical meaning of momenta. The full energy of the system expressed through coordinates and momenta, i.e. viewed as a function $H : T^*M \to \mathbb{R}$ on the cotangent bundle (or a time-dependent family of functions $H_t : T^*M \to \mathbb{R}$ if the system is not conservative) is called the *Hamiltonian* of the system. The dynamics is then defined by the Hamiltonian equations $\dot{z} = X_{H_t}(z), z \in T^*M$, where the Hamiltonian vector field X_{H_t} is determined by the equation $i(X_{H_t})\omega = -dH_t$, and in the canonical (p, q)-coordinates has the form

$$X_{H_t} = \sum_1^n -\frac{\partial H_t}{\partial q_i}\frac{\partial}{\partial p_i} + \frac{\partial H_t}{\partial p_i}\frac{\partial}{\partial q_i}.$$

The flow of the vector field X_{H_t} preserves ω, i.e. $X_{H_t}^*\omega = \omega$. The isotopy generated by the vector field X_{H_t} is called *Hamiltonian.*

More generally, Hamiltonian dynamics can be defined on any $2n$-dimensional manifold endowed with a *symplectic*, i.e. a closed non-degenerate differential 2-form ω. According to a theorem of Darboux any such form admits local *canonical coordinates* $p_1, \dots, p_n, q_1, \dots, q_n$ in which it can be written as $\omega = \sum\limits_1^n dp_i \wedge dq_i$. Diffeomorphisms preserving ω are called *symplectomorphisms.* Symplectomorphisms which can be included in a time dependent Hamiltonian flow are called *Hamiltonian*. When $n = 1$ a symplectic form is just an area form, and symplectomorphisms are area preserving transformations. In higher dimensions symplectomorphisms are volume preserving but the subgroup of symplectomorphisms represents only a small part of the group of volume preserving diffeomorphisms.

The projectivized cotangent bundle PT^*M serves as the phase space in *geometric optics*. It can be interpreted as the *space of contact elements* of the manifold M, i.e. the space of all tangent hyperplanes to M. The form pdq does not descend to PT^*M but its kernel does, and hence the space of contact elements carries a canonical field of hyperplanes tangent to it. This field turns out to be completely non-integrable. It is called a *contact structure*. More generally, a *contact structure* on a $(2n+1)$-dimensional manifold is a completely non-integrable field of tangent hyperplanes ξ, where the complete non-integrability can be expressed by the Frobenius condition $\alpha \wedge (d\alpha)^{\wedge n} \neq 0$ for a 1-form α (locally) defining ξ by the Pfaffian equation $\alpha = 0$. Though at first glance symplectic and contact geometries are quite different, they are in fact tightly interlinked and it is useful to study them in parallel. In particular, with any contact manifold (V, ξ) one can canonically associate a symplectic manifold (SV, ω_ξ), called the *symplectization* of (V, ξ), which is defined as follows. Let $N\xi \subset T^*V$ be the rank 1 conormal bundle to $\xi \subset TM$, and let SV be the total space of $N\xi$ with the 0-section removed. The contact condition for ξ is equivalent to the fact that the canonical symplectic form $d(pdq)$ on T^*V restricts to a symplectic form ω_ξ on SV. In case of a co-oriented ξ the term "symplectization" is usually referred to the half S_+V of SV consisting of 1-forms which define the given co-orientation of ξ. Any contactomorphism $f\colon (V, \xi) \to (V, \xi)$, i.e. a diffeomorphism $S\colon (V, \xi) \to (V, \xi)$ which preserves the contact structure, canonically lifts to a symplectomorphism $Sf\colon (SV, \omega_\xi) \to (SV, \omega_\xi)$ given by the formula $Sf = T^*f|_{SV}$. An important property of symplectic and contact structures is the following stability theorem due to Moser [50] in the symplectic case and to Gray [30] in the contact one:

Theorem 13.2.1 *Let ω_t, $t \in [0, 1]$, be a family of symplectic (resp. contact) structures on a manifold X which coincide outside of a compact set. In the symplectic case suppose, in addition, that $\omega_t - \omega_0 = d\theta_t$, $t \in [0, 1]$, where θ_t has a compact support. Then there exists an isotopy $h_t : X \to X$ with compact support which starts at the identity $h_0 = \mathrm{Id}$ and such that $h_t^*\omega_t = \omega_0$.*

Maximal integral (i.e. tangent to ξ) submanifolds of a $(2n+1)$-dimensional contact manifold (V, ξ) have dimension n and are called *Legendrian*. Their symplectic counterparts are n-dimensional submanifolds L of a $2n$-dimensional symplectic manifold (W, ω) which are isotropic for ω, i.e. $\omega|_L = 0$. They are called *Lagrangian* submanifolds. Here are two important examples of Lagrangian submanifolds. A diffeomorphism $f\colon W \to W$ of a symplectic manifold (W, ω) is symplectic if and only if its graph $\Gamma_f = \{(x, f(x));\ x \in W\} \subset (W \times W, \omega \times (-\omega))$ is Lagrangian. A 1-form θ on a manifold M viewed as a section of the cotangent bundle T^*M is Lagrangian if and only if it is closed. For instance, if $H_1(M) = 0$ then Lagrangian sections are graphs of differentials of functions, and hence the intersection points of a Lagrangian with the the 0-section are critical points of the function. A general Lagrangian submanifold corresponds to a *multivalued function*, called the *front* of the Lagrangian manifold. Given a submanifold $N \subset M$ (of any codimension), the set of all hyperplanes tangent to it in TM is a Legendrian submanifold of the space of contact elements PT^*M.

13.3 Poincaré's Dream and Arnold's Conjecture

Symplectic topology was born from Henri Poincaré's dream that Hamiltonian systems must have some special qualitative properties just because the phase flow of a Hamiltonian system preserves the symplectic form.

In particular, Poincaré's study of periodic orbits in the so-called restricted 3-body problem led him to the following statement, now known as the "last geometric theorem of H. Poincaré": *any area preserving transformation of an annulus* $S^1 \times [0, 1]$ *which rotates the boundary circles in opposite directions should have at least two fixed points.* Poincaré, see [55], provided many convincing arguments why the statement should be true, but the actual proof was found by G.D. Birkhoff[6] in 1913, only after Poincaré's death. Birkhoff's proof was purely 2-dimensional and further development of Poincaré's dream of what is now called *symplectic topology* had to wait till the 1960s when V. I. Arnold [3] formulated a number of conjectures formalizing this vision of Poincaré. These conjectures played an important role in the creation and development of symplectic topology. In particular, one of Arnold's conjectures stated that the number of fixed points of a Hamiltonian diffeomorphism is bounded below by the minimal number of critical points of a function on the symplectic manifold (which is positive only if the manifold is closed). For instance, for the 2-dimensional torus Arnold's conjecture predicted 4 fixed points in the non-degenerate case, and 3 fixed points without the non-degeneracy assumption. The statement looks very close to the statement of Poincaré's geometric theorem, and indeed it is easy to deduce the annulus case from the case of the 2-torus, but it seems impossible to adapt Birkhoff's annulus argument to the torus case.

13.4 Basic Symplectic Problems and Gromov's Alternative

Around the same time when Arnold formulated his conjectures there was an explosion of surprising results discovered on the flexible side of Mathematics. Several such results were proven in symplectic geometry, mainly by Gromov, and it was not at all clear whether any rigidity in symplectic geometry exist at all beyond dimension 2.

Here is a list of some problems (including Arnold's conjectures) which were considered in 1960–70s and were completely open at that time.

P1-abs. Which smooth manifolds admit symplectic or contact structures?

P1-rel. When can a symplectic or contact structure on a neighborhood of the boundary ∂M of a smooth manifold be extended to M?

P2-abs. Given two symplectic or contact structures on a manifold, when are they homotopic as symplectic or contact structures (and in the symplectic case possibly with an additional condition that the cohomology class of the symplectic form is preserved in the homotopy). According to Gray's and Moser's theorems this is equivalent for closed manifolds (together with the cohomological condition) to the existence of an isotopy between the structures.

P2-rel. When are two symplectic or contact structure on a manifold with boundary (e.g. the $2n$-ball) which coincide near the boundary homotopic via a homotopy fixed on a neighborhood of the boundary?

P3. Arnold's fixed point conjecture.

P4. Arnold's Lagrangian intersection conjecture. In particular, given a closed Lagrangian $L \subset T^*M$ which is Hamiltonian isotopic to the 0-section, the number of intersection points $\#L \cap M$ of L with the 0-section M is bounded below by the minimal number of critical points of a smooth function on M.

P5. The Lagrangian embedding problem, and in particular:

P5-$\mathbb{C}^n$. Are there embedded closed exact Lagrangian submanifols of $\mathbb{C}^n$? A Lagrangian submanifold L in an exact symplectic manifold $(X, d\alpha)$ with a chosen primitive α is called *exact* if the closed form $\alpha|_L$ is exact. If $H^1(X) = 0$ this condition is independent of the choice of the primitive α.

P5-nearby. Given a closed manifold M, are there closed exact Lagrangian submanifolds in $(T^*M, d(pdq))$ besides those which are Hamiltonian isotopic to the 0-section? The negative answer to this problem is known as *nearby Lagrangian conjecture* (which was formulated by V. I. Arnold later on around 1986).

P6. The Legendrian isotopy problem: are there obstructions for Legendrian isotopy beyond the formal ones? Here and below by *formal* obstructions we mean those which persist if we decouple maps and its derivatives, e.g. in the Legendrian case replace a Legendrian embedding of Λ into a contact manifold (Y, ξ) by a pair (ϕ, Φ_t), where $\phi : \Lambda \to Y$ is any smooth embedding, and $\Phi_t : T\Lambda \to TY$ a homotopy of injective homomorphisms connecting $\Phi_0 = d\phi$ with an isotropic homomorphism $\Phi_1 : T\Lambda \to \xi$.

P7. Are there obstructions for symplectic embeddings beyond the volume and bundle constraints?

P8. What is the C^0-closure of the group of symplectic or contact diffeomorphisms in the group of all diffeomorphisms of a symplectic or contact manifold?

One of Gromov's earlier results, [31], was that Problems P1-abs and P2-abs have an h-principle type answer for open manifolds. In other words, e.g. in the symplectic case:

Any almost symplectic structure (i.e. a non-degenerate but not necessarily closed 2-form) is homotopic on an open manifold to a symplectic form in any given cohomology class, and two symplectic forms homotopic as almost symplectic structures are homotopic as symplectic ones.

Gromov proved around the same time many other flexibility results in symplectic geometry. For instance, he proved, see [32], that Lagrangian *immersions* abide an h-principle (this was also proved independently by Lees [43]). Moreover, this is also the case for ϵ-Lagrangian *embeddings* (i.e. embeddings whose tangent planes deviate from Lagrangian directions by an angle $<\epsilon$). A remarkable h-principle was proven by Gromov for *iso-symplectic* and *iso-contact* embeddings.[1] For instance, in the symplectic case, Gromov proved that *if* (M, ω) *and* (N, η) *are two symplectic manifolds such that* $\dim N \geq \dim M + 4$ *then any smooth embedding* $f : M \to N$ *which pulls back the cohomology class of the form* η *to the cohomology class of* ω, *and whose differential* df *is homotopic to a symplectic bundle isomorphism, can be* C^0*-approximated by an iso-symplectic embedding* $\widetilde{f} : M \to N$, *i.e.* $\widetilde{f}^*\eta = \omega$. Gromov also proved the following alternative concerning Problem P8.

Theorem 13.4.1 *Either the group of symplectic (resp. contact) diffeomorphisms of a manifold* M *is* C^0*-closed in the group of all diffeomorphisms, or its* C^0*-closure coincides with the group of volume preserving (resp. all) diffeomorphisms.*

Though the proof of this alternative appeared only in Gromov's book [34], he explained to me a sketch of the proof already in 1971.

In fact, it was clear that there are a lot of links and inter-connections between all these problems. If the Gromov alternative had a flexible resolution (meaning that all volume preserving diffeomorphisms could be approximated by symplectic), this would imply that Arnold's fixed point conjecture is wrong in dim > 2. Indeed, it is easy to show that in dimensions >2 it is wrong for volume preserving, rather than symplectic diffeomorphisms.

In turn, an h-principle type answer to Problems P1-rel and P2-rel would imply the flexible resolution of the Gromov alternative, as the following argument (taken from [21]) shows. Consider, for instance, the contact case. Let (M, ξ) be a $(2n+1)$-dimensional contact manifold, and $f : M \to M$ a diffeomorphism isotopic to the identity which is somewhere not contact. Let us first note that according to Gromov's h-principle for open contact manifolds, the analogs of Problems P1-rel and P2-rel do have positive answers for neighborhoods of discs of positive codimension. Assuming that both Problems P1-rel and P2-rel have positive answers as well, we consider a small triangulation of M and inductively modify f by a C^0-small isotopy to make it contact on neighborhoods of k-skeleta of the triangulation, $k = 0, \dots, 2n + 1$. Suppose that we already constructed a diffeomorphism $f_{k-1} : M \to M$ preserving ξ on a neighborhood of the $(k-1)$-skeleton C_{k-1}. Take a neighborhood $G_\sigma \supset \sigma$ of a k-simplex σ. Consider a contact structure on $\mathcal{O}p\,(\partial G_\sigma \cup \sigma) \subset G_\sigma$ which coincides with ξ on $\mathcal{O}p\,\partial G_\sigma$ and with $(f_{k-1})_*\xi$ on $\mathcal{O}p\,\sigma$.[2] Using P1-rel we can then extend it as a contact structure on G_σ in the same formal homotopy class relative to ∂G_σ as ξ. Hence, P2-rel together with the Gray-Moser argument then implies the existence of a diffeomorphism $g_\sigma : M \to M$ compactly supported in G_σ such that $g_\sigma^*\xi = \widetilde{\xi}$. Note that one can arrange that the supports of g_σ and $g_{\widehat{\sigma}}$ are disjoint if $\sigma \cap \widehat{\sigma} = \varnothing$. Then the composition g of the diffeomorphisms g_σ for all k-simplices σ is C^0-small provided that the triangulation and the neighborhoods G_σ have small diameters. But then the

[1] An embedding $f : (M_0, \omega_0) \to (M_1, \omega_1)$ is called *iso-symplectic* if $f^*\omega_1 = \omega_0$, and an embedding $(M_0, \xi_0) \to (M_1, \xi_1)$ is iso-contact if its symplectization is iso-symplectic.

[2] We use Gromov's notation $\mathcal{O}p\ A$ for an unspecified neighborhood of A.

diffeomorphism $f_k := g \circ f_{k-1}$ is C^0-close to f and preserves the contact structure ξ on a neighborhood of the k-skeleton C_k. Continuing by induction we construct a contactomorphism f_{2n+1} which is C^0-close to f. But this contradicts to the C^0-closedness of the group of contact diffeomorphisms.

Hence, the resolution of the Gromov alternative became an existential question for symplectic *topology*. The rigid resolution of the alternative was a necessary condition for survival of symplectic topology as a subject.

13.5 Emergence of Symplectic Rigidity

Though I proved in 1979 Arnold's fixed point conjecture for all 2-dimensional surfaces [17], this was not yet a major breakthrough in symplectic rigidity because the result was purely 2-dimensional. The series of breakthroughs happened in the early 1980s. I proved a theorem about the combinatorial structure of wave fronts which implied the rigid resolution of the Gromov alternative in the symplectic case, see [26]. Then, in 1982, C. Conley and E. Zehnder proved Arnold's fixed point conjecture P3 for $2n$-dimensional tori. As it was explained above, their proof implied the rigid resolution of the symplectic Gromov alternative. In 1983 Bennequin showed that Problem P2-abs does not have an h-principle type answer even for contact structures on S^3 [4]. He also showed that not all diffeomorphisms of a contact 3-sphere can be approximated by contact diffeomorphisms, thus giving the rigid resolution of the 3-dimensional contact Gromov alternative.

13.6 Advent of Holomorphic Curves

The true new era of symplectic topology started with the publication of Gromov's paper [33]. Searching for tools to establish symplectic rigidity, and, in particular, to define invariants of symplectic manifolds, Gromov turned his attention to holomorphic curves which were known to be an important tool in Algebraic Geometry. However, the environment of integrable complex structures was too rigid to be useful in Symplectic Geometry. Gromov realized that Bers-Vekua's pseudo-analytic functions could be interpreted as J-holomorphic curves for a not necessarily integrable almost complex structure J. This turned out to be precisely the right tool to deal with in Symplectic Topology.

Let us recall that an almost complex structure on a $2n$-dimensional manifold W is a complex structure on its tangent bundle, i.e. an anti-involution $J: TW \to TW$, $J^2 = -\mathrm{Id}$. A map $f: (W_1, J_1) \to (W_2, J_2)$ between two almost complex manifolds of real dimension $2n_1 = \dim W_1$ and $2n_2 = \dim W_2$ is called *holomorphic* if the differential $df : TW_1 \to TW_2$ is complex linear, i.e.

$$\bar{\partial} f := \frac{1}{2}(df \circ J_1 - J_2 \circ df) = 0. \tag{13.1}$$

Written in local coordinates, Eq. (13.1) is a system of $2n_1n_2$ equations with respect to $2n_2$ unknown functions, and hence it is overdetermined unless $n_1 = 1$, i.e. when (W_1, J_1) is a Riemann surface. Respectively, when $n_1 > 1$ then for generic non-integrable J_1 or J_2 there are no holomorphic maps $(W_1, J_1) \to (W_2, J_2)$, even locally. On the other hand, when $n_1 = 1$ Eq. (13.1) is an elliptic equation with the same principal symbol as the standard Cauchy-Riemann equation in the integrable case. Hence, with appropriate boundary conditions (e.g. for closed holomorphic curves, or holomorphic curves with boundaries in totally real submanifolds) this is a Fredholm problem, and assuming certain transversality we get finite-dimensional moduli spaces of solutions.

At the time when Gromov was working on his theory of J-holomorphic curves there was already known an example of a remarkable application of an elliptic PDE in topology: Simon Donaldson [13] spectacularly applied in 4-dimensional topology moduli spaces of solutions of another elliptic problem, the so-called anti-self-dual Yang-Mills equation. Gromov's idea was to realize a similar scheme in symplectic topology using instead moduli spaces of holomorphic curves.

The starting problem in this scheme was to ensure compactness properties for the corresponding moduli spaces, i.e. to prove in the holomorphic curve setup an analog of Uhlenbeck's compactness theorem [63] in the Yang-Mills theory. Gromov proved a far-going generalization of the Schwarz lemma from complex analysis which allowed him to control derivatives of a holomorphic map in terms of the diameter of its image. Combining this lemma with an ingenious use of hyperbolic geometry of Riemann surfaces Gromov proved that

Theorem 13.6.1 *Given a sequence $f_n : (S, j) \to (M, J)$ of holomorphic curves in a closed almost complex manifold such that the area* $\mathrm{Area}(f_n(S))$ *is uniformly bounded, there exists a subsequence converging to a nodal holomorphic curve.*

Gromov spectacularly demonstrated how holomorphic curves can be successfully used for proving a wide range of remarkable results in symplectic topology. Here are a few examples.

13.6.1 Non-existence of Exact Lagrangians in $\mathbb{C}^n$

A symplectic manifold (X, ω) is called *Liouville* if the symplectic form ω is exact and it is fixed a primitive λ, called a *Liouville form*, of ω. A Lagrangian submanifold L in a Liouville manifold (X, λ) is said to be *exact* if the form $\lambda|_L$ is exact. Note that if $H_1(X) = 0$ then exactness is independent of the choice of the Liouville form λ.

Theorem 13.6.2 *There are no closed exact Lagrangian submanifolds in* $\mathbb{R}^{2n}$.

Gromov's ingenious strategy for proving this theorem was to consider an inhomogeneous $\overline{\partial}$-equation $\overline{\partial} f = c$ for maps $f : (D, \partial D) \to (\mathbb{C}^n, L)$ of the 2-disc with contractible boundary $f(\partial D) \subset L$, where c is a $(1, 1)$-form with constant coefficients. One then checks that for $c = 0$ all solutions are constant, and hence the moduli space of solutions is diffeomorphic to L itself, while for sufficiently large c there could not be any solution at all because solutions have to be harmonic functions, and hence satisfy an a priori derivative bound in terms of the diameter of L. In view of Gromov's compactness this leads to a conclusion that for a certain c there exists a bubbled solution where the bubble has to be a non-constant holomorphic disc with boundary in L. But the symplectic area of such a disc is positive which contradicts the exactness hypothesis.

The beautiful argument used by Gromov in this proof was replicated many times in later years by several authors. In particular, it is the starting point in a remarkable theorem of Abouzaid that for certain exotic n-spheres their cotangent bundles are not symplectomorphic (though all of them are diffeomorphic, as it follows from Smale's h-cobordism theorem) to T^*S^n, see [1].

13.6.2 Gromov's Non-squeezing Theorem

Theorem 13.6.3 *Suppose that* $0 < r < R$. *Then there are no symplectic embeddings* $D^{2n}(R) \to D^2(r) \times \mathbb{R}^{2n-2}$. *Here we assume that* $\mathbb{R}^{2k}$ *is endowed with the standard symplectic structure and denote by* $D^{2k}(r)$ *the ball of radius* r *in* $\mathbb{R}^{2k}$.

Gromov's argument used holomorphic curves in the following way. Suppose there exists a symplectic embedding $D^{2n}(R) \to D^2(r) \times \mathbb{R}^{2n-2}$. Symplectically embed $D^2(r) \times \mathbb{R}^{2n-2}$ into $S^2 \times \mathbb{R}^{2n}$, endowed with the symplectic form $\omega = \sigma_a \oplus \omega_0$, where σ_a is the area form on the 2-sphere S^2 of area $a \in (\pi r^2, \pi R^2)$, and ω_0 is the standard symplectic structure on $\mathbb{R}^{2n-2}$. Then $D^2(r) \times \mathbb{R}^{2n-2}$ symplectically embeds into $S^2 \times \mathbb{R}^{2n-2}$, and hence by assumption there exists a symplectic embedding $h : D^{2n}(R) \to S^2 \times \mathbb{R}^{2n-2}$. There exists an almost complex structure J on $S^2 \times \mathbb{R}^{2n}$ which is compatible with ω and coincides with the push-forward $h_* i$ of the standard complex structure on the ball $D^{2n}(R) \subset \mathbb{C}^n$. Gromov's theory of holomorphic curves then yields a J-holomorphic sphere S in $S^2 \times \mathbb{R}^{2n-2}$ in the homology class of $S^2 \times p$, $p \in \mathbb{R}^{2n-2}$, which passes through the image $h(0)$ of the center of the ball $D^{2n}(R)$. Then

$$\pi R^2 > a = \int_S \omega \geq \int_{S \cap h(D^{2n}(R))} \omega = \text{Area}(h^{-1}(S)). \tag{13.2}$$

But the curve $h^{-1}(S)$ passes through the center of $D^{2n}(R)$ and it is a properly embedded into $D^{2n}(R)$ holomorphic curve for the standard complex structure on

$\mathbb{C}^n$. But then it is a minimal surface, and hence by the isoperimetric inequality for minimal surfaces we have

$$\text{Area}(h^{-1}(S)) \geq \pi R^2, \tag{13.3}$$

which contradicts (13.2).

This theorem for the first time established the existence of specifically symplectic (e.g. different from the volume) invariants, and in particular, implied the symplectic rigidity theorem (i.e. the C^0-closedness of the group of symplectomorphisms in the group of all diffeomorphisms). One can also deduce from Theorem 13.6.3 that for any $n > 1$ there is a symplectic structure on $\mathcal{O}p\,\partial D^{2n}$ which does not extend to D^{2n}, while there are no formal obstructions for that. Recently it was shown, see Sect. 13.7.1 below, that there are no non-formal obstructions for the extension of contact structures. Gromov's paper also implied the contact rigidity. However, the full details of this argument were written only recently, see [51].

Slightly modifying the original Gromov definition we define $\text{swidth}(U, \omega)$ of a $2n$-dimensional symplectic manifold (U, ω), or as it is now usually called the *Gromov width* as

$$\text{swidth}(U) := \sup\{\pi r^2;\ D^{2n}(r) \text{ symplectically embeds into } (U, \omega)\}.$$

Thus, $\text{swidth}(D^2(r) \times D^{2n-2}(R)) = \pi r^2$ if $r \leq R$, and one can similarly prove that $\text{swidth}(S^2(a) \times \mathbb{R}^{2n-2}) = a$, where we denote by $S^2(a)$ the 2-sphere of area a, assume that the symplectic structure on $\mathbb{R}^{2n}$ is standard, and that the product is endowed with the split symplectic structure. It is interesting to note that this Gromov rigidity result coexists with the h-principle type observation of Polterovich (see [25]) that $\text{swidth}(T^2(a) \times \mathbb{R}^{2n-2}) = \infty$. In a similar vein, there is a result of Latschev-McDuff-Schlenk, see [41]: *the* 4*-torus admits an embedding of the* 4*-ball of full volume, and hence,* $\text{swidth}(T^2(a) \times T^2(b)) = \sqrt{2ab}$.

13.6.3 Packing Inequalities

Gromov's theory of holomorphic curves also implied that *for any almost complex structure J on $\mathbb{C}P^n$ which is tamed by the standard (Fubini-Study) symplectic form one has a J-holomorphic sphere in the class of the generator of $H_2(\mathbb{C}P^n)$ passing through any two points.* He then applied this result (cf. the proof of the non-squeezing theorem) to show that *if there is a symplectic embedding of two disjoint balls $D^{2n}(r_1)$ and $D^{2n}(r_2)$ into $D^{2n}(1)$ then $r_1^2+r_2^2 \leq 1$.* Indeed, if the balls $D^{2n}(r_1)$ and $D^{2n}(r_2)$ embed into $D^{2n}(1)$, then they also embed into $\mathbb{C}P^n$ with the symplectic area of the generator of $H_2(\mathbb{C}P^n)$ slightly bigger than π. Then one can choose an almost complex structure on $\mathbb{C}P^n$ tamed by the standard symplectic form and equal to the push-forward of the standard complex structure on the images of the balls. Finally, using the monotonicity theorem one concludes that the holomorphic sphere in the generator class passing through the centers of these balls have symplectic area $\geq \pi(r_1^2 + r_2^2)$, and the required inequality follows.

This result opened the whole new subject of *symplectic packing inequalities* with the most remarkable results proven in the 4-dimensional case. For instance, let us denote

$$v_k := \sup \frac{k\text{Volume}(D^4(r))}{\text{Volume}(D^4(1))},$$

where the supremum is taken over all r such that the disjoint union

$$\underbrace{D^4(r) \sqcup \cdots \sqcup D^4(r)}_{k}$$

symplectically embeds into $D^4(1)$. Gromov's result implies that $v_2 \leq \frac{1}{2}$. In the work of Karshon, Traynor, McDuff-Polterovich, and Biran [5, 48, 62], there were computed the precise values of v_k for all k. It turns out that

k	1	2	3	4	5	6	7	8
v_k	1	1/2	3/4	1	20/25	24/25	63/64	288/289

and $v_k = 1$ for $k \geq 9$.

Inspired by Gromov's definition of invariants of symplectic domains, Helmut Hofer defined in [38] a remarkable invariant of a Hamiltonian symplectomorphism, called nowadays the *Hofer norm* (a related invariant was defined by Claude Viterbo in [65]).

To connect Hofer's and Gromov's definitions let us consider a Hamiltonian diffeotopy $h_t\colon D^{2n} \to D^{2n}$, $t \in [0, 1]$, from the identity to $h_1 = h$. The diffeotopy is generated by a family of Hamitonian functions $H_t\colon D^{2n} \to \mathbb{R}$ equal to 0 on ∂D^{2n}. Let $\Gamma_{H_t} \subset D^{2n} \times [0, 1] \times \mathbb{R}$ be the graph of H_t:

$$\Gamma_{H_t} = \{u = H_t(x);\ x \in D^{2n}, t \in [0, 1]\} \subset (D^{2n} \times \mathbb{R}^2, \omega + dt \wedge du),$$

where ω is the standard symplectic form on D^{2n}. Let us assume for a moment that H_t is positive on the interior of the ball and consider the domain

$$U_{H_t} = \{(x, t, u) |\ 0 \leq u \leq H_t(x), t \in [0, 1], x \in D^{2n}\}.$$

Choosing a different Hamiltonian path $\widetilde{h}_t$ connecting Id with h, we observe that if the two paths are homotopic through paths with positive Hamiltonian functions then the corresponding domains U_{H_t} and $U_{\widetilde{H}_t}$ are symplectomorphic. Hence, the symplectic invariants of the domain U_{H_t}, e.g. its Gromov width, are invariants of the Hamiltonian diffeomorphism h (or more precisely, of its lift to the universal cover of the group of Hamiltonian diffeomorphisms).

Hofer got around the positivity constraint by defining his conjugation invariant norm $||h||$ for any Hamiltonian diffeomorphism $h\colon D^{2n} \to D^{2n}$ as

$$||h|| = \inf(\max H_t(x) - \min H_t(x)),$$

where the max and min are taken over all $(x, t) \in D^{2n} \times [0, 1]$, and the infimum is taken over all Hamiltonians H_t with $H_t|_{\partial D^{2n}} = 0$ generating h. One can think of $||h||$ as the Gromov width of a smallest box $D^{2n} \times [0, 1] \times [m, M]$ containing the graph Γ_{H_t}. The non-degeneracy of this norm, i.e. the fact that a symplectomorphism which can be generated by an arbitrary C^0-small Hamiltonian is equal to the identity, is parallel to Gromov's non-squeezing theorem. Hofer's norm can be equivalently defined by the formula $||h|| = \inf\limits_{H_t} \int\limits_0^1 ||H_t||_{C^0} dt$, where $||H_t||_{C^0} = \max\limits_{x \in D^{2n}} H_t(x) - \min\limits_{x \in D^{2n}} H_t(x)$ is the C^0-norm on the space of functions on D^{2n} equal to 0 on ∂D^{2n}, and the infimum is again taken over all Hamiltonians H_t with $H_t|_{\partial D^{2n}} = 0$ generating h. In this formulation we see that Hofer's norm is just the path-length norm on the group $\mathcal{H}$ of Hamiltonian diffeomorphisms corresponding to the Finsler metric given by the C^0-norm on the Lie algebra of the group $\mathcal{H}$. A remarkable theorem of Buhovsky and Ostrover, see [9], asserts that *any conjugation invariant Finsler (pseudo-)norm on the group of Hamiltonian diffeomorphisms that is generated by an invariant norm on the Lie algebra which is continuous with respect to the C^∞-topology, is either identically zero or equivalent to the Hofer norm.*

Hofer's norm, later generalized to all symplectic manifolds by Lalonde-McDuff [40], generates a bi-invariant metric on the group of Hamiltonian symplectomorphisms, which plays an important role in Hamiltonian Dynamics.

13.6.4 4-Dimensional Applications

As Gromov demonstrated in his seminal paper, holomorphic curves are especially useful in 4-dimensional symplectic geometry due to the *positivity of the intersection* property. As in the integrable case, transversely intersecting holomorphic curves in an almost complex 4-manifold intesect *positively*. Gromov sketched an argument to show that even in the singular case the analogy with the integrable case should hold. It turned out that the issue is quite subtle, and was settled in a series of papers of McDuff [47] and Micalleff–White [49]. Here are some of remarkable applications proven by Gromov.

Theorem 13.6.4

(i) Let J be any almost complex structure on $\mathbb{C}P^2$ tamed by the standard symplectic form on $\mathbb{C}P^2$. Then through any two distinct points there is a unique J-holomorphic sphere in the homology class of the generator of

$H_2(\mathbb{C}P^2)$, and any two such spheres intersect at one point. All these spheres are embedded.

(ii) Let (X, ω) be a symplectic 4-manifold with an almost complex structure J tamed by ω. Suppose that there exists an embedded J-holomorphic sphere $S \subset X$ with $S \cdot S = 1$, and there are no embedded J-holomorphic spheres with self intersection equal to -1. Then (X, ω) is symplectomorphic to $\mathbb{C}P^2$ with the standard symplectic form.

Theorem 13.6.4 (ii) was improved by McDuff [46]: *without assuming an absence of (-1)-curves one can conclude that X is symplectomorphic to $\mathbb{C}P^2$, possibly blown up at a few points.*

A corollary of Theorem 13.6.4 (ii) is that there exists a unique standard at infinity symplectic structure on $\mathbb{R}^4$, where uniqueness is understood up to symplectomorphism fixed at infinity. Another spectacular corollary is that the group of compactly supported symplectomorphisms of the standard symplectic $\mathbb{R}^4$ is contractible. This implies that the space of symplectic forms on $\mathbb{R}^4$standard at infinity is homotopy equivalent to the group of all compactly supported diffeomorphisms. Note, however, that nothing is currently known about the topology of this group.[3]

C.H. Taubes found a link, see [60, 61], between the Seiberg-Witten gauge theory and Gromov's theory of holomorphic curves. His result in combination with Theorem 13.6.4 (ii) implied that the uniqueness up to symplectomorphism result also holds for symplectic structures of a fixed total volume on $\mathbb{C}P^2$.

A lot of other great results on the rigidity side of symplectic topology were achieved in the years following Gromov's discovery. However, there were also significant advances on the flexible side of symplectic topology. We discuss some of them in the next section.

13.7 Successes on the Flexible Side of Symplectic Topology

13.7.1 Overtwisted Contact Structures

Gromov's h-principle for contact and symplectic structures on open manifolds reduced Problems P1-abs and P2-abs for closed manifolds to the extension problems P1-rel and P2-rel for the balls. Gromov non-squeezing result implied that an h-principle type answer cannot hold for Problem P1-rel in any symplectic dimension $2n \geq 4$. On the other hand, in the contact case Bennequin's theorem [4] in dimension 3 and later results, see

[3] Added in proof: Recently Tadayuli Watanabe found non-trivial elements in higher homotopies of this diffeomorphism group; see Some exotic nontrivial elements of the rational homotopy groups of Diff(S^4). Tadayuki Watanabe, Arxiv 1812.02448.

[64], implied that Problem P2-rel cannot have a positive h-principle type answer in any contact dimension $2n - 1 \geq 3$. On the other hand it was already known that P1-rel in contact dimension $2n - 1 = 3$ abides an h-principle.[4]

It was a surprising discovery in 1989 for the case $2n-1 = 3$, see [18], and in 2014 for all higher dimensions, see [7], that all problems P1-abs, P1-rel, P2-abs, P2-rel abide an h-principle for a certain subclass of contact structures, called *overtwisted* contact structures. In particular, if an odd-dimensional manifold M admits a stable almost complex structure (i.e. if the stabilized tangent bundle $TM \oplus \varepsilon^1$ admits a complex structure) then it also admits a genuine contact structure.

13.7.2 Donaldson's Hyperplane Sections

We already mentioned above Gromov's h-principle for iso-symplectic embeddings in codimension >2. Applying holomorphic curve techniques it is not difficult to construct counter-examples to a similar h-principle in codimension 2. However, Simon Donaldson, using his theory of *almost holomorphic sections* of complex line bundles over almost complex symplectic manifolds, proved among other remarkable results the following

Theorem 13.7.1 (Donaldson [14]) *For any closed $2n$-dimensional symplectic manifold (M, ω) with an integral cohomology class $[\omega] \in H^2(M)$ and a sufficiently large integer k there exists a codimension 2 symplectic submanifold $\Sigma \subset M$ which represents the homology class Poincaré dual to $k\omega$. Moreover, the complement $M \setminus \Sigma$ has the homotopy type of an n-dimensional cell complex (as it is the case for complements of hyperplane sections in complex projective manifolds).*

13.7.3 Guth's Symplectic Embeddings

Let us denote by $P(r_1, \dots, r_n)$ the polydisc $\{|z_1| \leq r_1, \dots, |z_n| \leq r_n\} \subset \mathbb{C}^n$, where we assume $r_1 \leq r_2 \leq \dots \leq r_n$. In the case $n = 2$ if $P(r_1, r_2)$ symplectically embeds into $P(R_1, R_2)$ then Theorem 13.6.3 implies that $r_1 \leq R_1$. We also have the volume constraint $r_1 r_2 \leq R_1 R_2$. Many people tried to prove that for a similar embedding problem for high-dimensional polydiscs there are additional constraints on intermidiate radii besides the width and volume constraints. However, Larry Guth proved the following remarkable result on the flexible side, which showed that the room for additional constraints is very limited.

[4]It seems that the proof of this result appeared in print only in [18], but it was, probably, known to W.P. Thurston, who used a very close argument in his proof of the h-principle for codimension 1 foliations.

Theorem 13.7.2 (Guth [35]) *There exists a constant $C(n)$ depending on the dimension n such that if $C(n)r_1 \leq R_1$ and $C(n)r_1 \dots r_n \leq R_1 \dots R_n$ then the polydisc $P(r_1, \dots, r_n)$ symplectically embeds into $P(R_1, \dots, R_n)$.*

13.7.4 Bounds on the Number of Double Points of Exact Symplectic Immersions

While the results confirming Arnold's conjecture on the intersection of two Lagrangian submanifolds remain one of the centerpieces of *rigid* symplectic topology, its analog concerning lower bounds for the number of double points of a Lagrangian immersion turned out to be wrong. For instance, an "Arnold type" conjecture predicts that the minimal number $s(L)$ of transverse double points of an exact Lagrangian immersion of an orientable n-dimensional closed manifold L into $\mathbb{R}^{2n}$ should satisfy the bound $s(L) \geq \frac{1}{2}\mathrm{rank} H_*(L)$, which for $L = T^n$ gives $s(L) \geq 2^{n-1}$.

However, it turns out (see [15]; for $n = 2$ the result is due to D. Sauvaget [56]) that this conjecture is wrong:

Theorem 13.7.3 *Let L be a closed n-dimensional orientable manifold. If the complexified tangent bundle $T^*(L) \otimes \mathbb{C}$ is trivial*[5] *then*

$$s(L) \begin{cases} = -\frac{1}{2}\chi(L), & \text{if } n \text{ is even and } \chi(L) < 0; \\ \leq \frac{1}{2}\chi(L) + 2, & \text{if } n \text{ is even and } \chi(L) \geq 0; \\ \leq 2, & \text{if } n \text{ is odd}; \\ = 1, & \text{if } n = 3, 7. \end{cases}$$

Here $\chi(L)$ is the Euler characteristic of L. For instance, *any* 3*-manifold admits a Lagrangian immersion into* $\mathbb{R}^6$ *with exactly* 1 *double point.*

It is interesting to contrast this theorem with the following rigidity result of Ekholm and Smith, [16]:

Theorem 13.7.4 *If a closed orientable $2k$-manifold $L, k > 2$, with $\chi(L) \neq -2$ admits an exact Lagrangian immersion into $\mathbb{R}^{4k}$ with one transverse double point and no other self-intersections, then L is diffeomorphic to the sphere.*

[5] This is a necessary and sufficient condition for existence of a Lagrangian *immersion*, as it follows from Gromov-Lees h-principle, [32, 43], for Lagrangian immersions.

13.7.5 Loose Legendrian Knots and Lagrangian with Conical Singularity

One of the applications of holomorphic curves techniques in contact topology was the introduction of new invariants of Legendrian knots, such as *Legendrian homology algebra*, [10, 20]. There is a certain operation on a Legendrian, called *stabilization*, see [11, 19, 52], which can be performed in a neighborhood of any point of the Legendrian knot, and which kills the Legendrian homology algebra. Moreover, it was known for 1-dimensional Legendrian knots, see [27], that formally isotopic Legendrian knots become genuinely Legendrian isotopic after sufficiently many stabilizations.

Emmy Murphy in her 2012 PhD dissertation [52] proved that for Legendrian knots of dimension > 1 formal Legendrian isotopy implies the genuine one already for once stabilized knots. A Legendrian knot L in a contact manifold (M, ξ) is called *loose* if it is Legendrian isotopic to a stabilization of another Legendrian knot.

Theorem 13.7.5 ([52]) *Any two formally isotopic loose Legendrian knots of dimension > 1 can be connected by a genuine Legendrian isotopy.*

Given a symplectic manifold (X, ω) we say that $L \subset M$ is a *Lagrangian submanifold with an isolated conical point* if it is a Lagrangian submanifold away from a point $x \in L$, and there exists a symplectic embedding $f : B_\varepsilon \to X$ such that $f(0) = x$ and $f^{-1}(L) \subset B_\varepsilon$ is a Lagrangian cone. Here B_ε is the ball of radius ε in the standard symplectic $\mathbb{R}^{2n}$. Note that this cone is automatically a cone over a Legendrian sphere in the sphere ∂B_ε endowed with the standard contact structure given by the restriction to ∂B_ε of the Liouville form $\lambda_{\mathrm{st}} = \frac{1}{2}\sum\limits_1^n (p_i dq_i - q_i dp_i)$.

It turns out that Lagrangians with 1 conical point exhibit a great deal of flexibility.

Theorem 13.7.6 ([24]) *Let L be an n-dimensional, $n > 2$, closed manifold such that the complexified tangent bundle $T^*(L \setminus p) \otimes \mathbb{C}$ is trivial. Then L admits an exact Lagrangian embedding into $\mathbb{R}^{2n}$ with exactly one conical point. In particular, a sphere admits a Lagrangian embedding into $\mathbb{R}^{2n}$ with one conical point for each $n > 2$.*

The analogous statement for $n = 2$ is wrong, as it follows from Bennequin's inequality, [4].

13.7.6 Construction of Symplectic Cobordisms

While existence problem for symplectic structures on closed manifolds remains widely open, the situation drastically changes when one allows symplectic structures to have one conical singularity. We say that a 2-form ω on the punctured $2n$-ball $D^{2n} \setminus 0$ is a *symplectic form on D^{2n} with a conical singularity at* 0 if it can be written as $d(r^2\alpha)$, where α is a contact form on $S^{2n-1} = \partial D^{2n}$ and r is the radial coordinate.

We will call $(S^{2n-1}, \zeta = \{\alpha = 0\})$ the *link* of the singularity. Note that if the contact structure ζ is standard, then the form extends to a non-singular symplectic form to D^{2n}.

Theorem 13.7.7 ([23]) *Let X be a closed manifold of dimension $2n > 4$ that admits an almost complex structure on $X \setminus p$, $p \in X$. Let $a \in H^2(X)$ be any cohomology class. Then for any closed symplectic $2n$-dimensional manifold (Z, ω), the connected sum $X\#Z$ admits a symplectic form with a conical singularity at p in the cohomology class $a + C[\omega]$ for a sufficiently large constant $C > 0$.*

13.8 Basic Problems: Where They Stand Now?

We finish this article with a brief survey of the current status of basic problems discussed in Sect. 13.4.

Problem P1-abs (Existence of Symplectic and Contact Structures)
Gromov's theorem fully answered this question for open manifolds. In the 3-dimensional contact case Problem P1-abs was positively answered by J. Martinet, [45], and R. Lutz, [44]. The paper [7] gave an h-principle type necessary and sufficient condition for higher-dimensional closed contact manifolds. In the closed contact symplectic case the problem is open and there are no counter-examples to the following *flexible conjecture*: any closed manifold of dimension $2n > 4$ which has an almost complex structure J and a cohomology class $a \in H^2(M)$ with $a^n \neq 0$ admits a symplectic structure ω with $[\omega] \in a$, and which is compatible with an almost complex structure homotopic to J. In dimension 4 this conjecture was proven to be wrong by Taubes, [59], with the help of Seiberg-Witten theory. For example, the connected sum of 3 copies of $\mathbb{C}P^2$ does not admit a symplectic structure, while it has an almost complex structure and a cohomology class $a \in H^2(M)$ with $a^2 \neq 0$.

P1-rel (Extension of Symplectic and Contact Structures)
In the contact case there is an h-principle type answer in all dimensions [7]. In the symplectic case Gromov's theory of holomorphic curves provides counter-examples to the flexible conjecture in all dimensions. Unfortunately, there are no known invariants beyond the formal ones which would serve as obstructions to the extension of a symplectic structure.

P2-abs (Homotopy of Symplectic and Contact Structures on Closed Manifolds)
In both the symplectic and contact cases there are obstructions for the homotopy between two formally homotopic structures on a given closed manifolds. In both cases invariants distinguishing the structures are based on Gromov's theory of holomorphic curves.

P2-rel (The Relative Version of P2-abs)
There are contact structures on the ball which coincide near the boundary but are not homotopic (e.g. see [64]). In the symplectic case the problem is open. In the 4-dimensional case Gromov proved that any symplectic structure on the ball D^4 which is standard near ∂D^4 is symplectomorphic to the standard one. However, nothing is known about the topology of the group $\mathrm{Diff}(D^4, \partial D^4)$, and hence it is unknown whether there could exist a symplectic structure on D^4 which is standard near ∂D^4 and which is not homotopic to the standard one.

P3 (Arnold's Fixed Point Conjecture)
Arnold's fixed point conjecture is proven for general symplectic manifolds in its minimal form, i.e. that assuming all fixed points are non-degenerate, the number of fixed points is bounded below by the rank of homology [28]. However, more optimistic conjectures (e.g. bounding the number of non-degenerate points of a symplectomorphism in terms of the minimal number of critical points of a Morse function on the manifold) are still open.

P4 (Arnold's Lagrangian Intersections Conjecture)
If a Lagrangian $L \subset T^*M$ is Hamiltonian isotopic to the 0-section M and transversely intersects M then the most optimistic Arnold conjecture for the lower bound for $\#L \cap M$ is equal to Morse(M), the minimal number of critical points of a Morse function function on M. But the best known estimate is $\#L \cap M \geq$ stabMorse(M), where stabMorse(M) is the *stable* Morse number, i.e. the minimal number of critical points of a Morse function $F : M \times \mathbb{R}^N \to \mathbb{R}$ of the form $F(x, y) = Q(y) + \varepsilon(x, y), x \in M, y \in \mathbb{R}^N$, where Q is a non-degenerate quadratic form and ε is a compactly supported function; see [42].

P5 (Lagrangian Embedding Problem)
A lot of progress was achieved towards understanding the topology of Lagrangian embeddings, beginning with Gromov's theorem about the absence of closed exact Lagrangian submanifolds in $\mathbb{R}^{2n}$. The progress is especially remarkable in dimension 4. A theorem of H. Whitney, see [66], implies[6] that the 2-torus is the only orientable closed surface which admits a Lagrangian embedding into $\mathbb{R}^4$ and that non-orientable closed surfaces of Euler characteristics non-divisible by 4 do not admit Lagrangian embeddings in $\mathbb{R}^4$. On the other hand all non-orientable surfaces of Euler characteristic $-4k, k \geq 1$, were known to admit such an embedding, see e.g. [29]. Finally the classification of all possible Lagrangian surfaces in $\mathbb{R}^4$ was completed by Shevchishin [57], who proved that there is no Lagrangian embedding of the Klein bottle into $\mathbb{R}^4$. In a recent paper of Dimitroglou Rizell et al. [12], it was shown that all Lagrangian tori in $\mathbb{R}^4$ are Lagrangian isotopic.

[6]Whitney proved that the normal Euler number of a closed surface embedded into $\mathbb{R}^4$ is equal to 0 in the orientable case, and is congruent to $2\chi \pmod 4$ in the non-orientable one. For Lagrangian surfaces normal and tangential Euler numbers differ only by sign, which implies the claim.

P5-nearby (Nearby Lagrangian Conjecture)
The nearby Lagrangian conjecture is still open despite a significant progress. It is fully proven for T^*S^2 [36], and T^*T^2 [12]. For all other cases the best known result [2], states that for any exact Lagrangian $L \subset T^*M$ its projection to M is a simple homotopy equivalence.

P6 (Legendrian Isotopy Problem)
The Legendrian algebra defined by Chekanov in [10] and independently in the framework of Symplectic Field Theory by Givental, Hofer and the author in [20, 22] proved itself to be an effective invariant of Legendrian knots. For 1-dimensional knots an alternative invariant in the framework of Heegaard homology theory was defined in [54].

P7 (Equidimensional Symplectic Embeddings)
Beginning with Gromov's non-squeezing theorem a lot of progress was achieved in the theory of symplectic embeddings, especially in dimension 4; see Hutchings's survey [39]. In higher dimensions there is still a large gap between known negative and positive results.

P8 (C^0-Symplectic Topology)
The resolution of Gromov's alternative in favor of rigidity suggested the notion of a symplectic *homeomorphism* as the one which could be C^0-approximated by symplectic diffeomorphisms. In turn, this yields a natural notion of a topological symplectic manifold. Most of the basic problems of symplectic C^0-topology are still open. For instance, it is still unknown whether a $2n$-dimensional closed topological symplectic manifold must have a cohomology class $a \in H^2(M)$ with $a^n \neq 0$. However, the subject continues to develop with many promising non-trivial results, both on the rigid and flexible sides of this theory; see e.g. [8].

Symplectic rigidity and flexibility continue their development towards each other in search of the ultimate boundary.

Acknowledgement Yakov Eliashberg was partially supported by NSF grant DMS-1505910.

References

1. M. Abouzaid. On the wrapped Fukaya category and based loops. *J. Symplectic Geom.*, 10(1):27–79, 2012.
2. M. Abouzaid and T. Kragh. Simple Homotopy Equivalence of Nearby Lagrangians. *J. Sympl. Geom.*, 2016.
3. V. I. Arnold. Sur une propriété topologique des applications globalement canoniques de la mécanique classique. *C. R. Acad. Paris*, 261:3719–3722, 1965.
4. D. Bennequin. Entrelacements et équations de Pfaff. In *Third Schnepfenried geometry conference, Vol. 1 (Schnepfenried, 1982)*, volume 107 of Astérisque, pages 87–161. Soc. Math. France, Paris, 1983.
5. P. Biran. A stability property of symplectic packing. *Invent. Math.*, 136(1):123–155, 1999.

6. G. D. Birkhoff. Proof of Poincaré's geometric theorem. *Trans. Amer. Math. Soc.*, 14(1):14–22, 1913.
7. M. S. Borman, Y. Eliashberg, and E. Murphy. Existence and classification of overtwisted contact structures in all dimensions. *Acta Math.*, 215(2):281–361, 2015.
8. L. Buhovsky, V. Humilière, and S. Seyfaddini. A C^0-counterexample to the Arnold conjecture. *Invent. Math.*, 213:759–809 (2018).
9. L. Buhovsky and Y. Ostrover. On the uniqueness of Hofer's geometry. *Geom. Funct. Anal.*, 21(6):1296–1330, 2011.
10. Y. Chekanov. Differential algebra of Legendrian links. *Invent. Math.*, 150(3):441–483, 2002.
11. K. Cieliebak and Y. Eliashberg. *From Stein to Weinstein and back*, volume 59 of American Mathematical Society Colloquium Publications. American Mathematical Society, Providence, RI, 2012. Symplectic geometry of affine complex manifolds.
12. G. Dimitroglou Rizell, E. Goodman, and A. Ivrii. Lagrangian isotopy of tori in $S^2 \times S^2$ and $\mathbb{C}P^2$. *Geom. Funct. Anal.*, 26:1297–1358 (2016).
13. S. K. Donaldson. An application of gauge theory to four-dimensional topology. *J. Differential Geom.*, 18(2):279–315, 1983.
14. S. K. Donaldson. Symplectic submanifolds and almost-complex geometry. *J. Differential Geom.*, 44(4):666–705, 1996.
15. T. Ekholm, Y. Eliashberg, E. Murphy, and I. Smith. Constructing exact Lagrangian immersions with few double points. *Geom. Funct. Anal.*, 23(6):1772–1803, 2013.
16. T. Ekholm and I. Smith. Exact Lagrangian immersions with a single double point. *J. Amer. Math. Soc.*, 29:1–59, 2015.
17. Y. Eliashberg. *Estimates on of the number of fixed points of area-preserving transformations of surfaces*. VINITI, Syktyvkar University, 1979.
18. Y. Eliashberg. Classification of overtwisted contact structures on 3-manifolds. *Invent. Math.*, 98(3):623–637, 1989.
19. Y. Eliashberg. Topological characterization of Stein manifolds of dimension > 2. *Internat. J. Math.*, 1(1):29–46, 1990.
20. Y. Eliashberg. Invariants in contact topology. In *Proceedings of the International Congress of Mathematicians, Vol. II (Berlin, 1998)*, number Extra Vol. II, pages 327–338, 1998.
21. Y. Eliashberg. Recent advances in symplectic flexibility. *Bull. Amer. Math. Soc.*, 52:1–26, 2015.
22. Y. Eliashberg, A. Givental, and H. Hofer. Introduction to Symplectic Field Theory. *Geom. and Funct. Anal.*, pages 560–673, 2000. Special volume.
23. Y. Eliashberg and E. Murphy. Making cobordisms symplectic. arXiv:1504.06312.
24. Y. Eliashberg and E. Murphy. Lagrangian caps. *Geom. Funct. Anal.*, 23(5):1483–1514, 2013.
25. Y. Eliashberg and L. Polterovich. Unknottedness of Lagrangian surfaces in symplectic 4-manifolds. *Internat. Math. Res. Notices*, (11):295–301, 1993.
26. Y. M. Eliashberg. A theorem on the structure of wave fronts and its application in symplectic topology. *Funkt. Anal. i Pril.*, 21(3):65–72, 96, 1987.
27. D.B. Fuchs and S. Tabachnikov. Invariants of Legendrian and transverse knots in the standard contact space. *Topology*, 36:1025–1053, 1997.
28. K. Fukaya and K. Ono. Arnold conjecture and Gromov-Witten invariant for general symplectic manifolds. In *The Arnoldfest (Toronto, ON, 1997)*, volume 24 of Fields Inst. Commun., pages 173–190. Amer. Math. Soc., Providence, RI, 1999.
29. A.B. Givental. Lagrangian imbeddings of surfaces and unfolded Whitney umbrella. *Funkt. Anal. i Pril.*, 20:35–41, 1986.
30. J.W. Gray. Some global properties of contact structures. *Ann. of Math. (2)*, 69:421–450, 1959.
31. M. Gromov. Stable mappings of foliations into manifolds. *Izv. Akad. Nauk SSSR Ser. Mat.*, 33:707–734, 1969.
32. M. Gromov. A topological technique for the construction of solutions of differential equations and inequalities. *Proc. Int. Congress Math.*, 2:221–225, 1970.
33. M. Gromov. Pseudoholomorphic curves in symplectic manifolds. *Invent. Math.*, 82(2):307–347, 1985.

34. M. Gromov. *Partial differential relations*, volume 9 of Ergebnisse der Mathematik und ihrer Grenzgebiete (3). Springer-Verlag, Berlin, 1986.
35. L. Guth. Symplectic embeddings of polydisks. *Invent. Math.*, 172(3):477–489, 2008.
36. R. Hind. Lagrangian spheres in $S^2 \times S^2$. *Geom. and Funct. Anal.*, 14(2):303–318, 2004.
37. M. W. Hirsch. Immersions of manifolds. *Trans. Amer. Math. Soc.*, 93:242–276, 1959.
38. H. Hofer. On the topological properties of symplectic maps. *Proc. Roy. Soc. Edinburgh Sect. A*, 115(1–2):25–38, 1990.
39. M. Hutchings. Recent progress on symplectic embedding problems in four dimensions. *Proc. Nat. Acad. Sci.*, 108(20):8093–8099, 2010.
40. F. Lalonde and D. McDuff. The geometry of symplectic energy. *Ann. of Math. (2)*, 141(2):349–371, 1995.
41. J. Latschev, D. McDuff, and F. Schlenk. The Gromov width of 4-dimensional tori. arXiv:1111.6566.
42. F. Laudenbach and Sikorav J.-P. Persistance d'intersection avec la section nulle au cours d'une isotopie hamiltonienne dans un fibré cotangent. *Invent. Math.*, 27:349–358, 1985.
43. J. A. Lees. On the classification of Lagrange immersions. *Duke Math. J.*, 43:217–224, 1976.
44. R. Lutz. Structures de contact sur les fibrés en cercles en dimension 3. *Ann Inst. Fourier*, 27:1–15, 1977.
45. J. Martinet. Forme de contact sur les variétés de dimension 3. *Lect. Notes in Math.*, 209:142–163, 1971.
46. D. McDuff. The structure of rational and ruled symplectic 4-manifolds. *J. Amer. Math. Soc.*, 3(3):679–712, 1990.
47. D. McDuff. Singularities and positivity of intersections of J-holomorphic curves. In *Holomorphic curves in symplectic geometry*, volume 117 of Progr. Math., pages 191–215. Birkhäuser, Basel, 1994. With an appendix by Gang Liu.
48. D. McDuff and L. Polterovich. Symplectic packings and algebraic geometry. *Invent. Math.*, 115(3):405–434, 1994. With an appendix by Yael Karshon.
49. M. J. Micallef and B. White. The structure of branch points in minimal surfaces and in pseudoholomorphic curves. *Ann. of Math. (2)*, 141(1):35–85, 1995.
50. J. Moser. On the volume elements on a manifold. *Trans. Amer. Math. Soc.*, 120:286–294, 1965.
51. S. Mueller and P. Spaeth. Gromov's alternative, contact shape, and C^0-rigidity of contact diffeomorphisms. arXiv:1310.0527.
52. E. Murphy. Loose Legendrian embeddings in high dimensional contact manifolds. arXiv:1201.2245.
53. J. Nash. C^1 isometric imbeddings. *Ann. of Math. (2)*, 60:383–396, 1954.
54. P. Ozsváth, Z. Szabó, and D. Thurston. Legendrian knots, transverse knots and combinatorial Floer homology. *Geom. and Topol.*, 12:941–980, 2008.
55. H. Poincaré. Sur une théorème de géométrie. *Rend. Circ. Mat. Palermo*, 33:375–507, 1912.
56. D. Sauvaget. Curiosités lagrangiennes en dimension 4. *Ann. Inst. Fourier (Grenoble)*, 54(6):1997–2020, 2004.
57. V. V. Shevchishin. Lagrangian embeddings of the Klein bottle and combinatorial properties of mapping class groups. *Izvestiya RAN*, 73(4):797–880, 2009.
58. S. Smale. The classification of immersions of spheres in Euclidean spaces. *Ann. of Math. (2)*, 69:327–344, 1959.
59. C. H. Taubes. The Seiberg-Witten invariants and symplectic forms. *Math. Res. Letters*, 1:809–822, 1994.
60. C. H. Taubes. SW $\Rightarrow$ Gr: from the Seiberg-Witten equations to pseudo-holomorphic curves. *J. Amer. Math. Soc.*, 9(3):845–918, 1996.
61. C. H. Taubes. Gr $\Longrightarrow$ SW: from pseudo-holomorphic curves to Seiberg-Witten solutions. *J. Differential Geom.*, 51(2):203–334, 1999.
62. L. Traynor. Symplectic packing constructions. *J. Differential Geom.*, 41(3):735–751, 1995.
63. K. K. Uhlenbeck. Connections with L^p bounds on curvature. *Comm. Math. Phys.*, 83(1):31–42, 1982.

64. I. Ustilovsky. Infinitely many contact structures on S^{4m+1}. *Int Math Res Notices*, 14:781–791, 1999.
65. C. Viterbo. Symplectic topology as the geometry of generating functions. *Mathematische Annalen*, 292(4):685–710, 1992.
66. H. Whitney. On the topology of differentiable manifolds. In *Lectures in Topology*. Mich. Univ. Press, 1940.

Chapter 14
Flat Affine, Projective and Conformal Structures on Manifolds: A Historical Perspective

William M. Goldman

Abstract This historical survey reports on the theory of locally homogeneous geometric structures as initiated in Ehresmann's 1936 paper *Sur les espaces localement homogènes*. Beginning with Euclidean geometry, we describe some highlights of this subject and threads of its evolution. In particular, we discuss the relationship to the subject of discrete subgroups of Lie groups. We emphasize the classification of geometric structures from the point of view of fiber spaces and the later work of Ehresmann on infinitesimal connections. The *holonomy principle*, first isolated by W. Thurston in the late 1970's, relates this classification to the representation variety $\mathsf{Hom}(\pi_1(\Sigma), G)$. We briefly survey recent results in flat affine, projective, and conformal structures, in particular the tameness of developing maps and uniqueness of structures with given holonomy.

1991 Mathematics Subject Classification: 57M05 (Low-dimensional topology); 20H10 (Fuchsian groups and their generalizations)

14.1 Introduction

On 23 October 1935, at the Geneva conference "Quelques questions de Geométrie et de Topologie," Charles Ehresmann [56] initiated the study of geometric structures *modeled on a homogeneous space* (X, G), or *locally homogeneous geometric structures* on manifolds. Here G is a Lie group and X a homogeneous space, representing a geometry in the sense of Klein's Erlanger program. A geometric structure is defined by an atlas of coordinate charts mapping into X with coordinate changes locally defined by transformations in G. We call such a structure *a* (G, X)-*structure,* and a manifold equipped with such a structure a (G, X)-*manifold.* A (G, X)-manifold M inherits all of the local geometry of X invariant under G.

W. M. Goldman (✉)
Department of Mathematics, University of Maryland, College Park, MD, USA
e-mail: wmg@math.umd.edu; http://www.math.umd.edu/~wmg

S. G. Dani, A. Papadopoulos (eds.), *Geometry in History*,
https://doi.org/10.1007/978-3-030-13609-3_14

These ideas were heavily influenced by Sophus Lie, Felix Klein, Henri Poincaré and Élie Cartan among others. Lie and Klein recognized how a group-theoretic viewpoint unified the disparate *classical geometries:* for them, a *geometry* consists of the properties of a space X upon which a group G acts transitively by symmetries *of that geometry.* Transitivity of the action means that the local geometries at any pair of points are equivalent. For example for *Euclidean geometry,* X is Euclidean space $\mathbb{E}^n$ and G is its group of isometries. Poincaré introduced the fundamental group $\pi_1(\Sigma, x_0)$ of a topological space Σ, consisting of loops based at a fixed (but arbitrary) base point $x_0 \in \Sigma$. Cartan introduced a general notion of *development* along paths, which corresponds to parallel transport of infinitesimal objects (for example, tangent vectors and frames) along paths. Development for an Ehresmann structure M modeled on a geometry (G, X) defines a homomorphism $\pi_1(M) \longrightarrow G$ compatible with a local homeomorphism $\widetilde{M} \longrightarrow X$.

Ehresmann begins with Riemannian manifolds of constant curvature, which he calls *Clifford-Klein space forms.* Such manifolds are locally modeled on Euclidean space $\mathbb{E}^n$, the sphere S^n, or hyperbolic space H^n, depending on whether the curvature is zero, positive or negative, respectively. Indeed, for these geometries, a (G, X)-structure is completely equivalent to a Riemannian metric of constant sectional curvature. The key property upon which he focuses is that any two points in such a space possess open neighborhoods which are isometric, that is, they have the *same local geometries.* He considers the more general situation of a manifold X with a transitive left action of a Lie group G; choosing a point $x \in X$,

$$
\begin{aligned}
G &\longrightarrow X \\
g &\longmapsto g(x)
\end{aligned}
$$

maps G (with its simply transitive group of left-multiplications) G-equivariantly to X. Furthermore this map passes down to a isomorphism of left G-spaces $G/\mathsf{Stab}(G, x) \longrightarrow X$, where

$$
\mathsf{Stab}(G, x) := \{g \in G \mid g(x) = x\}
$$

is the *stabilizer* of x in G. In modern parlance, X is a *homogeneous space* of G.

He then defines a *locally homogeneous space* to be a manifold M (having the same dimension as X) which is *locally modeled* on the G-invariant geometry of X. Specifically, M is covered by open neighborhoods, *coordinate patches,* U (which Ehresmann calls "elementary neighborhoods") equipped with homeomorphisms, *coordinate charts,* $U \xrightarrow{\psi} X$. The coordinate charts transfer the local G-invariant geometry of X to U.

Coordinate patches U and U' with corresponding charts ψ and ψ' respectively define possibly competing geometries on the intersection $U \cap U'$. Thus we require that ψ and ψ' define the same local geometry on $U \cap U'$: that is, each $p \in U \cap U'$ possesses an open neighborhood $V \subset U \cap U'$ such that $\psi'|_V = g \circ \psi|_V$ for some $g \in$

G. If we require that G acts effectively on X, then g will be uniquely determined. Furthermore g only depends on the connected component of $U \cap U'$ containing p.

This is what he calls a *locally homogeneous space of Lie,* to distinguish it from a *homogeneous space of Lie,* and he observes that every homogeneous space is locally homogeneous. The main question addressed in the paper is to what extent the converse holds.

He begins with the observation that a locally homogeneous space is a real analytic manifold. Although at the time, the global notion of a Lie group had not been popularized (Chevalley's book [35] would not be published for at least another decade), Ehresmann spends some time clarifying the relation between local groups of transformations on X.

In [56], Ehresmann calls M a *Clifford form* of X.

Ehresmann structures modeled on a Lie group G and its group of left-translations arise from discrete subgroups of G, at least under the assumption that the developing map is a covering space. Certainly when M is compact, such a (G, X)-structure corresponds to a discrete subgroup $\tilde{\Gamma} \subset \tilde{G}$ and an isomorphism $M \cong \tilde{\Gamma}\backslash\tilde{G}$. (For more information see [75].)

In the literature, "locally homogeneous spaces" sometimes refer to biquotients $\Gamma\backslash G/H$, since "homogeneous space" may refer to the quotient G/H. (Here $\Gamma \subset G$ is a discrete subgroup and $H \subset G$ is a closed subgroup, so that G/H is Hausdorff. Furthermore Γ is assumed to act *properly* on G/H, which, unless H is compact, is a nontrivial assumption on Γ.) If, in addition, Γ is assumed to act freely on G/H, then the double coset space $\Gamma\backslash G/H$ admits the natural structure of a (G, X)-manifold M, where $X = G/H$. Under various completeness assumptions (see below), "locally homogeneous" in our sense will imply that the geometric manifold is indeed a double coset space.

When H is compact, X carries a G-invariant Riemannian metric which is necessarily *geodesically complete,* and the Hopf Rinow theorem implies that a *closed* (G, X)-manifold is a double coset space in the above sense. These basic examples are particularly tame, although nonetheless extremely rich.

Acknowledgements I am grateful to Yves Benoist and Athanase Papadopoulous for their careful reading of the manuscript and many useful suggestions. I also wish to thank Thomas Delzant for useful comments on the literature.

14.2 Euclidean Manifolds

The most familiar geometry is *Euclidean geometry.* Euclidean geometry includes relations between points, lines, planes, and measurements such as distance, angle, area and volume. A more sophisticated aspect of Euclidean geometry is the theory of harmonic functions and Laplace's equation. The key property is that these objects, and the relations between them, are invariant under the transitive action of the isometry group $G = \mathsf{Isom}(\mathbb{E}^n)$ of Euclidean space $\mathbb{E}^n$.

The model space $X = \mathbb{E}^n$ is defined as the vector space $\mathbb{R}^n$ (or more accurately the affine space A^n, where the special significance of the additive identity $\mathbf{0} \in \mathbb{R}^n$ is removed). The isometry group is generated by the *group* of translations (identified as the vector space $\mathbb{R}^n$), and the group $\mathsf{O}(n)$ of orthogonal linear automorphisms of $\mathbb{R}^n$. Thus $G \cong \mathbb{R}^n \rtimes \mathsf{O}(n)$ and $X = \mathbb{E}^n$ identifies with the homogeneous space G/H where $H := \mathsf{Stab}(G, \mathbf{0}) = \mathsf{O}(n)$ is the stabilizer of the origin $\mathbf{0} \in \mathbb{R}^n$.

14.2.1 Riemannian Geometry

Euclidean structures are *flat Riemannian structures,* that is, Riemannian structures whose curvature tensor vanishes. We adopt the viewpoint that the Riemannian structure is the geometry defined by a Riemannian metric (tensor), which then leads to notions of *speed* and *length* of smooth curves, and finally the structure of a metric space.

The key point is that for $X = \mathbb{E}^n$, a positive inner product on the associated vector space $\mathbb{V} = \mathbb{R}^n$ extends to a $G = \mathsf{Isom}(\mathbb{E}^n)$-invariant metric tensor on X. (The G-invariant tensor is uniquely determined up to scaling by a nonzero constant.) This infinitesimal structure makes X into a metric space upon which G acts isometrically.

If M is a manifold, then an Ehresmann structure on M modeled on Euclidean geometry is essentially equivalent to a metric space locally isometric to a Euclidean structure. (We say "essentially" because the distance function is determined up to scaling by positive constant.) Equivalently, this is just a Riemannian metric which is *flat,* that is, one whose Riemann curvature tensor vanishes. Other Ehresmann structures can be defined in a similar way, using an infinitesimal form (*Cartan connections*) such that an object generalizing the curvature tensor vanishes. (Compare Sharpe [129].)

However, the theory of Euclidean manifolds really goes back much earlier, to crystallography and the theory of regular tilings of Euclidean space. Once the abstract notion of a *group of transformations* was formulated, late nineteenth-century crystallographers such as Schoenflies and Fedorov classified *crystallographic groups,* namely symmetry groups of tilings of $\mathbb{E}^3$ by compact polyhedra. These are the mathematical abstractions of *crystals*. In arbitrary dimension a *Euclidean crystallographic group* is a subgroup $\Gamma \subset \mathsf{Isom}(\mathbb{E}^n)$ acting properly discontinuously on $\mathbb{E}^n$ with compact quotient (equivalently, a compact fundamental domain). A compact flat Riemannian manifold M (that is, a *Euclidean manifold*) determines a crystallographic group. The Hopf-Rinow theorem (see Sect. 14.6 below) implies that the universal covering space $\widetilde{M}$ is isometric to $\mathbb{E}^n$, and the group $\pi_1(M)$ of deck transformations acts properly and isometrically on $\mathbb{E}^n$. Conversely, if $\Gamma \subset \mathsf{Isom}(\mathbb{E}^n)$ is discrete, then it acts properly and isometrically on $\mathbb{E}^n$. If, furthermore, Γ is torsionfree, it acts freely on $\mathbb{E}^n$ and the quotient $M := \Gamma \backslash \mathbb{E}^n$

is a manifold. In particular M identifies with a double coset space $\Gamma\backslash G/H$. Since $H = \mathsf{O}(n)$ is compact, the homogeneous space $\Gamma\backslash G$ is compact if and only if the locally homogeneous space $\Gamma\backslash G/H$ is compact. Crystallographic groups, then, are just discrete subgroups $\Gamma \subset \mathsf{Isom}(\mathbb{E}^n)$ which are *cocompact,* that is, when $\Gamma\backslash\mathsf{Isom}(\mathbb{E}^n)$ is compact. In the case that $G = \mathsf{Isom}(\mathbb{E}^n)$, this is equivalent to Γ being a *lattice* in G, namely a discrete subgroup such that $\Gamma\backslash G$ has finite Haar measure. Thus the classification of crystallographic groups is equivalent to the classification of lattices in $\mathsf{Isom}(\mathbb{E}^n)$. (See Milnor [120] and the references cited there for an excellent exposition of these ideas and their histroical motivation.)

14.2.2 The Bieberbach Theorems

In 1911, Bieberbach proved a *Structure Theorem* for crystallographic groups. Namely, the *linear holonomy* $\Gamma \xrightarrow{\mathbb{L}} \mathsf{O}(n)$ defined by the (constant) derivative of the isometry $\gamma \in \Gamma$ has finite image. Its kernel consists of all translations in Γ, and $\mathsf{Ker}(\mathbb{L}) = \Gamma \cap \mathbb{R}^n$ is a lattice Λ in $\mathbb{R}^n$ (the additive group spanned by a basis of $\mathbb{R}^n$). The geometric version is that *a compact Euclidean manifold* admits a finite covering space whose total space is a *flat torus* $\mathbb{R}^n/\Lambda$.

He also proved a *Rigidity Theorem* and a *Finiteness Theorem* for Euclidean manifolds. Euclidean manifolds are *rigid* in the following sense: Every isomorphism $\Gamma_1 \rightarrow \Gamma_2$ of crystallographic groups extends to an *affine* automorphism of $\mathbb{E}^n$ conjugating Γ_1 to Γ_2. Observe that the rigidity is up to *affine equivalence,* not *Euclidean isometry.* While *isometry* classes of marked Euclidean n-manifolds comprise a deformation space with rich geometry $\big($identifying with $\mathsf{GL}(\mathbb{R}^2)/\mathsf{O}(2)\big)$, the deformation space of affine equivalence classes of marked Euclidean n-manifolds is a point.

Finally any n admits finitely many isomorphism classes of crystallographic subgroups $\Gamma \subset \mathsf{Isom}(\mathbb{E}^n)$. For $n = 2$, only the torus and Klein bottle have Euclidean structures. For $n = 3$ only six orientable 3-manifolds admit Euclidean structures.

These three theorems provide a satisfactory qualitative picture of Euclidean structures on closed manifolds. Compare Wolf [145], Raghunathan [126], and Thurston [135].

By the Rigidity Theorem above, it seems natural to consider more general *affine crystallographic groups,* namely discrete subgroups $\Gamma \subset \mathsf{Aff}(\mathsf{A}^n)$ such that Γ acts properly on A^n, and the quotient $\Gamma\backslash\mathsf{A}^n$ is compact. (Our notation emphasizes *context:* A^n denotes the *affine space* underlying $\mathbb{E}^n$: that is, A^n "is" $\mathbb{E}^n$, but without the special structure defined by the Euclidean inner product. Similarly A^n "is" $\mathbb{R}^n$, but without the special structure given by the additive identity $\mathbf{0} \in \mathbb{R}^n$.)

14.2.3 Affine Crystallographic Groups

In dimension three, analogues of all three of Bieberbach's theorems fail for affine crystallographic groups: the image of the linear holonomy is generally infinite, there are generally infinitely many affine isomorphism types in a given topological type, and there are infinitely many topological types (Auslander [7]). Auslander and Markus [6] construct 3-dimensional flat *Lorentzian* manifolds which are *geodesically complete:* these are quotients $M := \Gamma\backslash\mathsf{A}^n$ by discrete subgroups $\Gamma \subset G = \mathsf{Aff}(\mathsf{A}^n)$ which act properly on A^n with compact quotient. The situation is now much more tricky, since M is a biquotient $\Gamma\backslash G/H$, where $G = \mathsf{Aff}(\mathsf{A}^n)$. However, since $H = \mathsf{Stab}(G, \mathbf{0}) = \mathsf{GL}(\mathbb{R}^n)$ is nocompact, generally discrete subgroups of G will *not* act properly on $X = \mathsf{A}^n$.

A structure theorem analogous to the Bieberbach's theorem may hold in this context, but presently is not known in general. This is the famous "Auslander Conjecture," since it was erroneously claimed in Auslander [8]. The assertion is that the fundamental group (or affine holonomy group) Γ is necessarily virtually solvable. This was proved in dimension three by Fried-Goldman[63], and Abels–Margulis–Soifer [2] have proved this in all dimensions < 6.

The Auslander Conjecture implies the following Structure Theorem: If $\Gamma \subset \mathsf{Aff}(\mathsf{A}^n)$ is an affine crystallographic group, then there exists a subgroup $G \subset \mathsf{Aff}(\mathsf{A}^n)$ such that

- G has finitely many connected components; $\Gamma \subset G$ is a lattice;
- The identity component G^0 acts simply transitively on A^n.

The last condition means that G inherits a left-invariant complete affine structure, and a finite-sheeted covering space of the complete affine manifold $M = \mathsf{A}^n/\Gamma$ identifies with the homogeneous space $\Gamma\backslash G^0$. The group G replaces the group of translations in the Euclidean (Bieberbach) case. For details, see the first section of [63]; we call Γ a *crystallographic hull,* but generally it is not unique.

When M is not required to be compact, then many examples are now known where Γ is not virtually solvable. The first ones were constructed in the late 1970's by Margulis [114, 115], where Γ is a nonabelian free group, and $n = 3$. For more information, see Abels [1], Charette-Drumm-Goldman-Morrill [31], Fried-Goldman [63], Milnor [123] and [76].

The three-dimensional examples found by Auslander [7] and Auslander-Markus [6] have special significance. Such a 3–manifold M^3 is a 2-torus bundle over S^1, and is a mapping torus of a linear automorphism of a flat torus T^2. We can identify T^2 as $\mathbb{R}^2/\mathbb{Z}^2$, and the automorphism corresponds to $A \in \mathsf{GL}(2, \mathbb{Z})$. The monodromy A is periodic if and only if M^3 is a Euclidean manifold, in which case the fundamental group Γ is a classical crystallographic group. When A is parablolic, then Γ is nilpotent and nonabelian, its crystallographic hull is the 3-dimensional Heisenberg group Nil, and M^3 is a nilmanifold.

The most interesting case arises when A is hyperbolic. In that case, the crystallographic hull G is the semidirect product $\mathbb{R}^2 \rtimes \mathbb{R}$ where $\mathbb{R}$ acts on $\mathbb{R}^2$ by the hyperbolic one-parameter group

$$\begin{aligned} \mathbb{R} &\longrightarrow \mathsf{GL}(\mathbb{R}^2) \\ t &\longmapsto \begin{bmatrix} e^t & 0 \\ 0 & e^{-t} \end{bmatrix}. \end{aligned}$$

Γ is a cocompact lattice in G, and the quotient $M^3 = \Gamma\backslash G$ is a 3-dimensional *solvmanifold.* We denote this group by Sol. Geometrically G is the identity component of the group of Lorentzian isometries of flat Minkowski 2-space, and this interpretation easily yields the flat Lorentzian structure on M.

14.3 Geometrization of 3-Manifolds

Ehresmann's viewpoint set the context for Thurston's geometrization program for 3-manifolds, and revolutionized the subject.

Every closed 2-manifold Σ admits a Riemannian metric of *constant curvature,* and hence a (G, X)-structure where X is a model space of constant curvature (the 2-sphere S^2, Euclidean space $\mathbb{E}^2$, or the hyperbolic plane H^2) and $G = \mathsf{Isom}(X)$. Which geometry is supported is determined by the topology of Σ: if $\chi(\Sigma) < 0$ (respectively $\chi(\Sigma) = 0$, $\chi(\Sigma) > 0$), then Σ admits hyperbolic structures (respectively, Euclidean structures, spherical structures). The deformation spaces of these structures (equivalent to the Teichmüller spaces of Σ by the *Uniformization Theorem*), are a powerful tool for understanding the topology of Σ.

In 1976, Thurston proposed that 3-manifolds possess a suggestive natural structure in terms of *canonical decompositions* into pieces which have locally homogeneous Riemannian structures. There are eight local models for such Riemannian structures, including the three constant curvature geometries (spherical, Euclidean and hyperbolic) as well as certain product and local-product geometries (such as $S^2 \times S^1$, $H^2 \times S^1$, the Heisenberg group and the solvable group $\mathsf{Isom}(\mathbb{R}^{1,1})$ and the unit tangent bundle $T_1(H^2) \cong \mathsf{PSL}(2, \mathbb{R})$). As these are metric structures, the Hopf-Rinow theorem implies that the developing maps are tame, so (at least when one passes to a simply-connected model space X) the structures are all quotient structures by discrete subgroups of G.

The tools for the decomposition existed at the time, due to earlier work of Seifert, Dehn, Kneser, Milnor, Haken, Waldhausen, Jaco, Shalen, Johannsen and many others; Thurston realized that these topological results gave an intimate and suggestive relationship between topology and differential geometry, *in dimension three.* The importance of these insights cannot be overestimated. See Scott [128], Bonahon [25] and Thurston [135] for further details.

Three of the geometries correspond to the Euclidean manifolds, nilmanifolds and solvmanifolds above. Namely, Euclidean geometry lives on quotients of flat 3-tori by finite groups. *Nilgeometry* lives on quotients of the Heisenberg group, and is defined as the geometry of a left-invariant metric. *Solvgeometry* lives on quotients of the solvable Lie group Sol described above, and is defined as the geometry of a left-invariant metric on Sol.

14.4 Ehresmann Structures

Now we return to Ehresmann's vision, as outlined in his paper [56] and the later paper [57], in the context of locally homogeneous structures which are *not necessarily* Riemannian.

In the later paper [57], he associates to a (G, X)-structure on M a fiber bundle with structure group G, fiber X and a section corresponding to the developing map. More generally, this structure corresponds to what is now called a *Cartan connection* on M, and the locally homogeneous structures described in [56] are precisely those Cartan connections which are *flat.* A flat Cartan connection is one for which the *curvature* vanishes. Sharpe [129] is a particularly readable exposition of this general theory. He calls a *Cartan geometry* a Cartan connection, and a flat one *Klein geometry.*

The local triviality of these structures implies that the study of such structures is essentially topological, and in particular closely related to the fundamental group and the universal covering space. Specifically, suppose M is a connected (G, X)-manifold with basepoint $p_0 \in M$. Let $U \subset M$ a coordinate patch containing p_0, with a coordinate chart $U \xrightarrow{\psi} X$. Let $\tilde{M} \xrightarrow{\Pi} M$ denote the corresponding universal covering space with covering group $\pi = \pi_1(M, p_0)$. Then ψ extends to a unique map $\tilde{M} \xrightarrow{\mathsf{dev}} X$ which is compatible with the (G, X)-atlas; it is a (G, X)-map, a morphism in the category whose objects are (G, X)-manifolds. As the restrictions of dev to coordinate patches are locally compositions of coordinate charts with transformations from G, the *developing map* dev is a local real-analytic diffeomorphism. (Since the action of G on X is real-analytic, a (G, X)-atlas determines a unique real-analytic structure.) Furthermore the group π of deck transformations acts by (G, X)-automorphisms of $\tilde{M}$, and therefore defines a homomorphism $\pi \xrightarrow{\rho} G$ such that

$$\begin{array}{ccc} \tilde{M} & \xrightarrow{\mathsf{dev}} & X \\ \gamma \big\downarrow & & \big\downarrow \rho(\gamma) \\ \tilde{M} & \xrightarrow{\mathsf{dev}} & X \end{array}$$

commutes.

This process of development originated with Élie Cartan and generalizes the notion of a developable surface in $\mathbb{E}^3$. If $S \hookrightarrow \mathbb{E}^3$ is an embedded surface of zero Gaussian curvature, then for each $p \in S$, the exponential map at p defines an isometry of a neighborhood of 0 in the tangent plane T_pS, and corresponds to rolling the tangent plane $\mathsf{A}_p(S)$ on S without slipping. In particular every curve in S starting at p lifts to a curve in T_pS starting at $0 \in T_pS$. Élie Cartan calls this the *development* of the surface (along the curve). For a Euclidean manifold M, this globalizes to a local isometry of the universal covering $\widetilde{M} \longrightarrow \mathbb{E}^2$, the *developing map* of the geometric structure. For general Ehresmann structures M, the developing map is a local homeomorphism $\widetilde{M} \xrightarrow{\mathsf{dev}} X$ from the universal covering space into X, which locally respects the geometric structure. Furthermore it is equivariant with respect to a representation $\pi_1(S) \xrightarrow{\rho} G$ (the *holonomy representation*) corresponding to the action of deck transformations of $\pi_1(S)$ on $\widetilde{M}$. The corresponding pair (dev, ρ) is unique up to the action of $g \in G$ given by

$$(\mathsf{dev}, \rho) \longmapsto (g \circ \mathsf{dev}, \mathsf{Inn}(g) \circ \rho).$$

The metric structure is actually subordinate to the affine connection, as this notion of development really only involves the construction of *parallel transport.*

Later this was incorporated into the notion of a *fiber space,* as discussed in the 1950 conference [133]. The collection of coordinate changes of a (G, X)-manifold M defines a fiber bundle $\mathcal{E}_M \longrightarrow M$ with fiber X and structure group G. The fiber over $p \in M$ of the associated principal bundle

$$\mathcal{P}_M \xrightarrow{\Pi_{\mathcal{P}}} M$$

consists of all possible germs of (G, X)-coordinate charts at p. The fiber over $p \in M$ of $\mathcal{E}_M$ consists of all possible *values* of (G, X)-coordinate charts at p. Assigning to the germ at p of a coordinate chart $U \xrightarrow{\psi} X$ its value

$$x = \psi(p) \in X$$

defines a mapping

$$(\mathcal{P}_M)_p \longrightarrow (\mathcal{E}_M)_p.$$

Working in a local chart, the fiber over a point in $(\mathcal{E}_M)_p$ corresponding to $x \in X$ consists of all the different germs of coordinate charts ψ taking $p \in M$ to $x \in X$. This mapping identifies with the quotient mapping of the natural action of the stabilizer $\mathsf{Stab}(G, x) \subset G$ of $x \in X$ on the set of germs.

For Euclidean manifolds, $(\mathcal{P}_M)_p$ consists of all *affine orthonormal frames,* that is, pairs (x, F) where $x \in \mathbb{E}^n$ is a point and F is an orthonormal basis of the tangent space $T_x\mathbb{E}^n \cong \mathbb{R}^n$. For an affine manifold, $(\mathcal{P}_M)_p$ consists of all *affine frames:* pairs (x, F) where now F is *any* basis of $\mathbb{R}^n$.

The coordinate atlas/developing map defines a section of $\mathcal{E}_M \to M$ which is transverse to the two complementary foliations of $\mathcal{E}_M$:

- As a section, it is necessarily transverse to the foliation of $\mathcal{E}_M$ by fibers;
- The nonsingularity of the coordinate charts/developing map implies this section is transverse to the horizontal foliation $\mathfrak{F}_M$ of $\mathcal{E}_M$ defining the flat structure.

The differential of this section is the *solder form* of the corresponding Cartan connection.

14.4.1 Properties of the Developing Map

Ehresmann [56] proves several basic facts about the development/holonomy pair:

Suppose that M is compact and $\pi_1(M)$ is finite. Then

- X must be compact and $\pi_1(X)$ is finite;
- The universal covering $\tilde{M}$ of M is (G, X)-isomorphic to the universal covering of X.

He defines a structure to be *normal* if and only if the developing map is a covering space. Structures on closed manifolds with finite fundamental group are normal.

14.4.2 Hierarchy of Structures

One may pass between different local models. We may define a *category of homogeneous spaces,* whose objects are pairs (G, X) where G is a Lie group and X is a manifold with a transitive action of G. A morphism $(G, X) \longrightarrow (G', X')$ is defined by a pair of maps $h : G \longrightarrow G'$ and $f : X \longrightarrow X'$, where h is a homomorphism and f is a local diffeomorphism which is h-equivariant, that is, for all $g \in G$,

$$\begin{array}{ccc} X & \xrightarrow{f} & X' \\ g\downarrow & & \downarrow h(g) \\ X & \xrightarrow{f} & X' \end{array}$$

commutes. Such a morphism induces a mapping from (G, X)-manifolds to (G', X')-manifolds.

Particularly interesting is the case when h is a local isomorphism of Lie groups. In this case the pseudogroups defined by (G, X) and (G', X') are identical, and the two categories of locally homogeneous structures identify. In this case Ehresmann calls X' a *Klein form* of X.

Here is another point of view concerning morphisms $(G, X) \longrightarrow (G', X')$. There is a unique (G', X')-structure on X such that $X \xrightarrow{f} X'$ is a (G', X')-map. Since the transformations of X defined by G are f-related to transformations of G', the action of G on X preserves this structure. In particular, for a given homogeneous space (G, X), a morphism $(G, X) \longrightarrow (G', X')$ is equivalent to a G-invariant (G', X')-structure on X.

In the special case that X is a Lie group and G is the group of left-multiplications, we see that a left-invariant (G', X')-structure on G is equivalent to a representation $G \longrightarrow G'$, together with an open orbit in X' which has discrete isotropy.

In many cases, the classification of geometric structures on a fixed topology proceeds by showing that the structures can be refined to certain subgeometries.

A particularly interesting and nontrivial example is Fried's classification of similarity structures on closed manifolds [61], whereby a compact manifold modeled on Euclidean similarity geometry is either a Euclidean manifold, or a finite quotient of a Hopf manifold. (See also Reischer-Vaisman [138] for a much different proof of the classification of closed similarity manifolds). This was first announced by Kuiper [103], but he implicitly assumed that the developing map was a covering-space onto its image.)

14.4.3 The Ehresmann-Weil-Thurston Holonomy Principle

Fundamental in the deformation theory of locally homogeneous (Ehresmann) structures is the following principle, first observed in this generality by Thurston [134]:

Theorem 14.4.1 *Let X be a manifold upon which a Lie group G acts transitively. Let M be a compact (G, X)-manifold with holonomy representation $\pi_1(M) \xrightarrow{\rho} G$.*

(1) *Suppose that ρ' is sufficiently near ρ in the representation variety $\mathsf{Hom}(\pi_1(M), G)$. Then there exists a (nearby) (G, X)-structure on M with holonomy representation ρ'.*
(2) *If M' is a (G, X)-manifold near M having the same holonomy ρ, then M' is isomorphic to M by an isomorphism isotopic to the identity.*

Here the topology on marked (G, X)-manifolds is defined in terms of the atlases of coordinate charts, or equivalently in terms of developing maps, or developing sections. In particular one can define a *deformation space* $\mathsf{Def}_{(G,X)}(\Sigma)$ whose points correspond to equivalence classes of marked (G, X)-structures on Σ. One might *like* to say the holonomy map

$$\mathsf{Def}_{(G,X)}(\Sigma) \xrightarrow{\mathsf{hol}} \mathsf{Hom}\big(\pi_1(\Sigma), G\big)/\mathsf{Inn}(G)$$

is a local homeomorphism, with respect to the quotient topology on $\mathsf{Hom}\big(\pi_1(\Sigma), G\big)/\mathsf{Inn}(G)$ induced from the classical topology on the $\mathbb{R}$-analytic set $\mathsf{Hom}\big(\pi_1(\Sigma), G\big)$. In many cases this is true (see below) but misstated in [75]. However, Kapovich [89] and Baues [12] observed that this is not quite true, because local isotropy groups acting on $\mathsf{Hom}\big(\pi_1(\Sigma), G\big)$ may not fix marked structures in the corresponding fibers.

In any case, these ideas have an important consequence:

Corollary 14.4.2 *Let M be a closed manifold. The set of holonomy representations of (G, X)-structures on M is open in $\mathsf{Hom}(\pi_1(M), G)$ (with respect to the classical topology).*

One can define a space of flat (G, X)-bundles (defined by a fiber bundle $\mathcal{E}_M$ having X as fiber and G as structure group) and the foliation $\mathcal{F}$ transverse to the fibration $\mathcal{E}_M \longrightarrow M$, that is *foliated bundles* or *flat bundles*. The foliation $\mathcal{F}$ is equivalent to a reduction of the structure group of the bundle from G with the classical topology to G with the discrete topology. This set of flat (G, X)-bundles over Σ identifies with the quotient of the $\mathbb{R}$-analytic set $\mathsf{Hom}(\pi_1(\Sigma), G)$ by the action of the group $\mathsf{Inn}(G)$ of inner automorphisms action by left-composition on homomorphisms $\pi_1(\Sigma) \to G$.

Conversely, if two nearby structures on a compact manifold M have the same holonomy, they are equivalent. The (G, X)-structures are topologized as follows. Let $\Sigma \longrightarrow M$ be a marked (G, X)-manifold, that is, a diffeomorphism from a fixed model manifold Σ to a (G, X)-manifold M. Fix a universal covering $\widetilde{\Sigma} \longrightarrow \Sigma$ and let $\pi = \pi_1(\Sigma)$ be its group of deck transformations. Choose a holonomy homomorphism $\pi \xrightarrow{\rho} G$ and a developing map $\widetilde{\Sigma} \xrightarrow{\mathsf{dev}} X$.

In the nicest cases, this means that under the natural topology on flat (G, X)-bundles $(X_\rho, \mathcal{F}_\rho)$ over M, the holonomy map hol is a local homeomorphism. Indeed, for many important cases such as hyperbolic geometry (or when the structures correspond to geodesically complete affine connections), hol is actually an embedding.

14.4.4 Historical Remarks

Thurston's holonomy principle has a long and interesting history.

The first application is the theorem of Weil [144] that the set of *discrete embeddings* of the fundamental group $\pi = \pi_1(\Sigma)$, of a closed surface Σ, in $G = \mathsf{PSL}(2, \mathbb{R})$ is open in the quotient space $\mathsf{Hom}(\pi, G)/G$. Indeed, a discrete embedding $\pi \hookrightarrow G$ is exactly a holonomy representation of a *hyperbolic structure* on Σ. The corresponding subset of $\mathsf{Hom}(\pi, G)/G$ is called the *Fricke space* $\mathfrak{F}(\Sigma)$ of Σ, and will be discussed more fully in Sect. 14.7.3. Weil's results are clearly and carefully expounded in Raghunathan [126], (see Theorem 6.19), and extended in Bergeron-Gelander [24].

In the context of $\mathbb{C}\mathsf{P}^1$-structures, this is due to Hejhal [81, 82]; see also Earle [55] and Hubbard [85]. This venerable subject originated with conformal mapping and the work of Schwarz, and closely relates to the theory of second order (Schwarzian) differential equations on Riemann surfaces. In this case, where $X = \mathbb{C}\mathsf{P}^1$ and $G = \mathsf{PSL}(2, \mathbb{C})$, we denote the deformation space $\mathsf{Def}_{(G,X)}(\Sigma)$ simply by $\mathbb{C}\mathsf{P}^1(\Sigma)$. See Dumas [52] and Sect. 14.8.1 below.

Thurston sketches the intuitive ideas for Theorem 14.4.1 in his notes [134]. The first detailed proofs of this fact are given in Lok [112], Canary-Epstein-Green [29], and Goldman [73] (the proof in [73] was worked out with M. Hirsch, and was independently found by A. Haefliger). The ideas in these proofs may be traced to Ehresmann [57], although he didn't express them in terms of moduli of structures. Corollary 14.4.2 was noted by Koszul [100], Chapter IV, §3, Theorem 3; compare also the discussion in Kapovich [90], Theorem 7.2.

14.5 Example: One Real Dimension

We illustrate these ideas in dimension 1, and classify geometric structures modeled on the real projective line $\mathbb{R}\mathsf{P}^1$, that is $\mathbb{R}\mathsf{P}^1$-manifolds. One-dimensional geometry (in our narrow *locally homogeneous* sense) reduces to projective geometry (where $X = \mathbb{R}\mathsf{P}^1$ and $G = \mathsf{PGL}(2, \mathbb{R})$). Let Σ be a compact connected 1-manifold (that is, a circle). Denote the deformation space $\mathsf{Def}_{(G,X)}(\Sigma)$ of $\mathsf{PGL}(2, \mathbb{R})$-equivalence classes of marked $\mathbb{R}\mathsf{P}^1$-structures on Σ by $\mathbb{R}\mathsf{P}^1(\Sigma)$. Denote the universal covering group of $\mathsf{SL}(2, \mathbb{R})$ by $\widetilde{\mathsf{SL}(2, \mathbb{R})})$; say that two elements $a, b \in \widetilde{\mathsf{SL}(2, \mathbb{R})})$ are *equivalent* if a is conjugate to b or b^{-1}.

The classification of $\mathbb{R}\mathsf{P}^1$-manifolds is due to Kuiper [106] and the following succinct description is due to Goldman [69].

Theorem 14.5.1 *The deformation space $\mathbb{R}\mathsf{P}^1(\Sigma)$ identifies with the space of equivalence classes of nontrivial elements of the universal covering group $\widetilde{\mathsf{SL}(2, \mathbb{R})})$.*

In other words,

$$\mathbb{R}\mathsf{P}^1(\Sigma) = \left(\widetilde{\mathsf{SL}(2, \mathbb{R})}) \setminus \{1\}\right) / \sim .$$

This space is a non-Hausdorff space containing several copies of $\mathbb{R}$, one corresponding to the lifts of an elliptic one-parameter subgroup, and others corresponding to cosets of a hyperbolic one-parameter subgroup. We describe the corresponding structures in detail below in Sect. 14.5.2.

14.5.1 Noncompact Manifolds

Let X be a 1-dimensional homogeneous space, and G its transitive group of automorphisms. A connected 1-manifold M is homeomorphic to either $\mathbb{R}$ or $S^1 \approx \mathbb{R}/\mathbb{Z}$. If $M \approx \mathbb{R}$, then it is simply connected, and a structure modeled on X is just an immersion $M \looparrowright X$. If X is not already simply connected, replace it by its universal cover $\tilde{X}$. Since $\tilde{X} \approx \mathbb{R}$, then a structure on M is an immersion, which must be an embedding. Such an embedding corresponds to a monotone function $\mathbb{R} \longrightarrow \mathbb{R}$. By choosing compatible orientations on M and X, we may assume that this monotone function is increasing. Such an increasing function is determined (up to the appropriate relation of isotopy) by the endpoints of the closure, that is a pair (a, b) where

$$-\infty \leq a < b \leq \infty.$$

14.5.2 Compact Manifolds

Now consider the case M is compact. In that case $M \approx S^1$, which we realize topologically as the quotient space of a closed interval $[a, b] \subset \mathbb{R}$ by the equivalence relation defined by identifying its two endpoints a, b. Denote the common image of the endpoints by $p_0 \in M$. The total space for the universal covering $\widetilde{M}$ is the quotient space of $[a, b] \times \mathbb{Z}$ by the equivalence relation defined by:

$$(b, n) \sim (a, n + 1)$$

for $n \in \mathbb{Z}$. The group $\pi_1(M)$ is the cyclic group $\langle \mu \rangle \cong \mathbb{Z}$ acting on $\widetilde{M}$ by:

$$[a, b] \times \mathbb{Z} \xrightarrow{\mu^m} [a, b] \times \mathbb{Z}$$

$$(u, n) \longmapsto (u, n + m)$$

where $u \in [a, b]$ and μ denotes the generator of $\pi_1(M)$.

Now we construct a developing map $\widetilde{M} \xrightarrow{\mathsf{dev}} X$. The developing map is determined by two pieces of information:

- Its restriction to $[a, b] \subset \widetilde{M}$ (corresponding to the subset $[a, b] \times \{0\} \subset [a, b] \times \mathbb{Z}$), which is an immersion

$$[a, b] \overset{f}{\looparrowright} X;$$

- A holonomy transformation $\eta = \rho(\mu) : X \to X$ such that $\eta f(b) = f(a)$.

Then f extends to the developing map by defining:

$$\mathsf{dev}(u, n) := \eta^n f(u)$$

for an arbitrary element $[(u, n)] \in \widetilde{M}$.

As above, it is convenient to lift f to the universal covering $\tilde{X} \approx \mathbb{R}$ and using a diffeomorphism $\widetilde{M} \approx \mathbb{R}$, identify f with a monotone function $\mathbb{R} \longrightarrow \mathbb{R}$.

14.5.3 Euclidean Manifolds

The first example arises when f is the embedding of $[a, b]$ as the unit interval $[0, 1] \subset \mathbb{E}^1$ and η is unit translation. Then M identifies naturally with the quotient $\mathbb{R}/\mathbb{Z}$. Its natural structure is that of a compact flat Riemannian manifold of total length 1.

More generally, for any $l > 0$, the quotient $\mathbb{R}/l\mathbb{Z}$ (where η is translation by l) is a Euclidean manifold of length l. Different values of l give non-isometric Euclidean structures, but homotheties define isomorphisms as *affine manifolds:*

$$\begin{aligned} \mathbb{R}/\mathbb{Z} &\xrightarrow{\cong} \mathbb{R}/l\mathbb{Z} \\ [x] &\longmapsto [lx]. \end{aligned}$$

Also observe that these structures are *homogeneous:* the group of translations acts transitively on M. Indeed, this defines a bi-invariant (Euclidean) geometric structure on the circle group.

14.5.4 Incomplete Affine Structures

Any $\lambda > 1$ generates a lattice inside the multiplicative group $\mathbb{R}_+$, which acts affinely on A^1. The quotient $\mathbb{R}_+/\langle\lambda\rangle$ also defines an affine structure on M, which is not a Euclidean structure since dilation by λ is not an isometry. Explcitly, take f to be a diffeomorphism onto the interval $[1, \lambda] \subset \mathbb{R} \approx \mathsf{A}^1$, so that dev is a diffeomorphism of $\widetilde{M}$ onto $(0, \infty) = \mathbb{R}_+ \subset \mathsf{A}^1$.

Like the preceding example, this affine structure is also bi-invariant with respect to the natural Lie group structure on $\mathbb{R}_+/\langle\lambda\rangle$.

Observe that, since the exponential map

$$\begin{aligned} \mathbb{R} &\longrightarrow \mathbb{R}_+ \\ x &\longmapsto e^x, \end{aligned}$$

converts addition (translation) to multiplication (dilation), it defines a diffeomorphism between two quotients

$$\mathbb{R}/l\mathbb{Z} \longrightarrow \mathbb{R}_+/\langle\lambda\rangle$$

where $l := \log(\lambda)$. This map also defines a (non-affine) analytic isomorphism between the corresponding Lie groups.

These structures are *geodesically incomplete,* and in fact model *incomplete closed geodesics* on affine manifolds. Namely, the geodesic on A^1 defined by

$$t \longmapsto 1 + t(\lambda^{-1} - 1)$$

begins at 1 and in time

$$t_\infty := 1 + \lambda^{-1} + \lambda^{-2} + \cdots = (1 - \lambda^{-1})^{-1} > 0$$

reaches 0. It defines a closed incomplete closed geodesic $p(t)$ on M starting at $p(0) = p_0$. The lift

$$(-\infty, t_\infty) \xrightarrow{\tilde{p}} \widetilde{M}$$

satisfies

$$\mathsf{dev}\,\tilde{p}(t) = 1 + t(\lambda^{-1} - 1),$$

which uniquely specifies the geodesic $p(t)$ on M. It is a geodesic since its velocity $p'(t) = (\lambda^{-1} - 1)\partial_x$ is constant (parallel). However $p(t_n) = p_0$ for

$$t_n := \frac{1 - \lambda^{-n}}{1 - \lambda^{-1}} = 1 + \lambda^{-1} + \cdots + \lambda^{1-n}$$

and as viewed in M, seems to go "faster and faster" through each cycle. By time $t_\infty = \lim_{n\to\infty} t_n$, it seems to "run off the manifold:" the geodesic is only defined for $t < t_\infty$. The apparent paradox is that $p(t)$ has *zero acceleration:* it would have "constant speed" if "speed" were only defined.

The model space X is the real projective line $\mathbb{R}\mathsf{P}^1$ and the structure group G is the group $\mathsf{PGL}(2, \mathbb{R})$ of collineations of $\mathbb{R}\mathsf{P}^1$. The fixed reference topology Σ is the *circle,* which we understand as a the identification space of a closed interval $[a, b]$ with its two endpoints a, b identified. We identify the universal covering $\widetilde{\Sigma} \longrightarrow \Sigma$ as the quotient of the Cartesian product $[a, b] \times \mathbb{Z}$, with identifications

$$(b, n) \sim (a, n + 1)$$

for $n \in \mathbb{Z}$. The group $\pi_1(\Sigma) \cong Z$ of deck transformations acts by:

$$(x, n) \longmapsto (x, n + m)$$

for $m \in \mathbb{Z}$. An $\mathbb{R}\mathsf{P}^1$-structure on Σ is defined by the restriction $\mathsf{dev}|_{[a,b]}$.

Choose a transformation $\gamma \in G$ to serve as the generator of the holonomy group. Specifically, define the presumptive holonomy homomorphism ρ by:

$$\mathbb{Z} \cong \pi_1(\Sigma) \xrightarrow{\rho} G$$
$$m \longmapsto \gamma^m.$$

Any immersion $f : [a, b] \looparrowright X$ such that $f(b) = \gamma f(a)$ extends to a ρ-equivariant immersion $\widetilde{\Sigma} \longrightarrow X$.

If $G \cong \mathbb{R}$, left-invariant structures form three equivalence classes, corresponding to the three conjugacy classes of one-parameter subgroups in G':

- An elliptic one-parameter subgroup acts simply transitively on all of X'. The deck transformation τ is a discrete subgroup of this one parameter group.
- A parabolic one-parameter subgroup acts simply transitively on the open interval between a point $x' \in X'$ and its image $\tau(x')$.
- A hyperbolic one-parameter subgroup acts simply transitively on the open interval between two points $x', y' \in X'$.

Of these structures, the last two are affine structures, since $X \setminus \{x'\}$ is an affine line and its stabilizer in G' is the affine group. For example, taking x', y' to be lifts of $0, \infty \in \mathbb{R}\mathsf{P}^1$ and g to be scalar multiplication by $\lambda > 1$, we obtain a structure, which we call a *Hopf structure*

$$\mathbb{R}_+/\lambda^n \mid n \in \mathbb{Z}, \tag{14.5.1}$$

since it is a special case of the construction of Hopf manifolds below.

In 1953, Kuiper [106] classified all such structures. In particular he found structures which are not homogeneous. These occur only if the holonomy group is parabolic or hyperbolic. Let $g \in G$ be either parabolic or hyperbolic, and let τ be the positive generator of the center of G' as above. Let $n > 0$. Choose a point $x \in X$ *not* fixed under g. Let $J \subset X'$ be a positively oriented interval going from x to $\tau^n g(x)$. A homeomorphism $[0, 1] \xrightarrow{h} J$ extends to the $\mathbb{Z}$-equivariant homeomorphism defined by:

$$\mathbb{R} \xrightarrow{\tilde{h}} X'$$
$$t \longmapsto \tau^n h(t - n),$$

(where $n = \lfloor t \rfloor$), which is a developing map for an $\mathbb{R}\mathsf{P}^1$-structure on the closed 1-manifold $\mathbb{R}/\mathbb{Z}$.

Kuiper [106] showed these comprise all the equivalence classes of structures. In particular the developing map is always a homeomorphism of the universal covering $\tilde{M}$ to either:

- all of X' (any structure with trivial or elliptic holonomy, or an inhomogeneous structure with parabolic or hyperbolic holonomy);
- the lift of the complement of one point (homogeneous structures with parabolic holonomy);
- the lift of the complement of two points (homogeneous structures with hyperbolic holonomy).

14.6 Geodesics on Affine Manifolds

14.6.1 Geodesic Completeness

The next examples we discuss are *affine structures.* In this case the affine group G' preserves the Euclidean connection on affine space (although not the Euclidean metric). These are manifolds with local *affine geometry.* A smooth vector field along a smooth curve γ has a well-defined *covariant derivative,* which is another vector field along γ. The *acceleration* γ'' of γ is the covariant derivative of the *veclocity vector field* γ', and γ is a *geodesic* if it has zero acceleration. If M is a manifold with an affine structure, and $(x, v) \in TM$ is a tangent vector, then there exists $\epsilon > 0$, and a unique geodesic $\gamma : (-\epsilon, \epsilon) \longrightarrow M$ with $\gamma(0) = x$ and $\gamma'(0) = v$.

If M is a Euclidean manifold with its underlying affine structure (or more generally if (M, g) is a Riemannian manifold with its Levi-Civita connection), then geodesic completeness is equivalent to the more intuitive notion of *metric completeness* of the associated metric space. This is the *Hopf-Rinow theorem,* and plays a fundamental role in controlling the developing map of flat structures.

As compact metric spaces are complete, a compact Riemannian manifold is geodesically complete. This also follows from the fact that the geodesic local flow of a compact Riemannian manifold reduces to local flows on the *energy hypersurfaces*

$$E_R(M) := \{(v, x) \in TM \times M \mid v \in T_x M, g(v, v) = R\},$$

which are compact. The complete integrability of these vector fields on $E_R(M)$ (for $R > 0$) implies geodesic completeness of (M, g).

14.6.2 Hopf Manifolds

In 1948, H. Hopf [84] constructed a compact complex manifold which is not Kähler. His construction also yields compact affine manifolds which are *geodesically incomplete.* Indeed, the examples in Sect. 14.5.4 above are the simplest case of Hopf's construction.

An affine manifold is *complete* if it is geodesically complete in the above sense, that is, for every initial location and velocity (x, v) there is a geodesic $\mathbb{R} \xrightarrow{\gamma} M$ with $\gamma(0) = x$ and $\gamma'(0) = v$. Auslander and Markus [5] showed that completeness is equivalent to bijectivity of the developing map. That is, a complete affine manifold is affinely isomorphic to a quotient of $\mathbb{R}^m$ by a discrete group of affine transformations acting freely and properly.

14.7 Surfaces and 3-Manifolds

14.7.1 Affine structures on surfaces

For a detailed survey of this subject, see Baues [14]. The first results are due to Kuiper [107], who listed the affine structures on 2-dimensional tori which are *convex,* that is, the developing map is an embedding onto a convex domain $\Omega \subset \mathsf{A}^2$. Either the structure is complete (which Kuiper also calls "normal"), in which case $\Omega = \mathsf{A}^2$, or Ω is either a half-plane or a quadrant.

The classification was completed in the 1970's by, independently, Nagano-Yagi [124] and Arrowsmith-Furness [64] (see also the classification of Klein bottles in [4]). In the remaining (*nonconvex*) cases after Kuiper, M is a quotient of the complement of a point in A^2. These are special cases of *radiant affine structures,* which we now describe.

14.7.1.1 Radiant Affine Structures

Affine manifolds which are quotients of the complement of a point $p \in \mathsf{A}^n$ have special properties, which deserve special attention. Necessarily p is fixed under the affine holonomy group $\Gamma \subset \mathsf{Aff}(\mathsf{A}^n)$. By applying a translation, we may conveniently assume that p is the origin $\mathbf{0} \in \mathbb{R}^n$. Since the stabilizer $\mathsf{Stab}\big(\mathsf{Aff}(\mathsf{A}^n), \mathbf{0}\big) = \mathsf{GL}(\mathbb{R}^n)$, the *affine* holonomy group is actually *linear.*

Such affine structures are called *radiant* in [62], since they are characterized by the existence of a *radiant vector field* which generates a homothetic flow (that is, scalar multiplications on the vector space $\mathbb{R}^2$).

Nagano-Yagi [124] observed that on a closed radiant affine manifold M, the developing image $\mathsf{dev}(\widetilde{M})$ is disjoint from the set $\mathsf{Fix}(\Gamma)$ of fixed points of Γ. Thus every radiant vector field on M is nonsingular. A purely topological consequence is that $\chi(M) = 0$. Another topological consequence is that M cannot have parallel volume (see Sect. 14.8.3.1 below), and therefore the first Betti number $\beta(M) \geq 1$.

In his 1960 unpublished lecture notes, L. Markus observed that the known examples of compact affine manifolds which are geodesically complete are precisely the known examples of compact affine manifolds with parallel volume. Validity of this observation, which Markus didn't really conjecture, has become known as the *Markus conjecture.*

These manifolds can be understood by a *suspension* construction of mapping tori of automorphisms of the compact $\mathbb{R}\mathsf{P}^1$-manifolds discussed in Sect. 14.5. This is due to Benzécri [23], who proved that, any $\mathbb{R}\mathsf{P}^n$-manifold M admits a double covering $\hat{M}$ such that the Cartesian product $\hat{M} \times S^1$ admits a radiant affine structure, where the radiant flow is the flow in the S^1-factor.

For example, all radiant affine 2-manifolds arise in this way. Furthermore the affine holonomy group contains a linear expansion of $\mathbb{R}^2$. The simplest example is the Hopf manifold described above.

A radiant affine manifold (M, ξ) is a radiant suspension if and only if the flow ρ suspension, and therefore is either a Seifert 3-manifold covered by a product $F \times S^1$, where F is a closed surface, a nilmanifold or a hyperbolic torus bundle (Fig. 14.1).

In dimensions 1 and 2 all closed radiant manifolds are radiant suspensions. Together with Kuiper's list of convex structures, these comprise all closed affine 2-manifolds, since a closed surface M admits an affine structure if and only if $\chi(M) = 0$ (Benzécri [22]). That is, either M is diffeomorphic to a Klein bottle (in which case its orientable double covering is diffeomorphic to a torus) or M is diffeomorphic to a torus. (Benzécri's theorem inspired Milnor's generalization [121] to flat oriented rank two vector bundles over surfaces; see [76] for a more detailed account of these developments.)

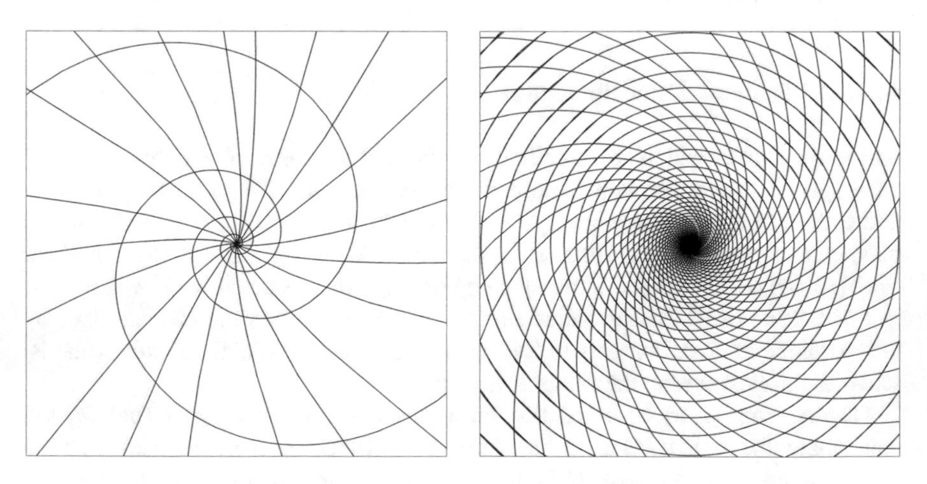

Fig. 14.1 Incomplete complex-affine structures on the 2-torus are radiant suspensions. The two examples depicted are suspensions of rotations (isometries) of the circle

14.7.2 Complete Affine Structures

The complete structures on T^2 are all *affine Lie groups,* that is, an affine structure on a Lie group invariant under both left- and right-multiplications. For example, the Euclidean structures are all quotients $\mathbb{R}^2/\Lambda$, where $\Lambda \subset \mathbb{R}^2$ is a lattice. The other structures are obtained by *polynomial deformations* of Euclidean structures, namely the diffeomorphism

$$\mathbb{R}^2 \xrightarrow{F_\epsilon} \mathbb{R}^2$$
$$(x, y) \longmapsto (x + \epsilon y^2, y),$$

conjugates translation by (s, t) to the affine transformation

$$(x, y) \longmapsto (x + 2\epsilon yt + (s + \epsilon^2 t^2), y + t)$$

and the quotient space

$$M = \mathbb{R}^2 / F_\epsilon \Lambda F_\epsilon^{-1}$$

is a non-Euclidean complete affine torus. Figure 14.2 illustrates a Euclidean torus and a polynomial deformation.

Baues [13] showed that the deformation space of complete affine structures on T^2 is homeomorphic to $\mathbb{R}^2$ (see also Baues-Goldman [15]), with the origin $(0, 0)$ corresponding to the single equivalence class corresponding to Euclidean structures. The action of $\mathsf{Mod}(T^2) = \mathsf{GL}(2, \mathbb{Z})$ on this deformation space identifies with the usual linear action of $\mathsf{GL}(2, \mathbb{Z})$ on $\mathbb{R}^2$. This action is highly chaotic: it is topologically mixing, and every continuous invariant function is constant.

Complete affine structures on *compact* 3-manifolds were classified by Fried-Goldman [63]. A closed 3-manifold admits a complete affine structure if and only if it is finitely covered by a 2-torus bundle over the circle; in other words, it is

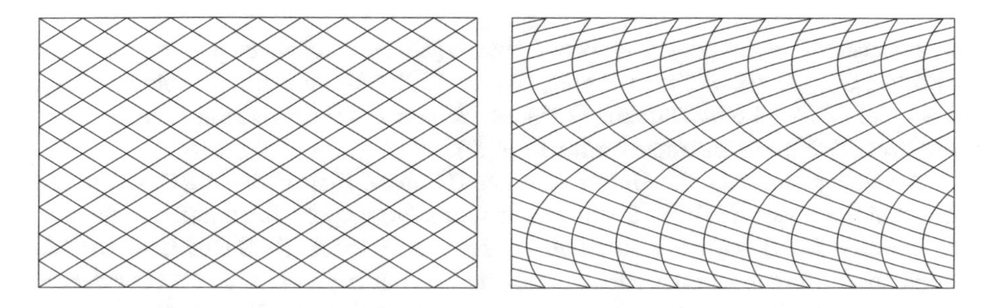

Fig. 14.2 Complete affine structures on the 2-torus

Euclidean, Heisenberg, or Sol in the Thurston geometrization [25, 128, 135]. For more information, see Abels [1], Baues [12, 13] and Goldman [75, 76].

Noncompact complete affine 3-manifolds are considerably more complicated, in the light of Margulis's discovery [115] of proper isometric actions of nonabelian free groups on Minkowski $2+1$-space. The structure is much better understood now, and they all arise from a Schottky-group construction using *crooked planes* invented by Drumm [50, 51]. We refer to [26, 32–34, 41, 48, 49] for more current information.

14.7.3 Hyperbolic Structures on Surfaces

A paradigm for this theory is the classification of hyperbolic structures on surfaces.

The Fricke space $\mathfrak{F}(\Sigma)$ embeds in $\mathsf{Hom}(\Gamma, G)/G$, where

$$G = \mathsf{Isom}(\mathsf{H}^2) \cong \mathsf{SO}(2,1),$$

and indeed defines a connected component in this space. As noted above, openness follows from Corollary 14.4.2.

Closedness is more special. If Γ is not virtually nilpotent and G is semisimple, then the discrete embeddings $\Gamma \hookrightarrow G$ form a *closed* subset of $\mathsf{Hom}(\Gamma, G)$ in the classical topology. (A proof of this well-known statement can be found in Goldman-Millson [77], although it was known much earlier.) In this special case, it is originally due to Chuckrow [45]. In general closedness follows from Kazhdan-Margulis uniform discreteness [92], see Chapter VIII of [126] , §4.12 of Kapovich [90], or §4.1 of Thurston [135].

It remains to see that Fricke space $\mathfrak{F}(\Sigma)$ is connected. In general the discrete embeddings $\Gamma \hookrightarrow G$ fall into many connected components. Counting the components is an interesting and difficult general problem. In this particular case, one can use the direct hyperbolic-geometry parametrization of $\mathfrak{F}(\Sigma)$ by Fenchel-Nielsen coordinates, whereby

$$\mathfrak{F}(\Sigma) \approx \mathbb{R}^{6g-6+3b}$$

is connected, completing the proof that $\mathfrak{F}(\Sigma)$ is a connected component of $\mathsf{Hom}(\Gamma, G)/G$. (See, for example, Buser [28], Abikoff [3], Ratcliffe [127], Theorem 9.7.4, §4.6 of Thurston [135], or Wolpert [147] for accessible accounts of the Fenchel-Nielsen parametrization of $\mathfrak{F}(\Sigma)$.

Another proof uses the Uniformization Theorem. First identify $\mathfrak{F}(\Sigma)$ with the Teichmüller space $\mathfrak{T}(\Sigma)$ of Σ (uniformization). Now apply Teichmüller's theorem to identify $\mathfrak{T}(\Sigma)$ with the unit ball in the vector space $Q(M)$ of holomorphic quadratic differentials on a Riemann surface M homeomorphic to Σ. (For details see Hubbard [86], Theorem 7.2.1.) Alternatively, following Wolf [146], identify $\mathfrak{T}(\Sigma)$ with all of $Q(M)$ using the Hopf differentials of harmonic maps from M to an arbitrary Riemann surface homeomorphic to Σ.

The components of $\mathsf{Hom}(\pi, G)/G$ were classified in [74] in terms of the Euler class. In particular $\mathfrak{F}(\Sigma)$ identifies with the component [74] *maximizing* this characteristic class. In terms of the foliated (G, X)-bundle X_ρ associated to a representation ρ, this result means that the necessary topological conditions for X_ρ to admit a transverse section are sufficient. See Goldman [75, 76] for further details and discussion.

Toledo [137] considered surface group representations when G acts on a Hermitian symmetric space X. He defined a characteristic number which includes the Euler class of flat $\mathsf{PSL}(2, \mathbb{R})$-bundes when $X = \mathsf{H}^2$. In particular Toledo's invariant is bounded by topological invariants of M. Representations maximizing Toledo's invariant have many properties of Fuchsian representations, in particular forming connected components consisting entirely of discrete embeddings. In a different direction, when G is a split $\mathbb{R}$-form, Hitchin [83] found components which naturally contain compositions of Fuchsian $\mathsf{SL}(2, \mathbb{R})$-representations with the *Kostant principal representation* $\mathsf{SL}(2, \mathbb{R}) \to G$. When $G = \mathsf{PGL}(3, \mathbb{R})$, the component identifies with the deformation space of convex $\mathbb{R}\mathsf{P}^2$-surfaces [42], as discussed in Sect. 14.8.2 below. Hitchin proved that these components are topologically open cells. Labourie [110] characterized Hitchin's representations dynamically, and proved that they are all quasi-isometric (and hence discrete) embeddings. (From a somewhat different viewpoint, these representations were also studied by Fock-Goncharov [59, 60] who found coordinates on these components.) This subject, sometimes called "higher Teichmüller theory," is surveyed in Burger-Iozzi-Wienhard [27] (with background expounded in Labourie [111]), to which we refer the reader for further details.

14.8 Projective and Conformal Structures

Finally we discuss Ehresmann structures modeled on compact homogeneous spaces, such as the sphere and projective space. Although this subject dates back to the nineteenth century, in the context of second order linear differential equations on Riemann surfaces and conformal mapping, many mysteries remain, and the subject is fundamental in the broader hierarchy of geometries. We then discuss real-projective structures on surfaces, for which a complete classification is known [42]. We then briefly discuss several results about flat conformal structures and real-projective structures in higher dimensions.

14.8.1 Projective Structures on Riemann Surfaces

The rich subject of *projective structures on Riemann surfaces* promises to be fundamental in the theory of Ehresmann locally homogeneous structures. When the underlying Riemann surfaces are allowed to vary, the resulting Ehresmann structures are called $\mathbb{C}\mathsf{P}^1$*-manifolds.* I find it rather striking that although the algebraic theory of the character variety is less pathological, the geometric theory is exceedingly profound and difficult. The parametrization of the deformation space $\mathbb{C}\mathsf{P}^1(\Sigma)$ as an affine bundle whose underlying vector bundle is $T^*\mathfrak{T}(\Sigma)$ is the (holomorphic) cotangent bundle of the Teichmüller space $\mathfrak{T}(\Sigma)$ is rather "soft" but the geometric theory of $\mathbb{C}\mathsf{P}^1$-manifolds is extremely subtle, involving some of the most technically difficult aspects of the "modern" theory of hyperbolic 3-manifolds and Kleinian groups (see Marden [113]). We only concentrate on the properties of the holonomy mapping, referring to the excellent survey article [52] by David Dumas. See also Gunning [78, 79] for background.

Gallo-Kapovich-Marden [65] answered a question first raised by Gunning [80]:

Theorem *Let* Σ *be a closed orientable surface of* $\chi(\Sigma) < 0$. *Denote the deformation space of marked* $\mathbb{C}\mathsf{P}^1$*-structures on* Σ *by* $\mathbb{C}\mathsf{P}^1(\Sigma)$. *The image of the holonomy mapping*

$$\mathbb{C}\mathsf{P}^1(\Sigma) \xrightarrow{\mathsf{hol}} \mathsf{Hom}\big(\pi_1(\Sigma)$$

consists of equivalence classes of representations ρ *for which:*

- ρ *lifts to a representation* $\pi_1(\Sigma) \xrightarrow{\tilde{\rho}} \mathsf{SL}(2, \mathbb{C})$;
- *the image* Γ *of* ρ *fixes no point on hyperbolic space* H^3, *fixes no point in the boundary* $\partial\mathsf{H}^3$, *and leaves invariant no geodesic in* H^3.

The first condition means that ρ lies in the connected component of $\mathsf{Hom}\big(\pi_1(\Sigma), G\big)$ containing the trivial representation (Goldman [74]). The second condition means that ρ is *nonelementary,* and is equivalent to numerous other conditions. For example, it is equivalent to the real Zariski closure of Γ being $\mathsf{PSL}(2, \mathbb{C})$ or conjugate to $\mathsf{PGL}(2, \mathbb{R})$. Another equivalent condition is that the image of ρ is *unbounded* (having noncompact closure) and non-solvable. Yet another condition is that the holonomy group Γ is *not amenable.*

Although W. Thurston announced this and communicated the outline of the proof to the author in the late 1970's, many details were missing. The full proof (following Thurston's outline) was completed by Gallo-Kapovich-Marden [65]. (An incorrect proof, but with an extremely interesting approach, can be found in [91].) See also Kamishima-Tan [88].

The injectivity of the holonomy mapping is also quite fundamental and mysterious. Goldman [72], using ideas inspired by the Thurston parametrization (see Sect. 14.8.4 below), computed the inverse image $\mathsf{hol}^{-1}(\mathfrak{F}_\Sigma)$ in terms of a *grafting construction,* first developed by Hejhal [82] and Maskit [118] (Theorem 5) and Sullivan-Thurston [132].

The main result is that, over the inverse image of the quasi-Fuchsian subset of $\mathsf{Hom}\big(\pi_1(\Sigma), G\big)$, the holonomy map hol is a covering space and the fiber admits an explicit topological description in terms of grafting. In this case, X decomposes into two subdomains Ω_+ and Ω_- along their common boundary which is the *limit*

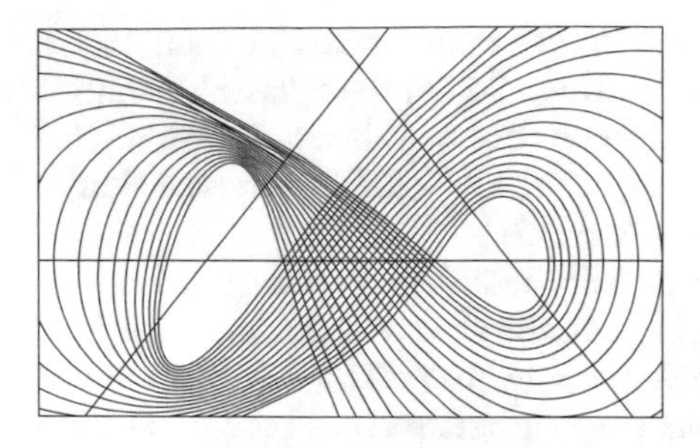

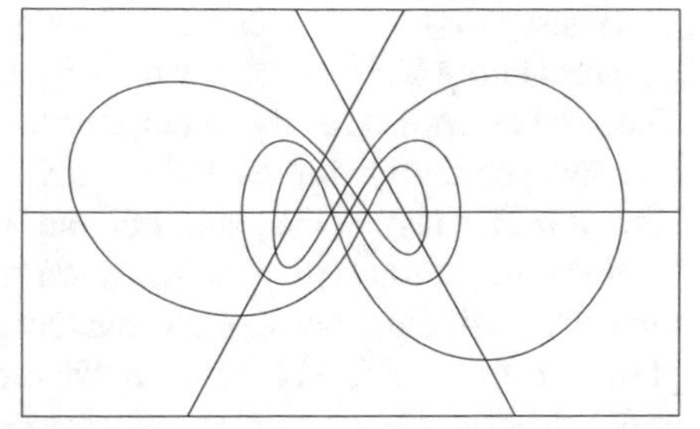

Fig. 14.3 Developing maps of $\mathbb{R}\mathsf{P}^2$-structures which are not covering spaces of their images, due to Sullivan-Thurston and Smillie

set of the holonomy group Γ. The geometric manifold M with this holonomy then admits a corresponding decomposition $M = M_+ \cup M_-$, and under the assumption that the holonomy homomorphism is an isomorphism $\pi_1(M) \xrightarrow{\cong} \Gamma$, one of $M_\pm$ is a union of annuli.

However, as pointed out by M. Kapovich and S. Choi, the proof of a key lemma (Theorem 2.2) of Goldman [72] is flawed. (A similar problem can be found in Faltings [58]). See Choi-Lee [44] for a corrected proof and extensive discussion. One would like to control the developing map by decomposing the geometric manifold into open submanifolds modeled on holonomy-invariant subdomains $\Omega \subset X$ where the holonomy Γ preserves a complete Riemannian structure g_Ω. However, even if M is compact, the induced metric on $\mathsf{dev}^{-1}(\Omega)$ may be incomplete. One needs a sharper argument involving the asymptotics of Γ, as in Kuiper [105]. (Indeed, the Sullivan-Thurston-Smillie examples discussed in Sect. 14.8.2 and depicted in Fig. 14.3 provide counterexamples to Theorem 2.2 of [72].)

Shinpei Baba's work [9–11] describes $\mathbb{C}\mathsf{P}^1$-structures with Schottky holonomy in terms of a similar grafting construction. Although developing maps for general $\mathbb{C}\mathsf{P}^1$-structures are intractable, under the assumption of Schottky holonomy, Baba obtains sharp results on decomposing the developing map into basic pieces.

14.8.2 Real-Projective Structures on Surfaces

When $X = \mathbb{R}\mathsf{P}^2$ and

$$G = \mathsf{Aut}(X) \cong \mathsf{PGL}(3, \mathbb{R}) \cong \mathsf{SL}(3, \mathbb{R})$$

is its group of collineations, we call the resulting Ehresmann structures $\mathbb{R}\mathsf{P}^2$-manifolds. We denote the deformation space $\mathsf{Def}_{(G,X)}(\Sigma)$ simply by $\mathbb{R}\mathsf{P}^2(\Sigma)$. Curiously, the case when $\chi(\Sigma) = 0$ has a much more complicated general picture than when $\chi(\Sigma) < 0$.

When $\chi(\Sigma) = 0$, then Σ admits affine structures, discussed in Sect. 14.7.1. The remaining $\mathbb{RP}^2$-structures on T^2 (and the Klein bottle) were classified in [68]. The new examples involve a surgery construction (due to Sullivan-Thurston [132] and independently Smillie [131]) analogous to the $\mathbb{CP}^1$-grafting construction of Hejhal [82], Maskit [118] and Sullivan-Thurston [132].

Here one starts with a collineation γ of $\mathbb{RP}^2$ described by a diagonal 3×3-matrix with distinct positive eigenvalues. The coordinate axes define the three fixed points of γ on $\mathbb{RP}^2$, which lie outside the developing image. The coordinate planes define invariant lines, whose complements are γ-invariant affine patches in $\mathbb{RP}^2$. Corresponding to the maximum (respectively minimum) eigenvalue are two affine patches in which γ acts by a linear expansion (respectively linear contraction). There corresponds a pair of Hopf manifolds with $\mathbb{RP}^2$-structures, which can be glued together along closed geodesics (corresponding to the coordinate plane with the middle eigenvalue) to form new $\mathbb{RP}^2$-surfaces with the "same" holonomy representation. (See [72, 132] for details.) The Sullivan-Thurston-Smillie example of a pathological developing map is illustrated in Fig. 14.3.

About 10 years later, the combined work [42] of the author [66] and Suhyoung Choi's dissertation [37, 38] (and its extensions in [39, 40]) completely described the deformation space $\mathbb{RP}^2(\Sigma)$. Choi shows that if Σ is closed and $\chi(\Sigma) < 0$, then any $\mathbb{RP}^2$-surface decomposes *canonically* into annuli bounded by closed geodesics and convex surfaces with totally geodesic boundary. $\mathbb{RP}^2(\Sigma)$ is a countable disjoint union of copies of the deformation space of *convex* $\mathbb{RP}^2$-structures (shown in [66] to be a cell of dimension $-8\chi(\Sigma)$). The components are parametrized by a discrete invariant involving the multicurve on Σ controlling Choi's convex decomposition.

Much more recently Choi observed that these grafted $\mathbb{RP}^2$-structures with Schottky $\mathsf{SO}(2,1)$-holonomy *compactify* Margulis spacetimes [41].

14.8.3 Incomplete Affine Structures on Closed 3-Manifolds

The classification of incomplete affine structures in dimension 3 is largely unkown, except under rather strong assumptions on the fundamental group. Smillie's work [131] on closed affine manifolds with abelian holonomy was generalized by Fried-Goldman-Hirsch [62] to nilpotent holonomy, and leads to a classification of closed 3-manifolds with nilpotent fundamental group. Serge Dupont [54] gave a beautiful classification of affine structures on hyperbolic 3-manifolds, which we briefly describe below.

14.8.3.1 Parallel Volume

An affine manifold has *parallel volume* if and only if its linear holonomy preserves volume (up to sign). Equivalently the linear holonomy has determinant ± 1. Another equivalent condition is the existence of a coordinate atlas whose coordinate changes preserve volume.

The obstruction to parallel volume is the class in $H^1(M;\mathbb{R})$ defined by the homomorphism

$$\begin{aligned}\pi_1(M) &\longrightarrow \mathbb{R}\\ \gamma &\mapsto \log|\det L(\gamma)|.\end{aligned}$$

When the first Betti number vanishes, every affine structure must admit *parallel volume.* Then the results of Sect. 14.8.3.1 below apply to give nonexistence of affine structures on certain closed 3-manifolds.

Theorem 14.8.1 (Smillie) *Let M be a closed affine manifold with a parallel exterior differential k-form which has nontrivial de Rham cohomology class. Suppose $\mathcal{U}$ is an open covering of M such that for each $U \in \mathcal{U}$, the affine structure induced on U is radiant. Then* $\dim\mathcal{U} \geq k$*; that is, there exist $k+1$ distinct open sets*

$$U_1, \ldots, U_{k+1} \in \mathcal{U}$$

such that the intersection

$$U_1 \cap \cdots \cap U_{k+1} \neq \emptyset.$$

(Equivalently the nerve of $\mathcal{U}$ has dimension at least k.)

A published proof of this theorem can be found in Goldman-Hirsch [67].

Using these ideas, Carrière, d'Albo and Meignez [30] have proved that a nontrivial Seifert 3-manifold with hyperbolic base cannot have an affine structure with parallel volume. This implies that the 3-dimensional Brieskorn manifolds $M(p,q,r)$ with

$$p^{-1} + q^{-1} + r^{-1} < 1$$

admit no affine structure whatsoever. (Compare Milnor [122].)

14.8.3.2 Hyperbolicity

The opposite of geodesic completeness is *hyperbolicity* in the sense of Vey [140] and Kobayashi [94, 95], which is equivalent to the following notion: Say that an affine manifold M is *completely incomplete* if there exists no affine map $\mathbb{R} \longrightarrow M$, that is, M admits *no* complete geodesic.

Similarly, an $\mathbb{RP}^n$-manifold is *completely incomplete* if there exists no projective map $\mathbb{R} \longrightarrow M$. As noted by the author (see Kobayashi [95]), the combined results of Kobayashi [95], Wu [148], and Vey [141] imply:

Theorem *Let M be a closed hyperbolic affine manifold. Then M is a quotient of a sharp convex cone.*

In particular M is radiant. Moreover M fibers over S^1 (which implies that $\chi(M) = 0$ and $b_1(M) > 0$).

For projective manifolds, taking the radiant suspension of a hyperbolic projective structure yields a radiant affine structure, which one easily sees to be is hyperbolic. Applying the above theorem implies that M is a quotient of a sharp convex cone.

This striking characterization of hyperbolicity uses *intrinsic metrics* in the category of affine and projective manifolds, developed by Vey [140] and Kobayashi [94, 95]. Their constructions were inspired by the intrinsic metrics of Carathéodory and Kobayashi for complex manifolds.

Denote by I the open unit interval $\big(-1, 1\big)$ and

$$g_I := \frac{4}{(1-u^2)^2} du^2$$

its *Poincaré metric.*

For projective manifolds M, one defines a "universal" pseudo-metric $M \times M \xrightarrow{d_M} \mathbb{R}$ such that affine (respectively projective) maps $I \to M$ are distance non-increasing with respect to g_I.

The definition of d_M enforces the triangle inequality by taking the infimum of g_I-distances over sequences $x_0 = x, x_1, \ldots, x_m = y$ where x_i and x_{i+1} are "close" in the following sense: there are projective maps $I \xrightarrow{f_i} M$ such that $x_i = f_i(a_i)$ and $x_{i+1} = f_i(b_i)$ for $-1 < a_i < b_i < 1$. Then define $d_M(x, y)$ as the infimum over all such sequences (f_i, a_i, b_i) of

$$\sum_{i=0}^{m-1} d_I(a_i, b_i),$$

where d_I is the distance function on the Riemannian 1-manifold $\big(I, g_{[-1,1]}\big)$. That is, $d_M(x, y)$ is the infimum of

$$\int_a^b f\big(\gamma'(t)\big)dt$$

over all piecewise C^1 paths $[a, b] \xrightarrow{\gamma} M$ with $\gamma(a) = x$, $\gamma(b) = y$.

This function has an infinitesimal form, defined by a nonnegative upper-semicontinuous function $TM \xrightarrow{\phi} \mathbb{R}$. For affine manifolds, completeness is equivalent to $f \equiv 0$.

Following Kobayashi and Vey, M is *projectively hyperbolic* if and only if d_M is a *metric,* that is, if $d_M(x, y) > 0$ for $x \neq y$. Then d_M is a *Finsler metric* and equals the Hilbert metric on the convex domain $\widetilde{M}$.

When M is affine, then Vey [141] proves that M is a quotient of a sharp convex cone. In that case there is (in addition to the Hilbert metric), a natural Riemannian metric introduced by Vinberg [142], Koszul [98, 99, 101, 102] and Vesentini [139]. In particular Koszul and Vinberg observe that this Riemannian structure is the covariant differential $\nabla\omega$ of a closed 1-form ω. In particular ω is everywhere nonzero, so by Tischler [136], M fibers over S^1.

14.8.3.3 Hessian Manifolds

Hyperbolic affine manifolds are closely related to *Hessian manifolds.* If ω is a closed 1-form, then its covariant differential $\nabla\omega$ is a symmetric 2-form. Since closed forms are locally exact, $\omega = df$ for some function; in that case $\nabla\omega$ equals the *Hessian* $d^2 f$. Koszul [101] showed that hyperbolicity is equivalent to the existence of a closed 1-form ω whose covariant differential $\nabla\omega$ is positive definite, that is, a Riemannian metric. More generally, Shima [130] considered Riemannian metrics on an affine manifold which are locally Hessians of functions, and proved that such a closed *Hessian* manifold is a quotient of a convex domain, thus generalizing Koszul's result.

14.8.3.4 Hyperbolic Torus Bundles

Although the class of affine structures on closed 3-manifolds with *nilpotent* holonomy are understood, the general case of *solvable* holonomy remains mysterious. However, Serge Dupont [54] completely classifies affine structures on 3-manifolds with solvable *fundamental group.* (Compare also Dupont [53].) In terms of the Thurston geometrization, these are the geometric 3-manifolds modeled on Sol, that is, 3-manifolds finitely covered by *hyperbolic torus bundles:* mapping tori (suspensions) of hyperbolic elements of $\mathsf{GL}(2, \mathbb{Z})$. Dupont shows that all such structures arise from left-invariant affine structures on the corresponding Lie group G, which is the semidirect product of $\mathbb{R}^2$ by $\mathbb{R}$, where $\mathbb{R}$ acts on $\mathbb{R}^2$ as a unimodular hyperbolic one-parameter subgroup (explicitly, G is isomorphic to the identity component in the group of Lorentz isometries of the Minkowski plane).

Two structures are particularly interesting for the behavior of geodesics in light of the results of Vey [141]. A properly convex domain $\Omega \in \mathsf{A}^n$ is said to be *divisible* if Ω admits a discrete group Γ of projective automorphisms acting properly on Ω such that Ω/Γ is compact. (Equivalently, the quotient space Ω/Γ by a discrete subgroup $\Gamma \subset \mathsf{Aut}(\Omega)$ is compact and Hausdorff.) Vey proved that a divisible domain is a cone. However, dropping the properness of the action of Γ on Ω allows counterexamples: the *parabolic cylinder*

$$\Omega := \{(x, y) \in \mathsf{A}^2 \mid y > x^2\}$$

is a properly convex domain which is not a cone, but admits a group Γ of automorphisms such that Ω/Γ is compact but not Hausdorff.

Now take the product $\Omega \times \mathbb{R} \subset \mathsf{A}^3$. The author [70] found a discrete subgroup $\Gamma \subset \mathsf{Aff}(\mathsf{A}^3)$ acting properly on $\Omega \times \mathbb{R}$ such that:

- The quotient $M = (\Omega \times \mathbb{R})/\Gamma$ is a hyperbolic torus bundle (and in particular compact and Hausdorff);
- $\Omega \times \mathbb{R}$ is not a cone.

Clearly $\Omega \times \mathbb{R}$ is not properly convex, showing that Vey's result is sharp. The Kobayashi pseudometric degenerates along a 1-dimensional foliation of M, and defines the hyperbolic structure transverse to this foliation discussed by Thurston [134], Chapter 4.

14.8.4 Flat Conformal Structures in Higher Dimensions

Flat conformal structures generalize $\mathbb{C}\mathsf{P}^1$-structures, under the identification of $\mathbb{C}\mathsf{P}^1$ with S^2 with its usual conformal structure. In general the group of conformal transformations of S^n is the group $\mathsf{PO}(n+1, 1)$, where $S^n \hookrightarrow \mathbb{R}\mathsf{P}^{n+1}$ embeds as a quadric invariant under the projectivized Lorentz group $\mathsf{PO}(n+1, 1)$. See Matsumoto [119] for an excellent survey.

Flat conformal structures arise in Riemannian geometry. Specifically, a Riemannian manifold (M, g) is *conformally flat* if and only if every point $p \in M$ possesses an open neighborhood U and a smooth coordinate chart $U \xrightarrow{\psi} \mathbb{E}^n$ such that the Riemannian structure on U induced by g is conformally equivalent to the Euclidean structure. (Sometimes this condition is called *locally conformally flat.*) When $n > 2$, a flat conformal structure in our sense is then equivalent to a conformal equivalence class of conformally flat Riemannian structures. See Kulkarni [108, 109] for more background.

Kuiper [104] initiated the subject of flat conformal structures, and in [105], classified those with abelian fundamental group. Kulkarni [108] defined a connected sum operation between flat conformal manifolds. The construction is based on the fact that a conformal inversion interchanges the two components of the complement of a hypersphere in S^n. In other words, the inside and the outside of a hypersphere in S^n are *conformally equivalent.* Thus a closed 3-manifold need not be *geometric* in Thurston's sense to admit a flat conformal structure. However, as shown by the author [71], 3-manifolds with nilgeometric and solvgeometric structures do *not* admit flat conformal structures. These were the first examples of 3-manifolds *without* flat conformal structures.

14.8.5 Real Projective Structures in Higher Dimensions

In a series of papers [16–20], Yves Benoist developed a vast theory of convex $\mathbb{R}\mathsf{P}^n$-structures on compact manifolds. (See Benoist [21] for a survey.) In particular he analyzed the boundary and showed that strict convexity is equivalent to hyperbolicity in various contexts.

All of these studies involve the projectively invariant *Hilbert metric* on a properly convex domain. When the domain is bounded by a quadratic, this metric is just the hyperbolic metric in the Beltrami-Klein projective model of hyperbolic space. See Marquis [117], and in general the collection [125] for surveys of Hilbert geometry on such manifolds. Recently Benoist's theory of convex $\mathbb{R}\mathsf{P}^n$-structures on compact manifolds has been extended to the analog of finite volume hyperbolic manifolds. In particular we mention the work of Cooper-Long-Tillmann [46] on cusped $\mathbb{R}\mathsf{P}^n$-manifolds, as well as Choi [36], Choi-Lee-Marquis [43] and Marquis [116].

Kapovich [91] gave examples of convex $\mathbb{R}\mathsf{P}^n$-structures on compact negatively curved Riemannian manifolds which admit no locally symmetric Riemannian metric.

Which closed 3-manifolds admit $\mathbb{R}\mathsf{P}^3$-structures (that is, $\mathbb{R}\mathsf{P}^3$*-manifolds*) is an interesting and difficult question.

Unlike in the case of flat conformal structures, the topology of $\mathbb{R}\mathsf{P}^3$ precludes any inversion such as the Steiner inversion facilitating the Kulkarni connected-sum operation.

(Indeed, the two components of the complement of a projective hyperplane in projective space are not even *topologically* equivalent.) In this direction, Weiqiang Wu [149] showed that any compact $\mathbb{R}\mathsf{P}^n$-structure bounded by a sphere on its convex side must be a disc—as noted above, this rigidity phenomenon is evidently absent for flat conformal manifolds. (Compare also Dupont [53].) In particular it seems notoriously difficult to construct an $\mathbb{R}\mathsf{P}^3$-structure on a connected sum. In this vein, Cooper-Goldman [47] showed that the connected sum $\mathbb{R}\mathsf{P}^3 \# \mathbb{R}\mathsf{P}^3$ fails to admit an $\mathbb{R}\mathsf{P}^3$-structure; as of yet we know very few obstructions for a 3-manifold *not* to admit a flat projective structure.

14.8.6 Complex Projective Structures in Higher Dimensions

In a different direction, Klingler [93] classified (holomorphic) projective structures on complex surfaces, following earlier work by Vitter [143] and Kobayashi-Ochiai [87, 96, 97]. Every closed $\mathbb{C}\mathsf{P}^2$-manifold is finitely covered by a manifold of one of the following types:

- the complex projective plane $\mathbb{C}\mathsf{P}^2$;
- complex hyperbolic manifolds;
- *complex solvmanifolds,* that is, homogeneous spaces $\Gamma\backslash G$ where G is a 4-dimensional (real) Lie group with left-invariant complex structure and $\Gamma \subset G$ is a lattice;

- Hopf manifolds $\mathbb{C}^2 \setminus \{0\}/\Gamma$, where Γ is a cyclic group of linear expansions;
- elliptic surfaces over $\mathbb{C}\mathsf{P}^1$-manifolds, that is, holomorphic fibrations by elliptic curves over a Riemann surface with a projective structure.

These two latter classes are affine structures.

Acknowledgements The author gratefully acknowledges research support from NSF Grants DMS1065965, DMS1406281, DMS1065965 as well as the Research Network in the Mathematical Sciences DMS1107367 (GEAR).

References

1. Herbert Abels, Properly discontinuous groups of affine transformations: a survey, *Geom. Dedicata* **87** (2001), no. 1–3, 309–333. MR 1866854
2. Herbert Abels, Gregory Margulis, and Gregory Soifer, The Auslander conjecture for dimension less than 7, arXiv:1211.2525.
3. William Abikoff, *The real analytic theory of Teichmüller space*, Lecture Notes in Mathematics, vol. 820, Springer, Berlin, 1980. MR 590044
4. D. K. Arrowsmith and P. M. D. Furness, Flat affine Klein bottles, *Geometriae Dedicata* **5** (1976), no. 1, 109–115. MR 0433358 (55 #6334)
5. L. Auslander and L. Markus, Holonomy of flat affinely connected manifolds, *Ann. of Math.* (2) **62** (1955), 139–151. MR 0072518 (17,298b)
6. ———, Flat Lorentz 3-manifolds, *Mem. Amer. Math. Soc.* No. **30** (1959), 60. MR 0131842 (24 #A1689)
7. Louis Auslander, Examples of locally affine spaces, *Ann. of Math.* (2) **64** (1956), 255–259. MR 0080957
8. ———, The structure of complete locally affine manifolds, *Topology* **3** (1964), no. suppl. 1, 131–139. MR 0161255
9. Shinpei Baba, A Schottky decomposition theorem for complex projective structures, *Geom. Topol.* **14** (2010), no. 1, 117–151. MR 2578302
10. ———, Complex projective structures with Schottky holonomy, *Geom. Funct. Anal.* **22** (2012), no. 2, 267–310. MR 2929066
11. ———, 2π-grafting and complex projective structures, I, *Geom. Topol.* **19** (2015), no. 6, 3233–3287. MR 3447103
12. Oliver Baues, Varieties of discontinuous groups, *Crystallographic groups and their generalizations* (Kortrijk, 1999), Contemp. Math., vol. 262, Amer. Math. Soc., Providence, RI, 2000, pp. 147–158. MR 1796130 (2001i:58013)
13. ———, Deformation spaces for affine crystallographic groups, *Cohomology of groups and algebraic K-theory*, Adv. Lect. Math. (ALM), vol. 12, Int. Press, Somerville, MA, 2010, pp. 55–129. MR 2655175 (2011f:57066)
14. ———, The deformations of flat affine structures on the two-torus, *Handbook of Teichmüller theory*, Vol. IV, European Mathematical Society, 2014, pp. 461–537.
15. Oliver Baues and William M. Goldman, Is the deformation space of complete affine structures on the 2-torus smooth?, *Geometry and dynamics*, Contemp. Math., vol. 389, Amer. Math. Soc., Providence, RI, 2005, pp. 69–89. MR 2181958 (2006j:57066)
16. Yves Benoist, Automorphismes des cônes convexes, *Invent. Math.* **141** (2000), no. 1, 149–193. MR 1767272 (2001f:22034)
17. ———, Convexes divisibles. II, *Duke Math. J.* **120** (2003), no. 1, 97–120. MR 2010735 (2004m:22018)

18. ______, Convexes divisibles. I, *Algebraic groups and arithmetic*, Tata Inst. Fund. Res., Mumbai, 2004, pp. 339–374. MR 2094116 (2005h:37073)
19. ______, Convexes divisibles. III, *Ann. Sci. École Norm. Sup.* (4) **38** (2005), no. 5, 793–832. MR 2195260 (2007b:22011)
20. ______, Convexes divisibles. IV. Structure du bord en dimension 3, *Invent. Math.* **164** (2006), no. 2, 249–278. MR 2218481 (2007g:22007)
21. ______, A survey on divisible convex sets, *Geometry, analysis and topology of discrete groups*, Adv. Lect. Math. (ALM), vol. 6, Int. Press, Somerville, MA, 2008, pp. 1–18. MR 2464391 (2010h:52013)
22. J.-P. Benzécri, *Variétés localement affines*, Seminaire de Topologie et Géom. Diff., Ch. Ehresmann (1958–60) (1959), no. 7, 1–34.
23. Jean-Paul Benzécri, Sur les variétés localement affines et localement projectives, *Bulletin de la Société Mathématique de France* **88** (1960), 229–332.
24. N. Bergeron and T. Gelander, A note on local rigidity, *Geom. Dedicata* **107** (2004), 111–131. MR 2110758 (2005k:22015)
25. Francis Bonahon, Geometric structures on 3-manifolds, *Handbook of geometric topology*, North-Holland, Amsterdam, 2002, pp. 93–164. MR 590044
26. Jean-Philippe Burelle, Virginie Charette, Todd A. Drumm, and William M. Goldman, Crooked halfspaces, *Enseign. Math.* **60** (2014), no. 1–2, 43–78. MR 3262435
27. Marc Burger, Alessandra Iozzi, and Anna Wienhard, Higher Teichmüller spaces: from SL(2, $\mathbb{R}$) to other Lie groups, *Handbook of Teichmüller theory*. Vol. IV, IRMA Lect. Math. Theor. Phys., vol. 19, Eur. Math. Soc., Zürich, 2014, pp. 539–618. MR 3289711
28. Peter Buser, *Geometry and spectra of compact Riemann surfaces*, Progress in Mathematics, vol. 106, Birkhäuser Boston, Inc., Boston, MA, 1992. MR 1183224
29. R. D. Canary, D. B. A. Epstein, and P. Green, Notes on notes of Thurston, *Analytical and geometric aspects of hyperbolic space* (Coventry/Durham, 1984), London Math. Soc. Lecture Note Ser., vol. 111, Cambridge Univ. Press, Cambridge, 1987, pp. 3–92. MR 903850
30. Yves Carrière, Francoise Dal'bo, and Gaël Meigniez, Inexistence de structures affines sur les fibrés de Seifert, *Math. Ann.* **296** (1993), no. 4, 743–753. MR 1233496 (94h:57031)
31. Virginie Charette, Todd Drumm, William Goldman, and Maria Morrill, Complete flat affine and Lorentzian manifolds, *Geom. Dedicata* **97** (2003), 187–198, Special volume dedicated to the memory of Hanna Miriam Sandler (1960–1999). MR 2003697
32. Virginie Charette, Todd A. Drumm, and William M. Goldman, Affine deformations of a three-holed sphere, *Geom. Topol.* **14** (2010), no. 3, 1355–1382. MR 2653729
33. ______, Finite-sided deformation spaces of complete affine 3-manifolds, *J. Topol.* **7** (2014), no. 1, 225–246. MR 3180618
34. ______, Proper affine deformations of the one-holed torus, *Transform. Groups* **21** (2016), no. 4, 953–1002. MR 3569564
35. Claude Chevalley, *Theory of Lie Groups. I*, Princeton Mathematical Series, vol. 8, Princeton University Press, Princeton, N. J., 1946. MR 0015396
36. Suhyoung Choi, *The convex real projective orbifolds with radial or totally geodesic ends: a survey of some partial results*, ArXiv:1601.06952.
37. ______, Convex decompositions of real projective surfaces. I. π-annuli and convexity, *J. Differential Geom.* **40** (1994), no. 1, 165–208. MR 1285533 (95i:57015)
38. ______, Convex decompositions of real projective surfaces. II. Admissible decompositions, *J. Differential Geom.* **40** (1994), no. 2, 239–283. MR 1293655 (95k:57016)
39. ______, The convex and concave decomposition of manifolds with real projective structures, *Mém. Soc. Math. Fr. (N.S.)* (1999), no. 78, vi+102. MR 1779499 (2001j:57030)
40. ______, The decomposition and classification of radiant affine 3-manifolds, *Mem. Amer. Math. Soc.* **154** (2001), no. 730, viii+122. MR 1848866 (2002f:57049)
41. Suhyoung Choi and William M. Goldman, Topological tameness of margulis spacetimes, ArXiv:104.5308, to appear in *Amer. J. Math.*
42. ______, The classification of real projective structures on compact surfaces, *Bull. Amer. Math. Soc. (N.S.)* **34** (1997), no. 2, 161–171. MR 1414974 (97m:57020)

43. Suhyoung Choi, Gye-Seon Lee, and Ludovic Marquis, Deformations of convex real projective manifolds and orbifolds, ArXiv:1605.02548.
44. Suhyoung Choi and Hyunkoo Lee, Geometric structures on manifolds and holonomy-invariant metrics, *Forum Math.* **9** (1997), no. 2, 247–256. MR 1431123 (98a:53052)
45. Vicki Chuckrow, On Schottky groups with applications to kleinian groups, *Ann. of Math.* (2) **88** (1968), 47–61. MR 0227403
46. D. Cooper, D. D. Long, and S. Tillmann, On convex projective manifolds and cusps, *Adv. Math.* **277** (2015), 181–251. MR 3336086
47. Daryl Cooper and William Goldman, A 3-manifold with no real projective structure, *Ann. Fac. Sci. Toulouse Math.* (6) **24** (2015), no. 5, 1219–1238. MR 3485333
48. Jeffrey Danciger, François Guéritaud, and Fanny Kassel, Geometry and topology of complete Lorentz spacetimes of constant curvature, *Ann. Sci. Éc. Norm. Supér.* (4) **49** (2016), no. 1, 1–56. MR 3465975
49. ———, Margulis spacetimes via the arc complex, *Invent. Math.* **204** (2016), no. 1, 133–193. MR 3480555
50. Todd A. Drumm, *Fundamental polyhedra for Margulis space-times*, ProQuest LLC, Ann Arbor, MI, 1990, Thesis (Ph.D.)–University of Maryland, College Park. MR 2638637
51. ———, Fundamental polyhedra for Margulis space-times, *Topology* **31** (1992), no. 4, 677–683. MR 1191372
52. David Dumas, Complex projective structures, *Handbook of Teichmüller theory*. Vol. II, IRMA Lect. Math. Theor. Phys., vol. 13, Eur. Math. Soc., Zürich, 2009, pp. 455–508. MR 2497780
53. Serge Dupont, Variétés projectives à holonomie dans le groupe $\mathrm{Aff}^+(\mathbf{R})$, *Crystallographic groups and their generalizations* (Kortrijk, 1999), Contemp. Math., vol. 262, Amer. Math. Soc., Providence, RI, 2000, pp. 177–193. MR 1796133
54. ———, Solvariétés projectives de dimension trois, *Geom. Dedicata* **96** (2003), 55–89. MR 1956834
55. Clifford J. Earle, On variation of projective structures, *Riemann surfaces and related topics: Proceedings of the 1978 Stony Brook Conference* (State Univ. New York, Stony Brook, N.Y., 1978), Ann. of Math. Stud., vol. 97, Princeton Univ. Press, Princeton, N.J., 1981, pp. 87–99. MR 624807
56. Charles Ehresmann, Sur les espaces localement homogènes, *L'Ens. Math* **35** (1936), 317–333.
57. ———, Les connexions infinitésimales dans un espace fibré différentiable, in Thone [133], pp. 29–55. MR 0042768
58. Gerd Faltings, Real projective structures on Riemann surfaces, *Compositio Math.* **48** (1983), no. 2, 223–269. MR 700005
59. V. V. Fock and A. B. Goncharov, Moduli spaces of convex projective structures on surfaces, *Adv. Math.* **208** (2007), no. 1, 249–273. MR 2304317
60. Vladimir Fock and Alexander Goncharov, Moduli spaces of local systems and higher Teichmüller theory, *Publ. Math. Inst. Hautes Études Sci.* (2006), no. 103, 1–211. MR 2233852
61. David Fried, Closed similarity manifolds, *Comment. Math. Helv.* **55** (1980), no. 4, 576–582. MR 604714 (83e:53049)
62. David Fried, William Goldman, and Morris W. Hirsch, Affine manifolds with nilpotent holonomy, *Comment. Math. Helv.* **56** (1981), no. 4, 487–523. MR 656210 (83h:53062)
63. David Fried and William M. Goldman, Three-dimensional affine crystallographic groups, *Adv. in Math.* **47** (1983), no. 1, 1–49. MR 689763
64. P. M. D. Furness and D. K. Arrowsmith, Locally symmetric spaces, *J. London Math. Soc.* (2) **10** (1975), no. 4, 487–499. MR 0467598 (57 #7454)
65. Daniel Gallo, Michael Kapovich, and Albert Marden, The monodromy groups of Schwarzian equations on closed Riemann surfaces, *Ann. of Math.* (2) **151** (2000), no. 2, 625–704. MR 1765706
66. William Goldman, Convex real projective structures on compact surfaces, *J. Differential Geom* **31** (1990), no. 3, 791–845.
67. William Goldman and Morris W. Hirsch, The radiance obstruction and parallel forms on affine manifolds, *Trans. Amer. Math. Soc.* **286** (1984), no. 2, 629–649. MR 760977 (86f:57032)

68. William M. Goldman, *Affine manifolds and projective geometry on surfaces*, senior thesis, Princeton University, 1977.
69. ______, *Discontinuous groups and the Euler class*, Ph.D. thesis, University of California, Berkeley, 1980.
70. ______, Two examples of affine manifolds, *Pacific J. Math.* **94** (1981), no. 2, 327–330. MR 628585 (83a:57028)
71. ______, Conformally flat manifolds with nilpotent holonomy and the uniformization problem for 3-manifolds, *Trans. Amer. Math. Soc.* **278** (1983), no. 2, 573–583. MR 701512
72. ______, Projective structures with Fuchsian holonomy, *J. Differential Geom.* **25** (1987), no. 3, 297–326. MR 882826
73. ______, Geometric structures on manifolds and varieties of representations, *Geometry of group representations* (Boulder, CO, 1987), Contemp. Math., vol. 74, Amer. Math. Soc., Providence, RI, 1988, pp. 169–198. MR 957518
74. ______, Topological components of spaces of representations, *Invent. Math.* **93** (1988), no. 3, 557–607. MR 952283 (89m:57001)
75. ______, Locally homogeneous geometric manifolds, *Proceedings of the International Congress of Mathematicians*. Volume II, Hindustan Book Agency, New Delhi, 2010, pp. 717–744. MR 2827816
76. ______, Two papers which changed my life: Milnor's seminal work on flat manifolds and bundles, *Frontiers in complex dynamics: In celebration of John Milnor's 80th birthday*, Princeton Math. Ser., vol. 51, Princeton Univ. Press, Princeton, NJ, 2014, pp. 679–703. MR 3289925
77. William M. Goldman and John J. Millson, Local rigidity of discrete groups acting on complex hyperbolic space, *Invent. Math.* **88** (1987), no. 3, 495–520. MR 884798
78. R. C. Gunning, *Lectures on Riemann surfaces*, Princeton Mathematical Notes, Princeton University Press, Princeton, N.J., 1966. MR 0207977
79. ______, Special coordinate coverings of Riemann surfaces, *Math. Ann.* **170** (1967), 67–86. MR 0207978
80. ______, Affine and projective structures on Riemann surfaces, *Riemann surfaces and related topics: Proceedings of the 1978 Stony Brook Conference* (State Univ. New York, Stony Brook, N.Y., 1978), Ann. of Math. Stud., vol. 97, Princeton Univ. Press, Princeton, N.J., 1981, pp. 225–244. MR 624816
81. Dennis A. Hejhal, Monodromy groups and linearly polymorphic functions, *Discontinuous groups and Riemann surfaces* (Proc. Conf., Univ. Maryland, College Park, Md., 1973), Princeton Univ. Press, Princeton, N.J., 1974, pp. 247–261. Ann. of Math. Studies, No. 79. MR 0355035
82. ______, Monodromy groups and linearly polymorphic functions, *Acta Math.* **135** (1975), no. 1, 1–55. MR 0463429
83. N. J. Hitchin, Lie groups and Teichmüller space, *Topology* **31** (1992), no. 3, 449–473. MR 1174252
84. H. Hopf, Zur Topologie der komplexen Mannigfaltigkeiten, *Studies and Essays Presented to R. Courant on his 60th Birthday, January 8, 1948*, Interscience Publishers, Inc., New York, 1948, pp. 167–185. MR 0023054
85. John H. Hubbard, The monodromy of projective structures, *Riemann surfaces and related topics: Proceedings of the 1978 Stony Brook Conference* (State Univ. New York, Stony Brook, N.Y., 1978), Ann. of Math. Stud., vol. 97, Princeton Univ. Press, Princeton, N.J., 1981, pp. 257–275. MR 624819
86. John Hamal Hubbard, *Teichmüller theory and applications to geometry, topology, and dynamics. Vol. 1*, Matrix Editions, Ithaca, NY, 2006, Teichmüller theory, With contributions by Adrien Douady, William Dunbar, Roland Roeder, Sylvain Bonnot, David Brown, Allen Hatcher, Chris Hruska and Sudeb Mitra, With forewords by William Thurston and Clifford Earle. MR 2245223
87. Masahisa Inoue, Shoshichi Kobayashi, and Takushiro Ochiai, Holomorphic affine connections on compact complex surfaces, *J. Fac. Sci. Univ. Tokyo Sect. IA Math.* **27** (1980), no. 2, 247–264. MR 586449

88. Yoshinobu Kamishima and Ser P. Tan, Deformation spaces on geometric structures, *Aspects of low-dimensional manifolds*, Adv. Stud. Pure Math., vol. 20, Kinokuniya, Tokyo, 1992, pp. 263–299. MR 1208313
89. Michael Kapovich, *Deformation spaces of flat conformal structures*, *Questions and Answers in General Topology* Special Issue **8** (1990), 253–264.
90. ______, *Hyperbolic manifolds and discrete groups*, Progress in Mathematics, vol. 183, Birkhäuser Boston, Inc., Boston, MA, 2001. MR 1792613
91. ______, Convex projective structures on Gromov-Thurston manifolds, *Geom. Topol.* **11** (2007), 1777–1830. MR 2350468
92. D. A. Každan and G. A. Margulis, A proof of Selberg's hypothesis, *Mat. Sb. (N.S.)* **75 (117)** (1968), 163–168. MR 0223487
93. Bruno Klingler, Structures affines et projectives sur les surfaces complexes, *Ann. Inst. Fourier (Grenoble)* **48** (1998), no. 2, 441–477. MR 1625606
94. Shoshichi Kobayashi, Intrinsic distances associated with flat affine and projective structures, *J. Fac. Sci. Univ. Tokyo* **24** (1977), 129–135.
95. ______, *Projectively invariant distances for affine and projective structures*, Banach Center Publ., vol. 12, PWN, Warsaw, 1984. MR 961077 (89k:53043
96. Shoshichi Kobayashi and Takushiro Ochiai, Holomorphic projective structures on compact complex surfaces, *Math. Ann.* **249** (1980), no. 1, 75–94. MR 575449
97. ______, Holomorphic projective structures on compact complex surfaces. II, *Math. Ann.* **255** (1981), no. 4, 519–521. MR 618182
98. J.-L. Koszul, Domaines bornés homogènes et orbites de groupes de transformations affines, *Bull. Soc. Math. France* **89** (1961), 515–533. MR 0145559 (26 #3090)
99. ______, Ouverts convexes homogenes des espaces affines, *Mathematische Zeitschrift* **79** (1962), no. 1, 254–259.
100. ______, *Lectures on groups of transformations*, Notes by R. R. Simha and R. Sridharan. Tata Institute of Fundamental Research Lectures on Mathematics, No. 32, Tata Institute of Fundamental Research, Bombay, 1965. MR 0218485 (36 #1571)
101. ______, Déformations de connexions localement plates, *Ann. Inst. Fourier (Grenoble)* **18** (1968), no. fasc. 1, 103–114. MR 0239529 (39 #886)
102. Jean-Louis Koszul, Variétés localement plates et convexité, *Osaka J. Math.* **2** (1965), 285–290. MR 0196662 (33 #4849)
103. N. H. Kuiper, Compact spaces with a local structure determined by the group of similarity transformations in E^n, *Nederl. Akad. Wetensch., Proc.* **53** (1950), 1178–1185 = Indagationes Math. 12, 411–418 (1950). MR 0039248
104. N.H. Kuiper, On conformally-flat spaces in the large, *Ann. of Math.* (2) **50** (1949), 916–924. MR 0031310 (11,133b)
105. ______, *On compact conformally Euclidean spaces of dimension* > 2, *Ann. of Math.* (2) **52** (1950), 478–490. MR 0037575
106. ______, Locally projective spaces of dimension one., *The Michigan Mathematical Journal* **2** (1953), no. 2, 95–97.
107. ______, *Sur les surfaces localement affines*, Géométrie différentielle. Colloques Internationaux du Centre National de la Recherche Scientifique, Strasbourg, 1953, 1953, pp. 79–87.
108. R. Kulkarni, On the principle of uniformization, *J. Diff. Geom* **13** (1978), 109–138.
109. ______, Conformal structures and Möbius structures, *Conformal geometry* (Bonn, 1985/1986), Aspects Math., E12, Vieweg, Braunschweig, 1988, pp. 1–39. MR 979787 (90f:53026)
110. François Labourie, Anosov flows, surface groups and curves in projective space, *Invent. Math.* **165** (2006), no. 1, 51–114. MR 2221137
111. ______, *Lectures on representations of surface groups*, Zurich Lectures in Advanced Mathematics, European Mathematical Society (EMS), Zürich, 2013. MR 3155540
112. Walter L. Lok, *Deformations of locally homogeneous spaces and kleinian groups*, Ph.D. thesis, Columbia University, 1984.

113. A. Marden, *Outer circles*, Cambridge University Press, Cambridge, 2007, An introduction to hyperbolic 3-manifolds. MR 2355387
114. GA Margulis, Complete affine locally flat manifolds with a free fundamental group, *Journal of Soviet Mathematics* **36** (1987), no. 1, 129–139.
115. Gregory Margulis, Free properly discontinuous groups of affine transformations, *Dokl. Akad. Nauk SSSR* **272** (1983), no. 4, 937–940.
116. Ludovic Marquis, Espace des modules marqués des surfaces projectives convexes de volume fini, *Geom. Topol.* **14** (2010), no. 4, 2103–2149. MR 2740643
117. ———, Around groups in Hilbert geometry, *Handbook of Hilbert geometry*, IRMA Lect. Math. Theor. Phys., vol. 22, Eur. Math. Soc., Zürich, 2014, pp. 207–261. MR 3329882
118. Bernard Maskit, On a class of Kleinian groups, *Ann. Acad. Sci. Fenn. Ser.* A I No. **442** (1969), 8. MR 0252638
119. Shigenori Matsumoto, Foundations of flat conformal structure, *Aspects of low-dimensional manifolds*, Adv. Stud. Pure Math., vol. 20, Kinokuniya, Tokyo, 1992, pp. 167–261. MR 1208312
120. J. Milnor, *Hilbert's problem 18: on crystallographic groups, fundamental domains, and on sphere packing, Mathematical developments arising from Hilbert problems* (Proc. Sympos. Pure Math., Northern Illinois Univ., De Kalb, Ill., 1974), Amer. Math. Soc., Providence, R. I., 1976, pp. 491–506. Proc. Sympos. Pure Math., Vol. XXVIII. MR 0430101
121. John Milnor, On the existence of a connection with curvature zero, *Commentarii Mathematici Helvetici* **32** (1958), no. 1, 215–223.
122. ———, On the 3-dimensional Brieskorn manifolds $M(p, q, r)$, *Knots, groups, and 3-manifolds* (Papers dedicated to the memory of R. H. Fox), Princeton Univ. Press, Princeton, N. J., 1975, pp. 175–225. Ann. of Math. Studies, No. 84. MR 0418127 (54 #6169)
123. ———, On fundamental groups of complete affinely flat manifolds, *Advances in Mathematics* **25** (1977), no. 2, 178–187.
124. Tadashi Nagano and Katsumi Yagi, The affine structures on the real two-torus. I, *Osaka Journal of Mathematics* **11** (1974), no. 1, 181–210.
125. Athanase Papadopoulos and Marc Troyanov (eds.), *Handbook of Hilbert geometry*, IRMA Lectures in Mathematics and Theoretical Physics, vol. 22, European Mathematical Society (EMS), Zürich, 2014. MR 3309067
126. M. S. Raghunathan, *Discrete subgroups of Lie groups*, Springer-Verlag, New York-Heidelberg, 1972, Ergebnisse der Mathematik und ihrer Grenzgebiete, Band 68. MR 0507234
127. John G. Ratcliffe, *Foundations of hyperbolic manifolds*, second ed., Graduate Texts in Mathematics, vol. 149, Springer, New York, 2006. MR 2249478
128. Peter Scott, The geometries of 3-manifolds, *Bulletin of the London Mathematical Society* **15** (1983), no. 5, 401–487.
129. Richard W Sharpe, *Differential geometry: Cartan's generalization of klein's erlangen program*, vol. 166, Springer, 1997.
130. Hirohiko Shima, *The geometry of Hessian structures*, World Scientific Publishing Co. Pte. Ltd., Hackensack, NJ, 2007. MR 2293045 (2008f:53011)
131. John Smillie, *Affinely flat manifolds*, Ph.D. thesis, University of Chicago, 1977.
132. Dennis Sullivan and William Thurston, Manifolds with canonical coordinate charts: some examples, *Enseign. Math* **29** (1983), 15–25.
133. Georges Thone (ed.), *Colloque de topologie (espaces fibrés), Bruxelles, 1950*, Masson et Cie., Paris, 1951. MR 0042768
134. William P. Thurston, *The geometry and topology of three-manifolds*, unpublished notes, Princeton University Mathematics Department, 1979.
135. ———, *Three-dimensional geometry and topology. Vol. 1*, Princeton Mathematical Series, vol. 35, Princeton University Press, Princeton, NJ, 1997, Edited by Silvio Levy. MR 1435975
136. D. Tischler, On fibering certain foliated manifolds over S^1, *Topology* **9** (1970), 153–154. MR 0256413
137. Domingo Toledo, Representations of surface groups in complex hyperbolic space, *J. Differential Geom.* **29** (1989), no. 1, 125–133. MR 978081

138. Izu Vaisman and Corina Reischer, Local similarity manifolds, *Ann. Mat. Pura Appl.* (4) **135** (1983), 279–291 (1984). MR 750537
139. Edoardo Vesentini, Invariant metrics on convex cones, *Ann. Scuola Norm. Sup. Pisa Cl. Sci.* (4) **3** (1976), no. 4, 671–696. MR 0433228 (55 #6206)
140. Jacques Vey, *Sur une notion d'hyperbolicité des variétés localement plates*, Ph.D. thesis, Université Joseph-Fourier-Grenoble I, 1969.
141. ———, Sur les automorphismes affines des ouverts convexes saillants, *Annali della Scuola Normale Superiore di Pisa-Classe di Scienze* **24** (1970), no. 4, 641–665.
142. È. B. Vinberg, The theory of homogeneous convex cones, *Trudy Moskov. Mat. Obšč.* **12** (1963), 303–358. MR 0158414 (28 #1637)
143. Al Vitter, Affine structures on compact complex manifolds, *Invent. Math.* **17** (1972), 231–244. MR 0338998
144. André Weil, On discrete subgroups of Lie groups, *Ann. of Math.* (2) **72** (1960), 369–384. MR 0137792 (25 #1241)
145. Joseph A. Wolf, *Spaces of constant curvature*, third ed., Publish or Perish, Inc., Boston, Mass., 1974. MR 0343214
146. Michael Wolf, The Teichmüller theory of harmonic maps, *J. Differential Geom.* **29** (1989), no. 2, 449–479. MR 982185
147. Scott A. Wolpert, *Families of Riemann surfaces and Weil-Petersson geometry*, CBMS Regional Conference Series in Mathematics, vol. 113, Published for the Conference Board of the Mathematical Sciences, Washington, DC; by the American Mathematical Society, Providence, RI, 2010. MR 2641916 (2011c:32020)
148. H. Wu, Some theorems on projective hyperbolicity, *J. Math. Soc. Japan* **33** (1981), no. 1, 79–104. MR 597482 (82j:53061)
149. Weiqiang Wu, *On embedded spheres of affine manifolds*, ProQuest LLC, Ann Arbor, MI, 2012, Thesis (Ph.D.)–University of Maryland, College Park, posted to ArXiv:math.110.3541. MR 3103703

Chapter 15
Basic Aspects of Differential Geometry

Marc Chaperon

Abstract This is a very partial description of differential geometry as elaborated by Élie Cartan and expressed in a suitable language by Charles Ehresmann.

This is a very partial description of differential geometry as elaborated by Élie Cartan and expressed in a suitable language by Charles Ehresmann. I am entirely responsable for the selection of materials and for the mistakes, if any.

The framework is that of smooth[1] (finite dimensional) manifolds and maps, whose definition is taken for granted—most of the notions we consider "pass" without any problem to the real analytic and (replacing $\mathbb{R}$ by $\mathbb{C}$) complex and/or Banach categories. The kth derivative of a map f is denoted by $D^k f$ as in [10]. Paths are defined on intervals.

15.1 Jets

Introduced by Ehresmann [14], curiously almost absent from [11, 12], they are at the very beginning of modern differential geometry, as they generalize Taylor expansions to maps between manifolds. Recall the *Faà di Bruno formula* giving the kth derivative of the composed map of two C^k maps between open subsets of Banach spaces:

$$\frac{1}{k!}D^k(g \circ f)(x)v^k = \sum D^{|p|}g(f(x))\left(\frac{1}{p_1!}\left(\frac{1}{1!}D^1 f(x)v^1\right)^{p_1}, \dots, \frac{1}{p_k!}\left(\frac{1}{k!}D^k f(x)v^k\right)^{p_k}\right),$$

[1] That is C^∞ or "smooth enough", the word being implicit when nothing is specified.

M. Chaperon (✉)
Institut de Mathématiques de Jussieu (IMJ-PRG), Université de Paris VII (Denis Diderot), Paris, France
e-mail: marc.chaperon@imj-prg.fr

S. G. Dani, A. Papadopoulos (eds.), *Geometry in History*,
https://doi.org/10.1007/978-3-030-13609-3_15

where x lies in the definition domain of $g \circ f$, the vector v in the ambient Banach space, $v^k := (\overbrace{v, \dots, v}^{k \text{ times}})$ and the sum is on all $p = (p_1, \dots, p_k) \in \mathbb{N}^k$ with $\sum j\, p_j = k$, setting $|p| = \sum p_j$.

This formula is obtained by "composition of kth order Taylor expansions" [8]. Its author, born in Alessandria in 1825, was an officer in the Italian Royal Army before studying mathematics in Paris under the supervision of Cauchy and Le Verrier and taking up the position of Professor of Mathematics at the university of Turin. He was beatified in 1988, one century after his death, for his work as a social reformer, most notably the foundation of the Minim Sisters of St. Zita. Also a musician, he had been ordained in 1876.

For each integer k, two C^k maps f and g, defined in the neighbourhood of a point a in a manifold M, taking their values in a manifold N, have the same kth *order jet* at a, denoted $j_a^k f = j_a^k g$, when they take the same value b at a and there exist local charts $\varphi : (M, a) \to \mathbb{R}^n$ and $\psi : (N, b) \to \mathbb{R}^p$ such that $\psi \circ f \circ \varphi^{-1}$ and $\psi \circ g \circ \varphi^{-1}$ have the same kth order Taylor expansion at $\varphi(a)$; fortunately for this definition, the Faà di Bruno formula implies that such is then the case for *all* local charts φ and ψ at a and b respectively.

Let $J^k(M, N)$ be the set of kth order jets $j_a^k f$ of maps of M into N. If M, N are open subsets U, V of $\mathbb{R}^n, \mathbb{R}^p$ respectively, $J^k(U, V)$ identifies to the open subset $U \times V \times J^k(n, p)$ of the finite dimensional vector space

$$J^k(\mathbb{R}^n, \mathbb{R}^p) = \mathbb{R}^n \times \mathbb{R}^p \times J^k(n, p) := \mathbb{R}^n \times \prod_{j=0}^{k} L_s^j(\mathbb{R}^n, \mathbb{R}^p),$$

where $L_s^j(\mathbb{R}^n, \mathbb{R}^p)$ is the space of symmetric j-linear maps of $(\mathbb{R}^n)^j$ into $\mathbb{R}^p$ and $L_s^0(\mathbb{R}^n, \mathbb{R}^p) := \mathbb{R}^p$; indeed, $j_a^k f$ is then naturally identified to $\left(a, \left(D^j f(a)\right)_{0 \le j \le k}\right)$, and this identification is bijective as every $(a, b_0, \dots, b_k)$ in $U \times V \times J^k(n, p)$ is of the form $j_a^k f$ for $f(x) = \sum_0^k \frac{1}{j!} b_j (x - a)^j$.

In the general case, it follows from the Faà di Bruno formula that $J^k(M, N)$ is endowed with a smooth manifold structure by the *natural charts* $\Phi_{\varphi,\psi}^k$ associated to pairs of local charts φ of M and ψ of N as follows:

- the definition domain $\mathbf{dom}\, \Phi_{\varphi,\psi}^k$ of $\Phi_{\varphi,\psi}^k$ is the set of $j_a^k f$ with $a \in \mathbf{dom}\, \varphi$ and $f(a) \in \mathbf{dom}\, \psi$,
- the chart $\Phi_{\varphi,\psi}^k$ is given by the formula

$$\Phi_{\varphi,\psi}^k(j_a^k f) := j_{\varphi(a)}^k(\psi \circ f \circ \varphi^{-1}),$$

  implying that the transition maps are $\Phi_{\varphi_1,\psi_1}^k \circ (\Phi_{\varphi,\psi}^k)^{-1} = \Phi_{\varphi_1 \circ \varphi^{-1}, \psi_1 \circ \psi^{-1}}$
- its range $\mathbf{im}\, \Phi_{\varphi,\psi}^k$ therefore is $J^k(\mathbf{im}\, \varphi, \mathbf{im}\, \psi)$.

Examples and "Derived Products" The manifold $J^0(M, N)$ is of course identified to $M \times N$ by the diffeomorphism $j_a^0 f \mapsto \big(a, f(a)\big)$.

The set of all $j_0^1 f \in J^1(\mathbb{R}, N)$ is a submanifold, the *tangent bundle* TN of N: each natural chart $\Phi^1_{\mathrm{id}_{\mathbb{R}},\psi}$ is an adapted chart for TN and restricts to the chart $T\psi : j_0^1\gamma \mapsto \big(\psi \circ \gamma(0), (\psi \circ \gamma)'(0)\big)$; moreover, $J^1(\mathbb{R}, N)$ is identified to $\mathbb{R} \times TN$ by the map $j_t^1\gamma \mapsto \big(t, j_0^1(\gamma \circ \tau_{-t})\big)$, where $\tau_{-t}(x) = x + t$. One calls $j_0^1(\gamma \circ \tau_{-t})$ the *velocity* $\dot\gamma(t)$ of the path γ at time t (the knowledge of this velocity includes that of the position $\gamma(t)$, but not that of the time t).

Symmetrically, the set of all $j_a^1 f \in J^1(M, \mathbb{R})$ with $f(a) = 0$ is a submanifold, the *cotangent bundle* T^*M of M: each natural chart $\Phi^1_{\varphi,\mathrm{id}_{\mathbb{R}}}$ is an adapted chart for T^*M and restricts to the chart $T^*\varphi : j_a^1 f \mapsto \Big(\varphi(a), D(f \circ \varphi^{-1})\big(\varphi(a)\big)\Big)$; moreover, $J^1(M, \mathbb{R})$ is identified to $T^*M \times \mathbb{R}$ by the map $j_a^1 f \mapsto \big(j_a^1(\tau_{f(a)} \circ f), f(a)\big)$. One calls $j_a^1(\tau_{f(a)} \circ f)$ the *differential* $d_a f$ of f at a (its knowledge includes that of a, but not of $f(a)$).

The natural charts endow $J^k(M, N)$ with much more than just a manifold structure, since the projections $j_a^k f \mapsto a$ ("source projection"), $j_a^k f \mapsto f(a)$ ("target projection") and $j_a^k f \mapsto j_a^\ell f$, $0 \le \ell < k$, are fibrations, as we shall now see.

15.2 Submersions and Fibrations

A map $\begin{array}{c} E \\ \downarrow \pi \\ B \end{array}$ between manifolds is a *submersion* when "it is locally *in E* the projection onto the first factor of a product": for every $a \in E$, there exist an open subset U of $\mathbb{R}^n$, an open subset V of $\mathbb{R}^r$, a local chart $\tilde\varphi$ of E at a and a local chart φ of B at $\pi(a)$ such that $\mathbf{im}\,\tilde\varphi = U \times V$, $\mathbf{im}\,\varphi = U$ and $\varphi \circ \pi = \mathrm{pr}_1 \circ \tilde\varphi$, where $\mathrm{pr}_1 : U \times V \to U$ denotes the projection onto the first factor. One then calls $\tilde\varphi$ a *fibred chart* of the submersion *over* φ.

Similarly, π is a *locally trivial fibration* when "it is locally *in B* the projection onto the first factor of a product": for every $b \in B$, there exist a local chart φ of B at b, a manifold F and a diffeomorphism $\tilde\varphi$ of $\pi^{-1}(\mathbf{dom}\,\varphi)$ onto $\mathbf{im}\,\varphi \times F$ such that $\varphi \circ \pi = \mathrm{pr}_1 \circ \tilde\varphi$, where $\mathrm{pr}_1 : \mathbf{im}\,\varphi \times F \to \mathbf{im}\,\varphi$ is the projection onto the first factor.

One can avoid the use of φ via an equivalent definition: for every $b \in B$, there exist an open subset $\Omega \ni b$ of B and a diffeomorphism h of $\pi^{-1}(\Omega)$ onto $\Omega \times F$ such that $\pi|_{\pi^{-1}(\Omega)}$ is the first component of the *local trivialisation* h of π.

Clearly (taking local charts of F) a fibration is a submersion and (by the very definition of a submanifold) the *fibres* $\pi^{-1}(b)$ of a submersion are submanifolds. When π is a fibration, one calls E (the *total space* of) a *fibre bundle over B* (called its *base space*) with *projection* π.

When F is an open subset of $\mathbb{R}^r$, the diffeomorphism $\tilde\varphi$ in the definition of a fibre bundle (which determines φ) is a chart of E. A *vector bundle* is defined by an atlas of such charts $\tilde\varphi$ with $F = \mathbb{R}^r$ (or a vector space), such that the transition maps

$\tilde{\varphi}_1 \circ \tilde{\varphi}^{-1}$ are linear with respect to F ("atlas of vector bundle"). Il follows that the fibres $E_b = \pi^{-1}(b)$ are endowed with a structure of vector space isomorphic to F. Replacing "linear" and "vector" by "affine", on gets the notion of an *affine bundle*, whose fibres are affine spaces.

Sections With the previous notation, a *smooth section of the submersion* π *over the open subset* U *of* B is a smooth map σ of U into $\pi^{-1}(U)$ such that $\pi \circ \sigma = \mathrm{id}_U$; if $U = B$, it is called a *section* of π. In the same way as a map is determined by its graph, a section is determined by its *image* $\sigma(U)$, that is a submanifold (it appears as a graph in the fibred charts $\tilde{\varphi}$). It is therefore natural—hence the terminology—to consider that a smooth section of π over U is a submanifold meeting each fibre of $\pi|_{\pi^{-1}(U)}$ at a unique point and *transversally* (see the sequel).

The Case of Jets Il is immediate that the projections $\pi_k^\ell : J^k(M, N) \to J^\ell(M, N)$ defined for $\ell \leq k$ by $\pi_k^\ell(j_a^k f) = j_a^\ell f$ are fibrations, whose typical fibre F is the vector space $\prod_{\ell<j\leq k} L_s^j(\mathbb{R}^n, \mathbb{R}^p)$: just take $\tilde{\varphi} = \Phi_{\varphi,\psi}^k$ and $\varphi := \Phi_{\varphi,\psi}^\ell$ in the definition. Similarly, taking $\tilde{\varphi} = \Phi_{\varphi,\psi}^k$ and $\varphi = \varphi$ (resp. $\varphi := \psi$) in the definition of a submersion, one sees that the source projection $s_k : j_a^k f \to a$ and the target projection $b_k : j_a^k f \to f(a)$ are submersions. By the Faà di Bruno formula,

- this defines on $J^1(M, N)$ a vector bundle structure with base space $J^0(M, N) = M \times N$, projection π_1^0 and typical fibre $L(\mathbb{R}^n, \mathbb{R}^p)$,
- thus the tangent bundle TN is a vector bundle over N with typical fibre $\mathbb{R}^p = L(\mathbb{R}, \mathbb{R}^p)$, and the cotangent bundle T^*M a vector bundle over M with typical fibre $\mathbb{R}^{n*} = L(\mathbb{R}^n, \mathbb{R})$,
- for $k > 1$, the fibre bundle $J^k(M, N)$ is an *affine* bundle with typical fibre $L_s^k(\mathbb{R}^n, \mathbb{R}^p)$ over $J^{k-1}(M, N)$,
- for $\ell < k \leq 2\ell + 1$, the space $J^k(M, N)$ is endowed by the charts $\Phi_{\varphi,\psi}^k$ with an *affine* bundle structure over $J^\ell(M, N)$,
- such is not the case for $k > 2\ell + 1$, the transition maps between natural charts being polynomial of degree at least 2 with respect to the typical fibre, *but*
- if N is a vector space, $J^k(M, N)$ is endowed for $0 \leq \ell < k$ with a structure of affine bundle over $J^\ell(M, N)$ (*vector* bundle if $\ell = 0$) by the charts $\Phi_{\varphi,\mathrm{id}_N}^k$.

The fibre $T_a M$ of TM over $a \in M$ is the *tangent space of* M *at* a.

Though it is a *vector* space, it should be pictured genuinely tangent to M at a when M is a submanifold of $\mathbb{R}^d$: indeed, $T_a M$ is obtained by looking at M through a microscope centred at a, taken as the origin of the *affine* space $\mathbb{R}^d$.

The fibre $T_a^* M$ of T^*M identifies naturally to the dual space $(T_a M)^*$, the duality form being $\big(\dot{\gamma}(a), d_a f\big) \mapsto (f \circ \gamma)'(a)$.

The source projection $s_k : j_a^k f \to a$ and the target projection $b_k : j_a^k f \to f(a)$ are in fact *fibrations*, whose typical fibres are respectively the set $J_0^k(\mathbb{R}^n, N)$ of all $j_0^k f \in J^k(\mathbb{R}^n, N)$ and the set $J^k(M, \mathbb{R}^p)_0$ of all $j_a^k f \in J^k(M, \mathbb{R}^p)$ with $f(a) = 0$.

The proof is the same as for the tangent and cotangent bundles: to each chart φ of M one can associate the diffeomorphism $\tilde{\varphi}$ of $s_k^{-1}(\mathbf{dom}\,\varphi)$ onto $\mathbf{im}\,\varphi \times J_0^k(\mathbb{R}^n, N)$ mapping $j_a^k f$ to

$(\varphi(a), j_0^k(f \circ \varphi^{-1} \circ \tau_{-\varphi(a)}))$; similarly, to each chart ψ of N is associated the diffeomorphism $\tilde{\psi}$ of $b_k^{-1}(\mathbf{dom}\,\psi)$ onto $\mathbf{im}\,\psi \times J^k(M, \mathbb{R}^p)_0$ mapping $j_a^k f$ to $(\psi \circ f(a), j_a^k(\tau_{\psi \circ f(a)} \circ \psi \circ f))$.

Examples of Sections For every smooth map f of an open subset U of a manifold M into a manifold N, the map $a \mapsto j_a^k f$ is a section $j^k f$ of the source projection $J^k(M, N) \to M$ over U, the kth *order jet of* f, clearly a section of the source projection $J^k(U, N) \to U$; such sections are called *holonomic*.

A section of the tangent bundle $TM \to M$ over U is called a *vector field* on U (at every point a of U one grows a vector $X_a \in T_a U = T_a M$).

For each smooth real function f on an open subset U of M, the map $df : a \mapsto d_a f$ is a section of the cotangent bundle $T^*M \to M$ over U or, equivalently, a section of the cotangent bundle $T^*U \subset T^*M$; a section of the cotangent bundle $T^*U \to U$ is called a "field of covectors" or *Pfaffian form* (or *differential form of degree* 1, or *differential* 1*-form*, or 1*–form*) on U.

More generally to each smooth map $f : M \to N$ is associated the map Tf of TM in TN defined by $Tf(\dot{\gamma}(a)) = \dot{\overline{f \circ \gamma}}(a)$; its restriction $T_a f$ to each fibre $T_a M$ is a *linear* map into $T_{f(a)}M$ ("linear map tangent to f at a"): this is expressed by calling Tf a *homomorphism of vector bundles*.

Of course, $T_a f$ is identified to $j_a^1 f$. In the seventies, some authors [11, 12] would replace for example $j^2 f$ by $T(Tf)$, but the ensuing inflation of dimensions and redondance are unreasonable.

Infinitesimal Characterisation of Submersions, Vertical and Horizontal Spaces and Sections It follows easily from the inverse mapping theorem that a smooth map $\begin{array}{c} E \\ \downarrow \pi \\ B \end{array}$ between manifolds is a submersion in the neighbourhood of $a \in E$ if and only if the tangent linear map $T_a\pi$ is onto; therefore, π is a submersion if and only if $T_a\pi$ is onto for every $a \in E$.

For each $a \in E$, setting $b = \pi(a)$, the tangent space at a to the fibre $\pi^{-1}(b)$ of the submersion π is the kernel $\mathbf{ker}\, T_a\pi$; it is called the *vertical space* $\mathcal{V}_a$ of π at a; in the case of a vector bundle, it therefore identifies to the vector space E_b; for an affine bundle, it is identified to the underlying vector space $\vec{E}_b$ of the fibre.

We can now characterise the smooth sections σ of the submersion π over an open subset U of B as submanifolds: they are the submanifolds W of $\pi^{-1}(U)$ that meet each fibre $\pi^{-1}(b)$ with $b \in U$ at a unique point a, such that the tangent space $T_a W$ is *horizontal*, i.e., a complement in $T_a E$ of the vertical space $\mathcal{V}_a$; in other words, $\pi|_W$ is a diffeomorphism of W onto U and the corresponding section σ is the composed map of $(\pi|_W)^{-1}$ and the inclusion $W \hookrightarrow \pi^{-1}(U)$.

Remarks In the case of the tangent bundle, one should therefore imagine the fibres $T_a M$ as being vertical, transversal to M (identified to the zero section). This somewhat contradicts the geometric intuition of submanifolds in $\mathbb{R}^d$, for which $T_a M$ lies along M, but one must understand that by identifying each $T_a M$ to the affine subspace so obtained, one gets a *very bad* representation of TM: in the case where M is a curve in $\mathbb{R}^3$, for example, the surface of $\mathbb{R}^3$ so obtained admits M

as a cuspidal line at points where the curve is "truly spatial", i.e., with nonegative curvature and torsion, even though these are the least singular points of the surface lying in M.

Similarly, the geodesics of a surface S in Euclidean space $\mathbb{R}^3$ are the parametrised curves γ with values in S whose acceleration $\gamma''(t)$ is *normal* to the surface for every t, whereas the second derivative $\ddot{\gamma}(t)$ is *horizontal* for the Levi-Civita connection (see the sequel). One has to get used to it...

Worse: the *rank* of a fibre bundle is the dimension of its fibre, i.e., the *corank* of its projection.

More Fibre Bundles The datum of a basis ("reference frame") $(e_1, \ldots, e_n)$ of a real vector space E is equivalent to that of the isomorphism $(x^1, \ldots, x^n) \mapsto x^1 e_1 + \cdots + x^n e_n$ of $\mathbb{R}^n$ onto E. An essential object, introduced (in a different language) by Élie Cartan, is the *frame bundle* of a manifold M of dimension n, whose fibre over $a \in M$ is the set of (linear) *isomorphisms* A_a of $\mathbb{R}^n$ onto $T_a M$; therefore, it is a dense open subset of the vector bundle over M (generalising TM) consisting of all $j_0^1 f \in J^1(\mathbb{R}^n, M)$, and obviously a fibre bundle whose typical fibre is the linear group $\mathrm{GL}_n(\mathbb{R})$ (L_n in Ehresmann's notation): this can be seen by restricting the natural charts $\Phi_{\mathrm{id}_{\mathbb{R}^n}, \varphi}$ of $J^1(\mathbb{R}^n, M)$.

This frame bundle, denoted by $\mathrm{Isom}(M \times \mathbb{R}^n, TM)$ in [12] (this is a little misleading, as it might make one believe that the sphere of dimension 2 is *parallelisable* in the sense given hereafter), is naturally endowed with the action $(B, A_a) \mapsto A_a \circ B^{-1}$ of $\mathrm{GL}_n(\mathbb{R})$, which is free and transitive in each fibre: this is expressed by calling it a *principal bundle* with *structural group* $\mathrm{GL}_n(\mathbb{R})$.

Ehresmann's "regular infinitesimal structures" are "principal subbundles of the frame bundle".

For example, the datum of a Riemannian metric on M (i.e., a scalar product in each tangent space $T_a M$, depending smoothly on a in the sense that the real function which to $v \in TM$ associates its scalar square is smooth) is equivalent to the datum of the subbundle of the frame bundle consisting of those A_a which map the canonical basis of $\mathbb{R}^n$ to an orthonormal basis for the scalar product in $T_a M$. This is a principal bundle whose structural group is the orthogonal group O_n, the *orthonormal frame bundle* of the Riemannian manifold. The scalar product on $T_a M$ is the image of the standard Euclidean scalar product on $\mathbb{R}^n$ by any of those "orthonormal frames" A_a.

Similarly, given a closed subgroup H of $\mathrm{GL}_n(\mathbb{R})$, the datum of a principal subbundle of the frame bundle, with structural group H, is equivalent to the datum, for each $a \in M$, of *one* of the frames A_a, the others being determined by the action of H. The "structure" preserved (or defined) by H is then transferred to $T_a M$ by any of the A_a's.

If one wishes frames A_a to depend smoothly on a, one must stay at the local level: otherwise, one would get an isomorphism of the trivial vector bundle $M \times \mathbb{R}^n$ onto TM, an isomorphism that does *not* exist [24] in the case of manifolds as respectable as the sphere of dimension 2: they are not *parallelisable*.

For each A_a, the n components of A_a^{-1} (coordinate functions in the frame A_a) are linear forms on $T_a M$; they constitute the "coframe" mentioned by Élie Cartan and Ehresmann; given a section of the frame bundle under study over the open U of M, i.e., for each $a \in U$, the choice of *one* frame A_a in the fibre, the components of $a \mapsto A_a^{-1}$ are therefore Pfaffian forms on U.

15.3 Pfaffian Systems and Systems of Partial Differential (in)Equations

The space $J^k(M, N)$ is not only a fibre bundle in many ways: for $k > 0$, it is also endowed with a canonical Pfaffian system, easy to understand when $M = \mathbb{R}^n$ and $N = \mathbb{R}^p$.

A section σ of the source projection of $J^k(\mathbb{R}^n, \mathbb{R}^p) = \mathbb{R}^n \times \prod_0^k L_s^j(\mathbb{R}^n, \mathbb{R}^p)$ over an open subset U of $\mathbb{R}^n$ is a map of U into $J^k(\mathbb{R}^n, \mathbb{R}^p)$ that writes $\sigma(x) = \big(x, y_0(x), \ldots, y_k(x)\big)$; clearly, it is holonomic (i.e. of the form $j^k f$) if and only if, modulo the canonical identification of $L\left(\mathbb{R}^n, L^j(\mathbb{R}^n, \mathbb{R}^p)\right)$ to $L^{j+1}(\mathbb{R}^n, \mathbb{R}^p)$ familiar in differential calculus, $Dy_j(x) = y_{j+1}(x)$ for $0 \le j < k$ for all $x \in U$.

Let us express this viewing σ as the submanifold $W = \sigma(U)$: if one writes $z = (x, y_0, \ldots, y_k)$ the points of $J^k := J^k(\mathbb{R}^n, \mathbb{R}^p)$, the section is holonomic if and only if, at every point z of W, the tangent space $T_z W$ (in other words, the image of $D\sigma(x)$) is contained in the subspace $\mathcal{K}_z^k = \mathcal{K}_z^k(\mathbb{R}^n, \mathbb{R}^p)$ of $T_z J^k \simeq J^k$ defined by the equations

$$dy_j = y_{j+1}\, dx \quad \text{pour} \quad 0 \le j < k, \tag{15.1}$$

i.e. consisting of those vectors $\delta z = (\delta x, \delta y_0, \ldots, \delta y_k)$ such that, modulo the canonical identification just mentioned, $\delta y_j = y_{j+1}\, \delta x$ for $0 \le j < k$; here, $y_{j+1}\, \delta x$ is the interior product ("contraction") of y_{j+1} by δx, i.e., the symmetric j-linear map $(\delta x_1, \ldots, \delta x_j) \mapsto y_{j+1}(\delta x, \delta x_1, \ldots, \delta x_j)$.

One calls (15.1) the *canonical Pfaffian system* or *Cartan system* (or *canonical contact structure*) of $J^k(\mathbb{R}^n, \mathbb{R}^p)$; equivalently, one can give the same name to the field of vector subspaces ("plane field") $z \mapsto \mathcal{K}_z^k$, that can be seen geometrically as the sub-vector bundle $\mathcal{K}^k = \mathcal{K}^k(\mathbb{R}^n, \mathbb{R}^p)$ of $T J^k \simeq J^k \times J^k$ union of the subsets $\{z\} \times \mathcal{K}_z^k$.

One can see that, for each $z \in J^k$, the "plane" $\mathcal{K}_z^k$ is the closure[2] of the union of all $T_z W$ when W varies among the holonomic sections through z; using the natural charts, this yields the following fact: given now two manifolds M and N, one defines a *Pfaffian system* $\mathcal{K}^k(M, N)$ on $J^k(M, N)$, i.e. a sub-vector bundle of the tangent bundle $T J^k(M, N)$, by the fact that its fibre over $z \in J^k(M, N)$ is the closure in

---

[2]One has to "catch" also the vertical vectors for the projection onto J^{k-1}.

$T_z J^k(M, N)$ of the union of the tangent spaces at z to holonomic sections through z. Naturally,

- it is called the *canonical Pfaffian system* or *Cartan system* (or *canonical contact structure*) of $J^k(M, N)$,
- one has $T_z\Phi\big(\mathcal{K}_z(M, N)\big) = \mathcal{K}_{\Phi(z)}(\mathbb{R}^n, \mathbb{R}^p)$ for every natural chart Φ of $J^k(M, N)$ and every jet $z \in \mathbf{dom}\,\Phi$, implying that $\mathcal{K}^k(M, N)$ is indeed a sub-vector bundle of $TJ^k(M, N)$.

The reader would have understood that a *Pfaffian system* on a manifold V can be defined as a sub-vector bundle $\mathcal{P}$ of the tangent bundle TV.

In "real life", we are going to see that the notion can be more complicated: the manifold V may have singular points, the dimension of the fibre $\mathcal{P}_z$ may vary at some points $z \in V$, etc.

An *integral manifold* of $\mathcal{P}$ is a submanifold W of V verifying $T_zW \subset \mathcal{P}_z$ for every $z \in W$; in this langage, a section of the source projection of $J^k(M, N)$ is holonomic if and only if, seen as a submanifold, it is an integral manifold of the Cartan system—which admits other integral manifolds, for example the fibres of the projection onto $J^{k-1}(M, N)$.

Example If $\dim N = 1$, the Cartan system $\mathcal{K}^1(M, N)$ is a field of hyperplanes, authentic contact structure in today's restrictive sense, and its integral manifolds of dimension n are called *Legendre submanifolds*, a terminology due to V.I. Arnold. In particular, (15.1) consists of one equation, and the Pfaffian form $\alpha = dy_0 - y_1\,dx$ on $J^1(\mathbb{R}^n, \mathbb{R})$ is a *contact form*, meaning that $d\alpha_z$ induces a nondegenerate bilinear form on $\mathcal{K}_z^1 = \mathbf{ker}\,\alpha_z$; according to a theorem of Darboux [8], up to diffeomorphism, all contact forms in dimension $2n + 1$ are *locally* equal to α.

Systems of Partial Differential Equations A system of q partial differential equations of degree k in p unknown functions of n variables is written in a condensed way as $F(j_x^k y) = 0$, where F is a map of an open subset of $J^k(\mathbb{R}^n, \mathbb{R}^p)$ into $\mathbb{R}^q$, the variable is $x \in \mathbb{R}^n$ and the unknown function y (with values in $\mathbb{R}^p$). A solution f of the system defined in an open subset of $\mathbb{R}^n$ is identified to $j^k f$, i.e. to a holonomic section of the source projection $J^k(\mathbb{R}^n, \mathbb{R}^p) \to \mathbb{R}^n$ over U that takes its values in $E = F^{-1}(0)$ or, in other words, to an integral manifold of the canonical contact structure contained in E and projecting diffeomorphically onto U.

A system of partial differential equations therefore identifies to a Pfaffian system, provided the name is given to the pair consisting of (15.1) and of the equation $F(z) = 0$. To use our first definition, one should take as a manifold V the smooth part of E (for analytic F, this makes sense) and as a Pfaffian system $\mathcal{P}_z := \mathcal{K}_z \cap T_zV$, a "fibre bundle" whose rank may have an unfortunate propension to jump (for example, if $k = n = p = q = 1$, it may well happen that $\mathcal{K}_z = T_zV$ at some points, which should be excluded from V if one is looking for a genuine sub-vector bundle).

Of course, all this extends to the case where E is a submanifold of codimension q of $J^k(M, N)$, not necessarily defined globally by q real equations.

For $k = p = q = 1$, it is fruitful to first forget the projection $J^1 \to J^0$ and consider the "geometric solutions" of the equation, i.e. the Legendre submanifolds contained in E, whether they are or not sections of the source projection. They sometimes have a physical meaning: for example, caustics are the projections into J^0 of such geometric solutions. This case, whose local theory goes back to the nineteenth century, still gives rise to new global developments.

Systems of Partial Differential Inequations The spaces of jets also serve as the framework of the homotopy principle or h–principle [18], introduced by Gromov (following Thom [25]) in his thesis as an astounding abstraction of Smale's classification of immersions. The idea is dual to what has just been done: in the case of immersions of a manifold M into a manifold N, one considers in $J^1(M, N)$ the open subset Ω consisting of jets of immersions, i.e. $j^1_a f$ such that $T_a f$ is injective. Given two immersions f_0, f_1 of M into N, the question is whether they are *regularly homotopic*, i.e. whether there exists a smooth path $[0, 1] \ni t \mapsto f_t$ joining them *in the space of immersions*; in other words, one wonders whether there exists a path of holonomic sections $j^1 f_t$ of $J^1(M, N) \to M$ joining $j^1 f_0$ to $j^1 f_1$ and such that all these sections take their values in Ω. Naturally, the same problem can be posed for various subsets Ω of various $J^k(M, N)$'s; the *homotopy principle* (when it is true) states that the question admits a positive answer if and only if this is the case *forgetting the contact structure but not the source projection*, meaning that one can join the two holonomic sections by a path in the set of not necessarily holonomic sections with values in Ω. With time, this has become astonishingly simple [15], back to Thom in fact (see Laudenbach's comment of [25] in [26]).

15.4 Connections

Here again, Ehresmann did a good job. The problem is that a submersion $\begin{array}{c} E \\ \downarrow \pi \\ B \end{array}$ does not allow even locally the unique lifting of paths, except when it is a local diffeomorphism at every point (in which case, if it is a fibration, one calls it a *covering*): if $\tilde{\varphi}$ is a fibred chart of π, with image $U \times V$, over a chart φ of B, then, for every path γ with values in $\mathbf{dom}\,\varphi$, any path $\tilde{\gamma}$ with values in $\mathbf{dom}\,\tilde{\varphi}$ of the form $\tilde{\gamma}(t) = \tilde{\varphi}^{-1}\big(\varphi \circ \gamma(t), f(t)\big)$ with $\mathbf{dom}\,\tilde{\gamma} = \mathbf{dom}\,\gamma$ is a *lift(ing)* of γ, meaning that $\pi \circ \tilde{\gamma} = \gamma$; therefore, even if one imposes to $\tilde{\gamma}$ a given value $a \in \pi^{-1}\big(\gamma(t_0)\big)$ for $t = t_0$, there are many possible choices f, none of which is a priori better than the others. The datum of a connection suppresses this indeterminacy and provides (at least locally) a unique lifting $\tilde{\gamma}$ of γ such that $\tilde{\gamma}(t_0) = a$.

For example, if E is the frame bundle of B (or a principal subbundle), a connection allows one to obtain along γ a *moving frame* $\tilde{\gamma}(t)$, well determined by its value at t_0. If the connection is better than the others, so will be this moving frame.

Definition A *connection* on the submersion π is a *field of horizontal spaces*, i.e., a Pfaffian system $\mathcal{H}$ on E such that $\mathcal{H}_a$ is, for every $a \in E$, a complementary subspace in $T_a E$ of the vertical space $\mathcal{V}_a = \mathbf{ker}\, T_a\pi = T_a\left(\pi^{-1}(a)\right)$; in other words, $T_a\pi|_{\mathcal{H}_a}$ is an isomorphism onto $T_{\pi(a)}B$.

The datum of $\mathcal{H}_a$ is equivalent to that of the projection of $T_a E$ onto $\mathcal{V}_a$ parallel to $\mathcal{H}_a$, used by Dieudonné [12] to define a connection; it can be denoted $\mathbf{v} \mapsto \mathbf{v}_V$ (*vertical component* of the tangent vector $\mathbf{v}$). The unique lifting ("horizontal lifting") $\tilde{\gamma}$ announced will be defined by the initial condition and by the fact that the derivative $\dot{\tilde{\gamma}}(t)$ is *horizontal* for every t, which writes (notation of [23])

$$\frac{D\tilde{\gamma}}{dt} := \dot{\tilde{\gamma}}(t)_V = 0. \tag{15.2}$$

Indeed, the connection $\mathcal{H}$ "reads" as follows in a fibred chart $\tilde{\varphi}$ of E over φ, with image the open $U \times V$ of $\mathbb{R}^n \times \mathbb{R}^r$: for every $a \in \mathbf{dom}\,\tilde{\varphi}$, if $\tilde{\varphi}(a) = (x, y)$, the image of $\mathcal{H}_a$ by $T_a\tilde{\varphi}$ is the graph of a linear map $-\Gamma(x, y)$ of $\mathbb{R}^n$ into $\mathbb{R}^r$: this defines the *Christoffel map* $\Gamma : U \times V \to L(\mathbb{R}^n, \mathbb{R}^r)$ of the connection $\mathcal{H}$ in the fibred chart $\tilde{\varphi}$, and it is smooth because $\mathcal{H}$ is; the equation $\mathbf{v}_V = 0$ expressing that $\mathbf{v} \in TE$ is horizontal therefore writes $\delta y + \Gamma(x, y)\delta x = 0$, where $\big((x, y), (\delta x, \delta y)\big) = T\tilde{\varphi}(\mathbf{v})$. Hence, if γ is a path in $\mathbf{dom}\,\varphi$ and $x(t) := \varphi \circ \gamma(t)$, a lifting $\tilde{\gamma}(t) = \tilde{\varphi}^{-1}\big(x(t), y(t)\big)$ of γ with values in $\mathbf{dom}\,\tilde{\varphi}$ is horizontal if and only if the path $t \mapsto y(t)$ verifies the differential equation

$$y'(t) + \Gamma\big(x(t), y(t)\big)x'(t) = 0$$

expressing (15.2); this enables one to use Cauchy's theorem on differential equations to obtain the local existence and uniqueness of the lifting $\tilde{\gamma}$ taking a given value at time t_0. Its *global* existence is ensured for example when π is *proper*, i.e., when $\pi^{-1}(K)$ is compact for every compact K of B: indeed, in that case, the solution $\tilde{\gamma}$ of (15.2) can not "go to infinity" at time $t \in \mathbf{dom}\,\gamma$. Let us deduce from this a fundamental result in differential topology:

Theorem (Ehresmann) *If the submersion π is proper, then it is a fibration.*

Proof For every $b \in B$, there exist an open subset $\Omega \ni b$ of B and a connection $\mathcal{H}$ on $\pi|_{\pi^{-1}(\Omega)}$: to see it, cover the compact manifold $\pi^{-1}(b)$ by the domains of finitely many fibred charts $\tilde{\varphi}_j$ and take $\Omega = \bigcap \mathbf{dom}\,\varphi_j$, where the φ_j's are the charts of B defined by the $\tilde{\varphi}_j$'s; restricting the $\tilde{\varphi}_j$'s, we may assume $\mathbf{dom}\,\varphi_j = \Omega$ for every j, so that the $\mathbf{dom}\,\tilde{\varphi}_j$'s form a finite cover of $\pi^{-1}(\Omega)$ and that there exists [11] a smooth partition of unity θ_j subordinate to this cover; for each j, there is a connection $\mathcal{H}_j$ on $\pi|_{\mathbf{dom}\,\tilde{\varphi}_j}$, for example that whose Christoffel map in the fibred chart $\tilde{\varphi}_j$ is identically zero; denoting by $\mathbf{v} \mapsto \mathbf{v}_{j,V}$ the corresponding projection, one can then take the connection $\mathcal{H}$ whose projection $T_a E \to \mathcal{V}_a$ is defined by $\mathbf{v}_V := \sum_j \theta_j(a)\mathbf{v}_{j,V}$ for each $a \in \pi^{-1}(\Omega)$ (as usual, the sum is on those j's such that $a \in \mathbf{dom}\,\tilde{\varphi}_j$).

Restricting Ω, one may assume that there exists a chart φ of B with $\mathbf{dom}\,\varphi = \Omega$ such that $\varphi(\Omega)$ is an open ball of centre $0 = \varphi(b)$ in $\mathbb{R}^n$. Thus, for $y \in \Omega$, one defines a path $\gamma_y : [0,1] \to \Omega$ joining y to b by $\gamma_y(t) := \varphi^{-1}\big((1-t)\varphi(y)\big)$; for all $x \in \pi^{-1}(y)$, the path γ_y admits a unique horizontal lift $\tilde{\gamma}_x : [0,1] \to E$ such that $\tilde{\gamma}_x(0) = x$, and the map $x \mapsto \tilde{\gamma}_x(1)$ of $\pi^{-1}(y)$ in $\pi^{-1}(b)$, called *parallel transport from time* 0 *to time* 1 *along the path* γ_y *for the connection* $\mathcal{H}$, is obviously bijective (its inverse is obtained by lifting $t \mapsto \gamma_y(1-t)$); as solutions of differential equations depend smoothly on initial conditions and parameters, the map $x \mapsto \tilde{\gamma}_x(1)$ is a diffeomorphism, and so is the map h of $\pi^{-1}(\Omega)$ onto $\Omega \times \pi^{-1}(b)$ given by $h(x) := \big(\pi(x), \tilde{\gamma}_x(1)\big)$, that is the required local trivialisation.

Remarks Conversely, a fibration with compact fibres is obviously proper. As in the definition of a fibration, if one wants the typical fibre to be unique up to diffeomorphism, then B must be assumed connected.

This very robust theorem holds, with the same proof, in the Banach framework. Proceeding as in the first part of the proof, one can see that a submersion defined on a paracompact manifold (as in real life) admits a connection, which can be used in the second part of the proof, Ω being the domain of any chart φ vanishing at b whose image is a ball.

An Example The contact structure $\mathcal{K}^1(M,\mathbb{R})$ is a connection for the fibration $\pi : j_a^1 f \mapsto d_a f$ of $J^1(M,\mathbb{R})$ onto T^*M. We shall return to it in the section on curvature.

15.5 Integral of Differential Forms, Pullbacks, Exterior Derivative

Direct Images of Paths, Curvilinear Integral, Pullback of Functions and 1-Forms Let g be a smooth map of a manifold M into a manifold N. The (direct) *image* under g of a path γ in M is the path $g_*\gamma := g \circ \gamma$ in N; similarly, the *inverse image* (or *pullback*) by g of a real function f on N is the real function $g^* f := f \circ g$ on M.

When γ is defined on a segment $[t_0, t_1]$ (γ is then called an *arc*), the (curvilinear) *integral along* γ *of a Pfaffian form* α on M is by definition

$$\int_\gamma \alpha := \int_{t_0}^{t_1} \alpha_{\gamma(t)}\big(\dot{\gamma}(t)\big)\, dt,$$

where $\alpha_{\gamma(t)} \in T^*_{\gamma(t)}M = (T_{\gamma(t)}M)^*$ denotes the value of α at $\gamma(t)$. This integral is invariant under parameter changes : if $\varphi : [s_0, s_1] \to [t_0, t_1]$ verifies $\varphi(s_j) = t_j$, then $\int_{\gamma\circ\varphi} \alpha = \int_\gamma \alpha$; when α is the differential df of a real function f on M, since

$df_{\gamma(t)}\big(\dot\gamma(t)\big) = (f\circ\gamma)'(t),$

$$\int_\gamma df = f\big(\gamma(t_1)\big) - f\big(\gamma(t_0)\big) \quad \text{(mean value formula).} \qquad (15.3)$$

A Pfaffian form α is determined by the integrals $\int_\gamma \alpha$.

Indeed, for every $x \in M$ and every $\mathbf{v} \in T_xM$, there exists an arc $\gamma : [0, 1] \to M$ such that $\dot\gamma(0) = \mathbf{v}$ (take a chart φ of M such that $\varphi(a) = 0$ and a path of the form $\varphi\circ\gamma(t) = \theta(t\,\varphi_*\mathbf{v})\,t\,\varphi_*\mathbf{v}$, where $\varphi_*\mathbf{v} = T_x\varphi(\mathbf{v})$ and $\theta : \mathbf{im}\,\varphi \to [0, 1]$ is C^∞ with compact support, equal to 1 near 0). If $\gamma_\varepsilon : [0, 1] \to M$ is given by $\gamma_\varepsilon(t) := \gamma(\varepsilon t)$, then $\lim_{\varepsilon\to 0} \varepsilon^{-1}\int_{\gamma_\varepsilon}\alpha = \lim_{\varepsilon\to 0}\int_0^1 \alpha_{\gamma(\varepsilon t)}\big(\dot\gamma(\varepsilon t)\big)\,dt = \alpha_{\gamma(0)}\big(\dot\gamma(0)\big) = \alpha_x\mathbf{v}$.

The *pullback by g of a Pfaffian form β* on N is the Pfaffian form $g^*\beta$ on M such that $\int_\gamma g^*\beta = \int_{g_*\gamma}\beta$ for every arc γ in M; it is given by the formula

$$(g^*\beta)_x = \beta_{g(x)} \circ T_xg.$$

For $f : M \to \mathbb{R}$, the chain rule, in intrinsic terms

$$T(f\circ g) = (Tf)\circ Tg,$$

therefore writes $g^*df = d(g^*f)$.

Differential Forms, Their Integral on Parametrised Rectangles and Their Pullbacks A *differential form of degree k* or *differential k-form*, or *k-form* α on a manifold M is a field of *alternate* k-linear forms $\alpha_x : (T_xM)^k \to \mathbb{R}$, i.e., a smooth section of the vector bundle $\bigwedge^k T^*M$ over M whose fibre over $x \in M$ is the space $L^k_{\text{alt}}(T_xM, \mathbb{R})$ of alternate k-linear forms on T_xM; an atlas of this vector bundle consists (naturally) of the *natural charts* $\bigwedge^k T^*\varphi : \alpha_x \mapsto \big(\varphi(x), (T_x\varphi)_*\,\alpha_x\big) \in \mathbf{im}\,\varphi \times L^k_{\text{alt}}(\mathbb{R}^n, \mathbb{R})$, where φ is a chart of M with values in $\mathbb{R}^n$ (the tangent linear map $T_x\varphi$ therefore maps T_xM onto $T_{\varphi(x)}\mathbb{R}^n = \mathbb{R}^n$), $\alpha_x \in L^k_{\text{alt}}(T_xM, \mathbb{R})$ and $(T_x\varphi)_*\,\alpha_x(\mathbf{v}_1, \ldots, \mathbf{v}_k) := \alpha_x\big((T_x\varphi)^{-1}\mathbf{v}_1, \ldots, (T_x\varphi)^{-1}\mathbf{v}_k\big)$ for $\mathbf{v}_1, \ldots, \mathbf{v}_k \in \mathbb{R}^n$.

For every smooth map $\rho : [0, 1]^k \to M$, the *integral* of α along the *parametrised rectangle* ρ of dimension k is by definition

$$\int_\rho \alpha := \int_{[0,1]^k} \alpha_{\rho(t)}\big(\partial_1\rho(t), \ldots, \partial_k\rho(t)\big)\,dt$$

(integral with respect to Lebesgue measure), where $\partial_j\rho(t) \in T_{\rho(t)}M$ is the partial derivative of ρ with respect to the jth factor and $\alpha_{\rho(t)} \in L^k_{\text{alt}}(T_{\rho(t)}M, \mathbb{R})$ denotes the value of α at $\rho(t)$.

A k-form α is determined by the integrals $\int_\rho \alpha$.

Indeed, for $x \in M$ and $\mathbf{v}_1, \ldots, \mathbf{v}_k \in T_xM$, there exists (same proof as for $k = 1$, replacing $t\varphi_*\mathbf{v}$ by $\sum t_j\varphi_*\mathbf{v}_j$) a parametrised rectangle $\rho : [0, 1]^k \to M$ such that $\partial_j\rho(0) = \mathbf{v}_j$ for every j; if

$\rho_\varepsilon : [0, 1]^k \to M$ is given for $0 < \varepsilon \leq 1$ by $\rho_\varepsilon(t) := \rho(\varepsilon t)$, then $\lim_{\varepsilon \to 0} \varepsilon^{-k} \int_{\rho_\varepsilon} \alpha = \alpha_x(\mathbf{v}_1, \ldots, \mathbf{v}_k)$ as for $k = 1$.

Given a smooth map $g : M \to N$ between manifolds, the *pullback by g of a k-form β on N* is the k-form $g^*\beta$ on M such that $\int_\rho g^*\beta = \int_{g_*\rho} \beta$ for every parametrised rectangle ρ of dimension k in M, using the notation $g_*\rho := g \circ \rho$; it is given by the formula

$$(g^*\beta)_x = \beta_{g(x)} \circ (T_x g)^k,$$

where $(T_x g)^k(\mathbf{v}_1, \ldots, \mathbf{v}_k) := (T_x g(\mathbf{v}_1), \ldots, T_x g(\mathbf{v}_k))$ for $\mathbf{v}_1, \ldots, \mathbf{v}_k \in T_x M$.

The Exterior Derivative That of a Pfaffian form α on M is the 2-form $d\alpha$ on M such that

$$\int_\rho d\alpha = \int_{\partial\rho} \alpha \tag{15.4}$$

for every C^2 parametrised rectangle $\rho : [0, 1]^2 \to M$, where $\partial\rho$ is the *oriented boundary* of ρ, obtained by concatenation of the paths $[0, 1] \ni s \mapsto \rho(s, 0)$, $[0, 1] \ni s \mapsto \rho(1, s)$, $[0, 1] \ni s \mapsto \rho(1 - s, 1)$ and $[0, 1] \ni s \mapsto \rho(0, 1 - s)$; it is given par

$$d\alpha_{\rho(t)}\big(\partial_1\rho(t), \partial_2\rho(t)\big) = \partial_1\big(\alpha_{\rho(t)}\partial_2\rho(t)\big) - \partial_2\big(\alpha_{\rho(t)}\partial_1\rho(t)\big). \tag{15.5}$$

More generally, for each $k \geq 1$, the *exterior derivative of a k-form α* on M is the $(k + 1)$-form $d\alpha$ on M verifying (15.4) for every parametrised rectangle ρ of dimension $k + 1$, setting

$$\int_{\partial\rho} \alpha := \sum_{i=1}^{k+1} (-1)^{i+1} \Big(\int_{\partial\rho_i^1} \alpha - \int_{\partial\rho_i^0} \alpha \Big),$$

where the "faces" $\partial\rho_i^j$ of ρ are the parametrised rectangles of dimension k defined by

$$\partial\rho_i^j(s) := \rho\big((s_\ell)_{\ell<i}, j, (s_\ell)_{\ell\geq i}\big), \ s = (s_1, \ldots, s_k) \in [0, 1]^k, \ j = 0, 1;$$

the identity (15.5) is the particular case $k = 1$ of the formula

$$d\alpha_{\rho(t)}\big(\partial_1\rho(t), \ldots, \partial_{k+1}\rho(t)\big) = \sum_{i=1}^{k+1} (-1)^{i+1} \partial_i \left(\alpha_{\rho(t)}\Big(\big(\partial_\ell\rho(t)\big)_{\ell<i}, \big(\partial_\ell\rho(t)\big)_{\ell>i} \Big) \right), \tag{15.6}$$

valid when ρ is a C^2 map with values in M defined on an open subset or an "open subset with corners" of $\mathbb{R}^k$, for example $[0, 1]^k$.

This formula follows from (15.4), the mean value formula and the Fubini theorem. indeed, if one alleviates notation by setting for example $\alpha\big(\partial_1\rho(t), \dots, \partial_k\rho(t)\big) := \alpha_{\rho(t)}\big(\partial_1\rho(t), \dots, \partial_k\rho(t)\big)$, then

$$\int_{\partial\rho_i^1}\alpha - \int_{\partial\rho_i^0}\alpha = \int_{[0,1]^k}\Big(\alpha\Big(\partial_{s_j}\rho\big((s_\ell)_{\ell<i}, 1, (s_\ell)_{\ell\geq i}\big)\Big)_{1\leq j\leq k} - \alpha\Big(\partial_{s_j}\rho\big((s_\ell)_{\ell<i}, 0, (s_\ell)_{\ell\geq i}\big)\Big)_{1\leq j\leq k}\Big)\,ds$$

$$= \int_{[0,1]^k}\int_0^1 \partial_\tau\alpha\Big(\partial_{s_j}\rho\big((s_\ell)_{\ell<i}, \tau, (s_\ell)_{\ell\geq i}\big)\Big)_{1\leq j\leq k}\,d\tau\,ds$$

$$= \int_{[0,1]^{k+1}} \partial_i\alpha\Big(\big(\partial_\ell\rho(t)\big)_{\ell<i}, \big(\partial_\ell\rho(t)\big)_{\ell>i}\Big)\,dt,$$

where $t := \big((s_\ell)_{\ell<i}, \tau, (s_\ell)_{\ell\geq i}\big)$. Naturally, the "miracle" is that the right-hand side of (15.6) depends only on the $\partial_j\rho(t)$'s: this can be checked in a chart, which reduces the problem to the case where M is an open subset U of $\mathbb{R}^n$, and using the fact that, then, $\partial_i\partial_\ell\rho = \partial_\ell\partial_i\rho$. Indeed, in that case, α identifies to its second component $U \to L^{k+1}_{\text{alt}}(\mathbb{R}^n, \mathbb{R})$ and $d\alpha : U \to L^{k+1}_{\text{alt}}(\mathbb{R}^n, \mathbb{R})$ is given by $d\alpha(x)(\mathbf{v}_1, \dots, \mathbf{v}_{k+1}) = \sum_{i=1}^{k+1}(-1)^{i+1}D\alpha(x)(\mathbf{v}_i)\big((\mathbf{v}_\ell)_{\ell<i}, (\mathbf{v}_\ell)_{\ell>i}\big)$, $x \in U$, $\mathbf{v}_1, \dots, \mathbf{v}_{k+1} \in \mathbb{R}^n$.

This definition of the exterior derivative is not too intrinsic, but it shows that a k-form is meant to be integrated on objects of dimension k, exterior derivation appearing as the dual ("coboundary") of the "oriented boundary" ∂ via the Stokes formula (15.4)—which generalises (15.3) and yields easily the other "Stokes formulae". Whitney even *constructed* the theory of differential forms out of it [29].

It follows at once from the definitions of the pullback and the exterior derivative that

$$d(g^*\beta) = g^*d\beta \tag{15.7}$$

for every smooth map $g : M \to N$ between manifolds and every differential form β on N.

Moreover, for every differential k-form α on M,

$$dd\alpha = 0. \tag{15.8}$$

Indeed, the integral of $dd\alpha$ on every parametrised rectangle $\rho : [0, 1]^{k+2} \to M$ is zero since by definition $\displaystyle\int_\rho dd\alpha = \sum_{i=1}^{k+2}(-1)^{i+1}\Big(\int_{\partial\rho_i^1}d\alpha - \int_{\partial\rho_i^0}d\alpha\Big) = \sum_{i=1}^{k+2}(-1)^{i+1}\Big(\int_{\partial\partial\rho_i^1}\alpha - \int_{\partial\partial\rho_i^0}\alpha\Big)$, in other words

$$\int_\rho dd\alpha = \sum_{i=1}^{k+2}(-1)^{i+1}\sum_{j=1}^{k+1}(-1)^{j+1}\Big(\int_{\partial(\partial\rho_i^1)_j^1}\alpha - \int_{\partial(\partial\rho_i^1)_j^0}\alpha - \int_{\partial(\partial\rho_i^0)_j^1}\alpha + \int_{\partial(\partial\rho_i^0)_j^0}\alpha\Big),$$

a sum where "each face of dimension k of ρ appears twice and with opposite signs" as

$$\partial(\partial\rho_i^\ell)_j^m = \partial(\partial\rho_j^m)_{i-1}^\ell, \quad 1 \leq j < i \leq k+2, \quad \ell, m \in \{0, 1\}.$$

If $k = 1$, these faces correspond to the edges of the cube $[0, 1]^3$.

A differential form β is *closed* when $d\beta = 0$; it is *exact* when it is the exterior derivative $\beta = d\alpha$ of a differential form, called a *primitive* of β and obviously unique up to the addition of a closed form (when one adds two sections α and β of a vector bundle E, it is of course fibrewise addition, i.e., $(\alpha + \beta)(x) = \alpha(x) + \beta(x)$ in E_x); the formula (15.8) therefore means that *every exact form is closed.*

15.6 Flows, Lie Derivative and Lie Bracket

Flows and Lie Derivative To every smooth vector field X on the manifold M is associated its *flow* or *one-parameter* (pseudo)*group* g_X^t, defined as follows: for every $a \in M$, the map $t \mapsto g_X^t(a)$ is the path in M that is the *maximal* solution of the differential equation $\dot{x} = X(x)$ ("integral curve of X ") passing through a at time $t = 0$. Here, *maximal* means "defined on an *interval* as large as possible."

Note As the integral curves are *parametrised*, they are not merely one-dimensional integral manifolds.

By the theory of differential equations, the domain of $g_X : (t, a) \mapsto g_X^t(a)$ is an open subset of $\mathbb{R} \times M$ and g_X is as smooth as X; clearly, $g_X^s\big(g_X^t(a)\big) = g_X^{s+t}(a)$ when the left-hand side makes sense or, equivalently, for $a \in \mathbf{dom}(g_X^t) \cap \mathbf{dom}(g_X^{s+t})$; in particular, since $g_X^0 = \mathrm{id}_M$, each g_X^t is a diffeomorphism of the open subset $\mathbf{dom}\, g_X^t \subset M$ onto the open subset $\mathbf{dom}\, g_X^{-t}$, and $(g_X^t)^{-1} = g_X^{-t}$.

If X has compact support, the solutions of $\dot{x} = X(x)$ cannot "go to infinity in finite time"; therefore, $\mathbf{dom}\, g_X = \mathbb{R} \times M$ and g_X is a smooth action of the additive group $\mathbb{R}$ on M, meaning that $t \mapsto g_X^t$ is a homomorphism of $\mathbb{R}$ into the group of diffeomorphisms of M onto itself; in that case, X (or its flow) is said to be *complete.*

The *Lie derivative* of a tensor field τ on M (here, a differential form of degree k or, as a little further, a vector field) with respect to X is by definition

$$\mathcal{L}_X\tau := \frac{d}{dt} g_X^{t\,*}\tau\Big|_{t=0}, \tag{15.9}$$

that is a tensor field of the same nature as τ; for example, the Lie derivative of a real function f on M is the real function on M which is the (interior) *product* or *contraction* $df(X)$ of df by X:

$$\mathcal{L}_X f = df(X) : x \mapsto d_x f(X_x).$$

For $k > 0$, the Lie derivative of a differential k-form α on M verifies the *Cartan formula*

$$\mathcal{L}_X \alpha = d(\alpha X) + (d\alpha)X, \tag{15.10}$$

where αX and $(d\alpha)X$ denote the *interior products* (or *contractions*) $x \mapsto \alpha_x X_x$ and $x \mapsto (d\alpha_x)X_x$ of α and $d\alpha$ by X, a notation introduced when we wrote the Cartan system of $J^k(\mathbb{R}^n, \mathbb{R}^p)$.

Though Élie Cartan undoubtfully knew and used the Cartan formula [4], it took some time for the Lie derivative—as for many primitive notions—to be recognised as such and it is Henri Cartan who wrote (15.10) under this form. One can, if one really wants to, take it as an intrinsic but incomprehensible definition of the exterior derivative.

Its proof is very easy: for all $x \in M$ and $(\mathbf{v}_1, \dots, \mathbf{v}_k) \in T_x M$, there exists $\rho_1 : (\mathbb{R}^k, 0) \to (M, x)$ such that $\mathbf{v}_j = \partial_j \rho_1(0)$ for $1 \le j \le k$, and one can take $\rho(t) := g_X^{t_1} \circ \rho_1(t_2, \dots, t_{k+1})$ and $t = 0$ in (15.6). Here is an important application:

Poincaré Lemma *Every closed differential form α of degree $k \geq 1$ on M is* locally *exact: each $a \in M$ has an open neighbourhood Ω such that $\alpha|_\Omega$ is exact.*

Indeed, if Ω is the domain of a chart φ vanishing at a whose image is a ball B of $\mathbb{R}^n$, let X be the vector field on Ω that is the pullback by φ of the radial field $Y_y := y$ on B; for every $x \in \Omega$, the points $g_X^t(x) = \varphi^{-1}\big(e^t \varphi(x)\big)$ with $t \le 0$ are well defined and, by (15.10), since $d\alpha = 0$,

$$\alpha_x = (g_X^{0*}\alpha)_x = (g_X^{0*}\alpha)_x - \lim_{t \to -\infty} (g_X^{t*}\alpha)_x = \int_{-\infty}^0 \frac{d}{dt}(g_X^{t*}\alpha)_x \, dt = \int_{-\infty}^0 (g_X^{t*}\mathcal{L}_X\alpha)_x \, dt$$

$$= \int_{-\infty}^0 \left(g_X^{t*} d(\alpha X)\right)_x dt = \int_{-\infty}^0 d\left(g_X^{t*}(\alpha X)\right)_x dt = \left(d \int_{-\infty}^0 g_X^{t*}(\alpha X)\, dt\right)_x,$$

where the last integral is in each fibre (one can find it more secure to work in the chart φ and take as variable $s = e^t$).

The de Rham Cohomology For $k > 0$, the quotient of the vector space of closed forms of degree k on M by the vector space of exact forms of degree k is the kth de Rham cohomology space $H^k(M, \mathbb{R})$; as every alternate k-linear form on a space of dimension $< k$ is zero, $H^k(M, \mathbb{R}) = \{0\}$ for $k > \dim M$; one denotes by $H^0(M, \mathbb{R})$ the space of locally constant functions on M and $H^\bullet(M, \mathbb{R}) := \bigoplus_{k \ge 0} H^k(M, \mathbb{R})$.

Pullback of Vector Fields, Lie Brackets Given a smooth map $h : M \to N$ between manifolds, a *pullback* of a vector field Y on N by h, if it exists, is a vector field X on M such that h "maps the integral curves of X onto those of Y", meaning that $h \circ g_X^t = g_Y^t \circ h$; as this relation holds for $t = 0$, it is equivalent to the one obtained by differentiating it with respect to time, which yields $T_x h(X_x) = Y_{h(x)}$ for every $x \in X$; one therefore sees that if h is *etale*, i.e., if all the $T_x h$'s are isomorphisms, then Y has a unique pullback by h, denoted by h^*Y and given by the formula

$$(h^*Y)_x = (T_x h)^{-1} Y_{h(x)}.$$

The formula (15.9) therefore has a meaning when τ is a vector field Y on M, and

$$\mathcal{L}_X Y = [X, Y]$$

is the *Lie bracket of the vector fields X and Y*, such that

$$\mathcal{L}_{[X,Y]} f = \mathcal{L}_X \mathcal{L}_Y f - \mathcal{L}_Y \mathcal{L}_X f$$

for every real function f on M (as $\mathcal{L}_X \mathcal{L}_Y f = \mathcal{L}_{\mathcal{L}_X Y} f + \mathcal{L}_Y \mathcal{L}_X f$ by "derivation of a product").

The *Jacobi identity* $\big[[X, Y], Z\big] + \big[[Y, Z], X\big] + \big[[Z, X], Y\big] = 0$ follows, making the C^∞ vector fields on M an archetypical Lie algebra.

By the formula for the derivation of a product and (15.9), for every choice of the function f, the tensor field τ, the vector fields X, Y and the differential form α of degree $k > 0$ on M, one has

$$\begin{aligned} \mathcal{L}_X(f\tau) &= (\mathcal{L}_X f)\tau + f\mathcal{L}_X\tau \\ \mathcal{L}_X(\alpha Y) &= (\mathcal{L}_X\alpha)Y + \alpha\mathcal{L}_X Y \\ &= d(\alpha X)Y + (d\alpha)XY + \alpha[X, Y]. \end{aligned} \tag{15.11}$$

If φ is a chart of M with values in $\mathbb{R}^n$, setting $X_\varphi(x) := T_x\varphi(X_{\varphi^{-1}(x)}) \in T_x\mathbb{R}^n = \mathbb{R}^n$ for every vector field X on M and every $x \in \mathbf{im}\,\varphi$, one has

$$[X, Y]_\varphi(x) = DY_\varphi(x)X_\varphi(x) - DX_\varphi(x)Y_\varphi(x). \tag{15.12}$$

15.7 Some Applications of the Cartan Formula

Infinitesimal Contact Transformations Let α be a *contact form* on a manifold V—recall that this means that $T_x M = \mathbf{ker}\,\alpha_x \oplus \mathbf{ker}\,d\alpha_x$ for every $x \in V$; let $\mathcal{K}$ be the associated *contact structure* $\mathcal{K}_x := \mathbf{ker}\,\alpha_x$. An *infinitesimal contact transformation* or *Lie field* for $\mathcal{K}$ is a vector field X on V whose flow $g^t := g^t_X$ preserves $\mathcal{K}$, meaning that $T_x g^t(\mathcal{K}_x) = \mathcal{K}_{g^t(x)}$ for every $(t, x) \in \mathbf{dom}\,g_X$: this is expressed by calling the maps g^t *contact transformations* or (local) *automorphisms* of $\mathcal{K}$.

Theorem (Libermann) *Under these hypotheses, a Lie field X is determined by its* Hamiltonian $-\alpha X$ *with respect to α, and every C^2 real function F on V is the Hamiltonian of a C^1 Lie field X_F. In particular, if α is C^∞, the map $F \mapsto X_F$ is an isomorphism of $C^\infty(V, \mathbb{R})$ onto the space of C^∞ Lie fields for $\mathcal{K}$, an isomorphism whose inverse is $X \mapsto -\alpha X$.*

Indeed, X is a Lie field if and only if its flow g^t verifies $(g^{t*}\alpha)_x = \mu_t(x)\alpha_x$ for every $x \in \mathbf{dom}\,g^t$, which (after derivation with respect to t) writes $\mathcal{L}_X\alpha = \lambda\alpha$, where λ is a real function on V; by

(15.10), the relations between X and $F := -\alpha X$ are therefore expressed for each $x \in V$ by the two equations

$$-\alpha_x X_x = F(x) \tag{15.13}$$

$$-d_x F + d\alpha_x X_x = \lambda(x)\alpha_x; \tag{15.14}$$

if $X_x = Y_x + Z_x$ in the decomposition $T_x V = \mathcal{K}_x \oplus \mathbf{ker}\, d\alpha_x$, (15.13) determines Z_x knowing $F(x)$ and vice versa since $\alpha_x|_{\mathbf{ker}\, d\alpha_x}$ is an isomorphism; as for (15.14), it may be written as

$$-d_x F|_{\mathbf{ker}\, d\alpha_x} = \lambda(x)\alpha_x|_{\mathbf{ker}\, d\alpha_x}$$

$$d_x F|_{\mathcal{K}_x} = (d\alpha_x\, Y_x)|_{\mathcal{K}_x};$$

the first equation determines $\lambda(x)$ knowing $d_x F|_{\mathbf{ker}\, d\alpha_x}$ and vice versa, and the second yields Y_x knowing $d_x F|_{\mathcal{K}_x}$ and vice versa, as the nondegenerate bilinear form $d\alpha_x|_{(\mathcal{K}_x)^2}$ induces the isomorphism $\mathbf{v} \mapsto (d\alpha_x \mathbf{v})|_{\mathcal{K}_x}$ of $\mathcal{K}_x$ onto its dual.

Having always [9] attributed this result to Sophus Lie, I nearly asked who was that Bermann the first time it was rightly [20] credited to Paulette Libermann in my presence. The result implies that the group of automorphisms of $\mathcal{K}$ is huge, the vector fields X_F with F compactly supported being complete.

Application: Local Theory of First Order Partial Differential Equations Under these hypotheses, given $F : V \to \mathbb{R}$, let $E := F^{-1}(0)$. Two preliminary observations:

(i) as there is no nondegenerate alternating bilinear form on a space of odd dimension, V is of odd dimension $2n + 1$;
(ii) an integral manifold W of $\mathcal{K}$ is of dimension at most n; indeed, if $\iota : W \hookrightarrow V$ is the inclusion, the relation $\iota^*\alpha = 0$ expressing that W is integral implies that $\iota^* d\alpha = d(\iota^*\alpha) = 0$, i.e., that each tangent space $T_x W$ is included in its orthogonal for the nondegenerate bilinear form $d\alpha_x|_{(\mathcal{K}_x)^2}$, hence the inequality $\dim T_x W \leq 2n - \dim T_x W$; the integral manifolds of dimension n are the *Legendre manifolds* of $\mathcal{K}$.

For every $x \in E$,

(iii) the previous proof shows that $X = X_F$ vanishes at x if $d_x F = 0$, since then $Y_x = Z_x = 0$;
(iv) it follows from (15.13)–(15.14) and the antisymmetry of $d\alpha_x$ that $d_x F(X_x) = 0$; hence, X_F is tangent at x to E for $d_x F \neq 0$ (F is a submersion in an open neighbourhood U of x, therefore $U \cap E$ is a submanifold of codimension 1 with tangent space $\mathbf{ker}\, d_x F$ at x);
(v) it follows from (15.13) that X_x belongs to $\mathcal{K}_x$.

Assertions (iii)–(iv) imply that one has $g_X^t(E \cap \mathbf{dom}\, g_X^t) \subset E$ for every t; assertion (v), together with the fact that the maps g_X^t preserve $\mathcal{K}$, therefore yields the following facts:

(vi) for every integral manifold $W_0 \subset E$ of $\mathcal{K}$ and every $a \in W_0$ with $X_a \notin T_a W_0$, there exists an open subset $\Omega \ni (0, a)$ of $\mathbb{R} \times W_0$ such that the map $j : \Omega \to E$

defined by $j(t,x) := g_X^t(x)$ is a diffeomorphism onto an integral manifold W of $\mathcal{K}$, which therefore verifies $\dim W = \dim W_0 + 1$;

(vii) this imposes $\dim W_0 < n$ by (ii); hence, a *geometric solution* of the *generalised partial differential equation* E, i.e., a Legendre manifold L contained in E, verifies $X_x \in T_x L$ for every $x \in L$;

(viii) if $\dim W_0 = n - 1$ (one then calls (E, W_0) a *generalised Cauchy problem, well-posed at* a), then W is a geometric solution of E;

(ix) conversely, by (vii), every geometric solution W of E is obtained in this fashion in the neighbourhood of each $a \in W$ where X_a is nonzero (just take for W_0 a hypersurface of W passing through a with $X_a \notin T_a W_0$); this proves the local existence and uniqueness of the solution of a generalised Cauchy problem.

If $V = J^1(\mathbb{R}^n, \mathbb{R})$, $\mathcal{K} = \mathcal{K}^1(\mathbb{R}^n, \mathbb{R})$ and, denoting by $(t, x) \in \mathbb{R} \times \mathbb{R}^{n-1}$ the points of $\mathbb{R}^n$, the equation E is of the form $\partial_t y = g(t, x, y, \partial_x y)$, a well-posed classical Cauchy problem is the datum of the value $y_0(x)$ of the unknown function for $t = 0$; this does determine the generalised Cauchy datum given by $W_0 = \big\{\big(0, x, y_0(x), g(0, j_x^1 y_0), Dy_0(x)\big)\big\} \subset E$, which defines at each of its points a well-posed problem whose local generalised solutions W are holonomic sections of the source projection $J^1(\mathbb{R}^n, \mathbb{R}) \to \mathbb{R}^n$ contained in E, jets of order 1 of the local solutions of the Cauchy problem.

Hamiltonian Vector Fields on a Symplectic Manifold Paulette Libermann is not foreign [20] to their intrinsic definition. A symplectic manifold is the pair consisting of a manifold V and a *symplectic form* on V, i.e., a *closed* 2-form ω such that every $\omega_x \in L^2_{\text{alt}}(T_x V, \mathbb{R})$ is *nondegenerate* (the dimension of V must therefore be even). A vector field X on V is *symplectic* when its flow $g^t = g_X^t$ preserves ω, meaning that $g^{t\,*}\omega = \omega$ in $\mathbf{dom}\, g^t$ for every t (the maps g^t are therefore *symplectic transformations* of ω). As this relation is verified if $t = 0$, this amounts to saying that $0 = \frac{d}{dt} g^{t\,*}\omega = g^{t*}\mathcal{L}_X\omega$ for every t, i.e., that $\mathcal{L}_X\omega = 0$; since ω is closed, it follows from (15.10) that this is the case if and only if the Pfaffian form ωX is *closed.*

When it is exact, $\omega X = dH$, one says that X is *Hamiltonian* and that the function H is a *Hamiltonian* of X; il determines X, and each real function H on V is the Hamiltonian of a unique Hamiltonian vector field X_H: indeed, for each $x \in V$, the equation $\omega_x \mathbf{v} = d_x H$ has a unique solution $\mathbf{v} \in T_x V$ since ω_x is nondegenerate. The group of (global) symplectic transformations of ω therefore is huge too, since it contains the maps $g_{X_H}^t$ with H compactly supported.

As $\mathcal{L}_{X_H} H = dH(X_H) = \omega(X_H, X_H) = 0$, the flow of $X = X_H$ preserves H, meaning that $H\big(g_X^t(x)\big) = H(x)$ for every $(t, x) \in \mathbf{dom}\, g_X$ ("conservation of energy"); one also calls H a *first integral* of X_H. Since $\mathcal{L}_{X_H} K = dK(X_H) = \omega(X_K, X_H) = -\mathcal{L}_{X_K} H$ for all real functions H and K on V, the *Poisson bracket* $\{H, K\} := \mathcal{L}_{X_H} K$ ("Poisson parentheses") is antisymmetric; this yields the (trivial but quite useful) Hamiltonian version of a theorem by Emmy Noether: if "X_K is an infinitesimal symmetry of H", meaning that H is a first integral of X_K, then K is a first integral of X_H. The Poisson bracket lifts to functions the Lie bracket of vector fields in the sense that $X_{\{H,K\}} = [X_H, X_K]$; it (therefore) satisfies the Jacobi identity, endowing $C^\infty(V, \mathbb{R})$ with a Lie algebra structure if ω is C^∞.

Similarly, if α is a contact form, the Lie bracket of Lie fields can be lifted to real functions by (the inverse of) the isomorphism $X \mapsto \alpha X$, the bracket so obtained being called the *Lagrange bracket*, it seems.

In the "concrete" case, studied since Lagrange at least [21], V is the cotangent bundle ("phase space") T^*M of a manifold ("configuration space") M, endowed with its *canonical symplectic structure* ω_M, unique 2-form on T^*M whose pullback by the projection $J^1(M, \mathbb{R}) \to T^*M$ is the exterior derivative of the canonical contact form $dy_0 - y_1\, dx$ defining $\mathcal{K}^1(M, \mathbb{R})$.

15.8 Curvature

Curvature of a Connection If $\mathcal{H}$ is a connection on a submersion $\begin{smallmatrix} E \\ \downarrow \pi, \\ B \end{smallmatrix}$ every vector field X on an open subset $U \subset B$ lifts to a unique *horizontal* vector field $\tilde{X}$ on $\pi^{-1}(U)$, given by $\tilde{X}_a = (T_a\pi|_{\mathcal{H}_a})^{-1} X_{\pi(a)}$. A remarkable fact of Nature is that, if Y is another vector field on U, the vertical component $([\tilde{X}, \tilde{Y}]_a)_V$ of the Lie bracket $[\tilde{X}, \tilde{Y}]$, at each point $a \in \pi(U)$, depends only on $\tilde{X}_a, \tilde{Y}_a \in \mathcal{H}_a$, i.e., on $X_{\pi(a)}, Y_{\pi(a)} \in T_{\pi(a)}B$; hence one can define an alternate bilinear map $R_a : T_{\pi(a)}B \times T_{\pi(a)}B \to \mathcal{V}_a$, the *curvature tensor of* $\mathcal{H}$ *at* a, by the formula

$$R_a(X_{\pi(a)}, Y_{\pi(a)}) := ([\tilde{X}, \tilde{Y}]_a)_V. \tag{15.15}$$

If Γ is the Christoffel map of $\mathcal{H}$ in a fibred chart $\tilde{\varphi}$ of π over the chart φ of B, it follows from (15.12) that

$$R_a(\mathbf{v}_1, \mathbf{v}_2) = D\Gamma(z)\big(\mathbf{x}_2, -\Gamma(z)\mathbf{x}_2\big)\mathbf{x}_1 - D\Gamma(z)\big(\mathbf{x}_1, -\Gamma(z)\mathbf{x}_1\big)\mathbf{x}_2\,, \quad \text{where} \quad \begin{cases} z := \tilde{\varphi}(a) \\ \mathbf{x}_j := T_{\pi(a)}\varphi(\mathbf{v}_j), \end{cases} \tag{15.16}$$

which proves our "fact of Nature" (see the next paragraph for a nicer argument).

When E is a vector bundle over B, the identification of $\mathcal{V}_a$ to the fibre $E_{\pi(a)}$ makes R_a into an element of $L^2_{\text{alt}}(T_{\pi(a)}B, E_{\pi(a)})$; in particular, if $E = TB$, one is in the perhaps more familiar situation where R_a takes its values in $T_{\pi(a)}B$.

For a general vector bundle, when $\mathcal{H}$ is *linear*, i.e., when the parallel transport from one time to another along any path is (which amounts to saying that the Christoffel maps $\Gamma(x, y)$ in the charts of the vector bundle are linear in y), it follows from (15.16) that the curvature R_a depends linearly on a viewed as an element of $E_{\pi(a)}$; setting $b = \pi(a)$, $R_a(\mathbf{v}, \mathbf{w})$ therefore is the value at $(a, \mathbf{v}, \mathbf{w}) \in E_b \times T_bB \times T_bB$ of a *trilinear* map R_b with values in E_b; if $E = TB$, the familiar monster of Riemannian geometry [23] is the *quadrilinear* form $(T_bB)^4 \ni (a, \mathbf{v}, \mathbf{w}, \mathbf{h}) \mapsto R_b(a, \mathbf{v}, \mathbf{w}) \cdot \mathbf{h}$ (scalar product).

If E is an affine bundle, R_a takes its values in the vector space $\vec{E}_{\pi(a)}$ underlying the fibre. More generally, when E is a principal bundle with structural group G, the datum of a enables one to identify $E_{\pi(a)}$ to G by the inverse of the bijection $G \ni g \mapsto ga$, and therefore identify $\mathcal{V}_a$ to the *Lie algebra* of G (the tangent space $\mathfrak{g} := T_1G$ of G at 1) by the inverse of the differential at 1 of the previous bijection; in this identification, one therefore has $R_a \in L^2_{\text{alt}}(T_{\pi(a)}B, \mathfrak{g})$.

"Curvature" of a Pfaffian System If $\mathcal{P}$ is a Pfaffian system on a manifold V (that is, a sub-vector bundle of the tangent bundle TV, the stupid cases of TV and its zero section being excluded), one can replace in the previous construction the "concrete" vertical space $\mathcal{V}_a$ by its "abstract" version

$$\nu\mathcal{P}_a := T_aV/\mathcal{P}_a$$

(which defines a vector bundle $\nu\mathcal{P}$ over V, the *normal bundle* of $\mathcal{P}$) and denote by $\mathbf{v} \mapsto \mathbf{v}_\nu$ the canonical projection $T_aV \to \nu\mathcal{P}_a$. The previous fact of Nature generalises: one defines the *"curvature tensor" $R_a \in L^2_{\text{alt}}(\mathcal{P}_a, \nu\mathcal{P}_a)$ of the Pfaffian system $\mathcal{P}$ at $a \in V$* by the formula

$$R_a(X_a, Y_a) := ([X, Y]_a)_\nu, \tag{15.17}$$

where X, Y vary among the *sections of the vector bundle $\mathcal{P}$* over open subsets $U \ni a$ of V (vector fields on U verifying $X_x, Y_x \in \mathcal{P}_x$ or, equivalently, $(X_x)_\nu = (Y_x)_\nu = 0$ for every x).

To prove our "fact of Nature", one can consider locally $\mathcal{P}$ as a connection (see the proof of the Frobenius theorem hereafter) and use (15.16) or, in a more elegant way, remark that if one multiplies for example Y by a real function f defined near a, (15.11) yields $[X, fY]_a = f(a)[X, Y]_a + \mathcal{L}_x f(a)Y_a$ and therefore $([X, fY]_a)_\nu = f(a)([X, Y]_a)_\nu$ since $(Y_a)_\nu = 0$, hence $([X, fY]_a)_\nu = ([X, Y]_a)_\nu$ if $f(a) = 1$.

Proposition *For every integral manifold W of $\mathcal{P}$, the curvature tensor R_a is identically zero on $T_aW \times T_aW$ for all $a \in W$.*

Indeed, if X, Y are vector fields on a neighbourhood of a in W, it is easy to extend them locally to sections $\bar{X}, \bar{Y}$ of $\mathcal{P}$ defined in the neighbourhood of a in V; by definition, $\bar{X}_a = X_a, \bar{Y}_a = Y_a$ and, moreover, $[\bar{X}, \bar{Y}]_a = [X, Y]_a \in T_aW \subset \mathcal{P}_a$ since, near a, the flow of $\bar{X}$ coincides on W with that of X. It follows that $R_a(X_a, Y_a) = R_a(\bar{X}_a, \bar{Y}_a) = ([\bar{X}, \bar{Y}]_a)_\nu = ([X, Y]_a)_\nu = 0$, hence the proposition since (X_a, Y_a) can be any pair of vectors tangent to W at a.

Definition An integral element of $\mathcal{P}$ at $a \in V$ is a plausible candidate to be the tangent space at a of an integral manifold of $\mathcal{P}$, i.e., a vector subspace I_a of $\mathcal{P}_a$ such that $R_a|_{I_a \times I_a} = 0$.

The *Cartan-Kähler theorem* for Pfaffian systems [1, 7, 12, 22] asserts that, in the analytic case, every "generic" integral element I_a of $\mathcal{P}$ is indeed of the form $I_a = T_a W$ for at least one (analytic) integral manifold W of $\mathcal{P}$. This statement is more Cartan than Kähler [2]; it is astounding that Élie Cartan, from three examples, could have the idea of so general a result and *see* how to "corner" the required integral manifold. Here are two extreme examples where this general result is not needed.

Example 1: Completely Integrable Pfaffian Systems They are those Pfaffian system $\mathcal{P}$ such that $R_a = 0$ for every $a \in V$ (in other words, $\mathcal{P}_a$ is an integral element). For example, the Pfaffian system $\mathcal{V}$ defined by the vertical spaces of a submersion is completely integrable (and completely integrated, the fibres being integral manifolds). A completely integrable connection is sometimes said to be *flat* since its curvature zero everywhere.

Frobenius theorem *If a Pfaffian system $\mathcal{P}$ on V is completely integrable, there does exist, for every $a \in V$, an integral manifold W of $\mathcal{P}$ such that $T_a W = \mathcal{P}_a$* (hence, for dimensional reasons, $T_x W = \mathcal{P}_x$ for every $x \in W$ if W is connected)*; moreover, this integral manifold is* locally unique*: if W' is another one, there exists an open neighbourhood U of a in V such that $W \cap U = W' \cap U$* (in words, W and W' have the same *germ*[3] at a).

Hence, the relation "there exists a connected integral manifold of $\mathcal{P}$ containing a and a'" between points a, a' of V is an equivalence relation, whose equivalence classes are called the *leaves* of the *foliation* of V defined by $\mathcal{P}$; they inherit from their definition a structure of connected manifold (injectively immersed) of the same dimension as the $\mathcal{P}_a$'s, but they are not (embedded) submanifolds in general. Even for $\dim \mathcal{P}_a = 1$ ("line field", always completely integrable since the R_a's are alternate), the global study of foliations is a very difficult subject to which, after Ehresmann and Reeb, contributed Haefliger, Bott, Novikov, Thurston among others and, in the case of line fields, all the great names of dynamical systems since Poincaré. Indeed, the theory includes the study of the *orbits* of a vector field X on V (considering the line field $x \mapsto \mathbb{R}X_x$ on the open subset of V where X does not vanish), which are the images of its integral curves.

Local structure of the foliation defined by a completely integrable Pfaffian system For every $a \in V$, there exist open subsets $U \subset \mathbb{R}^n$, $U' \subset \mathbb{R}^p$ and a chart ("plaque family") ψ of V with $a \in \mathbf{dom}\,\psi$ and $\mathbf{im}\,\psi = U \times U'$ such that the leaves of the foliation of $\mathbf{dom}\,\psi$ defined by $\mathcal{P}$ are the subsets $\psi^{-1}(U \times \{y_0\})$ with $y_0 \in U'$; each of these local leaves ("plaques") is obviously contained in one of the leaves of the global foliation, but this global leaf can come back and cut $\mathbf{dom}\,\psi$ following other plaques, whose union can even be dense in $\mathbf{dom}\,\psi$: for example, if α is an irrational

[3] In the beginning, Ehresmann used the word *jet*, little recommendable in this case except in the analytic framework.

number, all the orbits of the constant vector field $X_x := (1, \alpha) \in \mathbb{R}^2 = T_x\mathbb{T}^2$ on the torus $\mathbb{T}^2 = \mathbb{R}^2/\mathbb{Z}^2$ are dense.

Proof à la Dieudonné [10] of the Frobenius theorem and of the existence of plaque families Let φ be an arbitrary chart of V at a; composing it with a translation and a permutation of coordinates, one can assume that it takes its values in $\mathbb{R}^n \times \mathbb{R}^p$, that $\varphi(a) = 0$ and that $T_a\varphi(\mathcal{P}_a)$ is horizontal, i.e., complementary of the vertical space $\{0\} \times \mathbb{R}^p$ of the projection $\pi : (x, y) \mapsto x$. Restricting $\mathbf{dom}\,\varphi$, it follows that *all* the spaces $\mathcal{H}_{\varphi(z)} := T_z\varphi(\mathcal{P}_z)$ are horizontal; therefore, there exists a Christoffel map $\Gamma : \mathbf{im}\,\varphi \to L(\mathbb{R}^n, \mathbb{R}^p)$, such that $\mathcal{H}_{(x,y)}$ is the graph of $-\Gamma(x, y)$ for every $(x, y) \in \mathbf{im}\,\varphi$. The integral manifolds of maximal dimension of $\mathcal{P}$ in $\mathbf{dom}\,\varphi$ are the images by φ^{-1} of those of the connection $\mathcal{H}$ so defined, which integral manifolds are *locally* the graphs of solutions $y = f(x)$ of the "total differential equation"

$$\frac{dy}{dx} + \Gamma(x, y) = 0; \tag{15.18}$$

if such a solution f takes the value y_0 at 0, then, for every $x \in \mathbb{R}^n$ such that the segment $[0, x]$ is contained in $\mathbf{dom}\,f$, it follows that $f(tx)$ is for $0 \le t \le 1$ the value $R^t(x, y_0)$ at time t of the solution of the differential equation $\dfrac{dy}{dt} + \Gamma(tx, y)x = 0$ equal to y_0 at $t = 0$. As $R^t(x, y_0)$ exists for every t if $x = 0$, the theory of differential equations [8] tells us that there are open balls $U \subset \mathbb{R}^n$ and $U' \subset \mathbb{R}^p$ centred at 0 such that, for $x \in U$, the map $y_0 \mapsto R^1(x, y_0)$ is a diffeomorphism of U' onto an open subset of $\mathbb{R}^p$; in other words, $h : (x, y_0) \mapsto \big(x, R^1(x, y_0)\big)$ is a diffeomorphism of $U \times U'$ onto an open subset of $U \times \mathbb{R}^p$.

We now just have to check that the unique candidate $f : x \mapsto R^1(x, y_0)$ for every $y_0 \in U'$, to be in U the solution of (15.18) equal to y_0 for $x = 0$ is indeed a solution of (15.18): one will get the plaque family $\psi := h^{-1} \circ \varphi$ and, for $y_0 = 0$, the Frobenius theorem. Now, differentiating with respect to x the identity $\frac{\partial}{\partial t} f(tx) + \Gamma\big(tx, f(tx)\big)x = 0$ and using (15.16), one can see that $t \mapsto tDf(tx)$ and $t \mapsto -t\Gamma\big(tx, f(tx)\big)$ verify the same differential equation on $[0, 1]$ and take the same value $0 \in L(\mathbb{R}^n, \mathbb{R}^p)$ at $t = 0$; therefore, they are equal, hence the required result for $t = 1$.

Remarks For line fields, this is just the theory of "time dependent" differential equations. The construction performed in general (before the final verification, which uses curvature) is a local version of the proof of Ehresmann's theorem. The vanishing of curvature is imposed by the symmetry of the second derivative of solutions of (15.18). Dieudonné's proof works in infinite dimensions as well.

Example 2: Fields of Hyperplanes and Contact Structures If α is a nowhere vanishing Pfaffian form on V and $\mathcal{K}_z := \mathbf{ker}\,\alpha_z$, the curvature at z of the Pfaffian system $\mathcal{K}$ identifies to $-d\alpha_z|_{\mathcal{K}_z}$ by the isomorphism of $\nu\mathcal{K}_z = T_zV/\mathcal{K}_z$ onto $\mathbb{R}$ induced by α_z.

Indeed, for all local sections X, Y of the vector bundle $\mathcal{K}$ in the neighbourhood of z, one has $\alpha X = \alpha Y = 0$, and therefore $\alpha_z[X_z, Y_z] = -d\alpha_z(X_z, Y_z)$ by (15.11).

A contact structure therefore is “completely non integrable”, its curvature being at every point a nondegenerate bilinear form.

The canonical contact structure $\mathcal{K} = \mathcal{K}^1(M, \mathbb{R})$ of $J^1(M, \mathbb{R})$ is a connection on the trivial fibre bundle $J^1(M, \mathbb{R}) = T^*M \times \mathbb{R}$ over T^*M; therefore, it has an intrinsic “Christoffel map”: denoting the points of T^*M by $x = (q, p)$ ($p \in T_q^*M$), as in mechanics, and by $z = (q, p, y)$ those of $J^1(M, \mathbb{R})$, each $\mathcal{K}_z$ is defined by the equation $dy = p\,dq$; hence, it is the graph of the linear form $p\,dq$ on $T_x(T^*M)$; the Pfaffian form $\lambda = \lambda_M$ on T^*M given by $\lambda_x = p\,dq$ is called the *Liouville form* of T^*M.

The curvature of the connection $\mathcal{K}^1(M, \mathbb{R})$ on the trivial fibre bundle $J^1(M, \mathbb{R}) = T^*M \times \mathbb{R}$ over T^*M identifies therefore to the 2-form $d\lambda_M$ on T^*M: one obtains again the canonical symplectic form $\omega_M = -d\lambda_M$ of T^*M.

Remarks To obtain Hamilton’s equations under their historical form, one has the choice between our sign conventions and those of [21], namely $\omega_M = d\lambda_M$ and $\omega_M X_H = -dH$.

Every etale map g between open subsets of M lifts to the map T^*g of $T^*\,\mathbf{dom}\,g$ onto $T^*\,\mathbf{im}\,g$ given by $T^*g(q, p) := \big(g(q), p \circ (T_q g)^{-1}\big)$, which is obviously symplectic (it preserves the Liouville form); if X is a vector field on M, each $T^*g_X^t$ is the time t of the flow of the *Hamiltonian* vector field with Hamiltonian $K(q, p) = pX_q$; the first integrals of classical mechanics obtained by applying the “Hamiltonian Noether theorem” are in general such K’s.

Given a Pfaffian system $\mathcal{P}$ on V, let $\mathcal{P}^\perp$ be the sub-vector bundle of T^*V whose fibre over x consists of those $\xi \in T_x^*V$ which vanish on $\mathcal{P}_x$. For each $a \in V$, there exist r sections $\alpha_1, \dots, \alpha_r$ of $\mathcal{P}^\perp$ over an open subset U containing a such that $\alpha(x) := \big(\alpha_1(x), \dots, \alpha_r(x)\big)$ is a basis of $\mathcal{P}_x^\perp$ for every $x \in U$; in other words, $\alpha(x)$ induces an isomorphism of $\nu\mathcal{P}_x$ onto $\mathbb{R}^r$ that, as for $r = 1$, identifies R_x to $-d\alpha(x)|_{\mathcal{P}_x^2} = -\big(d\alpha_1(x), \dots, d\alpha_r(x)\big)|_{\mathcal{P}_x^2} \in L_{\mathrm{alt}}(\mathcal{P}_x, \mathbb{R}^r)$.

It follows from Thom’s transversality lemma that “almost every” Pfaffian form on a manifold of odd dimension is a contact form off a smooth hypersurface, see for example [9]; likewise, the exterior derivative of “almost every” Pfaffian form on a manifold M of even dimension is symplectic off a hypersurface, necessarily nonempty if M is compact without boundary.

In contrast, it is clear that, apart from those defined by a submersion and line fields, completely integrable Pfaffian systems *almost never* occur. Why devote so much effort to such improbable objects? An answer is that they appear in a rather robust way (despite a certain loss of regularity under perturbations) in the case of the stable and unstable foliations of an Anosov diffeomorphism—hence, it seems, Novikov’s initial interest in the subject; another answer, very present in Élie Cartan’s work, is that the most symmetric objects often are the most beautiful and the most useful; here is an illustration, assuming some knowledge of de Rham cohomology:

The "Gauss-Manin" Connection Associated to a Proper Submersion, and Monodromy One can associate to every proper submersion $\begin{array}{c}E\\ \downarrow\pi\\ B\end{array}$ the vector bundle $H^\bullet E$ over B on $\mathbb{K} = \mathbb{R}$ or $\mathbb{C}$ whose fibre over b is the cohomology space $H^\bullet(E_b, \mathbb{K})$. To see that it is indeed a vector bundle endowed with a canonical flat linear connection $\mathcal{H}$, we are going to construct (assuming E paracompact...) a vector bundle atlas $\{\tilde{\varphi}\}_{\varphi\in\Phi}$ such that, denoting by $\mathcal{H}_\varphi$ the linear flat connection on $\mathbf{dom}\,\tilde{\varphi}$ whose Christoffel map in the chart $\tilde{\varphi}$ is[4] $\Gamma = 0$, the connections $\mathcal{H}_\varphi$ and $\mathcal{H}_\psi$, coincide on $\mathbf{dom}\,\tilde{\psi} \cap \mathbf{dom}\,\tilde{\varphi}$ for $\varphi, \psi \in \Phi$; therefore, the local connections $\mathcal{H}_\varphi$ do define a global flat linear connection $\mathcal{H}$ on $H^\bullet E$.

In this construction, Φ is the atlas of B consisting of those charts whose image is an open ball centred at 0 in $\mathbb{R}^n$. A connection on π being chosen, the proof of Ehresmann's theorem shows that there exists for each $\varphi \in \Phi$ a trivialisation $h_\varphi : \pi^{-1}(\mathbf{dom}\,\varphi) \to \mathbf{dom}\,\varphi \times F_\varphi$ of π over $\mathbf{dom}\,\varphi$; for $b \in \mathbf{dom}\,\varphi$, the canonical injection $i_b : E_b \hookrightarrow \pi^{-1}(\mathbf{dom}\,\varphi)$ induces an *isomorphism* i_b^* of $H^\bullet(\pi^{-1}(\mathbf{dom}\,\varphi), \mathbb{K})$ onto $H^\bullet(E_b, \mathbb{K})$, which "reads" modulo h_φ as the isomorphism j_b^* of $H^\bullet(\mathbf{dom}\,\varphi \times F_\varphi, \mathbb{K})$ onto $H^\bullet(\{b\} \times F_\varphi, \mathbb{K})$ defined by $j_b : \{b\} \times F_\varphi \hookrightarrow \mathbf{dom}\,\varphi \times F_\varphi$ [the inverse isomorphism is p_b^*, where $p_b(x, y) := (b, y)$, as every closed differential form α on $\mathbf{dom}\,\varphi \times F_\varphi$ such that $j_b^*\alpha = 0$ is exact: to see it, just apply our proof of Poincaré's lemma to the vector field X on $\mathbf{dom}\,\varphi \times F_\varphi$ whose image by $\varphi \times \mathrm{id}_{F_\varphi}$ admits the flow $(x, y) \mapsto \big(b + e^t(x - b), y\big)$]. One can therefore associate to φ the chart $\tilde{\varphi}$ of $H^\bullet E$ over φ, with image $\mathbf{im}\,\varphi \times H^\bullet(\pi^{-1}(\mathbf{dom}\,\varphi), \mathbb{K})$, given by $\tilde{\varphi}(b, c) := \big(\varphi(b), (i_b^*)^{-1}c\big)$, $c \in H^\bullet(E_b, \mathbb{K})$. Il is easy to check that one gets in this fashion the required vector bundle atlas and flat connection.

A subtle feature of the construction is that the fibre bundle $H^\bullet E$ and the connection are $\mathbb{K}$–analytic when π is, whereas the local trivialisations h_φ are *not*—they are obtained using partitions of unity.

For $b \in B$, parallel transport along each loop γ in B, with base point b, defines an automorphism of $H^\bullet E_b$ since the connection is linear; as it is flat, this automorphism depends only on the homotopy class of γ; this defines a homomorphism of the fundamental group $\pi_1(B, b)$ into the group of automorphisms of $H^\bullet E_b$, called *monodromy*.

Torsion, Levi-Civita Connection and Variants The *torsion* $\tau_a \in L_{\mathrm{alt}}(T_aM, T_aM)$ at $a \in M$ of a linear connection on a manifold M (i.e., on its tangent bundle) can be defined quickly as follows: for every parametrised surface $\sigma : (\mathbb{R}^2, 0) \to (M, a)$, one has $\tau_a\big(\partial_1\sigma(0), \partial_2\sigma(0)\big) = D_2\partial_1\sigma(0) - D_1\partial_2\sigma(0)$, where $D_1\partial_2\sigma(s,t) := \frac{D}{\partial s}\frac{\partial}{\partial t}\sigma(s,t)$ and $D_2\partial_1\sigma(s,t) := \frac{D}{\partial t}\frac{\partial}{\partial s}\sigma(s,t)$. For each Riemannian metric on M, there exists a unique linear connection *without torsion* ("symmetric") on M that is *Riemannian*, i.e., such that the parallel transport from time s at time t along any path γ in M is an isometry of $T_{\gamma(s)}M$ onto $T_{\gamma(t)}M$: it is called the *Levi-Civita*

---

[4]More simply, $\tilde{\varphi}$ is a plaque family of the foliation defined by $\mathcal{H}_\varphi$, which therefore is born "integrated"; by the way, Élie Cartan named *infinitesimal connection* what we call a connection; the problem is to "connect" two nearby fibres E_b, $E_{b'}$—for (infinitesimal) connections with nonzero curvature, however, the result depends, even locally, on the arc from b to b' along which parallel transport is taken.

connection. The absence of torsion allows for example an intrinsic proof of the fact that the critical points of the action functional $\frac{1}{2}\int_0^1 \|\dot{\gamma}(t)\|^2\,dt$ on the space of paths γ with fixed endpoints $\gamma(0)$, $\gamma(1)$ in M are the *geodesics*, solutions of the equation $\frac{D}{dt}\dot{\gamma}(t) = 0$.

Since parallel transport for the Levi-Civita connection preserves the scalar product, it induces a parallel transport of orthonormal frames of the tangent spaces of M, which is the parallel transport of a connection on the bundle of orthonormal frames; this connection is *principal*, meaning that parallel transport preserves the action of the structural group.

15.9 As a Conclusion

Of course, I have barely touched the subject, my only ambition being to provide some access to the ideas of Ehresmann and his master Élie Cartan. The work of the latter is not yet finished, as each generation tries to cast some light on it. A first rate contribution in that respect was Charles Ehresmann's introduction of fibre bundles, jets and connections, but also pseudogroups and groupoids, now again very popular [17, 27, 28] in spite of their ugly name (to say nothing of the horrible *algebroids*, direct from a bad science fiction film).

Typical examples The diffeomorphisms between open subsets of a manifold M form a pseudogroup, and even a groupoid if it is forbidden to compose two of them when the domain of the second is not exactly the image of the first; the germs at points of M of such local diffeomorphisms form a groupoid (one can compose a germ f at a and a germ g at b only if $b = f(a)$), and so do their jets of order k. When M is endowed with an additional structure, for example a Riemannian metric or a symplectic form or a contact structure, the (jets or germs of) local diffeomorphisms preserving this structure form a sub-pseudogroup or a sub-groupoid of the previous one. The Riemannian example of an otherwise round sphere with a bump in the neighbourhood of a point shows that this pseudogroup or groupoid can be rather irregular, well apt to detect local symmetries ignored by the group of global isometries of our sphere onto itself, in general trivial. A fundamental object in foliation theory is the holonomy groupoid generalising monodromy.

In the works of Lie or Élie Cartan, "groups" were quite often pseudogroups—which appear already when one considers the flow of a non-complete vector field (similarly, what plays the role of a one-parameter group for time-dependent vector fields is a "groupoid with two parameters", which shows that many scientists manipulate groupoids without being aware of it!). The emphasis on abstract groups, which, according to the dogma, act only on themselves until they are represented, partially rejected into darkness Lie's original *groups*, i.e., pseudogroups of transformations that cannot always be abstracted from the space on which they act [3, 5].

To Élie Cartan, as I said, one goes back all the time: for example, the algorithmic "equivalence method" in [16] is a recent avatar of his "equivalence problem" [6, 19]. The equivalence problem is to find criteria for two structures to be locally equivalent up to local coordinate changes; of course, the langage of manifolds, used throughout this article, is coordinate-free, so that "coordinate change" means "diffeomorphism" (true problems cannot depend on the choice of coordinates).

Similarly, his theory of involution goes on to inspiring Malgrange [22] after Kuranishi and many others [1, 12], such as Ehresmann, whose jets allow an intrinsic formulation of the prolongations of a differential system.

It should also be time to go back to Ehresmann before his beautifully concise texts become inaccessible; thus, my proof of his most famous theorem is the original one [13], so elliptic that many people replaced it by arguments far less elegant and natural. Science progresses to a large extent because its actors do not really understand the work of their predecessors and make it into something else, sometimes more interesting than the original, but there are limits...

References

1. R. L. Bryant, S. S. Chern, R. B. Gardner, H. L. Goldschmidt, P. A. Griffiths. *Exterior Differential Systems*, Mathematical Sciences Research Institute Publications, Springer-Verlag, Berlin, 1991
2. É. Cartan, Sur l'intégration des systèmes d'équations aux différentielles totales, *Ann. Sci. Éc. Norm. Sup.*, 3e série, **16** (1901), 241–311
3. ——, Sur la structure des groupes infinis de tranformations, *Ann. Sci. Ec. Norm. Sup.* **21** (1904), 153–206
4. ——, *Leçons sur les invariants intégraux.* Hermann, Paris, 1922
5. ——, *La structure des groupes infinis*, Séminaire de Math., 4e année, 1936–37, G, (polycopié)
6. ——, *Les problèmes d'équivalence*, Selecta, Gauthier-Villars, Paris, 1937, 113–136
7. ——, *Les systèmes différentiels extérieurs et leurs applications géométriques*, Hermann, Paris, 1948
8. M. Chaperon, *Calcul différentiel et calcul intégral*, deuxième édition. Dunod, Paris, 2008
9. ——, Géométrie différentielle et singularités desystèmes dynamiques. *Astérisque* **138–139**, 1986
10. J. Dieudonné, *Foundations of Modern Analysis.* Academic Press, New York, 1960
11. ——, *Éléments d'Analyse*, tome 3. Gauthier-Villars, Paris, 1970
12. ——, *Éléments d'Analyse*, tome 4. Gauthier-Villars, Paris, 1971
13. C. Ehresmann, *Les connexions infinitésimales dans un espace fibré différentiable*, Colloque Top. Bruxelles 1950, 29–
14. ——, Les prolongements d'une variété différentiable, *C. R. Acad. Sc. Paris* **233** (1951), pp. 598, 777 et 1081
    ——, Structures locales et structures infinitésimales, ibidem, **234**, 1952, p. 587
15. Y. Eliashberg, N. Mishachev, *Introduction to the h-principle.* Graduate Studies in Mathematics, **48**. American Mathematical Society, Providence, RI, 2002
16. R. B. Gardner, *The method of equivalence and its applications*, Society for Industrial and Applied Mathematics, Philadelphia, Pa., 1989
17. M. Golubitsky, I. Stewart, Nonlinear dynamics of networks: the groupoid formalism, *Bull. Amer. Math. Soc.* **43** (2006), 305–364
18. M. Gromov, *Partial differential relations.* Springer-Verlag, Berlin, 1986

19. P. Libermann, *Sur le problème de l'équivalence de certaines structures infinitésimales.* Thèse, Université de Strasbourg, 21 mai 1953. Ann. Mat. Pura Appl. **36**, 1954, pp. 27–120
20. ——, *Sur les automorphismes infinitésimaux des structures symplectiques et des structues de contact.* Colloque de géométrie différentielle globale, 1958, pp. 37–59, Gauthier-Villars, Paris, 1959
21. ——, C.-M. Marle, Géométrie symplectique, bases théoriques de la mécanique. *Publications Mathématiques de l'Université Paris VII*, **21**, 1986–87
English translation par B. E. Schwarzbach: *Symplectic geometry and analytical mechanics.* Mathematics and its Applications, **35**. D. Reidel Publishing Co., Dordrecht, 1987
22. B. Malgrange, *Systèmes différentiels involutifs*, Panoramas et Synthèses **19**, Société mathématique de France, 2005
23. J. Milnor, *Morse Theory.* Princeton University Press, 1963
24. ——, *Topology from the differentiable viewpoint.* University of Virginia Press, 1969
25. R. Thom, Remarques sur les problèmes comportant des inéquations différentielles globales, *Bull. Soc. Math. Fr.*, **87**, 1959, pp. 455–461.
26. ——, *Œuvres mathématiques complètes, Vol. 1*, Documents mathématiques, **15**, Soc. Math. Fr., 2017.
27. A. Weinstein, Symplectic groupoids and Poisson manifolds, *Bull. Amer. Math. Soc.* **16** (1987), 101–104
28. ——, Groupoids: unifying internal and external symmetry, *Notices Amer. Math. Soc.* **43** (1996), 744–752
29. H. Whitney, *Geometric integration theory*, Princeton University Press, 1957

Chapter 16
The Global Study of Riemannian-Finsler Geometry

Katsuhiro Shiohama and Bankteshwar Tiwari

To the Memory of Marcel Berger

Abstract The aim of this article is to present a comparative review of Riemannian and Finsler geometry. The structures of cut and conjugate loci on Riemannian manifolds have been discussed by many geometers including H. Busemann, M. Berger and W. Klingenberg. The key point in the study of Finsler manifolds is the non-symmetric property of its distance functions. We discuss fundamental results on the cut and conjugate loci of Finsler manifolds and note the differences between Riemannian and Finsler manifolds in these respects. The topological and differential structures on Riemannian manifolds, in the presence of convex functions, has been an active field of research in the second half of twentieth century. We discuss some results on Riemannian manifolds with convex functions and their recently proved analogues in the field of Finsler manifolds.

AMS Classification: 53C60, 53C22, 53C70, 51H25

16.1 Introduction

The origin of Finsler geometry can be traced back to Riemann's 1854 Habilitation address "Uber die Hypothesen, welche der Geometrie zu grunde liegen" (On the Hypotheses which lie at the Foundations of Geometry), where he remarked: '... *The*

K. Shiohama
Institute of Information Science, Fukuoka Institute of Technology, Fukuoka, Japan
e-mail: k-siohama@fit.ac.jp

B. Tiwari (✉)
Centre for Interdisciplinary Mathematical Sciences, Institute of Science, Banaras Hindu University, Varanasi, India

S. G. Dani, A. Papadopoulos (eds.), *Geometry in History*,
https://doi.org/10.1007/978-3-030-13609-3_16

next case in simplicity includes those manifoldness in which the line-element may be expressed as the fourth root of a quartic differential expression. The investigation of this more general kind would require no really different principles, but would take considerable time and throw little new light on the theory of space, especially as the results cannot be geometrically expressed, I restrict myself, therefore, to those manifoldness in which the line-element is expressed as the square root of a quadratic differential expression...', translation by William Kingdon Clifford [35]. Later on, the geometry where the metric is the square root of a quadratic differential form, got well recognized as Riemannian geometry. The general case was initiated by Paul Finsler in 1918 in his thesis written under the supervision of Carathéodory. It was said by S.S. Chern that *Finsler Geometry is just Riemannian Geometry without the Quadratic Restriction* [11]. In this article, we are interested in Global Finsler Geometry considered as an intrinsic metric geometry. We often refer to Riemannian geometry for our development of global Finsler geometry. One of the basic differences between Riemannian and Finsler geometry is the possible asymmetry of distance functions. It turns out that in certain contexts Finsler geometry is more natural than Riemannian geometry, and closer to real world. Here is an example. On a slope of the earth's surface we may consider the "distance" in terms of time taken to traverse it. Consider a person walking from the bottom of a hill to its top. In this context, the "distance" will be larger from the bottom to the top, than from the top to the bottom. This example has been emphasized by Herbert Busemann, one of the most prominent promoters of Finsler geometry. Busemann's collected works were published in a 2-volume set by Springer Verlag, see [7]. Later Makoto Matsumoto explicitly showed that such metric is actually a Finsler metric, see [28].

Let us be more specific. A Finsler metric on a smooth manifold is a smoothly varying family of Minkowski norms on the tangent spaces, rather than a family of inner products in the case of a Riemannian metric. It turns out that every Finsler metric induces an inner product, one in each direction of a tangent space at each point of the manifold. Thus, a Finsler metric associates to the manifold a family of inner products parametrized by the tangent spaces of the manifold (instead of being parametrized by the manifold, in the case of a Riemannian metric). However, the perpendicularity between two tangent vectors does not make sense on a Finsler manifold. Thus, it seems difficult to talk about the angle between two tangent vectors on such a manifold. In the mathematical literature, several kinds of connections were defined on a Finsler manifold. Some of the well-known connections were introduced by J.L. Synge, J. H. Taylor, L. Berwald, E. Cartan, H. Rund, H. Hashiguchi and S.S. Chern and others. In Riemannian geometry, the Levi-Civita connection is the canonical connection. It is torsion free and metrical. There is no connection in Finsler geometry which is both torsion free and metrical. There are different connections which have their own importance. The Chern connection is important from two points of view: firstly when the Finsler metric corresponds to a Riemannian metric, it reduces to the Levi-Civita connection, and secondly, it solves the problem of equivalence in Finsler geometry. This connection is torsion free but not metrical.

On Finsler manifolds, geometric objects are two-sided; viz., forward and backward, arising from the asymmetry of the distance function. The study of the cut locus and the conjugate locus of Riemannian and Finsler manifolds is important for the development of global Finsler geometry. In this article we give an overview of some aspects of global Riemannian geometry, developed in the very beginning of the last century, and of extensions of the Riemannian results on the cut locus and conjugate locus to Finsler manifolds. Among others, the cut locus is most important in the study of global Riemannian geometry. We discuss pointed Blaschke-Finsler manifolds in connection with the Rauch conjecture on the cut locus and the conjugate locus of a compact simply connected Riemannian manifold. It should be emphasized that *convex sets and convex functions defined on a Finsler manifold are independent of the non-symmetric property of the distance function*. Hence, the notion of convexity is common to both Riemannian and Finsler geometries.

The comparison theorems of Rauch, Berger and Toponogov play essential roles in the study of complete Riemannian manifolds of non-negative sectional curvature. However, we do not use these comparison theorems here in our study of Finsler manifolds. Following the ideas from Busemann [8], we discuss several topics on Finsler manifolds with non-symmetric distance functions. They are (1) the cut locus, (2) the conjugate locus and (3) convex sets, including the Whitehead convexity theorem, (4) convex functions, and (5) Busemann functions. We also discuss Busemann functions on both complete Riemannian and Finsler manifolds.

The article is organized as follows. Definitions and notation are set up in Sect. 16.2. The forward cut locus and the forward conjugate locus and their fundamental properties, including the classical Whitehead convexity theorem are discussed in Sect. 16.3. A detailed discussion on cut locus and conjugate locus, including the classical results due to Klingenberg and Berger, which are very important in this article, are developed in Sect. 16.4. We discuss in Sect. 16.4, the well-known Blaschke problem on compact Finsler manifolds in connection with the Rauch conjecture [34]. We discuss the simplest case of a pointed Blaschke manifold. Berger initiated the study of compact simply connected even-dimensional Riemannian manifolds of positive sectional curvature whose diameter is minimal [2, 3]. Omori [31] discussed compact manifolds with minimal diameter with real analytic metric. In Sect. 16.5, we discuss the properties of Busemann functions and convex functions on complete non-compact Riemannian and Finsler manifolds. Finally, we summarize Riemannian and Finsler results on convex functions. Some of these results have already been announced in [36] and [22]. For the basic tools in Riemannian and Finsler geometry we refer to [1, 4, 8, 9, 11, 12, 24, 37].

The authors would like to express their sincere thanks to Professor N. Innami, Professor C.S. Aravinda and Professor Athanase Papadopoulos for reading and giving their valuable comments that improved this article.

16.2 Definitions and Preliminaries

We first give the definitions of Riemannian and Finsler metrics on a smooth manifold and discuss an important relation between them. The other notions that we present in this section are concerned with the non-symmetric properties of the distance function.

16.2.1 Riemannian and Finsler Metrics

Let M be a smooth manifold of dimension $n \geq 2$ and at each point $x \in M$, let g_x be a dot product on the tangent space $T_x M$ to M. For smooth vector fields X, Y defined in a neighborhood U of x in M, if the function $g(X, Y) : U \to \mathbb{R}$ defined as $x \mapsto g_x(X(x), Y(x))$ is smooth, then g is called a Riemannian metric, and the pair (M, g) is called a Riemannian manifold.

The tangent bundle $TM := \cup_{x \in M} T_x M$ over M is a smooth $2n$-manifold. Let $F : TM \to \mathbb{R}$ be a continuous function such that:

(1) F is smooth on $TM \setminus \{0\}$ (regularity);
(2) $F(x, cu) = cF(x, u)$ for all $c > 0$ and for all $(x, u) \in TM$ (positive homogeneity);
(3) $g_{ij}(x, u) := \frac{1}{2}\frac{\partial^2 F^2(x,u)}{\partial u^i \partial u^j}$ is a positive definite matrix for all $(x, u) \in TM$ (strong convexity).

The pair (M, F) is called a Finsler manifold and F its fundamental function. The positive homogeneity and the strong convexity of F lead us to the following facts:

Lemma 16.2.1 (See [1]) *Let (M, F) be a Finsler manifold and $x \in M$. If $u, z, w \in T_x M$ and if u is a non-zero vector, then we have*

(1) $g_{(x,u)}(z, w) = \frac{\partial^2 F^2(x,u+sz+tw)}{2\partial t \partial s}|_{(0,0)}$;
(2) $g_{(x,u)}(u, u) = F^2(x, u)$;
(3) $g(x, tu) = g(x, u)$ *for all* $t > 0$.

16.2.2 Intrinsic Distances and Geodesics

Let (M, F) be a Finsler manifold of dimension ≥ 2. For a smooth curve $c : [a, b] \to (M, F)$ the length $L(c)$ is given by

$$L(c) := \int_a^b F(c(t), c'(t))\, dt, \quad \text{where } c'(t) = \frac{dc}{dt}.$$

The reversed curve of c, viz. $t \mapsto c(a+b-t)$, $t \in [a,b]$, is denoted by c^{-1}. The length of c^{-1} is in general different from that of c:

$$L(c^{-1}) = \int_a^b F(c^{-1}(t), (c^{-1})'(t))\, dt.$$

The intrinsic distance $d(x, y)$ from a point $x \in M$ to a point $y \in M$ is defined by

$$d(x, y) := \inf\{L(c) \mid c \text{ is a smooth curve from } x \text{ to } y\}.$$

We note that in general $d(x, y) \neq d(y, x)$. The indicatrix $\Sigma_x \subset T_x M$ at a point x is the set of all unit vectors with respect to F:

$$\Sigma_x := \{u \in T_x M \mid F(x, u) = 1\}.$$

The *reversibility constant* $\lambda(C)$ *of a compact set* $C \subset M$ is defined by

$$\lambda(C) := \sup\left\{\frac{F(x, u)}{F(x, -u)} \mid x \in C,\ u \in T_x M \setminus \{0\}\right\}. \tag{16.2.1}$$

We then have

$$\lambda(C)^{-1} d(x, y) \le d(y, x) \le \lambda(C) d(x, y), \text{ for all } x, y \in C.$$

Let U be an open subset of a Finsler manifold (M, F). Let νU be the space of smooth vector fields on U and $\nu U^+ \subset \nu U$ be the subset of nowhere vanishing vector fields. For $V \in \nu U^+$ and for all $X, Y \in \nu U$ define a trilinear form $\langle \cdot\,, \cdot\,, \cdot \rangle_V$ by $\langle X, Y, Z\rangle_V = \frac{1}{4}\frac{\partial^3}{\partial r \partial s \partial t} F^2(V + rX + sY + tZ)|_{r=s=t=0}$, which is a symmetric $(0, 3)$ tensor, called the Cartan tensor. The Cartan tensor is a non-Riemannian quantity. It is easy to show that a Finsler metric reduces to a Riemannian metric if and only if its Cartan tensor vanishes. An affine connection ∇^V is a map $\nabla^V : (X, Y) \in \nu U \times \nu U \to \nabla^V_X Y \in \nu U$, linear in Y (not necessarily linear in X) and satisfying the following conditions $\nabla^V_X(fY) = f\nabla^V_X Y + X(f)Y$ and $\nabla^V_{fX} Y = f\nabla^V_X Y$ for all $f \in C^\infty U$ and $X, Y \in \nu U$.

Theorem 16.2.1 (See Rademacher [33]) *Let (M, F) be a Finsler manifold, $U \subset M$ an open set and $V \in \nu U^+$, then there is a unique affine connection ∇^V associated with V, called the Chern connection, satisfying the following conditions:*

(1) ∇^V *is torsion free, that is,* $\nabla^V_X Y - \nabla^V_Y X = [X, Y]$ *for all* $X, Y \in \nu U$.
(2) ∇^V *is almost metrical, that is,*

$$X g_V(Y, Z) = g_V(\nabla^V_X Y, Z) + g_V(Y, \nabla^V_X Z) + 2\langle \nabla^V_X V, Y, Z\rangle_V \text{ for all } X, Y, Z \in \nu U.$$

Using the connection ∇^V, we introduce the covariant derivative $\frac{\nabla^V}{dt}$ along a smooth curve $c : [a, b] \to M$. For a vector field X along the curve c with tangent vector field c', define $\frac{\nabla^V}{dt} X(t) = \nabla^V_{c'} X(t)$. If the vector fields V and c' along c coincide, we also write $\frac{\nabla^V}{dt} X(t) = \frac{\nabla}{dt} X(t)$.

Let $\gamma : [0, 1] \to (M, F)$ be a smooth curve on a Finsler manifold (M, F). Then γ is said to be a forward geodesic if $\frac{\nabla}{dt}\gamma'(t) = 0$, for all $t \in [0, 1]$. In the local coordinates, if $\gamma'(t) = \frac{dx^i}{dt}\frac{\partial}{\partial x^i}$ and $\Gamma^k_{ij}(x, y)$ are components of the Chern connection (see [1, 33]), then forward geodesics are the solutions of the second order non-linear differential equations

$$\frac{d^2x^k}{dt^2} + \Gamma^k_{ij}(x, y)\frac{dx^i}{dt}\frac{dx^j}{dt} = 0.$$

A vector field $V \in \nu U$ is said to be a geodesic vector field if $\nabla^V_V V = 0$, that is, if all the flow lines of V are geodesics.

Proposition 16.2.1 (See Rademacher [33]) *Let V be a nowhere-vanishing geodesic vector field defined on an open subset $U \subset M$. Denote by $\overline{\nabla}$, the Levi-Civita connection of the Riemannian manifold (U, g_V). Then $\nabla^V_X V = \overline{\nabla}_X V$, for all vector fields X; in particular, the vector field V is also a geodesic vector field for the Riemannian manifold (U, g_V).*

16.2.3 The Exponential Map and Geodesic Completeness

A forward geodesic $\gamma_u : [0, h) \to (M, F)$ with the initial conditions $\gamma_u(0) := x$ and $\dot{\gamma}_u(0) := u$ is the solution of a non-linear second order differential equation with smooth coefficients. Let $\Omega_x \subset T_x M$ be the star-shaped domain with respect to the origin of $T_x M$, such that

$$\Omega_x = \{u \in T_x M \mid \gamma_u(1) \text{ is defined}\}. \tag{16.2.2}$$

We then define the exponential map $\exp_x : \Omega_x \to (M, F)$ at x by

$$\exp_x u := \gamma_u(1), \quad u \in \Omega_x.$$

We say that (M, F) is *forward geodesically complete* if $\Omega_x = T_x M$ at some point $x \in M$. It then follows that $\Omega_y = T_y M$ for all $y \in M$. The classical Hopf-Rinow theorem states that any two points on a forward geodesically complete (M, F) are joined by a forward minimizing geodesic.

A forward (resp. backward) Cauchy sequence $\{q_j\}_{j=1}^{\infty}$ is defined by the condition that for every $\varepsilon > 0$ there exists an integer N_ε such that

$$d(q_j, q_k) < \varepsilon \text{ (resp. } d(q_k, q_j) < \varepsilon), \text{ for all } N_\varepsilon < j < k.$$

We say that (M, F) is *forward complete* (resp. *backward complete*) if every forward (resp. backward) Cauchy sequence converges.

Remark 1 If $F(p, u) = F(p, -u)$ holds for all $(p, u) \in TM$, then all the completeness conditions as above are equivalent to each other.

Proposition 16.2.2 (See [1]) *If a Finsler manifold (M, F) is forward geodesically complete, then (M, F) is forward complete.*

16.3 Forward Cut Locus and Forward Conjugate Locus

Let (M, F) be a forward geodesically complete Finsler manifold, i.e. at each point $x \in M$ the exponential map $\exp_x : T_xM \to (M, F)$ is defined on the whole tangent space.

16.3.1 Forward Cut Locus and Forward Conjugate Locus

The forward cut locus and forward conjugate locus to a point $x \in (M, F)$, denoted by $C(x)$ and $J(x)$ respectively, are subsets of M and have a significant role in the study of global differential geometry of Finsler manifolds. In particular, the forward cut locus to a point $x \in (M, F)$ *equipped with the equivalence relation provided by the exponential map* contains all the topological information of (M, F). We will define them shortly. Let $\gamma_u : [0, a] \to (M, F)$, for a unit vector $u \in \Sigma_x$, be a unit speed geodesic with $\gamma_u(0) := x$, $\dot{\gamma}_u(0) = u$. Define a function $i_x : \Sigma_x \to \mathbb{R}$ by

$$i_x(u) := \sup\{s > 0 \mid t = d(x, \gamma_u(t)), \text{ for all } t \in (0, s)\}.$$

The point $\gamma_u(i_x(u))$ is called the *forward cut point to x along γ_u*, $u \in \Sigma_x$. In the case where $i(x) = \infty$, we call x *a forward pole of M*. The forward injectivity radius function at x is defined by

$$i(x) := \inf\{i_x(u) \mid u \in \Sigma_x\}.$$

Let $v \in \Sigma_x$ be a unit vector with $g_u(u, v) = 0$. Here we employ the Riemannian metric g in (16.3.4) defined on $U_x \setminus \{x\}$ which will be described in the next section. We define the Jacobi field $Y_{u,v} : [0, a] \to TM$ along γ_u such that $Y_{u,v}(0) = 0$, $\nabla_{\frac{\partial}{\partial t}} Y_{u,v}(0) = v$. Namely, it is defined by

$$Y_{u,v}(t) := d(\exp_x)_{tu} tv, \quad v \in \Sigma_x, \quad g_u(u, v) = 0, \quad t \in [0, a]. \tag{16.3.3}$$

The forward first conjugate point along γ_u is defined as follows: Let $c_x : \Sigma_x \to (0, \infty)$ be a function defined by

$$c_x(u) := \sup\{\, s > 0 \mid \det(d(\exp_x)|_{tu}) \neq 0, \text{ for all } t \in (0, s)\}, \ u \in \Sigma_x.$$

In other words, a non-trivial Jacobi field Y along γ_u, with $Y(0) = 0$, exists such that $Y(c_x(u)) = 0$ and $Y(t) \neq 0$ for all $t \in (0, c_x(u))$. The point $\gamma_u(c_x(u))$ is called the *forward first conjugate point to x along* γ_u.

The forward tangential cut locus $\widetilde{C}(x) \subset T_xM$ and the forward first tangential conjugate locus $\widetilde{J}(x) \subset T_xM$ are defined by

$$\widetilde{C}(x) := \{i_x(u)u \mid u \in \Sigma_x\}, \quad \widetilde{J}(x) := \{c_x(u)u \mid u \in \Sigma_x\},$$

and their exponential images are the forward cut locus and forward conjugate locus to x respectively; they are denoted, respectively, by

$$C(x) := \exp_x \widetilde{C}(x), \quad J(x) := \exp_x \widetilde{J}(x).$$

The domain containing the origin of T_xM and bounded by $\widetilde{C}(x)$ is denoted by $\widetilde{U}_x$. Clearly we have $\partial\widetilde{U}_x = \widetilde{C}(x)$, and $\widetilde{U}_x \subset T_xM$ is the maximal domain on which $\exp_x$ is an embedding and denote $\exp_x \widetilde{U}_x$ by U_x. We observe from the definition of the cut locus to a point $x \in (M, F)$ that $C(x)$ contains all the topological information of M. In fact $M \setminus C(x)$ is just an open disk and the identification structure of $\widetilde{C}(x)$ via the exponential map defines the manifold.

16.3.2 Geodesic Polar Coordinates

We define geodesic polar coordinates around an arbitrary fixed point $x \in M$. Let $\varphi : \Sigma_x \times (0, i_x) \to (M, F)$ be defined by

$$\varphi(u, t) := \exp_x tu, \quad 0 < t < i_x.$$

The map φ is a diffeomorphism of $S^{n-1} \times (0, i_x)$ via identification of the indicatrix with the unit sphere $S^{n-1} \subset T_xM$ through the central projection. Property (3) in Lemma 16.2.1 defines a Riemannian metric g_u along γ_u. Let ξ be a radial vector field on U_x, i.e., $\xi(y) := d\exp_{tu}(u)$, $u \in \Sigma_x$, $y = \exp_x(tu)$, $0 < t < i_x$. Thus we have a smooth Riemannian metric g on $U_x \setminus \{x\}$ defined by,

$$g(y) := \bigcup_{y \in U_x} g(y, \xi(y)), \ y \in U_x \setminus \{x\}. \tag{16.3.4}$$

The polar coordinates centered at x are defined using φ. All the F-geodesics emanating from x are identified with geodesics as a Riemannian manifold (U_x, g).

The well-known first and second variation formulas along a geodesic γ_u with $u \in \Sigma_x$ are valid for (U_x, g). Thus we know that $q := \gamma_u(c_x(u))$ is a conjugate point to x along γ_u if and only if there is a non-trivial Jacobi field Y along γ_u such that $Y(0) = Y(c_x(u)) = 0$. If a unit speed geodesic $\sigma : [0, a] \to (M, F)$ admits a conjugate pair in its interior, then there is a 1-parameter variation $\alpha : (-h, h) \times [0, a] \to M$ along σ with $\alpha(\varepsilon, 0) = x$ and $\alpha(\varepsilon, a) = \sigma(a)$ for all $\varepsilon \in (-h, h)$, such that all of its variational curves have lengths less than a. This means that

$$c_x(u) \geq i_x(u), \quad \text{for all } u \in \Sigma_x, \quad \text{for all } x \in M. \tag{16.3.5}$$

We observe that if $\sigma : [0, a] \to (M, F)$ is a minimizing geodesic and if $\sigma(a)$ is conjugate to $\sigma(0)$ along σ, then $\sigma(a)$ is the cut point to $\sigma(0)$ along it.

16.3.3 The Whitehead Convexity Theorem

We define three kinds of convex sets on a complete Finsler manifold (M, F).

Definition 16.3.1 A set $V \subset M$ is by definition *convex* if any pair of points $x, y \in V$ is joined by a unique minimizing geodesic whose image is contained entirely in V. The existence of a convex ball centered at every point on (M, F) is stated in the Whitehead Convexity Theorem 16.3.1 below. Let $B(x, \delta(x))$ be a convex $\delta(x)$-ball around x. A closed set $V \subset M$ is called *locally convex* if every $x \in V$ has the property that $V \cap B(x, \delta(x))$ is convex. A set $V \subset M$ is called *totally convex* if every geodesic joining two points in V is contained entirely in V.

In the definition of locally convex sets above, the property of being closed is crucial; for, every open set would be locally convex. If two points x and y in a convex set U are joined by a non-minimizing geodesic, then the latter is not necessarily contained in U. For example, a closed hemi-sphere in the standard sphere $\mathbf{S}^n$ is locally convex and an open hemi-sphere is convex. $\mathbf{S}^n$ itself is the only totally convex subset of itself. Every sublevel set $\varphi^{-1}(-\infty, a]$ of a convex function $\varphi : (M, F) \to \mathbb{R}$ defined on a complete Finsler manifold (M, F) is totally convex.

J.H.C. Whitehead investigated the injectivity radius in [44] and the convexity radius in [43]. We describe here some of his results:

Let $\mathfrak{U} := \cup_{x \in M} \widetilde{U}_x \subset TM$. The natural projection $\Pi : TM \to M$ and the exponential map together define a smooth map $(\Pi, \exp) : \mathfrak{U} \to M \times M$ by:

$$(\Pi, \exp)(x, u) := (\Pi u, \exp_{\Pi u} u) \in M \times M.$$

The image $(\Pi, \exp)(x, u)$ of $(x, u) \in \mathfrak{U}$ is the pair of initial and end points of the geodesic $\gamma_u : [0, 1] \to (M, F)$. Clearly, the zero section $O \subset \mathfrak{U}$ has the property: $d(\Pi, \exp)|_O =$ Identity; hence we have a small neighborhood $\Omega \subset TM$ around the zero section and a small neighborhood $U(\Delta) \subset M \times M$ around the diagonal Δ of

$M \times M$ such that

$$d(\Pi, \exp)_\Omega : \Omega \to U(\Delta) \text{ is a diffeomorphism.}$$

This fact means that any pair of points sufficiently close to each other is joined by a unique minimizing forward geodesic (compare [44]). If $C \subset M$ is a compact set, then there exists a number $\alpha(C) > 0$ such that if $x, y \in C$ satisfy $d(x, y) < \alpha(C)$, then there is a unique minimizing geodesic joining x to y. Summing up, we have:

Proposition 16.3.1 *Let $C \subset M$ be a compact set. Then for every point $x \in C$, its injectivity radius $i(x)$ is bounded below by a positive number $\alpha(C)$, i.e., $i(x) \geq \alpha(C)$ for all $x \in C$.*

Theorem 16.3.1 (The Whitehead Convexity Theorem [43, 44]) *There exists for every point $x \in (M, F)$, a positive number $\delta(x)$ such that if $r \in (0, \delta(x))$, then a forward metric r-ball $B(x, r) := \{y \in M \mid d(x, y) < r\}$ has the property that any pair of points $y, z \in B(x, r)$ is joined by a unique minimizing geodesic whose image is contained entirely in the $B(x, r)$.*

Proof Let $C \subset M$ be a compact set and fix a small number $a > 0$. Using the notations as in the last subsection, we consider a $4n - 1$ dimensional smooth manifold

$$\Lambda_{C,a} := \{(u, v, t) \in \Sigma_x \times \Sigma_x \times (0, a], \mid x \in C, \quad g_u(u, v) = 0, \quad t \in (0, a]\}.$$

For an arbitrary fixed point $x \in C$, we shall employ the Riemannian metric g in (16.3.4). We observe that the maps $\Lambda_{C,a} \to TM$:

$$(u, v, t) \to Y_{u,v}(t), \quad (u, v, t) \to \nabla_{\frac{\partial}{\partial t}} Y_{u,v}(t)$$

are smooth and uniformly bounded on $\Lambda_{C,a}$. We employ here the Riemannian connection ∇ induced through the Riemannian metric g in (16.3.4). We then have from the construction of $Y_{u,v}$,

$$\frac{d}{dt} g_u(Y_{u,v}, \nabla_{\frac{\partial}{\partial t}} Y_{u,v})(t)|_{t=0} = 0,$$

and

$$\frac{d^2}{dt^2} g_u(Y_{u,v}, \nabla_{\frac{\partial}{\partial t}} Y_{u,v})(t)|_{t=0} = 1.$$

Then there exists a constant $0 < \beta(C) \leq 1$ independent of the choice of points on C, such that

$$g_u(Y_{u,v}, \nabla_{\frac{\partial}{\partial t}} Y_{u,v})(t) > 0, \quad (u, v, t) \in \Lambda_{C,\beta(C)}.$$

Let $\lambda(C) > 1$ be the reversibility constant for C. Let $B(x, r) \subset M$ be the forward metric r-ball around an arbitrary fixed point $x \in C$. If $y, z \in B(x, r)$, then the triangle inequality implies

$$d(y, z),\ d(z, y) < (1 + \lambda(C))r.$$

Let $\sigma : [0, d(y, z)] \to (M, F)$ be a minimizing geodesic with $\sigma(0) = y$, $\sigma(d(y, z)) = z$. We then observe that

$$d(x, \sigma(t)) < (\lambda(C) + 3)r/2, \quad \text{for all } t \in [0, d(y, z)] \text{ and for all } x \in C.$$

Therefore if $r < 2\alpha(C)/(\lambda(C) + 3)$, then every pair of points in $B(x, r)$ is joined by a unique minimizing geodesic whose image lies entirely in $B(x, r)$.

We next take an arbitrary pair of points $y, z \in B(x, r)$ with $r \in (0, 2\alpha(C))/(\lambda(C) + 3)$. Let $\tau_t : [0, \ell_t] \to (M, F)$ be the unique minimizing geodesic with $\tau_t(0) := x$, $\tau_t(\ell_t) := \sigma(t)$, $t \in [0, d(y, z)]$. Here we set $\ell_t := d(x, \sigma(t))$. The 1-parameter family of geodesics $\{\tau_t\}_{t\in[0,d(y,z)]}$ form a geodesic variation, along each τ_t, $t \in [0, d(y, z)]$. Clearly $t \mapsto \ell_t$ is a smooth function and the second variation formula implies that

$$\frac{d^2}{d\ell^2}\ell_t = g_u(Y_{u,v}(\ell_t), \nabla_{\frac{\partial}{\partial t}} Y_{u,v}(\ell_t)) > 0, \quad \text{for all } (u, v, t) \in \Lambda_{C,r},$$

if r satisfies $r < \beta(C)$. We conclude the proof by setting

$$\delta(x) := \min\left\{\frac{2\alpha(C)}{\lambda(C) + 3}, \frac{2\beta(C)}{\lambda(C) + 3}\right\}.$$

□

Remark 2 We define the convexity radius function $\delta : (M, F) \to \mathbb{R}$ by:

$$\delta(x) := \sup\{r > 0 \,|\, \text{every forward ball } B(y, t) \text{ contained in } B(x, r) \text{ is convex}\}.$$

A relation between the injectivity radius and the convexity radius was first discussed by Berger.

Proposition 16.3.2 (Berger's Remark: Oral Communication) *Let (M, g) be a compact Riemannian manifold. Let $\delta : (M, g) \to \mathbb{R}$ and $i : (M, g) \to \mathbb{R}$ be the convexity radius and the injectivity radius functions respectively. If $i(M)$ and $\delta(M)$ be the infimum of i and δ over M respectively, then we have*

$$\frac{1}{2}i(M) \geq \delta(M). \tag{16.3.6}$$

Proof Let (M, g) be a compact Riemannian manifold. Suppose that (16.3.6) does not hold. We then have $2\delta(M) > i(M)$.

Choose points $x, y \in M$ such that $i(M) = d(x, y) = i(x) = i(y)$. Let $\gamma : [0, i(M)] \to M$ be a minimizing geodesic with $\gamma(0) = x$, $\gamma(i(M)) = y$. Then, Klingenberg's Lemma (see Proposition 16.4.3) implies that either y is conjugate to x along γ, or γ extends to a simple closed geodesic such that $\gamma(0) = \gamma(2i(M)) = x$.

We first suppose that $\gamma : [0, 2i(M)] \to M$ is a simple closed geodesic. Since $y \notin B(x, \delta(M))$ and $x \notin B(y, \delta(M))$, the midpoints $\gamma(i(M)/2)$ and $\gamma(3i(M)/2)$ between x and y are contained in $B(x, \delta(M))$, which is a contradiction. In fact, convexity of $B(x, \delta(M))$ means that $\gamma[i(M)/2, 3i(M)/2] \subset B(x, \delta(M))$.

Suppose next that y is conjugate to x along γ. Let $\varepsilon \in (0, \frac{2\delta(M)-i(M)}{2})$, and extend γ to both sides and set $x' := \gamma(-\varepsilon)$, $q' := \gamma(i(M) + \varepsilon)$. Since $x', y' \in B(\gamma((i(M)/2), \delta(M))$ and $\gamma|_{[-\varepsilon, i(M)+\varepsilon]}$ is not minimizing, we have a unique minimizing geodesic joining x' to y' whose image lies in $B(\gamma(i(M)/2), \delta(M))$. Consequently, $\gamma[-\varepsilon, i(M) + \varepsilon]$ is not contained in $B(\gamma(i(M)/2), \delta(M))$, a contradiction. This completes the proof. □

Problem 16.3.1 *Is there any relation between the forward convexity radius and the forward injectivity radius on a compact Finsler manifold ?*

The non-symmetric property of the distance function on (M, F) makes it difficult to address Problem 16.3.1. In particular, it is not clear whether we can find a simple closed geodesic $\gamma : [0, 2i(M)] \to (M, F)$ if $i(\gamma(0)) = i(M, F)$ and $q = \gamma(i(M, F))$ is not conjugate to p along γ. This problem is discussed in Sect. 16.4.2.

Example 16.3.1 It should be remarked that Proposition 16.3.2 is optimal. In fact, for every rank-one symmetric space of compact type, equality holds in (16.3.6). A simple example is given here. Let $T^2 := S^1(a) \times S^1(b)$ with $0 < a < b$ be a flat torus whose fundamental domain is rectangular with edge length $2a\pi$ and $2b\pi$, then $i(T^2) = a\pi$ and $\delta(T^2) = \frac{1}{2}a\pi$.

Lemma 16.3.1 *The convexity radius function $\delta : (M, F) \to \mathbb{R}$ is locally Lipschitz.*

Proof Let $\lambda(C)$ be the reversibility constant of a compact set $C \subset M$ as defined in (16.2.1). We take two arbitrary points $x, y \in C$ sufficiently close to each other. We then observe that $B(y, \delta(x) - d(x, y)) \subset B(x, \delta(x))$ implies that $\delta(y) \geq \delta(x) - d(x, y)$ and similarly, $B(x, \delta(y) - d(y, x)) \subset B(x, \delta(x))$ implies $\delta(x) \geq \delta(y) - d(y, x)$. We conclude the proof by noting that

$$\frac{|\delta(x) - \delta(y)|}{d(x, y)}, \ \frac{|\delta(x) - \delta(y)|}{d(y, x)} \leq \lambda(C), \quad \text{for all } x, y \in C.$$

□

Using this notion of convexity radius function, Theorem 16.3.1 states that the forward distance function from a point on the manifold to $B(x, r)$ for $r \in (0, \delta(x))$ is convex.

16.4 The Properties of Cut Locus and Conjugate Locus

From now on, for simplicity, we shall only discuss forward geodesics, forward cut locus, forward conjugate locus, etc. The case where backward aspects are needed is rare. We shall state basic properties of cut locus and conjugate locus which have been discussed in Riemannian geometry by Whitehead, Myers, Klingenberg, Berger, Omori and many others. Rademacher has discussed in [33] the Finsler version of some results in the proof of the classical sphere theorem. We summarize them here.

16.4.1 Foots of Closed Sets

Let $C \subset M$ be a closed set and let $x \in M \setminus C$. A point $y \in C$ is called a *foot of x on C* if y satisfies $d(x, y) = d(x, C) := \inf\{d(x, y) \mid y \in C\}$. Let

$$B^{-1}(C, r) := \{x \in M \mid d(x, C) < r\} \text{ and } S^{-1}(C, r) := \{y \in M \mid d(y, C) = r\}$$

be the backward metric r-ball and the backward metric r-sphere around a closed set $C \subset M$ respectively. Let $\overline{B}^{-1}(C, r)$ be the closure of $B^{-1}(C, r)$. If a point $x \in M \setminus C$ has more than one feet, then x belongs to the backward cut locus to C.

With this notation the following Lemma (16.4.1) shows a common property of a foot of a point on a closed set in Riemannian and Finsler geometries. We show how the triangle inequalities are employed with a non-symmetric distance function.

Lemma 16.4.1 *Let $C \subset M$ be a closed set. Take a point $x \in M \setminus C$ and a positive number r such that $d(x, C) > r$. Let $y \in S^{-1}(C, r)$ be a foot of x on $S^{-1}(C, r)$ and $\gamma : [0, a] \to (M, F)$ is a unit speed minimizing geodesic with $\gamma(0) = x$, $\gamma(a) = y$. Then its extension reaches a point on C at $\gamma(a + r)$, which is the unique foot of y on C.*

Proof Let $z_1, z \in C$ be feet of x, y on C respectively. Let $\tau : [0, r] \to (M, F)$, $\sigma : [0, b] \to (M, F)$ be unit speed minimizing geodesics such that $\tau(0) = y$, $\sigma(0) = x$ and $\tau(r) = z$, $\sigma(b) = z_1$. Here we set $b := d(x, z_1) = d(x, C)$. Clearly, $\sigma[0, b]$ intersects $S^{-1}(C, r)$ at a point, say, $y_1 := \sigma[0, b] \cap S^{-1}(C, r)$. We then observe that $d(x, y_1) \geq d(x, S^{-1}(C, r)) = d(x, y)$ and $d(y_1, z_1) \geq d(y, C) = r$. The triangle inequality then implies $d(x, C) = d(x, y_1) + d(y_1, z_1) \geq d(x, S^{-1}(C, r)) + r = a + r$ and $d(x, C) \leq d(x, y) + d(y, z) = a + r$. We therefore have $d(x, y) + d(y, z) = d(x, z) = a + r$ and $d(y_1, z_1) = r$. We then assert that x, y and z belong to a minimizing geodesic. In fact, let $B(y, \delta(y))$ be a convex ball centered at y and take arbitrary points $x' \in \gamma(a - \delta(y), a) \cap B(y, \delta(y))$ and $z' \in \tau(0, r)$. The triangle inequality again implies that $d(x', z') = d(x', y) + d(y, z')$, and hence the uniqueness of minimizing geodesic joining two points in $B(y, \delta(y))$ implies that y is an interior point of the minimizing geodesic joining x' to z'. Therefore

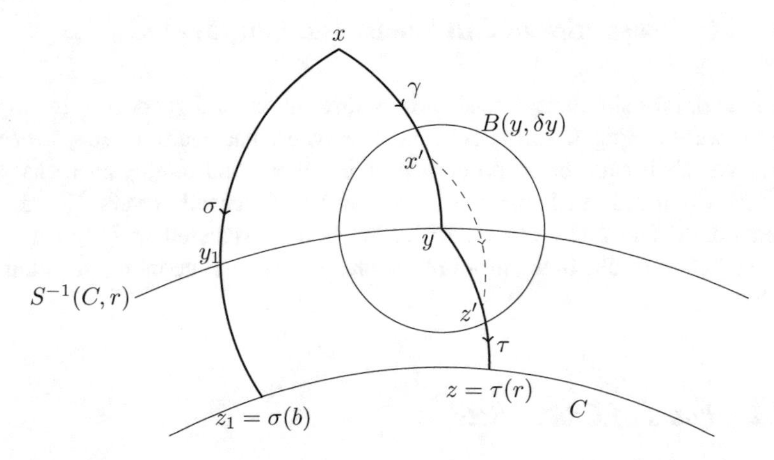

Fig. 16.1 Foot of a point x on the closed set C

$\gamma(a + r) = \tau(r) = z$. In particular, z and z_1 are unique feet of y and y_1 of C respectively (see Fig. 16.1). □

Let $C \subset (M, F)$ be a closed, connected and locally convex set. We then find an open set $U(C)$ of M and a strong deformation retract C of $U(C)$. In fact, we choose a countable open cover $\{U_i\}_{i=1,2,...}$ of C such that for each $i = 1, 2, \ldots$, the closure $\overline{U_i}$ of U_i is compact. Let $\delta_i := \delta(\overline{U_i})$ be the convexity radius of $\overline{U_i}$ and set

$$U(C) := \bigcup_{i=1}^{\infty} B^{-1}(\overline{U_i}, \delta_i).$$

If $x \in U(C)$, then we get a number i and a unique foot $f(x)$ of x on $U_i \cap C$. Suppose there is another foot $f'(x)$ of x on $U_i \cap C$. Clearly we have

$$f(x), f'(x) \in B(x, \delta_i) \cap C.$$

There is a unique minimizing geodesic $T(f(x), f'(x))$ joining $f(x)$ to $f'(x)$ and belonging to $B(x, \delta_i)$. The convexity of the distance function from x to $T(f(x), f'(x))$ implies that its minimum is attained at an interior point of $T(f(x), f'(x))$. This is a contradiction to the fact that $T(f(x), f'(x)) \subset C$. The above discussion may be summarized as follows:

Proposition 16.4.1 *A closed locally convex set $C \subset (M, F)$ admits an open set $U(C)$ of M such that C is a strong deformation retract of $U(C)$. The retraction is given by the foot map $f : U(C) \to C$.*

Remark 3 If the local convexity of a closed set $C \subset (M, F)$ is not assumed, then every open set $V \supset C$ may admit a point $x \in V$ with more than one foot on C. The backward cut locus $\mathrm{Cut}^{-1}(A) \subset (M, F)$ of a closed set $A \subset (M, F)$ is defined as follows:

$$\mathrm{Cut}^{-1}(A) := \left\{ x \in (M,F) \;\middle|\; \begin{array}{l} \text{if } f(x) \text{ is a foot of } x \text{ on } A, \text{then } T(x, f(x)) \text{ is not} \\ \text{properly contained in any } T(y, f(y)),\ y \in (M,F) \setminus A, \\ \text{where } f(y) \text{ is a foot of } y \text{ on } A. \end{array} \right\}$$

This means that x belongs to the backward cut locus to C, and hence the foot map $f : V \to C$ may fail to be a retraction map of C. Let $\mathrm{Cut}^{-1}(C) \subset (M, F)$ be the backward cut locus to C. We then have $d(\mathrm{Cut}^{-1}(C), C) \geq 0$, and equality holds if and only if there is a sequence $\{x_j\}_{j=1,2,\ldots} \subset \mathrm{Cut}^{-1}(C)$ such that $\lim_{j\to\infty} d(x_j, C) = 0$. Let $C \subset (M, F)$ be a compact set whose boundary ∂C is a smooth hypersurface of (M, F). Here we do not assume C to be locally convex. From the smoothness assumption on ∂C, we deduce that $d(\mathrm{Cut}^{-1}(C), C) > 0$. Let $U(C) \supset C$ be an open set as defined above. Then C has a strong deformation retract and its retraction map is the foot map $f : U(C) \to C$.

16.4.2 The Klingenberg Lemma

We now summarize the important properties of cut loci on complete Riemannian and Finsler manifolds [20].

Proposition 16.4.2 *Let (M, F) be a complete Finsler manifold and $x \in M$ an arbitrary fixed point such that $C(x) \neq \emptyset$. If $y \in C(x)$ and if $\gamma : [0, d(x, y)] \to (M, F)$ is a minimizing geodesic with $\gamma(0) = x$, and $\gamma(d(x, y)) = y$, then one of the following holds:*

(1) *y is conjugate to x along γ, or*
(2) *there exists another minimizing geodesic $\sigma : [0, d(x, y)] \to (M, F)$ such that $\sigma(0) = x$ and $\sigma(d(x, y)) = y$.*

Proposition 16.4.3 *Let (M, g) be a compact Riemannian manifold and $x \in M$, $y \in C(x) \subset (M, g)$ satisfy (2) in Proposition 16.4.2. If*

$$d(x, y) = i(x) = d(x, C(x)), \tag{16.4.7}$$

then there are exactly two distinct minimizing geodesics $\gamma, \sigma : [0, i(x)] \to (M, g)$ such that $\gamma(0) = \sigma(0) = x$, $\gamma(i(x)) = \sigma(i(x)) = y$, and $\dot\gamma(i(x)) = -\dot\sigma(i(x))$.

Remark 4 It turns out that Klingenberg's result in Proposition 16.4.3 that $\dot\gamma(i(x)) = -\dot\sigma(i(x))$ does not hold for a Finsler manifold, in general. This phenomenon shows an essential difference between Riemannian and Finsler geometry.

Proof of Proposition 16.4.2 Suppose that (1) does not hold. Thus $y \in C(x)$ is not conjugate to x along γ, and hence there is a small open set $O(y) \subset T_xM$ around

y such that $\exp_x|_{O(y)} : O(y) \to (M, F)$ is an embedding. Let $\{\varepsilon_j\}_{j=1,2,\ldots}$ be a decreasing sequence of positive numbers converging to 0 and $q_j := \gamma(d(x, y)+\varepsilon_j)$. For each j, let $\tau : [0, a_j] \to (M, F)$ be a minimizing geodesic such that $\tau_j(0) = x, \tau_j(a_j) = q_j$ for $j = 1, 2, \ldots$. Clearly $d(x, q_j) = a_j$ and $\lim_{j\to\infty} a_j = d(x, y)$. Then $\gamma[0, d(x, y) + \varepsilon]$ is not minimizing for all $\varepsilon > 0$ and hence $\dot{\gamma}(0) \neq \dot{\tau}_j(0)$ for all j. Choosing a subsequence, we find a limit of $\{\dot{\tau}_j(0)\}_j$ and the limit geodesic $\tau : [0, d(x, y)] \to (M, F)$ such that $\tau(0) = x, \tau(d(x, y)) = \gamma(d(x, y))$ satisfying $d(x, y)\dot{\tau}(0) \notin O(y)$. This proves that (2) holds. □

Lemma 16.4.3 below describes certain conditions under which (1) holds.

The following lemma due to Berger [3] is extended to Finsler manifolds. The proof employs the Riemannian metric in the geodesic polar coordinates as defined in Sect. 16.3.2, and the first variation formula, and is omitted here.

Lemma 16.4.2 *Let $x, y \in (M, F)$. Let $V \subset M$ be an open set around y such that the distance function $d(x, .) : V \to \mathbb{R}$ from x attains a local maximum at y. Then there exists for every vector $u \in T_yM$, $u \neq 0$ a minimizing geodesic $\sigma : [0, d(y, x)] \to (M, F)$ such that $\sigma(0) = x$, $\sigma(d(y, x)) = y$ and*

$$g(u, -\dot{\sigma}(d(x, y))) \geq 0. \tag{16.4.8}$$

16.4.3 The Berger-Omori Lemma

The following important lemma was first proved by Berger for even-dimensional compact Riemannian manifolds of positive sectional curvature with minimal diameter. Then Omori [31] proved it for any compact Riemannian manifold with minimal diameter. It is summarized in [4]. We prove the Riemannian version under certain weaker assumptions. The Berger-Omori Lemma can be extended to Finsler manifolds, and this has been carried out in [20].

For any vector $u \in TM$ we denote by $\|u\|$ the Riemannian norm of u.

Lemma 16.4.3 *Let (M, g) be a complete Riemannian manifold and $C(x) \neq \emptyset$ for a point $x \in M$. If $y \in C(x)$ satisfies $d(x, y) = i(x)$ and if there exist two distinct minimizing geodesics $\gamma_0, \gamma_1 : [0, i(x)] \to M$ such that $\gamma_0(0) = \gamma_1(0) = x$ and $\gamma_0(i(x)) = \gamma_1(i(x)) = y$ and such that $\dot{\gamma}_0(i(x))$ and $\dot{\gamma}_1(i(x))$ are linearly independent, then there exists a one-parameter family of minimizing geodesics γ_t : $[0, i(x)] \to M$, $\gamma_t(0) = x$, $\gamma_t(i(x)) = y$, $0 \leq t \leq 1$ such that*

$$\dot{\gamma}_t(i(x)) = \frac{(1-t)\dot{\gamma}_0(i(x)) + t\dot{\gamma}_1(i(x))}{\|(1-t)\dot{\gamma}_0(i(x)) + t\dot{\gamma}_1(i(x))\|}, \quad 0 \leq t \leq 1. \tag{16.4.9}$$

Proof Let $\rho := \delta(x)/2$ and $\rho' := i(x) - \rho$. We then have

$$B(\gamma_j(\rho'), \rho) \subset B(x, i(x)), \quad j = 0, 1,$$

and also

$$\{\bar{B}(\gamma_0(\rho'), \rho) \cup \bar{B}(\gamma_1(\rho'), \rho)\} \cap \partial B(x, i(x)) = \{y\}.$$

Again, let $\widetilde{U}_y \subset T_yM$ be the maximal domain in T_yM on which the exponential map at y is an embedding. We already know that $\widetilde{C}(y) = \partial\widetilde{U}_y$. Set

$$\mathcal{S}_j := (\exp_y|_{U_y})^{-1}(\partial B(\gamma_j(\rho'), \rho)), \quad j = 0, 1.$$

Then $\mathcal{S}_j$ is a smooth hypersurface in T_yM. Clearly, $c_j := (\exp_y|_{U_y})^{-1}(\gamma_j[\rho', i(x)]) \subset T_yM$ is a straight line segment, and normal to $\mathcal{S}_j$ at $O \in T_yM$. Using the assumption for γ_0 and γ_1, we choose a small number $0 < r_j < \rho$ and a vector $\xi_j := -r_j\dot{\gamma}_j(i(x)) \in T_yM$ such that $B(\xi_j, r_j) \subset (\exp_y|_{U_y})^{-1}(B(\gamma_j(\rho'), \rho))$, $j = 0, 1$. In fact, both $\partial B(\xi_j, r_j)$ and $(\exp_y|_{U_y})^{-1}(\partial B(\gamma_j(\rho'), \rho))$ are smooth hypersurfaces in T_yM and have the same tangent spaces at the origin. We then have

$$\exp_y\{\overline{B}(\xi_0, r_0) \cup \overline{B}(\xi_1, r_1)\} \cap \partial B(x, i(x)) = \{y\}.$$

Let $\xi_\lambda := (1 - \lambda)\xi_0 + \lambda\xi_1 \in T_yM$ for $\lambda \in [0, 1]$. Since the figures are all in a Euclidean space, it is clear that

$$B(\xi_\lambda, \|\xi_\lambda\|) \subset B(\xi_0, r_0) \cup B(\xi_1, r_1), \quad \forall\lambda \in [0, 1], \tag{16.4.10}$$

$$(\exp_y|_{U_y})(\partial B(\xi_\lambda, \|\xi_\lambda\|)) \cap \partial B(x, i(x)) = \{y\}. \tag{16.4.11}$$

Choosing $h_\lambda > 0$ sufficiently small, we deduce from the inclusion relation in T_yM that

$$B(\exp_y h_\lambda\xi_\lambda, \|h_\lambda\xi_\lambda\|) \subset \exp_y(B(\xi_\lambda, \|\xi_\lambda\|)).$$

Finally, the triangle inequality implies

$$d(x, \exp_y h_\lambda\xi_\lambda) + d(\exp_y h_\lambda\xi_\lambda, y) \geq i(x).$$

Setting $\sigma_\lambda : [0, \ell_\lambda] \to (M, g)$ a minimizing geodesic with

$$\sigma_\lambda(0) = x, \quad \sigma_\lambda(\ell_\lambda) = \exp_y h_\lambda\xi_\lambda,$$

we observe from the above triangle inequality that

$$d(\sigma_\lambda(\ell_\lambda), \sigma_\lambda(i(x)) = i(x) - \ell_\lambda \leq \|h_\lambda\xi_\lambda\|.$$

If $\sigma_\lambda(i(x)) \neq y$, then

$$\sigma_\lambda(i(x)) \in \bar{B}(\sigma_\lambda(\ell_\lambda), \|h_\lambda\xi_\lambda\|).$$

On the other hand, $\sigma_\lambda(i(x)) \in \partial B(x, i(x))$ means that the point $\sigma_\lambda(i(x))$ stays outside $\exp_y(B(\xi_0, \|\xi_0\|) \cup B(\xi_1, \|\xi_1\|)) \setminus \{y\}$, which is a contradiction. Thus we have $\sigma_\lambda(i(x)) = y$ for all $\lambda \in [0, 1]$. □

Remark 5 To discuss the Berger-Omori Lemma in the context of Finsler manifolds, we need to consider T_yM as a normed space and introduce a new idea for the proof of the Finsler version; see Theorem 4.2 in [20] for details.

16.4.4 The Rauch Conjecture

The classical Rauch conjecture [34] predicts that the cut locus $\widetilde{C}(x) \subset T_xM$ and the conjugate locus $\widetilde{J}(x) \subset T_xM$ to a point x on a compact and simply connected Riemannian n-manifold (M, g) have a point in common in its tangent space T_xM:

$$\widetilde{C}(x) \cap \widetilde{J}(x) \neq \emptyset, \quad \text{for all } x \in M. \tag{16.4.12}$$

The conjecture is true for all metrics on S^2 and for all compact rank-one symmetric spaces. Also it is true for a complete noncompact Riemannian 2-manifolds homeomorphic to $\mathbb{R}^2$. Weinstein [42] settled this conjecture negatively in general, by proving that if M is compact and not homeomorphic to S^2, then there exists a metric g and a point $x \in M$ such that

$$\widetilde{C}(x) \cap \widetilde{J}(x) = \emptyset.$$

We next discuss the results obtained in [21] on complete Finsler n-manifolds. Let (M, F) be a connected and geodesically complete Finsler n-manifold, $n \geq 2$. It is elementary to see that $i_x(u) \leq c_x(u)$, in general, and $i_x(u) = c_x(u) < \infty$ holds for some $u \in \Sigma_x$ if and only if the Rauch conjecture is valid at x. If $\widetilde{C}_x = \emptyset$ (or equivalently, $C(x) = \exp_x \widetilde{C}_x = \emptyset$), then $\gamma_u : [0, \infty) \to (M, F)$ is a ray for all $u \in \Sigma_x$. Such a point is called a *forward pole* of (M, F). An ellipsoid with foci at $x, y \in (M, F)$ and radius $r > d(x, y)$ is denoted by $E_{xy}(r) \subset M$ and

$$E_{xy}(r) := \{z \in M | d(x, z) + d(z, y) = r\},$$

$$B_{xy}(r) := \{z \in M | d(x, z) + d(z, y) < r\}.$$

Notice that $B_{xx}(2r) \neq B(x, r)$ follows if d is not symmetric. We further define the function $F_{xy} : (M, F) \to \mathbb{R}$ by

$$F_{xy}(z) := d(x, z) + d(z, y), \quad z \in M.$$

Notice also that $F_{xy} \neq F_{yx}$ and $E_{xy}(r) \neq E_{yx}(r)$.

Theorem 16.4.1 (Compare Theorem 1; [21]) *Let (M, F) be an n-dimensional compact Finsler manifold. Assume that there is a point $x \in M$ satisfying*

$$\widetilde{C}(x) \cap \widetilde{J}(x) = \emptyset. \tag{16.4.13}$$

Then for any point $y \in M \setminus \{x\}$ there exist at least two distinct geodesics emanating from x and ending at y.

Proof If there exist two distinct minimizing geodesics emanating from x and ending at y, there is nothing to prove.

If $y \in C(x)$ and if there exists a unique minimizing geodesic $\tau : [0, d(x, y)] \to (M, F)$ with $\tau(0) = x$ and $\tau(d(x, y)) = y$, then every extension $\tau|_{[0,d(x,y)+\varepsilon]}$ of τ beyond y is not minimizing for all $\varepsilon > 0$. The uniqueness of τ implies that if $\tau_\varepsilon : [0, d(x, \tau(d(x, y)+\varepsilon)] \to (M, F)$ is a minimizing geodesic with $\tau_\varepsilon(0) = x$ and $\tau_\varepsilon(d(x, \tau(d(x, y)+\varepsilon))) = \tau(d(x, y)+\varepsilon)$, then $\tau_\varepsilon \neq \tau|_{[0,d(x,y)+\varepsilon]}$. The uniqueness of a minimizing geodesic from x to y then implies that $\lim_{\varepsilon \to 0} \tau_\varepsilon = \tau$. Therefore $\exp_x : T_xM \to (M, F)$ is not bijective in any open set of $d(x, y) \cdot \dot{\tau}(0)$, and hence y is conjugate to x along τ. Thus we have $d(x, y) \cdot \dot{\tau}(0) \in TC(x) \cap TJ(x)$, a contradiction to (16.4.13). Thus we observe that there are at least two distinct minimizing geodesics from x to y for all $y \in C(x)$. We may therefore suppose that there exists a unique minimizing geodesic from x to y and that $y \notin C(x)$.

The construction of a non-minimizing geodesic joining x to y is achieved by using a technique developed in [21]. There exists a cut point $x_0 \in C(x)$ such that,

$$B_{xy}(\ell) \cap C(x) = \emptyset, \ x_0 \in E_{xy}(\ell) \cap C(x), \ \ell := F_{xy}(x_0). \tag{16.4.14}$$

Let $\sigma : [0, d(x_0, y)] \to (M, F)$ be a minimizing geodesic with $\sigma(0) = x_0$, $\sigma(d(x_0, y)) = y$. We assert that *there exist exactly two distinct minimizing geodesics $\gamma_j : [0, d(x, x_0)] \to (M, F)$, $j = 1, 2$, with $\gamma_j(0) = x$, $\gamma_j(d(x, x_0)) = x_0$ such that one of the composed geodesics $\gamma_1 * \sigma$ or $\gamma_2 * \sigma$ forms a geodesic joining x to y*. Here, $\gamma_j * \sigma : [0, \ell] \to (M, F)$ is a broken geodesic given by

$$\gamma_j * \sigma(t) = \begin{cases} \gamma_j(t), & t \in [0, d(x, x_0)], \\ \sigma(t - d(x, x_0)), & t \in [d(x, x_0), \ell]. \end{cases}$$

For the proof of the above assertion, we argue by deriving a contradiction, assuming the contrary. Suppose that there are more than two distinct minimizing geodesics from x to x_0. If σ is a fixed minimizing geodesic from x_0 to y, we choose two minimizing geodesics γ_1 and γ_2 such that both $\gamma_1 * \sigma$ and $\gamma_2 * \sigma$ are broken geodesics with a corner at x_0. We then find points $z \in \sigma(0, d(x_0, y)]$ and $y_j \in \gamma_j([0, d(x, x_0))$ lying in a convex ball around x_0. The short cut principle then implies that

$$d(y_j, z) < d(y_j, x_0) + d(x_0, z), \quad j = 1, 2,$$

and hence

$$\begin{aligned} F_{xy}(z) &= d(x, z) + d(z, y) < (d(x, y_j) + d(y_j, z)) + d(z, y) \\ &< d(x, y_j) + (d(y_j, x_0) + d(x_0, z)) + d(z, y) \\ &= d(x, x_0) + d(x_0, y) = F_{xy}(x_0) = \ell. \end{aligned}$$

We therefore have $z \in B_{xy}(\ell)$, and in particular, $z \notin C(x)$. This implies that $\sigma(0, d(x_0, y)) \subset \exp_x(U_x)$. From (16.4.13), there exists an open set $\Omega_i \in T_x M$ of $d(x, x_0) \cdot \dot{\gamma}_j(0)$ for $j = 1, 2$ such that $\exp_x|_{\Omega_j} : \Omega_j \to (M, F)$ is an embedding. The lifting of $\sigma(0, d(x_0, y))$ along the diffeomorphism $\exp_x|_{U_x} : U_x \to (M, F) \setminus C(x)$ forms distinct curves joining $d(x, x_0) \cdot \dot{\gamma}_j(0)$ to $d(x, y) \cdot \dot{\sigma}(0)$, $j = 1, 2$. However this is impossible by the uniqueness of $d(x, y) \cdot \dot{\sigma}(0)$.

As shown in the proof of the above assertion, one of $\gamma_1 * \sigma$ and $\gamma_2 * \sigma$ forms a geodesic emanating from x and ending at y; it may be noted that since this geodesic passes through a forward cut point x_0 to x, it is not minimizing. □

As an application of Theorem 16.4.1, we deduce the following:

Theorem 16.4.2 *Let (M, F) be a compact Finsler n-manifold, $n \geq 2$ and $\lambda = \lambda(M)$ be the reversibility constant of M, as defined in (16.2.1). If $x \in M$ is such that the Rauch conjecture does not hold, then there exists a geodesic loop $\gamma_x : [0, 2\ell_x] \to (M, F)$ with $\gamma_x(0) = \gamma_x(2\ell_x) = x$ such that*

$$(1 + \lambda^{-1})i(x) \leq L(\gamma_x).$$

Proof The same technique as involved in the proofs of Theorem 16.4.1 is employed here. For an arbitrary fixed point $y \notin C(x)$ with $y \neq x$, we choose a point $x_0(y) \in C(x)$ such that, setting $\ell(y) := d(x, x_0(y)) + d(x_0(y), y)$,

$$B_{xy}(\ell(y)) \cap C(x) = \emptyset, \quad x_0(y) \in E_{xy}(\ell(y)) \cap C(x).$$

By Theorem 16.4.1, there exists a geodesic $\gamma_{xy} : [0, \ell(y)] \to M$ such that $\gamma_{xy}(0) = x$, $\gamma_{xy} d(x, x_0(y)) = x_0(y)$ and $\gamma_{xy}(\ell(y)) = y$. From the construction we see that $L(\gamma_{xy}[0, d(x, x_0(y))]) = d(x, x_0(y))$ and $L(\gamma_{xy}[d(x, x_0(y)), \ell(y)]) = d(x_0(y), y)$. Taking a sequence $\{y\} \subset M \setminus C(x)$ converging to x, we find a geodesic loop $\gamma_x : [0, \ell_x] \to M$ as the limit of $\{\gamma_{xy}\}$, where $\gamma_x = \lim_{y \to x} \gamma_{xy}$ and $\ell_x = \lim_{y \to x} \ell(y)$. We then observe that $\gamma_x(d(x, x_0(x))) \in C(x)$ and that

$$L(\gamma_{xy}[0, d(x, x_0(y))]) \geq i(x).$$

From the triangle inequality we have,

$$\begin{aligned} L(\gamma_{xy}[d(x, x_0(y)), \ell(y)]) &= d(x_0(y), y) \geq \lambda^{-1} d(y, x_0(y)) \\ &\geq \lambda^{-1}[d(x, x_0(y)) - d(x, y)] \geq \lambda^{-1}(i(x) - d(x, y)). \end{aligned}$$

Therefore, by letting $y \to x$ we get, $\ell_x \geq (1 + \lambda^{-1})i(x)$. This proves Theorem 16.4.2. □

16.4.5 Poles

The original Rauch conjecture was considered on compact and simply connected Riemannian manifolds. We discuss it on complete non-compact Riemannian and Finsler manifolds admitting poles. For the discussion of the Rauch conjecture on complete non-compact manifolds, we need the notion of poles. The Rauch conjecture is valid for a point $x \in (\mathbb{R}^2, g)$ of a complete noncompact Riemannian 2-manifold homeomorphic to a plane, if $C(x) \neq \emptyset$. In fact, $C(x)$ for every point $x \in (\mathbb{R}^2, g)$ carries the structure of a tree. A cut point $y \in C(x)$, $x \in (\mathbb{R}^2, g)$ is called an *endpoint* of $C(x)$ if $C(x) \setminus \{x\}$ is connected. The cut loci of Riemannian 2-manifolds have been discussed by many authors; for instance, see [29, 30, 44] and [41]. Let $x \in (\mathbb{R}^2, g)$ be a point such that $C(x) \neq \emptyset$. Then there is a point $y \in C(x)$ which is an endpoint of $C(x)$. If $\gamma : [0, d(x, y)] \to (\mathbb{R}^2, g)$ is a minimizing geodesic with $\gamma(0) = x$, $\gamma(d(x, y)) = y$, then y is conjugate to x along γ, and hence $\widetilde{C}(x) \cap \widetilde{J}(x)$ contains $d(x, y)\dot{\gamma}(0)$, if $C(x) \neq \emptyset$.

Let (M, F) be a geodesically complete Finsler n-manifold. A unit speed forward geodesic $\gamma : [0, \infty) \to (M, F)$ is by definition a *forward ray* if every subarc $\gamma|_{[a,b]}$, $0 \leq a < b < \infty$ of γ is minimizing. A point $x \in (M, F)$ is called a *forward pole* if $C(x) = \emptyset$. Clearly, $\exp_x : T_xM \to (M, F)$ is a diffeomorphism if and only if x is a forward pole. A backward geodesic $\gamma^{-1}(-\infty, 0] \to (M, F)$ is called a *backward ray* if

$$d(\gamma^{-1}(s), \gamma^{-1}(t)) = t - s, \quad \text{for all } 0 > t > s > -\infty.$$

A point $y \in (M, F)$ is called a *backward pole* if every backward geodesic σ^{-1} : $(-\infty, 0] \to (M, F)$ with $\sigma^{-1}(0) = y$ is a backward ray.

The relation between the Rauch conjecture and poles on complete noncompact Riemannian n-manifolds has been discussed in [21]. We have the following relation between the Rauch conjecture and poles on complete non-compact Finsler manifolds. The proof is essentially contained in [21] and is omitted here.

Proposition 16.4.4 *Let (M, g) and (M, F) be complete Riemannian and Finsler manifolds respectively.*

(1) *If (M, g) admits a pole and if $x \in M$ is not a pole, then the Rauch conjecture is valid at x.*
(2) *If (M, F) admits a backward pole, then for $y \in M$ either the Rauch conjecture holds at y or y is a forward pole.*

16.4.6 The Continuity of the Injectivity Radius Function

We discuss the continuity of injectivity radius functions on complete Finsler manifolds. The compactification $[0, \infty] := [0, \infty) \cup \{\infty\}$ of the half line is employed here.

Lemma 16.4.4 *Let (M, F) be a complete Finsler manifold. The injectivity radius function $i : M \to [0, \infty]$ is continuous at every point $x \in M$ where $i(x) < \infty$.*

Proof Let $x \in M$ and $\{x_j\}_{j=1,2,\dots} \subset M$ be such that $\lim_{j\to\infty} x_j = x$. Let $\{y_j\}_{j=1,2,\dots} \subset M$ be chosen such that $d(x_j, y_j) = i(x_j) =: \ell_j$ for all $j = 1, 2, \dots$. Let $\gamma_j : [0, \ell_j] \to (M, F)$ be a minimizing geodesic with $\gamma_j(0) = x_j$ and $\gamma_j(\ell_j) = y_j$ for all $j = 1, 2, \dots$. In view of Proposition 16.4.2, by choosing a subsequence if necessary, it suffices to consider the following two cases:

Case 1. Assume that y_j is conjugate to x_j along γ_j. Setting $v_j := \ell_j \dot\gamma_j(0)$, we have

$$\det(d(\exp_{x_j})_{v_j}) = 0, \quad \text{for all } j = 1, 2, \dots \tag{16.4.15}$$

Thus we observe that $\lim_{j\to\infty} y_j = y$ is a conjugate point to x along γ, where γ is defined by the limit: $v = \dot\gamma(0) := \lim_{j\to\infty} \dot\gamma_j(0)$, and hence $i(x) \le F(x, v) = \lim_{j\to\infty} i(x_j)$.

Case 2. We now assume that there exist minimizing geodesics $\gamma_j : [0, \ell_j] \to (M, F)$ emanating from x_j and ending at y_j such that $i(x_j) = d(x_j, y_j)$ and $y_j \in C(x_j)$ is not conjugate to x_j along γ_j for all $j = 1, 2, \dots$. From Proposition 16.4.2, we get that there are exactly two minimizing geodesics $\gamma_j, \sigma_j : [0, \ell_j] \to (M, F)$ such that $\gamma_j(0) = \sigma_j(0) = x_j$ and $\gamma_j(\ell_j) = \sigma_j(\ell_j) = y_j$. Choosing a subsequence, if necessary, we get limit geodesics $\gamma := \lim_{j\to\infty} \gamma_j$ and $\sigma := \lim_{j\to\infty} \sigma_j$ together with $\ell := \lim_{j\to\infty} \ell_j$. Clearly γ and σ are distinct minimizing geodesics emanating from x and ending at y, and hence $y \in C(x)$. Therefore

$$d(x, y) = \lim_{j\to\infty} \ell_j \ge i(x).$$

This proves the lower semi-continuity of the injectivity radius function i.

We conclude the proof in this case by showing that

$$\lim_{j\to\infty} \ell_j \le \ell.$$

Suppose to the contrary that there exists a point x and a sequence $\{x_j\}$ converging to x such that

$$\lim_{j\to\infty} \ell_j > \ell. \tag{16.4.16}$$

Using (16.4.16) we choose a sufficiently small positive number

$$\varepsilon := \lim_{j\to\infty} (\ell_j - \ell)/2.$$

Let $U_j := U_{x_j} \subset T_{x_j}M$ be the domain such that $\partial U_j = \widetilde{C}(x_j)$. There exists a large number j_0 such that,

$$\bar{B}_j(O, \ell+\varepsilon) \subset U_j, \quad \text{for all } j > j_0.$$

Here we set $\bar{B}_j(O, r) \subset T_{x_j}M$ an r-ball centered at the origin $O \in T_{x_j}M$. Then $\exp_{x_j}|_{\bar{B}_j(O,\ell+\varepsilon)} : \bar{B}_j(O, \ell+\varepsilon) \to B(p_j, \ell+\varepsilon)$ is a smooth embedding. From the continuity of $\Pi : TM \to M \times M$, it follows that

$$\exp_x|_{U_x} : \bar{B}(O, \ell+\varepsilon) \to B(p, \ell+\varepsilon)$$

is an embedding, and hence $\ell := i(x) \geq \ell + \varepsilon$, a contradiction. □

For the proof of the continuity of the injectivity radius function, where M is non compact, we now only need to prove that it is continuous at any point where $i(x) = \infty$. This is achieved by the following:

Lemma 16.4.5 *Let (M, F) be a complete non-compact Finsler manifold. Then the injectivity radius function $i : (M, F) \to [0, \infty]$ is continuous at any point $x \in M$ where $i(x) = \infty$.*

Proof Let $\{x_j\}_{j=1,2,\cdots}$ be a sequence of points converging to x. We then prove that

$$\lim_{j\to\infty} i(x_j) = \infty.$$

Suppose contrary that there exists a sequence of points $\{x_j\}$ converging to x such that $\lim_{j\to\infty} i(x_j) < \infty$.

The same notations as in the previous Lemma 16.4.4 will be used. Let $y := \lim_{j\to\infty} y_j$, where y_j, for every j, is conjugate to x_j along γ_j. We observe that y is conjugate to x along $\gamma := \lim_{j\to\infty} \gamma_j$. However this is a contradiction to $i(x) = \infty$.

Now suppose that for each j the point y_j is not conjugate to x_j along γ_j. We then have two minimizing geodesics $\gamma_j, \sigma_j : [0, \lim_{j\to\infty} \ell_j] \to (M, F)$ joining x_j to y_j. If $\gamma := \lim_{j\to\infty} \gamma_j$ and $\sigma := \lim_{j\to\infty} \sigma_j$, then γ and σ are distinct minimizing geodesics from x to $y := \lim_{j\to\infty} y_j$. Therefore $y \in C(x)$, contradicting to $C(x) = \emptyset$. □

16.4.7 Pointed Blaschke Manifolds

The Riemannian Blaschke manifolds have been fully investigated by Berger and his colleagues and the findings are summarized in [4]. Instead of setting down the curvature assumption, a certain restriction on the diameter and injectivity radius of a compact Finsler manifold is proposed in this respect. Let (M, F) be a compact Finsler n-manifold. We have discussed the Finsler version of the fundamental properties of cut locus and conjugate locus. The diameter $d(M)$ of (M, F) is defined by

$$d(M) := \max \{d(x, y) \mid x, y \in M\}.$$

The injectivity radius $i(M)$ of (M, F) is defined by

$$i(M) := \min \{i(x) \mid x \in M\}.$$

Definition 16.4.1 A Finsler manifold (M, F) is called a *Blaschke Finsler manifold* if

$$d(M) = i(M), \tag{16.4.17}$$

and (M, F) is called a *pointed Blaschke manifold with a base point at $x \in M$* if

$$i(x) = \max \{d(x, y) \mid y \in M\}. \tag{16.4.18}$$

Such a pointed Blaschke manifold with a base point x is denoted by $(M, F : x)$.

We refer to [4] for a discussion on Riemannian Blaschke manifolds. Clearly (16.4.18) holds at each point of M if (16.4.17) is satisfied. A classical result by Berger and Klingenberg states that if (M, g) is a compact simply connected Riemannian manifold whose sectional curvature ranges over $[\frac{1}{4}, 1]$, then $i(M, g) \geq \pi$; see [2, 3], and [23]. Moreover M is homeomorphic to $\mathbf{S}^n$ if $d(M, g) > \pi$, and isometric to one of the symmetric spaces of compact type if $d(M, g) = i(M, g) = \pi$.

We set, for simplicity, $\ell := i(x)$ for a pointed Blaschke Finsler manifold $(M, F : x)$. Then every cut point $y \in C(x)$ has the property that $d(x, y) = \ell$ and that y is the farthest point from x. Therefore the assumptions in Proposition 16.4.2 and Lemma 16.4.3 are satisfied. Let $y \in C(x)$ and set

$$\Gamma_{xy} := \{\gamma : [0, \ell] \to (M, F : x) \mid \gamma(0) = x, \ \gamma(\ell) = y\},$$

and further set

$$A_{xy} := \{\dot{\gamma}(\ell) \mid \gamma \in \Gamma_{xy}\}.$$

We then observe from Lemma 16.4.1 that

$$\partial B(x, r) = \partial B^{-1}(C(x), \ell - r), \ \text{for all } r \in (0, \ell). \tag{16.4.19}$$

The discussion on pointed Blaschke Finsler manifolds is divided into two cases, according to whether the manifold is simply connected or non-simply connected. We first discuss a simpler case, where, roughly speaking, $(M, F : x)$ does not satisfy the Rauch conjecture at x.

Lemma 16.4.6 *Assume that the Rauch conjecture is not valid at the base point $x \in (M, F : x)$, i.e., $\widetilde{C}(x) \cap \widetilde{J}(x) = \emptyset$. Then every point $y \in C(x)$ is joined to x by exactly two distinct minimizing geodesics*

$$\gamma, \sigma : [0, \ell] \to (M, F : x), \ \gamma(0) = \sigma(0) = x, \ \gamma(\ell) = \sigma(\ell) = y,$$

such that y is not conjugate to x along them.

Lemma 16.4.6 is a direct consequence of Lemma 16.4.3 of Berger-Omori. Its proof is omitted. We observe from Lemma 16.4.6 that there exists a fixed-point free involution ψ on Σ_x such that $\psi(\dot{\gamma}(0)) = \dot{\sigma}(0)$. Clearly we have

$$\exp_x \ell u = \exp_x \ell\psi(u) \in C(x), \quad \text{for all} \quad u \in \Sigma_x.$$

Summing up the above discussion we have the following topological conclusion.

Theorem 16.4.3 *Let $(M, F : x)$ is a pointed Blaschke-Finsler manifold with base point x. If the Rauch conjecture is not valid at the base point $x \in (M, F : x)$, then we have*

(1) *the cut locus to x is a smooth hypersurface and diffeomorphic to the quotient space $\Sigma_x/\{\psi : \psi^2 = \mathrm{Id.}\}$, i.e., the cut locus is homeomorphic to a real projective space;*
(2) *the universal cover $\widetilde{M}$ of M is homeomorphic to $\mathbf{S}^n$ and M is homeomorphic to the real projective space;*
(3) *the fundamental group of M is cyclic of order two.*

Remark 6 The assumption in Theorem 16.4.3 is too strong. In fact we prove $\widetilde{C}(x) \cap \widetilde{J}(x) = \emptyset$ if $(M, F : x)$ satisfies

$$\widetilde{C}(x) \setminus \widetilde{J}(x) \neq \emptyset.$$

In the Riemannian case, γ and σ together form a simple closed geodesic loop at x, and $\widetilde{J}(x)$ is a 2ℓ-sphere and $J(x) = \{x\}$. In view of Theorem 16.4.3, we observe that if $(M, F : x)$ is a simply connected pointed Blaschke Finsler manifold, then the Rauch conjecture is valid at x. Moreover, if $(M, F : x)$ is simply connected, we have

$$\widetilde{C}(x) = \widetilde{J}(x).$$

The Berger-Omori Lemma 16.4.3 implies that A_{xy} is a convex set. Moreover the multiplicity of the conjugate point y to x along a minimizing geodesic $\gamma : [0, \ell] \to$

M with $\gamma(0) = x$, $\gamma(\ell) = y$ is independent of the choice of $\gamma \in \Gamma_{xy}$. Since the dimension of the convex sets A_{xy} is lower semi-continuous on $C(x)$, it is constant on $C(x)$. Therefore the rank of the exponential map $\exp_x$ at each point of $C(x)$ is constant. Hence the implicit function theorem implies that $\exp_x \widetilde{C}(x) = C(x)$ is a compact smooth submanifold of M. Thus the set of all points on minimizing geodesics belonging to Γ_{xy} forms a k-dimensional submanifold homeomorphic to $\mathbf{S}^k$, where k is the dimension of A_{xy}, $y \in C(x)$. It follows from the relation (16.4.19) that $B(x, r)$ for $r \in (0, \ell)$ is simply covered by $(k-1)$-dimensional spheres and its quotient space is nothing but $C(x)$. We still have much more discussion to complete this case.

16.5 Busemann Functions and Convex Functions

16.5.1 Busemann Functions

We discuss forward rays and forward Busemann functions on complete non-compact Finsler manifolds.

The definition of a Busemann function is found in §22 of [8]. A forward Busemann function $F_\gamma : (M, F) \to \mathbb{R}$ for a ray $\gamma : [0, \infty) \to (M, F)$ is defined as follows:

$$F_\gamma(x) := \lim_{t\to\infty} (t - d(x, \gamma(t))), \quad x \in M.$$

A backward ray $\gamma : (-\infty, 0] \to (M, F)$ and a backward Busemann function for the backward ray γ are similarly defined. A *super Busemann function* $F_x : (M, F) \to \mathbb{R}$ *at* x is defined by

$$F_x(y) := \sup\{ F_\gamma(y) \mid \gamma \text{ is a forward ray with } \gamma(0) = x\}, \quad y \in M.$$

Clearly, the function $t \to (t - d(x, \gamma(t)))$ is monotone increasing in t and bounded above by $d(x, \gamma(0))$. Thus F_γ is well defined, for $t - d(x, \gamma(t))$ converges uniformly on a compact set and F_γ is locally Lipschitz continuous. A unit speed forward ray $\sigma : [0, \infty) \to (M, F)$ is by definition *asymptotic to* γ if there exists a sequence of unit speed minimizing geodesics $\{\sigma_j : [0, \ell_j] \to (M, F)\}_{j=1,2,\dots}$ such that $\lim_{j\to\infty} \dot\sigma_j(0) = \dot\sigma(0)$, $\sigma_j(\ell_j) = \gamma(t_j)$ for a monotone divergent sequence $\{t_j\}$. The asymptotic relation is in general neither symmetric nor transitive. If (M, g) is a complete and simply connected Riemannian manifold of non-positive sectional curvature, then the asymptotic relation between two rays $\alpha, \beta : [0, \infty) \to (M, g)$ satisfies the following inequality:

$$d(\alpha(t), \beta(t)) \le d(\alpha(0), \beta(0)), \quad \text{for all } t \ge 0.$$

Therefore only in this case the asymptotic relation is an equivalence relation.

The sequence of points $\{\sigma_j(0)\}_{j=1,2,...}$ cannot be chosen to be a point $\sigma(0)$, as is seen in the following example.

Example 16.5.1 Let $\mathcal{F} \subset \mathbb{R}^3$ be a rotation surface of parabola in a Euclidean 3-space. Let $\{(r, \theta) \mid r > 0, \theta \in [0, 2\pi)\}$ be the geodesic polar coordinate system around the pole $(0, 0)$, and $\gamma_\theta : [0, \infty) \to \mathcal{F}$ for $\theta \in [0, 2\pi]$ be the meridian $\gamma_\theta(r) := (r, \theta)$, $r \geq 0$. We observe that all the meridians are asymptotic to each other. In fact, let $\theta_0 \in [0, 2\pi)$ be an arbitrary fixed number and $\{\theta_j\}_{j=1,2,...} \subset [0, 2\pi)$ be a monotone sequence with $\lim_{j\to\infty} \theta_j = \theta_0$. Let $\{r_j\}_{j=1,2,...}$ be a monotone decreasing sequence of positive numbers with $\lim_{j\to\infty} r_j = 0$. If we set $\gamma_j(t) := \gamma_{\theta_j}(t + r_j)$, $t > 0$, then γ_j for each $j = 1, 2, ...$ is asymptotic to γ_0, and hence so is $\gamma_{\theta_0} = \lim_{j\to\infty} \gamma_j$.

Assuming that a ray $\sigma : [0, \infty) \to (M, F)$, $y := \sigma(0)$ is asymptotic to another ray $\gamma : [0, \infty) \to (M, F)$, $x := \gamma(0)$, we say that σ is a *maximal asymptotic ray to* γ *if* σ *is not properly contained in any ray which is asymptotic to* γ. We also say that a ray is *maximal* if and only if it is not properly contained in any ray. A long-standing open problem proposed by Busemann in [6] is stated as follows:

Problem 16.5.1 *Is a maximal asymptotic ray a maximal ray?*

This problem was solved in the negative by Innami in [18] by exhibiting an example of a surface in $\mathbb{R}^3$ on which there is a maximal asymptotic ray which is not a maximal ray.

The local Lipschitz property (1) in Proposition 16.5.1 of F_γ implies that it is differentiable almost everywhere. Then (6) in Proposition 16.5.1 shows that F_γ is differentiable at an interior point of some asymptotic ray to γ. Let $\sigma(0)$ be the initial point of a maximal asymptotic ray to γ. If there exists a unique asymptotic ray to γ passing through $\sigma(0)$, we may view $\gamma(0)$ and $\gamma(\infty)$ as being conjugate pair along γ, (and this corresponds to (1) in Proposition 16.4.2). Otherwise, there exists another ray $\sigma_1 : [0, \infty) \to (M, F)$ which is asymptotic to γ. Therefore we may view the set of all the initial points of rays asymptotic to γ as the cut locus to a point at infinity obtained by $\gamma(\infty)$, (and this corresponds to (2) in Proposition 16.4.2). If F_γ attains its minimum at a point $x \in (M, F)$, then there exists for every unit vector $u \in \Sigma_x$ a ray $\sigma : [0, \infty) \to (M, F)$ asymptotic to γ such that $g_u(u, \dot\sigma(0)) \geq 0$. This corresponds to Lemma 16.4.2.

16.5.2 Properties of Busemann Functions on (M, F)

We denote by $F_\gamma^{-1}(\{a\})$ and $F_\gamma^{-1}(-\infty, a]$ the a-level set and the a-sublevel set of F_γ respectively. The basic properties of Busemann functions are stated in §§22 and 23 of [8] and those on complete Finsler and Riemannian manifolds (M, F) are summarized in [40] and [39] as follows:

Proposition 16.5.1 (Properties of Busemann Functions) *Let* $\gamma : [0, \infty) \to (M, F)$ *be a forward ray and* $F_\gamma : (M, F) \to \mathbb{R}$ *a Busemann function for* γ*. We then have:*

(1) F_γ *is locally Lipschitz.*
(2) *A level set* $F_\gamma^{-1}(\{a\})$ *for* $a \in F_\gamma(M)$ *is obtained by*

$$F_\gamma^{-1}(\{a\}) = \lim_{t\to\infty} S^{-1}(\gamma(t), t-a).$$

(3) *If* $a, b \in F_\gamma(M)$ *satisfies* $a \le b$*, then*

$$F_\gamma^{-1}(-\infty, a] = \{y \in F_\gamma^{-1}(-\infty, b] \mid d(y, F_\gamma^{-1}(\{b\}) \ge b - a\},$$

and

$$\begin{aligned} F_\gamma^{-1}(\{a\}) &= \{y \in F_\gamma^{-1}(-\infty, b]) \mid d(y, F_\gamma^{-1}(\{b\})) = b - a\} \\ &= S^{-1}(F_\gamma^{-1}(\{b\}), b-a) \cap F_\gamma^{-1}(-\infty, b]. \end{aligned}$$

(4) *A unit speed geodesic* $\sigma : [0, \infty) \to (M, F)$ *is a forward ray asymptotic to* γ *if and only if*

$$F_\gamma \circ \sigma(t) = t + F_\gamma \circ \sigma(0), \quad \text{for all } t \ge 0.$$

(5) *If* $x \in M$ *and* $a \in F_\gamma(M)$ *satisfy* $a > F_\gamma(x)$*, and if* $\sigma : [0, \ell] \to (M, F)$ *is a unit speed minimizing geodesic with* $\sigma(0) = x$ *such that* $\sigma(\ell)$ *is a foot of* x *on* $F_\gamma^{-1}(\{a\})$*, then the extension* $\sigma : [0, \infty) \to (M, F)$ *of* σ *is a forward ray asymptotic to* γ*.*
(6) F_γ *is differentiable at a point* $x \in M$ *if* x *is an interior of some ray asymptotic to* γ*.*

A detailed proof of Proposition 16.5.1 on Riemannian manifolds was given in [39] and the same proof technique for the Finsler case is seen in [40]. The proof is omitted here.

In the pioneering works [16] and [10], the authors have proved that a Busemann function on a complete and non-compact Riemannian manifold (M, g) is strongly convex if its sectional curvature is positive (see [16]) and convex if its sectional curvature is non-negative (see [10]). In particular, every super Busemann function is a convex exhaustion if its sectional curvature is non-negative. If the minimum set of a super Busemann function has non-empty boundary, then the negative of the distance function on the minimum set to the boundary is convex, and hence attains its minimum. Thus by iterating this, a totally convex compact totally geodesic submanifold without boundary, called a soul of M, is found in the minimum set. The well-known Cheeger–Gromoll structure Theorem (see [10]) states that a complete non-compact Riemannian manifold is homeomorphic to the normal bundle over the

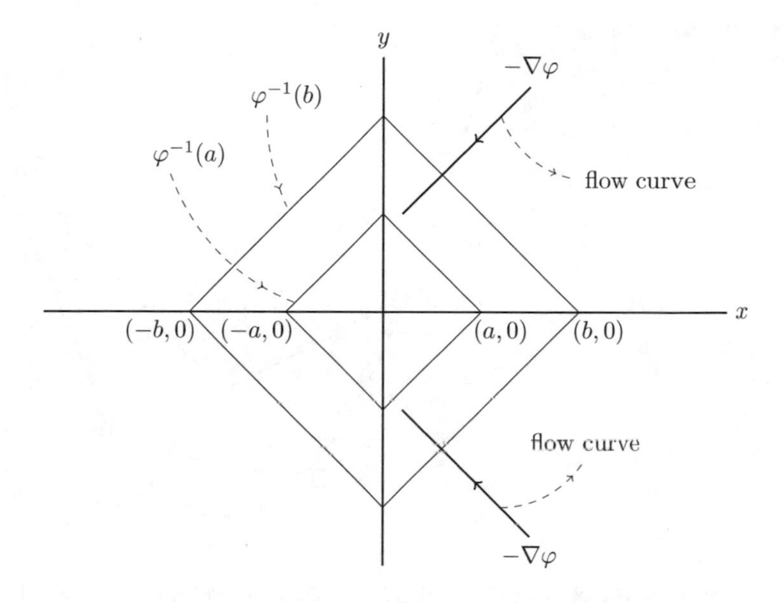

Fig. 16.2 Flows of $-grad(\phi)$

soul in M. If the sectional curvature is positive, the soul is a point, and hence M is diffeomorphic to $\mathbb{R}^n$. The Sharafutdinov construction [38] of flow curves along the negative of the subgradient of a Busemann function gives a distance non-increasing correspondence between two such flow curves. This was employed by Perelmann [32] for the proof of the famous soul conjecture.

A simple example is seen here (Fig. 16.2).

Example 16.5.2 Let $\varphi(x, y) := |x| + |y|$ for $(x, y) \in \mathbb{R}^2$ be a convex function. It is clear that the distance function is monotone non-increasing along two flow curves of $-\text{grad}(\varphi)$.

We do not know however, if anologues of the above results stated in [10, 16] and in Example 5.2 are valid on complete Finsler manifolds with positive (or non-negative) flag curvature (Flag curvature in Finsler geometry is an analogue of sectional curvature in Riemannian geometry; for details see [1]).

16.5.3 Convex Functions

A function $\varphi : (M, F) \to \mathbb{R}$ is said to be *convex* if along every geodesic $\gamma : [a, b] \to (M, F)$, the restriction $\varphi \circ \gamma : [a, b] \to \mathbb{R}$ is convex:

$$\varphi \circ \gamma((1 - \lambda)a + \lambda b) \leq (1 - \lambda)\varphi \circ \gamma(a) + \lambda\varphi \circ \gamma(b), \quad 0 \leq \lambda \leq 1 \tag{16.5.20}$$

Fig. 16.3 Graph of a convex function

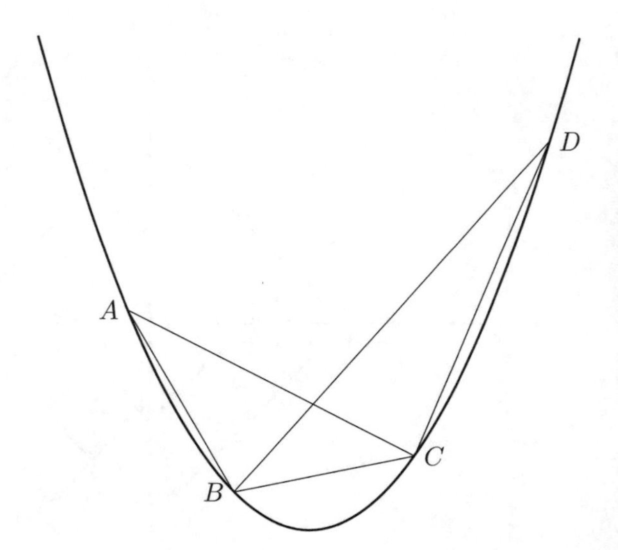

If the inequality in the above (16.5.20) is strict for all γ and for all $\lambda \in (0, 1)$, φ is called *strictly convex*, and *strongly convex* if the second order difference quotient, namely $\{\varphi o \gamma(h) - \varphi o \gamma(-h) - 2\varphi o \gamma(0)\}/h^2$ is positive for all γ and all $\lambda \in (0, 1)$. In the special case where equality in (16.5.20) holds for every γ and for every $\lambda \in [0, 1]$, the function is called an *affine function*. If a non-trivial convex function φ is constant on an open set, then it assumes its minimum on the open set and the number of components of a level set $\varphi^{-1}(\{a\})$, $a > \inf_M \varphi$, is equal to that of the boundary components of the minimum set of φ. A convex function φ is said to be *locally non-constant* if it is not constant on any non-empty open set of M. From now on, *we always assume that our convex functions are locally non-constant.*

The slope inequality of a one variable convex function is elementary, useful and employed throughout this section. Let $f : (\alpha, \beta) \to \mathbb{R}$ be a convex function. Let $\alpha < a < b < c < d < \beta$ and $A := (a, f(a))$, $B := (b, f(b))$, $C := (c, f(c))$ and $D := (d, f(d))$ be points on the graph of f (see Fig. 16.3). The slope inequality is expressed as

$$\text{slope}(AB) \leq \text{slope}(AC) \leq \text{slope}(BC) \leq \text{slope}(BD) \leq \text{slope}(CD). \quad (16.5.21)$$

We observe from (16.5.21) that the right and left derivatives $f'_+(t)$ and $f'_-(t)$ of f exist and $f'_+(t) \geq f'_-(t)$ for all $t \in (\alpha, \beta)$, and the equality holds if and only if f is differentiable at t.

The topology of Riemannian manifolds admitting locally nonconstant convex functions has been investigated in [14] and [13]. The topology of complete Alexandrov surfaces admitting locally nonconstant convex functions has been studied in [27] and [25]. The classification of Busemann G-surfaces admitting convex functions has been obtained in [17]. The isometry groups of complete Riemannian manifolds admitting strictly convex functions have been discussed in [45]. There are several extensions of convex functions, such as peakless functions introduced by Busemann [8], uniformly locally convex filtrations [46], etc. The splitting theorem for Alexandrov surfaces admitting affine functions has been established in [26]. Also in [27], the condition for compact Alexandrov surfaces to admit locally non-constant convex functions has been studied. A detailed discussion of convex sets on Riemannian manifolds of non-negative curvature is carried out in [5]. It is emphasized that the notion of convexity makes sense, irrespective of whether the distance function is symmetric or not. Hence the extended notion of convex functions will be discussed on Finsler manifolds.

We define the ends of a non-compact manifold M and proper maps on M, which are useful for the investigation of topology of manifolds admitting convex functions.

Definition 16.5.1 Let M be a noncompact manifold and $C_1 \subset C_2 \subset M$ be compact sets. Then, each component of $M \setminus C_2$ is contained in a unique component of $M \setminus C_1$. An end of X is by definition an element of the inverse limit system $\{$components of $M \setminus C$; C is compact$\}$ directed by the inclusion relation.

For example, if M is a compact manifold from which k points are removed, then M has k ends. $\mathbb{R}^1$ has two ends. For $n > 1$, $\mathbb{R}^n$ has one end, for it is homemorphic to S^n with one point removed. A cylinder $\mathbf{S}^{n-1} \times \mathbb{R}$ has two ends, for it is homeomorphic to S^n from which two points are removed.

Definition 16.5.2 A map $f : M \to \mathbb{R}$ is said to be *proper* if $f^{-1}(K)$ is compact for all compact set $K \subset \mathbb{R}$ and f is said to be an *exhaustion* of M if $f^{-1}(-\infty, a]$ is compact for all $a \in f(M)$.

We shall review the topology of geodesically complete Finsler manifolds admitting locally non-constant convex functions. The following Propositions are basic and important facts and have already been established in Riemannian geometry, (see [14] and [13]).

Without assuming the continuity of a convex function on (M, F), we have the following proposition. For its proof see [14, 36]:

Proposition 16.5.2 *Any convex function on* (M, F) *is locally Lipschitz.*

16.5.4 Riemannian and Finslerian Results on Convex Functions

The assumption on a convex function to be locally non-constant, as was introduced, is necessary. For, we can construct on every noncompact manifold a complete Riemannian metric and a non-trivial smooth convex function whose minimum set contains a non-empty open set. Therefore the existence of such a non-trivial convex function gives no restriction on the topology of a manifold. We first discuss the level sets of a locally nonconstant convex function.

Proposition 16.5.3 *[compare Proposition 2.3 in [14]] Let $\varphi : (M, F) \to \mathbb{R}$ be a convex function and $a > \inf_M \varphi$. Then the a-level set $\varphi^{-1}(\{a\})$ is a topological submanifold of dimension $n - 1$.*

Proposition 16.5.4 *Let $C \subset (M, F)$ be a closed locally convex set. Then there exists a totally geodesic submanifold W of M such that $W \subset C$ and its closure is C.*

Proof Since C is locally convex, every point $x \in C$ admits a convex set $B(x, r) \cap C$ for some $r \in (0, \delta(x))$. If y is a point in $B(x, r) \cap C$, then $\gamma_{xy} : [0, d(x, y)] \to (M, F)$ and $\gamma_{xy} : [0, d(y, x)] \to (M, F)$ are contained entirely in $B(x, r) \cap C$. Clearly, its interior is a totally geodesic submanifold of (M, F) of dimension at least one, contained in $B(x, r) \cap C$. Thus x is contained in a $k(x)$-dimensional totally geodesic submanifold which is contained entirely in $B(x, r) \cap C$ such that $k(x)$ is maximal dimension of all such totally geodesic submanifolds in $B(x, r) \cap C$. Setting $k = \max_{x \in C} k(x)$, we have a k-dimensional totally geodesic submanifold, say, $W(x)$ of (M, F) contained in $B(x, r) \cap C$. Suppose $W(x) \cap B(x, r) \subsetneqq C \cap B(x, r)$. We then find a point $z \in B(x, r) \cap (C \setminus W(x))$. Clearly, we have

$$d(z, \overline{W(x)}) \geq 0 \text{ and } \dot{\gamma}_{xz}(0) \text{ is transversal to the tangent space } T_x W(x) \text{ at } x.$$

Thus we find a small open set $\Omega \subset W(x)$ and a family of minimizing geodesics emanating from points on Ω and ending at z, whose initial vectors are transversal to $TW(x)$. We thus get a cone consisting of minimizing geodesics

$$\gamma_{yz} : [0, d(y, z)] \to B(x, r) \cap C, \quad y \in \Omega,$$

which is contained entirely in C and forms a totally geodesic submanifold of dimension $k + 1$, a contradiction.

Let $W = \cup_{x \in C} W(x) \subset C$. Again the transversality argument with $W(x)$ implies that W is a smooth totally geodesic submanifold of maximal dimension in C.

We finally prove that the closure $\overline{W}$ of W coincides with C. To prove this, suppose that there exists a point $z \in C \setminus \overline{W}$. We then find a point $y \in \overline{W}$ such that $d(z, y) = d(z, \overline{C})) > 0$. Let $T_y \overline{W} \subset T_y M$ be the linear subspace obtained as the limit $\lim_{j \to \infty} T_{y_j} W$; $y_j \in W$, $\lim_{j \to \infty} y_j = y$. If $\dot{\gamma}_{zy}(d(z, y))$ is transversal to $T_y \overline{W}$. Then the above argument shows the existence of a $(k+1)$-dimensional totally geodesic smooth submanifold in C, which is a contradiction. Therefore, we have $\dot{\gamma}_{zy}(d(z, y)) \in T_y \overline{W}$, and hence $\gamma_{zy}(0, d(z, y)) \subset W$. This proves $C = \overline{W}$. □

16.5.5 Level Set Configurations

An elementary observation based on the slope inequality (16.5.21) gives the following simple fact on a locally non-constant convex function $\varphi : (M, F) \to \mathbb{R}$ (and $\varphi : (M, g) \to \mathbb{R}$).

If there exists a compact level $\varphi^{-1}(\{a\}) \subset (M, g)$, then so are all the other levels. If $\varphi^{-1}(\{a\}) \subset (M, F)$ is compact, then so are $\varphi^{-1}(\{b\})$ for all $b \geq a$.

The proof is sketched as follows: Suppose there is a non-compact level $\varphi^{-1}(\{c\}) \subset (M, F)$. There is a sequence of minimizing geodesics emanating from an arbitrary fixed point $x \in \varphi^{-1}(\{a\})$ and ending at points $y_j \in \varphi^{-1}(\{c\})$, $j = 1, 2, \ldots$ where $\{y_j\}_{j=1,2,\ldots}$ is a sequence of points with $\lim_{j\to\infty} d(x, y_j) = \infty$. We then choose a ray obtained as the limit of these minimizing geodesics, along which φ must be bounded above by c. This means that this ray is contained in $\varphi^{-1}(\{a\})$, a contradiction.

Lemma 16.5.1 *Let $\varphi : (M, F) \to \mathbb{R}$ be a locally non-constant convex function. Let $\varphi^{-1}(\{a\})$ be a compact level. Then, $\varphi^{-1}[a, b]$ for a fixed $b > a$ is homeomorphic to the product $\varphi^{-1}(\{a\}) \times [a, b]$.*

Proof The slope inequality 16.5.21 plays an important role in this. We first choose two numbers $a_0 \in (\inf_M \varphi, a)$ and $a_{k+1} > b$ and let $\delta := \delta(\varphi^{-1}[a_0, a_k])$ be the convexity radius over $\varphi^{-1}[a_0, a_k]$. Take a sequence of real numbers

$$a_{k+1} > a_k := b > a_{k-1} > \cdots > a_1 := a > a_0,$$

such that for each integer $i = 2, \cdots, k+1$ and for each point $x \in \varphi^{-1}(\{a_i\})$, we find a unique foot $f(x)$ of x on $\varphi^{-1}(\{a_{i-2}\})$.

Let $x_k \in \varphi^{-1}(\{a_k\})$ be an arbitrary point. We then find a unique point $x_{k+1} \in \varphi^{-1}(\{a_{k+1}\})$ such that x_k belongs to the interior of $T(x_{k+1}, f(x_{k+1}))$, where we denote by $T(x_{k+1}, f(x_{k+1}))$ the unique foot of x_{k+1} on $\varphi^{-1}(\{a_{k-1}\})$.

The uniqueness of feet implies that there is a homeomorphism between $\varphi^{-1}(\{a_k\})$ and $\varphi^{-1}(\{a_{k-1}\})$ via the correspondence $x_k \mapsto x_{k-1}$. Thus we have a homeomorphism between $\varphi^{-1}[a_{k-1}, a_k]$ and $\varphi^{-1}(\{a_k\}) \times [a_{k-1}, a_k]$ through the feet.

By iteration, we have for an arbitrary fixed point $x_k \in \varphi^{-1}(\{a_k\})$, a sequence of points and a minimizing geodesics

$$\{x_{k+1}, x_k, \cdots, x_1\} \text{ and } \{T(x_{k+1}, f(x_{k+1})), T(x_k, f(x_k)), \cdots, T(x_2, f(x_2))\}$$

which satisfies the conditions

$$d(x_i, x_{i-1}) > d(x_i, f(x_{i+1})), \quad i = k, \cdots, 2.$$

The right derivative of $\varphi o T(x_{k+1}, \ldots, x_1)$ is monotone increasing (this is evident from the Figs. 16.4 and 16.5). The slope inequality then implies that the right and left derivatives of $\varphi \circ T(x_{i+1}, f(x_{i+1}))$ at x_i are larger than those of $\varphi \circ T(x_i, f(x_i))$ at x_i. Therefore, if φ is restricted to the union of broken geodesics

$$T(x_k, x_{k-1}, \ldots, x_1) := T(x_k, x_{k-1}) \cup T(x_{k-1}, x_{k-2}) \cup \cdots \cup T(x_2, x_1), \qquad (16.5.22)$$

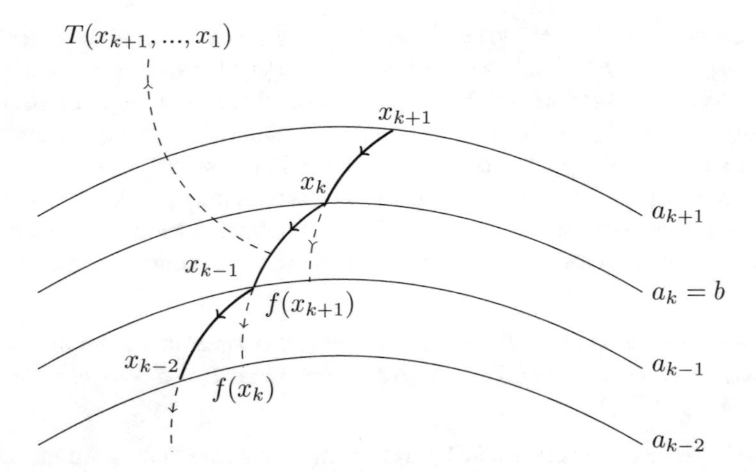

Fig. 16.4 Broken geodesics through levels of a convex function-I

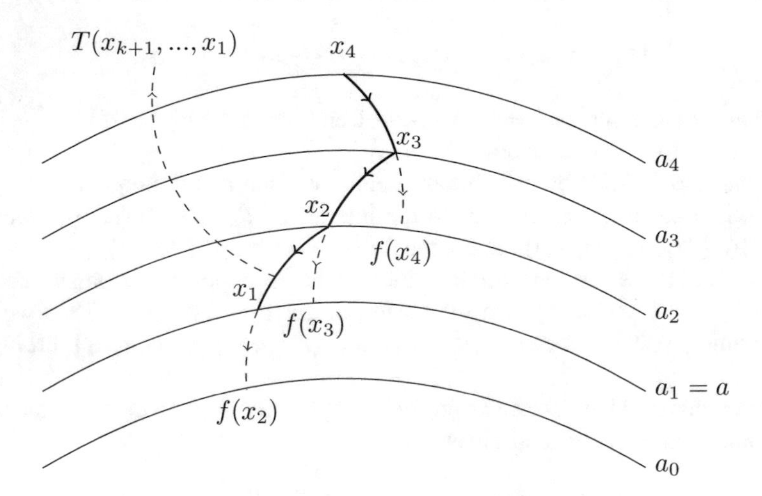

Fig. 16.5 Broken geodesics through levels of a convex function-II

then, it is monotone and convex. Clearly, the right and left derivatives at every point of $T(x_k, x_{k-1}, \ldots, x_1)$ are bounded above by a negative number $\mu = \mu(a_0, a, \delta)$; here μ is defined by

$$-\mu := \frac{a - a_0}{\max\{d(x, \varphi^{-1}(\{a_0\}))|x \in \varphi^{-1}(\{a\})\}}.$$

This means that the length of $T(x_k, x_{k-1}, \ldots, x_1)$ is bounded above by $\frac{(b-a)}{\mu(a_0,a,\delta)} < 0$. This completes the proof (Figs. 16.4 and 16.5). □

Let $\varphi : (M, g) \to \mathbb{R}$ be a locally non-constant convex function on a complete Riemannian manifold. The Sharafutdinov construction of flow curves along $-\text{grad}(\varphi)$ implies that the diameter function $t \mapsto \text{diam}(\varphi^{-1}(\{t\}))$ is monotone non-decreasing. Here we set

$$\text{diam}(\varphi^{-1}(\{t\})) := \sup\{d(x, y) \mid x, y \in \varphi^{-1}(\{t\}) \subset (M, g)\}. \tag{16.5.23}$$

The monotone property of $t \mapsto \text{diam}(\varphi^{-1}(\{t\}))$ may be roughly explained as follows:

Let $C \subset (M, g)$ be a closed convex set and $x, y \in M \setminus C$ be taken sufficiently close to C such that there exist unique foot $f(x), f(y)$ of x, y on C respectively. We observe that $d(x, y) \geq d(f(x), f(y))$. This shows that the diameter function has everywhere non-negative derivative (see [13, 14]). We therefore get that if φ admits a level that is compact, then so are all the others.

However we do not know if the monotone property of the diameter function is valid for Finsler manifolds. Irrespective of whether the diameter function is monotone or not, we get from Lemma 16.5.1 the following

Theorem 16.5.1 (See Theorem 1.1 in [36]) *Let (M, F) be a complete Finsler manifold and $\varphi : (M, F) \to \mathbb{R}$ be a locally non-constant convex function whose level sets are all compact. Then we have the following:*

(1) *If $\varphi^{-1}(\{c\})$ is connected for some $c > \inf_M \varphi$, then there exists a homeomorphism $H : \varphi^{-1}(\{c\}) \times (\inf_M \varphi, \infty) \to M$ such that*

  (a) *$H(x, t) \in \varphi^{-1}(\{t\})$ for all $(x, t) \in \varphi^{-1}(\{c\}) \times (\inf_M \varphi, \infty)$.*

  (b) *If $a, b \in \varphi(M)$, $a < b$, we then have $H(x, [a, b]) = T(x_k, x_{k-1}, \ldots, x_1)$ as defined in (16.5.22).*

(2) *If φ attains its infimum, say $m := \inf_M \varphi$, then M is homeomorphic to the normal bundle over $\varphi^{-1}(\{m\})$ in M.*

(3) *If there is a disconnected level, then φ attains its minimum $m = \inf_M \varphi$, and $\varphi^{-1}(\{m\})$ is a compact totally geodesic smooth hypersurface with trivial normal bundle. Moreover, M is homeomorphic to $\varphi^{-1}(\{m\}) \times \mathbb{R}$.*

Remark 7 Without the assumption of the existence of a compact level of $\varphi : (M, g) \to \mathbb{R}$, all the above statements are still valid in the Riemannian case. However we do not yet know this in the Finsler case.

16.5.6 Properness of Exponential Maps

The slope inequality of convex functions along geodesics leads us to the properness of the exponential maps on manifolds with convex functions. Clearly the exponential map $\exp_x : T_xM \to M$ at each point on a complete and simply connected

Riemannian manifold M gives a diffeomorphism, and hence it is proper. The proof is sketched as follows:

First of all, let (M, g) be a complete non-compact Riemannian manifold of positive sectional curvature. Then a super Busemann function $F_x : (M, g) \to \mathbb{R}$ at a point $x \in M$ is a strictly convex exhaustion. Under this condition Gromoll and Meyer [16] proved that the exponential map $\exp_x : T_x M \to M$ is proper. In fact, suppose to the contrary that there is a compact set $K \subset M$ such that $\exp_x^{-1}(K)$ is non-compact. Then there exists a divergent sequence $\{u_j\}_{j=1,2,\ldots} \subset T_x M$ of vectors with $\lim_{j\to\infty} \|u_j\| = \infty$ such that $\exp_x u_j \in K$ for all $j = 1, 2, \ldots$. Thus we find a geodesic $\gamma : [0, \infty) \to (M, g)$ such that $\varphi \circ \gamma$ is bounded above, and hence it is constant. This is impossible, for φ is strictly convex. Hence the exponential map $\exp_x : T_x M \to M$ is proper. The properness of exponential map on Finsler manifold has recently been extended as follows:

Theorem 16.5.2 (See [22]) *If (M, F) is a geodesically complete non-compact Finsler manifold, and if $\varphi : (M, F) \to \mathbb{R}$ is a strictly convex exhaustion function, then the exponential map at each point of (M, F) is proper.*

Notice that the exhaustion property in Theorem 16.5.2 is needed for the conclusion to hold. For instance, let $\mathcal{F} \subset \mathbb{R}^3$ be a surface of revolution with profile curve $y = e^x$, $x \in \mathbb{R}$. Then the exponential map is not proper at any point of $\mathcal{F}$.

16.5.7 Number of Ends

The number of ends of complete Riemannian manifolds admitting locally non-constant convex functions is estimated by using the slope inequality (16.5.21), Lemma 16.5.1 and Theorem 16.5.1.

Theorem 16.5.3 (Ends of (M, g), [14]) *Let (M, g) be a connected geodesically complete Riemannian manifold admitting a locally nonconstant convex function φ.*

(1) *If φ has a noncompact level, then M has one end.*
(2) *If φ assumes its minimum and if it admits a compact level, then M has one end.*
(3) *If φ has a disconnected compact level, then M has two ends.*
(4) *If φ has a compact level and if its infimum is not attained, then M has two ends.*

However the ends of geodesically complete Finsler manifolds admitting locally non-constant convex functions have not been fully understood yet. We do not know any example of a convex function $\varphi : (M, F) \to \mathbb{R}$ with compact and non-compact levels simultaneously.

16.5.8 Isometry Groups

Let (H, g) be a Hadamard manifold, namely H is a complete and simply connected Riemannian manifold of non-positive sectional curvature. Then the distance function $d(x, .)$ from a fixed point $x \in H$ is convex, with a unique minimum point at x. A well-known classical theorem by Cartan states that if G is a compact subgroup of the isometry group $I(H)$ of H, then it has a common fixed point. In fact, if $x \in H$ is a fixed point, then the G-orbit $G(x)$ of x is compact, and hence there exists a unique smallest ball $B(y, r)$ with $G(x) \subset \bar{B}(y, r)$. Clearly $B(y, r)$ is invariant under the action of G, and hence the center y is fixed under the actions of G.

We finally discuss how the existence of a convex function on (M, g) and (M, F) influences the group of isometries on them. The splitting theorem for Riemannian manifolds admitting affine functions has been discussed in [19]. It is proved in [15] that if (M, g) is a complete Riemannian manifold with non-compact isometry group and if (M, g) admits a convex function without minimum whose levels are all compact, then (M, g) is isometric to the Riemannian product $N \times \mathbb{R}$, where N is a compact smooth manifold. In [10] Cheeger and Gromoll constructed the compact totally convex filtration obtained by a super Busemann function on a complete non-compact Riemannian manifold of non-negative sectional curvature. They proved:

Theorem 16.5.4 *A complete Riemannian manifold (M, g) of non-negative sectional curvature splits off isometrically as the product:*

$$M = \overline{M} \times \mathbb{R}^k,$$

where the isometry group $I(\overline{M})$ of $\overline{M}$ is compact and $I(M) = I(\overline{M}) \times I(\mathbb{R}^k)$.

Without assuming that the sectional curvature is non-negative, there are some results on the relation between the isometry groups and convex functions defined on (M, g) and on (M, F) respectively.

Theorem 16.5.5 (See [45]) *Let (M, g) be a complete Riemannian manifold admitting a strictly convex function $\psi : (M, g) \to \mathbb{R}$. We then have*

(1) *If ψ admits a minimum, then every compact subgroup G of the isometry group of (M, g) has a common fixed point.*
(2) *If ψ has a compact level and if it has no minimum, then the group of isometries of (M, g) is compact.*

Proof For the proof of (1), we denote by μ the Haar measure on G, normalized by $\int_G d\mu = 1$. We define a function $\Psi : (M, g) \to \mathbb{R}$ by $\Psi(x) := \int_G \psi(gx)d\mu(g)$, $x \in M$. Clearly, Ψ is strictly convex. Since G is compact, Ψ attains its minimum. The strict convexity of Ψ means that the minimum set of Ψ consists of a single point. It follows from the construction of Ψ that the minimum set is a common fixed point of G.

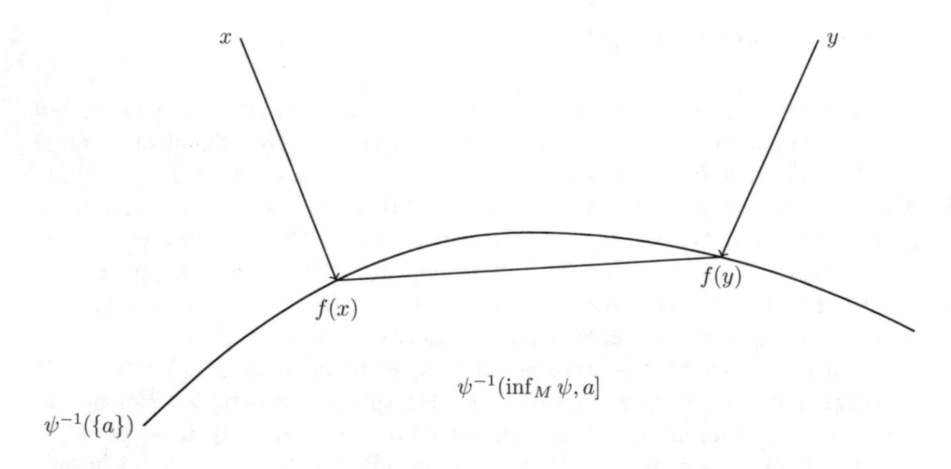

Fig. 16.6 Feet on a level set of a convex function

The strictly increasing property of the diameter function defined in (16.5.23) plays an important role for the proof of (2). This fact can intuitively be understood as follows:

Choose numbers $\inf_M \Psi < a < b$ such that $d(x, \psi^{-1}(\{a\}))$, for every $x \in \psi^{-1}(\{b\})$, is less than the convexity radius on the compact set $\psi^{-1}[a, b]$. If $x, y \in \psi^{-1}(\{b\})$ are sufficiently close to each other and if $f(x)$ and $f(y)$ are feet on $\psi^{-1}(\inf_M \Psi, a]$, we then have $d(x, y) > d(f(x), f(y))$, (see Fig. 16.6).

Roughly speaking, this is because of the angle property: $\measuredangle(x, f(x), f(y)) > \frac{\pi}{2}$ and $\measuredangle(y, f(y), f(x)) > \frac{\pi}{2}$. Here $\measuredangle(x, f(x), f(y))$ is the angle between two vectors at $f(x)$ tangent to minimizing geodesics joining $f(x)$ to x and $f(x)$ to $f(y)$. This infinitesimal version of the above observation will give the Sharafutdinov construction of the distance non-increasing strong deformation retract.

For the proof of (2), we argue by deriving a contradiction. Suppose that the isometry group $\mathbf{G}$ of M is non-compact. Then the orbit $\mathbf{G}(x)$ of an arbitrary point $x \in (M, g)$ forms an unbounded set. We know from Theorems 16.5.1 and 16.5.3 that M is homeomorphic to $\psi^{-1}(\{a\}) \times \mathbb{R}$, where $a := \psi(x)$. Since the diameter function $\mathrm{diam}_\psi(t)$ of ψ is strictly increasing, every isometry g_1 of (M, g) fixes each end of M. We may chose an element $g_1 \in \mathbf{G}$ so as to satisfy: $g_1 \circ \psi^{-1}(\{a\})$ is contained in $\psi[b, c]$, where $b - a$ is sufficiently large. Thus the diameter function of ψ satisfies $\mathrm{diam}_\psi(b) > \mathrm{diam}_\psi(a)$. We then choose a proper curve $\alpha : (\inf_M \psi, \infty) \to (M, g)$ such that $\psi \circ \alpha$ is strictly increasing and $\alpha[b, c]$ does not meet $g_1 \circ \psi^{-1}(\{a\})$ (see Fig. 16.7). It obviously follows that $g_1 \circ \alpha : (\inf_M \psi, \infty) \to (M, g)$ does not pass through any point of $\psi^{-1}(\{a\})$ and join the two ends of M, a contradiction. □

We know very little about the isometry groups of complete Finsler manifolds admitting convex functions. Proof of the following result can be found in [22].

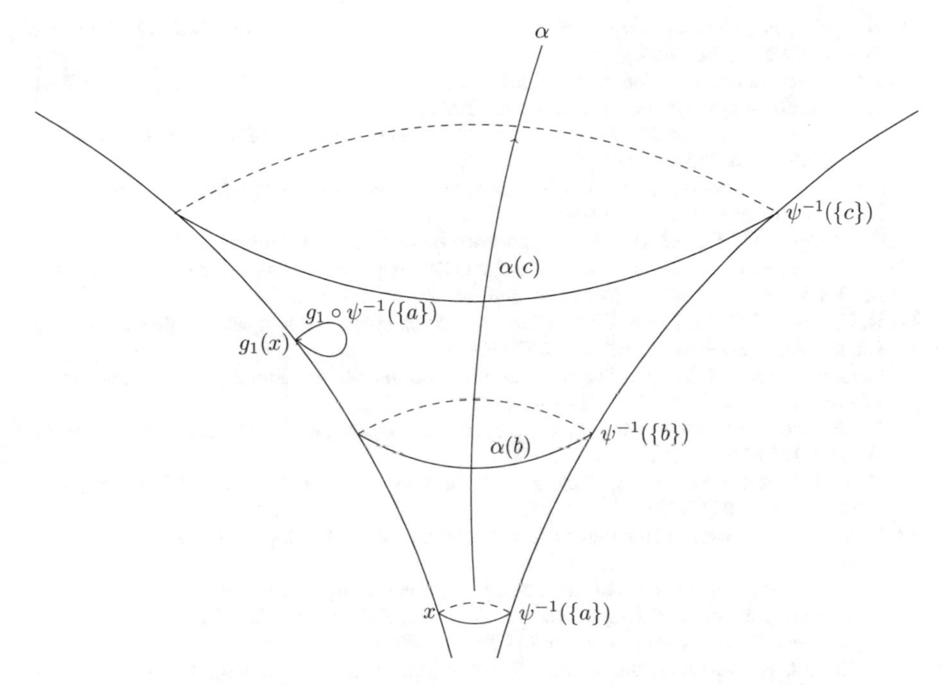

Fig. 16.7 Increasing diameter of a convex function

Theorem 16.5.6 (See [22]) *Let $\psi : (M, F) \to \mathbb{R}$ be a strictly convex exhaustion function. Then every compact subgroup of the group of isometries on (M, F) has a common fixed point.*

Acknowledgement The author's "Katsuhiro Shiohama" work was supported by JSPS KAKENHI Grant Number15K04864.

References

1. D. Bao, S.S. Chern and Z. Shen, *An Introduction to Riemannian-Finsler Geometry*, **200**, Graduate Texts in Math. Springer-Verlag, New York, 2000.
2. M. Berger, Sur les variétés a courbure positive de diamètre minimum, *Comment. Math. Helv.*, **35**, (1961), 28–34.
3. M. Berger, Sur quelques variétés riemannienne $\frac{1}{4}$-pincèes, *Bull. Soc. Math. France*, **88**, (1960), 57–71.
4. A. Besse, *Manifolds all of whose geodesics are closed*, Springer, Berlin (1978).
5. U. Burago and D. Zalgallar, Convex set in Riemannian spaces of non-negative curvature, *Russian Mathematical Surveys*, **32:3** (1977), 1–57 (Uspehki Mat. Nauk **32:3** (1977), 3–55.)
6. H. Busemann, *Recent Synthetic Differential Geometry*, Springer-Verlag, (1970).
7. H. Busemann, *Selected works*, 2 volumes, (ed. A. Papadopoulos), Springer Verlag, 2018.

8. H. Busemann, *The Geometry of Geodesics*, Pure and Applied Mathematics, **VI**, Academic Press Inc. New York, 1955.
9. J. Cheeger and D. Ebin, *Comparison Theorems in Riemannian Geometry,* AMS Chelsia Publ. Amer. Math. Soc. Provdence, Rhode Island, 2008.
10. J. Cheeger and D. Gromoll, On the structure of complete manifolds of non-negative curvature. *Ann. of Math.* **96**, 413–443 (1972).
11. S.S. Chern, Finsler Geometry is just Riemannian Geometry without the Quadratic Restrictions, *Notices of the AMS*, (1996), 959–963.
12. S.S. Chern, W.H. Chen and K.S. Lam, *Lectures on Differential Geometry,* 2005.
13. R. Greene and K. Shiohama, Convex functions on complete noncompact manifolds: Differentiable structure, *Ann. Scient. Ëc. Norm. Sup.* **14**, 357–367 (1981).
14. R. Greene and K. Shiohama, Convex functions on complete noncompact manifolds; Topological structure, *Invent. Math.* **63**, 129–157 (1981).
15. R. Greene and K. Shiohama, The isometry groups of manifolds admitting nonconstant convex functions, *J. Math. Soc. Japan,* **39** 1–16 (1987).
16. D. Gromoll and W. Meyer, On complete open manifolds of positive curvature. *Ann. of Math.* **90**, 75–90 (1969).
17. N. Innami, A classification of Busemann G-surfaces which possess convex functions, *Acta Math.* **148**, 15–29 (1982).
18. N. Innami, On the terminal points of co-rays and rays, *Arch. Math. (Basel)*, **45**, No.5, 468–470 (1985).
19. N. Innami, Splitting theorems of Riemannian manifolds, *Compositio Math.* **47** 237–247 (1982).
20. N. Innami, Y. Itokawa, T. Nagano and K. Shiohama, *Blaschke Finsler Manifolds and actions of projective Randers changes on cut loci,* Preprint 2017.
21. N. Innami, K. Shiohama and T. Soga, The cut loci, conjugate loci and poles in a complete Riemannian manifold, *GAFA*, **22**, 1400–1406 (2012).
22. Y. Itokawa, K. Shiohama and B. Tiwari, Strictly convex functions on complete Finsler manifolds, *Proc. Indian Acad. Sci. (Math. Sci.)* **126**, No. 623–627, (2016).
23. W. Klingenberg, Contributions to Riemannian geometry in the large, *Ann. of Math.* **69**, (1959), 654–666.
24. W. Klingenberg, *Riemannian Geometry,* de Gruyter Studies in Math. **1**, de Gruyter, Berlin (1982).
25. Y. Mashiko, A splitting theorem for Alexandrov spaces, *Pacific J. Math.* **204**, 445–458 (2002).
26. Y. Mashiko, Affine functions on Alexandrov surfaces, *Osaka J. Math.* **36** 853–859 (1999).
27. Y. Mashiko, Convex functions on Alexandrov surfaces, *Trans. Amer. Math. Soc.* **351**, no. 9, 3549–3567 (2006).
28. M. Matsumoto, A slope of a mountain is a Finsler surface with respect to a time measure, *J. Math. Kyoto Univ.*, **29**(1989), no. 1, 17–25.
29. S.B. Myers, Connections between differential geometry and topology I, *Duke Math. J.* **1**, 376–391 (1935).
30. S.B. Myers, Connections between differential geometry and topology II, *Duke Math. J.* **2**, 95–102 (1936).
31. H. Omori, A class of riemannian metrics on a manifold, *J. Differential Geom.*, **2**, (1968), 233–252.
32. G. Perelman, Proof of soul conjecture of Cheeger-Gromoll, *J.Diff. Geom.* **40**, (1994), 299–305.
33. H. Rademacher, *Nonreversible Finsler metrics of positive flag curvature*, Riemann-Finsler Geometry, MSRI Publications, **50**, (2004), 261–302.
34. H. Rauch, *Geodesics and Curvature in Differential Geometry in the Large*, Yeshiva University Press, New York (1959).
35. B. Riemann, *On the Hypotheses which lie at the Bases of Geometry*, Translated by William Kingdon Clifford, Nature, **VIII**, Nos. 183, 184, (1873), 14–17,36–37.
36. S. Sabau and K. Shiohama, Topology of complete Finsler manifolds admitting convex functions, *Pacific J. Math.* **276**, No 2, (2015), 459–481.
37. T. Sakai, *Riemannian Geometry*, Mathematical Monograph, **8** Amer. Math. Soc. (1996).

38. V.A. Sharafutdinov, The Pogorelov-Klingenberg theorem for manifolds homeomorphic to $\mathbf{R}^n$. *Sib. Math. J.*, **18** (1977), 915–925.
39. K. Shiohama, Topology of complete noncompact manifolds, *Geometry of Geodesics and Related Topics*, Advanced Studies in Pure Mathematics **3**, 432–450, (1984).
40. K. Shiohama, *Riemannian and Finsler Geometry in the Large,* Recent Advances in Mathematics, RMS-Lecture Notes Series **21**, (2015), 163–179.
41. K. Shiohama and M. Tanaka, *Cut loci and distance spheres on Alexandrov surfaces,* Round Table in Differential Geometry, Séminaire et Congrès, Collection SFM, no. 1, 553–560 (1996).
42. A. Weinstein, The cut locus and conjugate locus of a Riemannian manifold, *Ann. of Math.* (2), **36** 29–41, (1968).
43. J.H.C. Whitehead, Convex regions in the geometry of paths, *Quarterly Journal of Mathematics (Oxford)*, **3**, (1932), 33–42.
44. J.H.C. Whitehead, On the covering of a complete space by the geodesics through a point, *Ann. of Math.* (2) **36**, (1935), 679–704.
45. T. Yamaguchi, The isometry groups of Riemannian manifolds admitting strictly convex functions, *Ann. Sci. École Norm. Sup.* **15**, (1982), 205–212.
46. T. Yamaguchi, Uniformly locally convex filtrations on complete Riemannian manifolds, *Curvature and Topology of Riemannian Manifolds–Proceedings, Katata 1985*, Lecture Notes in Mathematics, **1201**, (1985) 308–318, Springer-Verlag.

Chapter 17
The Poincaré Conjecture and Related Statements

Valerii N. Berestovskii

Abstract The main topics of this paper are mathematical statements, results or problems related with the Poincaré conjecture, a recipe to recognize the three-dimensional sphere. The statements, results and problems are equivalent forms, corollaries, strengthenings of this conjecture, or problems of a more general nature such as the homeomorphism problem, the manifold recognition problem and the existence problem of some polyhedral, smooth and geometric structures on topological manifolds. Examples of polyhedral structures are simplicial triangulations and combinatorial simplicial triangulations of topological manifolds; so appears the triangulation conjecture, more exactly, the triangulation problem. Examples of geometric structures are Riemannian metrics that are locally homogeneous or have constant zero, positive or negative sectional curvature; more general structures are intrinsic or geodesic metrics with curvature bounded above or/and below in the sense of A.D. Alexandrov or with nonpositive curvature in the sense of H. Busemann.

MSC 2010: 57M40, 57R60, 57M35, 57R65, 57R05, 57R10, 52B70, 53B20, 51Fxx

17.1 Introduction

Poincaré introduced the notion of a (smooth) manifold for the needs of differential equation theory. For a deep study of manifolds he defined the fundamental group and the Betti numbers as predecessors of the homology groups. Much later, homology groups were defined under the influence of Emmy Noether, and higher homotopy groups were introduced by E. Cech and W. Hurewicz. As a result, there appeared an interpretation of the first integral homology group for manifolds as the quotient

V. N. Berestovskii (✉)
Sobolev Institute of Mathematics SB RAS, Novosibirsk, Russia

Novosibirsk State University, Novosibirsk, Russia
e-mail: berestov@ofim.oscsbras.ru

S. G. Dani, A. Papadopoulos (eds.), *Geometry in History*,
https://doi.org/10.1007/978-3-030-13609-3_17

group of the fundamental group by its commutator subgroup, and an interpretation of Betti numbers as ranks of rational or real homology groups.

The very first—erroneous—version of the Poincaré conjecture stated that a closed 3-manifold with trivial first homology group (in other words, with fundamental group equal to its commutator subgroup) must be homeomorphic to the three-dimensional sphere. However, very soon after, Poincaré gave the corrected version of his conjecture: a closed 3-manifold is homeomorphic to the 3-sphere if its fundamental group is trivial (the necessity of the last condition is obvious). This was after Poincaré discovered the famous homology 3-sphere with fundamental group of order 120.

Later on, there were many unsuccessful attempts to solve the Poincaré conjecture as well as very useful discoveries in 3-manifold theory. As examples, let us mention the spherical decomposition (the inverse operation to the operation of the connected sum) of any closed triangulated 3-manifold into a finite number of prime summands. Each summand, besides a 2-spherical bundle over the circle, is *irreducible*, i.e., any embedded 2-sphere in it bounds a 3-ball. In addition, the fundamental group of the initial manifold is a free product of fundamental groups of its summands. This decomposition is unique for orientable manifolds. Note that any closed 3-manifold with finite fundamental group is orientable.

In 1982 W. Thurston has conjectured that if M is a prime summand, then any piece of M of the following kind admits a locally homogeneous metric of a unique type among some eight types he indicated:

(1) M itself, if M is a 2-spherical bundle over the circle or if its fundamental group is finite;
(2) every open connected component of the decomposition of M via a finite family of embedded 2-tori (and possibly embedded projective planes or Klein bottles if M is nonorientable).

Thurston's conjecture implies that if the fundamental group of M is finite, then M admits a Riemannian metric of constant sectional curvature 1. Hence, a positive answer to Thurston's geometrization conjecture would imply the Poincaré conjecture.

G. Perelman proved the Thurston conjecture in three preprints, published in 2002 and 2003. For this, he used a version of R. Hamilton's program on the Ricci flow process using the decomposition of manifolds in this process, and his own results on Alexandrov's spaces with curvature bounded below.

In this survey, we formulate the Thurston geometrization conjecture without discussing later papers or results connected with its solution or with the Ricci flow. The reader can find more information about the conjecture and related topics in [177] (see also [162]).

Hamilton wrote his first paper on Ricci flows in [89], in the same year 1982 as [177]. A certain amount of information on these topics can be found in several preprints and papers [12, 24, 45, 49, 109, 133, 135–138, 151–153, 167]. The present survey may be considered as a complement to these texts.

Our main discussion concerning directly the Poincaré conjecture is about the early history of the subject, contained in [93, 99, 101, 155] and other papers which appeared before [177]. We introduce all the necessary notions or give necessary references to formulate the results, except for the first elementary notions of general topology, homology and homotopy theories. The latter are used rather moderately. We consider the fundamental group in detail, but do not discuss the techniques used in proofs of results.

In Sect. 17.2, we give a precise formulation of PC_3 (the Poincaré conjecture in dimension 3) and necessary definitions for this formulation.

In Sect. 17.3, we present some equivalent forms of PC_3 which admit simple statements. Let us mention especially the Bing characterization of the 3-sphere in Theorem 5 which supplies an equivalent to PC_3.

In Sect. 17.4, we quote another result of Bing: PC_3 is equivalent to the statement that any simplicial triangulation of a topological 4-manifold is combinatorial, or, in more common terms, it is a piecewise linear structure (PL-structure). Together with results of J.C.H. Whitehead about smoothing PL-structures on 4-manifolds, this implies that a topological 4-manifold admits a simplicial triangulation if and only if it admits PL and smooth structures. Then, by results of J. Cerf, there are one-to-one correspondences between these three structures on 4-manifolds; see Sect. 17.5.

It is known that any noncompact 4-manifold is smoothable. In Sect. 17.6, we discuss the smoothing question for closed simply connected oriented 4-manifolds. For any such manifold M, the intersection form $\omega = \omega_M$ is defined as an integral bilinear, symmetric, and unimodular form (on a free $\mathbb{Z}$-module). By the Whitehead theorem, the oriented homotopy type of a smooth manifold M is defined by ω_M. By M. Freedman's theorem, for any such form ω there exist exactly one (respectively, two) manifold M up to homeomorphism with $\omega_M = \omega$ if ω_M is even (respectively, odd); for at most one of them, $M \times S^1$ is smoothable. By Rochlin's Theorem 14 (respectively, Donaldson's Theorem 16), a smooth M has ω_M of a special kind if ω_M is even (respectively, definite). All this implies that the majority of manifolds M are nonsmoothable, hence are not triangulated. The smoothable M are not yet classified; possibly, the smoothable manifolds M are exactly connected sums of complex algebraic surfaces. Many of smoothable 4-manifolds, possibly all, admit infinitely many non-diffeomorphic smooth structures.

In Sect. 17.7, we define the homotopy groups, discuss their simple properties, and calculate the fundamental groups of 1-manifolds.

The main content of Sect. 17.8 is the Seifert-van Kampen theorem and some of its corollaries about the presentation of fundamental groups.

In Sect. 17.9 are given a classification of closed surfaces and presentations of their fundamental groups.

In Sect. 17.10 are discussed Heegaard splittings of closed 3-manifolds and corresponding finite presentations of their fundamental groups.

Section 17.11 contains two group-theoretic equivalent forms of the Poincaré conjecture and a mention of a third one by C.D. Papakyriakopoulos.

In Sect. 17.12 are considered Grushko's theorem, connected sums and prime decompositions of closed 3-manifolds and corresponding decompositions of fundamental groups into free products of fundamental groups of summands of prime decompositions. Any such summand, besides the 2-spherical bundles over the circle, is a closed irreducible 3-manifold. The Poincaré conjecture implies that no summand is a homotopy 3-sphere if the initial manifold is not a 3-sphere.

Any closed irreducible 3-manifold either contains a two-sided embedded projective plane P^2, hence is nonorientable, or has trivial second homotopy group. A closed irreducible 3-manifold is said to be *P^2-irreducible* if the first case does not occur. A theorem from Sect. 17.13 asserts that the universal covering of any such manifold M is either a compact homotopy 3-sphere (hence the 3-sphere by PC_3), or is noncompact and contractible (then M is called *aspherical*). Recent results imply that the universal covering of the last manifold is $\mathbb{R}^3$. The next theorem demonstrates on the ground of the loop theorem by J. Stallings that there exist contractible open 3-manifolds which are not homeomorphic to $\mathbb{R}^3$. The first example of such a manifold was discovered by Whitehead as a counterexample to his incorrect proof of the Poincaré conjecture.

In Sect. 17.14 are considered closed 3-manifolds with finite fundamental groups. Any such manifold is orientable; the Poincaré conjecture is equivalent to the statement that it is irreducible. The PC_3 also implies a surprising theorem of D.B.A. Epstein on the structure of compact non-orientable 3-manifolds with boundary and finite fundamental group and the fact that any closed 3-manifold with finite fundamental group is homeomorphic to a so-called *spherical space form*. Any such form is a Riemannian manifold with constant sectional curvature 1. J. Alexander was the first to discover that there exist non-homeomorphic forms with isomorphic cyclic fundamental groups (lense spaces).

Any nonorientable closed 3-manifold M has infinite first homology group $H_1(M, \mathbb{Z})$. In Sect. 17.15, we consider closed irreducible 3-manifolds M with infinite fundamental group and define incompressible surfaces in M. If M is P^2-irreducible, then its fundamental group $\pi_1(M)$ is without torsion. If $\pi_1(M)$ with the last condition contains no nontrivial free subgroup of finite index then $\pi_1(M)$ is free; then, using the Poincaré conjecture, one gets that M is a connected sum of 2-spherical bundles over the circle. A compact 3-manifold M is said to be *sufficiently large* if it contains a compact two-sided incompressible surface. This is so if $H_1(M, \mathbb{Z})$ is infinite. However, there are also closed sufficiently large 3-manifold with a finite $H_1(M, \mathbb{Z})$.

In Sect. 17.16 are discussed Haken 3-manifolds. A compact orientable 3-manifold is called a Haken manifold if it is irreducible and sufficiently large. Any continuous map between Haken manifolds, inducing an isomorphism of their fundamental groups, is homotopic to a homeomorphism. The universal covering manifold of a closed orientable Haken manifold is homeomorphic to $\mathbb{R}^3$. F. Waldhausen proposed the following *virtual Haken conjecture*: any compact irreducible 3-manifold with infinite fundamental group has a finite-sheeted covering by a Haken manifold. The homeomorphism problem has a positive algorithmic solution for Haken manifolds. This implies a positive solution to the problem of algorithmic knot classification.

Section 17.17 contains a formulation of the Jaco-Shalen-Johannson theorem on the torus decompositions of closed orientable Haken 3-manifolds. A smooth 3-manifold endowed with a smooth effective action of the Lie group $U(1)$ is called a Seifert fibration. A 3-manifold M is said to be atoroidal if any two-sided incompressible torus in M is boundary parallel in M. The JSJ-splitting theorem states that for any closed Haken manifold M^3 there exists a minimal finite (maybe empty) family W of tori such that any connected component of $M^3 - W$ is atoroidal or is a Seifert fibration; the family W is unique up to an isotopy of M^3.

In Sect. 17.18, we state the Thurston geometrization conjecture for closed orientable 3-manifolds indicating the relevant eight homogeneous Riemannian 3-manifolds. Let us note that recently I. Agol proved the virtual Haken conjecture (respectively, the Thurston conjecture on the virtual fibering over the circle) for closed aspherical (respectively, hyperbolic) 3-manifolds. As a consequence, the universal covering spaces of these manifolds are homeomorphic to $\mathbb{R}^3$. To prove the above two results, Agol used his proof of the D.T. Wise conjecture on the virtually special action of any word-hyperbolic group acting properly and cocompactly on a CAT(0) cube complex, and the Thurston geometrization conjecture.

The main result of Sect. 17.19 is a striking recent disproof by C. Manolescu of the (simplicial) triangulation conjecture for topological n-manifolds, $n \geq 5$. Together with the fact that there are nontriangulable closed simply connected 4-manifolds, we see that the triangulation conjecture is true only for $n = 1, 2, 3$. There is an epimorphism, ρ, called the Rochlin epimorphism, from the homology cobordism group Θ_3^H over $\mathbb{Z}_2$ of integral homology 3-spheres onto $\mathbb{Z}_2$. One gets the short exact sequence as in (17.4) and the corresponding Bockstein homomorphism $\beta : H^4(M^n, \mathbb{Z}_2) \rightarrow H^5(M^n, \ker(\rho))$ for any n-manifold M^n, $n \geq 5$. About 40 years ago D. Galewski and R. Stern proved that (1) the triangulation conjecture for $n \geq 5$ is true if and only if the sequence (17.4) splits, (2) a closed M^n, $n \geq 5$, is triangulable if an only if $\beta(k(M^n)) = 0$, where $k(M^n) \in H^4(M^n, \mathbb{Z}_2)$ is the Kirby-Siebenmann invariant of M^n which equals zero if and only if M^n admits a PL-triangulation, and they constructed for any $n \geq 5$ special n-manifolds N^n such that the triangulation conjecture for all $n \geq 5$ is true if and only if for at least one of the N^n, $\beta(k(N^n)) = 0$. Manolescu proved that (17.4) does not split and presented particular N^n with $\beta(k(N^n)) \neq 0$ for all $n \geq 5$.

In Sect. 17.20, using results by A.A. Markov and P.S. Novikov, we show that the homeomorphism problem is algorithmically unsolvable for closed n-manifolds in any dimension $n \geq 4$. Note that Seifert constructed the relevant manifolds in this respect.

Section 17.21 contains a discussion of the manifold recognition problem. A locally compact finite-dimensional metric space C is called an *absolute cone* if for any point $c \in C$, C is homeomorphic to an open cone over a compact base with vertex c. The de Groot absolute cone conjecture states that every n-dimensional absolute cone is homeomorphic to an open ball in $\mathbb{R}^n$. Assuming the Poincaré conjecture, C. R. Guilbault proved that the absolute cone conjecture is true for $1 \leq n \leq 4$ and false for $n \geq 5$. This permits easily to characterize topological n-manifolds for $1 \leq n \leq 4$. A *generalized n-manifold* (n-GM) is a locally compact, locally contractible, finite-dimensional separable metric space X with the local relative homology of $\mathbb{R}^n$, i.e., $H_*(X, X - \{x\}, \mathbb{Z})$ is isomorphic to $H_*(\mathbb{R}^n, \mathbb{R}^n - \{x\}, \mathbb{Z})$ for any $x \in X$. The space X is said to possess the *disjoint disk property* (DDP) if any two continuous maps of the disc B^2 into X can be approximated by arbitrary close maps with disjoint images. A connected space X is an n-manifold, $n \geq 5$, if and only if X is n-GM satisfying DDP and such that $i(X) = 1$, where $i(X) \in 1 + 8\mathbb{Z}$ is the integer F. Quinn obstruction (index). There is a well-known problem called the *Bing-Borsuk problem*: "Is it true that any connected locally compact locally contractible finite-dimensional topologically homogeneous metric space is a topological manifold?" W. Jakobsche proved that a positive answer to this problem in dimension 3 implies the Poincaré conjecture. A locally compact complete space M with intrinsic metric such that the shortest paths can be locally extended in a unique way is called a *Busemann G-space*. Up to now, answers to the following questions of Busemann are unknown: (1) Is it true that any Busemann G-space is finite-dimensional? (2) Is it true that any Busemann G-space M of finite dimension is a topological manifold? One can easily prove that M is locally contractible and topologically homogeneous. The first question has a positive answer if M contains at least one open ball U such that any two points in U are joined by a unique shortest path in M and this path lies in U. Any n-dimensional Busemann G-space is an n-GM. It is not difficult to show that small open balls in a Busemann G-space M are absolute cones over their bondary spheres. Then any n-dimensional Busemann G-space is a topological manifold for $1 \leq n \leq 4$. P. Thurston proved this for $n = 4$ not using PC_3.

In Sect. 17.22 we discuss triangulations and related questions on the canonical construction of Alexandrov's metrics with curvature ≤ 1 on simplicial complexes. The construction was suggested by the author of this survey. Alexandrov introduced his spaces with curvature $\leq K$ (respectively, $\geq K$) as a generalization of smooth Riemannian spaces with sectional curvature $\leq K$ (respectively, $\geq K$). Any such space can be characterized as a locally geodesic intrinsic metric space with a local possibility of isometric embedding of quadruples of points into three-dimensional Riemannian spaces with sectional curvatures $\leq K$ (respectively, $\geq K$). Any Busemann G-space with Alexandrov curvature bounded above or below is a Riemannian manifold. Any Alexandrov space of curvature ≤ 0 has also nonpositive curvature in Busemann's sense. P.D. Andreev proved that any Busemann G-space with nonpositive Busemann curvature is a topological manifold whose universal covering space is homemomorphic to a Euclidean space. By Alexandrov's definition, a K-region is a geodesic space which globally satisfies the curvature condition $\leq K$; much later, M. Gromov called it a CAT(K)-space. Any connected locally finite simplicial complex C admits a metric of 1-region; this implies a positive solution to a problem by Borsuk. For any point x in a locally compact Alexandrov space X with curvature $\leq K$ or $\geq K$, there are the associated space of directions $\Omega_x X$ and the tangent space $T_x X$ at x. One can show that there are topological n-spheres X for all $n \geq 5$ with a metric with curvature ≤ 1, such that at some point $x \in X$, $T_x X$ is homeomorphic to $\mathbb{R}^n$, while $\Omega_x X$ is not homeomorphic to S^{n-1}; if $\Omega_x S^4$ is homeomorphic to S^3 for each $x \in S^4$ and every metric on S^4 with curvature ≤ 1, then the Poincaré conjecture is true. Thus, there is a natural question for such metrics on S^4, whose positive solution would imply PC_3. Lytchak and Nagano gave in [115] an affirmative answer to this question *using the resolution of* PC_3!

We see post factum that the key results of 3-dimensional topology could be obtained by means of (generalized) Riemannian geometry and the corresponding analytical apparatus without numerous results obtained by topologists and algebraists in the twentieth century. So, the crucial theorems by Papakyriakopoulos, Whitehead and Shapiro, Stallings and Epstein on the Dehn lemma, the loop theorem and the sphere theorem were proved by W. Meeks III and S.T. Yau using minimal surfaces [129, 130]. Also Perelman proved the Poincaré conjecture in the realization of the Hamilton-Perelman program for the solution of Thurston's conjecture using the Ricci flow. However, the history (in particular, the history of mathematics) does not know the subjunctive mood, and the real centenary history culminating in the proof of the Poincaré conjecture once again demonstrates the unity of mathematics and the fruitfulness of ideas from mathematical physics. The same is true for the solution of the triangulation conjecture.

We also see that many interesting, even difficult, unsolved questions, directly or indirectly connected with the Poincaré conjecture, wait for a solution.

Finally, let us mention interesting popularization books [80] by Masha Gessen and [147] by Donal O'Shea on the Poincaré conjecture and events connected with its proof by Perelman as well as presentations of this proof. We warn the reader that both books contain mistakes and controversial statements, both mathematical as well as in other respects.

The text below is a revised version of the article [22].

The author is very obliged to professors A. Papadopoulos and S. Dani for many remarks which helped him improving the text and A. Lytchak for useful discussions.

17.2 The Poincaré Conjecture

We assume that the reader is familiar with the basic notions of general topology. If a topological space (X, τ) is homeomorphic to a topological space (Y, σ), we write $(X, \tau) \approx (Y, \sigma)$. The *Euclidean n-dimensional space* E^n is the space $\mathbb{R}^n$ (i.e., the n-fold Cartesian power of the real line $\mathbb{R}$) with the metric

$$\rho(x, y) = \sqrt{\Sigma_{k=1}^{n}(x_k - y_k)^2}$$

and the topology τ corresponding to this metric. We shall often denote a topological space (X, τ) (respectively, a metric space (X, ρ)) by X.

Definition 1 A metrizable separable topological space X is called an n-manifold with (possibly empty) boundary, if for any point $x \in X$ there exists an (open) neighborhood U_x of x such that $U_x \approx E^n$, or $U_x \approx (E^n)_+ = \{(x_1, \ldots, x_n) \in E^n : x_n \geq 0\}$ and there is no $U_x \approx E^n$. A point of the first (respectively second) kind is called an interior (respectively, boundary) point of X. The set of all interior (respectively, boundary) points of an n-manifold X is called the interior (respectively, the boundary) of X and is denoted by $\mathrm{Int}(X)$ (respectively, ∂X). A compact n-manifold without boundary is called a closed n-manifold.

Theorem 1 *The boundary ∂M^n of a (compact) n-manifold M^n is either empty or a (closed) $(n-1)$-manifold without boundary.*

Definition 2 Let $f, g : (X, \tau) \to (Y, \sigma)$ be continuous maps. The map f is said to be homotopic to the map g if there exists a continuous map (homotopy between f and g) $F : (X, \tau) \times I \to (Y, \sigma)$, where $I = [0, 1] \subset E^1$, such that $F(x, 0) = f(x)$, $F(x, 1) = g(x)$ for all $x \in X$. In this case we write $f \simeq g$.

The relation $\simeq$ is an equivalence relation on the set of continuous maps from (X, τ) to (Y, σ). The equivalence class of f is called the *homotopy class of* f and is denoted by $[f]$.

Definition 3 A path in a topological space X is a continuous map $p : I \to X$. The points $p(0)$ and $p(1)$ are called respectively the initial and terminal points of the path p. A path p is called a loop if $p(0) = p(1)$. A topological space X is said to be simply connected if for any two loops p and q in X there exists a homotopy $F : I \times I \to X$ such that $F(\cdot, 0) = p$, $F(\cdot, 1) = q$, and $F(\cdot, s)$ is a loop in X for all $s \in I$.

Any simply connected topological space X is path connected. This means that for any two points $x, y \in X$ there is a path p in X such that $p(0) = x$, $p(1) = y$. Any path connected topological space is connected. A topological manifold X is connected if and only if X is path connected.

Let (X, ρ) be a metric space, $x_0 \in X$, and r a positive real number. Recall that an *open (respectively, closed) ball*, or a *sphere of radius r with center x_0 in (X, ρ)* is the set of all $x \in X$ such that $\rho(x, x_0) < r$ (resp., $\rho(x, x_0) \leq r$) or $\rho(x, x_0) = r$. We use the following notation for these: $U(x_0, r)$ (respectively, $B(x_0, r)$) or $S(x_0, r)$.

Conjecture 1 PC_3. (The Poincaré Conjecture, 1904, [155]) Any closed simply connected topological 3-manifold X is homeomorphic to the unit sphere S^3 in E^4 centred at the origin.

17.3 Some Statements Related with the Poincaré Conjecture

Definition 4 A continuous map $f : X \to Y$ between topological spaces is called a homotopy equivalence if there exists a continuous map $g : Y \to X$ such that $g \circ f \simeq 1_X$, $f \circ g \simeq 1_Y$. Here 1_X is the identity map of X. In this case we write $f : X \simeq Y$. A topological space X is homotopy equivalent to a topological space Y if there is a homotopy equivalence $f : X \simeq Y$; we then write $X \simeq Y$.

Evidently, $\simeq$ is an equivalence relation between topological spaces (in general, this relation is more rough than the relation of homeomorphism). An equivalence class of a topological space X with respect to $\simeq$ is called the *homotopy type* (and is denoted by $[X]$). The properties of X being path connected or simply connected are invariants of the homotopy type. A topological space X is called *contractible* if X

has the homotopy type of a topological space reduced to one point. A topological space X is contractible if and only if $1_X \simeq i_x : X \to \{x\}$ for any point $x \in X$.

Theorem 2 PC$_3$ *is equivalent to the following statement* HS$_3$*: if X is a closed* 3-*manifold and $X \simeq S^3$ then $X \approx S^3$.*

Since the property of being simply connected is an invariant of homotopy type, and S^3 is simply connected, we have PC$_3$ $\Rightarrow$ HS$_3$. The converse statement is a consequence of the following Theorem 3.6 in [93].

Theorem 3 *A* 3-*manifold M is homotopy equivalent to the* 3-*sphere if and only if M is closed and simply connected.*

By a *simple closed curve* (respectively, 3-*cell*) we shall understand a subspace homeomorphic to S^1 (respectively, $B^3 = B(x, 1) \subset E^3$). A *fake* 3-*cell* will mean a compact contractible 3-manifold which is not homeomorphic to B^3.

Arguments as in the proof of Theorem 3 in [93] imply the following

Theorem 4 *The Poincaré conjecture is equivalent to the statement that there is no fake* 3-*cell.*

The sphere S^3 can be characterized as follows.

Theorem 5 *(R.H. Bing, 1958, [28]) Let M be a closed connected* 3-*manifold. Assume that each simple closed curve in M lies in a 3-cell in M. Then M is homeomorphic to S^3.*

Evidently, the converse of Theorem 5 is true. It is clear that the hypotheses of Theorem 5 imply that M is simply connected. Then this theorem follows from PC$_3$. In turn, it follows from this theorem that PC$_3$ can be obtained from the following statement.

Statement 1 *Let M be a closed (connected) simply connected* 3-*manifold. Then each simple closed curve in M lies in a 3-cell in M.*

Let us denote by PC$_n$ and HS$_n$ the statements obtained from PC$_3$ and HS_3 respectively by replacing 3 by n, a positive integer. HS_n is usually called the *generalized Poincaré conjecture*. It is enough easy to prove that PC$_n$ is true for $n = 1, 2$ and is false for $n \geq 4$; HS$_n$ is true for $n = 1, 2$.

In late 1950s and early 1960s, considerable progress was achieved due to the discovery that manifolds of higher dimensions can be investigated more easily than 3-manifolds. M.H.A. Newman proved in 1966, [142], the following

Theorem 6 HS_n *is true for $n > 4$.*

Remark 1 Smale proved this statement for combinatorial manifolds in 1961 [169]. Note that any homotopy n-sphere M^n, $n \geq 5$, has $H^4(M^n, \mathbb{Z}_2) = 0$, hence the Kirby-Siebenmann obstruction $k(M^n) \in H^4(M^n, \mathbb{Z}_2)$ vanishes, so M^n admits a combinatorial triangulation [107]. Related results were published by Stallings in 1960 [170], Wallace in 1961 [184], Yamasuge in 1961 [193], Zeeman in 1962 [194].

The case of $n = 4$ turned out to be much more difficult and the solution appeared in the paper by Freedman [70]. The following theorem is a partial result of his work.

Theorem 7 HS_4 *is true.*

17.4 Statements on Triangulations Related with the Poincaré Conjecture

The convex hull Δ_n of a set A_{n+1} consisting of $n+1$ points in E^m which do not lie in an $(n-1)$-plane of E^m is called *n-simplex* in E^m, $m \geq n$. The convex hull of any subset in A_{n+1} is called a *face* of the n-simplex Δ_n. An *n-simplex* in an abstract metric space is a subset isometric to an n-simplex in E^n; the faces of this abstract n-simplex are the inverse images of its isometric image in E^n.

A *geometric complex* is a metric space which is an union of a locally finite family of simplexes such that intersection of any two simplexes, when nonempty, is a face of each of them. T is called a *triangulation* of the geometric complex if T contains each face of each of its elements; the *i-skeleton* of a triangulation is the union of all its i-simplexes. A geometric complex endowed with a triangulation is called a *simplicial complex*.

If s is a simplex of a triangulation T of a geometric complex, then the *star* $st(s, T)$ is the union of all simplexes in T containing s. The *link* $lk(s, T)$ is the union of all simplexes s' in $st(s, T)$ such that $s \cap s' = \emptyset$. If s, s' are disjoint simplexes that are faces of the same simplex s'', then the *join* of s and s' is the union of all segments in s'' from points of s to points of s'. We will also use this term for arbitrary two subsets s, s' of a simplex s''.

A subset Y of a geometric complex C is called a *polyhedron* if C admits a triangulation T such that Y is a union of simplexes in T. A subset $Z \subset C$ is said to be *tame in C* if there exists a homeomorphism h of C onto itself such that $h(Z)$ is a polyhedron. A closed subset $X \subset C$ which is homeomorphic to a polyhedron but is not tame in C is said to be *wild*.

A map f of a geometric complex C into a geometric complex C' is *piecewise linear* if there is a triangulation T of C such that f is linear on each simplex of T. A map f of a simplicial complex C into a geometric complex C' is said to be *simplicial* if f maps each simplex s of C onto a simplex of C' and is linear on s. Geometric (respectively, simplicial) complexes C and C' are isomorphic if there is a bijection $f : C \to C'$ such that f and f^{-1} are piecewise linear (respectively, simplicial).

A *triangulated manifold* is a simplicial complex which is a manifold. A triangulated n-manifold M^n is a *piecewise linear* or *combinatorial manifold* if for any vertex v in T, $lk(v, T)$ is piecewise linearly homeomorphic to $\partial\Delta_n$.

For any topological manifold M^n, the following three natural questions arise.

Problem 1 Does M^n admit a triangulation? (*the triangulation conjecture*).

Problem 2 If a triangulation of M^n exists, is it unique? (*Hauptvermutung* (HV)). In more precise terms: Is it true that any two homeomorphic simplicial manifolds have isomorphic simplicial subdivisions?

Problem 3 Does a (non)-combinatorial triangulation of M^n exist? (*the (non)*-PL *-conjecture*).

It is clear that the PL-conjecture and the Hauptvermutung imply the absence of non-combinatorial triangulations.

$n = 1$. One can easily see that the PL-conjecture and the Hauptvermutung hold.

$n = 2$. T. Rado proved the existence of a PL-structure in 1924 [157]. B.V. Kerékjártó's classification in 1923 [105] implies the Hauptvermutung.

$n = 3$. Moise proved the PL-conjecture and Hauptvermutung in 1952 [134]; another proof was given by Bing [30].

$n = 4$. The work of Casson [2] in the 1980s provided counterexamples to the triangulation conjecture for closed 4-manifolds. We shall discuss such examples systematically in Sect. 17.6. Therefore, the triangulation conjecture (and hence the PL-conjecture) is false in the general case.

The following result is interesting in relation with the non-PL-conjecture.

Theorem 8 *The Poincaré conjecture* PC_3 *holds if and only if every triangulated* 4*-manifold is a combinatorial* 4*-manifold.*

The necessity is contained in Bing's Theorem 21 in [29]. The sufficiency is a consequence of the following Freedman suspension theorem in [70]: the suspension $S\Sigma^3$ over any closed simply connected 3-manifold Σ^3 is homeomorphic to S^4.

Recall that

Definition 5 The suspension SX over a topological space X is a quotient space $S(X) = (X \times I)/\sim$, where $(x, t) \sim (y, s)$ if $(x, t) = (y, s)$ or $t = s = 0$, or $t = s = 1$. In addition, if $p : X \times I \to SX$ is the quotient map, then two points $p(X \times \{0\})$ and $p(X \times \{1\})$ are said to be cone points of $S(X)$.

Remark 2 Another proof to Freedman's suspension theorem was given in [21] with essential usage of the main result from [26].

We shall continue this discussion of Problems 1, 2, and 3 later on in Sect. 17.19.

17.5 Smooth Manifolds

We assume that the reader is familiar with smooth (differentiable of class C^∞) manifolds (this term shall mean manifold without boundary, unless specified otherwise), smooth mappings between smooth manifolds, and diffeomorphisms. A topological manifold is *smoothable* if it admits at least one smooth structure, and is said to be *nonsmoothable* otherwise.

Analogously to the previous section, two natural problems arise.

Problem 4 The problem of smoothability of a given topological manifold M^n (*the smoothing conjecture*).

Problem 5 The problem of uniqueness of a smooth structure up to a diffeomorphism if this structure exists (*the conjecture on unique smoothing*).

Now let us consider relations with the previous section.

Cairns proved in [41] that any triangulable manifold M^n, $1 \leq n \leq 3$, admits a smooth structure. Together with the results of [134, 157], and [30] mentioned above, this implies that any topological manifold M^n, $1 \leq n \leq 3$, is smoothable. The corresponding smooth structure on M^n is unique up to a diffeomorphism [139, 140], and [189].

Theorem 9 ([40, 141, 186]) *Any smooth manifold M^n admits a unique natural* PL*-triangulation.*

Remark 3 Brouwer gave in [35] a proof of this theorem for closed manifolds using intuitionistic logic.

Theorem 10 *If M is a combinatorial* 4*-manifold, then M possesses a unique smooth structure compatible with its triangulation.*

In his note [42], S.S. Cairns claimed that the smooth structure from Theorem 10 exists, but his argument contains a gap. Much later, in [189] Whitehead gave a detailed proof for Cairns' claim, filling the gap in the proof of [42]. Another proof of the existence of such a smooth structure in [95] is based on the fact that the final group Γ^3 of the obstruction to smoothability of a 4-dimensional PL-manifold is trivial, due to the Smale theorem. The uniqueness of smooth structure follows from Cerf's results in [48]. Theorems 10 and 9 imply that PL- and smooth structures on a 4-manifold are in natural one-to-one correspondence with each other.

Now we know that there exists an uncountable set of different smooth structures on $\mathbb{R}^4$ [81] and [55]. Also there exist infinitely many homeomorphic but mutually non-diffeomorphic simply connected closed complex surfaces (the Dolgachev surfaces), [59, 72, 73]. Therefore, in general, the Hauptvermutung is false in dimension 4.

We get from Theorems 8, 9, and 10 the following

Corollary 1 *If the Poincaré conjecture* PC_3 *is true, then for any* 4*-manifold M^4 the following statements are equivalent:*

(1) M^4 admits a smooth structure;
(2) M^4 admits a combinatorial triangulation;
(3) M^4 admits a triangulation.

Moreover, a structure (in general, not unique) of each type on M^4 determines unique structures of the other types as above.

17.6 Closed Simply Connected 4-Manifolds and Smooth Structures

In this section, we denote by M^4 a closed simply connected oriented four-dimensional topological manifold. We need results by Dold from [56].

Set $\tilde{H}_2(M^4, \mathbb{Z}) = H_2(M^4, \mathbb{Z})/\,\mathrm{Tor}$, where Tor is the torsion subgroup (i.e., the subgroup of elements with finite order) of the two-dimensional singular integer homology group $H_2(M^4, \mathbb{Z})$. Let us define the integer symmetric bilinear intersection form

$$\omega = \omega_{M^4} : (a, b) \in \tilde{H}_2(M^4, \mathbb{Z}) \times \tilde{H}_2(M^4, \mathbb{Z}) \to \omega(a, b) \in \mathbb{Z}.$$

Using Poincaré duality, we identify the group $\tilde{H}_2(M^4, \mathbb{Z})$ with the group $\tilde{H}^2(M^4, \mathbb{Z}) = H^2(M^4, \mathbb{Z})/\,\mathrm{Tor}$ and set $\omega(a, b) = a \wedge b \in H^4(M, \mathbb{Z})$. Here $\wedge$ denotes the cohomology multiplication and $H^4(M^4, \mathbb{Z})$ is canonically identified with $\mathbb{Z}$ via an orientation of M^4. Since the multiplication $\wedge$ on $\tilde{H}^2(M^4, \mathbb{Z}) \otimes \tilde{H}^2(M^4, \mathbb{Z})$ is commutative, the form ω is bilinear and symmetric. The free abelian group $\tilde{H}^2(M^4, \mathbb{Z}) \cong \tilde{H}_2(M^4, \mathbb{Z})$ is a module over the ring $\mathbb{Z}$ whose rank is equal to the 2-dimensional Betti number $\beta_2(M^4)$. It follows from the definition of ω that this form is *unimodular*, i.e., the symmetric matrix of ω with respect to an arbitrary $\mathbb{Z}$-basis has determinant equal to ± 1. Clearly $\mathrm{rank}(\omega) = \beta_2(M^4)$.

If M^4 is a smooth oriented manifold, then one can realize cycles $z_1 \in c_1$, $z_2 \in c_2$ for any $c_1, c_2 \in \tilde{H}_2(M^4, \mathbb{Z})$ as smooth oriented surfaces in M^4 intersecting transversally at a finite set of isolated points (see [69], (E9) in Appendix E). Then $\omega(c_1, c_2)$ is equal to the index of intersection of the oriented cycles z_1 and z_2, i.e., to the algebraic sum of points in $z_1 \cap z_2$, where to each point we assign $+1$ or -1 according on whether the orientation of the space $Tz_1 \oplus Tz_2$ coincides with the orientation of TM^4 or not, respectively.

Definition 6 Let ω be an arbitrary integral symmetric bilinear unimodular form defined on a finitely generated free $\mathbb{Z}$-module μ. The form ω is said to be *even* (or of type II) if $\omega(x, x)$ is even for all $x \in \mu$, and *odd* (of type I) otherwise. The form ω is said to be *positive* (respectively, *negative*) definite if $\omega(x, x) > 0$ (respectively, $\omega(x, x) < 0$) for all $x \in \mu \setminus \{0\}$. The form ω is said to be definite if this form is positive or negative definite; otherwise ω is called indefinite. The signature $\sigma(\omega)$ of ω equals the dimension of the positive eigenspace minus the dimension of the negative eigenspace of the form ω on the real vector space $\mathbb{R} \otimes \mu$.

It is clear that the direct sum of two bilinear symmetric unimodular forms (more precisely, their sum on the direct sum of the corresponding $\mathbb{Z}$-modules) gives a bilinear symmetric unimodular form. The sum of two even or two odd forms is even; the sum of an odd and an even form is odd; the sum of two indefinite forms is indefinite; the sum of two definite forms can be definite or indefinite.

Theorem 11 ([132], Serre Theorem) *Let ω be an indefinite unimodular bilinear symmetric form of rank $r = p + n$ and signature $\sigma = p - n$ $(p, n > 0)$. Then*

(1) if ω is a form of type I, then $\omega \cong p(1) \oplus n(-1)$;
(2) if ω is a form of type II, then $\omega \cong aE_8 \oplus bU$, where $a = \frac{1}{8}b\sigma = \frac{1}{2}(r - |\sigma|)$ (note that $\frac{1}{8}\sigma \in \mathbb{Z}$). Here U is a form with matrix $\begin{pmatrix} 0 & 1 \\ 1 & 0 \end{pmatrix}$, E_8 is the form represented by the Cartan matrix of the exceptional Lie algebra E_8.

Consequently, each indefinite unimodular bilinear symmetric form is uniquely determined by its type, rank, and signature.

Theorem 12 ([132]) *The signature of a symmetric bilinear unimodular form of type II is divisible by* 8.

The complete classification of symmetric bilinear unimodular forms is unknown. When the rank increases, the number of different symmetric bilinear unimodular forms increases very rapidly. For example, even if we consider only positive definite forms of type II, we have one form of rank 8, 2 forms of rank 16, 24 forms of rank 24, at least 10^7 forms of rank 32, and at least 10^{51} forms of rank 40 [132].

The following is a classical theorem.

Theorem 13 ([187]) *Two oriented closed simply connected smooth* 4*-manifolds M_1 and M_2 are orientably homotopy equivalent if and only if $(\tilde{H}_2(M_1, \mathbb{Z}), \omega_{M_1})$ and $(\tilde{H}_2(M_2, \mathbb{Z}), \omega_{M_2})$ are isomorphic.*

An essential role in the further development of the theory of manifolds of dimension ≥ 4 was played by another classical theorem.

Theorem 14 ([85, 160]) *If M^4 is a closed smooth oriented spin* 4*-manifold, then the signature $\sigma(M^4) = \sigma(\omega_{M^4})$ is divisible by* 16.

Remark 4 A smooth oriented manifold M is said to be a spin manifold if M admits a spin structure. The existence of a spin structure is equivalent to the fact that the second Stiefel-Whitney class $w_2(M)$ of the tangent bundle TM vanishes [23]. If a manifold M^4 is simply connected (or, more generally, the group $H_1(M^4, \mathbb{Z})$ does not contain elements of the second order), then $w_2(M^4) = 0$ if and only if the form ω_{M^4} is even.

The following essential strengthening of Theorem 13 was proved by the Casson handle technique [85, 116].

Theorem 15 ([70]) *The oriented homeomorphism classes of closed simply connected oriented topological* 4*-manifolds M^4 are in one-to-one correspondence with the set of pairs*

$$\{([\omega], \alpha \in \{0, 1\}) : \textit{if } \omega \textit{ is even, then } \frac{1}{8}\sigma(\omega) = \alpha \bmod 2\},$$

where $[\omega]$ is the isomorphism class (over the ring $\mathbb{Z}$) of an integral symmetric bilinear unimodular form ω. Then $\omega = \omega_{M^4}$ and, if $\alpha = \alpha(M^4) = 0$ (respectively, 1*), then $M^4 \times S^1$ is smoothable (respectively, nonsmoothable).*

An essentially different proof of this theorem was given by Freedman and Quinn in [74]. Using the ideas and methods from mathematical physics, namely, the theory of gauge fields, S. Donaldson proved

Theorem 16 ([57]) *The definite intersection form ω_{M^4} of a simply connected smooth 4-manifold M^4 is standard, i.e., ω_{M^4} is equivalent to a form $\pm((1) \oplus \cdots \oplus (1))$.*

Corollary 2 *Any closed simply connected 4-manifold M^4 with definite form ω_{M^4} of type II (or of type I different from the standard form) is nonsmoothable.*

Corollary 3 *The form ω_{M^4} with matrix E_8 (the Cartan matrix of exceptional Lie algebra E_8) is even, the corresponding manifold $P := M^4$ (which exists by the Freedman theorem 15) is spin and $\sigma(\omega_P) = 8$; therefore, by the Rochlin Theorem 14, P is nonsmoothable.*

We need two notions from general topology. Let X and A be a topological space and a subset of it respectively. A point $a \in X$ is an interior point of A (relative to X) if there is a neighborhood U_a of a in X (i.e. an open subset in X which contains the point a) such that $U_a \subset A$. The set of all interior points of the set A (in X) we will denote by $\mathrm{Int}_X A$ or simply by $\mathrm{Int}\, A$. The set $\partial_X A$ or simply ∂A of all boundary points of A (in X) is defined as the set $X - (\mathrm{Int}_X(X - A))$.

Let M, M_1, and M_2 be connected closed n-manifolds, $n \geq 2$, and suppose there exist n-cells $B_i \subset M_i$, $i = 1, 2$, and topological embeddings $h_i : R_i := M_i - \mathrm{Int}\, B_i \to M$ with $h_1(R_1) \cap h_2(R_2) = h_1(\partial B_1) = h_2(\partial B_2)$ and $M = h_1(R_1) \cup h_2(R_2)$. We say that M is the *connected sum* of the manifolds M_1 and M_2 and we denote it by $M = M_1 \# M_2$. The operation of *spherical decomposition* which is inverse to the operation of connected sum can be described as follows. We take a PL-embedded dividing sphere S^{n-1} and replace its two-sided neighborhood $S^{n-1} \times I$ with $B^n \times S^0$ (they have the common boundary $S^{n-1} \times S^0$). Both operations are partial cases of *spherical surgery*. If M, M_1, and M_2 are oriented, we assume that $h_i : R_i \to M$ preserve orientations (this is equivalent to the condition that the homeomorphism identifying ∂B_1 with ∂B_2 reverses the induced orientations); in this case the operation is associative and commutative. There exists an operation of connected sum for smooth oriented manifolds [23] which we will use, but we give no precise definition here.

One can prove

Proposition 1 *If M, M_1, and M_2 are oriented closed simply connected 4-manifolds, and $M = M_1 \# M_2$, then $\omega_M = \omega_{M_1} \oplus \omega_{M_2}$, $\sigma(M) = \sigma(M_1) + \sigma(M_2)$.*

The following examples are taken from [71].

Example 1 $M = S^4$. Since $H_2(S^4, \mathbb{Z}) = 0$, then $\omega_{S^4} = \emptyset$ (empty set).

Example 2 Let CP^2 and $\overline{CP^2}$ be the complex projective plane with the standard orientation and respectively, the same plane with the inverse orientation. Then $H_2(CP^2, \mathbb{Z}) \cong \mathbb{Z}$, and a generator in $\mathbb{Z}$ is represented by the complex projective line $CP^1 \approx S^2$. Since any two (different) projective lines meet at a point, we have $\omega_{CP^2} = (1)$, $\omega_{\overline{CP^2}} = -(1)$, $\sigma(CP^2) = -\sigma(\overline{CP^2}) = 1$.

Example 3 $M = S^2 \times S^2$. Then $H_2(M, \mathbb{Z}) \cong \mathbb{Z} \oplus \mathbb{Z}$ is generated with $a = S^2 \times \{p\}$ and $b = \{p\} \times S^2$, where $\{p\}$ denotes a point in S^2. It is clear that a and b have zero self-intersection number, and they meet at a point. Hence $\omega_{S^2 \times S^2} \cong U$, $\sigma(S^2 \times S^2) = 0$.

Example 4 The Kummer surface: $M = K_3 = [z_0, z_1, z_2, z_3] \in CP^3$; $z_0^2 + z_1^2 + z_2^2 + z_3^2 = 0$, where $z_0 : z_1 : z_2 : z_3$ are homogeneous complex coordinates of a point in CP^3. The rank of $H_2(K_3)$ is equal to 22; the form ω_M is even, and $\sigma(M) = \sigma(\omega_M) = -16$. Hence the form ω_M is indefinite. By the classification of such forms given in Theorem 11, we have $\omega_{K_3} \cong -2E_8 \oplus 3U$.

Example 5 All non-degenerate algebraic surfaces V^d of degree d in CP^3 are diffeomorphic and their basic algebraic invariants are as follows:

(1) $\sigma(V^d) = -\frac{1}{3}d^3 + \frac{4}{3}d$;
(2) rank $H_2(V^d, \mathbb{Z}) = d^3 - 4d^2 + 6d - 2$;
(3) V^d is spin if and only if d is even.

These formulas explain Example 4.

It follows from Corollary 2, Proposition 1, Theorems 15, 11, and the above examples that smoothable closed simply connected (oriented) 4-manifolds (different from S^4) can be only among $n(S^2 \times S^2)\#mE_8$ and $nCP^2\#m\overline{CP^2}$, $n, m \geq 0$, $n + m \geq 1$. All these manifolds are spin. It is clear that all $nCP^2\#m\overline{CP^2}$ have smooth structures. What can one say about $n(S^2 \times S^2)\#mE_8$? The Rochlin Theorem 14 excludes the case of odd m. Considering the connected sum of several K_3 and $S^2 \times S^2$, we obtain by Examples 4 and 5 that $n(S^2 \times S^2)\#mE_8$ is smoothable if $3m \leq n$. This is equivalent to the inequality

$$b_2(M^4) \geq \frac{11}{8}|\sigma(M^4)|. \tag{17.1}$$

(Since the K_3-surface satisfies the equality, the coefficient $11/8$ cannot be replaced by a larger number.)

The assumption that the converse is also true, i.e., the inequality (17.1) holds for all smooth closed simply connected oriented 4-manifolds with even intersection form is known as the $\frac{11}{8}$-*conjecture*. This was suggested by Matsumoto in [123]. The $11/8$-conjecture is true for simply connected complex surfaces with even intersection form [15]. A positive solution to the conjecture would imply that any smooth simply connected 4-manifold is homeomorphic to a connected sum of algebraic surfaces. Also such a solution would give a classification of all closed simply connected topological 4-manifolds admitting smooth structures, or, what is equivalent by Corollary 1, simplicial triangulations. Now the best known result is the following inequality proved in [76] for smooth closed oriented spin manifolds with indefinite intersection form:

$$b_2(M^4) \geq \frac{10}{8}|\sigma(M^4)| + 2. \tag{17.2}$$

Note that if the intersection form is definite, a theorem of Donaldson implies the equality $b_2(M^4) = \sigma(M^4) = 0$ [58, 60].

A later consideration of the $11/8$-conjecture was continued in the paper [77]. Let us quote a phrase from its abstract: "We also show that the 'nilpotence phenomenon' explains why the Bauer-Furuta stable homotopy Seiberg-Witten invariants are not enough to prove $11/8$-conjecture'."

It is amusing that the smooth version of the corollary to the $11/8$-conjecture stated above, the so-called Smooth Decomposition Conjecture which asserts that any smooth simply connected 4-manifold is a result of *connected sum operation (from [23] for smooth manifolds)* of algebraic (hence complex) surfaces is not true: Gompf R.E. and Mrowka T.S. constructed in their paper [82] infinite families of smooth simply connected 4-manifolds that cannot be complex surfaces or connected sums (in the smooth category) of complex surfaces. Their examples are all homeomorphic to complex surfaces, either the Kummer surface or its nonspin version $3CP^2\#19\overline{CP^2}$, so this is really a smooth phenomenon.

Corollary 4 *A closed simply connected oriented 4-manifold M with odd intersection form ω_M is smoothable if and only if $M = nCP^2\#m\overline{CP^2}$, $n, m \geq 0$, $n + m \geq 1$. For each manifold M with this property there exists a unique closed simply connected 4-manifold M' with the form $\omega_{M'} = \omega_M$ which does not admit an orientation preserving homeomorphism onto M. Moreover, M' is orientably homotopic to M and nonsmoothable, hence M' admits no simplicial triangulation. Further, for each manifold M there exists a unique homeomorphic manifold M^- of this type, but it corresponds to the pair $(n^-, m^-) = (m, n)$; in particular, $M^- = M$ if and only if $n = m$.*

Corollary 5 *The manifold $nCP^2\#m\overline{CP^2}$ is homeomorphic to $n_1CP^2\#m_1\overline{CP^2}$ if and only if $(n, m) = (n_1, m_1)$ or $(n, m) = (m_1, n_1)$.*

Theorem 17 *The manifold $M = (n(S^2 \times S^2)\#m(K_3))\#(n_1CP^2\#m_1\overline{CP^2})$, where n, m, n_1, m_1 are nonnegative integers,* $\max(n, m) > 0$, $\max(n_1, m_1) > 0$, *is orientably homeomorphic to the manifold*

$$N = ((3m + n)CP^2\#(19m + n)\overline{CP^2})\#(n_1CP^2\#m_1\overline{CP^2}).$$

The proof is as follows. Proposition 1 and Examples 2–4 imply that $\sigma(M) = -16m + n_1 - m_1$, and the form ω_M is the direct sum of even and odd forms, hence it is odd and evidently indefinite. By Theorem 11, $M = n_2CP^2\#m_2\overline{CP^2}$, where

$$n_2 - m_2 = -16m + n_1 - m_1,$$

$$n_2 + m_2 = 22m + 2n + n_1 + m_1.$$

We solve this equation system with respect to n_2, m_2, and using the associativity and commutativity of the operation of connected sum in the case under consideration, we obtain the required result.

Remark 5 Theorems 17 and 35 demonstrate that, unlike the 3-dimensional case, the summands of connected sums of closed oriented (even simply connected) topological 4-manifolds are not uniquely determined in the general case. Apparently, the smooth manifolds M and N are not diffeomorphic, the manifold $P\#P$ is not triangulable, but $(P\#P) \times S^1$ is smoothable.

Theorem 18 ([71, 74]) *Each connected noncompact* 4*-manifold is smoothable. The manifold* $\mathbb{R}^4$ *admits uncountably many non-diffeomorphic smooth structures (in particular,* HV *is false for* $\mathbb{R}^4$*). There exists a smooth structure* s *on* $\mathbb{R}^4$ *such that for any other smooth structure* s_1 *on* $\mathbb{R}^4$, $(\mathbb{R}^4, s_1)$ *is diffeomorphic to an open submanifold in* $(\mathbb{R}^4, s)$.

Remark 6 It is known that any smoothable manifold of dimension greater than 4 admits only a finite number of smooth structures [107]. Also it is known that, for $n > 2$, there exist infinitely many topological manifolds which are homotopy equivalent to CP^n [183]. In contrast to this, in dimension 4 many (possibly all, as many experts assume) smoothable manifolds admit infinitely many different smooth structures [67] (and hence triangulations which have no isomorphic simplicial subdivision), and by Theorems 13, 15, there exist only two topological manifolds homotopically equivalent to CP^2. Thus dimension four is unique. This is possibly connected with the fact that we live in four-dimensional space-time.

17.7 Homotopy and Fundamental Groups

Let us recall some notions from general topology.

Definition 7 Let (X, τ) and $\sim$ be a topological space and an equivalence relation on X respectively. If $Y = X/\sim$ and $p : X \to Y$ are the corresponding quotient set and projection, then the family $\sigma = \{U \in Y : p^{-1}(U) \in \tau\}$ is a topology on Y. The topology σ and the topological space (Y, σ) are called respectively the quotient topology and the quotient (topological) space.

The quotient topology σ on the quotient set Y is characterized by the property that for any topological space Z and every map $g : (Y, \sigma) \to Z$, this map is continuous if and only if the composition $f = g \circ p : X \to Z$ is continuous.

In what follows we need the definition of the homotopy groups $\pi_n(X, x)$ for non-negative integer numbers n, of a topological space X and some fixed point $x \in X$.

$n = 0$. In fact $\pi_0(X, x)$ is not a group. This is the set of all path connected components of the space X. By definition, two points lie in a path connected component if there is a path in X joining them. Then we choose the path connected component of the fixed point x as distinguished point in $\pi_0(X, x)$.

Below $f : (X, A) \to (Y, B)$ denotes a continuous map $f : X \to Y$ such that $f(A) \subset B$ for $A \subset X$ and $B \subset Y$. By definition, maps $f, g : (X, A) \to (Y, B)$ are homotopic if there exists a homotopy $F : f \simeq g$ such that $F : (X \times I) \to (Y, B)$.

$n > 0$. As a set, $\pi_n(X, x)$ is the set of homotopy classes $[f]$ of all maps $f : (S^n, s) \to (X, x)$ with some fixed point $s \in S^n$. Let $I = [0, 1]$ and $I^n \subset E^n$ be the cube in E^n. Define an equivalence relation $\sim$ on I^n by $x \sim y$ if and only if $x = y$ or $x, y \in \partial I^n$. Then the quotient space $I^n/\!\sim$ is homeomorphic to S^n, therefore we can identify S^n with $I^n/\!\sim$ and take the point $p(\partial I^n)$ as distinguished point in $S^n = I^n/\!\sim$, where $p : I^n \to I^n/\!\sim$ is the quotient map. Hence we can assume that $\pi_n(X, x) = \{[f] \| f : (I^n, \partial I^n) \to (X, x)\}$. Then the product $[f] * [g]$ of elements $[f], [g] \in \pi_n(X, x)$ is $[f * g] \in \pi_n(X, x)$, where

$$(f * g)(t_1, t_2 \ldots, t_n) = \left\{ \begin{array}{ll} f(2t_1, t_2 \ldots t_n), & \text{if} \quad 0 \le t_1 \le \frac{1}{2}, \\ g(2t_1 - 1, t_2 \ldots t_n), & \text{if} \quad \frac{1}{2} \le t_1 \le 1. \end{array} \right\}$$

One can easily check that $[f * g]$ does not depend on a choice of the maps f, g in the corresponding homotopy classes.

Now we need to prove the group properties for this multiplication.

Associativity. $[(f * g) * h] = [f * (g * h)]$ or $F : (f * g) * h \simeq f * (g * h)$ for some $F : (I^n \times I, \partial I^n \times I) \to (X, x)$. One can take here the map F defined by the following formulas

$$F(t_1, t_2 \ldots, t_n, s) = \left\{ \begin{array}{ll} f(\frac{4t_1}{1+s}, t_2 \ldots t_n), & \text{if} \quad 0 \le t_1 \le \frac{1}{4}(1+s), \\ g(4t_1 - (1+s), t_2 \ldots t_n), & \text{if} \quad \frac{1}{4}(1+s) \le t_1 \le \frac{1}{4}(2+s), \\ h(\frac{4t_1}{2-s} - \frac{2+s}{2-s}, t_2 \ldots t_n), & \text{if} \quad \frac{1}{4}(2+s) \le t_1 \le 1. \end{array} \right\}$$

Similarly one can check that $[i_x]$, where $i_x(I^n) = \{x\}$, will be the unit, and for any $[f] \in \pi_n(X, x)$, $[f]^{-1} = [f-]$, where $(f-)(t_1, t_2, \ldots, t_n) = f(1 - t_1, t_2, \ldots, t_n)$.

In addition, the homotopy groups $\pi_n(X, x)$ are commutative for $n \ge 2$.

If the topological space X is path connected, then for any positive integer n and for any $x, y \in X$, the groups $\pi_n(X, x)$ and $\pi_n(X, y)$ are isomorphic.

The homotopy group $\pi_1(X, x)$ is called the *fundamental group*. The fundamental group and the higher homotopy groups $\pi_n(X, x)$, $n \ge 2$, were defined respectively by H. Poincaré and W. Hurewicz (the latter were defined around 1935).

Proposition 2 *A topological space is simply connected if and only if it is path connected and* $\pi_1(X, x) = 1$ *for any point* $x \in X$.

One can easily prove the following

Theorem 19 *If* X *and* Y *are homotopy equivalent path connected topological spaces, then for any points* $x \in X$, $y \in Y$, *and any positive integer* n, *the homotopy groups* $\pi_n(X, x)$, $\pi_n(Y, y)$ *are isomorphic.*

From Theorem 19 it follows

Corollary 6 *Any homotopy group* $\pi_n(X, x), n \geq 1$, *of a contractible topological space* X *is trivial.*

Up to homeomorphism, there are only two (separable) connected topological manifolds of dimension 1: the Euclidean straight line E^1 and the unit 1-dimensional sphere (circle) S^1, i.e. the set of complex numbers $\{z : |z| = 1\}$.

The contractibility of E^1 and Corollary 6 imply that all the homotopy groups of E^1 are trivial.

Definition 8 A continuous map $c : X \to Y$ between path connected topological spaces is called a covering (map) if for any point $y \in Y$ there is a neighborhood U_y such that $c^{-1}(U_y)$ is a disjoint union of open subsets in X, and the restriction of c to each of these subsets is a homeomorphism onto U_y. The space X is called a covering space of the space Y. A covering $c : X \to Y$ is universal if the space X is simply connected.

Example 6 The map $c : \mathbb{R} \to S^1$, where $c(t) = \exp(2\pi i t)$, is a universal covering.

One can easily prove the following lemma.

Lemma 1 *Let* $c : (Y, y) \to (X, x)$ *be a covering. Then for any continuous map* $p : (Z, z) \to (X, x)$ *of a simply connected topological space* Z *there is unique continuous map* $q : (Z, z) \to (Y, y)$ *such that* $c \circ q = p$. *(The map* q *is said to be a covering for* p.*) The converse statement is also true.*

Now one can easily prove

Proposition 3 *Let* $c : (Y, y) \to (X, x)$ *be a universal covering. Then for every path* $p : (I, 0) \to (X, x)$ *there is unique path* $q : (I, 0) \to (Y, y)$ *such that* $c \circ q = p$. *(The converse statement is also true.) Two loops* $p_1, p_2 : (I, \partial I) \to (X, x)$ *are homotopic if and only if for the paths* $q_1, q_2 : (I, 0) \to (Y, y)$, *which cover* p_1, p_2, *we have* $q_1(1) = q_2(1)$.

As a corollary of this proposition we obtain

Proposition 4 *For any path* $p : (I, 0) \to (S^1, 1)$ *there is a unique path* $q : (I, 0) \to (E^1, 0)$ *such that* $c \circ q = p$. *The path* p *is a loop if and only if* $q(1)$ *is a positive integer. Two loops* $p_1, p_2 : (I, \partial I) \to (S^1, 1)$ *are homotopic if and only if for the paths* $q_1, q_2 : (I, 0) \to (S^1, 1)$, *covering the paths* p_1, p_2, *we have* $q_1(1) = q_2(1)$. *The path* $q_1 * q_2$ *covers the loop* $p_1 * p_2$ *and* $(q_1 * q_2)(1) = q_1(1) + q_2(1)$.

Corollary 7 $\pi_1(S^1, 1)$ *is isomorphic to* $(\mathbb{Z}, +)$.

Similarly to Proposition 3 one can prove

Proposition 5 *Let* $c : (Y, y) \to (X, x)$ *be a covering. Then* $\pi_n(X, x)$ *is isomorphic to* $\pi_n(Y, y)$ *for any* $n \geq 2$.

Corollary 8 *All the homotopy groups* $\pi_n(S^1, 1)$ *are trivial for* $n \geq 2$.

Theorem 20 *For any continuous map $f : (X, x) \to (Y, y)$ and positive integer n, the formula $f_*([g]) = [f \circ g]$, where $[g] \in \pi_n(X, x)$, defines a homomorphism $f_* : \pi_n(X, x) \to \pi_n(Y, y)$. The homomorphism f_* depends only on the homotopy class $[f]$.*

Theorem 21 *Let X be a connected (separable) m-manifold. Then the fundamental group $\pi_1(X, x)$ is finite or countable. For any subgroup $H \subset \pi_1(X, x)$ there exists a covering map $c : (Y, y) \to (X, x)$ such that $c_*(\pi_1(Y, y)) = H$. In addition, $c_* : \pi_1(Y, y) \to \pi_1(X, x)$ is a monomorphism, and Y is a m-manifold determined by X and H up to a homeomorphism. In particular, there is a unique universal covering m-manifold $\tilde{X}$ of the manifold X.*

17.8 The Seifert-van Kampen Theorem

For calculating of the fundamental group, of great importance is the *Seifert-van Kampen theorem.*

Theorem 22 *Let X, U, V be a topological space together with open subspaces such that $X = U \cup V$, $x \in U \cap V$, and the spaces $U, V, U \cap V$ (hence X) are path-connected. Then the following square of homomorphisms induced by inclusions $i_U : U \cap V \to U$, $i_V : U \cap V \to V$, $j_U : U \to X$, and $j_V : V \to X$ in the diagram*

$$\pi_1(U \cap V, x) \xrightarrow{i_{U*}} \pi_1(U, x)$$

$$\downarrow i_{V*} \qquad\qquad \downarrow j_{U*}$$

$$\pi_1(V, x) \xrightarrow{j_{V*}} \pi_1(X, x)$$

is a pushout square, i.e. it has the following property: for any group G and any homomorphisms $\theta_U : \pi_1(U, x) \to G$ and $\theta_V : \pi_1(V, x) \to G$ such that $\theta_U \circ i_{U*} = \theta_V \circ i_{V*}$, there exists a unique homomorphism $\theta : \pi_1(X, x) \to G$ such that $\theta \circ j_{U*} = \theta_U$ and $\theta \circ j_{V*} = \theta_V$. In other terms, the group $\pi_1(X, x)$ is an amalgamated group product of the groups $\pi_1(U, x)$ and $\pi_1(V, x)$ with the subgroup $\pi_1(U \cap V, x)$; we write $\pi_1(X, x) = \pi_1(U, x) *_{\pi_1(U \cap V, x)} \pi_1(V, x)$. In particular, if the group $\pi_1(U \cap V, x)$ is trivial, then $\pi_1(X, x)$ is the free product $\pi_1(U, x) * \pi_1(V, x)$ of the groups $\pi_1(U, x)$ and $\pi_1(V, x)$.

Remark 7 In this theorem one can replace the open subspaces $U, V, U \cap V$ by their closed subsets $A, B, A \cap B$, if the inclusion maps $A \to U$, $B \to V$, and $A \cap B \to U \cap V$ are homotopy equivalences.

The *wedge product* $(X, x) * (Y, y)$ of topological spaces (with distinguished points) (X, x) and (Y, y) is the quotient space of the disjoint union of spaces X

and Y (even if X and Y do intersect, we assume that they are separated) relative to the equivalence relation: $u \sim v$ if and only if $u = v$, or $u = x$ and $v = y$, or $v = x$ and $u = y$.

Theorem 22 and Corollary 7 imply

Corollary 9 *The fundamental group of the wedge product of n circles* $(S^1, 1)$ *is isomorphic to the free product of n infinite cyclic groups.*

To explain Theorem 22, let us recall some notions from *combinatorial group theory*.

Definition 9 A subset $S \subset G$ of a group $(G, \cdot)$ is called a generator set of the group G if any $g \in G$ can be represented as a product of a finite number of elements from S or their inverses (any representation of this type is called a word in the alphabet S); any element from S is called a generator or a letter of the alphabet S. Assume now that S contains at most one of the elements $x, x^{-1} \in G$. If w is a word in the alphabet S, then we denote by w^{-1} the word obtained by rewriting the letters of w in the reverse order and multiplying the power of each letter of w by -1. The set of words in the alphabet S representing the unit element $e \in G$ will be called the set of units and will be denoted by E.

One can easily prove

Lemma 2 *If the words* w, w_1, *and* w_2 *are contained in* E, *then each word of the type* $u = w_1 w_2$ *or* $u = w^{-1}$ *or* $u = xwx^{-1}$, *or a word obtained from* w *by inserting or deleting* xx^{-1}, *where* $x \in S \cup S^{-1}$, *is also a word in* E.

Definition 10 A subset $R \subset E$ is said to generate E normally if any word in E can be obtained from the words of R using the composition of a finite number of operations described in Lemma 2. The set R is called the relation set (for the group G with the alphabet S). Any word in R is called a relation. In this case we write $G = \{S|R\}$, and the pair (S, R) is called a presentation of the group G. The group G is called free, or finitely generated, or finitely presented if it admits a presentation (S, R), where $R = \emptyset$, or S is finite, or S and R are finite, respectively.

Theorem 23 *Any pair* (S, R), *where* S *is a (finite or infinite) alphabet, and* R *is a set of words in the alphabet* S, *is a presentation of a certain (unique up to isomorphism) group* G. *Any group* G *has a presentation* (S, R).

In terms of presentations, we can formulate Theorem 22 in the following way.

Theorem 24 ([122]) *Suppose that, under the assumptions of Theorem 22, we are given the presentations*

$$\pi_1(U, x) = \{S_U|R_U\}, \quad \pi_1(V, x) = \{S_V|R_V\}, \quad \pi_1(U \cap V, x) = \{S_{U\cap V}|R_{U\cap V}\}.$$

Then the group $\pi_1(X, x)$ *has the presentation* $\pi_1(X, x) = \{S_U \cup S_V|R_U \cup R_V \cup R, \}$ *where* $R = \{w_U(z)(w_V(z))^{-1} : z \in S_{U\cap V}\}$, *and* $w_U(z)$ $(w_V(z))$ *is a representation of an element* $i_{U*}(z)$ $(\, i_{V*}(z))$ *in the alphabet* S_U *(respectively* S_V*).*

Obviously, $(\mathbb{Z}, +) = \{1|\emptyset\}$, i.e., this is a free group with one generator. From Theorem 24 and Corollary 9 we get

Corollary 10 *The fundamental group of the wedge product $W_n(S^1, 1)$ of n circles $(S^1, 1)$ is isomorphic to the group G with the presentation $G = \{a_1, \dots, a_n|\emptyset\}$, i.e., to the free group $\mathbb{F}_n$ with n generators (the free group of rank n).*

17.9 Closed Surfaces and Their Fundamental Groups

In this section, the topological types of closed surfaces (2-manifolds) are classified.

First of all, we have the simply connected *2-sphere* with distinguished point $x = (0, 0, 1)$:

$$S^2 = \{x \in E^3 : |x| = 1\} = S(0, 1) \subset E^3, \quad \pi_1(S^2, x) = \{1\}.$$

The projective plane P^2 is the space of orbits $\Gamma \setminus S^2$, where Γ is the 2-element group of motions of the space E^3 generated by the central symmetry c at zero, with projection p. Also let $b := [p(l_{xc(x)})]$, where $l_{xc(x)}$ is an arc of a large circle of S^2, passing from x to $c(x)$. Then, by Proposition 3, we have $\pi_1(P^2, p(x)) \cong \Gamma = \{c|c^2\}$.

The torus T^2 (respectively, the Klein bottle K) is the space of orbits $\Gamma_1 \setminus E^2$ (respectively, $\Gamma_2 \setminus E^2$), where Γ_1 (respectively, Γ_2) is generated by the parallel translations by vectors $a = (1, 0)$, $b = (0, 1)$ (respectively, by $d = b$ and f, the parallel translation by vector $\frac{1}{2}(1, 0)$ composed with the symmetry about x-axis). Let us set $a := [p_1([0a(0)])]$, $b := [p_1([0b(0)])]$, $d := [p_2([0d(0)])]$, $f := [p_2([0f(0)])]$ where $p_1 : E^2 \to P^2$ and $p_2 : E^2 \to K$ are the quotient maps. Then, $ab = ba$, $df = fd^{-1}$, and by Proposition 3,

$$\pi_1(T^2, p_1(0)) \cong \Gamma_1 = \{a, b|[a, b] = aba^{-1}b^{-1}\}, \quad \pi_1(K, p_2(0)) \cong \Gamma_2 = \{d, f|fdf^{-1}d\}.$$

The operation of connected sum is associative and commutative on the set of closed surfaces.

Theorem 25 ([122]) *Any closed surface is homeomorphic to one of the following surfaces: S^2, P^2, K, $S_g^2 := \#_{k=1}^g T^2(k)$, $P^2\#S_g^2$, $K\#S_g^2$, where $T^2(k) = T^2$, g is a positive integer, and any two of these surfaces are not homeomorphic to each other.*

Remark 8 It is well-known that S^2 and T^2 can be given by polynomial equations in E^3. This is true for all other orientable surfaces S_g^2, $g \geq 2$ [94]. Any non-orientable surface is not homeomorphic to any subspace in E^3; however, if we cut out any nonempty set from it, we get a subspace homeomorphic to a subset in E^3.

Theorem 26 ([122]) *The fundamental groups of the surfaces listed in Theorem 25 have the following presentations:*

$$\pi_1(S_g^2) = \left\{ a_1, b_1; \ldots; a_g, b_g | \prod_{k=1}^{g} [a_k, b_k] \right\},$$

$$\pi_1(P^2 \# S_g^2) = \left\{ a_1, b_1; \ldots; a_g, b_g | c^2 \prod_{k=1}^{g} [a_k, b_k] \right\},$$

$$\pi_1(K \# S_g^2) = \left\{ a_1, b_1; \ldots; a_g, b_g | f d f^{-1} d \prod_{k=1}^{g} [a_k, b_k] \right\}.$$

The proof can be done by induction with respect to g, if we restrict ourselves to the first case. Previously we have obtained the formula for $S_1^2 = T^2$. Now assume that our claim is true for $g \geq 1$. The surface S_{g+1}^2 can be obtained by gluing the (oriented) surfaces $\sigma_1 := S_g^2 - \operatorname{Int} B_2$, $\sigma_2 = S_1^2 - \operatorname{Int} B_2$ along their boundary circles s_1 and s_2 with the reversed induced orientations. One can easily see that the surface σ_2 is homotopic to the wedge product of two circles, then, by Corollary 9 we have $\pi_1(\sigma_2) = \{a, b\}$. In the same way, $\pi_1(\sigma_1) = \{a_1, b_1; \ldots; a_k, b_k\}$. Then $[s_2]^{-1} = [a, b]$, and (by induction with respect to g) $[s_1]^{-1} = \prod_{k=1}^{g} [a_k, b_k]$. Hence $[s_2] = [s_1]^{-1}$ in $S_{g+1}^2 = S_g^2 \# S_1^2$ and, by the Seifert- van Kampen theorem, we get

$$\pi_1(S_{g+1}^2) = \left\{ a_1, b_1; \ldots; a_g, b_g; a, b | \prod_{k=1}^{g} [a_k, b_k][a, b] \right\}.$$

The two other presentations from Theorem 26 are proved similarly.

Note that for closed surfaces, only the fundamental groups of S^2, P^2, and T^2 are commutative.

17.10 Heegaard Splittings and Fundamental Groups of Closed 3-Manifolds

We need the following definition to present the second description of closed orientable surfaces S_g^2, $g \geq 1$.

Definition 11 Let (M, ρ) be a metric space, $A \subset M$ a nonempty subset, $x \in M$, and r a real number. Then by definition, $\rho(x, A) = \inf_{a \in A} \rho(x, a)$ and similarly for open and closed balls and spheres, we can define

$$U(A, r) = \{x \in M : \rho(x, A) < r\},$$

$$B(A, r) = \{x \in M : \rho(x, A) \leq r\},$$

$$S(A, r) = \{x \in M : \rho(x, A) = r\}.$$

Let us consider the g-leafed rose given in cylindrical coordinates (r, θ, z) on E^3 by the relations $r_g = \{0 \leq r = \cos g\theta, z = 0\}$. It is clear that r_g is homeomorphic to the wedge product $W_g(S^1, 1)$. Let ε_g be a positive number such that the closed ball $B(p_0, \varepsilon_g)$ with center p_0, having the Cartesian coordinates $(\frac{1}{2}, 0, 0)$, does not intersect r_g.

One can prove that there exists a homeomorphism $f_g : S_g^2 \approx S(r_g, \varepsilon_g)$ of S_g^2, described in the previous section, with the following properties. Namely, f_g maps the loop a_k, $k = 1, \dots, g$, onto the loop

$$r = \cos g\theta, \quad z = \varepsilon_g; \quad \frac{2\pi k}{g} - \frac{\pi}{2g} \leq \theta \leq \frac{2\pi k}{g} + \frac{\pi}{2g},$$

and the loop b_k, $k = 1, \dots, g$, onto the loop, which homeomorphically projects onto the triangle in the xy-plane, whose two sides lie in the union of the (closed) half-planes

$$\theta = \frac{2\pi k}{g} \text{ and } \theta = \frac{2\pi k}{g} + \frac{\pi}{g},$$

and the third side is the image of the part of b_k, lying below the plane $z = 0$.

Hence f_g maps the common point p of all loops a_k, b_k; $k = 1, \dots, g$, to the point q with rectangular coordinates $(0, 0, \varepsilon_g)$. Let us denote by a_k, b_k their images $f_g(a_k), f_g(b_k)$; $k = 1, \dots, g$.

Example 7 Let $B_g^3 = B(r_g, \varepsilon_g)$, if $g \geq 1$, and B_0^3 be the unit closed ball $B^3 = B(O, 1) \subset E^3$. Then S_g^2 is the boundary of B_g^3, $g > 0$.

Proposition 6 *For any embedding $i_g : (S_g^2, q) \to (B_g^3, q)$, $g \geq 0$, the corresponding homomorphism $(i_g)_* : \pi_1(S_g^2, q) \to \pi_1(B_g^3, q)$ is surjective. Its kernel* $\ker(i_g)_*$ *is trivial if $g = 0$, and is normally generated by all loops b_k, $k = 1, \dots, g$, if $g \geq 1$. In addition, $\pi_1(B_g^3, q)$ is trivial if $g = 0$, and is a free group of rank g for $g \geq 1$.*

Let us prove this. It is clear that B_g^3 contracts in B_g^3 to $r_g = W_g S^1$ so that any loop a_k; $k = 1, \dots, g$, contracts to the leaf

$$r = \cos g\theta, \quad z = 0; \quad \frac{2\pi k}{g} - \frac{\pi}{2g} \leq \theta \leq \frac{2\pi k}{g} + \frac{\pi}{2g}.$$

Similarly, any loop b_k; $k = 1, \dots, g$, contracts in B_g^3 to a point. Now, all the claims of the Proposition follow from Corollary 9.

Theorem 27 ([93]) *Let B_g^3, $(B_g^3)_2$ be two disjoint copies of B_g^3, with boundaries S_g^2, $(S_g^2)_2$, $g \geq 0$, and $h_g : S_g^2 \to (S_g^2)_2$ be a homeomorphism. Define an* *equivalence relation* $\sim$ *on the disjoint union* $U = B^3_g \coprod (B^3_g)_2$ *by the condition* $x \sim y$ *if* $x = y$*, or* $x \in S^2_g$*,* $y \in (S^2_g)_2$*, and* $y = h_g(x)$*. Then the quotient space* $B^3_g \cup_{h_g} (B^3_g)_2 = U/\sim$ *is a closed orientable* 3*-manifold* M^3*. Conversely, each closed* 3*-manifold* M^3 *can be obtained in this way using a homeomorphism* h_g*,* $g \geq 0$.

Definition 12 The space B^3_g, $g \geq 0$, is called the handlebody of genus g. The handlebody of genus 1 is called the solid torus. The representation of a closed 3-manifold M^3 described in Theorem 27 is called a Heegaard splitting of genus g of M^3.

Example 8 For any homeomorphism $h_0 : S^2 \approx S^2$, the 3-manifold $(B^3)_1 \cup_{h_0} (B^3)_2$ is homeomorphic to S^3.

Example 9 The 3-sphere S^3 can be presented as the set of all points $(z_1, z_2) \in \mathbb{C}^2$ such that $|z_1|^2 + |z_2|^2 = 1$. The sets $B_1 = \{(z_1, z_2) \in S^3 || z_1| \geq |z_2|\}$, $B_2 = \{(z_1, z_2) \in S^3 || z_1| \leq |z_2|\}$ are homeomorphic to B^3_1, and $B^3_1 \cap B^3_2$ is homeomorphic to the torus $S^1 \times S^1 = T^2$. Thus we obtain a Heegaard splitting of genus 1 for S^3.

The following theorem is an immediate consequence of Proposition 6 and the Seifert-van Kampen theorem, applied to the triple $\{B^3_g, (B^3_g)_2, S^2_g\}$.

Theorem 28 *Let* $B^3_g \cup_{h_g} (B^3_g)_2$ *be a Heegaard spitting of a closed orientable* 3*-manifold* M^3*. For the embedding* $j_g : (S^2_g, q) \to (M^3, q)$ *the corresponding homomorphism* $(j_g)_* : \pi_1(S^2_g, q) \to \pi_1(M^3, q)$ *is surjective. In addition,*

$$\pi_1(M^3, q) = \{a_k, b_k; k = 1, \dots, g | b_k, [h_g^{-1}(b_{l,2})]; k, l = 1, \dots, g\},$$

where $[h_g^{-1}(b_{l,2})] \in \pi_1(S^2, q)$ *denotes a word in the alphabet* $\{a_k, b_k; k = 1, \dots, g, \}$ *representing this class, and* $b_{l,2}$ *denotes a loop in* $(S^2_g)_2 \subset (B^3_g)_2$*, similar to* b_l.

The following proposition is well known.

Proposition 7 *Any connected non-orientable n-manifold M admits a unique two-sheeted covering map* $p : M_0 \to M$*, where* M_0 *is orientable. The induced homomorphism* $p_* : \pi_1(M_0) \to \pi_1(M)$ *is a monomorphism and the group* $p_*(\pi_1(M_0))$ *is a normal subgroup of index* 2 *in* $\pi_1(M)$.

Theorems 25, 26, and Proposition 7 imply that the fundamental group $\pi_1(M)$ of any closed 3-manifold M is an order two extension of a finitely presented group $\pi_1(M_0)$ by the group $\pi_1(M)/p_*(\pi_1(M_0)) \cong \mathbb{Z}_2$. Thus we get the following exact sequence of groups and homomorphisms

$$\{1\} \to \pi_1(M_0) \stackrel{p_*}{\to} \pi_1(M) \to \pi_1(M)/p_*(M_0) \to \{1\}.$$

Any extension of a finitely presented group by a finitely presented group is itself finitely presented [14]. Therefore we obtain

Theorem 29 *The fundamental group of any closed 3-manifold is finitely presented.*

17.11 Group-Theoretic Statements, Equivalent to the Poincaré Conjecture

17.11.1 Splitting Homomorphisms

There are several ways to formulate the Poincaré conjecture in purely algebraic terms. Some of them involve the idea of *splitting homomorphism* [172].

Let $(B_g^3, (B_g^3)_2)$, $g \geq 1$, be a Heegaard splitting of a closed orientable 3-manifold M^3 and let $S_g^2 = B_g^3 \cap (B_g^3)_2$. We have the commutative diagram

$$\begin{array}{ccc} \pi_1(S_g^2) & \xrightarrow{\phi} & \pi_1(B_g^3) \\ \downarrow \phi_2 & & \downarrow i \\ \pi_1((B_g^3)_2) & \xrightarrow{i_2} & \pi_1(B_g^3) \times \pi_1((B_g^3)_2), \end{array}$$

where ϕ, ϕ_2 are induced by embeddings and i, i_2 are natural monomorphisms.

Definition 13 The map $\phi \times \phi_2 : \pi_1(S^2) \rightarrow \pi_1(B_g^3) \times \pi_1((B_g^3)_2)$ is called the splitting homomorphism associated with the given Heegaard splitting.

Lemma 3 *With the above notation, $\pi_1(M^3) = 1$ if and only if $\phi \times \phi_2$ is an epimorphism.*

In fact, it follows from Theorems 27 and 28 that all homomorphisms ϕ, ϕ_2, j_{g*} : $\pi_1(S_g^2) \rightarrow \pi_1(M^3)$ are surjective and $\ker j_{g*} = \ker\phi \cdot \ker\phi_2$, where $\ker\phi \cdot \ker\phi_2$ is generated by elements of $\ker\phi$ and $\ker\phi_2$. Therefore $\phi \times \phi_2$ is surjective if and only if $\ker\phi \cdot \ker\phi_2 = \pi_1(S_g^2)$.

By Proposition 6, $\pi_1(B_g^3)$ and $\pi_1((B_g^3)_2)$ are free groups of rank g. For each g we fix a free group F_g of rank g and consider the splitting homomorphism as the map $\phi \times \phi_2 : \pi_1(S^2) \rightarrow F_g \times F_g$. Such maps are *equivalent* if there exists a commutative diagram

$$\begin{array}{ccc} \pi_1(S_g^2) & \xrightarrow{\phi\times\phi_2} & F_g \times F_g \\ \downarrow \alpha & & \beta \times \beta_2 \downarrow \\ \pi_1(S_g^2) & \xrightarrow{\phi'\times\phi_2'} & F_g \times F_g \end{array}$$

with isomorphisms α, β, and β_2.

Lemma 4 ([97]) *Each map $\phi \times \phi_2 : \pi_1(S_g^2) \to F_g \times F_g$ with surjective ϕ and ϕ_2 is equivalent to the splitting homomorphism associated with a Heegaard splitting of a closed orientable* 3*-manifold.*

In [179] it is proved that any two Heegaard splittings of S^3 of the same genus are equivalent. A combination of this result with the above argument gives the following

Theorem 30 ([93]) *The Poincaré conjecture* PC_3 *holds if and only if for any positive integer g there exists a unique up to equivalence epimorphism*

$$\pi_1(S_g^2) \to F_g \times F_g$$

onto the direct product of two free groups of rank g.

17.11.2 The Mapping Class Group

In 1910, M. Dehn "proved" the theorem known now as

Lemma 5 (Dehn's Lemma) *Let M be a* 3*-manifold and $f : B^2 \to M$ a piecewise linear map such that for some neighborhood A of the boundary $S^1 = \partial B^2$ in B^2 the map $f|A$ is a topological embedding (i.e., a homeomorphism onto its image) and $f^{-1}(f(A)) = A$. Then $f|S^1$ can be extended to a topological embedding $g : B^2 \to M$.*

However, in 1929 H. Kneser discovered a serious gap in Dehn's proof and the problem remained open up to the middle of 1950s when Papakyriakopoulos [149] gave a correct proof for the Dehn lemma.

Definition 14 A homotopy $F : X \times I \to X$ between two homeomorphisms $f : X \approx X$ and $g : X \approx X$ is said to be an isotopy if $F(\cdot, s)$ is a homeomorphism of X onto itself for each $s \in I$. Homeomorphisms $f : X \approx X$, $g : X \approx X$ are isotopic if there exists an isotopy between them.

Theorem 30 admits another interpretation in terms of the *mapping class group* MC(S) of a closed orientable surface S which is defined as the group of classes of orientation-preserving homeomorphisms of S modulo homeomorphisms isotopic to the identity map. We denote by $< f > \in \mathrm{MC}(S)$ the class of homeomorphism which contains the homeomorphism $f : S \to S$. One can prove (see Theorem 13.1 in [93]) that each automorphism of the group $\pi_1(S)$ is induced by a homeomorphism of S. Two homeomorphisms of S inducing automorphisms of $\pi_1(S)$ which differ by an inner automorphism of $\pi_1(S)$ are (freely) homotopic and, by Epstein [65], are isotopic. Therefore MC(S) is isomorphic to the group of outer automorphisms of $\pi_1(S)$ (called the *homeotopy group* of the surface S).

Let (B_1, B_2) be a Heegaard splitting of a 3-manifold M, and $S = B_1 \cap B_2$. Choose a homeomorphism $h : S \to S$ which extends to a homeomorphism $H : B_1 \to B_2$ (h is determined up to conjugation by a homeomorphism of B_1), such that B_1 and h completely determine the splitting (B_1, B_2) of the manifold M. Let $K_i, i = 1, 2$, be the subgroup in MC(S) consisting of the mapping classes representing homeomorphisms f which can be extended to homeomorphisms of B_i onto itself. Then the Dehn lemma implies that

$$K_i = \{< f >: f_*(\ker \phi_i) = \ker \phi_i\},$$

where $\phi_i : \pi_1(S) \to \pi_1(B_i)$ is induced by the inclusion.

For every $g \geq 0$ there exists a *natural splitting* $(\overline{B}_1, \overline{B}_2)$ *of genus* g *of* S^3, where $\overline{B}_1 = B^3_g$, $\overline{B}_2 = S^3 - \text{Int}(\overline{B}_1)$, and S^3 is considered as the one-point compactification of E^3. Then, with the previous notation for the loops $a_k, b_k; k = 1, \ldots, g$, we get that $\ker \overline{\phi}_1$ is the normal subgroup generated by the loops $\{b_1, \ldots, b_g\}$, $\ker \overline{\phi}_2$ is the normal subgroup generated by $\{a_1, \ldots, a_g\}$, and $\overline{h}_*(a_i) = b_i, \overline{h}_*(b_i) = a_i$, $(< \overline{h}^2 > = 1)$.

Theorem 31 *The Poincaré conjecture is equivalent to the following statement: If* $k : \overline{S}^2_g \to \overline{S}^2_g$ *is a homeomorphism such that* $\ker \overline{\phi}_1 \cdot k_*(\ker \overline{\phi}_1) = \pi_1(\overline{S}^2_g)$, *then* $< k\overline{h}^{-1} > \in \overline{K}_1\overline{K}_2$.

Results by Nielsen and Thurston on mapping class groups can be found in [46].

Remark 9 There is also another paper [150] by Papakyriakopoulos with some group theoretic reductions of the Poincaré conjecture.

17.12 Connected Sums and Prime Decompositions of 3-Manifolds

In Sects. 17.12–17.17 we assume that all manifolds and their subspaces (respectively, maps) are polyhedra (respectively, piecewise linear) if the contrary is not stated explicitly. This is a natural assumption since any n-manifold $1 \leq n \leq 3$ admits a unique compatible PL-triangulation and smooth structure.

Example 10 The 2-sphere, the projective plane, and the Klein bottle with g handles, denoted respectively by S^2_g, P^2_g, and K_g, are respectively the g-multiple connected sum of the 2-sphere, the projective plane, and the Klein bottle with the torus S^2_1.

Lemma 6 ([93]) *The connected sum is a well defined associative and commutative operation in the category of oriented 3-manifolds and homeomorphisms preserving orientation.*

By the van Kampen theorem the inclusion maps $R_i \to M_i$ induce isomorphisms of fundamental groups if $n \geq 3$. If $A = h_1(R_1)$ and $B = h_2(R_2)$, then $A \cap B$ is homeomorphic to S^{n-1}, and hence is simply connected. Then the van Kampen theorem implies

Lemma 7 *If $M = M_1 \# M_2$ is a connected sum of n-manifolds and $n \geq 3$, then there is a unique natural isomorphism*

$$\pi_1(M) \cong \pi_1(M_1) * \pi_1(M_2).$$

A 3-manifold M is said to be *prime* if for any decomposition as $M = M_1 \# M_2$, M_1 or M_2 is the 3-sphere (then the other manifold is homeomorphic to M).

Theorem 32 ([93], Theorem 3.15) *Each compact 3-manifold can be represented as a connected sum of a finite number of prime manifolds.*

The proof of this theorem is based on Lemma 7, Theorem 29, and the following

Lemma 8 *Let G be a finitely generated group, $p(G) = \inf\{|A| : A$ generates $G\}$, and $G = G_1 * G_2$ with nontrivial groups G_1 and G_2. Then $p(G) = p(G_1) + p(G_2)$.*

In turn, this lemma immediately follows from the well-known Grushko theorem.

Theorem 33 *If $G_1 * G_2$ is a finitely generated group with nontrivial groups G_1 and G_2, F a finitely generated free group, and $\eta : F \to G_1 * G_2$ an epimorphism, then there exist free groups $F_i, i = 1, 2$, homomorphisms $\phi_i : F_i \to G_i$, and an isomorphism $\psi : F \to F_1 * F_2$ such that $(\phi_1 * \phi_2) \circ \psi = \eta$.*

For the proof of this theorem we refer the reader to [122]. A topological proof of the Grushko theorem is given in [171]. The proof of the following theorem, which is a converse to Lemma 7 in dimension 3 is based on this proof.

Theorem 34 ([93], Theorem 7.1) *Let M be a closed 3-manifold. If $\pi_1(M) \cong G_1 * G_2$, then $M = M_1 \# M_2$, where M_1 and M_2 are closed manifolds with $\pi_1(M_i) \cong G_i$, $i = 1, 2$.*

In [131] Milnor proved a uniqueness theorem for the prime decomposition of closed oriented 3-manifolds. A similar result is true in a more general case.

A prime decomposition $M = M_1 \# \ldots \# M_n$ of a 3-manifold M is said to be *normal* if some of the M_i can be $S^2 \times S^1$ only if M is orientable. By Lemma 3.17 in [93] any prime decomposition can be replaced with a normal decomposition.

Theorem 35 ([93], Theorem 3.21) *Let $M = M_1 \# \ldots \# M_n = M_1^* \# \ldots \# M_{n^*}^*$ be two normal prime decompositions of a closed 3-manifold M. Then $n = n^*$ and (up to a permutation) M_i is homeomorphic to M_i^* (in the oriented case the homeomorphisms preserve the orientation).*

17.13 Irreducible Closed and Open Contractible 3-Manifolds

A 3-manifold M is said to be *irreducible* if any 2-sphere S in M bounds a 3-cell in M. The well-known J.W. Alexander theorem in [5] implies that $\mathbb{R}^3$ and S^3 are irreducible under the assumptions set out at the beginning of Sect. 17.12.

However, a *topological* analogue of this statement is false for the (wild) 2-sphere S in $\mathbb{R}^3$ and S^3. Alexander constructed in [4] a topological (wild) 2-sphere S in S^3 such that one connected component of $S^3 - S$ is not simply connected (the so-called Alexander's horned sphere). One of the components of the complement of S in S^3 is (only topologically!) an open 3-cell, and the other D is not. Let $D' \cup S'$ be another copy of $D \cup S$ and $D \cup S$ be glued with $D' \cup S'$ along S and S' by a homeomorphism $h : S \approx S'$. R.H. Bing proved that the quotient space $(D \cup S) \cup_h (D' \cup S')$ is homeomorphic to S^3 [31]. It is clear that S bounds no (topological!) 3-cell in S^3.

Clearly, irreducible 3-manifolds are prime. The following lemma partially gives the converse statement.

Lemma 9 ([93], 3.13) *If M is a prime closed 3-manifold and M is not irreducible then M is a 2-spherical bundle over S^1, i.e., $M = S^1 \times S^2$ or $M = S^1 \times_\phi S^2$.*

Remark 10 In Lemma 9, $S^1 \times_\phi S^2$ denotes the quotient space of the direct product $I \times S^2$ with respect to the equivalence relation $\sim$, where $(t, s) \sim (t_1, s_1)$ if $(t, s) = (t_1, s_1)$ or $t = 0$, $t_1 = 1$, and $s_1 = \phi(s)$ for an orientation-reversing homeomorphism $\phi : S^2 \approx S^2$.

The following theorem gives a characterization of 2-spherical bundles over S^1.

Theorem 36 *M is a prime closed 3-manifold with a nontrivial free group $\pi_1(M)$ if and only if M is a 2-spherical bundle over S^1. In addition, $\pi_1(M) \cong (\mathbb{Z}, +)$.*

It is clear that $\pi_1(S^1 \times S^2) \cong \pi_1(S^1 \times_\phi S^2) \cong \pi_1(S^1) \cong (\mathbb{Z}, +)$ is a nontrivial free group. The necessity is proved in Theorem 5.2 of [93]. The following theorem generalizes this statement.

Theorem 37 *M is a closed 3-manifold with a free group $\pi_1(M)$ of rank $r > 0$ if and only if $M = \Sigma \# B_1 \# \ldots \# B_r$, where B_j is a 2-spherical bundle over S^1, and Σ is a homotopy 3-sphere.*

Note that the Σ can be dropped if M (or a covering of M) contains no fake 3-cell. This is true if and only if PC_3 holds.

Theorem 38 ([110, 131]) *Any closed orientable 3-manifold M^3 can be represented as a connected sum*

$$M^3 = (K_1 \# \ldots \# K_p) \# (L_1 \# \ldots \# L_q) \# (\#_1^r S^2 \times S^1) \tag{17.3}$$

of prime manifolds, where the K- and L-summands are closed irreducible 3-manifolds, the K-summands have infinite fundamental group and are aspherical (i.e., their universal coverings are contractible), the L-summands have finite *fundamental groups with homotopy 3-spheres as covering spaces, no L-summand is a 3-sphere; otherwise $M = L \approx S^3$. In this sum the summands are unique up to a permutation.*

Remark 11 Validity of the Poincaré conjecture implies that $\pi_1(L_i) \neq \{1\}$, $i = 1, \dots, q$.

In [188], the reader can find the following version of the *sphere theorem.*

Theorem 39 *For any orientable polyhedral 3-manifold M with $\pi_2(M) \neq 1$, there exists an embedded polyhedral 2-sphere in M which represents a nontrivial element of the group $\pi_2(M)$.*

Corollary 11 *Any closed orientable irreducible 3-manifold has trivial second homotopy group.*

Theorem 40 *The universal covering manifold $\overline{M}$ of a connected closed 3-manifold M with $\pi_2(M) = 0$ is either a compact homotopy 3-sphere or it is noncompact and contractible.*

A noncompact manifold (without boundary) is usually called an *open manifold.* The following theorem demonstrates that there are contractible open 3-manifolds which are not homeomorphic to E^3.

Theorem 41 ([93], Example 14.1) *Let W be an open 3-manifold represented as $W = \cup_{i=0}^{\infty} T_i$, where each T_i is a solid torus and for each $i \geq 0$*

(i) $T_i \subset \mathrm{Int}(T_{i+1})$
(ii) $\pi_1(\partial(T_{i+1})) \to \pi_1(T_{i+1} - T_i)$ *is a monomorphism,*
(iii) T_i *is contractible in* T_{i+1}.

Then W is a contractible open 3-manifold which is not homeomorphic to E^3.

The first example of such a manifold $W \subset E^3$ was given by Whitehead in [185] as a counterexample to his false statement in [190] that any open contractible 3-manifold is homeomorphic to E^3. This statement was part of a proof which Whitehead intended to give for the Poincaré conjecture, because if Σ^3 is a homotopy 3-sphere, then $\Sigma^3 - \{\text{point}\}$ is a contractible open 3-manifold.

Using a variant of Whitehead's technique, in [126] an uncountable family of pairwise nonhomeomorphic contractible open submanifolds W in E^3 was constructed. Each of these manifolds W satisfies the assumptions of Theorem 41, hence

(1) W is a contractible open 3-manifold;
(2) any compact subset in W can be topologically embedded in E^3.

Quite surprisingly, it was proved in [127] that these two properties imply that $W \times E^1$ is homeomorphic to E^4 and in [128] it was remarked that in fact $W \times E^1$ is combinatorially equivalent to E^4.

In [108], another uncountable family of pairwise nonhomeomorphic contractible open 3-manifolds W which admit no topological embedding in E^3 was constructed.

In the proof of Theorem 41 the following theorem called the *loop theorem* by Stallings [173] is used.

Theorem 42 *Let M be a 3-manifold with nonempty boundary, and F a connected 2-submanifold in ∂M. If N is a normal subgroup in $\pi_1(F)$ and* $\ker(\pi_1(F) \to \pi_1(M)) - N \neq \emptyset$, *then there exists an embedding $g : (B^2, \partial B^2) \to (M, F)$ such that $g(B^2) \cap \partial M = g(\partial B^2)$ and $[g|\partial B^2] \notin N$.*

17.14 Compact 3-Manifolds with Finite Fundamental Groups

Theorem 43 ([93]) *Any closed 3-manifold M with finite fundamental group $\pi_1(M)$ is orientable. Any such M is either irreducible, or can be represented as $M = \Sigma \# M_1$, where Σ is a homotopy 3-sphere which is not homeomorphic to S^3, and M_1 is a closed irreducible 3-manifold with $\pi_1(M_1) \cong \pi_1(M)$. In addition, M admits a finite covering by a homotopy 3-sphere.*

Corollary 12 *The Poincaré conjecture* PC_3 *is equivalent to the statement that any closed 3-manifold with finite fundamental group is irreducible.*

In [64] the following surprising theorem is proved.

Theorem 44 *Validity of the Poincaré conjecture* PC_3 *implies that any compact nonorientable 3-manifold with finite fundamental group is homeomorphic to $P^2 \times I$ with a finite number of open 3-cells removed, where P^2 is the projective plane.*

In the third chapter of [64] the question is investigated under what conditions a group can be a subgroup of the fundamental group of a 3-manifold M. For example, necessary and sufficient conditions are given for $\pi_1(M)$ to have a finite (nontrivial) subgroup and it is proved that this finite subgroup is isomorphic to the fundamental group of a closed 3-manifold [64, Corollary 8.7].

Let M be a closed 3-manifold with universal covering manifold S^3 (by the Poincaré conjecture, M is a closed 3-manifold with finite fundamental group). Then $M = \Gamma \backslash S^3$, where Γ is a finite group acting freely and topologically on S^3. The manifold M admits a piecewise linear structure [134], which can be smoothed [95]. Hence by Theorem 43, there is no loss of generality in supposing that Γ consists of smooth, orientation-preserving diffeomorphisms of the standard unit sphere S^3 in E^4. It is proved in [176] that $\Gamma \cong \pi_1(M)$ must be isomorphic to a subgroup $\Gamma_1 \subset SO(4)$, such that the standard linear action of Γ_1 on S^3 is free, under the assumption that the Smale conjecture (see below) is true. Moreover, *M is homotopy equivalent to $M_1 = \Gamma_1 \backslash S^3$*, a smooth Riemannian manifold of constant sectional curvature 1, a so-called *spherical Clifford-Klein form*.

The Smale conjecture is the assertion that the inclusion of the orthogonal group O(4) (respectively, SO(4)) into $\mathrm{Diff}(S^3)$ (respectively, $\mathrm{Diff}^+(S^3)$), the (orientation-preserving) diffeomorphism group of the 3-sphere with the

C^∞-topology, is a homotopy equivalence [168]. Later on A. Hatcher proved the Smale conjecture [90].

For the most part, the list of these subgroups Γ_1 was obtained by Hopf in [96]. A full classification of the groups $\Gamma_1 \subset \mathrm{SO}(4)$ and corresponding spherical Clifford-Klein forms is given in the books by Seifert and Threlfall [164] and Wolf [192].

Theorem 45

(1) A closed 3-manifold M is a homology 3-sphere (see Definition 21) if and only if M is orientable and $\pi_1(M)$ is a perfect group, i.e. $[\pi_1(M), \pi_1(M)] = \pi_1(M)$, where $[\pi_1(M), \pi_1(M)]$ is the commutant (the derived group) of $\pi_1(M)$.
(2) Any simply connected closed 3-manifold is a homology 3-sphere.
(3) Any (closed) 3-manifold which is a homotopy 3-sphere is a homology 3-sphere.
(4) Any simply connected closed 3-manifold is a homotopy 3-sphere.

As a corollary of Theorem 45 and the Seifert-van Kampen theorem, we get

Corollary 13 *A connected sum of two homology 3-spheres is a homology 3-sphere.*

At first Poincaré in [154] erroneously suggested the following

Conjecture 2 A closed 3-manifold, which is a homology 3-sphere, is simply connected and (hence?) homeomorphic to 3-sphere.

He soon found a mistake in his first assertion and constructed an example of a closed non-simply connected homology 3-sphere σ^3 with finite fundamental group $\pi_1(\sigma^3) = P_{120}$ of order 120, [155].

The Heegaard splitting of genus 2 for σ^3 is given in [93], Exercise 2.7. Here we only give the corresponding presentation of the group $\pi_1(\sigma^3)$:

$$\pi_1(\sigma^3) = \{a, b : abab^{-1}a^{-1}b^{-1}, aba^{-1}bab^{-1}\}.$$

Let us give another (more geometrical) description of σ^3. By definition, S^3 is the unit sphere in E^4, which can be identified with the associative non-commutative division algebra $\mathbb{H}$ of quaternions $q = x + y\mathbf{i} + z\mathbf{j} + w\mathbf{k}$ invented by W.R. Hamilton. Then S^3 is naturally identified with the set $Q = \{q : |q| = 1\}$ of all unit quaternions, which constitute a compact Lie group with respect to multiplication in $\mathbb{H}$. A *vector* (respectively, *scalar*) in $\mathbb{H}$ is a quaternion $q = x + y\mathbf{i} + z\mathbf{j} + w\mathbf{k}$ such that $x = 0$ (respectively, $y = z = w = 0$). The vectors v form the 3-dimensional real vector space $E^3 := V^3$.

For any unit quaternion $q \in Q$, the map $p(q) : V^3 \to V^3$, $p(q)(v) = qvq^{-1}$, is an isometric linear map of the space $V^3 = E^3$ onto itself, which preserves the orientation of E^3, or $p(q) \in \mathrm{SO}(3)$, where $\mathrm{SO}(3)$ is naturally identified with the Lie group of all orthogonal 3×3-matrices with determinant 1. In fact we obtain an epimorphism of the Lie groups $p : Q \to \mathrm{SO}(3)$ with kernel $\ker p = \{1, -1\}$. The epimorphism p is the two-sheeted universal covering map. The subgroup $I_{60} \subset \mathrm{SO}(3)$ of order 60 of all isometries in $V^3 = E^3$ which map a regular icosahedron (or dodecahedron) with center O onto itself, is the image of a finite group $P_{120} = p^{-1}(I_{60}) \subset Q$ of order 120, which is called the *binary icosahedral* or *binary dodecahedral group*.

Now $\sigma^3 = \mathrm{SO}(3)/I_{60} = Q/ P_{120}$ is the (homogeneous Riemannian) quotient space of left cosets of the group SO(3) (respectively, Q) by its finite subgroup I_{60} (respectively, P_{120}). Since $Q = S^3$ is simply connected, $\pi_1(\sigma^3) \cong P_{120}$.

We can add to Theorem 43 the following interesting Theorem 3.1 from [66].

Theorem 46 *Let M be a compact 3-manifold with finite $\pi_1(M)$. Then $\pi_1(M)$ is solvable if $\pi_1(M)$ is not isomorphic to the binary dodecahedron group P_{120} or to the direct product of P_{120} and a cyclic group whose order is coprime to* 120.

Corollary 14 *If M^3 is a (closed) homology 3-sphere with finite nontrivial fundamental group $\pi_1(M^3)$, then $\pi_1(M^3) \cong P_{120}$.*

We get from Theorems 43 and 53 the following

Corollary 15 *If a closed 3-manifold M with finite $\pi_1(M)$ contains the projective plane P^2, then $\pi_1(M) \cong \mathbb{Z}_2$.*

In the same manner one can consider the *binary tetrahedral group* P_{24} and the *binary octahedral group* P_{48} and the corresponding homogeneous Riemannian quotient spaces $\mathrm{SO}(3)/T_{12} = Q/P_{24}$ and $\mathrm{SO}(3)/O_{24} = Q/P_{48}$ with fundamental groups isomorphic to P_{24} and P_{48}, respectively. These closed 3-manifolds are not homology 3-spheres, because the groups P_{24} and P_{48} are solvable.

One can also consider a regular two-sided plane n-polygon in E^3, where $n \geq 3$, as a dihedron and its *dihedral* symmetry group $D_{2n} \subset \mathrm{SO}(3)$, as well as the *binary dihedral* group $\mathrm{BD}_{4n} = p^{-1}(D_{2n})$ and the corresponding homogeneous Riemannian space Q/BD_{4n}. Any element $e^{2\pi i/p}$ of the multiplicative group Q, where $p \geq 2$ is a positive integer, generates a finite cyclic subgroup e_p and determines the corresponding homogeneous Riemannian space Q/e_p. It is clear that Q/e_2 is the *real projective 3-space* P^3. All other spaces Q/e_p, where $p \geq 3$, are called *lens spaces* and denoted by $L_{p,1}$.

All the above mentioned finite subgroups Γ of the multiplicative Lie group $\mathbb{Q}$ determine homogeneous Riemannian spaces $Q/\Gamma = S^3/\Gamma$. Any other Riemannian manifold S^3/Γ is non-homogeneous [192].

Among the other nonhomogeneous manifolds S^3/Γ we mention only lens spaces $L_{p,q}$, where $p, q \geq 2$ are mutually coprime positive integers. By definition, $L_{p,q} = S^3/\Gamma_{p,q}$, where $\Gamma_{p,q}$ is a cyclic subgroup of order p in SO(4) generated by the diagonal complex (2×2)-matrix $\mathrm{diag}(e^{2\pi i/p}, e^{(2\pi i)q/p})$, and E^4 is understood as $\mathbb{C}^2$.

Theorem 47

(1) $L_{p,q}$ is homotopy equivalent to $L_{p',q'}$ if and only if $p = p'$.
(2) $L_{p,q}$ is homeomorphic to $L_{p',q'}$ if and only if $p = p'$ and $q \equiv \pm q' \mod p$, or $qq' \equiv \pm 1 \mod p$.

Corollary 16 *There exist homotopy equivalent but nonhomeomorphic lens spaces.*

Note that J.W. Alexander was the first to discover that there exist nonhomeomorphic lens spaces with isomorphic fundamental groups [3]. In [93] the reader can find another description of lens spaces in terms of Heegaard splittings of genus one. Apart from lens spaces one can describe in this way S^3, P^3, and the two above mentioned S^2-bundles over S^1.

17.15 Irreducible Closed 3-Manifolds with Infinite Fundamental Group

It follows from theorems on the prime (normal in the nonorienrable case) decomposition of a closed 3-manifold into a finite connected sum which were exposed in Sect. 17.12, that the investigation of these manifolds can be reduced to the case of closed irreducible 3-manifolds with infinite fundamental group. By Theorem 38, in the orientable case, these manifolds are aspherical K-summands. By Lemma 9, in the nonorientable case any prime summand in the corresponding decomposition is either irreducible or homeomorphic to $S^2 \times_\phi S^1$. First let us give information on infinite fundamental groups of 3-manifolds (mainly from [93]).

From Lemma 6.7 in [93] we get the following

Theorem 48 *Any non-orientable closed 3-manifold M has infinite first homology group $H_1(M, \mathbb{Z})$.*

Definition 15 A closed, possibly not connected, surface F in a closed 3-manifold M is said to be two-sided in M if there is an embedding $h : F \times I \to M$ with $h(x, \frac{1}{2}) = x$ for all $x \in F$.

Theorem 49 *Let M be a prime closed 3-manifold with infinite fundamental group $\pi_1(M)$. Then any element of finite order in $\pi_1(M)$ has order 2. If M contains no 2-sided projective plane P^2, then $\pi_1(M)$ is torsion free. In particular, if M is orientable, then M contains no embedded 2-sided P^2, and therefore $\pi_1(M)$ is torsion free.*

The first two statements of this theorem follow from Corollary 9.9 in [93]. The last statement follows from Theorems 53 and 34.

Theorem 50 ([93], Theorem 10.7) *Let M be a closed 3-manifold with $\pi_1(M)$ torsion free. If $\pi_1(M)$ contains a nontrivial free subgroup of finite index, then $\pi_1(M)$ is free (and the structure of M is as given in Theorem 37).*

Definition 16 A closed 3-manifold M is said to be P^2-irreducible if M is irreducible and M contains no two-sided P^2.

Theorem 51 ([93], Lemma 10.4) *Let M be a P^2-irreducible 3-manifold and $p : \tilde{M} \to M$ be a two-sheeted covering map. Then $\tilde{M}$ is P^2-irreducible.*

Proposition 8 *If M is a P^2-irreducible 3-manifold, then its universal covering manifold $\overline{M}$ is a homotopy 3-sphere if $\pi_1(M)$ is finite and is a contractible noncompact 3-manifold if $\pi_1(M)$ is infinite. For nonorientable M only the second possibility holds.*

Let us prove this. If M is orientable, this follows from Corollary 11 and Theorem 40. If M is nonorientable, then its two-sheeted orientable covering space $\tilde{M}$ is P^2-irreducible by Theorem 51 and we can apply the previous argument. The last statement now follows from Theorem 48.

Conjecture 3 The universal covering manifold of any connected P^2-irreducible 3-manifold with infinite group $\pi_1(M)$ is homeomorphic to E^3.

Definition 17 A compact connected surface F in a compact 3-manifold M is said to be incompressible in M if $\partial F \subset \partial M$ and neither of the following conditions hold:

(i) F is a sphere bounding a homotopy 3-cell in M,
(ii) there exists a 2-cell $D \subset M$ with $D \cap F = \partial D$ not contractible in F.

Theorem 52 ([93], Corollary 6.2) *If F is a 2-sided incompressible surface in a compact 3-manifold M, then $\ker(\pi_1(F) \to \pi_1(M)) = \{1\}$. Hence $\pi_1(M)$ has a 2-torsion if $F = P^2$.*

Theorem 53 ([93], Lemma 6.3) *Let M be an orientable 3-manifold containing a closed surface with odd Euler characteristic, and let F be a closed surface in M with a maximal odd Euler characteristic. Then F is incompressible in M. If $F = P^2$, then $\pi_1(M) \cong \mathbb{Z}_2 * G$ (it is possible that $G = \{1\}$).*

Theorem 54 ([93], Lemma 6.6) *A closed 3-manifold M contains a 2-sided non-separating incompressible closed surface if and only if the group $H_1(M, \mathbb{Z})$ is infinite.*

Definition 18 A compact 3-manifold M is said to be sufficiently large if M contains a compact 2-sided incompressible surface.

From Theorems 54 and 48 we get

Corollary 17 *Any closed 3-manifold M with infinite $H_1(M, \mathbb{Z})$ is sufficiently large. In particular, any nonrientable closed 3-manifold is sufficiently large.*

Remark 12 The converse of Corollary 17 does not hold; there exist closed orientable sufficiently large 3-manifolds with finite (or even trivial) $H_1(M, \mathbb{Z})$.

17.16 Haken Manifolds

Definition 19 A compact 3-manifold is called Haken if it is irreducible and contains a closed connected incompressible two-sided surface F which is not homeomorphic to P^2.

Proposition 9 *Any Haken 3-manifold M has infinite fundamental group $\pi_1(M)$.*

Let us prove this. Since M is irreducible, then in Definition 19, F cannot be S^2 or P^2. Hence $\pi_1(F)$ is infinite. Now the required statement follows from the first statement of Theorem 52.

Proposition 10 *Let us consider the following three conditions on a closed 3-manifold M:*

(i) M is P^2-irreducible and sufficiently large;
(ii) M is Haken;
(iii) M is irreducible and sufficiently large.

Then $(i) \Rightarrow (ii) \Rightarrow (iii)$. When M is orientable, the three conditions are equivalent.

Let us prove this. The first statement is evident. If M is orientable, then it contains no two-sided nonorientable closed surface (and any orientable closed surface in M is two-sided), hence M is P^2-irreducible, if it is irreducible.

Remark 13 F. Waldhausen and W. Jaco defined a Haken manifold to be a compact orientable irreducible sufficiently large 3-manifold [99, 181]. Proposition 10 shows that the term "orientable Haken manifold" has the usual meaning.

Theorem 55 ([99], Corollary X.8) *Let M and M' be compact orientable Haken manifolds and $f : (M, \partial M) \to (M', \partial M')$ be a map such that $f_* : \pi_1(M) \to \pi_1(M')$ is an isomorphism. Then f is homotopic to a homeomorphism (it is possible that the image $\partial M \times I$ is not contained in $\partial M'$). Moreover, if $f|\partial M \to \partial M'$ is a homeomorphism, then f is homotopic* rel∂M *to a homeomorphism h, i.e. there is a homotopy between f and h constant on ∂M.*

In [181] Waldhausen proved the following partial case of Conjecture 3.

Theorem 56 *The universal covering of any closed orientable Haken manifold is homeomorphic to E^3.*

In [182] Waldhausen proposed the following conjecture called the *virtual Haken Conjecture* (see also [61]), which, by Theorem 56, is stronger than Conjecture 3.

Conjecture 4 Let M be a compact irreducible 3-manifold with infinite fundamental group. Then M has a finite-sheeted covering which is Haken.

The following text, till the end of this section, is an excerpt from the book [125] by Matveev.

The question: "Does there exist an algorithm to decide whether or not two given 3-manifolds are homeomorphic?" in [88] is known as the *recognition problem for*

3-*manifolds*. The aim of Chapter 6 in [125] is to present a positive solution to this problem for Haken manifolds. This case is especially important because it implies a positive solution to the problem of algorithmic knots classification, one of the most intriguing problems of low-dimensional topology. Recall that a *knot* is a circle topologically embedded in S^3. Two knots are *equivalent*, if there exists an isotopy $S^3 \to S^3$ moving one knot to the other.

The history of the positive solution to the recognition problem for Haken 3-manifolds is very interesting. In 1962 Haken proposed an approach to a solution to this problem [88]. But this approach had a gap. Thanks to the efforts of several mathematicians, at the beginning of the seventies a crucial obstacle to the solution was found, and when in 1978 Hemion overcame it in [91], it was widely announced that the problem has been solved [102, 180]. Later on many topologists used extensively this result.

Nonetheless, in trying to understand the proof of this theorem in detail, Matveev discovered that the complete proof was not written anywhere. All the papers and even books [92, 102, 103, 180] devoted to this subject were written according to the same scheme: they contained nonformal description of Haken's approach, obstacles, Hemion's result, and the claim that these three ingredients give together the proof. At the same time, there was no article containing this proof. Matveev investigated this question and came to the following conclusions.

1. The statement that the recognition problem for Haken manifolds is algorithmically solvable is true.
2. There is another obstacle of similar nature that cannot be overcome by the same tools as the first one.
3. This obstacle can be overcome by using an algorithmic version of Thurston's theory of surface homeomorphisms that appeared only in 1995 [25].

17.17 The JSJ-Splitting Theorem for Orientable Haken Manifolds

In this section we discuss the JSJ-*splitting theorem* proved independently by Johannson [101] and Jaco and Shalen [98]. Another proof of this theorem for Haken manifolds together with its algorithmic version is given in Matveev's book [125]. We restrict ourselves to the case of orientable Haken manifolds.

Let $S^1 := \{z \in \mathbb{C} : |z| = 1\}$ be the multiplicative group and M a 3-manifold (with or without boundary) of class C^∞. A *smooth effective action of* S^1 on M is a C^∞-map $f : M \times S^1 \to M$ such that

(1) the map $t \in S^1 \longmapsto f(\cdot, t)$ is a monomorphism into the diffeomorphism group of M;
(2) for any $x \in M$ there exists $t \in S^1$ such that $f(x, t) \neq x$.

A 3-manifold M endowed with such a structure is called a *Seifert fibration* and is denoted by $\mathbb{S}$. M splits into the orbits of the group S^1, i.e., the sets $S_x = f(x, S^1)$, $x \in M$, which are circles, called *fibres*. A fibre S_x is said to be *regular* (respectively, *exceptional*) if $\Gamma_x := \{t \in S^1 : f(x,t) = x\} = \{1\}$ (respectively, $\Gamma_x \neq \{1\}$). Each exceptional fibre of the Seifert fibration is isolated; if $\mathbb{S}$ is compact, then the number of exceptional fibres is finite.

The quotient (orbit) space $\mathbb{B} = S^1 \backslash M$ of the Seifert fibration $\mathbb{S}$ obtained by identifying each fiber to a point is a 2-manifold with distinguished points, is called an *orbifold*. The corresponding quotient map $p : \mathbb{S} \to \mathbb{B}$ is called the *projection*. Note that $\partial\mathbb{S} \neq \emptyset$ if and only if $\partial\mathbb{B} \neq \emptyset$, and $\partial\mathbb{S} = p^{-1}(\partial\mathbb{B})$. Further, if $M = \mathbb{S}$ is compact then any connected component of $\partial\mathbb{S}$ is a foliated two-dimensional torus with regular fibres-circles.

The following statement is a combination of Lemma II.2.3 and Corollary II.2.4 in [98].

Theorem 57 *Let $\mathbb{S}$ be a compact connected Seifert fibration and $\partial\mathbb{S} \neq \emptyset$. Then $\mathbb{S}$ is irreducible and sufficiently large. Any component of $\partial\mathbb{S}$ is incompressible except if $\mathbb{S}$ is homeomorphic to the solid torus.*

Seifert fibrations were defined and classified by H. Seifert in [163]. The reader can find an English translation of this nice paper in [166]. See also the book [146] by Orlik. Useful information can be found also in [93] and [161].

Two surfaces C, C_1 in a 3-manifold M are said to be *parallel* if there is a 3-manifold $Q \subset M$ such that $\partial Q = C \cup C_1$ and Q is homeomorphic to $C \times I$. A surface C in M is called *boundary parallel* in M if there is a surface C_1 in ∂M which is parallel to C.

A 3-manifold M is said to be *atoroidal* if any two-sided incompressible torus $T = S_1^2$ in M is boundary parallel in M. In particular, a closed 3-manifold M is atoroidal if and only if M contains no two-sided incompressible torus; this is true if M is not a Haken manifold.

Theorem 58 ([98, 101]) *A compact orientable irreducible manifold M^3 with an infinite fundamental group is a Seifert fibration if and only if the group $\pi_1(M)$ contains an infinite cyclic normal subgroup.*

Let W be a two-sided $(m-1)$-manifold in a compact m-manifold. We denote by $\sigma_W(M)$ the m-manifold obtained by splitting M along W. There is a canonical "identifying map" $r : \sigma_W(M) \to M$ such that $r^{-1}(W)$ consists of two homeomorphic copies of W, and $r|r^{-1}(M - W)$ is one-to-one. This map will be denoted by $r_W(M)$.

Theorem 59 ([98, 101]) *Let M be a closed connected orientable Haken 3-manifold. Then there exists a (possibly nonconnected) surface $W \subset M$ with two-sided incompressible components, unique up to isotopy of the entire space such that*

(a) the components of W are tori;

(b) *any component of* $\sigma_W(M)$ *is either a Seifert fibration or an atoroidal manifold (both cases can coexist);*
(c) W *is a minimal two-sided 2-manifold in* M *with properties (a) and (b).*

In fact this theorem was proved for any compact orientable Haken 3-manifold. In this framework this result can be stated as follows. First note that any compact irreducible orientable 3-manifold with nonempty incompressible boundary is a Haken manifold.

Theorem 60 *Let* M *be a compact connected orientable irreducible 3-manifold with nonempty incompressible boundary* ∂M*, and* $DM \supset M$ *a closed 3-manifold (the double of* M*) such that* M *is an orbit space of* DM *relative to an involutive homeomorphism* $j : DM \to DM$ *with fixed point set* ∂M*. Then the statement for Theorem 59 for* DM *(instead of* M*) is the same, with the additional requirement that we consider only* j*-invariant surfaces* $W \subset DM$*. In addition, for the natural quotient map* $p : DM \to DM/j = M$*,*

(a) $p(W)$ *is a collection of tori or bands in* M *which are not parallel to the boundary;*
(b) *any component of* $\sigma_{p(W)}(M)$ *is either a Seifert fibration or an atoroidal 3-manifold, or an* I*-bundle over a compact surface.*

17.18 The Thurston Geometrization Conjecture

In order to formulate this conjecture we need the notion of *geometric structure*.

Definition 20 Let (N, ν) be a simply connected homogeneous Riemannian manifold, i.e., the group G of all isometries of the manifold (N, ν) is a Lie group which acts transitively on N, and let $\Gamma \subset G$ be a closed discrete subgroup which acts freely (without fixed points) on N. The orbit space $\Gamma\backslash N$ with the natural smooth structure and Riemannian metric μ such that the natural projection $(N, \nu) \to (\Gamma\backslash N, \mu)$ is a local isometry, is called a locally homogeneous Riemannian manifold of type (N, ν). A topological manifold M (possibly with boundary) admits a geometric structure (of finite volume) of type (N, ν) if $\operatorname{Int} M$ is homeomorphic to $\Gamma\backslash N$ for some pair $((N, \nu), \Gamma)$ (and $\operatorname{Vol}(\Gamma\backslash N, \mu) < \infty$).

Remark 14 Under the latter condition of Definition 20, $\pi_1(M) \cong \Gamma$.

Theorem 61 *Any compact surface* M^2 *admits a geometric structure. Here* $(N, \nu) = S^2 \subset E^3$ *if* M *is homeomorphic to* S^2 *or* P^2*;* $(N, \nu) = E^2$ *if* M^2 *is homeomorphic to* T^2 *or* K*; and* $(N, \nu) = H^2$ *(the Lobachevskii plane of curvature* -1*) in all other cases.*

Theorem 62 ([177]) *There are exactly eight* 3*-dimensional simply connected homogeneous Riemannian manifolds* (N, ν) *admitting a quotient manifold* $(\Gamma\backslash N, \mu)$ *of finite volume:*

(1) the sphere S^3 *of constant sectional curvature* 1*;*
(2) the Euclidean space E^3 *of constant sectional curvature* 0*,*
(3) the Lobachevskii space H^3 *of constant sectional curvature* -1*;*
(4) the direct metric product $S^2 \times E^1$*;*
(5) the direct metric product $H^2 \times E^1$*;*
and the following Lie groups with left invariant Riemannian metrics:
(6) the universal covering $\widetilde{\mathrm{SL}(2, \mathbb{R})}$ *of the Lie group* $\mathrm{SL}(2, \mathbb{R})$*;*
(7) Nil*, the nilpotent Heisenberg group of real upper triangular* (3×3)*-matrices with units on the main diagonal;*
(8) Sol*, the solvable Lie group which is the semidirect product of* R^* *and* $(R^2, +)$*, where the action of* $t \in R^*$ *on* $(R^2, +)$ *is given by the* (2×2)*-matrix* $\mathrm{diag}(t, t^{-1})$.

Theorem 63 ([161]) *Let* M *be a closed* 3*-manifold. Then*

(1) M *is a Seifert fibration if and only if* M *admits a geometric structure of type different from Sol and* H^3.
(2) M *admits a geometric structure of type Sol if and only if* M *has a finite-sheeted covering by a bundle* P *over* S^1 *whose fibres are tori and the gluing map is hyperbolic.*

Remark 15 The condition in (2) that the gluing map is hyperbolic means that $P \approx (T^2 \times I)/f$, where $f : T^2 \times \{0\} \approx T^2 \times \{1\}$ induces an isomorphism $f_* : \pi_1(T^2) \to \pi_1(T^2) \cong \mathbb{Z} \oplus \mathbb{Z}$ with matrix in $\mathrm{GL}(2, \mathbb{Z})$ with respect to a $\mathbb{Z}$-lattice which has no eigenvalues equal to 1.

Theorem 64 ([161]) *Any closed* 3*-manifold admits at most one geometric structure of one of the above types.*

Remark 16 This geometric structure is not unique. However, for the type H^3, by Mostow's theorem, the manifold (M^3, μ) is determined up to isometry by the group $\pi_1(M)$.

Conjecture 5 ([177]. (Thurston's Geometrization Conjecture)) Let a spherical decomposition (17.3) of an orientable closed 3-manifold M^3 be given. Then the summands L_j, $S^2 \times S^1$ and the interior part of any connected component of the torus decomposition $\sigma_{W_i} K_i$ from Theorem 59 of each K_i from (17.3), $i = 1, \dots, p$, admit a geometric structure of finite volume.

We need to give an explanation. If K_i is Haken, then it is possible to apply the JSJ theorem. If K_i is not Haken, then K_i has no incompressible torus, W_i is empty, and $\sigma_{W_i} K_i = K_i$. The following two corollaries are consequences of the Thurston geometrization conjecture.

Corollary 18 *Any closed* 3*-manifold with finite fundamental group is homeomorphic to* $\Gamma\backslash S^3$ *with a group* $\Gamma \subset \mathrm{SO}(4)$ *from Sect. 17.14. This implies the Poincaré conjecture.*

Corollary 19 *A closed orientable manifold* M^3 *admits a geometric structure of type* H^3 *if and only if* M^3 *is irreducible, atoroidal, is not a Seifert fibration, and whose fundamental group* $\pi_1(M^3)$ *is infinite.*

Agol proved the following long-awaited result.

Theorem 65 ([1], Theorem 9.1) *The virtual Haken conjecture is true for closed aspherical* 3*-manifolds.*

This theorem also gives a partial positive answer to Thurston's Question 16 in [177]. The following two theorems are consequences of Theorem 65.

Theorem 66 *The virtual Haken conjecture is true for closed orientable irreducible* 3*-manifolds with infinite fundamental group.*

Theorem 67 ([161]) *Let* M^3 *be a closed orientable irreducible non-Haken manifold with infinite fundamental group. Then* M^3 *is a Seifert fibration or is atoroidal. Consequently* M^3 *admits a geometric structure different from Sol.*

The next statement is a consequence of Theorems 65 and 56.

Corollary 20 *Conjecture 3 is true.*

Theorem 68 ([1], Theorem 9.2) *Let* M *be a closed hyperbolic* 3*-manifold. Then there is a finite-sheeted covering* $\tilde{M} \to M$ *such that* $\tilde{M}$ *fibers over the circle (i.e.,* $\tilde{M}$ *is a fibration over* S^1 *with a closed* 2*-surface* F *as the fiber).*

Theorem 68 gives a partial positive answer to Thurston's Question 18 in [177]. Theorems 65 and 68 were proved under the assumption of a positive answer to the Thurston geometrization conjecture and the following theorem which gave a positive answer to Problem 11.7 in the paper [86] by F. Haglund and Wise.

Theorem 69 ([1], Theorem 1.1) *Let* G *be a word-hyperbolic group acting properly and cocompactly on a* CAT(0) *cube complex* X*. Then* G *has a finite-index subgroup* H *acting specially on* X*.*

We shall consider CAT(0) spaces in Sect. 17.22. The (word-)hyperbolic groups were defined by Gromov in [83]. Special cube complexes were defined in [86] (see also [87]).

17.19 Triangulations of Topological Manifolds

In this section, we again discuss the triangulation conjecture (Problem 1). To begin with we note the known fact that any topological n-manifold is homotopy equivalent to an n-dimensional simplicial complex.

Let us recall some facts alluded to above. Any n-manifold is triangulable, and each triangulation is combinatorial if $0 \leq n \leq 3$. The positive solution to the Poincaré conjecture implies that (1) a 4-manifold is triangulable if and only if it is smoothable, (2) any triangulation is combinatorial. Any noncompact 4-manifold is triangulable. The "majority" of closed simply connected 4-manifolds are not triangulable. Many (or, possibly, all) triangulable 4-manifolds admit infinitely many triangulations which are not equivalent with respect to simplicial subdivision. The triangulation and non-PL conjectures, as well as the Hauptvermutung are false in dimension 4. The question of whether closed simply connected 4-manifolds with indefinite even intersection form are triangulable has not been completely solved.

R. Kirby and L. Siebenmann proved that M^n, $n \geq 5$, admits a piecewise linear structure if and only if the *Kirby-Siebenmann obstruction* $k(M^n) \in H^4(M^n, \mathbb{Z}_2)$ vanishes [107]. In particular, M^n admits a PL-structure if $H^4(M^n, \mathbb{Z}_2) = 0$. In dimensions $n \geq 5$ there exist manifolds with $k(M^n) \neq 0$; e.g. $M^n = T^{n-4} \times E_8$. Therefore, the PL-conjecture (Problem 3) is false in all dimensions $n \geq 4$.

Definition 21 A homology n-sphere for a positive integer n is a closed topological n-manifold M with integral homology groups isomorphic to the integral homology groups of S^n. Specifically, $H_0(M, \mathbb{Z}) \cong \mathbb{Z}$, $H_n(M, \mathbb{Z}) \cong \mathbb{Z}$, and $H_k(M, \mathbb{Z}) = 0$ if $0 < k < n$.

Remark 17 To obtain the notion of rational homology n-sphere one needs to replace the ring $\mathbb{Z}$ by the field $\mathbb{Q}$.

Example 11 Let Σ^{n-2}, $n \geq 5$ be a triangulated homology $(n-2)$-sphere with $\pi_1(\Sigma^{n-2}) \neq \{1\}$ (for $n = 5$ we can take as Σ^3 the Poincaré homology 3-sphere P^3 from Corollary 14). The suspension $S(\Sigma^{n-2})$ is not a manifold, but the double suspension $S(S(\Sigma^{n-2}))$ is a topological manifold homeomorphic to the sphere S^n, by the double suspension theorem of Edwards [62, 63], and Cannon [44]. A triangulation of $S(S(\Sigma^{n-2}))$ induced by Σ^{n-2} is not a combinatorial triangulation, since the link of any cone point v in $S(S(\Sigma^{n-2}))$ is $lk(v) = S(\Sigma^{n-2})$, which is not a manifold so is not a PL-sphere.

M. Kervaire proved that for any dimension $k \geq 4$ there exist smooth homology k-spheres Σ^k with $\pi_1(\Sigma^k) \neq \{1\}$ [106]. This result and Example 11 imply the following

Corollary 21 *For any manifold of dimension $n \geq 5$, the non-*PL*-conjecture holds, while the Hauptvermutung is false.*

To consider the triangulation conjecture for dimension ≥ 5, let us begin with some statements that do not involve new notions. If a simplicial complex is a topological n-manifold, then the link of each vertex is a homotopy $(n-1)$-sphere (which in general is not an $(n-1)$-manifold). Any orientable topological 5-manifold is triangulable (see Corollary 22 below). There exists a closed nonorientable topological 5-manifold M^5 such that all the topological n-manifolds, where $n \geq 5$, are triangulable if and only if M^5 is triangulable [78, 79]. An exact statement about this is given in Theorem 75.

Theorem 70 ([85, 159]) *Any smooth closed 3-manifold M^3 is diffeomorphic to the boundary of a smooth 4-manifold M^4 (with boundary) which is orientable if M^3 is orientable.*

If the manifold $M^3 = \Sigma^3$ is a homology 3-sphere, then M^4 is a spin 4-manifold (endowed with an arbitrary orientation). Then the signature $\sigma(M^4)$ is divisible by 8. From Rochlin's results it follows that $\rho(\Sigma^3) := \frac{1}{8}(\sigma(M^4))(\text{mod } 2) \in \mathbb{Z}_2$ depends only on Σ^3. This is called the *Rochlin invariant* of Σ^3.

Theorem 71 ([107]) *Every orientable topological 5-manifold may be simplicially triangulated if and only if there exists a* PL*-homology 3-sphere of Rochlin invariant* 1 *whose double suspension is homeomorphic to S^5.*

The Poincaré homology 3-sphere P^3 has Rochlin invariant 1 and $S(S(P^3)) \approx S^5$ by Example 11. Then Theorem 71 implies

Corollary 22 *Any orientable topological 5-manifold is triangulable.*

By Lemma 6 and Corollary 13, the set $\mathcal{H}$ of all smooth oriented homological 3-spheres Σ^3 is a commutative monoid with respect to the operation of connected sum; this set modulo the Σ^3 that bound smooth *acyclic* [56] 4-manifolds is a group Θ_3^H called the *homology cobordism group of homology 3-spheres*.

The Rochlin invariant ρ is a surjective homomorphism $\rho : \Theta_3^H \to \mathbb{Z}_2$. Thus we get the short exact sequence

$$1 \to \ker(\rho) \to \Theta_3^H \xrightarrow{\rho} Z_2 \to 1 \tag{17.4}$$

and the corresponding homomorphism called Bockstein homomorphism [56]

$$\beta : H^4(M^n, Z_2) \to H^5(M^n, \ker(\rho))$$

for any n-manifold $M^n, n \geq 5$.

Theorem 72 ([79, 124]) *A topological manifold M^n, $n \geq 5$, is triangulable if and only if $\beta(k(M^n)) = 0$.*

Theorem 73 ([79, 124]) *Each topological manifold M^n, $n \geq 5$, is triangulable if and only if there exists an element $\Sigma^3 \in \Theta_3^H$ of order two (i.e., $\Sigma^3 \# \Sigma^3$ bounds an acyclic 4-manifold) such that $\rho(\Sigma^3) = 1$.*

Theorem 74 ([158]) *The short exact sequence (17.4) splits if and only if $\beta(k(M^n)) = 0$ for any topological manifold M^n, $n \geq 5$.*

Let us consider the short exact sequence

$$0 \to Z_2 \xrightarrow{\times 2} Z_4 \to Z_2 \to 0.$$

The Bockstein homomorphism β associated to this sequence is the first Steenrod square Sq^1 on cohomology [174],

$$H^k(M^n, Z_2) \stackrel{Sq^1}{\to} H^{k+1}(M^n, Z_2).$$

Theorem 75 ([78, 79]) *For each $n \geq 5$, there exist topological n-manifolds M^n such that $0 \neq Sq^1(k(M^n)) \in H^5(M^n, Z_2)$. All the topological n-manifolds, where $n \geq 5$, are triangulable if and only if at least one of these manifolds is triangulable.*

A manifold M^n as in the first statement in Theorem 75 is said to be the *Galewski-Stern manifold*.

In 2013 Manolescu proved the following striking result.

Theorem 76 ([117]) *The short exact sequence (17.4) does not split.*

Then, by Theorems 72, 74, and 76,

Theorem 77 ([117]) *For every $n \geq 5$, there exist nontriangulable n-dimensional manifolds. In other words, the triangulation conjecture is false for every $n \geq 5$.*

Theorem 76 implies

Proposition 11 *If M_n, $n \geq 5$, is a topological manifold and $Sq^1(k(M^n)) \neq 0$, then $\beta(k(M^n)) \neq 0$.*

In particular no Galewski-Stern manifold M^n, $n \geq 5$ can be triangulated. To see explicit examples, it suffices to present an M^5. It will then follow that M^5, and hence $M^5 \times T^{n-5}$ for any $n \geq 5$, are nontriangulable.

Following [117], one can construct such manifold M^5 as follows.

By the positive answer to the Poincaré conjecture, Corollary 1, and Theorem 15, there is an orientable fake $CP^2\#\overline{CP^2}$, i.e., nontriangulable simply connected closed topological 4-manifold W with intersection form $Q = \omega_W = (+1) \oplus (-1)$. Since the form Q is isomorphic to $-Q$, then by Theorem 15, W admits an orientation-reversing homeomomorphism $f : W \to W$. Let M^5 be the mapping torus of f, i.e. the space obtained from $W \times I$ by gluing all points $(w, 0) \in W \times I$ with $(f(w), 1) \in W \times I$. We have $k(W) = 1 \in H^4(W, Z_2) = Z_2$, and therefore $Sq^1(k(M^5)) = k(W) \wedge u \neq 0$, where u is a nontrivial element of the cohomology group $H^1(S^1, Z_2) \cong Z_2$.

In [52], the authors applied Gromov's hyperbolization technique [83] to Freedman's E_8-manifold to show that there exist closed aspherical (i.e., with contractible universal covering) 4-manifolds that cannot be triangulated. In the paper [53], with the help of the hyperbolization technique applied to the Galewski-Stern manifolds, it is proved that for each $n \geq 6$ there exist nontriangulable closed aspherical n-manifolds. The question, whether such manifolds exist in dimension 5, remains open.

In dimensions $n \geq 6$ there also exist nontriangulable oriented n-manifolds.

In 1985 Casson constructed a new powerful integer invariant $\lambda : \mathcal{H} \to \mathbb{Z}$ [2, 85]. The invariant λ has the following properties:

(0) $\lambda(S^3) = 0$ and $\lambda(\mathcal{H})$ is not contained in a proper subgroup of the group $\mathbb{Z}$,
(4) $\lambda(-\Sigma^3) = -\lambda(\Sigma^3)$,
(5) $\lambda(\Sigma_1^3 \# \Sigma_2^3) = \lambda(\Sigma_1^3) + \lambda(\Sigma_2^3)$,
(6) $\lambda(\Sigma^3) = \rho(\Sigma^3) (\mathrm{mod}\ 2)$,
(7) if $\lambda(\Sigma^3) \neq 0$, then $\pi_1(\Sigma^3)$ has a nontrivial representation in the Lie group SU(2).

The other properties (1), (2), (3), (3') of λ are related to links (in particular, to knots) in $\Sigma^3 \in \mathcal{H}$ and to Dehn surgery along them [2, 85]. The invariant λ is determined uniquely up to a sign by the above mentioned properties.

It follows from (6) that the Rochlin invariant is determined by the Casson invariant. By properties (4), (6), (7), the Rochlin invariants of homotopy 3-spheres and mirror homology 3-spheres vanish. The Casson invariant is not invariant under the homology cobordisms. Therefore, unlike the Rochlin invariant, it is not defined on Θ_3^H. Note that Casson's invariant of an oriented Σ^3 can be thought of as the number of conjugacy classes of irreducible representations of $\pi_1(\Sigma^3)$ in SU(2) counted with signs.

Casson gave a topological definition of his invariant. An analytic definition of Casson's invariant is the subject of the article [175] by Taubes. Roughly speaking, Casson's invariant can be defined using gauge theory as an infinite-dimensional generalization of the classical Euler characteristic. This new definition of Casson's invariant requires some basic facts from gauge theory (connections, curvature, and covariant derivatives) which can be found in [69] and [113].

In the paper [68], A. Floer assigns to an oriented homology 3-sphere Σ^3 a $\mathbb{Z}_8$-graded homology group $I_*(\Sigma^3)$ whose Euler characteristic is twice Casson's invariant $\lambda(\Sigma^3)$. The definition uses a construction of instantons on $\Sigma^3 \times \mathbb{R}$. Later on, the group $I_*(\Sigma^3)$ was called the *instanton Floer homology group* of Σ^3.

Galewski-Stern [79] and Matumoto [124] reduced Problem 1 to a problem in 3+1 dimensions. The solution of this reduced problem is given in [117], and uses the Pin(2)-equivariant Seiberg-Witten Floer homology theory for homology 3-spheres.

Variants of the $U(1)$-equivariant Seiberg-Witten Floer homology theory for rational homology 3-spheres were constructed by Marcolli-Wang in [119], by Manolescu in [118], and Froshov in [75], as well as by Kronheimer-Mrowka in [112] for all three-manifolds. The Pin(2)-equivariant theory was first defined by Manolescu in [117] for rational homology 3-spheres. Later on a different construction of the Pin(2)-equivariant theory for all three-manifolds was given by Lin in [114]. Lin's construction provides an alternative proof of Theorem 77. All these papers presented definitions of the so-called *equivariant monopole Floer homology groups*.

For Theorem 76, it is enough to consider a smooth integral homology 3-sphere $Y = \Sigma^3$, and pick a Riemannian metric g on Y. There is a unique Spin^c- structure on (Y, g) denoted by $\mathfrak{s}$. Specifically, $\mathfrak{s}$ consists of a rank-2 Hermitian bundle $\mathbb{S}$ on

Y, together with a Clifford multiplication map $\mu : TY \xrightarrow{\cong} \mathfrak{su}(\mathbb{S}) \subset \mathrm{End}(\mathbb{S})$. Here $\mathfrak{su}(\mathbb{S})$ denotes traceless (tr $A = 0$), skew-adjoint ($A + A^* = 0$) endomorphisms of $\mathbb{S}$. Everything in the construction from [117] is Pin(2)-equivariant, because the Spin^c-structure is Spin. Here $\mathrm{Pin}(2) = U(1) \cup jU(1) \subset \mathbb{C} \cup j\mathbb{C} = \mathbb{H}$, $j^2 = -1, ij = ji$. The Seiberg-Witten equations are also Pin(2)-equivariant. Thus, Manolescu gives in [117] the definition of the Pin(2)-equivariant Seiberg-Witten Floer homology.

17.20 The Homeomorphism Problem for Closed n-Manifolds, $n \geq 4$

Now we present some results connected with the algorithmic insolubility of the homeomorphy problem for closed n-manifolds, $n \geq 4$, as was remarked by Markov in [120] and [121].

Theorem 78 *Any finitely presented group is isomorphic to the fundamental group of some closed n-manifold if $n \geq 4$.*

Let us prove this theorem.

Let $G = \{a_1, \dots, a_l | r_1, \dots, r_m\}$ be a presentation of a group G and $n \geq 4$ be any positive integer. First, let us consider the manifold $V^n = (S^{n-1} \times S^1)\# \dots \#(S^{n-1} \times S^1)$ (l copies). Obviously, $\pi_1(S^{n-1} \times S^1) \cong (\mathbb{Z}, +)$. Then by Lemma 7, $\pi_1(V^n)$ is a free group on l generators (loops with a base point $x_0 \in V^n$) which we denote by $\alpha_1, \dots, \alpha_l$.

Denote by $\rho_1, \dots, \rho_m$ loops in V^n, obtained from corresponding words $r_1, \dots, r_m$ via the substitution of α_i instead of $a_i, i = 1, \dots, l$. Using free homotopies (without fixed ends) of loops, we can assume that all loops $\rho_1, \dots, \rho_m$ are mutually disjoint circles S^1, embedded in V^n. Then there exist mutually disjoint closed neighborhoods U_j of $\rho_j, j = 1, \dots, m$, in V^n, homeomorphic to $S^1 \times B^{n-1}$, so the boundary ∂U_j is homeomorphic to $S^1 \times S^{n-2}$. Now, we can move homotopically every circle ρ_j in U_j to a circle on the boundary ∂U_j, preserving the same notation ρ_j for the moved circle. Next we remove every open set Int $U_j, j = 1, \dots, m$ and glue in place of it a copy of $B^2 \times S^{n-2}$ via a homeomorphism of its boundary $S^1 \times S^{n-2}$ to the boundary ∂U_j. This is possible, because ∂U_j is homeomorphic to $S^1 \times S^{n-2}$. As a result we get the desired closed n-manifold M^n with $\pi_1(M^n)$ isomorphic to G, because $B^2 \times S^{n-2}$ is simply connected, so any circle $\rho_j, j = 1, \dots, m$ is null-homotopic in M^n. To be accurate, instead of the very last sentence we need only use m times the Seifert-van Kampen theorem.

Novikov in [144] (see also [145]) proved the following

Theorem 79 *There exists a finitely presented group with undecidable word problem, i.e., such that it is impossible to find an algorithm for deciding the identity of group elements given by products of powers of generators.*

As a corollary of Theorems 78 and 79 we get

Corollary 23 *If we are given arbitrary two n-manifolds M_1 and M_2, where $n \geq 4$, then in general it is impossible to decide (algorithmically) whether they are homeomorphic or not.*

Finally, notice that Markov indicated in [121] that a manifold M^n, similar to the one given in the proof of Theorem 78, was described in exercises from [165].

Remark 18 In reality, the situation is much worse than it is stated in Corollary 23. Theorems 78 and 79 imply that in general there is no algorithm to solve the word problem for presentations of fundamental groups of closed topological manifolds in any dimension $n \geq 4$. But in fact to solve the homeomorphism problem it is necessary to overcome after the word problem several other problems, namely, (1) the conjugacy problem, (2) the isomorphism problem (for fundamental groups), after that the topological problems: (3) the homotopy equivalence, and only at the end (4) the homeomorphism problem. The passage from every mentioned problem to the next one is very difficult or maybe even impossible.

17.21 The Manifold Recognition Problem

In [43] Cannon posed the following problem: find a short list of topological properties that separate topological manifolds from topological spaces. He assumed that the manifolds can be characterized as generalized n-manifolds with general position property. For dimensions ≥ 5 he proposed the following disjoint disk property (DDP): any two continuous maps of the disc B^2 into the space can be approximated by arbitrary close maps with disjoint images.

A *generalized n-manifold* (n-GM) is a locally compact, locally contractible, finite-dimensional separable metric space X with local relative homology of E^n, i.e., $H_*(X, X - \{x\}, \mathbb{Z})$ is isomorphic to $H_*(E^n, E^n - \{x\}, \mathbb{Z})$ for any $x \in X$. Generalized manifolds naturally arise as sets of fixed points of group actions on manifolds, limits of sequences of manifolds, and boundaries of groups of negative curvature.

Theorem 80 ([191]) *Any generalized n-manifold is a topological n-manifold for $n = 1, 2$.*

By definition, a compact subset C of a topological space X is *cell-like* if for any neighborhood U of C in X, C contracts to a point in U. A proper map (i.e., a map such that the inverse image of each compact set is compact) $f : X \to Y$ between locally compact metric spaces is said to be *cell-like* if $f^{-1}(y)$ is cell-like for any $y \in Y$.

It is clear that any n-manifold is an n-GM, but the converse is false for $n > 2$. Let $f : M \to X$ be a proper cell-like surjective map defined on an n-manifold, and $\dim X < \infty$. Then X is an n-GM (by definition, X is a *resolvable* n-GM and the pair (M, f) is a *resolution* for X). The classical example of Bing's dog bone space

[27] demonstrates that, in general, X is not a manifold. Edwards shows in [63] (for details, see [51]) that a resolvable n-GM, $n \geq 5$, is an n-manifold if and only if it satisfies the DDP.

For a connected n-GM space X, the integer obstruction (index) $i(X) \in 1 + 8\mathbb{Z}$ was defined by Quinn [156]. This obstruction is defined locally and is locally constant; $i(X \times X') = i(X) \times i(X')$, and $i(X) = 1$ if and only if X is resolvable ($n \geq 4$). Hence for $n \geq 5$ a connected space X is an n-manifold if and only if X is an n-GM satisfying DDP and such that $i(X) = 1$. In [37] it is proved that for each $n \geq 6$ there exist generalized n-manifolds X with any index $i(X) \in 1+\mathbb{Z}$, including unresolvable. Since the Quinn index is locally constant, we obtain that any point of an unresolvable X has no neighborhood homeomorphic to a manifold.

A locally compact complete space M with intrinsic metric which satisfies the property that the shortest paths can be locally extended in a unique way is called a *Busemann G-space* [39]. Busemann G-spaces have many properties of regular Finsler manifolds. Up to now, answers to the following questions of Busemann from [39] are unknown.

Problem 6 Is it true that any Busemann G-space is finite-dimensional?

Problem 7 Is it true that any Busemann G-space M of finite dimension is a topological manifold?

One can easily prove that a Busemann G-space M is locally contractible and *topologically homogeneous*, i.e., for any $x, y \in M$ there exists a homeomorphism $f : M \to M$ such that $f(x) = y$. (Moreover, one can find such a homeomorphism f which is isotopic to the identity map). Therefore, a positive answer to Problem 7 would follow from a positive answer the well-known *Bing-Borsuk problem*

Problem 8 ([32]) Is it true that any connected locally compact locally contractible finite-dimensional topologically homogeneous metric space is a topological manifold?

The following theorem is of interest.

Theorem 81 ([100]) *If the answer to Problem 8 is positive in dimension* 3*, then the Poincaré conjecture holds.*

Problem 6 has a positive answer if M has at least one open nonempty ball which is geodesically convex with respect to the induced metric [16]. Any finite-dimensional Busemann G-space satisfies the property of domain invariance (D. Montgomery). The answer to Problem 7 is positive for dimensions $n = 1, 2$ [39], $n = 3$ [111], and $n = 4$ [178]. B. Krakus applied K. Borsuk's theorem on the characterization of the 2-sphere. P. Thurston subsequently proved the statement in four steps.

Step 1. M^n is an n-GM. In fact, the present author proved this result earlier (unpublished).

Step 2. M^4 admits arbitrary small topological embeddings of 2-spheres, 3-cells, and 3-spheres.

Step 3. $i(M^4) = 1$.
Step 4. Final.

Problem 9 Is it true that $i(M^n) = 1$ or DDP holds for M^n, $n \geq 5$?

There is another method to get a positive answer to Problem 7 for dimensions $1 \leq n \leq 4$.

Definition 22 A (locally) compact finite-dimensional metric space X is called an *absolute suspension* (respectively, an *absolute cone*) if for any two points $x_1, x_2 \in X$, $x_1 \neq x_2$, (respectively, for any $x \in X$) X is homeomorphic to a suspension (respectively, to an open cone over a compact base) with vertices x_1, x_2 (respectively, with vertex x).

Conjecture 6 ([54]) Every n-dimensional absolute suspension is homeomorphic to S^n.

Conjecture 7 ([54]) Every n-dimensional absolute cone is homeomorphic to U^n (open n-cell).

Theorem 82 ([84]) *The absolute cone conjecture 7 is true for all $1 \leq n \leq 4$ and false for $n \geq 5$, if the Poincaré conjecture holds.*

Using some isotopies of small balls in a Busemann G-space M, one can easily see that they are absolute cones over their boundary spheres (and hence have a local product structure with respect to their boundaries which are known to be generalized $(n-1)$-manifolds if M is n-dimensional). Then Theorem 82 implies

Corollary 24 *Any n-dimensional Busemann G-space is a topological manifold for $1 \leq n \leq 4$.*

Let Q be a polyhedron with triangulation T and $z \in Q$. Taking, if necessary, a subdivision of T, one can assume that z is a vertex of T. If Q is topologically embedded as a closed subset into a generalized 3-manifold X, then we say that the set $X - Q$ has a *free local fundamental group at* $z \in Q$ (briefly, 1-FLG at z) if for any sufficiently small neighborhood U of z there exists another neighborhood V of z such that $z \in V \subset U$ and if W is any connected open neighborhood of z in V, then for any nonempty connected component W' of $W - Q$ the image $\pi_1(W') \to \pi_1(U')$ (induced by inclusion) is a free group with $m - 1$ generators, where U' denotes the connected component of $U - Q$ containing W', and m the number of "components" of $\mathrm{st}(z) - \{z\}$ whose images intersect $\overline{W'}$. As usual, we simply say that $X - Q$ is 1-FLG in X if this space is 1-FLG at every point $z \in Q$.

We say that a map $f : K \to X$ from a compact 2-dimensional polyhedron K into a generalized 3-manifold X is *simplicial* if $f(K)$ is a polyhedron whose complement is 1-FLG in X and $f : K \to f(K)$ is simplicial with respect to certain triangulations of K and $f(K)$. A generalized 3-manifold X is said to satisfy the property of *relative simplicial approximation* (RSA) if for any map $f : B^2 \to X$ and any compact subpolyhedron $Q \subset B^2$ such that $f|Q$ is simplicial, and for each

$\varepsilon > 0$, there exists a simplicial map $F : B^2 \to X$ such that the natural distance $\mathrm{dist}(F, f)$ is less than ε and $F|Q = f|Q$.

Theorem 83 ([50, 104]) *Assume that the Poincaré conjecture is true. Then a generalized* 3*-manifold X is a* 3*-manifold if and only if X has the property RSA.*

17.22 Triangulations and Alexandrov Spaces with Curvature Bounded Above

A.D. Alexandrov introduced spaces of curvature $\leq K$ in [7] (cf. [8, 9, 34]). A smooth Riemannian manifold has curvature $\leq K$ (respectively, $\geq K$) in the sense of Alexandrov if and only if it has sectional curvature $\leq K$ (respectively, $\geq K$).

Let us introduce some definitions and notation. We denote by xy the distance between points x, y of a metric space M. A point y is said to lie between points x and z if y is different from x and z, and $xz = xy + yz$; we write (xyz) to denote such a triple. A quadruple of distinct points in M will be called *singular* if one of its points lies between two different pairs of the quadruple and the quadruple does not lie on a straight line, i.e., it does not embed isometrically into a Euclidean straight line. Otherwise the quadruple is called *standard*. We denote by $U(x, r)$ and $B(x, r)$ respectively the open and closed balls of radius r with center x; and by S_K a complete simply connected three-dimensional Riemannian manifold of constant sectional curvature K. A subset $[x, y] \subset M$ is said to be a *shortest path* or a *segment with endpoints* x, y if $[x, y]$ is isometric to a segment of the real line. A space M is called a *space with intrinsic metric* (respectively, a *geodesic space*) if any two points $x, y \in M$ can be joined by a path in M whose length is arbitrarily close (respectively, equal) to xy. A geodesic space M is said to have *Busemann nonpositive curvature* if for any five points $x, y, z, y_1, z_1 \in M$ such that $xy_1 = y_1y = \frac{1}{2}xy$, $xz_1 = z_1y = \frac{1}{2}xz$, the inequlity $y_1z_1 \leq \frac{1}{2}yz$ holds [39]. Gromov called such spaces *metrically convex* [83]. A detailed investigation of metrically convex spaces was undertaken by Papadopoulos in his book [148].

Theorem 84 ([16]) *Sufficiently small balls in a geodesic locally metrically convex space are geodesically and metrically convex. Moreover, any Busemann G-space with Busemann nonpositive curvature is finite-dimensional.*

Theorem 85 ([13]) *Any Busemann G-space with Busemann nonpositive curvature is a topological n-manifold for some positive integer n. In addition, its universal covering space is homeomorphic to the Euclidean space E^n.*

For an ordered triple (x, y, z) in M such that $xy + yz + zx < \frac{2\pi}{\sqrt{K}}$ if $K > 0$, we denote by $\gamma_K(xyz)$ the angle in the triangle Δ_K on S_K with sides of lengths xy, xz, and yz, which lies opposite to the side of length xz. For the shortest paths $k = [y, x]$ and $l = [y, z]$, we denote by $\underline{\gamma}(k, l)$ (respectively, $\overline{\gamma}(k, l)$) the lower (respectively, upper) limit $\underline{\lim}(\gamma_K(x'yz'))$ (respectively, $\overline{\lim}(\gamma_K(x'yz')))$ under the condition that

$x' \in k$, $z' \in l$, and $x' \to y$, $z' \to y$. These limits do not depend on the choice of K. The value $\underline{\gamma}(k,l)$ (respectively, $\overline{\gamma}(k,l)$) is called *lower (upper) angle between the shortest paths* k and l. If $\underline{\gamma}(k,l) = \overline{\gamma}(k,l)$, then their common value $\gamma(k,l)$ is called the *angle between the shortest paths* k and l. The upper angle $\overline{\gamma}$ between the shortest paths is nonnegative and satisfies the triangle inequality. However it can be zero even if the shortest paths k and l have only one common point. If we fix a point $y \in M$ and identify the shortest paths k and l with the starting point at y if $\overline{\gamma}(k,l) = 0$, we obtain a metric space $\omega_y(M)$. The metric completion $\Omega_y(M)$ of this space is called the *space of directions to* M *at* y.

In [17] and [19] it was announced and proved that a metric space M with local existence of shortest paths is a space of curvature $\leq K$ (respectively, $\geq K$) if and only if for any point $x \in M$ there exists $r = r(x) > 0$ ($r \leq \frac{\pi}{3\sqrt{K}}$ if $K > 0$) such that any standard (respectively, any) quadruple of points in $U(x,r)$ is isometric to a quadruple of points in $S_{K'}$, $K' \leq K$ (respectively, $K' \geq K$), where K' depends on the chosen quadruple. In a space of curvature $\leq K$ (respectively, $\geq K$), for any two shortest paths with a common endpoint the angle between them exists at this point.

An important notion of K-region introduced in [8] and later called a CAT(K)-space [34] is a special case of a geodesic space of curvature $\leq K$, which globally satisfies the properties from the previous paragraph. More precisely, in case $K \leq 0$ (respectively, $K > 0$) the space M is a K-region if and only if any two points $x, y \in M$ (respectively with $xy < \frac{\pi}{\sqrt{K}}$) can be joined by a unique shortest path and any standard quadruple of points x, y, z, w in M (respectively with the additional condition that for any triple from this quadruple, e.g., x, y, z, we have $xy+yz+zx < \frac{2\pi}{\sqrt{K}}$, etc.) is isometric to a quadruple of points in $S_{K'}$, $K' \leq K$, where K' depends on the chosen quadruple.

A geodesic space M is a 0-region if and only if for all points $p, q, r, m \in M$ satisfying equalities $qm = mr = \frac{1}{2}qr$, the *Bruhat-Tits*-CN-*inequality* [36] holds:

$$(pq)^2 + (pr)^2 \geq 2(mp)^2 + \frac{1}{2}(qr)^2.$$

Any 0-region is metrically convex.

In [6] it was proved that any simply connected geodesic space of locally (Busemann) nonpositive curvature has globally nonpositive curvature. It follows from this that any simply connected complete space of curvature $\leq K \leq 0$ is a K-region.

G. Perelman proved a rather wide generalization of V.A. Toponogov's theorem on comparison of angles in geodesic triangles in complete Riemannian manifolds of sectional curvature $\geq K$ [38]. This generalization is equivalent to the following statement: Any quadruple of points in a complete Alexandrov space of curvature $\geq K$ is isometric to a quadruple of points in $S_{K'}$, $K' \geq K$, where K' depends on the chosen quadruple.

Theorem 86 ([18]) *Any connected locally finite simplicial complex* C *admits a geodesic metric of* 1*-region.*

Since any two points in a 1-region of distance less than π are joined by a unique shortest arc, Theorem 86 gives a positive answer to Problem (10.2) from Borsuk's book [33]. One can describe briefly the metrization as such that any k-dimensional simplex of the first barycentric subdivision of a given triangulation T of the complex C is isometric to the regular k-dimensional simplex with edges of length $\pi/2$ in the unit k-dimensional sphere S_1^k.

In the proof and in the original construction of this metric the construction and properties of the *K-cone $C_K Y$ over a metric space Y* for a real number K was used [9, 18]. Let (Y, d) be a metric space and K a real number. If $K \leq 0$, then $C_K Y$ (as a set) is the quotient set of $[0, \infty) \times Y$ with respect to the equivalence relation $\simeq$ defined by $(t, y) \simeq (t', y')$ if $(t = t' = 0)$ or $(t = t' > 0$ and $y = y')$. If $K > 0$, then the *suspension* $S_K Y$ is the quotient set of $[0, \frac{\pi}{\sqrt{K}}] \times Y$ with respect to the equivalence relation $\simeq$ defined by $(t, y) \simeq (t', y')$ if $t = t' = 0$, or $t = t' = \frac{\pi}{\sqrt{K}}$, or $(t, y) = (t', y')$. The point of the cone (respectively, suspension) corresponding to $t = 0$ (or $t = \frac{\pi}{\sqrt{K}}$) is called the *vertex*. Let us denote by ty the equivalence class of a point (t, y). Set $d_\pi(y, y') = \min\{\pi, d(y, y')\}$. Let us define the distance xx' between the points $x = ty$ and $x' = t'y'$ in the following way:

$$(xx')^2 = t^2 + (t')^2 - 2tt' \cos(d_\pi(y, y'))$$

if $K = 0$,

$$\cosh(\sqrt{-K}xx') = \cosh(\sqrt{-K}t)\cosh(\sqrt{-K}t') - \sinh(\sqrt{-K}t)\sinh(\sqrt{-K}t')\cos(d_\pi(y, y'))$$

if $K < 0$; and $xx' \leq \frac{\pi}{\sqrt{K}}$,

$$\cos(\sqrt{K}xx') = \cos(\sqrt{K}t)\cos(\sqrt{K}t') + \sin(\sqrt{K}t)\sin(\sqrt{K}t')\cos(d_\pi(y, y')),$$

if $K > 0$. For $K > 0$, the cone $C_K Y$ can be obtained from the suspension by the restriction $0 \leq t \leq \frac{\pi}{2\sqrt{K}}$.

Theorem 87 ([9, 18]) *If the space Y is a 1-region, then $C_K Y$ is a K-region.*

The following is an analogue of this theorem:

Theorem 88 ([38]) *If the space Y is a complete space of curvature ≥ 1, then $C_K Y$ is a complete space of curvature $\geq K$.*

The following theorem was proved with the help of Theorem 87.

Theorem 89 ([11]) *The interior of each compact contractible n-dimensional piecewise linear manifold with boundary ($n \geq 5$) admits a metric of K-region for $K \geq 0$.*

This gives a new series of simple examples providing a negative answer to Gromov's question: *Are metrically convex geodesic manifolds homeomorphic to the*

Euclidean space? The first examples of this kind were constructed in [52] (also with usage of Alexandrov's metrics of nonpositive curvature).

The space $T_y(M) := C_0(\Omega_y(M))$ is called the *tangent space* or the *tangent cone to M at the point y* [9, 38]. The terms *space of directions* and *tangent space* can be explained by the fact that for each Riemannian manifold (M^n, μ), $n \geq 2$, endowed with its natural intrinsic metric and any $y \in M^n$, the spaces $\Omega_y(M^n)$ and $T_y(M^n)$ are isometric respectively to S_1^{n-1} and E^n.

Theorem 90 ([143]) *For any space M of curvature $\leq K$ and any point $y \in M$, $\Omega_y(M)$ (respectively, $T_y(M)$) is a metrically complete 1-region (respectively, 0-region).*

Here is an analogue of this result:

Theorem 91 ([38]) *For any finite-dimensional space M of curvature $\geq K$ and any point $y \in M$, $\Omega_y(M)$ (respectively, $T_y(M)$) is a complete (respectively, locally) compact space of curvature ≥ 1 (resp. ≥ 0).*

Problem 10 ([10]) Is it true that if M is a metrized manifold of dimension $n \geq 2$ which is an Alexandrov space of curvature $\leq K$, then the space of directions $\Omega_y(M)$ is homeomorphic to S^{n-1} for any point $y \in M$?

One can prove that the answer to this problem is affirmative for $n = 2, 3$.

Theorem 92 ([20]) *Let (Σ^{n-2}, T), $n \geq 5$, be a closed manifold with triangulation T which is a homology, but not homotopy $(n-2)$-sphere endowed with a metric d from Theorem 86. Then the double suspension $M = S_1(S_1((\Sigma^{n-2}, d)))$ (respectively, the cone $M = C_K(S_1((\Sigma^{n-2}, d)))$ for $K \leq 0$) is a 1-region (respectively, K-region) homeomorphic to S^n (respectively, to E^n), but the space of directions at any vertex of the second suspension (respectively, of the cone) is not a topological manifold, hence is not homeomorphic to S^{n-1}.*

Remark 19 The topological statements of this theorem follow from Example 11 and the first sentence after it. The same argument shows that the tangent cone $T_y(M)$ at any point $y \in M$ is homeomorphic to E^n.

Corollary 25 *The answer to Problem 10 is in the negative for all $n \geq 5$.*

Theorem 93 ([21]) *Let (Σ^3, T) be a three-dimensional closed manifold with triangulation T which is a homotopy 3-sphere endowed with a metric d from Theorem 86 so that (Σ^3, d) is a 1-region. Then the suspension $M = S_1((\Sigma^3, d))$ (respectively, the cone $M = C_K((\Sigma^3, d))$ for $K \leq 0$) is a 1-region (respectively, K-region) homeomorphic to S^4 (respectively, to E^4), and the space of directions at any vertex of the suspension (respectively, of the cone) is homeomorphic to Σ^3.*

Topological versions of this theorem follow from the papers [70] and [26].

Very recently, using the resolution of the Poincaré conjecture PC_3, A. Lytchak and K. Nagano proved as a part of Theorem 6.4 in [115] the following important

Theorem 94 *Problem 10 has affirmative answer for $n \leq 4$.*

Acknowledgement The author was supported by the Ministry of Education and Science of the Russian Federation (Grant 1.3087.2017/4.6).

References

1. Agol, I., The virtual Haken conjecture. With an appendix by Agol, D. Groves, J. Manning, *Doc. Math.* 18(2013), 1045–1087.
2. Akbulut, S., McCarthy, J.D., *Casson's invariant for oriented homology* 3*-spheres,* Mathematical Notes, Princeton University Press, Princeton, NJ, 36(1990).
3. Alexander, J.W., Note on two three-dimensional manifolds with the same group, *Trans. Amer. Math. Soc.* 20(1919), 339–342.
4. Alexander, J.W., An example of a simply connected surface bounding a region which is not simply connected, *Proc. Nat. Acad. Sci. U.S.A.* 10(1924), 8–10.
5. Alexander, J.W., The combinatorial theory of complexes, *Ann. of Math.* 31(1930), 292–320.
6. Alexander, S.B., Bishop, R.L., The Hadamard-Cartan theorem in locally convex spaces, *L'Enseign. Math.* 36(1990), 309–320.
7. Alexandrov, A.D., A theorem on triangles in metric spaces and its applications (Russian), *Tr. Mat. Inst. im. V.A. Steklova, Acad. Nauk SSSR*, 38(1951), 5–23.
8. Alexandrov, A.D., Über eine Verallgemeinerung der Riemannschen Geometrie (German), *Schriftenr. Inst. Mat. Deutsch. Acad. Wiss., H.* 1(1957), 33–84.
9. Alexandrov, A.D., Berestovskii, V.N., Nikolaev, I.G., Generalized Riemannian spaces, *Russian Math. Surveys* 41(1986), no. 3, 1–54.
10. Alexandrov, A.D., Berestovskii, V.N., Riemannian spaces, generalized, M. Hazewinkel (Managing ed.) *Encyclopaedia of mathematics*, Kluwer Academic Publishers, Dordrecht-Boston- London, 8(1992), 150–152.
11. Ancel, F.D., Guilbault, C.R., Interiors of compact contractible n-manifolds are hyperbolic ($n \geq 5$), *J. Differential Geom.* 45(1997), 1–32.
12. Anderson, M., Geometrization of 3-manifolds via the Ricci flow, *Notices Amer. Math. Soc.* 51(2004), no. 2, 184–193.
13. Andreev, P.D., Proof of the Busemann conjecture for G-spaces of nonpositive curvature, *St. Petersburg Math. J.* 26(2015), no. 2, 193–206.
14. Baer, R., Noethersche gruppen, *Math. Z.* 66(1956), 269–288.
15. Barth, W.P.; Hulek, K.; Peters, C.A.M., Van de Ven, A., *Compact complex surfaces,* Springer-Verlag, Berlin, Heidelberg, 2004.
16. Berestovskii, V.N., On the problem of a Busemann G-space to be finite-dimensional, *Siber. Math. J.* 18(1977), no. 1, 159–161.
17. Berestovskii, V.N., On spaces with bounded curvature, *Soviet Math. Dokl.* 23 (1981) 491–493.
18. Berestovskii, V.N., Borsuk's problem on the metrization of a polyhedron, *Soviet Math. Dokl.* 27 (1983), no. 1, 56–59.
19. Berestovskii, V.N., Spaces with bounded curvature and distance geometry, *Siber. Math. J.* 27 (1986), no. 1, 8–19.
20. Berestovskii, V.N., Manifolds with intrinsic metric of unilaterally bounded Alexandrov curvature (Russian), *Journal of Matematical Physics, Analysiz, Geometry* 1(1994), no. 1, 41–59.
21. Berestovskii, V.N., Pathologies in Alexandrov spaces with curvature bounded above, *Siberian Adv. Math.* 12(2002), no. 4, 1–18 (2003).
22. Berestovskii, V.N., Poincaré conjecture and related statements, *Russian Mathematics (Iz VUZ)* 51(2007), no. 9, 1–36.
23. Besse, A., *Géometrie Riemannienne en dimension 4. Séminaire Arthur Besse, 1978/79,* Cedic/Fernand Nathan, Paris, 1981.

24. Besson, G., *Preuve de le conjecture de Poincaré en déformant la métrique par la courbure de Ricci, d'après G. Perel'man,* Astérisque, Société Mathématique de France 307(2006).
25. Bestvina, M., Handel, M., Train-tracks for surface homeomorphisms, *Topology* 34(1995), no. 1, 109–140
26. Bestvina, M.; Daverman, R.J.; Venema, G.A.; Walsh, J.J., A 4-dimensional 1-LCC shrinking theorem. Geometric topology and geometric group theory (Milwaukee, Wi, 1997), *Topology Appl.* 110(2001), no. 1, 3–20.
27. Bing, R.H., A decomposition of E^3 *into points and tame arcs such that the decomposition space is topologically different from* E^3, *Ann. of Math.* (2) 65(1957), 484–500.
28. Bing, R.H., Necessary and sufficient conditions that a 3-manifold be S^3, *Ann. of Math.* (2) 68(1958), no. 1, 17–37.
29. Bing, R.H., Some aspects of the topology of 3-manifolds related to the Poincaré conjecture, *Lectures on modern mathematics II* (T.L. Saaty, editor), 93–128, John Wiley and Sons, New York, 1964.
30. Bing, R.H., An alternative proof that 3-manifolds can be triangulated, *Ann. of Math.* 69(1959) 37–65.
31. Bing, R.H., A homeomorphism between the 3-sphere and the sum of two solid horned spheres, *Ann. of Math.* 56(1952), no. 2, 354–362.
32. Bing, R.H., Borsuk, K., Some remarks concerning topologically homogeneous spaces, *Ann. Math.* 81(1965), 100–111.
33. Borsuk, K., *Theory of retracts,* PWN – Polish scientific publishers, Warszawa, 1967.
34. Bridson, M.R., Haefliger, A., *Metric spaces of non-positive curvature,* Springer-Verlag, Berlin, Heidelberg, 1999.
35. Brower, L.E.J., Zum Triangulationsproblem, *Nederl. Akad. Wetensch. Proc.* 42(1939), 701–706.
36. Bruhat, F., Tits, J., Groupes réductifs sur un corps local. I. Données radicielles valuées, *Inst. Hautes Études Sci. Publ. Math.* 41(1972), 5–251.
37. Bryant, R., Ferry, S., Mio, W., Weinberger, S., Topology of homology manifolds, *Ann. of Math.* (2) 143(1996), no. 3, 435–483.
38. Burago, Y., Gromov, M., Perelman, G., A.D. Alexandrov's spaces with curvature bounded below, *Russian Math. Surveys* 47(1992), no. 2, 1–58.
39. Busemann, H., *Geometry of geodesics,* Academic Press Inc., Publishers, New York, 1955.
40. Cairns, S.S., On the triangulation of regular loci, *Ann. of Math.* 35(1934), 579–587.
41. Cairns, S.S., Homeomorphisms between topological manifolds and analytic Riemannian manifolds, *Ann. of Math.* (2) 41(1940), 796–808.
42. Cairns, S.S., Introduction of a Riemannian Geometry on a triangulable 4-manifolds, *Ann. of Math.* (2) 45(1944), no. 2, 218–219.
43. Cannon, J.W., The recognition problem: what is a topological manifold, *Bull. Amer. Math. Soc.* 84(1978), 832–866.
44. Cannon, J.W., Shrinking cell-like decompositions of manifolds. Codimension 3, *Ann. of Math.* (2) 110(1979), no. 1, 83–112.
45. Cao, H.-D., Zhu, X.-P., A complete proof of the Poincaré and geometrization conjectures. –Application of the Hamilton-Perelman theory of the Ricci flow, *Asian J. Math.* 10(2006), no. 2, 145–492.
46. Casson, A.J., Bleier, S.A., *Automorphisms of surfaces after Nielsen and Thurston,* Cambridge Univ. Press Cambridge, New York, New Rochelle, Melbourne Sydney, 1988.
47. Cerf, J., Groupes d'automorphismes et groupes de difféomorphismes des variétés de dimension 3, *Bull. Soc. Math. France* 87(1959), 319–329.
48. Cerf, J., *Sur les difféomorphismes de la sphère de dimension trois* ($\Gamma_4 = 0$), Springer Lecture Notes in Math., No. 53, 1968.
49. Colding, T.H.; Minicozzi II, W.P., Estimates for the extinction time for the Ricci flow of certain 3-manifolds and a question of Perelman, *J. Amer. Math. Soc.* 18(2005), no. 3, 561–569.

50. Daverman, R.J., Thickstun, T.L., The 3-manifolds recognition problem, *Trans. Amer. Math. Soc.* 358(2006), 5257–5270.
51. Daverman, R.J., *Decompositions of manifolds,* Academic Press, Inc., Orlando, Fl., 1986.
52. Davis, M.W., Januszkiewicz, T., Hyperbolization of polyhedra, *J. Differential Geom.* 34(1991), no. 2, 347–388.
53. Davis, M.W., Fowler, J., Lafont, J.-F., Aspherical manifolds that cannot be triangulated, *Alg.& Geom. Topology* 14(2014), 795–803.
54. de Groot, J., On the topological characterization of manifolds, in: *General Topology and its Relations to Modern Analysis and Algebra, III* (Proc. Third Prague Topological Sympos., 1971), Academia Prague, 1972, 155–158.
55. De Michelis, S., Freedman, M.H.A., Uncountably many exotic $\mathbb{R}^4$'s in standard 4-space, *J. Diferential Geom.* 35(1992), no. 1, 219–255.
56. Dold, A., *Lectures on algebraic topology,* Springer-Verlag, Berlin-Heidelberg-New York, 1972.
57. Donaldson, S.K., An application of gauge theory to the topology of 4-manifolds, *J. Diferential Geom.* 18(1983), 269–316.
58. Donaldson, S.K., Connections, cohomology, and the intersection forms of 4-manifolds, *J. Diferential Geom.* 24(1986), 275–371.
59. Donaldson, S., Irrationality and h-cobordism conjecture, *J. Diferential Geom.* 26(1987), no. 1, 141–168.
60. Donaldson, S.K., The orientation of Yang-Mills moduli spaces and four manifold topology, *J. Diferential Geom.* 26(1987), 397–428.
61. Dunfield, N.M., and Thurston, W.P., The virtual Haken conjecture: Experiments and examples, *Geometry and Topology* 7(2003), 399–441.
62. Edwards, R.D., *Suspensions of homology spheres,* arXiv:[math]0610573.
63. Edwards, R.D., The topology of manifolds and cell-like maps, *Proceedings of the International Congress of Mathematicians (Helsinki, 1978),* Acad. Sci. Fennica, Helsinki, 1980, 111–127.
64. Epstein, D.B.A., Projective planes in 3-*manifolds, Proc. London Math. Soc.* 11(1961), no. 3, 469–484.
65. Epstein, D.B.A., Curves on 2-manifolds and isotopies, *Acta Math.* 115(1966), 83–107.
66. Evans, B. and Moser, L., Solvable fundamental groups of compact 3-manifolds, *Trans. Amer. Math. Soc.* 168(1972), 189–210.
67. Fintushel, R., Stern, R.J., Knots, links, and 4-manifolds, *Invent. Math.* 134(1998), no. 2, 363–400.
68. Floer, A., An instanton-invariant for 3-manifolds, *Commun. Math. Phys.* 118(1988), 215–240.
69. Freed, D.S., Uhlenbeck, K.K., *Instantons and four-manifolds,* Springer-Verlag, New York-Berlin- Heidelberg- Tokyo, 1984.
70. Freedman, M.H., The topology of four-dimensional manifolds, *J. Dif. Geom.* 17(1982), 357–453.
71. Freedman, M.H., Luo, F., *Selected applications of geometry to low-dimensional topology,* Marker lectures in the mathematical sciences. The Pennsylvania State University. University lecture series I. Amer. Math. Soc. Providence, Rhode Island, 1989.
72. Freedman, M.H.A., Morgan, J., On the diffeomorphism type of certain algebraic surfaces, *J. Differential Geom.* 27(1988), 297–369.
73. Freedman, M.H.A., Morgan, J., Algebraic surfaces and 4-manifolds: Some conjectures and speculations, *Bull. Amer. Math. Soc.* 18(1988), 1–19.
74. Freedman, M.H., Quinn, F., *Topology of* 4-*manifolds,* Princeton Univ. Press, Princeton, New Jersey, 1990.
75. Froshov, K.A., Monopole Floer homology for rational homology 3-spheres, *Duke. Math. J.* 155(2003), no. 3, 519–576.
76. Furuta, M., Monopole equation and the $\frac{11}{8}$-conjecture, *Math. Res. Lett.* 8(2001), 279–291.
77. Furuta M., Kametani M., Matsue H., Minami N., *Homotopy theoretical considerations of the Bauer-Furuta stable homotopy Seiberg-Witten invariant,* Geometry & Topology Monographs 10 (2007), 155–166.

78. Galewski, D.E., Stern, R.J., A universal 5-manifold with respect to simplicial triangulations, *Geometric topology (Proc. Georgia Topology Conf., Athens, GA, 1977)*, pp. 345–350.
79. Galewski, D.E., Stern, R.J., Classification of simplicial triangulations of topological manifolds, *Ann. Math.* (2) 111(1980), no. 1, 1–34.
80. Gessen, M., *Perfect rigour (A genious and the mathematical breakthrough of the century)*, Icon Books Ltd, London, 2011.
81. Gompf, R.E., An infinite set of exotic R^4's, *J. Diferential Geom.* 21(1985), no. 2, 283–300.
82. Gompf R.E., Mrowkai T.S., Irreducible 4-manifolds need not be complex, *Ann. of Math.* (2) 138 (1993), no. 1, 61–111.
83. Gromov, M., Hyperbolic groups, *Essays in Group theory* (Ed. S. Gersten), MSRI Publications, 1985, 72–263.
84. Guilbault, C.R., A solution to de Groot absolute cone conjecture, *Topology* 46(2007), 89–102.
85. Guillou, L., Marin, A., *À la recherche de la topologie perdue,* Birhäuser, Boston-Basel-Stuttgart, 1986.
86. Haglund, F., Wise, D.T., Special cube complexes, *Geom. Funct. Anal.* 17(2008), 1551–1620.
87. Haglund, F., Wise, D.T., A combination theorem for special cube complexes, *Ann. of Math.* 176(2012), 1427–1482.
88. Haken, W., Über das Homöomorphieproblem der 3-*Mannigfaltigkeiten. I. (German), Math. Z.* 80(1962), 89–120.
89. Hamilton, R.S., Three-manifolds with positive Ricci curvature, *J. Differential Geom.* 17(1982), 255–306.
90. Hatcher, A., A proof of the Smale conjecture, $\mathrm{Diff}(S^3) \cong O(4)$, *Ann. of Math.* 117(1983), 553–607.
91. Hemion, G., On the classification of knots and 3-dimensional spaces, *Acta of Math.* 142(1979), no. 1–2, 123–155.
92. Hemion, G., *The classification of knots and 3-dimensional spaces,* Oxford Science Publications. The Clarendon Press, Oxford University Press, New York, 1992.
93. Hempel, J., 3-*manifolds,* Ann. of Math. Studies, 86, Princeton Univ. Press, Princeton, 1976.
94. Hirsch, M.W., *Differential topology,* Corrected reprint of the 1976 original. Graduate texts in Mathematics, 33. Springer-Verlag, New-York, 1994.
95. Hirsch, M., Masur, B., *Smoothing piecewice linear manifolds,* Annals of Math. Studies, 80, Princeton Univ. Press, Princeton, 1974.
96. Hopf, H., Zum Clifford-Kleinschen Raumproblem, *Math. Ann.* 95(1925–26), 313–319.
97. Jaco, W., Heegaard splittings and splitting homomorphisms, *Trans. Amer. Math. Soc.* 144(1969), 365–379.
98. Jaco, W.J., and Shalen, P.B., Seifert fibred spaces in 3-manifolds, *Memoirs of Amer. Math. Soc.* 21(1979), no. 220.
99. Jaco, W., *Lectures on* 3-*manifold topology,* CBMS Regional Conference Series in Mathematics, 43. American Mathematical Society, Providence, R.I., 1980.
100. Jakobsche, W., The Bing-Borsuk conjecture is stronger than the Poincaré conjecture, *Fund. Math.* 106(1980), no. 2, 127–134.
101. Johannson, K., *Homotopy equivalences of* 3-*manifolds with boundaries,* Lect. Notes in Math. 761, Springer, Berlin, 1979.
102. Johannson, K., Topologie und Geometrie von 3-Mannigfaltigkeiten (German), *Jahresber. Deutsch. Math.-Verein.*, 86(1984), no. 2, 37–68.
103. Johannson, K., *Classification problems in low-dimensional topology,* Geometric and algebraic topology, 37–59, Banach Center Publ., 18, Warsaw, 1986.
104. Kavichcholi, A., Repovs, D., Tickstun, T., Geometric topology of generalized 3-manifolds (Russian), *Fundam. Prikl. Mat.* 11(2005), no. 4, 71–84.
105. Kerékjártó, B.V., *Vorlesungen über Topologie, I, Flächentopologie,* Springer, 1923.
106. Kervaire, M.A., Smooth homology spheres and their fundamental groups, *Trans. of Amer. Math. Soc.* 144(1969), 67–72.

107. Kirby, R.C., Siebenmann, L.C., *Foundational Essays on topological manifolds, smoothings, and triangulations,* Princeton Univ. Press and University of Tokyo Press, Princeton, New Jersey, 1977.
108. Kister, J.M. and McMillan, D.R., Jr., Locally euclidean factors of E^4 *which cannot be embedded in* E^3, *Ann. of Math.* 76(1962), 541–546.
109. Kleiner, B.; Lott, J., Notes on Perelman's papers, *Geom. Topol.* 12(2008), no. 5, 2587–2855.
110. Kneser, H., Geschlolossene Flächen in dreidimensionale Mannigfaltigkeiten, *Geom. Jahresber. Deutsch. Math.* 38(1929), 248–260.
111. Krakus, B., Any 3-dimensional G-space is a manifold, *Bull. Acad. Pol. Sci. Sér. Math. Astronom. Phys.* 16(1968), 285–291.
112. Kronheimer, P.B., Mrowka, T.S., *Monopoles and three-manifolds,* New Mathematical Monographs, vol 10, Cambridge University Press, Cambridge, 2007.
113. Lawson, H.B., *The theory of gauge fields in four dimension,* Amer. Math. Soc., Providence, RI., 1986.
114. Lin, F., The surgery exact triangle in Pin(2)-monopole Floer homology, *Algebraic & Geometric Topology* 17–5(2017), 2915–2960.
115. Lytchak, A., Nagano, K., *Topological regularity on spaces with an upper curvature bound,* ArXiv:1809.06183 [math.DG], 17 Sep 2018.
116. Mandelbaum, R., Four-dimensional topology; an introduction, *Bull. Amer. Math. Soc. (N.S.)* 2(1980), no. 1, 1–159.
117. Manolescu, C., Pin(2)-equivariant Seiberg-Witten-Floer homology and the triangulation conjecture, *J. of Amer. Math. Soc.* 29(2016), 147–176.
118. Manolescu, C., Seiberg-Witten-Floer stable homotopy type of three-manifolds with $b_1 = 0$, *Geom. Topol.* 7(2003), 889–932.
119. Marcolli, M., Wang, B.-L., Equivariant Seiberg-Witten Floer homology, *Comm. Anal. Geom.* 9(2001), no. 3, 451–639.
120. Markov, A.A., Jr., Insolvability of the problem of homeomorphy. (Russian), *Dokl. Akad. Nauk SSSR* 121(1958), 218–220.
121. Markov, A.A., Jr., Insolvability of the problem of homeomorphy. (Russian), *Proc. Internat. Congress Math. 1958*, pp. 300–306 Cambridge Univ. Press, New York, 1960.
122. Massey, W.S., *Algebraic topology: an introduction,* Reprint of the 1967 edition. Graduate text in Mathematics, 56. Springer-Verlag, New York-Heidelberg, 1977.
123. Matsumoto, Y., On the bounding genus of homology 3-spheres, *J. Fac. Sci. Univ. Tokyo Sect. IA Math.* 29(1982), 287–318.
124. Matumoto, T., Triangulation of manifolds, *Algebraic and geometric topology (Proc. Sympos. Pure Math., Stanford Univ., Stanford, Calif., 1976)*, Part 2, pp. 3–6, Proc. Sympos. pure Math., XXXII, Amer. Math. Soc., Providence, R.I., 1978.
125. Matveev, S., *Algorithmic Topology and Classification of* 3*-Manifolds,* Algorithms and Computation in Mathematics, Vol. 9. Springer-Verlag, Berlin, Heidelberg, New York, 2003.
126. McMillan, D.R., Jr., Some contractible open 3-manifolds, *Trans. Amer. Math. Soc.* 102(1962), 373–382.
127. McMillan, D.R., Jr., Cartesian products of contractible open manifolds, *Bull. Amer. Math. Soc.* 67(1961), 510–514.
128. McMillan, D.R., Jr., and Zeeman, E.C., On contractible open manifolds, *Proc. Camb. Phil. Soc.* 58(1962), 221–224.
129. Meeks, W.H. III, Yau, S.T., Topology of three-dimensional manifolds and the embedding problems in minimal surface theory, *Ann. of Math.* 112(1980), 441–484.
130. Meeks, W.H. III, Yau, S.T., The classical Plateau problem and the topology of three-dimensional manifolds. The embedding of the solution given by Douglas-Morrey and an analytic proof of Dehn lemma, *Topology* 21(1982), 409–442.
131. Milnor, J.W., A unique factorization theorem for 3*-manifolds, Amer. J. Math.* 74(1962), 1–7.
132. Milnor, J., Husemoller, D., *Symmetric bilinear forms,* Springer-Verlag, Berlin, Heidelberg, New York, 1973.

133. Milnor, J., Towards the Poincaré conjecture and the classification of 3-manifolds, *Notices of Amer. Math. Soc.* 50(2003), no. 10, 1226–1233.
134. Moise, E.E., Affine structures on 3-manifolds, V. The triangulation theorem and Hauptvermutung, *Ann. of Math.* 56(1952) 96–114.
135. Morgan, J.W., Recent progress on the Poincaré conjecture and the classification of 3-manifolds, *Bull. Amer. Math. Soc. (N.S.)* 42(2004), no. 1, 57–78.
136. Morgan, J.W., *100 years of topology; work stimulated by Poincaré's approach to classifying manifolds. The Poincaré conjecture,* Clay Math. Proc., 19. Amer. Math. Soc., Providence, RI, 2014.
137. Morgan, J.W., Fong, F. T.-H., *Ricci flow and geometrization of 3-manifolds,* University Lecture Series, 53. Amer. Math. Soc., Providence, RI, 2010.
138. Morgan, J.W., Tian, G., *The geometrization conjecture,* Clay Mathematics Monographs. Amer. Math. Soc., Providence, RI; Clay mathematics institute, Cambridge, MA, 2014.
139. Munkres, J.R., *Some applications to triangulation theorems (thesis),* University of Michigan, 1955.
140. Munkres, J.R., Obstructions to smoothing piecewise-differentiable homeomorphisms, *Ann. of Math.* 72(1960), 521–554.
141. Munkres, J.R., *Elementary differential topology,* Ann. of Math. Studies, No. 54, Princeton Univ. Press, Princeton, N.J., 1966.
142. Newman, M.H.A., The engulfing theorem for topological manifolds, *Ann. of Math.* 84(1966), 555–571.
143. Nikolaev, I.G., The tangent cone of an Alexandrov space of curvature $\leq K$, *Manuscripta Math.* 86(1995), 683–689.
144. Novikov, P.S., *On the algorithmic insolvability of the word problem in group theory. (Russian),* Trudy Mat. Inst. im. Steklov. no. 44 Izdat. Akad. Nauk SSSR, Moscow, 1955.
145. Novikov, P.S., *On the algorithmic insolvability of the word problem in group theory,* Amer. Math. Soc. Translations, Ser. 2, Vol. 9, pp. 1–122, AMS, Providence, R.I. 1958.
146. Orlik, P., *Seifert manifolds,* Lect. Notes in Math. 291, Springer, Berlin, 1972.
147. O'Shea Donal, *The Poincaré conjecture (In search of the shape of the Universe),* Penguin Books Ltd, London, 2007.
148. Papadopoulos, A., *Metric spaces, convexity and non-positive curvature.* Second edition, IRMA Lectures in mathematics and Theoretical Physics, 6. European Mathematical Society (EMS), Zürich, 2014.
149. Papakyriakopoulos, C.D., On Dehn's lemma and asphericity of knots, *Ann. of Math.* 66(1957), 1–26.
150. Papakyriakopoulos, C.D., A reduction of the Poincaré conjecture to group-theoretic conjectures, *Ann. of Math.* 77(1963), 250–305.
151. Perelman, G., *The entropy formula for the Ricci flow and its geometric applications,* arXiv.org/abs/math.DG/0211159.
152. Perelman, G., *Ricci flow with surgery on three-manifolds,* arXiv.org/abs/math.DG/0303109.
153. Perelman, G., *Finite extinction time for the solutions to the Ricci flow on certain three-manifolds,* arXiv.org/abs/math.DG/0307245.
154. Poincaré, H., Second complément à l'Analysis Situs, *Proc. London Math. Soc.* 32(1900), 277–308.
155. Poincaré, H., Cinquième complément à l'Analysis situs, *Rend. Circ. Mat. Palermo* 18(1904), 45–110. (See *Ouevres*, Tome VI, Paris, 1953, p. 498.)
156. Quinn, F., An obstruction to the resolution of homology manifolds, *Mich. Math. J.* 34(1987), 285–291.
157. Rado, T., Über den Begriff der Riemanschen Fläche, *Acta Litt. Scient. Univ. Szeged* 2(1925), 101–121.
158. Ranicki, A.A., On the Hauptvermutung, *The Hauptvermutung book*, K-monographs in Math., Kluwer Academic Publishers, 1996, 3–31.
159. Rochlin, V.A., Any three-dimensional manifold is a boundary of four-dimensional manifold (Russian), *Dokl. Acad. Nauk SSSR* 81(1951), 355–357.

160. Rochlin, V.A., New results in the theory of four-dimensional manifolds (Russian), *Dokl. Acad. Nauk SSSR* 84(1952), 221–224.
161. Scott, P., The geometries of 3-manifolds, *Bull. of the London Math. Soc.* 15(1983), no.5, 401–487.
162. Scott, P., The symmetry of intersection numbers in group theory, *Geometry and Topology* 2(1998), 11–29, Correction (ibid) (1998), 333–335.
163. Seifert, H., Topologie dreidimensionaler gefaserter Räume, *Acta Math.* 60(1933), 147–238.
164. Seifert, H., and Threlfall, W., Topologische Untersuchung der Diskontinuitäts-bereiche endlicher Bewegungsgruppen des dreidimensionalen sphärischen Raumes, *Math. Ann.* 104(1930–31), 1–70.
165. Seifert, H., and Threlfall, W., *Lehrbuch der Topologie,* Teubner, Leipzig, 1934.
166. Seifert, H., and Threlfall, W., *A textbook of topology,* Pure and Applied Mathematics 89, Academic press, 1980.
167. Shioya, T., Yamaguchi, T., Volume collapsed three-manifolds with a lower curvature bound, *Math. Ann.* 333(2005), 131–155.
168. Smale, S., review of the paper [47], *Math Reviews* 22(1961), 1218.
169. Smale, S., Generalized Poincaré's conjecture in dimensions greater than 4, *Ann. of Math.* (2) 74(1961), 391–406.
170. Stallings, J., Polyhedral homotopy spheres, *Bull. Amer. Math. Soc.* 66(1960), 485–488.
171. Stallings, J., A topological proof of Grushko's theorem on free products, *Math. Zeit.* 90(1965), 1–8.
172. Stallings, J., *How not to prove the Poincaré conjecture,* Ann. of Math. Studies, vol. 60, Princeton Univ. Press, 1966, 83–88.
173. Stallings, J., On the loop theorem, *Ann. of Math.* 72(1960), 12–19.
174. Steenrod, N.E., Epstein, D.B.A., *Cohomology operations,* Princeton, New Jersey, Princeton University Press, 1962.
175. Taubes, C.H., Casson invariant and gauge theory, *J. Differential Geom.* 31(1990), 547–599.
176. Thomas, C.B., Homotopy classification of free actions by finite groups on S^3, *Proc. London Math. Soc.* (3)40(1980), 284–297.
177. Thurston, W.P., Three-dimensional manifolds, Kleinian groups and hyperbolic geometry, *Bull. Amer. Math. Soc. (N.S.)*, 6(1982), no. 3, 357–381.
178. Thurston, P., 4-dimensional Busemann G-spaces are manifolds, *Differential Geom. and Appl.* 6(1996), no. 3, 245–270.
179. Waldhausen, F., Heegaard-Zerlegungen der 3-sphere, *Topology* 7(1968), 195–203.
180. Waldhausen, F., Recent results on sufficiently large 3-manifolds, *Algebraic and geometric topology (Proc. Sympos. Pure Math., Stanford Univ., Stanford, Calif., 1976)*, Part 2, pp. 21–38, Proc. Sympos. pure Math., XXXII, Amer. Math. Soc., Providence, R.I., 1978.
181. Waldhausen, F., On irreducible 3-*manifolds which are sufficiently large,* *Ann. of Math.* (2)87(1968), 56–88.
182. Waldhausen, F., The word problem in fundamental groups of sufficiently large irreducible 3-manifolds, *Ann. of Math.* (2)88(1968), 272–280.
183. Wall, C.T.C. (Ed. Ranicki, A.A.), *Surgery on compact manifolds, 2nd Ed.* Mathematical Surveys and Monographs 69. Amer. Math. Soc. Providence, Rhode Island, 1999.
184. Wallace, A., Modifications and cobounding manifolds, II, *Journal of Mathematics and Mechanics*, 10(1961), 773–809.
185. Whitehead, J.H.C., A certain open manifold whose group is unity, *Quart. J. Math., Oxford Ser.* 6(1935), 268–279.
186. Whitehead, J.H.C., On C^1-complexes, *Ann. of Math.* 41(1940), 809–824.
187. Whitehead, J.H.C., On simply connected 4-dimensional polyhedra, *Comment. Math. Helv.* 22(1949), 48–92.
188. Whitehead, J.H.C., On 2-spheres in 3-manifolds, *Bull. Amer. Math. Soc.* 64(1958), 161–166.
189. Whitehead, J.H.C., Manifolds with transverse fields in euclidean space, *Ann. of Math.* (2) 73(1961), 154–212.

190. Whitehead, J.H.C., *The mathematical works of J.H.C. Whitehead. Vol. II: Complexes and manifolds. Edited by I.M.James,* Pergamon Press, Oxford- New York- Paris, 1962.
191. Wilder, R.L., Topology of manifolds, *Amer. Math. Soc. Coll.*, 32(1949).
192. Wolf, J.A., *Spaces of constant curvature,* McGraw-Hill, Inc., New York, 1967.
193. Yamasuge, H., On Poincaré conjecture for M^5, *J. Math. Osaka City Univ.* 12(1961), 1–17.
194. Zeeman, E.C., *The Poincaré conjecture for* $n \geq 5$, Topology of 3-manifolds and related topics. Englewood Cliffs, NJ: Prentice Hall, 1962, 198–204.

Chapter 18
A Glimpse into the Problems of the Fourth Dimension

Valentin Poénaru

18.1 Introduction

The message of this short survey is that four-dimensional topology is very special indeed. Also, four dimensions is the place where, today, as far as topology of manifolds is concerned, more than anywhere else, there are still big questions waiting to be solved.

Some people might add that four dimensions correspond exactly to our ambient space-time, but this kind of argument I will not pursue here.

The first section of this short paper lists some basic questions and NO/YES answers to them (sometimes question marks) and, to a certain extent at least, it corresponds to my own idiosyncrasies. The next three sections, which explore the big abyss between DIFF and TOP in four dimensions (and only there, is there such an abyss), are more conventional.

My choice here was to use the Casson-Handles and Yang-Mills rather than, let us say, the gropes of Stanko-Freedman-Quinn [8] or the equations of Seiberg-Witten. For a longer survey, that would have been a better policy, but just to introduce those topics would have required a much longer paper. Of course, if complete proofs would have had to be supplied, it would have been simpler, or much simpler, with gropes and with Seiberg-Witten, [14].

I wish to thank the IHES for its friendly help and, in particular Cécile Gourgues for the typing and Marie-Claude Vergne for the drawings.

V. Poénaru (✉)
Professor Emeritus at the Université Paris Sud-Orsay, Orsay, France

S. G. Dani, A. Papadopoulos (eds.), *Geometry in History*,
https://doi.org/10.1007/978-3-030-13609-3_18

18.2 Geometrical Questions, and Answers

We will review now some simple geometrical questions, which should illustrate the special position of dimension four, inside the category of DIFF manifolds.

A smooth connected manifold is said to be geometrically simply connected (GSC) if it possesses a smooth handlebody decomposition with a unique handle of index zero and with its 1-handles and 2-handles in cancelling position, i.e. without 1-handles. (I am not sure who concocted the GSC concept, it may well have been Terry Wall [27].) I will be more explicit concerning the "handles in cancelling position". Remember that an n-dimensional handle of index λ is an n-ball factorized as $B^{\lambda} \times B^{n-\lambda}$. Here the $\partial B^{\lambda} \times B^{n-\lambda}$, respectively $B^{\lambda} \times \partial B^{n-\lambda}$ are called attaching zone, respectively lateral surface. The reason for this terminology is that, via an embedding of the attaching zone into the boundary of some n-manifold, the handle can be glued and/or attached and then the lateral surface appears as a piece of the boundary of the newly created n-manifold.

Elementary Morse theory tells us that, in the DIFF context, any connected manifold can be gotten by starting either with an n-ball (i.e. a handle of index $\lambda = 0$), in the compact case, or with the n-dimensional regular neighbourhood of an infinite tree, in the non-compact case, and then adding, successively, handles of index $\lambda = 1$, $\lambda = 2$, a.s.o. This kind of thing is called a handlebody decomposition. In the non-compact case one adds here the proviso that there should be no infinite accumulation at finite distance.

In this context, let us consider two handles $H^1 = B^1 \times B^{n-1}$, $H^2 = B^2 \times B^{n-2}$ of index $\lambda = 1$ and $\lambda = 2$ respectively. In the generic case, any connected component of

$$\{\text{the lateral surface } B^1 \times S^{n-2} \text{ of } H^1\} \cap \{\text{the attaching zone } \partial B^2 \times B^{n-2} \text{ of } H^2\}$$

is a product $B^1 \times B^{n-2}$, with $B^1 \subset S^1$ and $B^{n-2} \subset S^{n-2}$. We will denote by $H^2 \cdot H^1$ the number of these connected components. No orientations, no $\pm$ signs and no algebra are involved here.

With this, any handlebody decomposition comes with a matrix $H_j^2 \cdot H_i^1$ which is finite in the compact case and infinite in the non-compact case. It is called the ***geometric intersection matrix***. Here comes now a definition, for ***square*** matrices a_{ij} with entries in Z_+. We say that the matrix is of the ***easy id + nilpotent*** type if $a_{ji} = \delta_{ji} + \eta_{ji}$ where $\eta_{ji} > 0 \Rightarrow j > i$. Dually, it is of the ***difficult id + nilpotent*** type if $a_{ji} = \delta_{ji} + \eta_{ji}$ where now $\eta_{ji} > 0 \Rightarrow j < i$. Of course, the two notions are equivalent in the finite case.

By definition, for a handlebody decomposition, the 1-handles and 2-handles are in cancelling position if we can identify a family of 2-handles H_i^2 in bijection with the 1-handle H_i^1 such that the geometric intersection matrix $H_j^2 \cdot H_i^1$ is of the easy id + nil type. Then the 1-handles can be gotten rid of, by cancelling them with the 2-handles.

In formula (6) below, we will meet the open Whitehead manifold Wh^3. This can be proved not to be GSC, but it admits a handlebody decomposition with $H_j^2 \cdot H_i^1$ of the difficult id + nil type.

The proof of this fact does not exist in print, to the best of my knowledge, but I will give here a hint of how one gets it. Look first at the figure 2.3.1 in [22], p. 669. What one sees there is a disc with boundary the curve T_i, coming back so as to hit itself along a double line (the $g\,(M_2(g \mid D_i))$ in the drawing). All this is really a 2^{d} object, but one can also read it as a recipe for a handlebody decomposition of $S^1 \times D^2$, with geometric intersection matrix

$$H_1^2 \cdot H_1^1 = 1\,, \qquad H_1^2 \cdot H_2^1 = 2\,.$$

There are here one 0-handle, two 1-handles and one 2-handle, and the H_2^1 corresponds to the double line above.

In this 2^{d}-object, one picks up now a closed loop going through a piece of T_i and through H_2^1, and next one treats the loop in question like we just did with T_i, adding along it a disc which bites itself back. The higher 2^{d} object gotten this way is again a recipee for $S^1 \times D^2$, with this time, the matrix

$$H_1^2 \cdot H_1^1 = 1\,, \quad H_1^2 \cdot H_2^1 = 2\,, \quad H_2^2 \cdot H_2^1 = 1\,, \quad H_2^2 \cdot H_3^1 = 2\,.$$

The pattern continues indefinitely, and it generates an infinite matrix of the difficult id + nilpotent type. But the resulting 2^{d} object is certainly not Wh^3 and also some sacro-sancted local finiteness conditions get violated, in the way.

So the construction has to be modified, or rather enriched. After each one of the infinitely many steps, we thicken the result in 3-dimensional, by adding additional handles of index 1, 2 and 3. If one does this carefully, then what emerges is a PROPER (no accumulation at finite distance) handlebody decomposition for the Whitehead manifold Wh^3 which has a geometric intersection with a main difficult id + nilpotent part (and with some additional easy id + nilpotent contributions which do not interfere with the main difficult part). But the complete detailed version of this story is too long for the format of the present short survey. And, of course, one has to be a bit less cavalier than we have been, in the complete definition of the difficult id + nil.

The fact that a manifold is not GSC is closely related to it not being simply connected at infinity. More precisely, GSC implies the existence of an exhaustion by compact simply-connected complexes, which for open 3-manifolds implies $\pi_1^\infty = 0$. Obviously, GSC implies $\pi_1 = 0$ and here is the first

Question 1 Assume that the smooth manifold M^n is simply-connected. Is it then, also, GSC? Here is the complete answer to this question for compact M^n's. I will organize the yes/no answers to our question by decreasing dimensions.

The Case $n \geq 5$ The answer is YES. This is part of Smale's proof for his h-cobordism theorem, which includes the Poincaré Conjecture in dimensions $n \geq 5$.

There is here a little piece of dramatic history. About 1960, at the time when Steve Smale announced his result, John Stallings also announced his own proof for the Poincaré Conjecture (in dimension ≥ 7), via engulfing. Then, during the Arbeitstagung in Bonn, where both of them were lecturing, Stallings detected an error in Smale's original argument, exactly at the level of our present question

$$\pi_1 = 0 \xRightarrow{?} \text{GSC}\,.$$

At that moment, it looked as if Smale's proof was in shambles. But, after very few weeks, Smale came with a nice and clever argument, using handles of index two ***and*** three; thus he put his proof back, now on a solid footing. He had gotten the idea during the boat-trip on the Rhine, traditional part of the same Arbeitstagung. I have always admired Smale for his courage in front of adversity and for the beautiful way by which he rescued everything. I must add here that I was not present there at that Arbeitstagung, since I was living on the dark side of the Moon at that time, but I have reconstructed the story from what was told to me, a bit later. See here [24] for more on this story.

The Case $n = 4$ Here the answer is NO, since there exist compact bounded contractible 4-manifolds which are not GSC. This was proved by Andrew Casson. I will be a bit more specific here. Towards 1959, Barry Mazur and the present author [12, 16], independently of each other, had constructed smooth compact bounded contractible 4-manifolds M^4 such that

(a) $\pi_1\, \partial\, M^4$ admits non trivial representations into some permutation group. So, $\pi_1\, \partial\, M^4 \neq 0$ and hence $M^4 \underset{\text{TOP}}{\neq} B^4$, already in the topological category.
(b) Nevertheless, $M^4 \times [0, 1] \underset{\text{DIFF}}{=} B^5$.
(c) M^4 has a 2-spine: there is a contractible finite complex $K^2 \subset M^4$ such that the 4-dimensional regular neighbourhood $N^4(K^2) = M^4$.

Now, what Casson proved is that if a contractible compact smooth bounded 4-manifold X^4 is such that there is a non trivial homomorphism

$$\pi_1\, \partial\, X^4 \xrightarrow{\ \rho\ } \{\text{compact connected Lie group}\}\,,$$

then X^4 ***cannot*** be GSC. His argument is a beautiful mixture of discrete groups and Lie groups. Casson never published his proof, but I believe many people knew about it. I learned about it from Mike Freedman, many years ago.

The Case $n = 3$ Here our question is equivalent to the 3-dimensional Poincaré Conjecture and so, via the very celebrated work of G. Perelman, the answer is YES, [15].

When we move from the compact case to the one of open manifolds, asking the same question 1, then for $n \geq 5$ and $n = 3$ things are just like above, provided we add the assumption $\pi_1^\infty = 0$. When we move, next, to open 4-manifolds or to non compact n-manifolds with non-empty boundary, then except for obvious trivialities, there are no theorems nor even general clean conjectures. Notwithstanding this, the present author, in his work on low-dimensional manifolds and on geometric group theory, consistently met this issue $\pi_1 = 0 \overset{?}{\Longrightarrow}$ GSC and had to prove that various things ***are*** GSC, in the murky non compact realms just described, (see here, for instance [18] and [20]). So, we go now to the next

Question 2 Let K^2 be a contractible finite complex such that, for some n, the regular neighbourhood $N^n(K^2)$ makes sense. Is then $N^n(K^2) \underset{\text{DIFF}}{\overset{?}{=}} B^n$? Here are the answers, for increasing n's.

The Case $n = 3$ Again YES, since the question is now equivalent to the 3-dimensional Poincaré Conjecture, and so we can invoke once more Perelman's work.

The Case $n = 4$ The answer here is NO. In our previous discussion of Question 1, we have already met the Mazur-Po manifolds M^4. These are contractible, $M^4 \neq B^4$ and we have $M^4 = N^4(K^2)$.

The Case $n = 5$ Here we have a question mark, it is an open question, the answer is not known today. I will come back to this later.

The Case $n \geq 6$ The answer is YES, via Smale, just like for Question 1 above.

The question mark above corresponds to the following

Conjecture A If K^2 is contractible, then

$$N^5(K^2) \underset{\text{DIFF}}{=} B^5 .$$

This question is connected with another big open problem, the smooth 4-dimensional Poincaré Conjecture, and remember here that Michael Freedman has proved that any TOP manifold which has the homotopy type of S^4 is homeomorphic to S^4. This is the TOP 4-dimensional Poincaré Conjecture.

So, let then Σ^4 be a DIFF 4-dimensional homotopy sphere, and define

$$\Delta^4 \equiv \Sigma^4 - \text{int}\, B^4 , \tag{18.1}$$

where $B^4 \subset \Sigma^4$ is a smoothly embedded standard 4-ball. The DIFF 4-dimensional Poincaré Conjecture, an open question indeed, is equivalent to the question $\Delta^4 \underset{\text{DIFF}}{\overset{?}{=}} B^4$. There is here a particularly interesting special case, namely when Δ^4 embeds smoothly into S^4, $\Delta^4 \subset S^4$. The issue $\Delta^4 \underset{\text{DIFF}}{\overset{?}{=}} B^4$ is then the so-called smooth 4-dimensional Schoenflies problem.

The general DIFF Schoenflies problem is to describe, up to diffeomorphism, the smooth compact bounded n-manifolds Δ^n gotten by splitting S^n along a smooth embedding $S^{n-1} \subset S^n$. The celebrated work of Mazur (and besides [11] I suggest here the reference [19]), completed by Kervaire and Milnor [10], shows that if $n \neq 4$ then the Schoenflies n-ball is $\Delta^n \underset{\text{DIFF}}{=} B^n$. For $n = 4$, the only known fact, as of today, is that

$$\Delta^4_{\text{Schoenflies}} - \{\text{a boundary point}\} \underset{\text{DIFF}}{=} B^4 - \{\text{a boundary point}\},$$

which implies, of course, that $\Delta^4_{\text{Schoenflies}} \underset{\text{TOP}}{=} B^4$. This is also proved in Mazur's celebrated paper. But the full DIFF 4-dimensional Schoenflies problem is another big open question.

Assume now that, for the general smooth homotopy 4-ball Δ^4 from (18.1), we would also know that it is GSC. Then, Conjecture A would easily imply that $\Delta^4 \times I \underset{\text{DIFF}}{=} B^5$ and, if we would also have the DIFF 4-dimensional Schoenflies, we would then get that $\Delta^4 \underset{\text{DIFF}}{=} B^4$, i.e. $\Sigma^4 \underset{\text{DIFF}}{=} S^4$, the DIFF 4-dimensional Poincaré Conjecture.

Personally, I have strong reasons to believe the truth of both Conjecture A and the DIFF 4-dimensional Schoenflies, but I am less sure concerning the $\Delta^4 \in$ GSC, for a general DIFF homotopy 4-ball. It is here, I believe, that the DIFF 4-dimensional Poincaré Conjecture may fail, if it fails.

At this point, once we talk about geometric conjectures, I would like to end this discussion with the following old conjecture, due, I think, to the late Sir Christopher Zeeman, who incidentally happens to have been my cousin, and which would immediately imply the 3-dimensional Poincaré Conjecture:

Conjecture B Let K^2 be any finite complex which is contractible. Then $K^2 \times [0, 1]$ is collapsible.

Although the 3-dimensional Poincaré Conjecture is proved by now, this does not imply Conjecture B, which to the best of my knowledge is still open, with a meaning which is mysterious.

At this point, some historical comments have to be fitted in. Way back in the nineteen-fifties, and earlier too of course, people believed that, in topology, difficulties increased with dimension. And, since 3-dimensional was clearly very hard, one was afraid of the higher dimensions, the difficulties of which were believed to be stratospheric.

Then came Barry Mazur (and he was barely 18 at the time), with his big breakthrough in the Schoenflies problem [11]. This was a psychological breakthrough too, not only mathematical. People stopped being afraid of the high dimensions and now the road was open for the spectacular next successes of Steve Smale [23], John Stallings [25], Christopher Zeeman [28], and others, who showed the dimensions ≥ 5 were actually not that hard, after all.

It was in dimensions 3 and 4 that the real difficulties were hidden. Then Mike Freedman [8] and Simon Donaldson [4] clarified a lot dimension four while Bill Thurston [26] and Grisha Perelman [3, 15], to a large extent completely cleaned the field in dimension three. Our lists of questions above, reflect this historical development.

18.3 Casson Handles

Almost 50 years ago, Andrew Casson introduced a class of non compact DIFF 4-manifolds V^4 with non-empty boundary having all of them with the following general features:

$$\text{int}\, V^4 \underset{\text{DIFF}}{=} R^4_{\text{standard}}, \qquad \partial\, V^4 = S^1 \times \overset{\circ}{B^2}, \tag{18.2}$$

and before I will say some more things about them, let us contemplate (18.2). We all know that knots are loops, of which I will always think of as being thickened, which live in the boundary of B^4, i.e. things like

$$S^1 \times \overset{\cup}{B^2} \subset S^3 = \partial\, B^4 \text{ (null-framing is meant here)}. \tag{18.3}$$

Now, what we see displayed in (18.2) is such an $S^1 \times \overset{\circ}{B^2}$ but living not in $\partial\, B^4$ but at the infinity of R^4. I will call such a thing a sort of knot. And if a sort of knot like (18.2) can be extended to something like (18.3), then I will call it tame, otherwise it is wild. But here the distinction DIFF/TOP can be injected too, we will come back to this below. Changing the topic for a little while, there is a famous lemma of Whitney, for eliminating intersections of submanifolds of complementary dimensions and this was an essential tool for Smale's h-cobordism theorem. A good reference for Whitney's lemma is [7]. But the Whitney lemma fails in dimension four and Casson invented his handles as a way to circumvent this difficulty.

Their explicit construction is too long for the format of the present article, but a very good description of all Casson Handles can be found in [9]. Concerning these Casson Handles, there is a big theorem of Freedman [6]

Theorem C *For any Casson Handle V^4, we have*

$$V^4 \underset{\text{TOP}}{=} B^2 \times \overset{\circ}{B^2}. \tag{18.4}$$

This is the main step in Freedman's proof of $\Sigma^4 \underset{\text{TOP}}{=} S^4$ and also in his complete classification for closed simply-connected TOP 4-manifolds [6].

Via Theorem C, the sort of knot (18.2) is trivial, in the TOP category, i.e. it is also topologically tame. But then, using the kind of arguments which will be developed in the last section of this paper one can also show that there have to be Casson

Handles V^4 such that $V^4 \underset{\text{DIFF}}{\neq} B^2 \times \overset{\circ}{B}{}^2$. Such Casson Handles are sort of knots which have to be smoothly wild. Here is the argument. Assume that the Casson Handle V^4, above is actually a sort of knot which is smoothly tame.

It is a standard fact that any Casson Handle V^4 is properly homotopy equivalent with $B^2 \times \overset{\circ}{B}{}^2$. (This is, of course, immediately implied by Theorem C too.) But then, using the now classical knot theory, our smoothly tame V^4 is also smoothly trivial, hence $V^4 \underset{\text{DIFF}}{=} B^2 \times \overset{\circ}{B}{}^2$, contradiction!

We talked here about sort of knots, but then, of course there are also sort of links, living at the infinity of R^4 too. I have encountered a lot of such in my work, and I had to invest quite a big effort in showing that certain particular classes were smoothly tame; I needed this; see here [17], for instance. But then, notice that all these kind of issues wild versus tame in the context DIFF or TOP are unknown in classical knot theory, to which they are foreign. It would be very interesting, I think, to try to develop the appropriate quantum topology in this quite nonstandard context. I believe this would be of consequence for physics too.

Each Casson Handle V^4 is labeled by a certain kind of infinite tree. These trees may be organized in some sort of "moduli space" and I will only describe here, explicitly, the Casson handle V^4 which corresponds to an extremal point of this "moduli space" (with the tree reduced to a half-line).

We start with the embedding $T_1 \subset T_2$ of one solid torus into the interior of another one, which Fig. 18.1 suggests and which is obviously connected to the Whitehead link.

One can iterate infinitely many times the map $T_1 \subset T_2$, getting thus

$$T_1 \subset T_2 \subset T_3 \subset \dots \tag{18.5}$$

and the union

$$\text{Wh}^3 \equiv \bigcup_{n=1}^{\infty} T_n \tag{18.6}$$

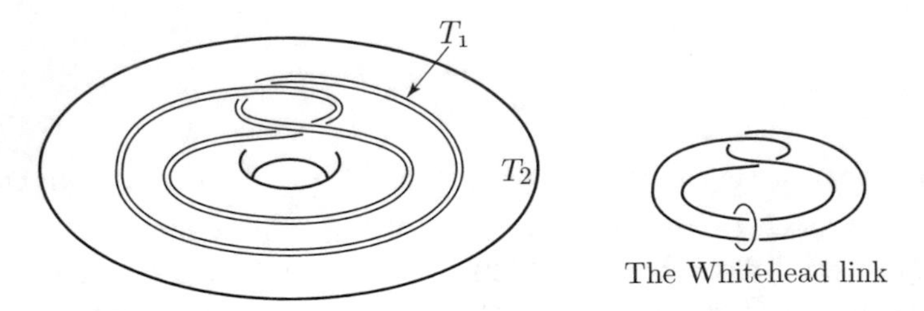

Fig. 18.1 The embedding of T_1 in T_2

is called the Whitehead manifold. This venerable object is an open contractible 3-manifold such that $\pi_1^\infty \text{Wh}^3 \neq 0$, i.e. $\text{Wh}^3 \underset{\text{TOP}}{\neq} R^3$ too. It was discovered by Henry (J.H.C.) Whitehead around 1934 as a counterexample to his attempted proof of the 3-dimensional Poincaré Conjecture. It is actually a counterexample in many different contexts, in particular in the context of geometric group theory.

It is very easy to show that $\text{Wh}^3 \times R$ which I will express as

$$\text{Wh}^3 \times R = \bigcup_{n=1}^{\infty} \mathring{T}_n \times (0,1) \tag{18.7}$$

is diffeomorphic to our standard R^4. With this, here is the simplest possible example of a Casson Handle

$$V^4 = \left(\mathring{T}_1 \times [0,1)\right) \cup \bigcup_{n=2}^{\infty} \mathring{T}_n \times (0,1)\,, \quad \partial V^4 = \mathring{T}_1 \times \{0\}\,, \tag{18.8}$$

and ***all*** Casson Handles admit descriptions of this type.

Some Comments

(A) We can continue (18.5) to the left, by a similar sequence

$$\ldots \subset T_{-2} \subset T_{-1} \subset T_0 \subset T_1 \tag{18.9}$$

and $K \equiv \bigcap_{n=1}^{-\infty} T_n$ is called the Whitehead continuum. With it, comes a dual description of the Whitehead manifold, namely

$$\text{Wh}^3 = S^3 - K\,, \tag{18.10}$$

an easily provable fact.

(B) Our Fig. 18.1 can also be read as a smooth injection $T_2 \xrightarrow{f} T_2$, with $fT_2 = T_1$ and this is a discrete non-hyperbolic dynamical system which deserves, I think, to be studied. I would, for instance, be personally curious to know what is its zeta-function, à la Artin-Mazur [1] looks like; I never found enough leisure to compute it, but it should not be too hard.

But there are also other dynamical aspects of the Whitehead manifold Wh^3, in particular connections with the Julia sets of quadratic maps (see [22]).

Our story, so far, has carried us through differential topology, topological topology (i.e. TOP manifolds) and through wild topology (à la R.H. Bing). And, next, we will meet the Standard Model of the elementary particles. It is this mixture which, among other things, makes four dimensions such a fascinating topic. But, before going on, this is an appropriate moment for another historical comment.

As already mentioned, Casson's motivation for inventing those handles which bear his name, was the dismal failure of Whitney's lemma, in dimension four.

And those handles were created as a means for circumventing that difficulty. But when he discovered that wild topology was propping up, Casson failed to take full advantage of his discovery. And so it was left for Mike Freedman to make the big breakthrough, the complete classification of closed simply-connected 4-manifolds, in the TOP category, via a big symphony of infinite processes. He married differential topology and the wild topology in the style of R.H. Bing. And all the spectacular results in high dimensions, could be extended to that context of TOP 4-dimensional manifolds too.

That made the issue of the quadratic intersection form for DIFF closed 4-manifolds become very acute. And now the tools came from a really unexpected direction, the physics of the Standard Model of elementary particles and fields. Sir Michael Atyiah [2] discovered that the non-linear Yang-Mills equations (non-abelian gauge theory) contained important mathematics, opening thus the way for the discoveries of Simon Donaldson, Karen Uhlenbeck, Clifford Taubes and the others.

This is the object of what comes next. And, for this section and the next two, there is also the reference [21].

18.4 Yang-Mills Theory, From a Topological Viewpoint

In what comes next, M^4 will be a closed DIFF 4-manifold, with $\pi_1 M^4 = 0$ and with a chosen orientation. Over M^4, we are also given a complex bundle $E \to M^4$ with structural group G and with a connection A. Here G is a compact Lie group and with this comes the curvature 2-form $F = F_A \in \Omega^2(M^4, \mathcal{L}G)$. We are now in dimension four and here, since $2 + 2 = 4$, for the Hodge operator we find that it maps 2-forms into 2-forms

$$\Omega^2 \xrightarrow{*} \Omega^2 ,$$

meaning that $* F \in \Omega^2$ too. If d_A is the covariant differential, we have the following automatic equation, always verified

$$d_A F = 0 \qquad \text{(Bianchi's theorem)}. \tag{18.11}$$

But then, there is also the following equation, which is certainly ***not*** automatic, the celebrated Yang-Mills equation

$$d_A * F = 0 \tag{18.12}$$

(written here without a source term), living at the core of the Standard Model of elementary particles. The "unknown quantity" in Eq. (18.12) is the connection A, and for our non-abelian $G = \mathrm{SU}(n)$, $n \geq 2$, Eq. (18.12) is non-linear.

If $G = U(1)$, then we can replace d_A by the mundane exterior differential d and the formulae (18.11) + (18.12) become the linear equations

$$dF = 0\,, \qquad d * F = 0\,. \tag{18.13}$$

But before we go on with topology, I feel like injecting here a bit of physics. If one wants to bring Maxwell's theory (electro-magnetism) into the framework of Einstein's special relativity, then one has to lump together the electric field $\vec{E}$ and the magnetic field $\vec{B}$ into the following 2-form

$$F = -E_x\, dt \wedge dx - \ldots + B_x\, dy \wedge dz + \ldots \in \Omega^2(R^4)\,, \tag{18.14}$$

with the "..." standing for circular permutations. With this meaning for F, the (18.13) are exactly Maxwell's equations (in the absence of currents and electric charges). The A is now the gauge potential, more classically it is the scalar potential and the vector potential lumped together. To be very specific here, the Bianchi part of (18.13) corresponds to Faraday's law of induction and to the non-existence of magnetic monopoles, while the second, Yang-Mills part corresponds to the laws of Gauss and Ampère (with Maxwell's "displacement field" added here too). The real meaning of $G = \mathrm{U}(1)$ is provided by quantum theory, and I will not discuss it here. Actually, quantum-theoretically, the A stands for the photon field.

Also, in the framework of Maxwell, the Hodge-star operation interchanges $\vec{E}$ and $\vec{B}$, i.e. $*\vec{E} = -\vec{B}$, a.s.o. This is the so-called electro-magnetic duality, with deep prolongations in string theory. It is a very hot topic these days. In a maybe less serious vein, I heard some people say that our world (the space-time) is 4-dimensional, so as to make possible the existence of Maxwell's equations.

The non-linear Eq. (18.12) for $G = \mathrm{SU}(2)$ and $G = \mathrm{SU}(3)$ correspond to the nuclear weak forces respectively to the strong forces. Actually, in our real world, $G = \mathrm{SU}(2)$ has to be replaced by $G = \mathrm{SU}(2) \times \mathrm{SU}(1)$, in what is called the electro-weak unification of Glashow, Salam and Weinberg.

Historically, Eq. (18.12), with $G = \mathrm{SU}(2)$ was discovered by Yang and Mills around 1954, in an attempt to describe nuclear forces by a nonlinear gauge theory inspired by Maxwell's equations. How, once combined with an appropriate symmetry-breaking mechanism, this eventually fitted into the Standard Model is a fascinating story on its own, but we cannot tell it here.

Anyway we close here our little excursion into physics. And going back to (18.12), notice that the infinite-dimensional group $\mathcal{G}$ of automorphisms of the bundle E, called the group of gauge transformations, acts on the whole theory.

For our present purposes, we will work with $G = \mathrm{SU}(2)$, and assume that the second Chern class $c_2(E) = -1$. Also, we will replace (18.12) by the simpler, so called self-dual Yang-Mills equation

$$F = *F \tag{18.15}$$

which, via Bianchi implies (18.12).

We consider now the moduli-space

$$\mathcal{M} = \{\text{self dual connections}\}/\mathcal{G}\,, \tag{18.16}$$

and it is a nontrivial fact that $\mathcal{M}$ is a non-void Hausdorff space. For a generic Riemannian metric on M^4, this $\mathcal{M}$ can be compactified into a 5-manifold $\overline{\mathcal{M}}$ which is smooth except for some isolated, very controlled singularities and which is such that $\partial\,\overline{\mathcal{M}} = M^4$. The condition $c_2(E) = -1$ is used here, and the key fact is a deep theorem of Karen Uhlenbeck, which states that from every sequence $A_1, A_2, \ldots \in \mathcal{M}$ which does not converge, one can extract a subsequence $A_{n_1}, A_{n_2}, \ldots \in \mathcal{M}$ such that: There is a point $p \in M^4$ such that the curvatures $F_{A_{n_1}}, F_{A_{n_2}}, \ldots$ get more and more concentrated around p, leaving a flat connection outside p. See here [5].

Now, any oriented M^4 has a quadratic intersection form $q(M^4)$ defined over Z, which via Poincaré duality is nondegenerate. Actually, any M^{4n} has such $q(M^{4n})$, and this is an important invariant.

Donaldson's big discovery was that the topology of $\overline{\mathcal{M}}$ contains important information concerning its boundary M^4, our smooth 4-manifold of interest. We are here in the DIFF context, remember.

Theorem D (Donaldson) *If $\pi_1 M^4 = 0$ and $q(M^4) > 0$, then $q(M^4)$ is diagonalizable, over Z. [4]*

In particular, this excludes $q(M^4) = E_8 + E_8$, which at the time of his discovery was a big novelty. The E_8 had already been excluded by a famous theorem of Rohlin, but not $E_8 + E_8$. All this concerns the DIFF case. For TOP, things are quite different.

It is via this theorem of Donaldson that one knows that some of the Casson Handles are smoothly wild.

Questions Are there criteria, coming let's say from some kind of quantum topology, for detecting Casson Handles which are smoothly wild? And are all Casson Handles smoothly wild?

18.5 Exotic R^4's

If X^n is a smooth manifold homeomorphic to R^n then, if also $n \neq 4$, it was known since a while that X^n is diffeomorphic to the standard R^n. This basic fact has been proved by Moise for $n \leq 3$ [13] and by Stallings for $n \geq 5$ [25]. Moreover it was also known, essentially via the work of Kervaire and Milnor [10] that, in dimensions

$n \neq 4$, the difference between DIFF and TOP (which can be there in dimensions $n \geq 5$) is completely controlled by the discrete invariants of algebraic topology, let us say by some "discrete quantum numbers". All the facts above, which were already well-known by, let us say 1970 or before, fail to be true when $n = 4$. This is so because we have here the following surprising result.

Theorem E (Freedman-Donaldson) *There exist open manifolds smooth X^4 s.t.*

$$X^4 \underset{\text{TOP}}{=} R^4 \qquad \textit{but} \qquad X^4 \underset{\text{DIFF}}{\neq} R^4 .$$

It is here that one can really perceive that there is a big mysterious chasm between dimension four and the other dimensions. In dimension four the discrete invariants of algebraic topology are of no use when one tries to compare DIFF and TOP. But we will not discuss this further here, and I will just sketch the proof of Theorem E, modulo the various other things already said. Notice, first the obvious fact that

$$B^4 + \{\text{two handles of index two added along the Hopf link}$$

$$\text{with null-framing}\} \underset{\text{DIFF}}{=} S^2 \times S^2 - \text{int}\, B^4 . \tag{18.17}$$

Figure 18.2 should illustrate this fact, with two dimensions less.

If in the context of formula (18.17) we replace the 2-handles by some Casson Handles, and if we also scrape away all the boundary then we get a smooth open manifold which I will generically denote by W^4_{Casson}. There is, a priori at least, a whole uncountable infinity of them. It follows from Theorem C that

$$W^4_{\text{Casson}} \underset{\text{TOP}}{=} S^2 \times S^2 - B^4 , \tag{18.18}$$

which should be, of course compared to (18.17). Here, the smooth embedding $B^4 \subset S^2 \times S^2$ factors through another smooth embedding

$$B^4 \subset \text{int}\, B^4_1 \subset B^4_1 \subset S^2 \times S^2 . \tag{18.19}$$

Fig. 18.2 $S^1 \times S^1 - \text{int}\, B^2$

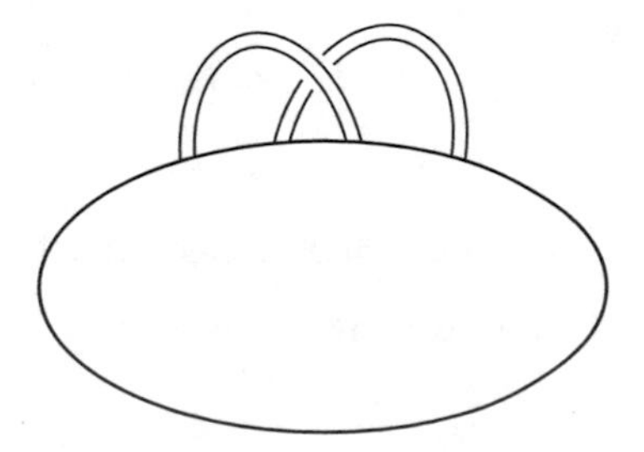

This allows us to introduce the following compact core C^4 of the W^4_{Casson}

$$C^4 \equiv S^2 \times S^2 - \operatorname{int} B_1^4 \subset W^4_{\text{Casson}} \,. \tag{18.20}$$

We have now the following

Lemma F (A. Casson) *For every W^4_{Casson}, there is a smooth embedding*

$$W^4_{\text{Casson}} \subset S^2 \times S^2 \,. \tag{18.21}$$

The proof uses ideas similar to those in Comment A, Sect. 18.2. So we can consider now

$$C^4 \subset W^4_{\text{Casson}} \subset S^2 \times S^2$$

and when we look, next, at the smooth open manifold

$$X^4 \equiv S^2 \times S^2 - C^4 \,, \tag{18.22}$$

then we can very easily see that it is contractible (since any non-trivial topology has gone with the core C^4) and also that its unique end is the standard $S^3 \times R$. This is so because the (18.19) is very nice and tame; we may actually assume that the $\partial B_1^4 \subset S^2 \times S^2$ has a collar $\partial B_1^4 \times (-\varepsilon, \varepsilon) \subset S^2 \times S^2$. It follows then, from the TOP 4-dimensional Poincaré Conjecture proved by M. Freedman [6], that $X^4 \underset{\text{TOP}}{=} R^4$.

We will show that, among these X^4's defined by (18.22), some have to be exotic, $X^4 \underset{\text{DIFF}}{\neq} R^4$.

We consider now the classical Kummer surface K^4, defined in CP^3 by

$$K^4 = \{z_0^4 + z_1^4 + z_2^4 + z_3^4 = 0\} \,,$$

in projective coordinates. What we need to know here is that K^4 is a smooth closed 4-manifold (which happens to be spin, too), coming with $\pi_1 K^4 = 0$ and with the following quadratic intersection form

$$q(K^4) = E_8 + E_8 + \begin{pmatrix} 0 & 1 \\ 1 & 0 \end{pmatrix} + \begin{pmatrix} 0 & 1 \\ 1 & 0 \end{pmatrix} + \begin{pmatrix} 0 & 1 \\ 1 & 0 \end{pmatrix} . \tag{18.23}$$

We see here occuring three times the $q(W^4_{\text{Casson}})$, namely the $\begin{pmatrix} 0 & 1 \\ 1 & 0 \end{pmatrix}$.

We have now

Lemma G (Casson) *We can find three* W^4_{Casson}*'s which come with smooth disjoined embeddings inside* K^4*, realizing exactly the* $\begin{pmatrix} 0 & 1 \\ 1 & 0 \end{pmatrix} + \begin{pmatrix} 0 & 1 \\ 1 & 0 \end{pmatrix} + \begin{pmatrix} 0 & 1 \\ 1 & 0 \end{pmatrix}$ *in (18.23).*

Here is what I can say concerning this. The Casson definition of the Casson Handles is an infinite process, which is sort of a very high powered version of the definition of the Whitehead manifold Wh^3. And Casson found a nice simple algebraic criterion which, when satisfied, allows us to construct a Casson Handle V^4 (never mind which), when we might need one. It is the Casson criterion which is behind Lemma G; see [9] for more details.

So, we have here three X^4's like in (18.22) and I will show that, if each of them comes with

$$X^4 \underset{\text{DIFF}}{=} R^4 , \tag{18.24}$$

then we get a contradiction with Donaldson's theorem.

In the context of Lemma F, let us consider the wild closed subset

$$k \equiv S^2 \times S^2 - W^4_{\text{Casson}} . \tag{18.25}$$

Here, we have two disjoined embeddings into $S^2 \times S^2$, namely

$$k \subset S^2 \times S^2 \supset C^4 . \tag{18.26}$$

Assuming from now on that each of our six X^4's is diffeomorphic to the R^4_{standard}, we can find inside each of them a smoothly embedded copy of the 3-sphere

$$S^3 \subset X^4 = S^2 \times S^2 - C^4 \subset S^2 \times S^2 , \tag{18.27}$$

which separates our wild closed set k from the infinity of X^4. Metaphorically speaking, a standard R^4 has a smooth horizon while, generally speaking an exotic R^4 has a fractal horizon. [I am touching here an issue which, I think, deserves to be investigated, namely the possible connection between exotic SMOOTH 4-dimensional structures and chaotic dynamics. Such ideas, partly suggested by John Hubbard, were behind [22].]

This $S^3 \subset X^4$ splits from X^4 a smooth compact bounded manifold $A^4 \subset X^4 \subset S^2 \times S^2$, which is such that

$$S^3 = \partial A^4 \qquad \text{and} \qquad k \subset \text{int}\, A^4 .$$

Inside our $S^2 \times S^2$ from (18.27) we also find that

$$S^3 \subset S^2 \times S^2 - k = W^4_{\text{Casson}} , \tag{18.28}$$

and now, our smooth S^3 splits from $W^4_{\text{Casson}} \subset K^4$ a smooth compact bounded manifold $D^4 \subset W^4_{\text{Casson}}$, which comes with

$$\{\text{the core } C^4 \text{ of } W^4_{\text{Casson}}\} \subset \text{int}\, D^4\,, \qquad \partial\, D^4 = S^3\,.$$

For each of the three individual X^4's coming from the three W^4_{Casson}'s in Lemma G, we find now the inclusions

$$C^4 \subset D^4 \subset K^4 \qquad \text{and} \qquad k \subset A^4 \subset S^2 \times S^2\,,$$

coming with a canonical diffeomorphism $\partial\, D^4 = \partial\, A^4$ (our S^3). Let Y^4 be the smooth closed simply-connected 4-manifold gotten by replacing each $D^4 \subset K^4$ by the corresponding A^4, an operation which makes sense, once

$$\partial\, D^4 = \partial\, A^4\,.$$

We clearly have $q(Y^4) = E_8 + E_8$, which contradicts Donaldson's theorem. The smooth manifold Y^4 cannot exist and so Theorem E is now proved.

The theorem in question is the first sign that the difference between DIFF and TOP, which is non-existant at category level in dimensions ≤ 3, and completely amenable to the discrete techniques of algebraic topology in dimensions ≥ 5, is a big wide open gap in dimension exactly four, totally un-understandable by algebraic topology. I think that understanding this gap is an important problem, and I tend to view the smooth 4-dimensional Schoenflies issue as a first small little step in that direction. But what we will eventually need is a new sort of quantum topology, or something like it, not yet available to us. This should allow us to make good sense of the gap DIFF/TOP in dimension four, as well let us say, as we understand that kind of issue today in dimensions higher than five, via the sophisticated methods of differential and algebraic topology, methods which seem powerless for our gap.

References

1. M. Artin and B. Mazur, On periodic points, *Ann of Math.* **2** (1965), pp. 82–99.
2. M.F. Atyiah, *Geometry of Yang-Mills Fields*, Lezioni Fermiane, Pisa (1979).
3. L. Bessières, G. Besson, M. Boileau, S. Maillot and J. Porti, Geometrization of 3-manifolds, *EMS Tracts in Math.* **13** (2010).
4. S.K. Donaldon, An application of gauge theory to four-dimensional topology, *J. Diff. Geom.* **18** (1983), pp. 278–315.
5. D.S. Freed and K.K. Uhlenbeck, *Instantons and four-manifolds*, Springer (1984).
6. M. Freedman, The Topology of four-dimensional manifolds, *J. Diff. Geom.* **17** (1983), pp. 279–315.
7. M.H. Freedman and F. Luo, *Selected applications of geometry to low-dimensional topology*, Marker Lectures, AMS (1990).

8. M.H. Freedman and F. Quinn, *Topology of 4-manifolds*, Princeton mathematical Series **39**, Princeton N.J. (1990).
9. L. Guillou and A. Marin, À la recherche de la topologie perdue. Birkhäuser 1986.
10. M. Kervaire and J. Milnor, Groups of homotopy spheres I, *Ann. of Math.* **77** (1963), pp. 504–537.
11. B. Mazur, On embeddings of spheres, *BAMS* **65** (1959), pp. 59–65.
12. B. Mazur, A note on some contractible 4-manifolds, *Ann. of Math.* **73** (1961), pp. 221–228.
13. E.E. Moise, Affine structures in 3-manifolds V, the triangulation theorem and Hauptvermutung, *Ann. of Math.* **55** (1952), pp. 96–114.
14. J. Morgan, The Seiberg-Witten Equations and applications to the topology of smooth four-manifolds, *Mathematical Notes* **44**, Princeton (1996).
15. J. Morgan and G. Tian, *Ricci flow and the Poincaré Conjecture*, Clay Math. Monographs **3**, AMS (2007).
16. V. Poénaru, Les décompositions de l'hypercube en produit topologique, *Bull. Soc. Math. de France* **88** (1960), pp. 113–129.
17. V. Poénaru, Processus infinis et conjecture de Poincaré en dimension trois, IV: Le théorème de non-sauvagerie lisse (The smooth tameness theorem) Part A, *Prépublications Orsay* **93–83** (1993), Part B, *Prépublications Orsay* **95–33** (1995).
18. V. Poénaru, Geometric Simple Connectivity and Low-Dimentional Topology, *Proceedings Steklov Inst. of Math.* **247** (2004), pp. 195–208.
19. V. Poénaru, What is ... an infinite swindle, *Notices AMS*, Vol. 54, n° 5, May 2007, pp. 619–622.
20. V. Poénaru, Geometric Simple Connectivity and finitely presented groups, arXiv:1404.4283 (math G.T.) (2014).
21. V. Poénaru and C. Tanasi, Introduzione alla geometria e alla topologia dei Campi di Yang-Mills, *Sopplemento dei Rendicondiconti del Circolo Matematico di Palermo*, S.2, N.13 (1986), pp. 1–55.
22. V. Poénaru and C. Tanasi, Representations of the Whitehead Manifold Wh^3 and Julia Sets, *Ann. Toulouse*, Vol. IV, n° 3 (1995), pp. 665–694.
23. S. Smale, On the Struture of Manifolds, *Amer. J. of Math.* **84** (1962), pp. 387–399.
24. S. Smale, The Story of the Higher Dimensional Poincaré Conjecture (What actually happened on the beaches of Rio). *Mathematical Intelligencer*, Vol. 12, n° 2 (1990).
25. J. Stallings, The piece-wise linear structure of Euclidean spaces, *Proc. Cambr. Math. Soc.* **58** (1962), pp. 481–488.
26. W. Thurston, *The geometry and topology of 3-manifolds*, Lecture Notes, Princeton University (1980).
27. C.T.C. Wall, Classifications of handlebodies, *Topology* **2** (1963), pp. 263–272.
28. E.C. Zeeman, The Poincaré Conjecture for $n \geq 5$, in *Topology of 3-manifolds and related topics*, M.K. Ford ed., Prentice-Hall (1962).

Chapter 19
Memories from My Former Life: The Making of a Mathematician

Valentin Poénaru

To Barry and Gretchen

I will give you now some snapshots from my former life, which ended very suddenly the 14th of August 1962 at 6 am.

As a very young adolescent, my first big interest was Philosophy, and my hero was Immanuel Kant, the Kant of the Critique of Pure Reason, but I also had a big interest in Ludwig Wittgenstein and in the Vienna Circle, Der Wiener Kreiss.

What for some might be the Bible, but that book I have never read, was for me in those very youthful years *Die Kritik der reiner Vernunft* and the *Tractatus*. And once we are talking about these kind of things, I maybe should tell you right now that I do not believe in God, more precisely, that I fail to understand what the whole question of His existence or nonexistence might mean. But then I do believe that mystical activity, that totally individualistic quest for transcendence, is something objective, on par with other objective things. But I should leave this topic for another time.

Pretty soon I got the idea that in order to start any serious philosophical meditations, I better understand how the world functions. And as far as that functioning of the world was concerned, for me it meant Physics. I can muse now in retrospect, why life sciences were not included here. I guess that, at least in those old days, I thought that life was an accident, while physics meant universality. And as far as human society and its organization were concerned, that might well be important for our everyday life, but so is good plumbing too, and it never counted among the topics of higher intellectual value which I cherished.

I may have slightly mollified some of these ideas, later on in my second life, then that universality of physics may have come into some questioning. For instance, our Universe may well be only one of many possible ones, all somehow existing out there, more or less independently of each other, in particular with different physical laws. For quite some time, in the past, I had believed that there was one and only one mathematically consistent system of equations for the world, waiting

V. Poénaru (✉)
Université Paris Sud-Orsay, Orsay, France

S. G. Dani, A. Papadopoulos (eds.), *Geometry in History*,
https://doi.org/10.1007/978-3-030-13609-3_19

to be discovered. In my second life, I have even worked with some physicist friends, for some years, on a small piece of those so elusive equations, but that is another story. Anyway, I now no longer think that this sacrosanct uniqueness makes that good sense. And I also realize now that, after all, the brain can think mathematics and understand the world.

The fact still remains that, even today, a Black Hole fascinates me much more than the most exquisitely beautiful animal or plant. It would not be appropriate to explain you here the rather technical reasons that make Black Holes so exciting for me. Besides, at the time when the present story unfolds, nobody had thought seriously about them, they were still in limbo for years to come.

Coming back to those really old days, I soon came to realize that one cannot get anywhere far in Physics without a solid amount of Mathematics. I really hated the silly maths which the school tried to teach me, I was totally unmotivated as far as it was concerned, and I am unable to function without motivation. But then, at the age of about eleven, I started to learn more serious Mathematics, all by my own, with only occasional bits of help from people who knew more.

You understand that for expository reasons which should be obvious, I am telling you here, in a linear order, various things which happened to me, all more or less simultaneously, mixed with each other and intermingled, during a certain span of time, between the ages of 11 and 14 or maybe 15, in one big lump.

So I successively learned Algebra, Analytic Geometry and Trigonometry, and then, at the age of about 13–14, Calculus. I got so excited when I understood Calculus that, I immediately started teaching it to my friend and classmate Gussi. We were sharing a double bench in our Romanian high-school class of fifty or so pupils, and our bench was located in the most remote corner of the classroom, so that we should not be disturbed by our teacher's voices, we liked to do our own business in peace. This way, Gussi also learned Calculus, he later became a professional mathematician, and as you shall see we even wrote quite a number of mathematical papers together. He actually ended his career as director of the Mathematical Institute in Bucharest, but that came after Ceauçescu's fall.

Our activity was not restricted to Maths or Physics (and Physics in those old days meant for us Newton, Lagrange and Maxwell) or Philosophy (this was in my case); about four times a week we left surreptitiously our high school, after having socialized with the other chums, and went to the boxing school instead. Our classes started at about 2 pm, but at 4 pm we jumped over the fence around the school yard, and went straight to the boxing school. There we were together with professional boxers too, and the training was tough. But after a few years of that, we both became very athletic, we were the strongest guys in our school. Sometimes it so happened that we quarreled and that we started fighting. His game was usually to try to keep me at a distance with his fists, while my game was to try to get through that, and close in on him, because I knew that if I had managed that, then I had won. The other pupils gathered in a big circle around us, like for a circus show. And nobody in that school ever dared challenge us.

There was another boy in our class, Michael, a small frail but very brilliant guy, who would later become a well-known stage director at the Bucharest theater. He had the idea that if we all, the whole class, joined together the Communist Youth organization we would be safe with the Communist system.

It was a serious thing, but a bit of a joke too. You have to understand that communism is a very verbose thing. Using the politically correct word in the correct place may have meant the difference between being highly praised for your good knowledge of marxism-leninism, as this was in my case; after all I had a philosophical background and all that was very easy for me, or when you did not manage to find those correct words or, even worse, you came out with something heretical, then you might even have had to face jail. So, quite naturally, in my high-school I was the propagit, which in communist jargon means the person in charge with propaganda and agitation. That is how it was, in those old days.

I have just told you how I learnt calculus at the age of about thirteen or fourteen. It so happens that there were some twelve boys, of the same age, in that country, who like me did the same. Their motivation might have been physics, like in my case, but not always. I will call them from now on "the bright boys" and they, or at least most of them, will appear big in this story of mine. And soon the bright boys got to know about each other's existence too.

This is a good time to take a break and make a parenthesis. Karl Marx once said that people do not understand the history they live through, and this is damn true. I have lived through this sudden burst of youthful mathematical enthusiasm, in a scientifically underdeveloped Balkan country, and I still do not fully understand how that could happen. And after my own crop of bright boys, the next generation in Romania came with a new crop, and the process was many times renewed. How was that possible? I have, of course some conjectures about that, having to do, for instance, with a transfer of interest from humanities, which was rather strong before the war, and then made impossible by communism, to something equally lofty, which was possible to do now. But I cannot see very well how, for such a conjecture one could find any shred of factual evidence. History, which after maths, physics and philosophy, is my next big topic of interest, is clearly a very difficult subject. When I have such a hard time in getting some understanding for events in my own life, then imagine how painstaking it should be to disentangle the more distant past. I love History, but I never could have become a real historian trying to prune through the evidence, written or other, which the past has left us. I certainly do not like old archives, that is certainly not something for me.

And then also, at least when talking about one's own past history, so often did I talk with some friend about some past event we shared, and more than often our memories did not quite match, sometimes ridiculously badly so. I strongly suspect that this same kind of mechanism concerns the more global history too.

Finally also, history can and should be done at various levels, at various scales of magnification, in time and/or in field of activity, a bit like the renormalization group, if you see what I mean. For instance, I am very fascinated by the history of recent or very recent Science, and the only sciences I know, are of course maths and physics. Just an example. A few hours before flying to the South Pacific I heard that Bill Thurston had died, I do not even know the details, here and now, in Fiji. Bill has left us a magnificent mathematical legacy but as for right now, large chunks of it only survive in the minds of a few of us. We should better hurry to do something about this, because when we will be gone it will be gone too. I know that various people will re-work and re-write things, make them rigorous by other means, and so on and so forth. But then something of the fantastic potentiality which is there in Bill's work, as it stands, will be lost in the process, maybe forever.

But I should not loose track of my own story, and I go back to it now. So, time went by and I had to start thinking about the University. When my parents heard that what I was planning to do was to study maths and physics or physics and maths, they got very worried. Very gently, my father told me that this was certainly the surest road to starvation. He thought there were only three reasonable lines of University studies with which one could make, afterwards, a comfortable living. These were medicine, law, and engineering, which included architecture too. How little did he grasp what was going on in that country of ours. Lawyers were already pariahs in the communist society, and medical doctors were to follow them in less than 10 years time. My father's dream was that I should become a dentist, like himself, and eventually take over his medical practice. I believe that this would have been about as unsuitable for me as it could have been, and with some sadness my father understood that too. I shudder even now, with horror, thinking of what would have happened if, instead of being myself the way I am, I would have been a traditionalist, would have followed in the steps of my father, and then today, if still alive, I would be an old sedated dentist, clearly feeling like a total failure, living in some dark God-forgotten corner of that Balkan country where I happened to have been born. But tradition, of any kind, is not my thing. I am sure, of course, that in certain appropriate conditions and for the appropriate people, tradition may be a good thing, but in my specific case it would clearly have meant sheer disaster. One cannot stick to tradition when the world around you is in upheaval, nor when your own ideals are too far removed from those of your parents or ancestors.

Also, when my father wanted me to become a dentist, little did he fancy, at that stage, that private medical practice was going to be abolished, like so many other things. And when, later on, he understood that, he died with a broken heart.

With these kind of thoughts being around, only about half of the bright boys went in for science. All of them went in for maths, except one who became a physicist. As I soon found out myself, maths studies in Romania were quite underdeveloped, but physics studies were even more so.

But then also, about half of the bright boys went into engineering. I met at least some of them, a few years later. They were by then broken and disillusioned. The bright light in their eyes was gone, their youthful ideals were shattered, and they were busy counting sacs of cement or doing other similarly exciting tasks. A slightly older mathematician, Ganea, a very close friend of mine, had been obliged by his parents to study engineering before going fully into maths. And he told me quite gory stories about that. Some examples:

Exam question: How does one proceed for paving a road?
Expected answer: With care.

Then, the students ask their professor questions about some more interesting topics. The professor says to them: You do not have to worry about these things, which are not for you. If ever such things become actual, then one will call an engineer from Germany to take over. Beautiful, isn't it? Ganea, the man with the blackest, often self-critical sense of humor which I ever knew, will occur big again in this story.

Few months before actually entering the University, in June 1951, I went there to attend a series of lectures on Abstract Algebra, "Rings and Ideals" it was called, geared for mathematically minded young people, and given by Moisil, a university professor who was soon going to play a big role in my life. Just before the first lecture started, another student, a tall intense looking young man with bushy eyebrows, came to me saying: I know who you are and I want us to become friends. So we did, and we stayed friends forever after. Sorin, the young man in question, is that unique one among the bright boys who became a physicist. Also, about a year after we first met, he introduced me to the mountains, which became then so important for me. In his later life, after many years spent at Trinity College in Dublin, Sorin finished his career as a professor at the University of Montpellier. He is a regular visitor of the CERN in Geneva, and brings me news from there. But this is no longer part of this story.

The first semester at the university was common for maths, physics and for training future high-school teachers. The three lines separated at the end of the semester. One of the first new fellow students which I met, was Samy Z., a very brilliant and quite aggressive guy. He brushed away my fears of not having gone into engineering and started to feed me with little problems to solve. He was more advanced than I was then, in point set topology and real variables. Every time he served me some little Fundamenta type problem, he told me that I was an idiot if I could not prove that. You might remember that Fundamenta is the Polish maths journal, famous between the two wars.

This changed my life. Instead of taking the overcrowded tram, I started to walk to the University and back home, a total trip of about an hour and a half. And while walking, I thought about Samy's problems, this taught me to work without paper and pencil, all in the head. I always managed to prove what Samy asked me to prove, and I remember only one of his little riddles, which took me more than the walk home to do. I thought about it during a good part of that afternoon, in the middle of a lively and noisy party, until I solved it. It went about like this. Try to cram in a plane as many two by two disjoint topological copies of the figure Y as you can; it is claimed that you cannot put more than countably many.

About a year later, Samy and I had our first mathematical paper published, a joint piece of work on Fundamenta type arguments; that was in the very beginning of 1953.

So now our whole little gang of bright boys, we found ourselves first year students, actually in the first semester, and all the courses were very lousy. Then we decided to attend some fourth year courses, the closest thing to a graduate school in Bucharest, in those days. I attended a course in set theory, à la Cantor, and another one in algebraic topology. To this one was also attached a little appendix, on group theory. The professor for these two joined courses was the same Moisil already mentioned.

Moisil was brilliantly clever, he had a rather broad knowledge of many mathematical topics and his main field was mathematical logic. But, at the same time he was quite superficial, in fact his knowledge was all quite shallow. He had a Balkan mentality and Balkan ambitions (avoid carefully the difficult deep questions and, instead, enjoy toying with those harmless and easy ones). He was also sparklingly witty and praised a nicely coined witty sentence above all things. For the sake of making a nice pun, he was willing to create for himself a lifelong enemy, and he had many such. Here is one of his many favorite aphorisms: Do not trust a man who neither smokes, nor drinks, nor has plenty of affairs with women, because only God knows what kind of secret tastes he might hide. Moisil himself certainly indulged big in the three pleasures above. He also told me once that, as a young man he went to a lot of parties, and in order to sharpen his mind, so as to be able to put up that brilliant conversation of his, just before the party he would read some G.B. Shaw; he trained his brain like for a sporting event of sorts.

He also loved to go to restaurants, surrounded by a crowd of followers; I soon became one of the youngest of them. Everybody paid for his or her food, and Moisil paid the wine for everybody, and plenty of wine there always was. His table talk was pure pleasure to listen to. With his round bald Churchillian head, his clever eyes and his deep voice he was quite a character. And the town was full of funny little stories concerning him, I guess some true and some not.

Between the two wars, in a country dominated by the extreme right, Moisil had been a well-known left-wing university professor, and he had been targeted by the Iron Guard, the violently vicious Romanian fascist party. So now he was an important member of the Communist party, with a lot of power. But Moisil was nice and kind and he helped me a lot, in difficult times; and he helped many other people too. He used his power to be good. He would soon become my boss, and one could not survive in Romania, in those days, without one.

Moisil's algebraic topology course was flying very low, and it was highly non-inspiring. In his little group theory course he rather assigned some research papers to various graduate students to read and report about. So, 1 day I learned about an item which was supposed to be next week's topic. The issue was this. If a real function respects addition and if it is also continuous, then quite trivially it has to be linear homogenous. But what I heard now was that there existed also discontinuous solutions for the corresponding functional equation. That was next week's topic, and a graduate student, who was soon to become a friend of mine, was reading a paper about it, preparing to make a report. But this was good food for my mathematical thinking walks, and before the week was over, I knew how to construct all those discontinuous solutions. I had never been exposed to linear algebra so I recreated from scratch everything which was necessary now, looking at something which I later recognized as being the vector space of the reals over the rationals, but that language was then unknown to me. But that did not matter; on the other hand, what I did know quite well, were transfinite ordinals, which were certainly needed for the construction. I was never to be exposed to that linear algebra, but some 2 years later I had become an expert on Hilbert and Banach spaces and I had also created my own private brand of theory of distributions. And so the standard linear algebra looked to me like a silly childish game, really Kindergarten stuff. I happily skipped the University courses on the topic, like I skipped most of them, as you shall see.

Anyway, when that next week came, I reported on my solution in the group theory class. There were two consequences of this, firstly I decided to go in for math the second term and not for physics, as Sorin was pressing me to do. So, it was a big forking point for me. Secondly, Moisil did then become my boss and me his protégé. And some 6 months later, when I risked to be expelled from the university for political reasons, his protective shield saved me.

In my high-school days, the communist youth organization was a bit of a joke, but at the university it was a deadly serious affair. Some of my fellow students were like hateful monsters, coming straight out of a book by Dostojevski, or so they seemed to me.

And just before the end of the first academic year, a political earthquake shook us all. I had always, since 1947 when the communists took over, felt Stalin's oppressive presence, physically, as if he were there. Now, in the last year of his life, Stalin got a big access of paranoia and wanted to purge thoroughly the whole communist world, at all levels. At the highest level, that meant a lot of political trials of communist leaders, in various communist countries, accused of this and of that, and more often than not, shot or hanged.

At my more modest level what this meant was that beginning with June 1952, every week there was sort of a political trial, inside the university, in the framework of the communist youth organization. No blood was actually shed, but the punishment was being excluded from that organization, with possible consequences being expelled from the university or even jail. Each of these shows, with its ritual of accusations, confessions and self-vilifications, started at 10 pm and lasted until the early morning. It so happens that at the very first trial already, I tried to defend a friend of mine, claiming that he should not be punished for what his father had done. And, for once actually, the father in question had done something, he had been part of an attempted military coup against the regime, and he was now in prison for life.

The only net result of my attempt to help my friend was that now I was myself targeted for the next trial. I had decided to stay home for the occasion, but they came and fetched me. When my turn came, it was four o'clock in the morning, I was exhausted, and I tried to shorten the procedure by simply returning my little red membership card, after which I left. This was taken very badly, I was declared, in my absence, an enemy of the people and this stuck to my secret police file for many years to come, with all sort of unpleasant consequences.

In those months, during the Summer and Fall of 1952, I lived through a sort of very fast re-enacting of the French Revolution, let us say the French revolution of the years 1791 to 1794, coming with its murderous blend of terror and fanaticism; those two items mix together better than one might think, even inside a single person's mind. No heads were actually rolling, but careers were destroyed and lives got broken. Those who had expelled me from the communist youth were themselves expelled a few weeks later; those who had expelled them were in their turn expelled too, and so it went on and on, until Stalin died, in the early months of 1953. Then, all of a sudden, everything calmed down, the big oppressive pressure had disappeared, and at least for the next few years one could start breathing a bit more freely.

My daily routine had also changed. I was now a second year student and, in the morning, when I would get close to the university, then I rarely went in, normally I would go straight to the next café, where I knew I would find my chums, the bright boys. With two exceptions, Complex Analysis and Lebesgue theory, all the courses were really bad, and we had decided to skip them all, except for those two.

We learned mathematics from each other, in the long and lively café discussions. We talked loud, we seldom waited for the other person to finish his sentence, but in those immensely intense exchanges of ideas, we grew up to become mathematicians. Of course, each of us also read mathematics on his own and thought on his own, but the two kinds of activities, alone or together with the group, complemented each other quite well. They were both necessary.

Our professors were so much out of touch with what was going on, mathematically speaking, that our discussions would have completely passed over their heads. We learned very little from them and had to do it all by ourselves. My boss Moisil being a logician, I still had heard about Gödel's incompleteness and about Turing machines, but I had to learn all that by myself. He certainly encouraged me to do that, and he was very happy when I lectured about these things in his seminar. For somebody with my background it was only natural that I should pick up these topics. But the other bright boys did not seem very interested in those things, so there was nobody with whom I could really talk about them, in that former life of mine.

Later on, in the second life, I was sometimes dreaming of what it must have meant to live in a mathematical climate, where let's say Hodge combined classical algebraic geometry with the electro-magnetism of Maxwell, about which he had first hand knowledge, where Atiyah became a student of Hodge and where in his own turn, Donaldson became a student of Atiyah. We of course, did not have access to such niceties, and we had to try to do the best we could with what we had. And each other's company was our most precious asset; we had no other.

About the end of 1952, beginning of 1953, Foias, Gussi and myself started a big piece of joint work, more than fifteen published papers, in all, on nonlinear partial differential equations. We certainly were not doing fore-front mathematics, we were after all still apprentices, but in the rather modest niche inside which we were operating, our team of three was quite respected by the colleagues in the US, in France, in Russia, or in Italy. Some of our correspondents and competitors were rather well-known mathematicians with high academic positions, while we were still second or third year undergraduate students. But at least as far as Foias and myself were concerned, we were after much bigger game than this.

I became very close friend with Foias during the Fall of 1952. He was coming from a small Transylvanian town and, in the beginning he felt very lost in the big capital, Bucharest. He was immensely proud of his family. His father had been the main medical doctor in the little town, and also its liberal mayor; the Liberals were one of the old political parties in Romania.

But in the stormy Fall of 1940 the Iron Guard took complete control of the country, for a while. The Iron Guard, the Romanian fascist party, was the most viciously murderous fascist party in Europe, according to most historians. Even the nazis thought of the Iron Guard as being too extreme and, somebody like let us say Mussolini and his gang of thugs, were nice little angels compared with them. They had their own brand of very chauvinistic ultra-nationalism, their very ugly form of antisemitism, dark orthodox religious creeds mixed with archaic rituals, and they were also violently anti-capitalistic; their word for capitalism was plutocracy, something which they thought was a Jewish avatar. They also thought that all town people were corrupt and that only the peasants were the pure, true children of God. When they did their bloody murders, they always had the name of God on their lips.

The liberals were anathema for the Iron Guard, and 1 day they stepped in through the main door of Foias's parents house, in order to shoot Foias's father. But Foias's father fled through the back door, jumped on his already prepared fast horse, and rode away to safety. By then also, Hitler had decided that the Iron Guard were really too crazy and he had the Romanian army, under Antonescu, the new boss of the country, put them down, temporarily. He always thought that they might come in handy for him later on.

Anyway, Foias's father now was safe, for a while. When the communists came to power in 1946–1947 he went to jail, where he stayed a long time, because he had been a Liberal mayor.

And then, in 1952 Foias's magnificent family castle collapsed, his beautiful mother ran away with a younger man. Foias was shattered, but he became a different, stronger person, in the process. His mother's eloping was a big scandal in the little town, and her parents did not want to see her, nor hear from her, any longer. And it was young Foias who managed to make peace between his grand-parents and their daughter, his mother. And he also stayed close to both of his, by now completely estranged parents.

Foias was a superb mathematician, and a few years later he started doing magnificent work in the theory of single operators in Hilbert space (he was never an operator algebra man) and on the Navier-Stokes equations of fluid dynamics. He was fascinated by the mathematics of turbulence.

My old friend Gussi, my very first pupil, and the third man in our mathematical trio, came from a very aristocratic and rich originally Greek family; actually a very large part of the high middle class in Romania was of Greek origin. Gussi's father, as a young ambassador to Spain, in the early forties, was the first ambassador since the times of Philip the second, to go down in the arena and fight the bull, like a torrero. He was still looking very youthful when Gussi and me we were in high-school and people usually thought he was the older brother. He was notoriously gay too. In 1947, when the communists were in power, he had a brilliant idea: he gave up all his positions, his fortune was already in the process of being confiscated, and he became a factory worker, which both saved him from jail, and also made that Gussi, now the son of a factory worker, was safe, for a while.

Among the three of us, Gussi was the best in gauging what was important in mathematics. He had an excellent mathematical taste and a very good nose for those important things. His advices have often been very useful to me, in my later career.

But Gussi also had a big problem. He never produced any mathematics, except for the joint work of our trio, nothing ever of his own, alone. One reason for this is simple and sad. He wanted a lot of things, like a motorcycle and other gadgets. For this he needed money and his salary was certainly not enough. So he gave private lessons, made translations and a lot of other such extra jobs. But then he never had the leisure, the time, nor the energy to do maths, other than reading and staying well informed about what was going on. But he did not do anything creative, and by the time he understood what was happening to him, it was too late. There is a second reason too: the very unpleasant conditions in which he and his parents were living, completely discouraged him.

The three of us also loved to do sports, we could not have lived without that. So, Gussi went big into rugby and he became one of the two best half-backs in the country. But let me tell you a little story which may amuse you. Gussi liked to use what he had learned in our boxing school in his normal life, when he thought it was necessary. And on his way to and from the rugby field, he had to go through some very ill-famed streets of Bucharest. He told me that there, if someone asked you the hour, you better hit him hard and fast, before he had the chance of doing that to you. After that, you might even tell him what time it was. That is how some sections of Bucharest used to be, and maybe still are. Actually, not so many years ago Jean-Pierre Serre was almost torn to pieces by a pack of wild, vagabond dogs without master, in the streets of that old town of mine.

Foias and myself went in for another sport. We ganged up with another bright boy, Igor, and as a new trio we did a lot of mountain climbing, actually rock-climbing together, and we did it big. You will hear much more about that, later on in this story. But I close now the long parenthesis, and go back to the mathematics of our trio.

Foias, Gussi and myself were deeply involved in functional analysis too. But, although I knew that in certain aspects of Hilbert space, for instance, I could navigate better than anybody else, I felt that all this was not really my road. My friend Foias, for instance, had started to play the following mathematical game, where I felt I could not be his equal. I will remind you first a Kindergarten fact. If T is an operator and $f(z)$ is a holomorphic function on the spectrum of T, then $f(T)$ makes good sense, and this is a nice useful little fact. Now, Foias had a very clever and slick way to make sense, under certain conditions, of $f(T)$ when $f(z)$ had singularities on the spectrum. And that, when it worked, had big consequences.

So, by 1955, while I was still a student in the last year at the university, I moved from analysis to algebraic topology, not like in the old Moisil lectures, but now for serious. And in those days I was completely lonely in Romania, in that area and I had nobody to talk to, about my new field of interest. Only a bit later did others join in, which made me quite happy. It did not take me long to discover that there was, in topology, a big open problem, which I decided was my kind of thing, the Poincaré Conjecture. And It turned out to be one of the most important mathematical problems that there was. And I started some very naive, clumsy first steps towards it.

When the Summer 1955 came, I had my university degree and I had a job too. Since my political file was so lousy, the University could not hire me as an assistant, and I only got a lesser job, called preparateur, I think it had no equivalent elsewhere. It implied the same teaching load as an assistant, with a lot more administrative tasks on top, and with only half the salary. In the meanwhile, I had also met Sanda.

She was a student in the Theater School and her boy-friend, Lucian, was an old friend of mine, from the high-school days. He was her fellow-student, and everybody in that Theater school thought that Lucian was a genius. And I think he was a genius. This very intense young man, made later on a number of movies, mostly in English, which were pure masterpieces. Also his Turandot, on the scene of a Paris theater was a historical event. He became one of the big stage directors of his generation.

But in those by now so old days, when I met Sanda, there was big tension in her relation with Lucian and they were moving towards a breaking point. I must confess that I took advantage of that crack in their relation, and started to court Sanda. Of course Lucian came to me and asked me to give up, and of course I promised him to do so, but then I could not keep my promise. By then, Sanda and I were in love. My friendship with Lucian did not survive these tumultuous events.

Sanda and I were now lovers, and I remember a little incident from the very beginning of our love affair. It was a little thing, but it was so full of further meaning, a harbinger of future things to come. Sanda and me we were sitting on a bench in the gardens of the Academy, and soon we noticed that some ten meters away, on the next bench, were sitting Lucian himself, immersed in deep amorous conversation with young Julie, whom I only knew by sight in those old days. And Sanda, rather jokingly, maybe not a very good taste joke, said to me: why don't you go and seduce her. Little did she know, and little did I know then, what Fate had in store for Julie and myself. This little incident must have happened in the Spring of 1955.

At the beginning of the next year, although our love was shining somehow less bright, Sanda and I decided to get married and we announced the news to our families, which then met. That was not a big success. My future mother-in-law started by saying that I was too young, while my father, after taking a glance at that prospective mother-in-law of mine, decided that he did not like the whole set-up. The only thing on which the two families could agree, was that they both opposed the marriage.

Now, we were both twenty-three, we ourselves had started to believe that the marriage idea was maybe not such a great one, but a parental opposition was a challenge which we felt had to be met, our honor was at stake. And so we did get married.

But let me come back, for an instant, to that business of being too young. Not so many years later, in my next life, another mother-in-law declared me to be too old. So, in a very short span of time, I must have gone from extreme youth to old age, without ever becoming an adult.

Be that as it might be, the marriage with Sanda was a big failure, and it is not hard to give at least a couple of good reasons why that was so. We did not like the same things, we did not like the same people, and above all, there was hardly anything which we enjoyed doing together. I think we were both nice and clever people, but we did not really fit. The love was not yet dead, but that was not enough.

I was living now in the world of the theater, which I got to know from the inside, and that had its own fascination. My wife was a stage director and an actress too, not at all easy to do both those things at the same time, but she was both gifted and had a lot of energy. My father-in-law was a famous actor. He was also a bit of a poet, he wrote biting little epigrams, which created many enemies for him, and which did not always do him a lot of good.

I maybe should have said all through here, theater and movie world, but the actor takes his or her pleasure in the theater, through the almost physical contact with the public, rather than in the movies, although that brings more fame and more money. An actor's job, which he or she does using the glandular system, or the brain, or a mixture of both, is to pass a certain kind of magical, invisible fluid to the public in front and make it vibrate.

For this magic to work, a certain kind of tension inside the actor's mind is necessary. It is a tension between some sort of shyness or inhibition, and the quite opposite urge to shout out loud one's strong emotions. A perfect balance of these two ingredients is necessary here.

And then, like in so many other things, there is also a question of both inborn talent and acquired technique. The first girl I was ever in love with, while I was still a teenager, another Sanda, also went to the Theater school. And at her first test, everybody thought she had a fantastic talent. So she might have had, but she never learned anything more, never developed, and she became a third class actress, with pitiful little roles.

Understanding all these things about how an actor functions, in relation to the public, was rather useful in my next life, when lecturing or teaching.

Through my not so happy years of marriage, relatively short years actually, from May 1956 to September 1958, I continued to work hard on the Poincaré conjecture And, quite naively I might add, I thought I had managed to prove it, by the Summer of 1957.

I wrote a longish paper, had it typed and then sent it to Georges de Rham in Switzerland. In those days, when I was writing to Georges I addressed him as "Très honoré Maître". It so happens, that not so many years later, but that was happening now in my second life, we became very close friends, although Georges was about 30 years older than I was. And, notwithstanding the fact that what will come right next is really part of my second, future life, I still feel like recording it here.

At the time when our friendship started, Georges was a big professor at the universities of Geneva and Lausanne, President of the International Mathematical Union, while I was just a young refugee. But he had a high opinion about what I had done by then in mathematics (since 1959), and he was very fond of me. He helped me a lot in my second life.

Georges was not only a great mathematician, he was also one of the big mountain climbers of his generation. He climbed during all his life, he feasted his 60th birthday by leading a very difficult climb in the Mont Blanc, and he kept doing hard rock-climbing into his very late sixties, or even early seventies. Since I was myself a very passionate climber, you will hear more about that later, Georges and I did a lot of rock-climbing together, in the Swiss and the French Alps.

And we talked a lot together, about mathematics and everything, in his palatial home in Lausanne with the sumptuous living-room decorated by his ancestor's armor plates, and by medieval cannon, but we also talked in various cafés of Lausanne or Geneva. Georges loved good wines and, as one says in French, he was a fin gourmet. I can still remember the sunny day, years ago, but still well inside my second life, when Henri Cartan, who knew that Georges and I were friends and who liked us both quite a lot, invited us for a beautiful lunch, in one of the best restaurants in Strasbourg, where the three of us had quite a good time together.

I learned a lot from Georges, both in mathematics, and about many other things. It was him, for instance, who introduced me to the Whitehead manifold, which later in my life was always in the back of my mind, as a mathematical scarecrow.

Georges had a young Polish girl-friend, she must have been about my age, but later on she died in a mountain accident.

One day, in one of the chats with Georges, I discovered something which amused me a lot and which I feel like recording here. I must start by telling you that, years before, as a child, I had been often with my rich godparents. My godmother was my mother's older sister, while her husband, my godfather, was a first cousin of her mother, my maternal grandmother. This was about as close as one was allowed to be, blood-wise, and still get married.

They did not have children of their own, and they treated me and my cousin Liliana, the daughter of my mother's younger sister, as their own children. Liliana was the same age as me and for years she was like my sister. I am sad that the stormy sea of life made that, later on, we became complete strangers. But that is too long a story for right now.

As a child, I spent the first half of each Summer, in the mansion of my godparents, on their estate in the country side, and then the second half in their very large villa, in a beautiful mountain resort.

And close to my godparents estate, in the next village, there was another, even more magnificent mansion. Talking 1 day with Georges, about this and that, I discovered that the magnificent mansion in question had belonged to his, by then dead, brother, who had had a Romanian wife. The world is small, indeed. It really sounds odd that a Swiss Marquess, member of the highly aristocratic HSP (Haute Société Protestante) should own an estate in Romania, neighboring the estate of my own godparents. And that his brother later became a good friend of mine, isn't all that very strange?

But please do not get things wrong, at this point. Georges himself was deeply democratic, in the noblest sense of the word, his two usual climbing mates and close friends, with whom I climbed myself often too, were Carlo, a gardener, employee of the Lausanne township and Apo, a railway worker. Georges was really a wonderful person. Except for its very beginning this long prentice about him really concerns my second life; but I felt like telling these things here. I move now back to my main story, where I had left it.

1958 was the black year of my life. One February night, in my dreams, I saw a sort of a mathematical monster, which I immediately understood to mean the complete collapse of my childish Poincaré proof. And the monster was certainly there, for good. Of course, I immediately informed de Rham, he was not yet Georges to me, about the disaster.

In April, my father died, and this was the single biggest blow which ever befell on me. I might not always have agreed with my father, but our love bond was of the strongest possible kind there may be. It took me years to get over my sorrow.

By then also, the relative liberalization which had followed Stalin's death came to an end, and as an aftermath of the crushing of Budapest by Soviet tanks in 1956, the communist system had become very tough again. And nasty rumors concerning what was going to happen next at the university, were in the air. And indeed, one black Friday, in the beginning of September 1958, a list of about thirty names was posted on the door of the university. And the people on that list were supposed to report the next Monday at human resources; in Romania that meant a branch of the secret police, the Securitate, located inside the university. It did not feel good at all to find myself on that list. About half of it were Jews, my friend Samy was included, the other half were people with various other political sins, like myself.

That next Monday, we were each informed in turn, that we were dismissed from our jobs at the university, each with his or her precise reason. Samy was supposedly a Zionist, and I was an enemy of the people.

Poor Samy! Every year we were given some forms to fill in, from the administration of the university. Among other things, there were two items to be filled in: citizenship and nationality, which in communist countries did not mean the same thing. I wrote "Romanian" in both places, but when Samy did the same, the forms were invariably returned to him with "Jewish nationality" added in red and "Romanian nationality" crossed out.

Samy managed to leave Romania and landed first in Italy, where he got some university degree and an Italian wife too. "Una magnifica ragazza", he wrote to me. But when he married, he had to promise that his children would be raised in the Roman-Catholic religion. So, Samy might well have been born in the ghetto, his first religion might well have been marxism, he ended up as father of six Catholic children. He also became a specialist in decoding the Papal texts, trying to find loop-holes for contraception. He is now professor at the Université de Montréal, in Canada and, these days, when we meet, we tend to speak Italian together, rather than Romanian. And he is not the only person from the old country with whom I use now that language.

When I became jobless, in September 1958, my already shaky marriage collapsed completely. Sanda and I decided to divorce; that took about 10 days and costed something like the equivalent of four or five euros. That is how it was in Romania in those days. It was a completely friendly affair, and we did stay friends and chatted from time to time about our respective problems; you will hear more about that.

Just before anybody else would have grabbed it, I moved fast into my father's former study, in my mother's flat.

Notwithstanding the situation, life felt now better than before. While I was married, I had often felt very lonely, my by now ex-wife was often until late in the theater, and we had no common friends. Now I felt like a free man, many of my various inhibitions had vanished, I became a very good dancer, something which I had never been before, and it looked as if a lot of girls wanted to see me. For the first time in my life, I started moving around feeling free with respect to obligations, social conventions, and all that. I was stepping now firmly on the ground, doing what I wanted to do and not what other people wanted me to do. All this was new for me and it was clearly a liberation.

But it was only a foretaste of that really big liberation, which came with my second life, when I felt like a wild stallion who had escaped from the stable, and was now running madly free through the big open fields. By then I was also loose from all those so many mental ropes which earlier had held my elbows back, preventing me from acting freely. The influence of my good friend Barry Mazur was certainly very important here too, but all this is certainly no longer part of the present story of my earlier life.

All this having been said, I still had to earn a living, in that Fall of 1958. So I started to give private lessons of maths. This was quite illegal and rather risky, but I earned now substantially more than my earlier meager university salary. Of course, some time and energy had to be spent on those lessons too. Of course also, according to the law I had to stand in line, periodically, at the unemployment office, and declare myself as a job-seeking person. For some reason, one had to start queuing there at 4 am. And indeed, I got some offers for various jobs: to sell apples in a state-owned supermarket, or to go and teach small children in a God-forgotten village in Northern Romania. I always politely refused these job offers and chose instead to stay with my black market activity of giving private lessons. This was the only way by which I could survive as a mathematician, and I have never thought that my own life was worth living without mathematics. The price was the risk involved, of course.

Through all these upheavals, totally oblivious of the, by a large, hostile outside world, I moved inside my mathematical universe, and I continued to work hard, in all my available time and with all my available energy. My favorite activity consisted in long walks through the parks, during dusk and afterwards, thinking deeply about mathematics, under the starry sky. And since my February Poincaré disaster, I had learned by now a lot of mathematics, mostly by banging my head against the walls, so to say. Sometimes I had to recreate, from scratch, various mathematical theories which the Bucharest libraries did not give me access to.

And I also felt now that at least inside the field which I had chosen, I had become an adult, seasoned mathematician and that I was no longer a beginner.

During my long mathematical walks I discovered that the monster which had killed my childish Poincaré attempt, was a mathematical object of some interest, behind which lay hidden a little treasury. So, I started digging and, by May–June

1959, I made a discovery: I found a smooth four-manifold which was a non-trivial factor of the five-dimensional cube.

I was jobless, but I had access to the outgoing mail-box of the Mathematical Institute. This was a big courtesy of its director Stoilov, my former teacher in complex analysis. So, I wrote a paper on that nontrivial factor, and I sent it to Jean Leray. He decided that this was indeed, an important thing, and with the expert advice of R.H. Bing from Texas, he thoroughly checked my paper, helped me re-write it completely, and had it published in France. When that paper appeared, it created quite a commotion in the mathematical world and I may say that, ever since, my road in life was open and smooth, and that I never had problems any more.

That paper had saved my life. I had a big admiration for Leray, reason for which I had sent him my paper, to begin with. And then, more technically speaking, as I saw things then, what I had done had some connection with things which he himself had done earlier. He helped me a lot and I am sad to say that, later on, our relations became ice-cold. But this was in my other life, and it is not part of this story, any longer.

I was not alone in my discovery. Barry Mazur, who was already a big mathematician did, independently and at the same time as myself, the same discovery. So we started to write to each other, and soon we got to meet and became best friends.

But in September 1959, I was still jobless. My paper had not yet seen the light of day, it had not yet made its impact, and my boss Moisil thought that if I did not fast get a job, some job, any job, then I would get into trouble with the authorities. After all, I was now a black marketeerer who made his living by performing an illegal activity, giving private lessons in maths. So, at the end of September 1959, I became a janitor at the Romanian Mathematical Society, an organization located inside the university, but catering for high-school teachers. Moisil was the big super-duper boss of that Society, that is how I got the job.

Never mind now the exact chronology, but several interesting things happened during my 13 months of joblessness or soon after, when I was a janitor. I feel I have to record them here.

Once I was a free man, free I really was, and I started to go every weekend, and in addition during long periods in Summer or Winter, to the mountains. And, as I have already told you earlier, we were three friends doing this together, Foias, Igor and myself. We went together rock-climbing and we did it the big way. I think rock-climbing, if one loves it, is one of the basic pleasures which a human being can have. If you hang on a vertical wall, on one finger and one toe, with many hundreds of meters of void behind you, if you still manage to climb higher, and if you do not care a damn about the danger, then this is like the most divinely delicious strong drink, the taste of which you can never forget, afterwards.

Quite a long time ago, the great Italian climber Emilio Comici, the man who had managed to climb for the first time the Cima Grande di Lavaredo, in the Dolomites, had said: “What for many is Death, for some of us is just a game”.

In my own case, all this came with a side benefit. I used to be quite clumsy. Now, when you hang on that finger and on that toe, then with the other hand you have to do something tricky, like planting a python or mounting a carabine. You better do that fast, you certainly cannot afford to be clumsy, this is now a matter of life and death. And so I learned not to be clumsy, at least when necessary.

The other climbers called us the ones in rags, because we always were in rags. Until 1 day, a state sponsored mountain club came to us and said: Listen guys, why don't you join us, we will give you boots, clothes, new ropes and equipment, food too. We will also pay your train tickets to the mountains. You will continue to do your climbing as you like, but from time to time you will have to give us a hand with the beginners. We did join the club. Until then we had climbed together, and there was always an issue about who should lead. Also, climbing in a team of three is not the most efficient thing. Now we were each of us three, leaders on our own. Each of us had a second, all by himself.

When the difficulty was up to fourth degree included, I always left my second, Sandu B. lead. When the difficulty was five or more, then it was me who was leading. This way he learned things and we were both happy.

Igor lived to become the best climber in the country. His technique was fantastic, he was wiry but very strong, very agile and supple, and then he also had nerves out of steel. He was not only a superb athlete, but also very gifted in many fields. He was certainly a full-fledged member of the group of bright boys already mentioned. It is through him that I got my first exposure both to Lie groups and to quantum field theory. But then also, because of his background of deep poverty and persecution in those very stormy and troubled times of the Romania of our youth, he was somehow a desperate man, who did not care a damn whether next instant he would be still alive or not. And all this allowed him to do things which nobody else could. For instance he did various hard climbs alone, freely and without any balaying, sometimes two the same day, one after the other. And in December 1962, but then I was already away, and this time with a second, he climbed in two successive days and nights, in full winter, the "Fissure Bleue", the hardest climb there was in Romania. It was close to impossible to do, even in Summer, and nobody before had ever dreamt of doing it in winter.

He eventually got back his job at the university, he had lost it at the same time as I did, but then he died in a stupid mountain accident, in February 1963.

The third man in our mountaineering trio, my old friend Foias together with whom I had also done all that maths earlier on, is an American today. He is a professor at a University somewhere in Texas. He is also staunchly Republican, of the most arch-conservative biblical type. Fortunately for our friendship, we see each other not more often than every 5 years or so. Would we live in the same town, I am not so sure how that friendship would fare.

At some time during my jobless period, the phone rang and a young woman informed me that comrade minister So-and-So wanted to talk to me. That was the Minister of Education and I got quite excited, thinking that I was getting my job back. The comrade minister invited me to come and see him, which I immediately did, of course. He received me very friendly and told me: You have been highly recommended to me, and I have a son who needs private maths lessons. The son in question was on the verge of entering the university and I started going once a week to give him a lesson. The comrade minister used to keep me a bit afterwards for snacks, and I think I must have drunk quite a number of bottles of good wine together with the minister. He always asked me very kindly about my affairs, and always wished me good luck. So much for the help. But I was very well paid, and that was that.

Let us stop here for an instant and think a bit. You may have noticed that in this little story which I have just told you, a cabinet minister asks happily and casually somebody to do for him a black market, totally illegal job. I think Romania is about the only place where such things can happen and the same comment may also apply to other instances in this story of mine too.

Some help actually did come, from an unexpected direction. There was a woman, let me call her comrade M., who was a very big shot. She had a young son who had a German nanny, a good friend of my own former German nanny. So, when years before, comrade M. had needed a dentist, she heard about my father, through her nanny, and she came to see him. My father had been a very charming and charismatic person, and I understand that comrade M. liked him quite a lot. When comrade M. heard that the son of that nice man, now dead, meaning myself, was in deep trouble, she sent a word for me to come and see her at her office at the Central Committee of the Party. So I did, and she promised to ask for an investigation by the central committee on my case. I never heard anything more about that, but I understand that what she actually did, was to get hold of my very damaging file and destroy it. That did help me a lot, later on.

When, as a result of the Leray paper I got, later on, a job at the mathematical institute of the Academy, then that kind of special Human Resources office which I have already mentioned before, had to create a brand new file for me. And that new file might still not have been such a fantastically good one either, but at least it was no longer deadly damaging.

The Romanian Mathematical Society, where I was now employed as a janitor, was located, as I said, inside the university. Igor, who as I have told you had also been expelled from his university job, was there as a junior secretary, and so was also our old friend from student days, Gunther Bach. My monthly salary would not have been enough for buying two pairs of normal shoes. I still had to continue with the private lessons, but now I was a normal working person and hence no longer in danger.

There was not a very hard working atmosphere in that Society of Maths. As soon as the boss, who was quite unpleasant, was away, which meant almost always, Igor, Bach and myself, went straight to the café, while the ladies, the secretaries, went to do their shopping. Incidentally, the boss of the society hated Moisil, and hence he hated me too. My duties as a janitor were varied and many, and I did them all very badly. The many letters which I was supposed to answer, I put straight in the waste-basket, any caller who came to ask for something from the Society was invariably told by me to come the next week, and when he did come again that next week, then the same thing started all over again. The letters which I was supposed to put stamps on and take to the post office, I never sent, the lamps which I was supposed to repair I left unrepaired, and so it went.

The only thing which I did happily, was to turn a big printing press with my arms, I thought that would be a good physical exercise. I may have been the world's worst janitor, ever.

I have described you a bit of Igor's short life. Our friend Gunther Bach, had always dreamt of going to Göttingen and doing big mathematics there, but he died of alcoholism in Romania, with his dream unfulfilled. He did not gauge either, that since Hitler, Göttingen was no longer the Göttingen of his dreams.

As I have already told you, the Mathematical Society was located inside the university, and I took a certain pride and pleasure in carrying, as part of my janitorial duties, big sacs with various junk on my back, under the noses of my former professors.

One day, in October 1959, in the corridors of the university, I happened to see L., a handsome black haired girl, a student in physics, whom I had briefly met some time before, at a party. I decided it was a good idea to invite her to go out with me. I also sensed, very fast, that she was going to refuse me flatly; but I was not leaving her any chance to do so. I engaged into a fast, interesting and funny conversation, where I did most of the talking myself, and after a short time the ice was melting and she very gladly accepted my offers.

From there on, an affair with L. started, then it grew and unfolded. She became, so to say, my official girl-friend. And for the time being that seemed like a happy situation for both of us.

In the early weeks of December 1959, I came rather close to that Death which I used to tease, without ever being afraid of it. One weekend of that early December, Igor and I went to the mountains, with ropes and all the gear, to do a certain climb in snow and ice. But the snow was very difficult, we advanced slower than anticipated, and by the time we were on top of the mountain, it was already pitch dark and we did not manage to find our way down. So we had to spend the night there on top, it was very cold, we had neither food nor water, except for the possibility to eat a bit of snow, and we only had a minimum of clothes with us. We also knew quite well that falling asleep meant certain death. So, we made a hole in the snow, just big enough to sit down, back against back, to keep each other a bit warm. We kept talking, during that long cold winter night, each of us watching carefully that the other one should not fall asleep. When the morning finally came, then our boots, which we had stupidly taken off so as to keep our feet inside our rucksacks, had to be softened with our hammers, but we were safe, and we got home that Monday evening. It took me about a week in order to get warmed up again, and on my toes I still have the scars of that night.

The next Tuesday morning, when we reappeared at the Society of Maths, everybody knew already that we had gotten lost in the mountains and they thought that we might have been already dead. But our boss was worried only about (I forget now which) little nonsensical objects, which were in our charge, and which he thought could get lost, in case we did not come back alive.

And soon after this, finally quite exciting little adventure, one afternoon in the same month of December, I was invited to a bridge party, where I finally met Julie.

Of course, we both knew very well about each other and we also both remembered that little quiproquo in the Academy gardens, even if she had not heard what Sanda had said. Among the guests was also Gelu, her future husband; he met Julie now for the first time too. He never liked me, but that might not be so surprising. There were other people there too, including my own uncle Dan, who knew very well Julie. He was a close scientific collaborator of her father and a very frequent visitor of their house.

Dan was a younger cousin of my mother, and this made him my uncle, by Romanian standards. He was actually the preferred and most beloved cousin of my mother, who had many cousins. Dan's own mother was Jewish and he had always thought of himself as being a Jew. He was accepted by the Jewish community as such. So was also his very elegant older sister, my aunt. Of course neither my uncle nor my aunt were religious; but then so few Romanian Jews were, in those days.

Dan was one of the kindest and sweetest persons I knew, well-beloved by everybody. There was nobody in the extended family to cater so well for the old aunts, like Dan did. He was also quite gay, a fact which my very shy mother never mentioned. But he made no secret about it, the whole town knew, and I knew too.

Through Dan, I got to know a lot of things about Julie's father, his boss, who was a well-known university professor, in virology. Both him and Dan had originally been trained as medical doctors. But Julie's father was also one of the big heads of a venerable, very many centuries old, half-secret international organization, with infinitely many ramifications, secret ceremonials, and big connections. He was, somehow, the certainly unofficial representative of the organization, within the Romanian communist party. He had happily survived, always in the highest positions, through the stormy years of the recent Romanian history. During the nazi times, although his wife, Julie's mother, was Jewish, actually like Dan, Julie thought of herself as being a Jew, he managed to be good friends with Antonescu, Hitler's stooge, the dictator of Romania during the war days.

And now he was a very big shot, member of the central committee of the communist party, president of the Romanian academy, an important member of the National assembly, the communist parliament, director of various institutes, and so on.

He and his family lived like princes of old. They had a big beautiful mansion, which included a large private collection of Far Eastern art, they had their own butler, chauffeur, cook, chambermaids, and so on and so forth. As I also learned from Julie later on, her father lived with the permanent fear of getting arrested and always had a little suitcase ready for the occasion. Do not believe, like many westerners do, that people in the kind of privileged situation which I have described, were necessarily communists. A far as Julie was concerned, for instance, she would invariably refer to the communists as "the pigs".

I also knew that entering that family implied joining the venerable organization and hence becoming a piece of a highly structured system, loosing one's freedom. This is, of course, exactly what Gelu, who married Julie a bit later, did. Later in time, he actually became himself president of the Romanian Academy, ambassador to Paris and Brussels, and so on and so forth. But I am a free man and I did not, and still do not want to have neither master nor God, for that matter.

My uncle Dan of course, would have been delighted if I, his nephew, would have gotten together with Julie. Incidentally, many years later Dan left Romania and moved to Italy, where during my second life I met him again. He died in Rome, in the early eighties.

At that famous bridge afternoon party, very soon some magical invisible current passed between me and Julie, and I well knew that it only depended on myself to let events happen and unfold, between her and me. But, as I have just said, I wanted to stay a free man and so, notwithstanding the big mutual attraction, which there obviously was between us two, I decided not to do anything.

I hardly saw Julie during the next few months and then, via Dan and other common friends, I knew that she was getting married to Gelu. And the day before that marriage, I got a phone call from Julie who asked me to come and join her in a neighboring public garden. So, we met that afternoon, and again I knew perfectly well that it only depended on myself to totally change the course of events. Again I decided to stay a free man and did not do anything. We had a long walk together, with a big exchange of lofty ideas. Later Julie referred to that afternoon as our trip to Africa. In fact she did ask me rather insistently to read a book, "Le Lion", by Joseph Kessel. I did that only much later, in my next life, and then I understood the metaphor she had in mind that afternoon, about me (King the lion), herself (the young girl) and Gelu (the massai warrior).

In the meanwhile, my paper was out, I became then a well-known mathematician, so Moisil, Stoilov and the other big bosses thought it was indecent that I should continue to be a janitor. As a little parenthesis, Moisil and Stoilov were not quite good bed-fellows, they disliked each other, life cannot always be linearly simple. But all that was immaterial for the matter at hand. Of course, the university could not take me back, they thought I could have politically contaminated the students. But in May 1960 I got a much nicer job, a purely research job at the Mathematical Institute of the Academy. The contribution of comrade M. who had destroyed my bad file was essential at this point. And when I left my janitor job for the new one, the very unpleasant boss of the math society looked like a tiger from whom his meat had been taken away.

The Mathematical Institute where I was working now, was a singularity in communist Romania. It was the creation of Stoilov, the only one among my old professors, of some reasonably higher mathematical quality. He had been my complex analysis teacher, you may remember.

Like Moisil, Stoilov had always been a left-wing intellectual, then at an early stage he joined the communists. He was high in their hierarchy and had a lot of power too. But, just like Moisil, he used his power trying to do good, and what those two really thought about communism in the depth of their mind, was always a mystery to me.

At Stoilov's Institute, everybody was free to work on what he or she liked, and as they liked. You could come when it suited you, if it suited you, not like in any of those regular usual jobs, which were certainly not my kind of thing. Of course, only people with very high motivation for mathematical research, were ever hired at this institute. And we were fewer than 20, maybe about 15 or 16.

Every month there was a little fight with the accountant, who did not want to pay us our salaries. According to her, being present at the institute for 8 h a day, something which we clearly never did, was the only thing that mattered, and as far as producing all that mathematics, about that she could not have cared less, and said so. Every month the director, Stoilov, had to force her to pay our salaries. There was nothing else like this, in communist Romania.

So, beginning with May 1960 I had the best job I could have wished for right then, perfectly well suited for me. My mathematics proceeded quite well, and I had a good number of mathematical friends, throughout the world, with most of them I could only communicate by letters, since I could not travel to the West. Among them was an English mathematician, Christopher Zeeman, later Sir Christopher, whose young Danish cousin, just 2 years after my second life had started, became my much beloved second wife, finally the good one.

The years 1959–1960 saw the glorious revolution of high-dimensional topology, lead by Jack Milnor, Michel Kervaire, Barry, Steve Smale, John Stallings, Chris Zeeman, and few others. I had not yet met any of these big mathematicians, it is only a few years later that I was getting to know all of them quite well, but I had access to their work, long before it was in print. I joined with great passion and excitation this movement, placing myself somewhere at the border between high dimensions and low dimensions.

Anyway, in the Spring of 1961, I could happily pursue both with my mathematics and with my climbing, the Sun was still bright, and life looked quite good, for the time being.

But I had a strong feeling that this situation depended of a very unstable balance, and that it was anything but lasting. My fears were quite right. Very few years after my departure from Romania, Ceauçescu the then dictator of the country, declared publicly that he wanted to have a bomb explode under the Mathematical Institute, which he claimed created disorder in his country. And he had the Institute closed down. Those who were still working there, at that time, became either unemployed or were given various not very pleasant jobs, maybe as a punishment for the too good time they had before. So my fears were certainly not pure paranoia. If I would not have gone out of Romania in time, horrible thought if any, then I would have been caught in that big disastrous crash. I do not know, under those hypothetical conditions, how I would fare to-day, if still alive. Anyway, occasionally, this is food for my nightmares.

But let me go back to my main story now. L. was a very nice, sweet, beautiful girl, but there was a little problem too. Although nothing of the kind was ever said between us, I knew she wanted very much to get married to me. Her parents wanted the same, and my own mother wanted that too. And I was not up to that, maybe crudely put, I was not enough in love with L. But all this was too much for me, and I started to feel like climbing up the walls.

So, one afternoon during the very early days of May 1961, I went to pay a visit to Julie, at her lab. She was by then a medical doctor, a lab-doctor. I had hardly seen her again since our trip to Africa, the day before her marriage, about a year earlier.

And now, when I met her again, the dam finally burst and things unfolded then very fast, all the big way through, between her and me. At the very beginning already, I had broken with L., quite abruptly and brutally, I am sorry to say. It so happened, by one of those accidents of chance, that L. met, face to face, with Julie and me, in one of those gardens. And what was going on between us two, was so crystal clear and so much beyond any shadow of doubt, that L. burst into tears, started crying and ran away. I still feel sorry now, for the way she felt hurt and humiliated that day. But the big irresistibly powerful torrent swept everything else away.

A very big, passionate love affair, between Julie and myself was now on. We were also anything but discrete, we displayed proudly and happily our love to the Sun and to the Moon, and so we were, by then, the big talk of the whole town, but we could not have cared less. We were madly in love with each other, and we moved inside our private magical garden.

At this point, another very big event, so important for my life ever after, occurred too. In those days, I was not entitled to travel to the West, although I had been by now invited several times as a speaker at various important conferences. But I was allowed to travel inside the communist bloc. So, in August 1961 I went to Prague, to a mathematical meeting. And Barry, who knew I was coming, joined that meeting too and so we finally met, for the first time. It was big friendship at first sight. We did not need to search out and discover who the other one was. We knew already plenty about each other, about our work, our interests and our thoughts. We became almost instantly good friends, so it stayed for ever after, our friendship only grew and developed. Barry IS my best friend. Of course this is by now largely part of the story of my second life.

As I have several times noticed, the world is very small, indeed. It so happens that L.'s mother was the accountant at one of the Institutes where Julie's father was the director. And about what I am going to tell you right now, I only heard later on, after my break with L. One day, before the break in question, L.'s mother had asked young Anna, a friend of mine who was also working at that same institute, to come and see her. And when she did, she scolded young Anna, asking her to stay away from me since, that is what L.'s mother said, I was promised to her daughter. She did not dream then, what was going to happen soon.

And since L.'s mother worked at that institute, at various official occasions, L. and Julie had actually met too. I knew these things from L., before the big irresistible torrent had swept our poor little relation away. And Julie who always had felt like a princess, could be very haughty, and she snubbed big little L. I do not think Julie knew then that L. was my girlfriend. But that does not matter, one way or another, now.

At the end of 1961, something like November 15th to December 15th, I was sent by my Institute for a scientific visit to Moscow. In those days I could speak some Russian, I was even able to lecture, in Russian, about my own work. My Russian was clumsy and stuffed with mistakes, but I was able to communicate in Russian. I wish I could still do that to-day; but I can only read it, and with difficulty.

I was coming from another communist country and that gave me a privileged position which a westerner, which I was not yet then, could never have had. Very fast, I could feel where the person who was in front of me stood with respect to the regime, and that person in front felt the same about me. So, I could speak freely with the Russians, in a way no Westerner could have done. And I never saw anywhere else as strong anti-communism as I saw in those Russian days of mine.

And I talked a lot with the Russian people, in the streets or in the restaurants, where some always gathered at my table. It was the Khrushchev period when they could not order much vodka. Since I was a foreigner I could, so I ordered for them, and we had a very good time together. These were really bygone days, when a full dinner on black caviar, which I very often had, costed the same price as a steak.

And this same mechanism of mutual fast recognition, followed by completely free talk, also worked with my mathematical colleagues, and some of them became my friends too.

I met a good number of the great luminaries, like Kolmogorov, Alexandrov, Pontryaguine, Shafarevitch, Postnikov; I had met already earlier Sobolev, and I knew him a bit. With more time available, there would be some interesting stories to be told concerning him too. I must confess that when I was introduced to Kolmogorov, I was very pleased when he immediately said he knew who I was, and in order to show that he actually did, he started quoting some of my mathematics. But I spent most of my time with the mathematicians of my age, Novikov, Arnold, Anosov, Cernavski and others. Novikov and I became quite good friends, and stayed so, afterwards.

We were all very young in those old days. I still remember how Novikov and Arnold talked very outspokenly against the regime, so outspokenly that it did not do them much good. I spent many hours in various Moscow cafés, together with Serguei Novikov and some other friends, talking about mathematics and everything else. Serguei was already working big on the invariance of the Pontryaguine classes and I remember how 1 day he flew to Leningrad (Saint Petersburg to-day) to speak to Rohlin, to whom he was very close. There was a rumor that Rohlin had proved the invariance of those classes. Serguei flew back to Moscow the next day with the news that Rohlin's supposed proof had a big hole, and he continued to work big, on his own. He was a very charismatic person and, in those 1961 days, possibly the most handsome young man I had ever met. And I also remember, how many years later, he impressed Milen with his fantastic knowledge of Scandinavian literature, which he seemed to know better than herself.

During my Moscow trip, but also later, during my second life, I heard from Serguei a lot of very interesting stories concerning the KGB, and also about the fates of the people whom the KGB interrogated. He thought that only very few of them managed to survive in one piece, without being permanently damaged. He mentioned Sakharov and Solzhenytsin among the rare happy exceptions. But most were destroyed in the process, and with quite many, something very strange happened. They were turned, they went on the side of their tormentors and joined them. They had gotten somehow to like them. Human mind is a complicated and strange thing.

Serguei thought that Shafarevitch was a case all by himself, in addition to these KGB incidences. As a very young mathematician, he got a big prize for his solution of the inverse Galois problem. But then, in later years, he or one of his students, found a big hole in that supposed proof. And Shafarevitch suppressed this fact, both publicly and in his own mind. And from there on, he started doing really weird things.

I was often invited to Serguei's home, where I was always very nicely received by his mother, Ludmila Keldysh, the Tartar princess, a very distinguished mathematician herself. I did not meet the father, Piotr Sergueevitch, of Word problem fame, an item in which I was myself involved then and quite for some time afterwards, in my second life. You may remember that the big discovery of Novikov father was the following item, which I will tell here in a popular form. He found a finitely presented group, the explicit rules of the game of which can be written down on one single page (or less), which you can easily feed into your lap-top, but which is such that no Turing machine can ever unravel all its mysteries, hence no imaginable computer, classical or quantum, can either, ever. This is something very close to my heart and which my mathematics several times touched.

Unfortunately, I could not meet Piotr Sergueevitch because at the time of my Moscow trip he was in a hospital, for a cure of alcoholic desintoxication.

My Moscow trip was one of those things which I could not easily forget. Then, in the Summer of 1962, Ludmila Keldysh, through her brother Mstislav Keldysh, Serguei's uncle, the then president of the Soviet academy, and a very big shot in the Soviet Union, started to make official arrangements for a very extended visit of mine to Moscow, beginning with the Fall of 1962. But Fate had decided otherwise.

Some time in January 1962, my friend Ganea told me that he had gotten the exit visa from Romania, and asked me what my own plans were. At that time, Jews like Ganea could be bought out of Romania, sometimes other people too. Ganea thought that it would be a very bad idea for me to continue to rot in that country where I was living then, and very helpfully, he went to the Jewish organization, of which his father had been the president earlier, and asked whether I could be added on their lists, as some sort of a honorary Jew, so that I could leave Romania that way. Since that turned out to be impossible, he came back with a half-joke: Why don't you just marry my mother, he said, and then come out with us? This was pure Ganea!

A more practical idea was that, if I wanted, he could organize for me to be bought out, just like he was being bought out then. Since I had not yet made up my mind and since he was leaving that next week, we devised the following plan. If and when, at some later time, I would decide to go, then I should write a mathematical letter to a common mathematical friend, Peter Hilton from Birmingham, a letter containing a subtle mathematical error which only somebody very familiar with my own work could detect as such, and which the censorship hence could not decode.

Although what will immediately follow now concerns the very beginning of my second life, I have to tell it here, because it is about how Ganea saved me from a very dangerous situation, in the beginning of September 1962. I had just left Romania, and how that came about I will tell you soon. I had landed in Stockholm and at the time of the little incident to be told about now, I had been there already for a few weeks. What happened during those weeks was quite eventful but I will not tell about that now.

I was now completely alone in Stockholm, living in the flat of a Swedish-American friend, himself away on vacation. I had plenty of money too, and there was no problem as far as that was concerned. But, in a nutshell, here was my situation, on a very precise day, that month of September: the bridges back to Romania were certainly burned, my provisional permit for staying in Sweden had expired, and although quite friendly, the Swedish police had made it very clear that they did not want me to ask for political asylum in their country and, finally, my visa for France had just been refused. The French police thought that I was a spy. So, at that very instant, it looked as if there was no corner of the planet, where I could safely be.

I sent a desperate telegram to Ganea, in Paris, asking for help. And Ganea went to Charles Ehresmann, who he knew wanted very much that I should settle in France. Ehresmann went to the minister of police and talked to him and then in less than 24 h I had my visa for Paris. This really was the end of my escape from Romania. My friendship with Ganea continued big, and I was very saddened when some years later he died of a very nasty cancer. But this is no longer part of the present story, to which we move back now.

At the very beginning of the Spring of 1962, Julie's parents were faced with the following situation. Their daughter who had already left her husband and was staying again with them, wanted to get a divorce and move in with her lover, meaning with me, while at the same time their son-in-law Gelu threatened with suicide. So Julie's father decided that he wanted to know more about me, before anything else. And it is Julie who told me what follows next.

In his position, Julie's father could use the Romanian secret police (the famous Securitate) as in a Western country one might use some private detectives. Few weeks later, the Securitate came with their answer: Comrade, they said, you cannot let your daughter go with that man, he is crazy! Julie's family strongly vetoed then the whole thing.

Julie was a very romantic person, and she often had said that if for some reason we two got separated, then she would go to a nunnery and become a nun, while I should join the Foreign Legion, an idea which I had sometimes toyed with.

Then, in May 1962 I sent that famous letter to Peter Hilton, and about that same time I got an invitation as a speaker at the International Congress of Mathematicians, in Stockholm, August 1962; that is a kind of event happening every fourth year. I had never been allowed to go to the West before and I did not think there would be any reason for them to let me go now. What I did not know then, was that the buying out system was finished, at least for non-Jews and also that via Ganea and Hilton, many western colleagues knew by then that I wanted to leave Romania.

Actually, at the time when I wrote that letter to Peter Hilton, certain things had maturated and crystalized in my mind too, and I was burning now with a big desire to go. I also knew that this step was crucial for my mathematics, and that was what mattered most for me, it was more important than anything else. Mathematics was and is, my big passion in life.

Julie's father had never liked me, but her mother Raia rather did. And Raia was a woman who yielded a lot of power, and she had very long arms. Several of her relatives were in high positions in various communist parties, in various countries, on both sides of the Iron Curtain. A cousin of her's for instance, was one of the big bosses of the French communist party. And here comes something which I understood only much later: Raia devised a plan for getting us all out of the dead end, namely to have me shipped away. Like Julie, Raia knew that if I got to the West, then I would never come back to Romania, and she also knew that I had an invitation for Stockholm. So, she put pressure on the man who was the head of the communist party organization of the Academy, that institution of which my own Institute depended, a man who had never met me, and asked him, or maybe forced him, to vouch for me, and guarantee that I would return, if allowed to go to Stockholm. So he did, and as a consequence he lost his job.

Then, to everybody's surprise, mine included, I did get that permission and visa too, to go to the conference in Stockholm. And about twenty people knew, because I had told them, that I would not return from Stockholm, but leave Romania forever. These twenty or so included, of course Julie, my very close friends and my mother. But they also included some others, of various ages and genders. I obviously had made a good choice in trusting those twenty people, they were all loyal to me, and nobody gave me in. I am also sorry to have to tell you that some relatives of mine were mortally offended forever, because I had not told them. But I had made a choice anyway.

And, maybe I was crazy, as that police report claimed. I was certainly recklessly incautious when I took with me on that plane to Stockholm, two big suitcases, supposedly for 10 days. They contained, among other things, my complete archives, dangerous stuff would I have been searched. And in those archives, basically by chance, mixed up with other papers, was my divorce certificate. Few weeks before, my ex-wife Sanda, who had just received it, had handed it casually to me, saying that I might need it. Sanda was not included in that list of twenty people who knew, and she did not have the foggiest idea that I would never return; like most people in town she thought that I was much too unpractical to be able to manage alone, all by myself, in a foreign land. When she said that I might find that certificate useful, what she had in mind was, undoubtedly, me and Julie. And what Sanda did not know and what myself I could not have dreamt then, was how badly that divorce certificate turned out to be actually needed, but now in that second life of mine.

The 14th of August 1962, at 6 am, I flew away from Romania. And then I got reborn for my next, second life.

I felt an urge to write all these things down, after a Fijian night when, with a mere fortnight before my eightieth birthday, the long-forgotten memories started bubbling in my mind, and the words to tell them started to get organized in my head too, just by themselves, with me as a mere spectator.

And after having told you all of this, I feel now like I am coming back from the land of the dead, from the dead times and from the dead by-gone world. I must confess that I feel a bit shaky.

A Little Side-Story

– In Lieu of a Short Story Inside the Bigger Tale –

In my youth, when I was eagerly reading those big writers of old, I loved their shorter stories inside the main stories. But, unlike Cervantes or Dostojevski, what I will offer you now, will be not something inside my main story, but a little aside, which you may take with you if you like to do so, or leave.

I have often been thinking about Winston Churchill. You see, I am not a historian, I have already told you that, and so I can indulge in playing with history, in a way no professional historian could do, for so obvious reasons.

You might remember that our Winston Churchill, was the grand-grand-grand-son of John Churchill, first Duke of Marlborough, who together with prince Eugenio di Savoia, the Austrian general, stopped the villain of Europe of those days, Louis the fourteenth of France, at Blenheim, about 300 years ago. You may also remember that Blenheim is located in the Black Forest, der Schwarzwald, that place where both the Rhine and the Danube have their sources so close to each-other.

The reason why I mention the old Duke now, is because our Churchill often identified himself with his illustrious ancestor and in more than one way, the first Duke had always loomed big in his mind.

Churchill has done a lot of mistakes and said a lot of stupid things during his life, that is the kind of things which most of us do. But there was one big historical instant when he rose above all of us, or at least above most of us. That was in the Summer of 1940, when our kind of world was crumbling and when, almost single-handedly and against all odds, Churchill stood between Hitler and that victory of Hitler's which seemed then so close at hand. Churchill decided that his island would not raise hands up, but rather fight to death, to the bitter end.

At that time, many of the other British politicians of the day, would have rather wanted to talk to Hitler, see what he wanted, maybe some arrangement or other could be found, that is what they hoped. Churchill strongly disagreed, he did not want to go on that slippery slope, that is what he said. And it was so fortunate that it was Churchill who was the British PM and not, let us say somebody like Lord Halifax, bringing with him his underling and accomplice Rap Butler. And in May 1940, when history bifurcated, it was almost by chance that it was Churchill who became the PM, and not the Halifax in question.

Churchill could not beat Hitler then, of course, but he could prevent him from winning, and as we all know, Hitler eventually lost his war. It is fair to say that Churchill has saved us all at that time.

I have also dreamt a lot about the same Churchill, at an earlier age, and during an earlier war. He was then the First Lord of the Admiralty, with at his side Jacky Fisher, the foremost fighting sailor since the days of Horatio Nelson, of Trafalgar fame. Fisher was Churchill's First Sea Lord; you see, I am very much at home with the complicated arcanes of the British hierarchy.

The two of them, Churchill and Fisher, devised a master plan to end the war fast, the Gallipoli campaign, sometimes referred to as the Dardanelles, and also called by those Brits of that day, imbued with classical culture, the Hellespontus.

Sadly for all of us, the Gallipoli campaign in question was a total failure. One of the many reasons for that failure, was that, by then Fisher was too old, in point of fact he was younger than I am now, but then almost everybody is so, and he got cold feet.

But if the Gallipoli campaign, which I so often have dreamt about, would have succeeded, then nobody to-day would have ever heard neither about Hitler, naziism and fascism, nor about Stalin and communism. You have understood by now, from my main story, that those beasts of the Apocalypse haunted so big the dark times of my childhood and youth.

And I sometimes also dream that I can break away from the prison of time, get together with Churchill and have a nice long chat with him, about history and other things, in front of a bottle of good champagne, his preferred kind of wine, which I do not dislike either. We might have a very good time together, or at least so I think.

Orsay, France
October 2012

Valentin Poénaru

2017: I thank Cécile Gourgues for typing this text.

Index

S. G. Dani, A. Papadopoulos (eds.), *Geometry in History*,
https://doi.org/10.1007/978-3-030-13609-3

Printed in the United States
By Bookmasters